Deontic Logic and Normative Systems

15th International Conference,
DEON 2020/2021

Deontic Logic and Normative Systems

15th International Conference,
DEON 2020/2021

Edited by

Fenrong Liu

Alessandra Marra

Paul Portner

and

Frederik Van De Putte

© Individual authors and College Publications 2021
All rights reserved.

ISBN 978-1-84890-352-4

College Publications
Scientific Director: Dov Gabbay
Managing Director: Jane Spurr

http://www.collegepublications.co.uk

Original cover design by Laraine Welch

All rights reserved. No part of this publication may be reproduced, stored in a retrieval system or transmitted in any form, or by any means, electronic, mechanical, photocopying, recording or otherwise without prior permission, in writing, from the publisher.

Contents

1   The Roles of Authority and Norm-Addressees in Deontic Puzzles
    Edgar Avendaño-Mejía and Yolanda Torres-Falcón

17 The Gentle Murder Paradox in Sanskrit Philosophy
    Kees van Berkel, Agata Ciabattoni, Elisa Freschi, Francesca Gulisano and Maya Olszewski

36 If You Want to Smoke, Don't Buy Cigarettes: Near-Anankastics, Contexts, and Hyper Modality
    Kees van Berkel, Dov Gabbay and Leendert van der Torre

56 The Varieties of Ought-Implies-Can and Deontic STIT Logic
    Kees van Berkel and Tim Lyon

77 Deontic Action Logics via Algebra
    Pablo F. Castro, Valentin Cassano, Raul Fervari and Carlos Areces

94 Sequent Rules for Reasoning and Conflict Resolution in Conditional Norms
    Agata Ciabattoni and Björn Lellmann

114 Proof Systems for the Logics of Bringing-It-About
    Tiziano Dalmonte, Charles Grellois and Nicola Olivetti

CONTENTS

133 The Original Position: A Logical Analysis
Thijs De Coninck and Frederik Van De Putte

151 Impossible and Conflicting Obligations in Justification Logic
Federico L. G. Faroldi, Meghdad Ghari, Eveline Lehmann and Thomas Studer

166 The Manchester Twins: Conflicts Between Directed Obligations
Stef Frijters and Thijs De Coninck

183 Reason-Based Deontic Logic
Alessandro Giordani

201 Axiomatizing Norms Across Time and the 'Paradox of the Court'
Daniela Glavaničová and Matteo Pascucci

219 How Deontic Logic Ought to Be: Towards a Many-Sorted Framework for Normative Reasoning
Valentin Goranko

239 A Defeasible Deontic Logic for Pragmatic Oddity
Guido Governatori, Silvano Colombo Tosatto and Antonino Rotolo

255 Is Free Choice Permission Admissible in Classical Deontic Logic?
Guido Governatori and Antonino Rotolo

272 Moral Principles: Hedged, Contributory, Mixed
Aleks Knoks

291 A Reduction in Violation Logic
Timo Lang

CONTENTS

308 Interpretive Normative Systems
Juliano Maranhão and Giovanni Sartor

323 A Logical Analysis of Freedom of Thought
Réka Markovich and Olivier Roy

339 Deontic Logic Based on Inquisitive Semantics
Karl Nygren

358 Input/Output Logic With a Consistency Check–the Case of Permission
Maya Olszewski, Xavier Parent and Leendert van der Torre

376 Term-Sequence-Dyadic Deontic Logic
Takahiro Sawasaki and Katsuhiko Sano

394 Forms and Norms of Indecision in Argumentation Theory
Daniela Schuster

414 Goal-Directed Decision Procedures for Input/Output Logics
Alexander Steen

427 Prioritized Defaults and Formal Argumentation
Christian Straßer and Pere Pardo

Editorial Preface

Fenrong Liu Alessandra Marra Paul Portner
Frederik Van De Putte

* * *

This volume contains the proceedings of DEON2020/2021, the 15th International Conference on Deontic Logic and Normative Systems that was organized by the Munich Center for Mathematical Philosophy at LMU Munich (Germany) on 21st-24th July, 2021. The biennial DEON conferences are designed to promote interdisciplinary cooperation amongst scholars interested in linking the formal-logical study of normative concepts, normative language and normative systems with computer science, artificial intelligence, linguistics, philosophy, organization theory, and law.

There have been fourteen preceding DEON conferences: Amsterdam 1991; Oslo 1994; Sesimbra 1996; Bologna 1998; Toulouse 2000; London 2002; Madeira 2004; Utrecht 2006; Luxembourg 2008; Fiesole 2010; Bergen 2012; Ghent 2014; Bayreuth 2016; and Utrecht 2018.

Special Focus of DEON2020/2021

In addition to the general themes of the DEON conferences, DEON2020/2021 had a special thematic focus on "Norms in Social Perspective". While deontic logic is still often approached as the logic of objective, agent-independent notions of obligation, permission, and prohibition, there is a growing body of literature on the role of groups, social interaction, and networks in our normative reasoning. The social perspective plays a role in our reasoning in at least three different ways. First, it determines the ways norms are adopted and updated. How individual and group agents relate to one another (e.g., via a social network or as part of an institution) is a crucial factor that helps govern which norms these agents endorse, and the way their norms change. Second, social interaction is essential for the way normative language is used and interpreted: deontic terms acquire meaning via social conventions; we use deontic terms to communicate our ethical stances to one another and to convince each other of the rationality of certain choices. Third, the social dimension is also essential in the way we evaluate norms: often, obligations cannot be properly explained in terms of a single normative code or one individual agent's attitudes. Instead, one needs to refer to the preferences, goals, and norms of several agents, and to lift these to the group level in order to specify what ought to be done.

Contributed Papers and Keynote Speakers

The contributed papers collected in this volume respond both to the general themes of DEON and to the special theme of DEON2020/2021. They cover foundational issues on the formal study of normativity and its application to a wide array of topics. These topics range from the fine-grained logical structure of obligations and permissions, their relation with agency, and their role in conditional constructions, to normative conflicts, the structure of normative language in argumentation, and the analysis of normative concepts in legal traditions as well as in Eastern and Western philosophies.

Our four keynote speakers were chosen with an eye on the special theme and on their outstanding contributions to law, philosophy, logic, and computer science: Marcia Baron (Indiana University, Bloomington), Emiliano Lorini (Université Paul Sabatier, Toulouse), Shyam Nair (Arizona State University, Tempe), and Sonja Smets (ILLC, University of Amsterdam). Titles and abstracts of their keynote addresses are presented below.

Marcia Baron, *Recklessness and Negligence in the Criminal Law*
Abstract: Criminal law theorists debate whether negligence should suffice for criminal liability. Put differently, should acting negligently ever be enough to supply the *mens rea*, or culpability, component, required for criminal conviction? Everyone agrees that recklessness should suffice, but there is disagreement about negligence. The debate is marred by unclarity about just how negligence differs from recklessness. It is not unusual to have the experience I recently had when arguing that negligence should suffice: someone in the audience says, "But that isn't really negligence! That example you just gave is really an example of recklessness." The disagreement is surprising because the parties to the debate all claim to be relying on the definitions of negligence and recklessness in the Model Penal Code. However, the definitions are dense and the implications are not always obvious. In addition, some parts of the definitions are often neglected by those offering summaries of what, according to the MPC, the difference between negligence and recklessness is. In my presentation, after first explaining the *mens rea* requirement so as to provide the context for the debate, I seek to shed light on what, according to the MPC definitions, the difference between negligence and recklessness is. I then indicate some implications for the controversy concerning whether negligence should suffice for criminal liability.

Emiliano Lorini, *Logics of Evaluation.*
Abstract: We present a family of logics of evaluation which clarify the relationship between knowledge, values and preferences of multiple agents in an interactive setting. Evaluation is a fundamental concept for understanding how an ethical agent's decision is affected by her values. We present complete axiomatics for these logics as well as a dynamic extension by the concept of value expansion. We show that value expansion indirectly affects the agents' preferences by inducing a preference upgrade operation.

Shyam Nair, *Reasons-Based Theories of Obligation and Optimality.*
Abstract: It is common in certain circumstances for philosophers and logicians to conflate what is obligatory or required with what is optimal or ought to be done. But it is a by now familiar thought that these notions should be separated. Terms such as 'must' and 'have to' are most naturally used to express obligations or requirements. Terms such as 'ought' and 'should' do not express requirements. Instead they are used to express a kind of optimality. Paul McNamara and other logicans have done much to improve our understanding of these notions (as well as various other related notions) within a broadly value- or preference-based deontic logic framework. But comparatively less work has been done exploring these issues in the broadly imperatival or reasons-based deontic logic tradition. A notable exception is a recent paper by Robert Mullins. This talk discusses various choice points and generalization of Mullins' framework. The aim is to highlight that there are many different frameworks that deserve our attention and provide a preliminary assessment of the costs and benefits of each.

Sonja Smets, *The Logical Dynamics of Social Norms.*
Abstract: The flow of information is what drives our information society of interconnected agents capable of reasoning, communication, and learning. In this context we are interested in the logical study of how information flows in social networks by focusing on the spread of behaviors, ideas, and the adoption of norms across a social network. With respect to the study of social norms, our aim is to design a logical model that can give an adequate description of agents being influenced to adopt a new norm. This refers to situations as described by C. Bicchieri in [1]: "if people believe that a sufficiently large number of others uphold a given norm, then, under the right conditions, they will conform to it". We will focus on these triggers of being persuaded to adopt a new norm, which are here stated in terms of a "sufficiently large number of others", or "enough people", upholding the norm. We will capture these triggers for adopting norms in a qualitative logical framework and model the diffusion process as well as the long-term informational evolution of our networks. For this presentation, which is based on ongoing work with A. Baltag and on the work presented in [2], we will make use of the tools of Dynamic Epistemic Logic as well as Modal Mu-Calculus.

[1] C. Bicchieri, The Grammar of Society, The Nature and Dynamics of Social Norms, Cambridge University Press, 2006.

[2] A. Baltag, Z. Christoff, R. K. Rendsvig, S. Smets, Dynamic Epistemic Logics of Diffusion and Prediction in Social Networks, Studia Logica, 107(3):489–531, 2019.

Format of DEON2020/2021

DEON2020/2021 was a special DEON conference in various respects. The complex scenario yielded by the COVID-19 pandemic forced us to reconsider the schedule and structure of the conference. With the hope of benefiting the DEON community, we decided to postpone the conference and have two rounds of submissions: in 2020, when the conference was originally scheduled, and in 2021.

In the 2020 round, 45 papers were submitted in total, out of which 18 were accepted for presentation, and 17 of these can be found in the proceedings. In the 2021 round, 19 papers were submitted, out of which another 8 were accepted for presentation and publication. As these numbers suggest, both calls for papers were very successful, especially given the precarious situation that so many members of our community found themselves in.

All papers underwent double-blind peer review by at least two Program Committee members. We made the final selections on the basis of the resulting scores and the reviews themselves, our own insight into the papers, and diversity considerations. Even though at points this made for tough choices, we are nevertheless very glad to present these papers as editors, and we are convinced that they will contribute to deontic logic and the study of normative systems just as much as previous DEON editions did.

Acknowledgments

Organizing a conference, especially in the given circumstances, is impossible without teamwork. We are very much indebted to everyone who made this possible. First of all, we wish to thank the invited and contributing speakers for providing the very content of the conference and the proceedings. Special thanks go to the Program Committee for their careful reviews of submitted papers, and for their constructive discussions and reconsideration in cases of disagreement. We also thank the local organizing team, especially Hannes Leitgeb, Norbert Gratzl, Ursula Danninger, and Karsten Maria Thiel for their help with the practical organization and fund-raising for DEON2020/2021. Thank you to Andreas Lüchinger for making these proceedings a concrete object. We also thank Jan Broersen and Jeff Horty, Chair and Vice Chair of the DEON Steering Committee, respectively, for their advice, practical wisdom and continuing goodwill. Finally we are indebted to College Publications, and to Dov Gabbay and Jane Spurr in particular, for their support in getting the proceedings published and available to the DEON community.

These proceedings were published thanks to financial support by Frederik Van De Putte's MSCA-IF project DYCODE (grant agreement ID:795329) and the Chair of Philosophy I at the University of Bayreuth.

DEON2020/2021 is funded by the Deutsche Forschungsgemeinschaft (DFG, GermanResearch Foundation) [446508593], which is gratefully acknowledged. Gefördert durch die Deutsche Forschungsgemeinschaft (DFG) – [446508593].

July 2021,   the Editors of DEON2020/2021

Committees

Chairs of the Program Committee
Fenrong Liu (Tsinghua University)
Alessandra Marra (MCMP/LMU Munich)
Paul Portner (Georgetown University)
Frederik Van De Putte (University of Rotterdam and Ghent University)

Local Organizing Committee
Alessandra Marra (MCMP/LMU Munich)
Norbert Gratzl (MCMP/LMU Munich)
Hannes Leitgeb (MCMP/LMU Munich)

Program Committee
Thomas Ågotnes (University of Bergen)
Maria Aloni (University of Amsterdam)
Christoph Benzmüller (Freie Universität Berlin)
Jan Broersen (Utrecht University)
Mark Brown (Syracuse University)
Ilaria Canavotto (University of Amsterdam)
Fabrizio Cariani (University of Maryland, College Park)
José Carmo (University of Madeira)
Ivano Ciardelli (MCMP/LMU Munich)
Roberto Ciuni (University of Padova)
Cleo Condoravdi (Stanford University)
Robert Demolombe (ONERA)
Huimin Dong (Zhejiang University)
Hein Duijf (Vrije Universiteit Amsterdam)
Federico Faroldi (Ghent University)
Stephen Finlay (University of Southern California)
Guido Governatori (CSIRO)
Davide Grossi (University of Groningen)
Andreas Herzig (CNRS, IRIT, Univ. Toulouse)
John Horty (University of Maryland)
Magdalena Kaufmann (University of Connecticut)
Piotr Kulicki (John Paul II Catholic University of Lublin)
Beishui Liao (Zhejiang University)
Juliano Maranhão (University of São Paulo)
Réka Markovich (University of Luxembourg)
Joke Meheus (Ghent University)
John-Jules Meyer (Utrecht University)
Robert Mullins (Queensland University)
Xavier Parent (TU Wien)
Gabriella Pigozzi (Université Paris-Dauphine)
Martin Rechenauer (MCMP/LMU Munich)
Antonino Rotolo (University of Bologna)

Olivier Roy (Universität Bayreuth)
Chenwei Shi (Tsinghua University)
Audun Stolpe (Norwegian Defence Research Establishment FFI)
Christian Strasser (Ruhr-University Bochum)
Allard Tamminga (University of Groningen)
Paolo Turrini (University of Warwick)
Peter Vranas (University of Wisconsin-Madison)
Malte Willer (University of Chicago)
Tomoyuki Yamada (Hokkaido University)
Leon van der Torre (University of Luxembourg)

DEON Steering Committee
Jan Broersen (Utrecht University) - Chair
John Horty (University of Maryland) - Vice Chair
Christoph Benzmüller (Freie Universität Berlin)
Cleo Condoravdi (Stanford University)
Melissa Fusco (Columbia University)
Beishui Liao (Zhejiang University)
Juliano Maranhão (University of São Paulo)
Alessandra Marra (MCMP/LMU Munich)
Paul McNamara (University of New Hampshire)
Joke Meheus (Ghent University)
Gabriella Pigozzi (Université Paris-Dauphine)
Paul Portner (Georgetown University)
Antonino Rotolo (University of Bologna)
Olivier Roy (Universität Bayreuth)
Leon van der Torre (University of Luxembourg)
Malte Willer (University of Chicago)

The Roles of Authority and Norm-Addressees in Deontic Puzzles

Edgar Avendaño-Mejía

Universidad Autónoma Metropolitana
Mexico
Mexico City

Yolanda Torres-Falcón

Universidad Autónoma Metropolitana
Mexico
Mexico City

Abstract

Attempts to solve classical deontic logic puzzles do not deem relevant the roles of normative authority and norm-addressees. However, drawing a line in imperative logic between the sets of actions validated by authority and the actions that agents can fulfill in different situations prevents apparently unrelated paradoxes from arising at all. This separation also establishes clearer criteria for deeming a result as paradoxical, rather than a vague appeal to intuitions. The old semantic distinction between norm validity and satisfaction is thus brought back and refined through this discussion: norm validity as a reflection of the will of authority; norm satisfaction as the effects of norm-following throughout possible worlds.

Keywords: authority, addressee, validity, satisfaction, agent, puzzle, norm, imperative, deontic.

1 Introduction

Authority and norm-addressees in the logic of norms The will of authority and the actions of norm-addressees are two main actors in the act of norm validation and norm fulfillment or satisfaction. However, this social aspect of the normative phenomenon is usually not deemed as necessary for a basic logic of norms; only propositional contents, the right normative operators and the usual logical connectives are considered to be basic.

It has even been said that the intentions and wishes of authority are not a matter of logic, even if they are necessary to determine whether some norm is correctly satisfying something commanded by authority:

To determine whether an imperative is 'separable' or 'inseparable', i.e. whether doing A alone produces something 'right' with respect to an imperative !$(A \land B)$ or not, it is necessary to examine the intentions and wishes of the authority that used the imperative, it is not a matter of logic. [2, p. 171]

However, the roles of authority and norm-addresses, when they are represented in a formal system through the semantic distinction between validity and satisfaction [1], offer two main reasons to consider the distinction not only relevant, but basic to the logic of norms: first, it solves many apparently unrelated puzzles usually thought to require different approaches to be solved; second, it offers clear grounds to deem a result in normative logic as paradoxical or puzzling, rather than some vague appeal to natural language intuitions. To this end, we propose to take the formal system described in Krister Segerberg's 1990 article 'Validity and satisfaction in imperative logic' [4] as a system formalizing both the logic of norm validation and the logic of norm satisfaction. [2]

One important remark about Segerberg's system is that it is described in terms of imperatives, not in terms of deontic concepts. Nonetheless, we argue that it throws light onto the normative phenomenon in general. We will comment on some of the counterarguments presented by Jörg Hansen in [2]. [3]

We will denote the formal system described in [4] VSL (Validity-Satisfaction Logic). Our purpose is to extend the analysis already offered there; whatever merits are found in the formal system described and the basis of the arguments developed should redirect the reader's attention to Segerberg's numerous works.

Validity and satisfaction in VSL In VSL, Segerberg's goal is to prevent the Ross Paradox, precisely the one which gave rise to this distinction between validity and satisfaction most convincingly. He acknowledges a couple of presuppositions that he needs in order to develop the intended interpretation of his formal system. These presuppositions also describe the main reason why this system represents so adequately the idea that authority and addressees are basic to the logic of norms. The first one has to do with maintaining a separation between the world of facts and the will and actions of authority and norm addressees [4].

[...] we believe that it is well to leave both the commanding authority (the commander, Ross's "imperator") and the subject (the agent) out of it [the world]. They have different roles to play, both having to do with changes in the world. The subject's is to act; he tries to manipulate the way in which

---

[1] Long set aside in the discussions on normative logic.

[2] This work takes a standpoint against the idea that we need to choose among truth, validity, satisfaction or some other semantic value for norms, in order to obtain one single true standard deontic logic. Deontic puzzles may require not only a variety of symbolic tools, but also a variety of semantic values.

[3] He puts together important challenges to those who try to develop logics for norm validity and norm satisfaction, but no analysis of the formal system to be described here is presented there, so we will try to answer some of his objections through Segerberg's system and also by refining the notion of satisfaction involved in deontic systems like SDL (Standard Deontic Logic).

[4] He calls them 'subjects' or 'agents'.

the world changes. The world is in one state one moment, in another state the next; but what the next state is may depend on the subject—on his will. [...] Ultimately it is change in the world that is the authority's concern too, but the ways of authority are indirect, proceeding via the subject. [4, p. 204]

The second presupposition is that the realm of norm validity is seen as a reflection of the will of authority and that this may be achieved by a special semantic device separate from the one describing the different possible situations in a model, which is standard in modal logic semantics. Although he doesn't intend to represent the subject performing any actions because of the complexity of the phenomenon, we suggest there is already a representation of the decisions of agents by way of the actions available to any norm-addressee throughout possible situations, thus taking the context of norm satisfaction as a reflection of norm addressees.

Here we are doing elementary logic and so shall not be able to do more than scratch the logical surface. In fact, we shall not even touch on the question of how to represent the subject performing any actions. However, we will represent the authority issuing commands. To this end we need to introduce a semantic device to keep track of the commands issued by the authority. [4, p. 204]

The way we are intending to represent authority and norm addressees doesn't allow us to identify and distinguish among separate individual subjects in their role as authority or addressee, but only to distinguish in a most general way among this two different roles. This means it's not relevant if authority is taken to be a singular subject, a group of people, a paper with rules written on it or even the customs of society which eventually may deem some action as normatively valid. It is only relevant to distinguish between the role played by those who validate norms and the very different role played by those who are supposed to follow those valid norms.

2   VSL's language, syntax and semantics

Since VSL is not a well-known system, we will take some pages to describe it, but only certain aspects of it for space reasons, pointing out specially where the notions of validity and satisfaction come up. We will use the same symbols and conventions which Segerberg uses originally, the same names for axioms, inference rules, semantic conditions, language symbols, etc.

2.1   VSL language

Definition 2.1 A *well-formed expression* is either a formula or a term. Every formula is either theoretical or practical, but not both.

(i) Every propositional letter is a theoretical formula.

(ii) $\perp$ is a theoretical formula.
   If A and B are formulas:

(iii) $A \to B$ is a formula: practical if A or B are practical, theoretical if both are theoretical.

(iv) $[\alpha]B$ is formula if α is term: it is theoretical if B is theoretical, it is practical if B is practical.

(v) δA is term if A is theorical formula.

If α and β are terms:

(vi) $\alpha + \beta$ is term.

(vii) $\alpha; \beta$ is term.

(viii) $!\alpha$ is practical formula.

(ix) There are no more well-formed expressions.

Theoretical formulas are those true or false formulas which do not contain formulas with imperatives at any point. Practical formulas are those which do contain at some point at least one formula with an imperative.[5] Imperative formulas are considered practical formulas and they don't have truth value, but norm-validity value; they are *prescriptive* norms in the sense that they are valid or invalid on the basis of a defined command set.[6] On the other hand, terms are expressions representative for actions and are of the form 'δA', where A is a theoretical formula and the term is read 'to bring it about that A is the case'.[7] They don't have truth value, neither is it relevant to them the truth value of the formulas which compose them. In this sense they are not actions already *taken* but actions *available* and in order to appear as part of a formula in a truth/falsity context, they need to be part of a modal-dynamic formula as in '$[\delta A]B$'. This modal-dynamic formulas '$[\delta A]B$' are read as usual in dynamic logic: 'bringing it about that A is the case *leads always to situations where* B is the case', where the 'always' is the reading of the brackets '[ ]'. Again, the complete formal definition of these symbols can be found in Segerberg's article.

2.2 Syntax

Axioms and inference rules The full list of axioms and inference rules can be found in [4, p. 206]. We remark that they are grouped in four categories: propositional, modal, action and imperative.

Propositional

(PA1) *Every instance of tautology of propositional calculus in the VSL language.*

(MP) *If A is a theorem and $A \to B$ is a theorem, then B is a theorem.*

---

[5] They don't simply describe facts, but depend at some extent on the will of authority.

[6] We could further add a definition for *descriptive* norms, commonly known as normative propositions, also based on this command sets which would have truth value. But it is the main goal of this text to keep things as simple as possible.

[7] There's also an article by Segerberg where he develops this operator [3]

Modal

(MA0) $[\alpha](A \to B) \to ([\alpha]A \to [\alpha]B)$ [8]

(MA1) $[\alpha](B \wedge C) \equiv ([\alpha]B \wedge [\alpha]C)$

(MA2) $[\alpha]\top$

(N) If A is a theorem, then $[\alpha]A$ is a theorem, for any α

(MR1) If $B \equiv C$ is a theorem, then $[\alpha]B \equiv [\alpha]C$ is also a theorem.

Action

(AA1) $[\delta A]A$

(AA2) $[\delta A]B \to ([\delta B]C \to [\delta A]C)$

(AA3) $[\alpha + \beta]C \equiv [\alpha]C \wedge [\beta]C$

(AA4) $[\alpha; \beta]C \equiv [\alpha][\beta]C$

(AR1) If $A \equiv B$ is a theorem, then $[\delta A]C \equiv [\delta B]C$ is also a theorem, given that A and B are theoretical formulas.

Imperative

(IA1) $(!\delta A \wedge !\delta B) \to !\delta(A \wedge B)$

(IA2) $!(\alpha; \beta) \to !\alpha$

(IA3) $!(\alpha; \beta) \to [\alpha]!\beta$

(IA4) $!\alpha \to ([\alpha]!\beta \to !(\alpha; \beta))$

(IR1) If $[\alpha]C \equiv [\beta]C$ is a theorem for every C, then $!\alpha \equiv !\beta$ is also a theorem.

2.3 VSL semantics

We offer a brief sketch for the semantics of VSL. Further reference for metasemantic proofs such as soundness and completeness is to be found in [4]. The semantics for VSL are defined in the familiar Kripke structures style, but we first define truth for theoretical formulas and realization for terms in a model. Only later can we define truth or validity for practical formulas.

Definition 2.2 A *VSL model* is a quintuple $\mathfrak{M} = <\mathbf{U}, \mathbf{A}, \mathbf{D}, \mathbf{P}, \mathbf{V}>$, such that:

(i) $\mathbf{U} \neq \emptyset$; $\mathbf{U}$ is a non-empty set. [9]

(ii) $\mathbf{A} \subseteq \mathbb{P}(\mathbf{U} \times \mathbf{U})$; the elements of $\mathbf{A}$ are sets of ordered pairs belonging to $\mathbf{U} \times \mathbf{U}$. [10]

(iii) $\mathbf{P} \subseteq \mathbb{P}(\mathbf{U})$; The elements of $\mathbf{P}$ are *sets* containing elements of $\mathbf{U}$. [11]

---

[8] This axiom is not stated explicitly in Segerberg's 1990 article, but it is assumed since he acknowledges that this system includes every theorem of the smallest *normal* modal logic [4, p. 211].

[9] Elements of **U** are usually interpreted as possible states or situations where different propositions may be the case.

[10] These are the actions of the model.

[11] We follow the traditional view of characterizing propositions extensionally as the set of

(iv) **D:P**$\longrightarrow$**A** ; **D** is a function from **P** to **A**. [12]

(v) **V:L**$\longrightarrow \mathbb{P}(\mathbf{U})$; V is the standard *valuation* function. [13]

Each one of this elements should fulfill certain conditions described in [4, pp. 208-9].

2.4 Truth, validity and satisfaction

Truth and satisfaction Truth and satisfaction are defined in [4, p. 209] through a definition of intension for formulas $\|A\|$ and for terms $\|\alpha\|$. We will here focus on the last three (IC5-7), the satisfaction conditions central to the argumentation of this text:

(IC5) $\|\delta A\| = D\|A\|$

(IC6) $\|\alpha + \beta\| = \|\alpha\| \cup \|\beta\|$

(IC7) $\|\alpha;\beta\| = \|\alpha\| \mid \|\beta\|$

Since each action δA is to be understood as a *set* of ordered pairs, where each pair represents a *transition* between possible worlds or situation, the notion of satisfaction is given in such terms, that is, what satisfies an action is not a state of affairs, but rather a transition between possible situations given by an ordered pair. The definition of satisfaction for terms is thus:

Definition 2.3 Given $\|\alpha\| \in \mathbf{A}$, $<x,y>$ satisfies α if and only if $<x,y> \in \|\alpha\|$.

Such a set $\|\alpha\|$ is defined according to the above definitions (IC5-7), so it ultimately relies on the **D** function of the defined VSL-model.

These conditions set the basis for the role of norm-addressees in satisfying the commands issued by authority. In a sense, they offer a way of answering the question: 'What changes in the world would count as performing which actions?', which would be a first step in answerring how to satisfy a certain imperative. The second step would be asking if the action in question is normatively valid, the criteria to answer it comes next.

Norm validity Lastly we get to the definition of norm validity, which is found in the semantics for practical formulas on [4, pp.209-212]. A command system Γ in a VSL-model $\mathfrak{M}$ is defined as:

$$\Gamma = \{\Gamma_x : x \in \mathbf{U}\}$$

where Γ_x is called a command set of x. Any command set Σ should fulfill conditions (C0-4) found in [4, p. 210]. We will emphasize only the first two:

(C0) $\Sigma \subseteq \mathbb{P}(\mathbf{U} \times \mathbf{U})$

(C1) if $DX, DY \in \Sigma$ then $D(X \cap Y) \in \Sigma$, for all $X, Y \in \mathbf{P}$.

---

situations or states where they are true.

[12] This is the δ-operator, which defines an action in terms of a transition that leads to a certain proposition being true.

[13] **L** is a set of propositional letters in VSL's language. Thus the function assigns a set of elements of **U** to each propositional letter from the language of VSL.

Condition (C0) is central since it states that command sets are given in the same terms as actions; as sets of ordered pairs of elements of **U**, sets which represent transitions between possible situations, so that only transitions are commanded and not directly propositions. Condition (C1) will be refered to when analyzing the paradox of conflicting oblgiations in the next section. It should also be noted that the conditions (C0-4) make it possible that there are empty command sets.

Norm validity is thus defined by recursion in (RC1-5) [4, p. 211]. We will only cite the most relevant one:

(RC5) $\Gamma \models_x !\alpha$ iff $\| \alpha \| \in \Gamma_x$.

The crucial definition (RC5) tells us that a commanded term is normatively valid whenever it is found in the command system of the world in question. This represents what is commanded by authority.

Other definitions We present important definitions not explicit in Segerberg's article.

Definition 2.4 A formula ϕ is *VSL-valid* if at any $x \in \mathbf{U}$ of any $\mathfrak{M}$, if ϕ is either true or normatively valid in x. That is, if either $x \in \| \phi \|$ or $\Gamma \models_x \phi$.

Definition 2.5 A formula ϕ is *logical consequence* of ξ, if there is no model $\mathfrak{M}$ and a situation $x \in \mathbf{U}$ of the model where: ϕ is either true or normatively valid in x and ξ is neither true nor normatively valid in x.

3 Reinterpreting three classic puzzles

We will discuss three main puzzles: the logical necessity of obligations, the paradox of conflicting obligations and Chisholm's paradox. The reasons for these particular choices are mainly to show how paradoxes with different formal sources may be approached through this lens.[14]

3.1 Logical Necessity of Obligations

This puzzle is a clash between a very intuitive idea about the contingency of norms and a very straightforward result of SDL (Standard Deontic Logic). The idea of contingency of norms may be expressed in a simple statement:

(1) There are possible worlds without norms.

The theorem of SDL in clash with (1) is:

(2) (ON) $O\top$ [15]

This theorem basically says that it is a logical truth that every tautology is obligatory. Since being a logical truth means being true at all possible worlds, it follows that in all possible worlds tautologies are obligatory, so none of those worlds is without norms.

---

[14] This analysis is part of a PhD thesis to be presented in the Autumn of 2020, under the tentative title: 'Validity and satisfaction in deontic logic', by the first author of this text. Eight more puzzles are analyzed in that text.

[15] Reading '⊤' as any tautology of propositional logic.

It is no mystery that the source of (ON) is the necessitation rule together with the tautologies of propositional calculus. But why is it deemed as paradoxical? What does it mean to say that according to this theorem 'tautologies *are* obligatory' and to say that 'there *are* possible worlds *without* norms'? Under what standard should we interpret there *being* norms or *not being* any norms. We should review the semantic definition of formulas with the form Oα in SDL.

The semantic conditions of truth in terms of Kripke semantics state that, for a formula Op to be true in a world x, the formula p has to be true in all the worlds which have the normative-acceptability relation with the world x. Given this definition, it should be obvious that the theorem has to be a logical truth, since tautologies are true in all possible worlds, no exception for the normatively acceptable ones. Thus the formula is satisfied throughout normatively acceptable possible worlds, it's modally-normatively satisfied.

But why then is the result paradoxical? It would be hard to argue against the possibility of worlds without complex beings, complex enough to state rules. Our own universe didn't have any normative authorities when life wasn't even possible, in that possible world there would be no norms. The crucial focus here is on the word *are* when saying 'there *are* no norms'. What do we mean by that? The most natural answer may be that norms haven't been stated or validated by anyone.[16] In that sense, of course, there are possible worlds where no norms have been validated, in the abscence of any normative authority necessary for that action.

SDL clearly fails to follow that simple intuition, but it correctly expresses a truth about the context of satisfaction [17] of the norm-validating and norm-following activities. Namely, that the criteria to consider an event as going according to some norm is trivially satisfied even in the absence of conscious agents capable of even understanding rules.[18] In this sense, from the viewpoint of norm-addressees, this theorem says that they shouldn't worry about obligations of making a tautology true, for anything they do or don't do will trivially satisfy that obligation. This may be the first intuitive answer one may think about when first trying to make sense of this puzzle.

VSL doesn't have this ambiguity problem, for it has separate criteria to deem a normative formula as valid and a theoretical formula as true or satisfied.

---

[16] In this discussion it should be prefered to talk about 'validating' a norm, instead of 'stating' one, for 'stating' may be interpreted as'uttering', which is not necessarily meant when talking about norm validity. A norm may be uttered and thus validated, but its logical consequences may have not been uttered and still be validated by their antecesor.

[17] The term 'context of satisfaction' is taken from a very interesting account of this concept in [5]

[18] This is argued in the same spirit in which propositions are said to be true even in possible worlds where there are no living creatures capable of uttering any sentence. Since uttering a true sentence and its being true can be considered different matters, commanding an imperative and its being satisfied can also be considered different matters. An event may occur which would make some sentence true, even if that sentence has never been uttered. In the same way, an event may occur which would count as satisfying some command, even if that command has never been uttered.

That is, the theorem is not logically valid in VSL, but the fact that anything would count as satisfying such a command is also preserved. This is how this same theorem would look in VSL:

(3) $(VON) !\delta\top$

This formula (VON) should be interpreted as saying:

(4) Bring it about that $\top$ (an instance of any tautology) is the case!

Clearly, any action or even any failure to act would satisfy such a command, in every possible world. There are only two conditions which the function **D** has to fulfill whenever it belongs to a model $\mathfrak{M}$ of VSL, given in [4, p. 209] as FD1 and FD2. These should hold for any X, Y$\in$**P**

(FD1) $DX \subseteq \{<x,y>: y \in X\}$

(FD2) If $<x,y> \in DX \Rightarrow y \in Y$, for all y, then $<x,z> \in DX \Rightarrow <x,z> \in DY$.

The relevant condition for this puzzle is clearly (FD1), since it states that the *image* of the function assigned to a proposition through the δ-operation should be contained in the set of worlds represented by the proposition affected by the operation. In other words, 'doing A' should always get you to a state where 'A' is the case. Since $\top$ is the case in every possible world, any set of ordered pairs $\| \alpha \| \in$ **A** will count as seeing to it that $\top$ is true, for any set of ordered pairs will take us to worlds where $\top$ is true. In other words, the way we defined what counts as doing a certain action makes it true that any action whatsoever will count as seeing to it that a tautology is true.

But that wouldn't be enough to deem VON as logically valid in VSL, for the semantics of formulas with the form $!\delta A$ are not based on what propositions are true in any possible world or situation of the model, but rather on command sets specific to each possible world. These command sets may be empty according to the semantic conditions in [4, pp. 209-211]; in that case $\| \delta\top \|$ wouldn't belong to the command set in question, thus it wouldn't be normatively valid in virtue of mere logic. Given (RC5), it would be necessary to have the specific formula VON in a command set of a certain world to deem it as valid, regardless of the fact that any action would count as $\delta\top$, thus distinguishing clearly between what norms agents fulfill by doing '$\delta\top$' (anything at all) and what the authority wills to be done (perhaps not particulary $\delta\top$).

3.2 Paradox of conflicting obligations

This paradox calls for attention to the intolerance of standard deontic systems towards normative conflict, making it escalate to a logical contradiction. Consider thus this two conflicting obligations:

(1) Op

(2) O¬p

It's not hard to prove that from (1) and (2) a contradiction is derivable in SDL. It's also commonly accepted that even a conflict of obligations as impossible to fulfill simultaneously as the one between (1) and (2) shouldn't be considered as serious and strong as a logical contradiction.

But we could ask again exactly what aspect of norms makes it so paradoxical that a conflict of norms derives in a logical contradiction. Let's introduce the points of view of authority and norm-addressees to clear out this question.

From the point of view of norm-addressees and specifically when considering the context of satisfaction of (1) and (2), trying to fulfill both obligations would amount to a contradiction. The very semantic conditions of SDL make it clear that making both true in a certain world would require that in all normatively adequate worlds both p and ¬p be true, which would cause contradictions in all such worlds. This supports the idea that a certain notion of satisfaction is the best way to interpretic deontic formulas in SDL.

From the point of view of authority, there's certainly some kind of tension (maybe a *rationality* tension) between two norms which can't be fulfilled, but it's not clear that this tension should amount to a contradiction. This point of view favojrs the idea which gives rise to the paradox: a conflict of obligations is not as strong as a logical contradiction.

But let's consider how VSL deals with this sort of conflict. The traslation of (1) and (2) would be:

(3) !δA

(4) !δ¬A

Where A is any theoretical formula of VSL. There should be no doubt that the normative conflict between (3) and (4), although stated in terms of imperatives, is completely analogous to that of (1) and (2). Does VSL allow a logical contradiction from (3) and (4)? The short answer is no, but the details are very interesting: although (3) and (4) don't result in a logical contradiction, they can't be both normatively valid in the same world; also, the system allows us to see that both commands can't be simultaneously fulfilled, so the system tells us something about the context of validity and also about the context of satisfaction of these commands in VSL.

Firstly, these formulas don't result in contradiction because these are imperatives, practical formulas which have no relevance in the factic description of any world of a VSL-model, this task is left to theoretical formulas. Axioms and rules of VSL back up this, since none of them is analogous to the axiom of SDL which gives rise to the paradox: Op$\rightarrow$ ¬O¬p.

Secondly, one consequence of rendering both formulas normatively valid is a trivialization of the norm-giving activity of authority. This is due to semantic condition for command sets C1. Given (C1), if both (3) and (4) belong to a command set, the following should also belong to that command set:

(5) !δ(A$\wedge$¬A)

This would amount to command an impossibility, just as in SDL Op and O¬p would amount to commanding an impossibility. This can happen in VSL according to its semantic conditions, not only can a command set be empty, but also the empty set could belong to a command set. However, the problem doesn't escalate to a logical contradiction, it just makes the command system of the model useless, as Segerberg proves in his Proposition 4.1 [4, p. 210]. This proposition says that, if $\emptyset \in \Gamma_x$, then $\Gamma_x = \mathbf{A}$ and if $<x,y> \in R$, for any $R \in \mathbf{A}$, then $\Gamma_y = \mathbf{A}$. That is, it deems the command set, and all the command sets with which it is related through an action, equal to the action set of the model, so everything would be normatively validated. But this is different from having a propositional contradiction, for this trivialization of norm validity would not make every theoretical formula a theorem. The system is at least able to contain the explosivity of normative contradiction within the domain of normative reasoning, without affecting the descriptive or propositional part of the system.

This view is in consonance with the intuition that there is some tension between (3) and (4) which should neither scalate to contradiction, nor should it be normatively or rationally indifferent for authority to command contradicting things.

Lastly, we notice that commands (3) and (4) could not possibly be *fulfilled* simultaneously. Because of the way the δ-operator is defined, there is no way to execute both δA and $\delta \neg A$, for it would require a world where A and $\neg A$ were true.[19] Norm-addressees would certainly backup the idea that fulfilling both commands would amount to a logical contradiction.

3.3 Chisholm's paradox

This paradox can't be solved as straightforwardly as the last two and that is why we chose it for exposition, to show some of the drawbacks of VSL and the distinction in question. Notwithstanding, some aspects of the paradox are very interesting under this light.

This paradox is usually taken to reveal SDL's lack of capacity to represent normative conditionals and also its questionable capacity to deal with normatively unacceptable worlds, where a logic of norms should nevertheless hold.[20] As we will see, VSL's modal-dynamic symbols add some expressive capacity in those areas, but it remains questionable how succesful it is in solving this puzzle since it also requires an additional semantic condition and an additional axiom to represent some important intuitions about this sentences.

Consider the following statements:[21]

(1) It ought to be that John goes to the assistance of his neighbours.

(2) It ought to be that if John goes to the assistance of his neighbours, then he tells them he is coming.

---

[19] But this leaves the possibility to fulfill both commands sequentially, first getting to a world where A is the case and then to one where it is not the case or viceversa: ($\delta A; \delta \neg A$) or ($\delta \neg A; \delta A$).

[20] This already hints to the relevance of distinguishing between validity and other semantic values, it leads to evaluate separately which norms have been violated or fulfilled and which norms are nevertheless valid or invalid.

[21] The formulation is taken from [1, p. 83]

(3) If John doesn't go to the assistance of his neighbours, then he ought not tell them he is coming.

(4) John does not go to their assistance.

The problem arises when different formalizations of this four sentences clash with at least one of the following intuitions about them: That (1)-(4) are mutually consistent and also logically independent.

The formalizations differ on the scope of the normative operator in sentences (2) and (3), precisely the sentences with conditionals involved. They can be symbolized with a wide scope (as in 'O(p→q)') or with a narrow scope (as in 'p→Oq'). Either way, and no matter which of the conditionals is symbolized in these different ways, the set of sentences will be either inconsistent or there will be logical dependency among some of the sentences.

Let's see how VSL could symbolize this set of sentences without using the classical conditional:

(5) !δA

(6) [δA]!δD

(7) [δ¬A]!δ¬D

(8) ¬A

The modal-dynamic ingredient added in VSL allows us to represent conditionals (2) and (3) in terms the worlds or situations where doing A or not doing A would take us (whether John fulfills or doesn't fulfill his obligation stated in (1)). This is one of the two important aspects where SDL fails, in its capacity to deal with less-than-ideal worlds, meanwhile VSL can tell us which command is valid in any of the two cases, even considering that (5) validates that δA is the normatively right way to go.

From the point of view of authority, this sentences are covering what authority wills to be done, whether the addressee decides to assist his neighbors or not: the commands that would hold as valid in any of the two cases are already foreseen by the command sets representing the will of authority. From the point of view of John, the norm-addressee, it is also clear which actions would satisfy which commands and the consequences of his choices.

Regarding the alleged consistency and independence of this sentences, both seem to be respected. It is to be noted that this particular modeling in VSL allows an inference when considering axiom AI4 [22]:

(AI4) !α → ([α]!β →!(α;β))

The inference seems harmless, since all we get is !(δA;δD). But it does reveal that the modal-dynamic brackets may not be rescuing the conditional spirit of the advice given in (6), for the result would read something like this: 'Assist your neighbors! Then tell them you are coming!', which is certainly bizarre since it shouldn't be a matter of timing or order of execution, it should

---

[22] [4, p. 206]

rather be a matter of a command being valid whenever some action should be taken. But (6) describes what happens when the action is taken, not when it is normatively valid; therefore, that couldn't be represent either, since the rules for well-formed expression only allow terms to be between modal-dynamic brackets, not imperatives or formulas of any kind.

We could stick to the traditional implication, leaving (5) and (8) as they are and changing only (6) and (7):

(6') $A \to !\delta D$

(7') $\neg A \to !\delta \neg D$

We would be following the narrow scope formalization, but then we could derive (6') from (8), using propositional logic and thus would be rendering the sentences not logically independent.

Another observation is that, if we stick to the formalization with brackets, there's yet another problem regarding (7) and (8). That is, it should be clear that from the original sentences (3) and (4) follows that John ought not tell his neighbors that he is coming, which is precisely the consequent in (3). But from (7) and (8) VSL doesn't allow us to infer the validity of '!$\delta \neg D$'. For this, an additional action axiom and an additional semantic condition for the function **D** would be needed:

(AA5) $A \to ([\delta A]B \to B)$

(FD3) For any $u \in U$ and $X \in P$, if $u \in X$, then $<u, u> \in DX$

The axiom means that if you are in a world where A is the case and doing A always takes you to a situation where B is the case, then you are already in a situation where B is the case. [23]. The condition FD3 reflects this on a semantic level by saying that any world where a proposition X holds should always be included as one of the possible transitions that leads to X. [24]

Further observations could be made, but the goal of showing the limitations of this distinction between authority and norm-addressees to solve this puzzle has already been met.

4 Counterarguments and further questions

Valdity as utterance An interesting argumentation line against the very idea of a logic of imperative validity is that, in trying to explain what it means to logically infer one imperative from other, we may be implying that the *existence* of an imperative somehow implies the *existence* of another imperative. Hansen cites [2, p. 153] many examples of different logicians warning us against the idea that a logic of imperative may be understood as the logic of the *existence*

---

[23] This may lead to say that whenever something is the case you may assume some action lead to it, which is rather questionable since actions are usually thought of in relation to some agent. It would be a matter of discussion how this affects the characterization of action in this system

[24] This is also an odd condition, since it is doubtful that any instance of $<u,u>$ should really count as a *transition* between situations.

of a command, its utterance, the action of stating or similar definitions. The warning is certainly helpful, but it hardly undermines the very idea of a logic of norm validity. Let us take the example from Aleksander Peczenik:

> The premiss 'love your neighbour' may be regarded as describing the fact that the authority – Jesus – has in fact said 'love your neighbor'. The imperative existed because it was uttered by Jesus. But the conclusion, for example, 'love Mr. X' does not describe anything which in fact has been said by Jesus. [25]

In this sense, to say that an imperative exists would mean something like the 'stating' or 'uttering' of a certain command in the right context. But why should we define norm validity like that? In VSL, for example, there are conditions to ensure some kind of rationality for command sets, the ones listed (C0-4). Condition C1, for example, says that if two different commands belong to the command set, then the **D** function of the intersection of both propositions also belong to the set, thus rendering it normatively valid. That the set of commands issued by an authority may be closed under logical consequence doesn't mean that authority is somehow 'silently uttering' commands, but rather that authority is committed to the validity of implicit norms, just as asserting propositions may committ a speaker to the implicit truths.

In general, we may take this kind of objection as a healthy warning against the identification between the kind of actions that may render some norm as valid (its utterance, for example), and the validity itself, which may be acquired in a variety of ways (by being written in some particular place, being uttered by some person, being the result of a certain social convention, being performed by a number of individuals through a long period of time, being implied by some other valid norm, etc.). Taking the warning seriously doesn't require to throw away altogether the concept of a logic of norm validity.

Imperatives and deontic concepts A more serious objection may be that it is doubtful that a logic of imperatives can contain a logic of deontic concepts without loss. Even by defining the obligation operator O exactly as the imperative ! in VSL, it is doubtful that the specific logic for the concept of permission would be adequately represented. This would affect the scope of the present arguments, so that we would only be arguing for the relevance of the distinction between authority and norm-addressees regarding the logic of imperatives and not for the logic of norms in general.

Moreover, an interesting problem arises when trying to define the concept of permission in terms of the distinction between validity and satisfaction. When the propositional content of an obligation is true, it's satisfied; when it's false, it's violated. If I have an obligation to pay taxes, then whenever it is true that I pay them, the obligation is certainly satisfied. The same may hold for imperatives, but it doesn't hold for permission, for if I have a permission to drive a car, were it true that I drive one doesn't satisfy it in any meaningful

---

[25] In a letter by Peczenik found in [6], cited in [2, p. 152]

sense and were it false, it wouldn't be in violation of the permission.

What, then, is to satisfy a permission? A possible answer may lie in comparing permissions to rights, in the sense that they are duties or *obligations* for authorities (usually norm-givers), thus inverting the directionality of responsibilty for norm satisfaction from norm-adreessees to norm-givers or authorities.

5 Conclusions

Although Chisholm's Paradox couldn't be solved with the aid of the distinction between authority and norm-addressees, the other two paradoxes were not only overcome, but it is clearer why they are considered as paradoxes to begin with: it depends on the semantic values we are assuming to evaluate the soundness of each result, either validity or satisfaction. This is where we intend to make a contribution, in showing that not only the Ross paradox calls for an analysis which makes a clear semantic distinction between norm validity and satisfacion, but other classical paradoxes could also be better understood and solved through this approach.

We also wish to address how the semantic distinction in question is also related to the points of view of authority and norm-addressees, suggesting that their relevance to a standard logic of norms may be more important than usually regarded. The system VSL may be too complex to be considered basic to the logic of norms, but its capacity to solve paradoxes calls for a detailed analysis of the aspects of this complexity that should be preserved in order to define a truly standard logic for normative reasoning.

The suggestion to add axiom (AA5) and semantic condition (FD3) is a small formal contribution to VSL. Hopefully, it adds to the suggestions of possible ways to make more positive contributions to the enrichment of systems of normative logic seeking resources from dynamic logic to solve the problems of conditionals in normative contexts and the logic of norms in less-than-ideal situations.

The correct interpretation of deontic formulas in standard systems such as SDL is problematic, but a refinement in the notion of norm satisfaction may help reivindicate the line of interpretation which leans towards *satisfaction* for this systems. The refinement here proposed would be the one we called 'normative-modal satisfaction' in Section 3.1: the fulfillment of a norm in a specific set of possible worlds and not just its truth value in the actual world.

The problem of defining a reasonable notion of satisfaction for permission in terms of inverting the directionality of responsibility among authority and norm-addressees was also brought to light through this view. It could also be argued that the notion of satisfaction can't make sense at all regarding the concept of permission, but it would be hard to explain the relevance of permissions in evaluating the overall fulfillment of a normative system which includes permissions.

References

[1] Gabbay, D., J. Horty, X. Parent, R. van der Meyden and L. van der Torre, *Handbook of deontic logic and normative systems* (2013).
[2] Hansen, J., "Imperative logic and its problems." College Publications, 2013 pp. 137–191.
[3] Segerberg, K., *Bringing it about*, Journal of philosophical logic (1989), pp. 327–347.
[4] Segerberg, K., *Validity and satisfaction in imperative logic.*, Notre Dame journal of formal logic **31** (1990), pp. 203–221.
[5] Stolpe, A., *Normative consequence: The problem of keeping it whilst giving it up*, in: International Conference on Deontic Logic in Computer Science, Springer, 2008, pp. 174–188.
[6] Walter, R., *Some thoughts on peczenik's replies to jrgensen's dilemma and how to face it (with two letters by a. peczenik)*, Ratio Juris **10** (1997), p. 392.

The Gentle Murder Paradox in Sanskrit Philosophy

Kees van Berkel[*,1]   Agata Ciabattoni[*,1]   Elisa Freschi[**,2]
Francesca Gulisano[*,3]   Maya Olszewski[*,1]

*Institute of Logic and Computation, TU Wien**
Vienna, Austria
*Department of Philosophy, University of Toronto**
Toronto, Canada

Abstract

For decades, the gentle murder paradox has been a central challenge for deontic logic. This article investigates its millennia-old counterpart in Sanskrit philosophy: the *śyena* controversy. We analyze three solutions provided by Mīmāṃsā, the Sanskrit philosophical school devoted to the analysis of normative reasoning in the Vedas, in which the controversy originated. We introduce axiomatizations and semantics for the modal logics formalizing the deontic theories of the main Mīmāṃsā philosophers Prabhākara, Kumārila, and Maṇḍana. The resulting logics are used to analyze their distinct solutions to the *śyena* controversy, which we compare with formal approaches developed within the contemporary field of deontic logic.

Keywords: Mīmāṃsā, Dyadic Deontic Logic, Instruments, Gentle Murder Paradox

---

1 Introduction

Introduced by Forrester [9], the *Gentle Murder Paradox* (GMP) is a well-known problem for monadic deontic logic [13,28], motivating the use of alternative systems employing dyadic deontic operators, e.g., [16,21]. The GMP in a nutshell: (i) x is obliged not to kill, (ii) if x kills, x is obliged to kill gently, (iii) gentle killing implies killing, and (iv) x will kill. Although intuitively consistent, the sentences (i)-(iv) lead to a contradiction in Standard Deontic Logic, implying x's obligation to kill. Originally, the GMP was introduced as a stronger Good Samaritan Paradox [24], but it is commonly taken as a variant of Chisholm's Paradox [6]. Under the former reading, (i)-(iv) imply conflicting obligations (i.e., a dilemma), inconsistent under normality of deontic operators. Under the latter reading, the GMP relates to challenges of reasoning with violations and

---

[1] {kees, agata, maya}@logic.at
[2] elisa.freschi@utoronto.ca
[3] f.gulisano2201@gmail.com

contrary-to-duty (CTD) obligations (i.e., the obligation (ii) is only in force if (i) is violated). In fact, the GMP has features of both paradoxes [17].

While the GMP was introduced to the deontic logic community only a few decades ago, a similar example has been thoroughly investigated in Sanskrit philosophy for more than two millennia. This is the renowned *śyena* controversy. The *śyena* is a one-day long ritual in which the Soma beverage is offered. Its putative result is the death of the sacrificer's enemy. Unlike animal sacrifices it does not involve violence in its performance, violence is only found in its result. The controversy is due to the fact that the *śyena* appears to be prescribed in the Vedas —the sacred texts of what is now known as "Hinduism"—, which also prohibits the performance of violence. The *śyena* controversy in short [4]:

(A) The one who desires to kill their enemy should sacrifice with the *śyena*

(B) One should not harm any living being

(C) Performing *śyena* implies causing someone's death

(D) Causing someone's death implies harming

With (A)-(D), the Vedas seem to provide contradicting commands concerning the performance of violence, a possibility which is ruled out by the (indisputable) claim that the Vedas are consistent.

The Sanskrit philosophical school of Mīmāṃsā—which flourished between the last centuries BCE and the 20th c. CE—paid exceptional attention to the controversy, explaining why the *śyena* should not be performed and why the sacred texts prescribing it are not contradictory. In general, the Mīmāṃsā school focused on the rational interpretation and systematization of the prescriptive portions of the Vedas. To reason with Vedic commands, and resolve seeming conflicts, the Mīmāṃsā developed a vast system of theories containing rigorous analyses of deontic concepts. Key to their enterprise was the formulation of general reasoning principles called *nyāya*s, and the distinction among elective duties (to be performed only if one wishes their specific result), fixed duties (to be performed no matter what), and prohibitions. The resulting theories, which have been extremely influential in Sanskrit philosophy, theology and law, provide an inexhaustible resource for deontic investigation, largely still unexplored.

Although all Mīmāṃsā authors agree that *śyena* should not be performed, they disagree on the reasons underlying it. In this article, we focus on the three main Mīmāṃsā authors: Kumārila, Prabhākara (both ca. 7th c. CE), and Maṇḍana (ca. 8th c. CE). They are known for their distinctive deontic theories, which give rise to different interpretations of Vedic commands. Likewise, their solutions to the *śyena* controversy are markedly distinct.

We provide three modal logics [5] describing the deontic theories of these

---

[4] (A) and (B) are direct translations from Sanskrit, whereas (C) and (D) are derived from Mīmāṃsā arguments about the *śyena*.

[5] The logics in this paper are intended to reason about commands as interpreted by the Mīmāṃsā. Since the Vedas are self-contained and immutable, new Vedic commands cannot

authors, whose rational, structured approach makes their accounts particularly suitable for formalization. The resulting logics are obtained by "extracting" Hilbert axioms out of translated and parsed Mīmāṃsā *nyāya*s and additional passages by the three authors. While the logics for Prabhākara and Kumārila are a modification of those presented in [7] and [20], Maṇḍana's logic is novel.

The main contributions of this paper are threefold: First, we develop a logic formalizing Maṇḍana's deontic theory. His account is particularly noteworthy due to its *deontic reduction*: the reduction of all commands of the Vedas (i.e., fixed and elective obligations, as well as prohibitions) to mere descriptive statements of instrumentality. For instance, according to Maṇḍana, an obligation to perform an action means the action is an instrument for attaining a certain result. The introduced logic reproduces Maṇḍana's reduction by adopting a PDL-like language [8,23], together with a modified Andersonean reduction of deontic modalities [1]. Second, we offer a consistent formalization of the *śyena* controversy as interpreted by Kumārila and Maṇḍana, faithful to the explanations found in Mīmāṃsā texts. Kumārila's formalization is achieved by introducing a neighbourhood semantics for its logic. Prabhākara's solution was formally analyzed in [7], however, the logic presented there contained only obligations. In [12,20] it was shown that obligations and prohibitions in Mīmāṃsā are not inter-definable and, hence, we extend Prabhākara's logic (and solution) with a prohibition operator. Third, we analyze and compare the three formal solutions to the *śyena* controversy and discuss their relations to approaches in contemporary deontic logic. In particular, the dual reading of the GMP is reflected in the different approaches to the *śyena* controversy: As for the Chisholm paradox, Prabhākara takes the *śyena* prescription as a contrary-to-duty obligation. Kumārila addresses the controversy by interpreting *śyena* as an elective sacrifice, to which he assigns no deontic force. As for the Good Samaritan Paradox, Maṇḍana endorses the view that there is a proper dilemma in the controversy, but addresses it through his reduction, arguing for a pragmatic rational-choice solution based on a cost-benefit analysis of (un)desirable outcomes.

Our work is the first study of the *śyena* controversy in Mīmāṃsā, the school in which the controversy originated. The interest in this controversy from the point of view of modern deontic logic is also testified by the recent work [14], where the *śyena* is analyzed from the perspective of the Navya-Nyāya, a different school of Sanskrit philosophy.[6]

2 Prabhākara and Kumārila

Prabhākara and Kumārila interpret Vedic prescriptions as proper commands with deontic force. Despite their shared view on fixed duties (to be read as obligations) and on prohibitions, Prabhākara and Kumārila disagree on the

---

be derived through Logic. Accordingly, our logics deal with commands on the derived level.

[6] The work [14], also relating the *śyena* to the GMP, was published while the present paper was under review.

reading of elective sacrifices, which are always conditioned on a desire.

Prabhākara's system is eminently deontic: Vedic statements are binding, independently from their conditions; hence, an elective sacrifice is also a type of obligation. The desire for a specific worldly result, necessarily mentioned as the condition of an elective ritual, only represents the requirement through which the eligible agents are identified, but it does not weaken the deontic force of the injunction. By contrast, for Kumārila, elective sacrifices are of a different type, not enjoining any deontic force, and can be omitted without risk. Still, an eligible agent—i.e., an agent who desires the expected result of the sacrifice—feels prompted to undertake the sacrifice due to its presence in the Vedas: such sacrifices represent a "guaranteed" method for obtaining the desired results. Hence, whereas Prabhākara sees elective sacrifices as conditional obligations, Kumārila sees them as a different type of Vedic command.

The two logics presented in this section will reflect this distinction. Since their only difference is the presence of elective sacrifices as a distinct deontic concept, the logic for Prabhākara will be a proper subset of Kumārila's. However, the distinction causes wholly different solutions to the *śyena* controversy. The logics are variants of the formalism introduced in [20], whose properties were extracted from a collection of general Mīmāṃsā reasoning principles (*nyāya*s, see Sect. 1). By adding the deontic operators for prohibitions $\mathcal{F}(\cdot/\cdot)$ and for injunctions prescribing elective duties $\mathcal{E}(\cdot/\cdot)$, the resulting logics extend the non-normal dyadic deontic logic bMDL. Introduced in [7] to formalize the deontic theory of Prabhākara, bMDL only contained a single deontic operator $\mathcal{O}(\cdot/\cdot)$ for obligations.

2.1 Deontic logics for Kumārila and Prabhākara

The languages $\mathcal{L}_{\mathsf{LPr}}$ for Prabhākara and $\mathcal{L}_{\mathsf{LKu}}$ for Kumārila are defined through the following BNF (with $\mathcal{X} \in \{\mathcal{O}, \mathcal{F}\}$ for $\mathcal{L}_{\mathsf{LPr}}$ and $\mathcal{X} \in \{\mathcal{O}, \mathcal{F}, \mathcal{E}\}$ for $\mathcal{L}_{\mathsf{LKu}}$):

$$\varphi ::= p \mid \neg\varphi \mid \varphi \vee \varphi \mid \boxed{\mathsf{u}}\varphi \mid \mathcal{X}(\varphi/\varphi) \qquad \text{with } p \in \mathsf{Atom}$$

Atom is the set of atomic propositions, $\neg$ and $\vee$ are primitive connectives, the others defined as usual. $\boxed{\mathsf{u}}\varphi$ reads "it is universally necessary that φ". The operators $\mathcal{O}(\varphi/\psi)/\mathcal{F}(\varphi/\psi)/\mathcal{E}(\varphi/\psi)$ read as "φ is obligatory/forbidden/enjoined by an injunction prescribing an elective ritual, given ψ".

Axiomatization. The properties of the operators $\mathcal{O}, \mathcal{F}$, and $\mathcal{E}$ are extracted from the following Mīmāṃsā principles (see [11] for details on how these principles were transformed into axioms for $\mathcal{O}$, and [20] for the remaining axioms):

(P1) "If the accomplishment of a task presupposes the accomplishment of another connected but different task, the obligation to perform the first task prescribes also the second one".

(P2) "Two actions that exclude each other cannot be prescribed simultaneously to the same group of eligible people under the same conditions".

(P3) "If two sets of conditions always identify the same group of eligible agents, then a command valid under the conditions in one of those sets is also enforceable under the conditions in the other set".

The two logics are described in Definition 2.1. In contrast with bMDL [7], we use S5 to characterize necessity $\boxed{U}$, instead of S4: Note that the concept of necessity is not explicitly defined by Mīmāṃsā authors in the context of deontic reasoning, and the choice of S4 in [7] was motivated by the simpler proof theory of this logic, with respect to S5. In this paper we use necessary statements mainly as global assumptions (assertions commonly recognised as describing "facts"); hence, any assumption defines an equivalence class of states sharing the same truths.

Using the corresponding universal modality of S5 makes the bMDL axiom $\Box((\psi \to \theta) \land (\theta \to \psi)) \land \mathcal{O}(\varphi/\psi) \to \mathcal{O}(\varphi/\theta)$ redundant (it is derivable by using axiom T and the congruence rule of S5), also in the versions for $\mathcal{E}$ and $\mathcal{F}$.

Definition 2.1 Prabhākara's logic LPr extends S5 with the following axioms:

$A_{\mathsf{LKu}}1.$ $(\boxed{U}(\varphi \to \psi) \land \mathcal{O}(\varphi/\theta)) \to \mathcal{O}(\psi/\theta)$

$A_{\mathsf{LKu}}2.$ $(\boxed{U}(\varphi \to \psi) \land \mathcal{F}(\psi/\theta)) \to \mathcal{F}(\varphi/\theta)$

$A_{\mathsf{LKu}}3.$ $\boxed{U}(\psi \to \neg\varphi) \to \neg(\mathcal{X}(\varphi/\theta) \land \mathcal{X}(\psi/\theta))$ for $\mathcal{X} \in \{\mathcal{O}, \mathcal{F}\}$

$A_{\mathsf{LKu}}4.$ $\boxed{U}(\varphi \to \psi) \to \neg(\mathcal{O}(\varphi/\theta) \land \mathcal{F}(\psi/\theta))$

$A_{\mathsf{LKu}}5.$ $(\boxed{U}((\psi \to \theta) \land (\theta \to \psi)) \land \mathcal{X}(\varphi/\psi)) \to \mathcal{X}(\varphi/\theta)$ for $\mathcal{X} \in \{\mathcal{O}, \mathcal{F}\}$

Kumārila's deontic logic LKu, extends LPr with the following axioms:

$A_{\mathsf{LKu}}6.$ $(\boxed{U}(\varphi \to \psi) \land \mathcal{E}(\varphi/\theta)) \to \mathcal{E}(\psi/\theta)$

$A_{\mathsf{LKu}}7.$ $\boxed{U}(\neg\varphi) \to \neg\mathcal{E}(\varphi/\psi)$

$A_{\mathsf{LKu}}8.$ $(\boxed{U}((\psi \to \theta) \land (\theta \to \psi)) \land \mathcal{E}(\varphi/\psi)) \to \mathcal{E}(\varphi/\theta)$

A derivation of φ in LKu (i.e., $\vdash_{\mathsf{LKu}} \varphi$) is defined as usual [3] (similarly for LPr).

Axioms $A_{\mathsf{LKu}}1$, $A_{\mathsf{LKu}}2$, $A_{\mathsf{LKu}}6$ are based on (P1) and correspond to the property of monotonicity. Axioms $A_{\mathsf{LKu}}3$, $A_{\mathsf{LKu}}4$ formally represent (P2) (found in Kumārila's *Tantravārtika* ad 1.3.3 [27]). Last, the Mīmāṃsā property (P3) is ensured by $A_{\mathsf{LKu}}5$, $A_{\mathsf{LKu}}8$.

Semantics We present a neighbourhood semantics (e.g., see [5]) for LPr and LKu (resp.), as defined along the lines of the one for bMDL in [7]:

Definition 2.2 An LPr-frame $\mathfrak{F}_{\mathsf{LPr}} = \langle W, R_{\boxed{U}}, \mathcal{N}_\mathcal{O}, \mathcal{N}_\mathcal{F} \rangle$ is a tuple where $W \neq \emptyset$ is a set of worlds, $R_{\boxed{U}} = W \times W$ is the universal relation, and $\mathcal{N}_\mathcal{X} : W \mapsto \wp(\wp(W) \times \wp(W))$ is a neighborhood function (for $\mathcal{X} \in \{\mathcal{O}, \mathcal{F}\}$). $\mathfrak{F}_{\mathsf{LPr}}$ satisfies:

(i) If $(X, Z) \in \mathcal{N}_\mathcal{O}(w)$ and $X \subseteq Y$, then $(Y, Z) \in \mathcal{N}_\mathcal{O}(w)$;

(ii) If $(X, Y) \in \mathcal{N}_\mathcal{X}(w)$, then $(\overline{X}, Y) \notin \mathcal{N}_\mathcal{X}(w)$ for $\mathcal{X} \in \{\mathcal{O}, \mathcal{F}\}$;

(iii) If $(X, Z) \in \mathcal{N}_\mathcal{F}(w)$ and $Y \subseteq X$, then $(Y, Z) \in \mathcal{N}_\mathcal{F}(w)$;

(iv) It cannot be the case that $(X, Z) \in \mathcal{N}_\mathcal{O}(w)$ and $(X, Z) \in \mathcal{N}_\mathcal{F}(w)$.

An LPr-*model* $\mathfrak{M}_{\mathsf{LPr}} = \langle W, R_{\boxed{U}}, \mathcal{N}_\mathcal{O}, \mathcal{N}_\mathcal{F}, V \rangle$ extends the LPr-frame by a *valuation* function V which maps propositional variables to subsets of W.

Definition 2.3 An LKu-frame $\mathfrak{F}_{\mathsf{LKu}} = \langle W, R_{\boxed{U}}, \mathcal{N}_\mathcal{O}, \mathcal{N}_\mathcal{F}, \mathcal{N}_\mathcal{E} \rangle$ is an LPr-frame extended with a neighbourhood function $\mathcal{N}_\mathcal{E} : W \mapsto \wp(\wp(W) \times \wp(W))$ s.t.:

(v) If $(X, Z) \in \mathcal{N}_\mathcal{E}(w)$ and $X \subseteq Y$, then $(Y, Z) \in \mathcal{N}_\mathcal{E}(w)$;
(vi) If $(X, Y) \in \mathcal{N}_\mathcal{E}(w)$, then $X \neq \emptyset$.

An LKu-*model* $\mathfrak{M}_{\mathsf{LKu}} = \langle \mathfrak{F}_{\mathsf{LKu}}, V \rangle$ is an LKu-frame with a valuation function V.

Note that (i), (iii), (v) express the property of monotonicity in the first argument of the deontic operators (cf. $\mathsf{A}_{\mathsf{LKu}}1$, $\mathsf{A}_{\mathsf{LKu}}2$, $\mathsf{A}_{\mathsf{LKu}}5$); (ii), (iv) correspond to the principle (P2) (cf. $\mathsf{A}_{\mathsf{LKu}}3$, $\mathsf{A}_{\mathsf{LKu}}4$), and (vi) expresses the self consistency of statements prescribing elective sacrifices (cf. $\mathsf{A}_{\mathsf{LKu}}6$).

Definition 2.4 Let $\mathfrak{M}_{\mathsf{LKu}}$ be an LKu-model and $\|\theta\| = \{w \in W \mid \mathfrak{M}_{\mathsf{LPr}}, w \models \theta\}$. We define the *satisfaction* of a formula $\varphi \in \mathcal{L}_{\mathsf{LKu}}$ at any w of $\mathfrak{M}_{\mathsf{LKu}}$ inductively:

$\mathfrak{M}_{\mathsf{LKu}}, w \models p$ iff $w \in V_{\mathsf{LPr}}(p)$, for any $p \in$ Atom
$\mathfrak{M}_{\mathsf{LKu}}, w \models \varphi \to \psi$ iff $\mathfrak{M}_{\mathsf{LKu}}, w \not\models \varphi$ or $\mathfrak{M}_{\mathsf{LKu}}, w \models \psi$
$\mathfrak{M}_{\mathsf{LKu}}, w \models \neg \varphi$ iff $\mathfrak{M}_{\mathsf{LKu}}, w \not\models \varphi$
$\mathfrak{M}_{\mathsf{LKu}}, w \models \boxdot \varphi$ iff   for all $w_i \in W$ s.t. $(w, w_i) \in R_{\boxdot}$, $\mathfrak{M}_{\mathsf{LKu}}, w_i \models \varphi$
$\mathfrak{M}_{\mathsf{LPr}}, w \models \mathcal{X}(\varphi/\psi)$ iff $(\|\varphi\|, \|\psi\|) \in \mathcal{N}_\mathcal{X}(w)$ for $\mathcal{X} \in \{\mathcal{O}, \mathcal{F}, \mathcal{E}\}$

Global truth and validity are defined as usual [3]. Note that *satisfaction* for $\mathfrak{M}_{\mathsf{LPr}}$-models is defined as for $\mathfrak{M}_{\mathsf{LKu}}$, without the clause for $\mathcal{N}_\mathcal{E}(w)$.

Theorem 2.5 (Soundness and completeness) *The logic* LKu *(*LPr*) is sound and complete with respect to the class of* LKu-*frames (*LPr $-$ *frames).*

Soundness and completeness are proven as usual. The latter is shown using the method of canonical models [5], generalized to the dyadic setting.

2.2 The solutions of Prabhākara and Kumārila

The sentences (A)-(D) comprising the *śyena* controversy (Sect. 1) are formalized in a similar way by the two authors. The only difference is their interpretation of (A), prescribing the *śyena* sacrifice: for Prabhākara this is a conditional obligation (A_P), whereas Kumārila interprets it as an elective sacrifice (A_K). Their formalization:

(A_P) $\mathcal{O}(\text{Śy}/\texttt{des\_kill})$ (B) $\mathcal{F}(\text{harm}/\top)$
     (C) $\boxdot(\text{Śy} \to \text{death})$
(A_K) $\mathcal{E}(\text{Śy}/\texttt{des\_kill})$ (D) $\boxdot(\text{death} \to \text{harm})$

Fig. 1 shows the models $\mathfrak{M}_P$ and $\mathfrak{M}_K$ demonstrating the mutual satisfiability of (A_P), (B), (C), (D) in LPr and of (A_K), (B), (C), (D) in LKu, respectively, and hence the consistency of the *śyena* controversy for both authors. That is, there is always at least one world in which no command is violated. (A command $\mathcal{O}(\phi/\psi)$ or $\mathcal{E}(\phi/\psi)$ is violated if ψ is satisfied, but ϕ is not. $\mathcal{F}(\phi/\psi)$ is violated when both ϕ and ψ are satisfied.)

The models $\mathfrak{M}_P$ and $\mathfrak{M}_K$, satisfying Def. 2.2 and 2.3, are defined as: $W^P = W^K = \{w_i \mid 1 \leq i \leq 8\}$ s.t. $\|\text{harm}\| = V(\text{harm}) = \{w_2, w_3, w_4, w_6, w_7, w_8\}$, $\|\text{kill}\| = V(\text{kill}) = \{w_2, w_3, w_6, w_7\}$, $\|\text{Śy}\| = V(\text{Śy}) = \{w_4, w_8\}$, $\|\texttt{des\_kill}\| = V(\texttt{des\_kill}) = \{w_5, w_6, w_7, w_8\}$ (with $V^P = V^K = V$), $\mathcal{N}^P_\mathcal{F}(w_i) = \mathcal{N}^K_\mathcal{F}(w_i) = \{(X, Y) \mid X \subseteq \{w_2, w_3, w_4, w_6, w_7, w_8\}, Y = W\}$, $\mathcal{N}^P_\mathcal{O}(w_i) = \mathcal{N}^K_\mathcal{E}(w_i) = \{(V, Z) \mid \{w_2, w_6\} \subseteq V, Z = \{w_5, w_6, w_7, w_8\}\}$ and $\mathcal{N}^K_\mathcal{O}(w_i) = \emptyset$.

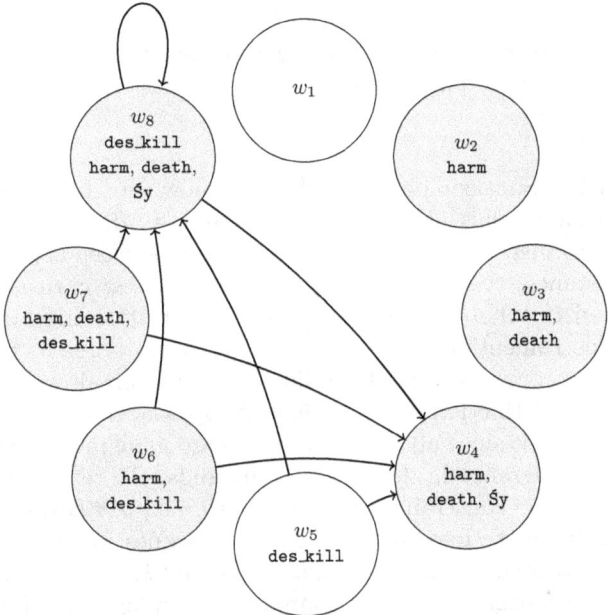

Fig. 1. depicts the models $\mathfrak{M}_P$ and $\mathfrak{M}_K$ satisfying the Śyena controversy (A_i)-(D) $(i \in \{P, K\})$: for $(\|\text{harm}\|, \|\top\|) \in \mathcal{N}_{\mathcal{F}}^i$ ($i \in \{P, K\}$). The worlds $w_i \in \|\text{harm}\|$ are coloured grey and for $(\|\text{Śy}\|, \|\text{des\_kill}\|) \in \mathcal{N}_{\mathcal{E}}^P(w_i) = \mathcal{N}_{\mathcal{O}}^K(w_i)$ (expressing A_P and A_K, resp.) the elements are indicated by arrows from each $w_i \in \|\text{des\_kill}\|$ to each $w_j \in \|\text{Śy}\|$. For Prabhākara w_1 is the only deontically acceptable world, while Kumārila also accepts w_5, as (A_K) has no deontic force.

From Kumārila's perspective, all the worlds that are not coloured grey—i.e., worlds where the prohibition (B) is not violated—are deontically acceptable, namely, no command with deontic force is violated. Kumārila's answer relies on the distinction between obligations and statements prescribing elective sacrifices, which are mutually independent: i.e., in case they conflict with a prohibition, elective sacrifices can be omitted without risk, thus avoiding to violate the prohibition. In contrast, in Prabhākara's logic the two neighbourhoods associated with (A_P) and (B) are not independent: i.e., condition (iv) of Def. 2.2 excludes the possibility that the same neighbourhood of a world represents both a prohibition and an obligation. However, since the eligibility conditions of the two commands do not exactly coincide, there is at least one world—i.e., w_1—in which no command is violated. That is, w_1 is the only deontically acceptable world from Prabhākara's point of view, the state in which one does not desire to kill one's enemy. Since desires are interpreted by Prabhākara as irreversible decisions—i.e., for Prabhākara the desire to kill amounts to a decision to kill—his solution is a case of CTD reasoning: the injunction to perform the *śyena* represents an obligation taking effect when a violation (the decision to cause a death) has occurred.

Following [7], the model $\mathfrak{M}_P$ also explains Prabhākara's claim that the Vedas do not impel one to perform the malevolent sacrifice *śyena*, they only say that it is obligatory. This claim that a Vedic obligation does not necessarily impel was wrongly considered meaningless, for instance, in [25].

3 Maṇḍana

Maṇḍana is a key Sanskrit philosopher, also known for his revolutionary approach to deontic modals. He reduces commands to descriptions of states of affairs, that is, to *instrumentality* relations holding between actions and desired results. For instance, "you are obliged to perform the *kārīri* ritual, if you desire rain" is reduced to "the *kārīri* is an instrument for attaining rain". Presently we interpret 'instrument' as an action sufficient to guarantee its result. Despite his revolutionary approach, Maṇḍana did not wish to break with the Mīmāṃsā tradition and its distinction among different types of duties. Still, his reduction may suggest that since all command-types are mere instrument statements there is also no difference in degrees of commands. To retain the distinction, Maṇḍana adopts two constraints involving *pāpa*, i.e., *bad karma*.

First, to individuate fixed duties, Maṇḍana argues for the universal desirability of their coveted result: the *reduction of bad karma*. For Maṇḍana, the reduction of bad karma is a desire shared by every agent. The introduction of this fixed desire, preserves the distinction between obligations (instruments that reduce bad karma) and elective duties (instruments serving specific desires). Second, to ensure the prohibitive strength of actions leading to undesirable outcomes, Maṇḍana argues that prohibited actions are instruments to outcomes whose undesirability is incommensurably greater than any desirable result. This universally undesired result is the *accumulation of bad karma*.

As will be shown at the end of this section, Maṇḍana's solution to the *śyena* controversy centers on the rationality of the agents involved. No rational agent would desire the small benefit of performing *śyena* in exchange for its accessory negative result, the accumulation of bad karma.

Related work. As Maṇḍana reasons about actions and outcomes, a PDL-like language [8,23] seems adequate. Actually, a minimal action-language suffices: i.e., negation and combination. Hence, we base our logic on [2]: a basic PDL-like language reducing action-modalities to action constants. The formalism in [2] is aimed at representing Von Wright's theory of instrumentality and hence appears particularly suitable. An alternative approach may be BDI logics [22], due to its connection to means-end reasoning (cf. [18]). However, they do not accommodate the required distinction between actions and outcomes. (Due to the role of desires in Maṇḍana's account, BDI-like extensions of our logic will be reserved for future work.)

To reason about bad karma, we adopt and enhance an Andersonean reduction to deontic logic [1]: φ is obligatory iff $\neg\varphi$ necessarily implies a sanction. The reduction was adapted by Meyer [23] to the deontic action setting: action Δ is obligatory iff all performances of its complement $\overline{\Delta}$ lead to a violation. Similarly, Maṇḍana can be seen as a reductionist of deontic reasoning: every

Vedic command is an instrumentality statement about actions leading to states of affairs, sanctions, and rewards. In deontic logic, the use of positive constants was introduced by Kanger [19]: φ is obligatory iff in the good state φ holds. However Kanger's approach takes φ as a necessary condition for the 'good state' whereas Maṇḍana takes φ as sufficient condition for 'reducing bad karma'.

3.1 The logic LMa: Language, Axioms and Semantics

We introduce the normal modal logic LMa equipped with action constants and karma constants. LMa captures Maṇḍana's intended reduction of norms to claims of instrumentality. We start by introducing an algebra of action $\mathcal{L}_{\mathsf{Act}}$ and the logical language $\mathcal{L}_{\mathsf{LMa}}$ into which these actions will be translated. Presently, a single-agent setting suffices. Let Act be a set of atomic actions δ (such as 'opening the window'). The action language $\mathcal{L}_{\mathsf{Act}}$ is defined as

$$\Delta ::= \delta \mid \overline{\Delta} \mid \Delta \cup \Delta \qquad \text{with } \delta \in \mathsf{Act}$$

The operator $-$ denotes the complement of an action, whereas $\cup$ is read as a disjunctive action. We use uppercase Greek letters $\Delta, \Gamma, \ldots$ to denote arbitrary actions. We define $\Delta \cap \Gamma = \overline{\overline{\Delta} \cup \overline{\Gamma}}$ as the the joint performance of actions.

The language $\mathcal{L}_{\mathsf{LMa}}$ for Maṇḍana is defined through the following BNF:

$$\varphi ::= p \mid \mathsf{d}_\delta \mid \mathsf{P} \mid \mathsf{R} \mid \neg\varphi \mid \varphi \vee \varphi \mid \boxed{\mathsf{s}}\varphi \mid \boxed{\mathsf{u}}\varphi$$

with $p \in \mathsf{Atom}$ and $\mathsf{d}_\delta \in \mathsf{Wit}_{\mathsf{Act}}$, where Atom is the set of atomic propositions, and $\mathsf{Wit}_{\mathsf{Act}}$ the set of atomic *constants* called 'action-witnesses'. The constant d_δ is to be read as a witness stating that 'the action δ has just been successfully performed'.[7] P is a constant reading 'bad karma is accumulated' and the constant R reads 'bad karma is reduced'. The unary operators $\boxed{\mathsf{s}}$ and $\boxed{\mathsf{u}}$ are interpreted as 'in all succeeding states it holds that' and 'it is universally necessary that', respectively. Their respective duals $\Diamond\!\!\!\!\!\!{\scriptstyle\mathsf{s}}$ and $\Diamond\!\!\!\!\!\!{\scriptstyle\mathsf{u}}$ are defined as usual.

The translation between $\mathcal{L}_{\mathsf{Act}}$ and action formulae in our object language $\mathcal{L}_{\mathsf{LMa}}$ is established through the following recursive definition:

- For all $\delta \in \mathsf{Act}$, $t(\delta) = \mathsf{d}_\delta$
- For all $\Delta \in \mathcal{L}_{\mathsf{Act}}$, $t(\overline{\Delta}) = \neg t(\Delta)$
- For all $\Delta, \Gamma \in \mathcal{L}_{\mathsf{Act}}$, $t(\Delta \cup \Gamma) = t(\Delta) \vee t(\Gamma)$

The upshot of the above translation is that it enables us to reason with actions on the object language level. The resulting versatility will prove useful in (i) defining a variety of modal operators (including instruments and commands) and (ii) axiomatizing action-properties. For instance, $\boxed{\mathsf{s}}(t(\Delta) \to \varphi)$ reads "at every successor state witnessing the successful performance of action Δ, the

---

[7] The logic LMa does not allow to keep track of action histories, only the last executed actions are known (cf. the presence of action witnesses). This is due to the absence of modalities referring to the past, which are not required in our present analysis of instruments.

state-of-affairs φ holds". When taken together with action, the modality $\boxed{\text{s}}$ can be seen as an indeterministic *execution operator*, in the spirit of propositional dynamic logic (PDL): "every successful execution of Δ, guarantees φ". See [2] for a discussion of this basic PDL-reductionist approach, and for a formal analysis of different notions of instrumentality.

Axiomatization. As for the previously introduced logics, also for Maṇḍana we want to avoid imposing any property that cannot be traced back to the Mīmāṃsā in general, and Maṇḍana in particular. For this reason, the proposed logic will be rather minimal. The Hilbert-style axiomatization of LMa is presented in Def. 3.1 (below) and justified accordingly: Both $\boxed{\text{u}}$ and $\boxed{\text{s}}$ are normal modal operators due to the S5 charactization of the former and $A_{\text{LMa}}1$ for the latter. $A_{\text{LMa}}2$ expresses a bridge axiom, stating that what holds universally, must also hold at any successor state. $A_{\text{LMa}}3$ conveys the Mīmāṃsā principle that whenever bad karma is attainable, it is also avoidable. (This principle is based on the Mīmāṃsā meta-rule according to which all commands need to be non-trivial and to prescribe something new, see [10].) $A_{\text{LMa}}4$ captures the same property for the reduction of bad karma, and $A_{\text{LMa}}5$ gives a central Mīmāṃsā principle: "if an action is executable, then it is executable in such a way that it does not trigger both the reduction and the increase of bad karma" ([29] ad 1.1.2), see Remark 3.9.

Definition 3.1 Maṇḍana logic LMa extends $\boxed{\text{u}}$-S5 with the following axioms:

$A_{\text{LMa}}1.\ \boxed{\text{s}}(\varphi \to \psi)$
$\quad\to (\boxed{\text{s}}\varphi \to \boxed{\text{s}}\psi)$

$A_{\text{LMa}}2.\ \boxed{\text{u}}\varphi \to \boxed{\text{s}}\varphi$

$A_{\text{LMa}}3.\ \Diamond P \to \Diamond \neg P$

$A_{\text{LMa}}4.\ \Diamond R \to \Diamond \neg R$

$A_{\text{LMa}}5.\ \Diamond t(\Delta) \to \Diamond(t(\Delta) \wedge (\neg P \vee \neg R))$

A derivation of $\varphi \in \mathcal{L}_{\text{LMa}}$ in LMa from a set $\Sigma \subseteq \mathcal{L}_{\text{LMa}}$ (i.e., $\Sigma \vdash_{\text{LMa}} \varphi$) is defined as usual [3]. When $\Sigma = \emptyset$, we say φ is an LMa-theorem, and write $\vdash_{\text{LMa}} \varphi$.

Semantics. We introduce a relational semantics for the logic LMa:

Definition 3.2 An LMa-frame $\mathfrak{F}_{\text{LMa}} = \langle W, \{W_\delta : \delta \in \text{Act}\}, W_P, W_R, R_{\boxed{\text{u}}}, R_{\boxed{\text{s}}} \rangle$ is a tuple with $W \neq \emptyset$ a set of worlds $w, v, u...$ etc. For every $d_\delta \in \text{Wit}_{\text{Act}}$, let $W_\delta \subseteq W$ be the set of worlds witnessing the successful performance of δ. Let $W_{\overline{\Delta}} = W \setminus W_\Delta$, and $W_{\Delta \cup \Gamma} = W_\Delta \cup W_\Gamma$. $W_P \subseteq W$ and $W_R \subseteq W$, are sets of worlds witnessing the accumulation and reduction of bad karma, resp. Last, $R_{\boxed{\text{s}}} \subseteq W \times W$ and $R_{\boxed{\text{u}}} = W \times W$ are binary relations s.t. the following holds:

(i) $R_{\boxed{\text{s}}} \subseteq R_{\boxed{\text{u}}}$;

(ii) $\forall w, v \in W((w, v) \in R_{\boxed{\text{s}}}$ and $v \in W_P)$ implies $\exists u((w, u) \in R_{\boxed{\text{s}}}$ and $u \notin W_P)$

(iii) $\forall w, v \in W((w, v) \in R_{\boxed{\text{s}}}$ and $v \in W_R)$ implies $\exists u((w, u) \in R_{\boxed{\text{s}}}$ and $u \notin W_R)$

(iv) $\forall w, v \in W((w, v) \in R_{\boxed{\text{s}}}$ and $v \in W_\Delta)$ implies $\exists u((w, u) \in R_{\boxed{\text{s}}}$ and $u \in W_\Delta \setminus W_R \cap W_P)$

An LMa-model is a tuple $\mathfrak{M}_{\text{LMa}} = \langle \mathfrak{F}_{\text{LMa}}, V \rangle$ where $\mathfrak{F}_{\text{LMa}}$ is an LMa-frame and V is a valuation function mapping atomic propositional symbols from Atom $\cup$ Wit$_{\text{Act}} \cup \{P\} \cup \{R\}$ to sets of worlds, such that the following conditions are satisfied: $V(\mathsf{d}_\delta) = \mathsf{W}_\delta$ for every $\mathsf{d}_\delta \in \mathsf{Wit}_{\mathsf{Act}}$, $V(\mathsf{P}) = \mathsf{W}_\mathsf{P}$, and $V(\mathsf{R}) = \mathsf{W}_\mathsf{R}$. (n.b. constants P, R and those from $\mathsf{Wit}_{\mathsf{Act}}$ have a fixed evaluation over frames). We use $\mathcal{C}^f_{\mathsf{LMa}}$ to refer to the entire class of LMa-frames.

The ⊡-modality behaves as a universal modality, hence its corresponding accessibility relation $R_{\boxed{\mathsf{U}}}$ is reflexive, symmetric and transitive (cf. Sect. 2). The purpose of ⊡ is to represent universally true statements, which should hold 'at every world'. The intended use of the ⊡-modality is to represent the possible outcomes of transitions triggered by actions. We have adopted a very general notion of the 'immediate successor' relation, by imposing no additional properties on this relation (cf. the absence of irreflexivity and asymmetry). We point out that there is no Mīmāṃsā characterization of time available to justify such properties. However, we do realize that, in general, these properties may be desirable in a temporal logic of action. Following [2], one can show that LMa can likewise be characterized by a subclass of LMa-frames including only asymmetric and intransitive *tree-like* frames (this is due to the fact that languages such as $\mathcal{L}_{\mathsf{LMa}}$ cannot force these additional frame properties; cf. [3]). For the purpose of this paper, a general notion of the immediate successor relation suffices.

Semantic evaluation of formulae φ from $\mathcal{L}_{\mathsf{LMa}}$ is defined accordingly:

Definition 3.3 Let $\mathfrak{M}_{\mathsf{LMa}}$ be an LMa-model and $w \in W$ of $\mathfrak{M}_{\mathsf{LMa}}$. We define the *satisfaction* of a formula $\varphi \in \mathcal{L}_{\mathsf{LMa}}$ in $\mathfrak{M}_{\mathsf{LMa}}$ at w inductively:

$\mathfrak{M}_{\mathsf{LMa}}, w \vDash \chi$	iff	$w \in V(\chi)$, for any $\chi \in \mathsf{Atom} \cup \mathsf{Wit}_{\mathsf{Act}} \cup \{\mathsf{P}\} \cup \{\mathsf{R}\}$
$\mathfrak{M}_{\mathsf{LMa}}, w \vDash \neg\varphi$	iff	$\mathfrak{M}_{\mathsf{LMa}}, w \nvDash \varphi$
$\mathfrak{M}_{\mathsf{LMa}}, w \vDash \varphi \vee \psi$	iff	$\mathfrak{M}_{\mathsf{LMa}}, w \vDash \varphi$ or $\mathfrak{M}_{\mathsf{LMa}}, w \vDash \psi$
$\mathfrak{M}_{\mathsf{LMa}}, w \vDash \boxed{\mathsf{U}}\varphi$	iff	for all $v \in W$ s.t. $(w,v) \in R_{\boxed{\mathsf{U}}}$, $\mathfrak{M}_{\mathsf{LMa}}, v \vDash \varphi$
$\mathfrak{M}_{\mathsf{LMa}}, w \vDash \boxed{\mathsf{S}}\varphi$	iff	for all $v \in W$ s.t. $(w,v) \in R_{\boxed{\mathsf{S}}}$, $\mathfrak{M}_{\mathsf{LMa}}, v \vDash \varphi$

The semantic clauses for the dual operators ⟨s⟩ and ⟨u⟩ as well as global truth, validity and semantic entailment are defined as usual (see [3]).

Theorem 3.4 *(Soundness) For all $\varphi \in \mathcal{L}_{\mathsf{LMa}}$ and $\Gamma \subseteq \mathcal{L}_{\mathsf{LMa}}$, if $\Gamma \vdash_{\mathsf{LMa}} \varphi$, then $\mathcal{C}^f_{\mathsf{LMa}}, \Gamma \vDash_{\mathsf{LMa}} \varphi$*

Proof. Soundness is proven as usual. Explicating the use of constants we prove axiom $\mathsf{A}_{\mathsf{LMa}}5$. Let $\mathfrak{M}_{\mathsf{LMa}}$ be an LMa-model with $w \in W$. Suppose $\mathfrak{M}_{\mathsf{LMa}}, w \vDash \langle s \rangle t(\Delta)$. Then $\exists v \in W$ s.t. $(w,v) \in R_{\boxed{\mathsf{S}}}$ and $\mathfrak{M}_{\mathsf{LMa}}, v \vDash t(\Delta)$. So $v \in W_\Delta$. By (iv) of Def. 3.2, $\exists u \in W$ s.t. $(w,u) \in R_{\boxed{\mathsf{S}}}$ and $u \in W_\Delta \setminus W_\mathsf{R} \cap W_\mathsf{P}$. So $\mathfrak{M}_{\mathsf{LMa}}, u \vDash t(\Delta)$ and $\mathfrak{M}_{\mathsf{LMa}}, u \nvDash \mathsf{R} \wedge \mathsf{P}$. Which gives $\mathfrak{M}_{\mathsf{LMa}}, w \vDash \langle s \rangle (t(\Delta) \wedge (\neg\mathsf{R} \vee \neg\mathsf{P}))$. □

Strong completeness is proven via canonical model construction, adjusted to the inclusion of constants. LMa-maximal consistent sets (MCS) are defined as usual, enjoying the usual properties. We define the following canonical model:

Definition 3.5 Let $\mathsf{M}^c = \langle \mathsf{W}^c, \{\mathsf{W}^c_{\mathsf{d}_\delta} | \mathsf{d}_\delta \in \mathcal{L}_{\mathsf{LMa}}\}, \mathsf{W}^c_\mathsf{P}, \mathsf{W}^c_\mathsf{R}, \mathsf{R}^c_{\boxed{\mathsf{U}}}, \mathsf{R}^c_{\boxed{\mathsf{S}}}, V^c \rangle$ be a canonical model, where W^c is the set of all LMa-MCSs ($\Gamma, \Sigma, \Phi...$) and:

- For all $\mathsf{d}_\delta \in \mathcal{L}_{\mathsf{LMa}}$ and $\Sigma \in \mathsf{W}^c$, $\Sigma \in \mathsf{W}^c_{\mathsf{d}_\delta}$ iff $\mathsf{d}_\delta \in \Sigma$

- For $\alpha \in \{\mathsf{P}, \mathsf{R}\}$, and all $\Sigma \in W^c$, $\Sigma \in W^c_\alpha$ iff $\alpha \in \Sigma$
- For $\alpha \in \{\boxed{\mathsf{s}}, \boxed{\mathsf{u}}\}$, and all $\Sigma, \Gamma \in W^c$, $(\Sigma, \Gamma) \in R^c_\alpha$ iff $\{\phi |\ [\alpha]\phi \in \Sigma\} \subseteq \Gamma$
- For all $\chi \in \mathsf{Atom} \cup \mathsf{ActWit} \cup \{\mathsf{P}\} \cup \{\mathsf{R}\}$, $V^c(\chi) = \{\Sigma | \chi \in \Sigma \in W^c\}$

The existence lemma and truth lemma are proven in [3, Sect. 4.2] (nb. LMa is a normal modal logic). We show that M^c belongs to the class of LMa-models, i.e., satisfying the properties of Def. 3.2.

Theorem 3.6 M^c *is an* LMa-*model*.

Proof. We demonstrate the LMa specific properties (ii) and (iv) (Def. 3.2). The proofs of the remaining properties are similar.

(ii) For all $\Sigma, \Gamma \in W^c$, $(\Sigma, \Gamma) \in R^c_{\boxed{\mathsf{s}}}$ with $\Gamma \in W^c_\mathsf{P}$, there exists a $\Theta \in W^c$ s.t. $(\Sigma, \Theta) \in R^c_{\boxed{\mathsf{s}}}$ and $\Theta \notin W^c_\mathsf{P}$. Assume the antecedent, we construct the set Θ. Let $\Theta^- = \{\neg \mathsf{P}\} \cup \{\phi | \boxed{\mathsf{s}}\phi \in \Sigma\}$. Suppose Θ^- is not LMa-consistent. Hence for some $\phi_1, .., \phi_n \in \Theta^-$, we have $\vdash_{\mathsf{LMa}} \phi \wedge ... \wedge \phi_n \to \mathsf{P}$. By LMa we have $\vdash_{\mathsf{LMa}} \boxed{\mathsf{s}}(\phi \wedge ... \wedge \phi_n \to \mathsf{P})$, which implies $\vdash_{\mathsf{LMa}} \boxed{\mathsf{s}}(\phi \wedge ... \wedge \phi_n) \to \boxed{\mathsf{s}}\mathsf{P}$, and so $\vdash_{\mathsf{LMa}} \boxed{\mathsf{s}}\phi \wedge ... \wedge \boxed{\mathsf{s}}\phi_n \to \neg\diamondsuit\neg\mathsf{P}$. By monotonicity of LMa, $\vdash_{\mathsf{LMa}} \boxed{\mathsf{s}}\phi \wedge ... \wedge \boxed{\mathsf{s}}\phi_n \wedge \diamondsuit\mathsf{P} \to \neg\diamondsuit\neg\mathsf{P}$. By assumption $\boxed{\mathsf{s}}\phi_1, ..., \boxed{\mathsf{s}}\phi_n, \diamondsuit\mathsf{P} \in \Sigma$ and MCS properties, we have $\neg\diamondsuit\neg\mathsf{P} \in \Sigma$. However, since Σ is a LMa-MCS we have $\diamondsuit\mathsf{P} \to \diamondsuit\neg\mathsf{P} \in \Sigma$, and thus $\diamondsuit\neg\mathsf{P} \in \Sigma$. Contradiction. Hence, Θ^- is LMa-consistent. Let Θ be the LMa-MCS extending Θ^- (Lindenbaum's lemma). By construction of M^c we obtain $(\Sigma, \Theta) \in R^c_{\boxed{\mathsf{s}}}$ and since $\neg\mathsf{P} \in \Theta^- \subseteq \Theta$ we have $\Theta \notin W^c_\mathsf{P}$.

(iv) For all $\Sigma, \Gamma \in W^c$, if $(\Sigma, \Gamma) \in R^c_{\boxed{\mathsf{s}}}$ with $\Gamma \in W^c_{t(\Delta)}$, then there exists a $\Theta \in W^c$ s.t. $(\Sigma, \Theta) \in R^c_{\boxed{\mathsf{s}}}$ and $\Theta \in W^c_{t(\Delta)} \setminus W^c_\mathsf{R} \cap W^c_\mathsf{P}$. Assume the antecedents, we construct such a Θ. Let $\Theta^- = \{t(\Delta)\} \cup \{\neg\mathsf{R} \vee \neg\mathsf{P}\} \cup \{\phi | \boxed{\mathsf{s}}\phi \in \Sigma\}$. Suppose Θ^- is LMa-inconsistent. Then there are $\phi_1, ..., \phi_n \in \Theta^-$ s.t. $\vdash_{\mathsf{LMa}} \phi_1 \wedge ... \wedge \phi_n \to \neg(t(\Delta) \wedge (\neg\mathsf{R} \vee \mathsf{P}))$. Hence, we have $\vdash_{\mathsf{LMa}} \boxed{\mathsf{s}}\phi_1 \wedge ... \wedge \boxed{\mathsf{s}}\phi_n \to \neg\diamondsuit(t(\Delta) \wedge (\neg\mathsf{R} \vee \mathsf{P}))$. By monotonicity, $\vdash_{\mathsf{LMa}} \boxed{\mathsf{s}}\phi_1 \wedge ... \wedge \boxed{\mathsf{s}}\phi_n \wedge \diamondsuit(t(\Delta)) \to \neg\diamondsuit(t(\Delta) \wedge (\neg\mathsf{R} \vee \mathsf{P}))$. Since $\boxed{\mathsf{s}}\phi_1, ..., \boxed{\mathsf{s}}\phi_n, \diamondsuit(t(\Delta)) \in \Sigma$, we get $\neg\diamondsuit(t(\Delta) \wedge (\neg\mathsf{R} \vee \mathsf{P})) \in \Sigma$. By inclusion of axiom $\diamondsuit t(\Delta) \to \diamondsuit(t(\Delta) \wedge (\neg\mathsf{R} \vee \neg\mathsf{P})) \in \Sigma$, we get a contradiction. Hence, Θ^- is consistent. Let Θ be the LMa-MCS extending Θ^-. By construction of Θ we get $(\Sigma, \Theta) \in R^c_{\boxed{\mathsf{s}}}$. Since $t(\Delta) \in \Theta$ we have $\Theta \in W^c_{t(\Delta)}$. Last, since $\neg\mathsf{R} \vee \mathsf{P} \in \Theta$ we get $\Theta \notin W^c_\mathsf{R} \cap W^c_\mathsf{P}$, hence $\Theta \in W^c_{t(\Delta)} \setminus W^c_\mathsf{R} \cap W^c_\mathsf{P}$. □

Corollary 3.7 *(Strong Completeness for* LMa*)* *For all* $\phi \in \mathcal{L}_{\mathsf{LMa}}$ *and* $\Gamma \subseteq \mathcal{L}_{\mathsf{LMa}}$, *we have: if* $\mathcal{C}^f_{\mathsf{LMa}}, \Gamma \models \phi$, *then* $\Gamma \vdash_{\mathsf{LMa}} \phi$.

3.2 Instrumentality and Mīmāṃsā properties

We introduce Maṇḍana's notion of instruments, his deontic reduction, and discuss important Mīmāṃsā properties and their rendering in Maṇḍana's logic.

Instruments and Maṇḍana's deontic reduction. Maṇḍana's program consists in reducing all deontic modalities to a uniform notion of instrumentality. Our uniform definition of instrumentality must satisfy the following Maṇḍana-criteria: First, (i) the instrument relation contains three components: (a) an action Δ, serving as the instrument; (b) a state-of-affairs φ, representing the outcome of Δ; and (c) a state-of-affairs χ defining the circumstances in which Δ functions as an instrument for bringing about φ. Second, (ii) the circumstances χ must be meaningful, that is, χ must be possible in the broadest sense. Last, the agent must have a *choice* to perform the action Δ when circumstances χ occur; (iii) Δ can be performed and (iv) Δ can be refrained from (for (ii–iv) see Śabara on *Mīmāṃsāsūtra* 6.1 in [27]). In short, we take $\mathcal{I}(\Delta/\varphi/\chi)$ to read "Δ is an instrument for guaranteeing φ in circumstances χ", which amounts to:

"(i) Whenever circumstances χ hold, performing Δ guarantees φ, (ii) χ is a possible circumstance, (iii) at χ, Δ is possible, and (iv) at χ, $\overline{\Delta}$ is possible."

The formal definition of instrumentality, based on (i)-(iv), is given in Def. 3.8.

Maṇḍana's reduction, that is, the reduction of commands to statements of instrumentality, is then obtained accordingly: prohibited and obligatory actions are defined in terms of those actions being instrumental to the outcome of bad karma (i.e., P) and the reduction of bad karma (i.e., R), respectively. Electives are those actions instrumental to outcomes that are neither P nor R.

Definition 3.8 Maṇḍana's notion of instruments in the logic LMa:

$\mathcal{I}(\Delta/\varphi/\chi)$:= (i) $\boxed{\mathsf{U}}(\chi \to \boxed{\mathsf{S}}(t(\Delta) \to \varphi)) \quad \wedge$
(ii) $\diamondsuit \chi \quad \wedge$
(iii) $\boxed{\mathsf{U}}(\chi \to \diamondsuit t(\Delta)) \quad \wedge$
(iv) $\boxed{\mathsf{U}}(\chi \to \diamondsuit \neg t(\Delta))$

Maṇḍana's reduction of obligations, prohibitions and elective sacrifices in LMa:

$\mathcal{O}(\Delta/\chi)$:= $\mathcal{I}(\Delta/\mathrm{R}/\chi)$
$\mathcal{F}(\Delta/\chi)$:= $\mathcal{I}(\Delta/\mathrm{P}/\chi)$
$\mathcal{E}(\Delta/\varphi/\chi)$:= $\mathcal{I}(\Delta/\varphi/\chi)$ with $\varphi \not\vdash_{\mathsf{LMa}} \mathrm{P}$ and $\varphi \not\vdash_{\mathsf{LMa}} \mathrm{R}$

(n.b. we differentiate actions (kill, sacrifice), from results (death, P, R), with a contrasting font style.)

Recall that for the Mīmāṃsā school, obligations, prohibitions and electives cannot be expressed in terms of one another [12,20]. Similarly, Maṇḍana adopts this irreducibility by limiting the result of the instruments corresponding to the three norm types. This property is preserved in Def. 3.8. In particular, we note that the result of an elective sacrifice cannot entail either of the results identified with obligations or prohibitions. We come back to this in Sect. 3.3, when we discuss Maṇḍana's solution to the *śyena* controversy. Furthermore, in LMa, the elective operator $\mathcal{E}$ has one additional argument. This is because the instrument notation is more expressive, and there are variables for both the eligibility condition (the desire) and the purpose (the object of the desire).

Remark 3.9 Now that we have defined instruments, let us go back to axiom $\mathsf{A}_{\mathsf{LMa}}5$ in Def. 3.1. Observe that it limits the interaction between actions and karma constants. In essence, $\mathsf{A}_{\mathsf{LMa}}5$ ensures that an action Δ cannot be both obligatory and prohibited, i.e., Δ cannot at the same time be an *instrument* for the reduction and accumulation of bad karma (n.b. the inconsistent action $\Delta \cap \overline{\Delta}$ is excluded from being an instrument by Def. 3.8). Nevertheless, we still allow for singular situations where we end up with both P and R after executing Δ, however $A_{\mathsf{LMa}}5$ guarantees that this must be the result of some other action Γ executed alongside Δ (one being a prohibition, the other an obligation).

We show below that the logic LMa is expressive enough to entail other principles that can be found in Mīmāṃsā (as LMa-theorems).

Contingency. For the Mīmāṃsā it is essential that actions in commands are meaningful (see Śabara on *Mīmāṃsāsūtra* 6.1, [27]). For an action to be meaningful, an agent must have the choice to perform it as well as refrain from performing it:

$$\mathcal{I}(\Delta/\varphi/\chi) \to \boxed{\scriptstyle \mathsf{U}}(\chi \to (\Diamond t(\Delta) \wedge \Diamond \neg t(\Delta))) \text{ for } \varphi \in \{\mathsf{P}, \mathsf{R}\} \text{ or } (\varphi \not\vdash \mathsf{P} \text{ and } \varphi \not\vdash \mathsf{R})$$

In deontic logic this property is known as the *contingency principle* [30, p. 11][1]. The above formula is an LMa-theorem, guaranteed solely by our definition of instruments ((iii) and (iv) in Def. 3.8). However, for obligations and prohibitions the property is also implied in association with axioms $A_{\mathsf{LMa}}3$ and $A_{\mathsf{LMa}}4$. That is,

$$\mathcal{I}(\Delta/\mathsf{R}/\chi) \equiv \mathcal{O}(\Delta/\chi) \equiv (\Diamond \chi \wedge \boxed{\scriptstyle \mathsf{U}}(\chi \to \boxed{\scriptstyle \mathsf{S}}(t(\Delta) \to \mathsf{R})) \wedge \boxed{\scriptstyle \mathsf{U}}(\chi \to \Diamond t(\Delta))$$

and,

$$\mathcal{I}(\Delta/\mathsf{P}/\chi) \equiv \mathcal{F}(\Delta/\chi) \equiv (\Diamond \chi \wedge \boxed{\scriptstyle \mathsf{U}}(\chi \to \boxed{\scriptstyle \mathsf{S}}(t(\Delta) \to \mathsf{P})) \wedge \boxed{\scriptstyle \mathsf{U}}(\chi \to \Diamond t(\Delta))$$

are LMa-theorems. These theorems demonstrate that condition (iv) of instruments (Def. 3.8) is *admissible* in the light of Maṇḍana's analysis. However, (iv) is still necessary to ensure meaningfulness of actions for elective duties; i.e., $\not\vdash_{\mathsf{LMa}} \mathcal{E}(\Delta/\varphi/\chi) \equiv (\Diamond \chi \wedge \boxed{\scriptstyle \mathsf{U}}(\chi \to \boxed{\scriptstyle \mathsf{S}}(t(\Delta) \to \varphi)) \wedge \boxed{\scriptstyle \mathsf{U}}(\chi \to \Diamond t(\Delta))$.

No impossible commands. Although the logic LMa does not adopt a D-axiom for deontic consistency, the following formula is in fact an LMa-theorem:

$$\vdash_{\mathsf{LMa}} \neg(\mathcal{F}(\Delta/\chi) \wedge \mathcal{F}(\overline{\Delta}/\chi))$$

The theorem corresponds to the Mīmāṃsā principle: "It is impossible that the Vedas tell you that you'll fall (i.e., be reborn in hell) both if you do something and if you don't do it" ([29, p. 32]). The quote refers to the impossibility of the Vedas giving contradictory instruments. The theorem is a direct consequence of the definition of instrumentality together with axiom $A_{\mathsf{LMa}}3$. In fact, we obtain a similar theorem for obligations from axiom $A_{\mathsf{LMa}}4$. Clearly, the scheme does not hold for elective duties (cf. Def. 3.8). Last, the logic LMa satisfies the Mīmāṃsā principle that obligations and prohibitions are strictly incompatible (even on the derived level). That is, the following formula is LMa-valid:

$$\neg(\mathcal{O}(\Delta/\chi) \wedge \mathcal{F}(\Delta/\chi))$$

Mīmāṃsā principles. The logics for Prabhākara and Kumārila are built upon principles (P1)-(P3) recalled in Sect. 2. A natural question to ask is whether these principles are preserved in Maṇḍana's reduction logic. Their reformulation in LMa is as follows (notice that (P1)-(P3) were postulated for commands in particular, not instruments in general):

(p1) $(\mathcal{I}(\Delta/\varphi/\chi) \wedge \boxed{\mathsf{u}}(t(\Delta) \to t(\Gamma))) \to \mathcal{I}(\Gamma/\varphi/\chi)$ such that $(\star)$

(p2) $(\mathcal{I}(\Delta/\varphi/\chi) \wedge \boxed{\mathsf{u}}(\varphi \to \neg\psi)) \to \neg\mathcal{I}(\Delta/\psi/\chi)$ such that $(\star)$

(p3) $(\mathcal{I}(\Delta/\varphi/\chi) \wedge \boxed{\mathsf{u}}(\chi' \equiv \chi)) \to \mathcal{I}(\Delta/\varphi/\chi')$ such that $(\star)$

$(\star)$ $\varphi \in \{\mathsf{P}, \mathsf{R}\}$ or $(\varphi \not\vdash \mathsf{P}$ and $\varphi \not\vdash \mathsf{R})$

(p1)-(p3) deal with instruments that are obligations, prohibitions and electives.

Principle (p1) is not an LMa-valid formula, (a counter-model is easily obtained), and for good reasons: instrumentality is a notion of sufficient cause. Maṇḍana knew this, but he had to somehow preserve the property expressed in (P1). He achieved this by explaining necessity as external to instruments: that is, Maṇḍana's account of *universally desirable outcomes* (i.e., R and P) ensures that no agent would, from a rational point of view, transgress such commands. Hence, although necessary conditions of instruments leading to reducing bad karma are themselves not recognized as instruments, from a meta point of view, no rational agent would refrain from performing them.

Principle (p2) is LMa-valid and it follows from Maṇḍana property (cf. Def. 3.8) that actions must be meaningful (thus leading to meaningful outcomes).

Last, (p3) is LMa-valid and follows from the fact that universal necessity is a normal modal operator.

3.3 Maṇḍana's solution

We utilize Maṇḍana's reduction and demonstrate that, when formalized in terms of *instrumentality*, the sentences (A)-(D) from Sect. 1 are satisfiable. That is, we show the consistency of Maṇḍana's solution to the *śyena* controversy by providing an LMa-model satisfying the following:

(A_M) $\mathcal{E}(\mathsf{\acute{S}y/death/des\_kill}) \equiv \mathcal{I}(\mathsf{\acute{S}y/death/des\_kill})$
$\equiv \boxed{\mathsf{u}}(\mathsf{des\_kill} \to \boxed{\mathsf{s}}(t(\mathsf{\acute{S}y}) \to \mathsf{death})) \wedge \diamondsuit \mathsf{des\_kill} \wedge$
$\boxed{\mathsf{u}}(\mathsf{des\_kill} \to \diamondsuit t(\mathsf{\acute{S}y})) \wedge \boxed{\mathsf{u}}(\mathsf{des\_kill} \to \diamondsuit \neg t(\mathsf{\acute{S}y}))$

(B_M) $\mathcal{F}(\mathsf{harm}/\top) \equiv \mathcal{I}(\mathsf{harm}/\mathsf{P}/\top) \equiv \boxed{\mathsf{u}}\diamondsuit t(\mathsf{harm}) \wedge \boxed{\mathsf{u}}\boxed{\mathsf{s}}(t(\mathsf{harm}) \to \mathsf{P})$

(C_M) $\boxed{\mathsf{u}}(t(\mathsf{\acute{S}y}) \to \mathsf{death})$

(D_M) $\boxed{\mathsf{u}}(\mathsf{death} \to t(\mathsf{harm}))$

A model satisfying (A_M)-(D_M) is defined as follows: $\mathfrak{M}_{\mathsf{LMa}} = \langle \mathfrak{F}_{\mathsf{LMa}}, V \rangle$ with $\mathfrak{F}_{\mathsf{LMa}} = \langle W, W_{\mathsf{\acute{S}y}}, W_{\mathsf{harm}}, W_{\mathsf{P}}, W_{\mathsf{R}}, R_{\boxed{\mathsf{u}}}, R_{\boxed{\mathsf{s}}} \rangle$ s.t.: $W = \{w_1, w_2, w_3\}$, $W_{\mathsf{\acute{S}y}} = W_{\mathsf{harm}} = W_{\mathsf{P}} = \{w_2\}$, $W_{\mathsf{R}} = \emptyset$, $V(\mathsf{des\_kill}) = \{w_0\}$, $V(\mathsf{death}) = \{w_2\}$, $R_{\boxed{\mathsf{u}}} = W \times W$, $R_{\boxed{\mathsf{s}}} = \{(w_1, w_2), (w_1, w_3), (w_2, w_2), (w_2, w_3), (w_3, w_2), (w_3, w_3)\}$. Note that $\mathfrak{M}_{\mathsf{LMa}}$ satisfies the properties in Def. 3.2. The model

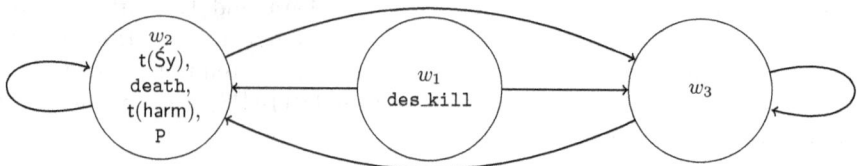

Fig. 2. *śyena* model in LMa, with the arrows representing the relation $R_{\boxed{s}}$.

is represented in Fig. 2.

For all $w \in \{w_0, w_1, w_2\}$, we have $\mathfrak{M}, w \models \mathcal{I}(\mathtt{harm}/\mathtt{P}/\top) \land \mathcal{I}(\mathtt{\acute{S}y}/\mathtt{death}/\mathtt{des\_kill}) \land \boxed{\mathtt{u}}(t(\mathtt{\acute{S}y}) \to \mathtt{death}) \land \boxed{\mathtt{u}}(\mathtt{death} \to t(\mathtt{harm}))$. The model shows that the *śyena* example is consistent. Furthermore, it illustrates that, given our assumptions, one cannot perform the *śyena* without accumulating bad karma P. It can easily be verified that $\mathcal{I}(\mathtt{\acute{S}y}/\mathtt{P}/\mathtt{des\_kill})$ is the case (conditions (ii)-(iv) of the definition of instrumentality (Def. 3.8) follow from A_M, and condition (i) follows from A_M, B_M and D_M). So, the *śyena* is an instrument for bad karma. In fact, in the logic LMa, that is on a *derived level* (see footnote 5), the *śyena* sacrifice is prohibited $\mathcal{F}(\mathtt{\acute{S}y}/\mathtt{des\_kill})$. By contrast, on the Vedic level *śyena* is not prohibited. Following Maṇḍana, in LMa something can be prohibited and elective at the same time, without it being inconsistent. Maṇḍana's reasoning for the *śyena* sacrifice is the following: from a state where one desires to kill their enemy, it is rationally preferable not to perform the *śyena*. Performing it would transgress the prohibition of harming a living being, with the result of accumulating bad karma. This is necessarily undesirable for anyone, as discussed in Maṇḍana:

> When it comes to pain and its cause, the one who is afflicted by them will always desire their removal. And the one who desires well-being desires to destroy the obstacle (bad karma) towards it. Therefore, the destruction of bad karma, a destruction which is the cause of what is desired, is always desired. (*Vidhiviveka* ad 2.8 [26])

4 Discussion of the three *śyena* solutions

We presented formal models that capture Prabhākara, Kumārila and Maṇḍana's responses to the *śyena* case. Here we compare the different solutions relating them to the history of deontic logic. Recall the main challenge facing the three authors: how to deal with seemingly conflicting prescriptions coming from a source that is assumed to be consistent. Prabhākara's solution is akin to CTD reasoning in deontic logic, which introduces (sub-ideality) levels to a normative system, not treating every norm on equal footing. We distinguish norms that hold primarily (possibly conditioned on circumstances) from norms that only arise in case of a norm violation, the latter being CTD obligations. In this case, the prohibition to commit violence is a primary norm, whereas the prescription of the *śyena* is an obligation that only comes into force once a violation has occurred: for Prabhākara, the intention to kill one's enemy amounts to violence. Here, we see a striking similarity with the most common interpretation of the GMP, namely Chisholm's paradox [6].

Although Prabhākara and Kumārila agree that the *śyena* case does not constitute a dilemma, they argue so on different grounds. For Kumārila, prohibitions do not interact with electives in a mutually conflicting way. In particular, as an elective sacrifice the *śyena* has no deontic force and is thus overturned by the Vedic prohibition to commit violence. Despite some shallow similarities with the deontic logic literature on priority orderings (e.g., [15])—i.e., obligations and prohibitions being of highest priority for Kumārila—and hierarchies of different norm systems (e.g., [4])—i.e., obligations and prohibitions forming a proper norm system in contrast to electives—we note that Kumārila's approach is different, in the sense that he assigns no deontic force to elective sacrifices whatsoever. They are mere sacrificial ways to attain one's end, without being compelling, eliminating the controversy altogether.

Maṇḍana preserves Kumārilas distinction between obligatory and elective sacrifices but offers a different solution: deontic modalities are just variations of a shared underlying structure, namely, instrument relations. In order to preserve the appealing distinction between the three norm types, Maṇḍana relates obligations and prohibitions to the reduction and accumulation of bad karma. Elective sacrifices are karma-independent. They might have indirect consequences on the reduction/accumulation of bad karma (e.g., the *śyena*), but their direct results are not karma-results. Maṇḍana argues that avoiding the accumulation of bad karma is *a priori* desired by all human beings, similarly its reduction. By reducing the Vedic norm system to notions of instruments and desires Maṇḍana does not yet resolve the problem, but transforms the seeming problem of conflicting norms to a problem of conflicting desires instead. What remains is a conflict between one's desire to kill an enemy and one's "rational" desire to avoid accumulating bad karma. In other words, Maṇḍana does not address the problem on the command level, but on the instrument level and, subsequently, solves it on the desire level. Interestingly, Maṇḍana is the only author that endorses the view that there is a real dilemma or conflict at stake in the *śyena* case. Nevertheless, he resolves the dilemma by arguing that avoiding the accumulation of bad karma is the highest of desires, which implies that no rational agent would ever perform the *śyena*. We find a priority ordering on the level of desires, consequently resolving the implied commands. There are striking similarities between Maṇḍanas approach and the Kanger-Anderson approach to deontic logic [1,19], by opting for a unifying approach reducing a variety of modalities to a single (alethic) modality together with the notions of sanction (accruing bad karma) and goodness (reducing bad karma). However, Maṇḍana's final solution to the *śyena* controversy is a decision-making problem that occurs on the meta-level, by making an appeal to rationality and undesirability.

Future work. Our interdisciplinary work only scratches the surface of the research opportunities offered by formal approaches to Mīmāṃsā reasoning. As illustrated in this work, these approaches can provide a better understanding of Mīmāṃsā texts, and may offer new stimuli for the deontic logic community.

Since the logics of the first two authors, Prabhākara and Kumārila, have been extensively studied elsewhere [7,20], further investigation of the logic of Maṇḍana and his reduction is planned. For instance, to simplify matters, in this work we took desires as regular terms of our object language. We plan to investigate the logical behaviour of desires as an intentional modality interacting with instruments and norms.

Acknowledgements. Work funded by the projects WWTF MA16-028 and FWF W1255-N23.

References

[1] Anderson, A. R. and O. K. Moore, *The formal analysis of normative concepts*, American Sociological Review **22** (1957), pp. 9–17.
[2] van Berkel, K. and M. Pascucci, *Notions of instrumentality in agency logic*, in: *International Conference on Principles and Practice of Multi-Agent Systems*, Springer, 2018, pp. 403–419.
[3] Blackburn, P., M. de Rijke and Y. Venema, *Modal logic*, Cambridge tracts in theoretical computer science (2004).
[4] Boella, G. and L. van der Torre, *Institutions with a hierarchy of authorities in distributed dynamic environments*, Artificial Intelligence and Law **16** (2008), pp. 53–71.
[5] Chellas, B. F., "Modal Logic," Cambridge University Press, Cambridge, 1980.
[6] Chisholm, R. M., *Contrary-to-duty imperatives and deontic logic*, Analysis **24** (1963), pp. 33–36.
[7] Ciabattoni, A., E. Freschi, F. A. Genco and B. Lellmann, *Mīmāṃsā deontic logic: Proof theory and applications*, in: H. De Nivelle, editor, *Automated Reasoning with Analytic Tableaux and Related Methods* (2015), pp. 323–338.
[8] Fischer, M. J. and R. E. Ladner, *Propositional dynamic logic of regular programs*, Journal of computer and system sciences **18** (1979), pp. 194–211.
[9] Forrester, J., *Gentle murder, or the adverbial samaritan*, J. of Philosophy (1984), pp. 193–197.
[10] Freschi, E., *The role of* paribhāṣās *in Mīmāṃsā: rational rules of textual exegesis*, Asiatische Studien/Études Asiatiques **72** (2018), pp. 567–595.
[11] Freschi, E., A. Ciabattoni, F. A. Genco and B. Lellmann, *Understanding prescriptive texts: Rules and logic as elaborated by the Mīmāṃsā school*, Journal of World Philosophies **2** (2017).
[12] Freschi, E. and M. Pascucci, *Deontic concepts and their clash in Mīmāṃsā: towards an interpretation*, Theoria (2021).
[13] Goble, L., *Murder most gentle: The paradox deepens*, Philosophical Studies: An International Journal for Philosophy in the Analytic Tradition **64** (1991), pp. 217–227.
[14] Guhe, E., *Killing gently by means of the śyena: The navya-nyāya analysis of vedic and secular injunctions (vidhi) and prohibitions (niṣedha) from the perspective of dynamic deontic logic*, Journal of Indian Philosophy (2021).
[15] Hansen, J., *Prioritized conditional imperatives: problems and a new proposal*, Autonomous Agents and Multi-Agent Systems **17** (2008), pp. 11–35.
[16] Hansson, B., *An analysis of some deontic logics*, in: *Deontic Logic: Introductionary and Systematic Readings* (1971), pp. 121–147.
[17] Hilpinen, R. and P. McNamara, *Deontic logic: A historical survey and introduction*, Handbook of deontic logic and normative systems **1** (2013), pp. 3–136.
[18] Hughes, J., P. Kroes and S. Zwart, *A semantics for means-end relations*, Synthese **158** (2007), pp. 207–231.

[19] Kanger, S., *New foundations for ethical theory*, in: *Deontic logic: Introductory and systematic readings*, Springer, 1971 pp. 36–58.
[20] Lellmann, B., F. Gulisano and A. Ciabattoni, *Mīmāṃsā deontic reasoning using specificity: a proof theoretic approach*, Artificial Intelligence and Law (2021).
[21] Lewis, D., *Semantic analysis for dyadic deontic logics*, in: *Logical Theory and Semantical Analysis* (1974), pp. 1–14.
[22] Meyer, J., J. Broersen and A. Herzig, *Bdi logics*, Handbook of Logics of Knowledge and Belief (2015), p. 453.
[23] Meyer, J.-J. C. et al., *A different approach to deontic logic: deontic logic viewed as a variant of dynamic logic.*, Notre Dame J. Formal Log. **29** (1988), pp. 109–136.
[24] Prior, A. N., *Escapism: The logical basis of ethics*, in: A. Melden, editor, *Essays in Moral Philosophy* (1958).
[25] Stcherbatsky, T. I., *Über die Nyāyakaṇikā des Vācaspatimiśra und die indische Lehre des kategorischen Imperativ*, in: W. Kirfel, editor, *Beiträge zur Literaturwissenschaft und Geistesgeschichte Indiens. Festgabe Hermann Jacobi zum 75. Geburtstag*, Kommissionsverlag F. Klopp, Bonn, 1926 pp. 369–380.
[26] Stern, E., editor, "Vidhiviveka of Maṇḍana Miśra with the commentary Nyāyakaṇikā of Vācaspati Miśra," forthcoming.
[27] Subbāśāstrī, editor, "Śrīmajjaiminipraṇitaṃ Mīmāṃsādarśanam," Ānandāśramamudrāṇālaya, Poona, 1929-1934.
[28] van der Torre, L. and Y. hua Tan, *Deliberate robbery, or the calculating samaritan*, in: *In Proceedings of the ECAI'98 Workshop on Practical Reasoning and Rationality (PRR'98)*, 1998.
[29] Viraraghavacharya, U. T., editor, "Seśvaramīmāṃsā-Mīmāṃsāpaduke, Seswara Mīmāṃsā and Mīmāṃsā paduka [by Veṅkaṭanātha]," Ubhaya Vedanta Granthamala, Madras, 1971.
[30] von Wright, G. H., *Deontic logic*, Mind **60** (1951), pp. 1–15.

If You Want to Smoke, Don't Buy Cigarettes: Near-Anankastics, Contexts, and Hyper Modality

Kees van Berkel[a][1] Dov Gabbay[b,c] Leendert van der Torre[c,d]

[a] *Technische Universität Wien, Vienna, Austria*
[b] *King's College London, London, United Kingdom*
[c] *University of Luxembourg, Esch-sur-Alzette, Luxembourg*
[d] *Zhejiang University, Hangzhou, China*

Abstract

In this discussion paper we are interested in anankastic conditionals such as "if you want to smoke you must buy cigarettes" and near-anankastic conditionals such as "if you want to smoke, you must not buy cigarettes." First, we discuss challenges to representing such conditionals in deontic logic, in particular in relation to the use of context. We do this through a discussion of the Tobacco shop scenario, an example dealing with ambiguity of certain deontic conditionals. Second, we illustrate how ambiguity of natural language can be formally represented through the use of hyper-modalities, using a minimal modal logic for (near-)anankastic conditionals. We illustrate how the hyper-modal setting can disambiguate such conditionals. As the Tobacco shop scenario suggests, in our formalism interaction between antecedent, consequent, and context can reduce ambiguity in the involved conditionals.

Keywords: anankastic conditionals, deontic logic, desire modalities, hyper modalities

1 Introduction

Natural language offers a wide assortment of sometimes ambiguous deontic expressions. For example, consider the sentence "if you want to smoke, you must buy cigarettes." This natural language sentence can be interpreted in at least two ways. On the one hand, we may say that the best means to smoke is to buy cigarettes. On the other hand, we may say that the most ideal way to satisfy your desire is to buy cigarettes (better than, say, stealing cigarettes). The former is a teleological interpretation of 'must' (i.e., referring to a goal), and the latter is called a deontic interpretation (i.e., referring to a

---

[1] Kees van Berkel acknowledges the projects WWTF MA16-028 and FWF W1255-N23. Leendert van der Torre acknowledges financial support from the Fonds National de la Recherche Luxembourg (INTER/Mobility/19/13995684/DLAl/van der Torre). Contact details: kees@logic.at, dov.gabbay@kcl.ac.uk, and leon.vandertorre@uni.lu.

duty). A modality such as 'must' has many different interpretations [9,15]. As a basic example, the utterance 'it must rain' may refer to an epistemic necessity which says that it cannot but rain, but it could also refer to an optimal ideal expressing that it ought to rain. Often, context may help to disambiguate. For instance, if it does not rain here and now, then we know 'must' cannot receive an epistemic necessity reading. Hence, different contexts will imply different logical reasoning with the same modality. For example, epistemic necessity may be an S5 modality, whereas deontic obligation is a KD modality. Likewise, the different interpretations of the conditional "if you want to smoke, you must buy cigarettes" will have distinct logical formalisations. Following [7], we call modalities that may receive different interpretations in different contexts *hyper-modalities*.

In this paper, we discuss hyper-modality through a discussion of a challenging example concerning *anankastic* and *near-anankastic* conditionals: the Tobacco shop scenario. The anankastic conditional "if you want to smoke, you must buy cigarettes" is one of the scenario's central premises. Von Wright is said to be the first to adopt the term 'anankastic' in his philosophy of agency [15], but thorough investigation of such conditionals began with the work of Sæbø [12]. In [12], Sæbø points out that the nature of 'desire' in the antecedent generates some challenges when it comes to the interpretation of the modal 'must' in the consequent, challenges which are particular to anankastic conditionals. Since then, alternative accounts have been proposed (e.g., [4,13,14]), properly introducing anankastics to the research agenda of deontic modality.

The contribution of this discussion paper is twofold. First, we argue that the analysis of deontic modalities—such as 'must'—in natural language expressions can bring new challenges to deontic logic. In particular, we discuss how the consequent plays an important role in evaluating an anankastic conditional, via interaction with the conditional's embedded context and antecedent. Second, we argue that logical techniques can help to bring some aspects traditionally referred to pragmatics, within the reach of logical analysis. That is, we develop a hyper-modal setting in which ambiguous (near-)anankastic conditionals—such as those in the Tobacco shop scenario—can be formally represented and which facilitates partial disambiguation of such conditionals.

The paper is structured as follows: In Section. 2, we discuss the Tobacco shop scenario and (near-)anankastic conditionals. Section. 3 contains a modal logic for the Condoravdi-Lauer analysis of (near-)anankastics and we formalise four types of such conditionals. Section. 4 extends this logic to the hyper-modal setting, internalising part of the pragmatics of interpreting (near-)anankastics. In Section. 5, we provide a hyper-modal analysis of the Tobacco shop scenario.

2 The Tobacco shop scenario

The development of deontic logic has been driven by deontic benchmark examples [11]. In this paper we are interested in the *Tobacco shop scenario*, a scenario which circulates since at least 2016 in various forms [5]:

Dr. Smoke wanders through the university's inner courtyard. Prof. Pragmatics notices a slight disturbance in Dr. Smoke's mood. She asks him: "what's on your mind?" Smoke shares with her his craving for a cigarette. Prof. Pragmatics replies "if that is so, then you must go to the tobacco shop!" At that moment, Prof. Restraint crosses the lawn and, by chance, catches Pragmatics' last remark. Restraint asks: "what is going on here?" Pragmatics: "He wants a cigarette!" Prof. Restraint looks surprised: "if that is the case," she exclaims, "then you surely should not go to the tobacco shop. "

The Tobacco shop scenario illustrates a scenario in which Dr. Smoke (henceforth, S) receives seemingly incompatible advice, conditional on his desire to smoke. At face value, we have two similar conditional premises (henceforth, P1 and P2) which share the same antecedent:

If S wants to smoke, then S must buy cigarettes. (P1)

If S wants to smoke, then S must not buy cigarettes. (P2)

Provided that the conditionals involved are of the same form ([2] makes a strong case for uniformity of conditionals), we have to accept the following inference:

If S wants to smoke, then S must buy cigarettes and S must not buy cigarettes. (1)

In any deontic logic that allows for factual detachment and monotonicity of its modal operators, while also adopting a deontic consistency axiom [8] a logical inconsistency of P1 and P2 arises in the light of S's actual desire:

S wants to smoke. (2)

Which gives us:

S must buy and not buy cigarettes. (3)

A consequence such as (3) is not just undesirable, it also does not seem to do justice to the nature of the involved conditionals. Premises P1 and P2 do not merely express conflicting obligations given a shared antecedent, they convey additional information: the relation between smoking and buying cigarettes is clearly of a different nature than the relation between smoking and not buying cigarettes. For instance, by looking at the antecedent and consequent of P1, we observe that buying cigarettes, as an activity, may serve as a means for smoking. This is not the case for P2. In fact, not buying may even prove obstructive to satisfying one's desire to smoke. We find that the interpretation of 'must' in P2 differs from the one adopted in P1: 'must' is a hyper-modality. In the case of P2, the consequent suggests the need for additional context in which the conditional must be embedded. Conditionals that relate statements of desire to statements of must, are called *(near-)anankastic conditionals.*

Many of the benchmark examples in deontic logic revolve around challenges of reasoning with conditionals in normative settings [8]. Likewise, we find that the Tobacco shop scenario focuses on a specific, yet ubiquitous, type of conditionals: anankastic and near-anankastic conditionals. In this section we will see that, whereas often only the antecedent is considered in evaluating the consequent [1], the consequent of a (near-)anankastic conditional plays an important role in evaluating the conditional, through interacting with the conditional's embedded context and antecedent.

2.1 Terminology

We first go through some terminological matters. The group of conditionals we are interested in share the following structure: 'if you want ϕ, you must ψ'. Depending on the disambiguation of the modalities 'desire' and 'must', but also on the relation between their internal structure ϕ and ψ, we may obtain different conditionals called *anankastic* and *near-anankastic* conditionals. We recall the terminology of Condoravdi and Lauer [4]:

$$\text{If } S \underbrace{\underbrace{\text{wants to}}_{\text{desire predicate}} \underbrace{[S] \text{ smoke}}_{\text{internal antec.}}}_{\text{antecedent}}, \text{ then } S \underbrace{\underbrace{\text{must}}_{\text{modal}} \underbrace{[S] \text{ buy cigarettes}}_{\text{prejacent}}}_{\text{consequent}}$$

An anankastic conditional transmits that the complement of 'must' functions as a necessary *precondition* for the realisation of the complement of 'desire'. See [4] for a discussion. A near-anankastic conditional, on the other hand, has the same general structure but lacks the relation between internal antecedent and consequent as one of necessary precondition.

Given the above distinction, we find that the two central premises P1 and P2 of the Tobacco shop scenario are, respectively, an anankastic and a near-anankastic conditional. Namely, P1 expresses a positive relation between the internal antecedent 'S smokes' and the prejacent 'S buys cigarettes', i.e., S's buying cigarettes is instrumental to S's smoking. In this particular case, the relation is a *best-means* relation which indicates that buying is an optimal means serving the goal of smoking. When 'must' is taken to refer to an optimal realisation of a goal, we say it receives a *teleological* reading. Premise P2 does not express such a relation (in fact, it hints at a relation to the contrary) and for that reason it is called a near-anankastic conditional. As a first observation we find that, in order to determine the nature of the conditional we must thus go into its *substructure*: to correctly interpret the conditional, we must (i) determine the relation between the four central components of a (near-)anankastic conditional and (ii) disambiguate the involved modalities 'must' and 'desire'. Interpreting the substructure of the conditional, subsequently, often depends on the context in which the statement occurs.

2.2 Two types of desire and two types of obligations

The right interpretation of the 'desire' expressed in a (near-)anankastic's antecedent, plays a central role in correctly interpreting the consequent (and vice versa). Condoravdi and Lauer [4] distinguish between two types of desires: *mere desires* and *action-relevant desires*. Mere desires are desires that are 'psychological facts' (nothing more), whereas action-relevant desires reflect the agent's goal and a corresponding intention to realise that goal. We refer to D^1

and D^2 as mere desires, respectively, action-relevant desires. Since the latter is related to action (i.e., a goal), it is subject to additional constraints. Hence, the two notions have a different logic which will influence the logical behaviour of conditionals in which they occur. In other words, 'desire' in a (near-)anankastic conditional is a *hyper modality* too. Disambiguation may give:

D^2 smoke $\Rightarrow O$ buy cigarettes. [2] (p1)

D^1 smoke $\Rightarrow O$ not buy cigarettes. (p2)

Similarly, we can distinguish different kinds of obligation. For instance, O^1 may denote a teleological 'must', whereas O^2 represents a deontic 'must':

D smoke $\Rightarrow O^1$ buy cigarettes. (p1)

D smoke $\Rightarrow O^2$ not buy cigarettes. (p2)

Given these possible readings of 'desire' and 'must' we already obtain four different interpretations of P1 and P2. We come back to this in Section. 3.

2.3 Various deontic contexts

We represent the *context* of a sentence Pi by Δ_i. The context expresses the conversational background in the light of which a sentence is uttered (cf. [9]). Such a context may contain facts, beliefs, desires, obligations, and what have you (from an agentive perspective we may call the context epistemic, in the sense that it expresses that which is *known* to the speaker of the sentence). Thus, we may take P1 and P2 as implicit renditions of:

$\Delta_1 \wedge D$ smoke $\Rightarrow O$ buy cigarettes. (p1)

$\Delta_2 \wedge D$ smoke $\Rightarrow O$ not buy cigarettes. (p2)

Contexts may change how we interpret the two conditionals and their involved modalities. In multi-agent scenarios, in which utterances come from different speakers, different contexts for the individual sentences are likely to occur. For instance, in the Tobacco shop scenario Prof. Restraint may know of Dr. Smoke's desire (or promise) to stop smoking, whereas Prof. Pragmatics does not. Usually, when the relevant context merely contains facts and common knowledge, it is left out of the conditional. Think of a case in which you need to apologise because you did not keep a promise. If it is common knowledge that "you must keep a promise," the conditional may be safely abbreviated to "if you break a promise, you should apologise." Unfortunately, often such common knowledge is falsely assumed and this may lead to ambiguity and miscommunication. In such cases, just as in the Tobacco shop scenario, we must ask for certain context to be made explicit. For instance, upon inquiry Prof. Pragmatics may recall that the tobacco shop is just around the corner, thus making the attainment of Smoke's goal most optimal. In other cases, looking at the content of a conditional's constituents, may help to reconstruct possible contexts and interpretations.

---

[2] In what follows, we write p1 and p2 to indicate alternative formal readings of P1 and P2.

To illustrate the above, Prof. Restraint may recall that Dr. Smoke also has a desire to be healthy, which would be unattainable in the light of smoking:

D smoke $\Rightarrow O$ buy cigarettes. (p1)

D healthy $\wedge D$ smoke $\Rightarrow O$ not buy cigarettes. (p2)

Given the additional context, the consequent in p2 above seems to suggest a priority for health over the desire to smoke. Here, the consequent provides additional information about the context too. If Dr. Smoke would buy cigarettes he would, given his desire to smoke, most likely start to smoke, thus compromising his health. The consequent seems to suggest that (i) the context contains an action-relevant desire and (ii) the antecedent contains a mere desire. Other contexts worth investigating are factual, normative, and intentional contexts.

2.4 Analysis: four observations concerning (near-)anankastics

In conditional logic, it is normally assumed that the context of a conditional is determined by the antecedent only [1]. One of the interesting aspects of the Tobacco shop scenario, is that this approach is no longer sufficient: P1 and P2 have the same antecedent but different consequents. In this analysis, interpreting conditionals such as P1 and P2 depends on the (mutual) interaction between antecedent, consequent, and context. Ambiguous sentences such as P1 and P2 may receive their correct interpretation through this interaction and additional context. Through disambiguation, the antecedents of the two conditional obligations may no longer be the same. Thus, we find that simple applications of aggregation to P1 and P2 are not always warranted for (cf. inference (1)) and consequently the pair of sentences is no longer inconsistent in standard deontic logic (cf. inference (3)).

In what follows, we adopt an explicit context to specify and investigate possible interactions between and interpretations of desire and obligation modalities. We have the following *general* representation of the Tobacco shop scenario:

$(\Delta_1, D$ smoke$) \Rightarrow O$ buy cigarettes. (p1)

$(\Delta_2, D$ smoke$) \Rightarrow O$ not buy cigarettes. (p2)

Still, the way in which we interpret these conditionals depends on the possible interpretations that can be assigned to the ambiguous modalities 'desire' and 'must'. Namely, in the above D and O represent hyper-modalities that may receive different logical interpretations depending on their appearance in the conditional (together with their corresponding context Δ_i). Several readings of D and O are possible, but in the present work we limit their possible interpretations to those discussed above, i.e., D^1, D^2, O^1, O^2.

We make four key observations about the logic of (near-)anankastics:

Role of consequent The consequents of P1 and P2 must inform us on their relation with their respective antecedents and contexts.

Ambiguity of modalities The semantic interpretations of 'desire' and 'must' may vary from context to context. The constituents of the conditional, together with its context, must aid in determining the appropriate interpretations of these ambiguous modalities.

Ambiguity of conditionals The semantic interpretation of a conditional, such as P1 and P2, will likewise vary with its context. This depends partially on resolving ambiguity of the modalities 'desire' and 'must'.

Aggregation rule The different nature of P1 and P2 suggests that the application of aggregation to P1 and P2 may not be warranted for.

Formal analysis of the Tobacco shop scenario must thus explain how the antecedent *together with* the consequent receive their interpretation. An immediate question would be: *can we still formally reason with (near-)anankastic conditionals, even if we cannot completely resolve ambiguity?* We come back to this in Section. 5 where we formally discuss the Tobacco shop scenario.

3 Condoravdi-Lauer and (near-)Anankastic conditionals

Anankastic conditionals have been extensively discussed in formal linguistics. Many authors developed non-standard ways to deal with the relation between desires and obligations in anankastic conditionals [9,13,14,4]. For example, in the setting of Kratzer [9], the obligations are based on a so-called ordering source, and this ordering source is updated by the desire in the antecedent (i.e., the restrictor analysis). Condoravdi and Lauer [4] provide an account that does not only address the compositionality problems that arises in previous approaches, their account is generalised to the inclusion of near-anankastic conditionals (dealing with several natural language examples of (near-)anankastics which previous accounts could not satisfactorily address). They argue that better results can be obtained by adopting a standard approach with counterfactual implications and standard dyadic, teleological obligations. Without taking a stance in this debate, since the Tobacco shop scenario deals with both types of conditionals, we will base our logic on the Condoravdi-Lauer approach [4]. In this section, we present the modal logic La, which allows us to formally represent four different interpretations of (near-)anankastic conditionals. In Section. 4, we will extend this account to a hyper-modality setting in order to reason with ambiguity in such conditionals.

3.1 A Modal logic for Anankastic Conditionals

The properties of desire and teleological modalities are taken from Condoravdi and Lauer [4], and we refer to their paper for an in-depth discussion of these properties and the various alternatives. The two kinds of desire, D^1 and D^2—i.e., mere-desire, respectively action-relevant desire—are differentiated by the property 'conjunction introduction' and a 'consistency' requirements (both are not valid for D^1, but are valid for D^2). For our purposes, we make a few modifications and simplifications: The teleological modality $O^1(\phi, \psi)$ reads 'ϕ holds when ψ is optimally realised'. We adopt a dyadic, deontic obligation $O^2(\phi, \psi)$ which reads 'ϕ is obligatory given ψ' and adopt a triadic conditional

$(\phi, \psi) \Rightarrow \theta$ expressing "given context ϕ, if ψ, then θ".[3] See [6] for motivating the use of ternary conditionals. A universal modality $\Box$ is used to represent facts rigid across contexts: $\Box \phi$ reads 'ϕ holds universally'. We do not formalise desires as priority rankings. The language $\mathcal{L}_{\mathsf{La}}$ is defined by the following BNF:

$$\phi ::= p \mid \neg \phi \mid \phi \vee \phi \mid D^1 \phi \mid D^2 \phi \mid O^1(\phi, \phi) \mid O^2(\phi, \phi) \mid (\phi, \phi) \Rightarrow \phi \mid \Box \phi$$

with $p \in \mathsf{Atoms}$. The connectives $\neg$ and $\vee$ are read as usual, and other connectives are obtained in the standard way. We use $\Diamond$ for the dual of $\Box$. The modalities are interpreted as discussed above.

We provide a Hilbert-style axiomatization for the logic of anankastics La.

Definition 3.1 The *Logic of anankastics* La extends the logic S5 for $\Box$ with:

A1. $(D^2 \phi \wedge D^2 \psi) \to D^2(\phi \wedge \psi)$ \hfill (C)

A2. $\mathsf{X}(\psi \wedge \theta, \phi) \to (\mathsf{X}(\psi, \phi) \wedge \mathsf{X}(\theta, \phi))$ for $\mathsf{X} \in \{O^1, O^2\}$ \hfill (M)

A3. $\mathsf{X}(\psi, \phi) \wedge \mathsf{X}(\theta, \phi)) \to \mathsf{X}(\psi \wedge \theta, \phi)$ for $\mathsf{X} \in \{O^1, O^2\}$ \hfill (C)

A4. $\neg D^2 \bot$ \hfill (P)

A5. $\neg(D^2 \phi \wedge D^2 \neg \phi)$ \hfill (D)

A6. $\Diamond \phi \to \neg(\mathsf{X}(\psi, \phi) \wedge \mathsf{X}(\neg \psi, \phi))$ for $\mathsf{X} \in \{O^1, O^2\}$ \hfill (D)

A7. $\Diamond \phi \to \mathsf{X}(\phi, \phi)$ for $\mathsf{X} \in \{O^1, O^2\}$ \hfill (Id)

A8. $\neg \mathsf{X}(\phi, \bot)$ for $\mathsf{X} \in \{O^1, O^2\}$ \hfill (F)

A9. $\Diamond(\phi \wedge \psi) \to (\phi, \psi) \Rightarrow (\phi \wedge \psi)$ \hfill (Id)

A10. $(\phi, \psi) \Rightarrow \theta \to (\Diamond(\phi \wedge \psi) \wedge \Diamond \theta)$ \hfill (F)

A11. $(\phi, \psi) \Rightarrow (\chi \wedge \theta) \to ((\phi, \psi) \Rightarrow \chi \wedge (\phi, \psi) \Rightarrow \theta)$ \hfill (M)

A12. $((\phi, \psi) \Rightarrow \chi \wedge (\phi, \psi) \Rightarrow \theta) \Rightarrow (\phi, \psi) \Rightarrow (\chi \wedge \theta)$ \hfill (C)

R1. Congruence rule: holds for $D^1, D^2, O^1, O^2,$ and $\Rightarrow$ (all arguments) \hfill (RE)

La-derivability and La-theorems are defined as usual [3].

In Def. 3.1, M and C denote monotonicity, respectively, conjunction introduction. P and D are consistency constraints on D^2, O^1, O^2. Id is identity for consistent formulae. F states that no obligation O^1, O^2 holds given $\bot$, and that the antecedent and consequent of $\Rightarrow$ are jointly consistent, respectively, consistent. Since La is a non-normal logic, we use neighbourhood semantics [3]:

Definition 3.2 An La-frame is a tuple $F = \langle W, \mathcal{N}_{D^1}, \mathcal{N}_{D^2}, \mathcal{N}_{O^1}, \mathcal{N}_{O^2}, \mathcal{N}_\Rightarrow \rangle$, where $W \neq \emptyset$ is a non-empty set of worlds $w, v, u, \ldots$ (possibly indexed) and $\mathcal{N}_i$ ($i \in \{D^1, D^2, O^1, O^2, \Rightarrow\}$) are neighbourhood functions such that:

- $\mathcal{N}_j : W \mapsto \mathcal{P}(\mathcal{P}(W))$ \hfill for $j \in \{D^1, D^2\}$
- $\mathcal{N}_k : W \mapsto \mathcal{P}(\mathcal{P}(W) \times \mathcal{P}(W))$ \hfill for $k \in \{O^1, O^2\}$

---

[3] Another way to look at $(\phi, \psi) \Rightarrow \theta$ is to take $\Rightarrow$ as a stereotypicality conditional, in line with the covert outer modal in [4]: 'in the most stereotypical ϕ and ψ worlds, θ holds'.

- $\mathcal{N}_{\Rightarrow}: W \mapsto \mathcal{P}(\mathcal{P}(W) \times \mathcal{P}(W) \times \mathcal{P}(W))$

F satisfies the following constraints, for all $w \in W$, and all $X, Y, Z, U \subseteq W$:

- (c_1) if $Z \in \mathcal{N}_{D^2}(w)$ and $Y \in \mathcal{N}_{D^2}(w)$, then $Z \cap Y \in \mathcal{N}_{D^2}(w)$;
- (m_2) $i \in \{O^1, O^2\}$, $(X \cap Y, Z) \in \mathcal{N}_i(w)$ implies $(Y, Z) \in \mathcal{N}_i(w)$ and $(X, Z) \in \mathcal{N}_i(w)$;
- (c_2) $i \in \{O^1, O^2\}$, $(X, Z) \in \mathcal{N}_i(w)$ and $(Y, Z) \in \mathcal{N}_i(w)$ implies $(X \cap Y, Z) \in \mathcal{N}_i(w)$;
- (p) $\emptyset \notin \mathcal{N}_{D^2}(w)$;
- (d_1) if $X \in \mathcal{N}_{D^2}(w)$, then $\overline{X} \notin \mathcal{N}_{D^2}(w)$;
- (d_2) $i \in \{O^1, O^2\}$, if $X \neq \emptyset$, $(Y, X) \in \mathcal{N}_i(w)$, then $(\overline{Y}, X) \notin \mathcal{N}_i(w)$;
- (id_1) $i \in \{O^1, O^2\}$, if $X \neq \emptyset$, then $(X, X) \in \mathcal{N}_i(w)$;
- (f_1) $i \in \{O^1, O^2\}$, if $X = \emptyset$, then $(Y, X) \notin \mathcal{N}_i(w)$;
- (id_2) if $X \cap Y \neq \emptyset$, then $(X, Y, X \cap Y) \in \mathcal{N}_{\Rightarrow}(w)$;
- (f_2) if $X \cap Y = \emptyset$ or $Z = \emptyset$, then $(X, Y, Z) \notin \mathcal{N}_{\Rightarrow}(w)$;
- (m_3) if $(X, Y, Z \cap U) \in \mathcal{N}_{\Rightarrow}(w)$, then $(X, Y, Z) \in \mathcal{N}_{\Rightarrow}(w)$ and $(X, Y, U) \in \mathcal{N}_{\Rightarrow}(w)$;
- (c_3) if $(X, Y, Z) \in \mathcal{N}_{\Rightarrow}(w)$ and $(X, Y, U) \in \mathcal{N}_{\Rightarrow}(w)$, then $(X, Y, Z \cap U) \in \mathcal{N}_{\Rightarrow}(w)$;

An La-model is a tuple $M = \langle F, V \rangle$ s.t. F is an La-frame and V is a valuation function assigning atoms $p \in \mathsf{Atoms}$ to sets of worlds; i.e. $V : \mathsf{Atoms} \mapsto \mathcal{P}(W)$.

Last, we semantically evaluate formulae of $\mathcal{L}_{\mathsf{La}}$ as usual:

Definition 3.3 Let M be an La-model, $w \in W$ and $\|\phi\| = \{v \in W | M, v \models \phi\}$:

- $M, w \models p$ iff $w \in V(p)$.
- $M, w \models \neg\phi$ iff $M, w \not\models \phi$
- $M, w \models \phi \vee \psi$ iff $M, w \models \phi$ or $M, w \models \psi$
- $M, w \models \mathsf{X}\phi$ iff $\|\phi\| \in \mathcal{N}_{\mathsf{X}}(w)$ with $\mathsf{X} \in \{D^1, D^2\}$
- $M, w \models \mathsf{Y}(\phi, \psi)$ iff $(\|\phi\|, \|\psi\|) \in \mathcal{N}_{\mathsf{Y}}(w)$ with $\mathsf{Y} \in \{O^1, O^2\}$.
- $M, w \models (\phi, \psi) \Rightarrow \theta$ iff $(\|\phi\|, \|\psi\|, \|\theta\|) \in \mathcal{N}_{\Rightarrow}(w)$.
- $M, w \models \Box\phi$ iff for all $v \in W$, $M, v \models \phi$

Satisfiability, validity, and model-validity are defined as usual [3]. (nb. The operator $\Box$ expresses model-validity.)

Comparing the axioms of Def.3.1 with the properties of Def.3.2, we see that La is highly modular. Consequently, completeness is obtained following the standard approach for neighbourhood semantics [3] (proofs are omitted):

Theorem 3.4 (SOUNDNESS AND COMPLETENESS) Let $\phi \in \mathcal{L}_{\mathsf{La}}$, and let C_{La} be *the class of* La*-frames:* $\mathcal{C}_{\mathsf{La}} \models \phi$ *iff* $\vdash_{\mathsf{La}} \phi$.

3.2 Anankastics and near-anankastics in the logic La

In this section we discuss formalisations of four teleological (near-)anankastic conditionals. We focus on those that play a role in the Tobacco shop scenario.

The anankastic conditional. Anakastic conditionals are identified by the fact that both the antecedent and consequent receive an action-relevant reading of desire [4]. Let $\Delta \subseteq \mathcal{L}_{\mathsf{La}}$ be the context representing the finite knowledge base of the speaker. The formalised anankastic conditional $(\Delta, D^2\phi) \Rightarrow_{ac} O^1\psi$ is informally interpreted as: "(i) all the most stereotypical worlds consistent with Δ in which $D^2\phi$ holds, are such that whenever all the addressee's known goals, including ϕ, are optimally realised, then ψ holds and (ii) the hypothesised goal ϕ is compatible with what the speaker knows Δ". This definition resonates the account provided in [4]. The first conjunct (i) expresses the teleological optimality of the prejacent with respect to the internal antecedent. The second conjunct (ii) captures the requirement that action-relevant desire must be realistic: i.e., the goal must be compatible with what is known. Given a context of utterance Δ, a speaker may know of some of the addressee's actual action-relevant desires, we let $\Sigma_\Delta^{D^2} = \{\theta | D^2\theta \in \Delta\}$ denote the set of the addressee's actual goals and call θ a goal whenever $D^2\theta$. In what follows, we slightly abuse notation and write Δ and $\Sigma_\Delta^{D^2}$ for the conjunction of formulae in Δ and $\Sigma_\Delta^{D^2}$, respectively. Let the (teleological) *anankastic conditional* (tac) be defined as: [4]

$$(\Delta, D\phi) \Rightarrow_{tac} O\psi := (\Delta, D^2\phi) \Rightarrow O^1(\psi, \Sigma_\Delta^{D^2} \wedge \phi) \wedge \Diamond(\Delta \wedge \phi) \qquad (4)$$

Applying (4) to premise P1 of the Tobacco shop scenario $(\Delta, D\mathsf{smoke}) \Rightarrow_{tac} O\mathsf{buy}$ gives us the following formal definition:

$$(\Delta, D^2\mathsf{smoke}) \Rightarrow O^1(\mathsf{buy}, \Sigma_\Delta^{D^2} \wedge \mathsf{smoke}) \wedge \Diamond(\Delta \wedge \mathsf{smoke}) \qquad (5)$$

Informally, (5) reads "in the most stereotypical worlds in which Δ and $D^2\mathsf{smoke}$ are the case, buy proves teleological optimal given the optimal realisation of the known goals $\Sigma_\Delta^{D^2}$ together with the goal of smoking". Let us look at some logical consequences of definition (4).

Conflicting and non-conflicting goals. Conflicting goals relate to a part of the compositionality problem of anankastics: the addressee's actual action-relevant desires should not matter in the analysis, *unless* these are known to the speaker, see [4]. In La, the issue is accounted for in the same manner as in [4]: only when action-relevant desires are known in Δ, they will be taken into consideration. There are two cases of possible conflict: First, when the desire $D^2\phi$ is incompatible with context Δ. For instance,

---

[4] On the left side of (4) we leave D and O underspecified, but the index '*tac*' on $\Rightarrow_{tac}$ tells us to interpret the involved modalities as D^2 and O^1 specified on the right side of (4).

suppose $\Delta_1 = \{D^2\text{health}, \Box(\text{smoke} \to \neg\text{health})\}$, then the anankastic $(\Delta_1, D\text{smoke}) \Rightarrow_{tac} O\text{buy}$ is not satisfiable (cf. c_1 p of Def.3.2). Second, when the goal ϕ in $D^2\phi$ is incompatible with Δ. For instance, when you know the shops are closed, and the only chance of smoking would be when the shops are open: given $\Delta_2 = \{\neg\text{open}, \Box(\text{smoke} \to \text{open})\}$, the second conjunct $\Diamond(\Delta_2 \wedge \text{smoke})$ of the anankastic conditional (4) becomes inconsistent.

Failure of strengthening of the antecedent (SA). In line with [4], there are two ways in which SA may fail: (i) through strengthening that makes the antecedent inconsistent with what is known, and (ii) through strengthening that selects other most stereotypical worlds. As an example, suppose we don't know whether the shops are open today. By later strengthening the antecedent with ¬open, we may obtain a different set of most stereotypical worlds. Failure of SA is guaranteed by the non-monotonic nature of the ⇒ modality and its consistency requirement (cf. f_3 of Def.3.2).

The teleological near-anankastic conditional. Near-anankastic conditionals come in different shapes, depending on what readings of the ambiguous 'desire' and 'must' modalities are assigned to the conditional's antecedent and consequent, respectively. Let the *teleological* near-anankastic conditional (tnc) be defined accordingly:

$$(\Delta, D\phi) \Rightarrow_{tnc} O\psi := (\Delta, D^1\phi) \Rightarrow O^1(\psi, \Sigma_\Delta^{D^2} \wedge D^1\phi) \tag{6}$$

The formal definition reads: "all the most stereotypical worlds consistent with Δ in which $D^1\phi$ holds are such that the optimal realization of all the addressee's known goals, together with $D^1\phi$, also realize ψ". The presence of $D^1\phi$ in $O^1(\psi, \Sigma_\Delta^{D^2} \wedge D^1\phi)$ is important: We take the mere-desire for ψ as a *cause* for the necessitated consequent. In contrast to (4) where optimality is conditioned on the realization of the antecedent's goal, we condition (6) on the desire itself. See [4] for a discussion. The second premise P2 of the Tobacco shop scenario can be assigned this form. Suppose we know that the addressee has an action-relevant desire for $D^2\text{health} \in \Delta$, that smoking is not healthy $\Box(\text{smoke} \to \neg\text{health})$, and that buying cigarettes together with a mere-desire to smoke will lead to smoking $\Box(\text{buy} \wedge D^1\text{smoke} \to \text{smoke})$. In that case, the antecedent $D^1\text{smoke}$ together with buy will lead to a conflict with the optimal realization of the addressee's known desire $D^2\text{health} \in \Delta$. We write,

$$(\Delta, D^1\text{smoke}) \Rightarrow O^1(\neg\text{buy}, \Sigma_\Delta^{D^2} \wedge D^1\text{smoke}) \tag{7}$$

For teleological near-anakastic conditionals we likewise have failure of SA. However, note that (6) is not subject to a realism condition due to the presence of a mere desire D^1 in the antecedent.

Deontic near-anankastics with action-relevant desires. We introduce deontic counterparts to the teleological (near-)anankastics. Following [4], deontic near-anankastics emerge when the conditional does not have a purpose reading (e.g., when 'not buying does not serve the purpose of 'smoking'), but a deontic reading of the consequent 'must'. The structure of the first deontic conditional, with an action-relevant reading, is similar to that of the anankastic conditional. The main difference is that in evaluating the hypothesized goal ϕ in $D^2\phi$, we are not concerned with what is deontically optimal given the realization of all the agent's action-relevant desires, but only with what is deontically implied when the goal ϕ is *actualized* given those stereotypical worlds in which Δ and $D^2\phi$ hold (cf. [4]). This is reflected in how the antecedent influences the consequent in (8). To illustrate, think of an agent with a desire to smoke who can either buy or steal cigarettes. Since stealing is forbidden, it must deontically be the case that if she smokes, then she bought the cigarettes. We formalize deontic near-anankastics (dac) with action-relevant desires accordingly:

$$(\Delta, D\phi) \Rightarrow_{dac} O\psi := (\Delta, D^2\phi) \Rightarrow O^2(\psi, \phi) \land \Diamond(\Delta \land \phi) \tag{8}$$

Since we are dealing with action-relevant desires, the realism clause is preserved.

Deontic near-anankastics with mere-desires. This conditional is similar to (6). The main difference is again that we are not concerned with what is deontically optimal given the realization of the agent's action-relevant desires, but only with what is deontically implied when the agent has the mere-desire expressed in the antecedent. That is, the occurrence of 'want' is not vacuous but the actual cause of the obligation (cf. [4]). We formalize deontic near-anankastics (dnc) with mere-desires as:

$$(\Delta, D\phi) \Rightarrow_{dnc} O\psi := (\Delta, D^1\phi) \Rightarrow O^2(\psi, D^1\phi) \tag{9}$$

An example of (9) would be when there is an obligation not to smoke. Then, in all deontically optimal worlds where you do not smoke, but desire to smoke, you do not buy cigarettes (since buying, together with a desire to smoke, stereotypically implies smoking). Perhaps less common, deontic conditionals with mere-desires also arise in CTD-scenarios in which desires are forbidden.

For both deontic near-anankastic conditionals (8) and (9) SA fails.

3.3 Ambiguity, (near-)anankastics, and pragmatics

The four readings show that in the consequent, different use is made of the context and the desire modality occurring in the antecedent. This interaction between consequent, antecedent, and context is reflected in the different interpretations of (near-)anankastic conditionals. As observed in Section. 2, there is no difference between the four types of conditionals when we look at "if you want ϕ, you must do ψ". Still, we can differentiate them through linguistic analysis. In particular, the four definitions are differentiated through (i) the role of the context and (ii) the interpretation of the involved hyper-modalities.

Note that we take the antecedent to do double duty: it serves as a "restrictor" of the modal operator, but also conditionalises the modal claim to an assumption. The Tobacco shop scenario illustrates that this is not only desirable, but even necessary. It is normally assumed that an if-clause either restricts an operator, or functions as a supposition. However, this would make P1 and P2 indistinguishable. The double duty of the antecedent is motivated by the fact that only when we consider the consequent as well, we can properly distinguish anankastics from near-anankastics. This is in line with [4].

So far we *assumed pragmatics*: that is, we assumed that we know with which interpretations of 'desire', 'must', and the conditional we are dealing, prior to formalization. Often, we don't have access to a determined interpretation and ambiguity remains. The question is, can we reason with such conditionals even though we don't have a definite interpretation? In the next section, we provide a hyper formalism that enables us to represent and reason with ambiguous conditionals and modalities. By internalizing part of the pragmatics, we may formally reduce possible interpretations through explicit interaction between context, antecedent, and consequent in logic.

4 Hyper modalities: interpreting (near-)anankastics

The two conditionals in the Tobacco shop scenario share the general structure 'if you desire ϕ, you must ψ'. We have argued that 'desire', 'must', and the involved conditional are ambiguous and may receive different readings. Such modalities are called *hyper modalities*. In the previous section, we discussed four possible readings of 'if you desire ϕ, you must ψ'. There, we used distinct modalities for the different readings of 'desire' (D^1 and D^2) and 'must' (O^1 and O^2), and more importantly we *assumed* access to the correct readings of these conditionals and their modalities, prior to their formalization.

We present a way to make ambiguity and interpretation part of the logic, for this we will use the hyper modality framework, as developed in [7]. We introduce the hyper modalities $\mathbb{D}$ and $\mathbb{O}$ to represent the ambiguous 'desire' and 'must'. Such hyper modalities may receive different semantic interpretations depending on their context of evaluation (but under other contexts ambiguity may persist). In Section. 5, we will deploy the formalism to disambiguate and reason with the involved modalities in the Tobacco shop scenario. We point out that the reader may temporarily skip this technical section and first consult the hyper-modal analysis of the Tobacco shop scenario in Section. 5.

4.1 Preliminaries: a brief introduction to hyper-modalities

Why do we need hyper-modalities? Such modalities occur in natural language: for example, "soon p will be true". The reading of 'soon' depends on time: In the 19th century 'soon' could have meant within a week, whereas nowadays 'soon' would mean within 24 hours. Another example, already discussed, is the context dependence of the meaning of 'must' which may refer (among others) to logical necessity, epistemic certainty, and deontic optimality. To represent the linguistic distinctions that may occur in certain contexts, we need to allow for 'must' (and other modalities) to have several semantic interpretations (e.g., S5 for epistemic certainty, but KD for deontic optimality). Hence, in contrast to standard modal logic approaches, we need modalities which do not have a fixed meaning, but *receive their meaning* through evaluation in a context.

4.2 From neighbourhood semantics to hyper modality semantics

Before moving to the multi-modal setting, we introduce the formalism by considering an example language with a single modality $\mathbb{M}$, for 'must'. Let $\mathcal{N}_{\mathbb{M}}$ be a neighbourhood function from worlds $w \in W$ to sets of subsets: $\mathcal{N}_{\mathbb{M}} : W \mapsto \mathcal{P}(W)$. Semantics of atoms and the connectives $\neg$ and $\vee$ are defined as usual, and for $\mathbb{M}$ we adopt $w \models \mathbb{M}\phi$ iff $\|\phi\| \in \mathcal{N}_{\mathbb{M}}(w)$.

We turn $\mathbb{M}$ into a hyper-modality if we allow for each world $w \in W$ an *option* of neighbourhoods functions $\mathcal{N}^1_{\mathbb{M}}(w), \mathcal{N}^2_{\mathbb{M}}(w), ..., \mathcal{N}^n_{\mathbb{M}}(w)$. We call these options the different *modes* of modality $\mathbb{M}$ and denote them by $\Psi^i(w,\mathbb{M}) = \mathcal{N}^i_{\mathbb{M}}(w)$ (for $i \in \{1, ..., n\}$). That is, $\Psi^i(w,\mathbb{M})$ denotes a possible mode for interpreting $\mathbb{M}$ at w. Let $\mathcal{M}odes$ be the set of modes Ψ^i for $i \in \{1, ..., n\}$. Since a modality may have various possible modes, we need a table function, $f : \mathcal{M}odes \mapsto \mathcal{P}(\mathcal{M}odes)$. So, for each $\Psi^i \in \mathcal{M}odes$, $f(\Psi^i)$ denotes the set of option modes $\{\Psi^k, ..., \Psi^l\}$ (with $1 \leq l \leq k \leq n$). Last, the semantic clause of a hyper modality $\mathbb{M}$ is relativized to the use of modes Ψ^i, denoted by $\models_{\Psi^i}$. We have, for all $w \in W$:

$$w \models_{\Psi^i} \mathbb{M}\phi \text{ iff for some mode } \Psi^j \in f(\Psi^i), \{v \mid v \models_{\Psi^j} \phi\} \in \mathcal{N}^i_{\mathbb{M}}(w) \quad (10)$$

Hence, $\mathbb{M}\phi$ is satisfiable at w at mode Ψ^i, whenever there is a mode $\Psi^j \in f(\Psi^i)$ (possibly several) for $\mathbb{M}$ such that $\{v \mid v \models_{\Psi^j} \phi\}$ is in the $\mathbb{M}$-neighborhood for mode Ψ^j. Note that modes are only relevant for evaluating hyper-modalities.

Let us consider an example. We formalize the utterance "it does not rain, but it must rain" as $\neg\mathbf{rain} \wedge \mathbb{M}\mathbf{rain}$. Let there be two modes for $\mathbb{M}$: for any w, let $\Psi^{deo}(w, \mathbb{M}) = \mathcal{N}^{deo}_{\mathbb{M}}(w)$ s.t. $\mathcal{N}^{deo}_{\mathbb{M}}(w)$ does not contain $\emptyset$ (i.e., Ψ^{deo} interprets $\mathbb{M}$ deontically by excluding inconsistencies) Let $\Psi^{epi}(w, \mathbb{M}) = \mathcal{N}^{epi}_{\mathbb{M}}(w)$ s.t. $\mathcal{N}^{epi}_{\mathbb{M}}(w)$ is restricted to all the sets containing the world w (i.e., Ψ^{epi} takes $\mathbb{M}$ as some sort of epistemic certainty: 'if ϕ is epistemically certain, ϕ must be true now'). Let $f(\Psi^{deo}) = f(\Psi^{epi}) = \{\Psi^{deo}, \Psi^{epi}\}$, which means that in both the deontic and epistemic mode, $\mathbb{M}\phi$ may be interpreted deontically, as well as epistemically. How do we evaluate $\neg\mathbf{rain} \wedge \mathbb{M}\mathbf{rain}$ at w? We need to pick a starting mode. Suppose it is Ψ^{deo}. Hence, $w \models_{\Psi^{deo}} \neg\mathbf{rain} \wedge \mathbb{M}\mathbf{rain}$ iff $w \not\models_{\Psi^{deo}} \mathbf{rain}$ and $w \models_{\Psi^{deo}} \mathbb{M}\mathbf{rain}$. The last conjunct in equivalent to $\{v \in W \mid v \models_{\Psi^i} \mathbb{M}\mathbf{rain}\} \in \mathcal{N}^i_{\mathbb{M}}(w)$ for some $\Psi^i \in \{\Psi^{epi}, \Psi^{deo}\}$. If $\Psi^i = \Psi^{epi}$, then $w \in \{v \in W \mid v \models_{\Psi^{epi}} \mathbf{rain}\}$, but $w \in \{v \in W \mid v \models_{\Psi^{deo}} \neg\mathbf{rain}\}$ (modes do not apply to atoms). We have a contradiction. Hence, the ambiguous $\mathbb{M}\mathbf{rain}$ cannot be interpreted epistemically given $\neg\mathbf{rain}$ (whether it can be interpreted deontically, remains to be determined). See [7] for further examples.

4.3 Interpreting (near-)anankastics using hyper-modalities

Since the conditional 'if you want ϕ, you must ψ' depends on the ambiguous 'want' and 'must', the conditional $\Rightarrow$ is likewise ambiguous. The hyper modalities that we will consider are thus 'desire' $\mathbb{D}$, 'must' $\mathbb{O}$ and 'conditional' $\Rightarrow$. We build our hyper modality setting on top of the La-neighbourhood semantics of Def.3.2. The hybrid language $\mathcal{L}_{\mathsf{LaH}}$ is defined through the following BNF:

$$\phi ::= p \mid \neg\phi \mid \phi \vee \phi \mid D^1\phi \mid D^2\phi \mid O^1(\phi,\phi) \mid O^2(\phi,\phi) \mid \Box\phi \mid (\phi,\phi) \Rightarrow \phi \mid \mathbb{D}\phi \mid \mathbb{O}\phi$$

with $p \in$ Atoms. The language La properly extends $\mathcal{L}_{\mathsf{LaH}}$ for the reason that, in evaluating a (near-)anankastic, we involve a context Δ which may contain information about unambiguous desires and obligations. Note that $\mathbb{O}\phi$ is monadic but will be interpreted dyadically in the hyper-setting, conditioning it on a given context (if there is no context we evaluate $\mathbb{O}\phi$ conditional on $\top$).

To facilitate readability, we write $\|\phi\|_{\Psi^i} = \{v \in W \mid v \models_{\Psi^i} \phi\}$ to indicate the presence of a mode Ψ^i for evaluating ϕ. Furthermore, we explicitly index the modes Ψ^i with the formula's arguments. As an example, $w \models_{\Psi^i} \mathbb{O}(\phi,\psi)$ means $\mathbb{O}(\phi,\psi)$ is evaluated with respect to the mode $\Psi^i_{\phi,\psi}(w,\mathbb{O})$. Most formulae will be evaluated with respect to what we call the *common* mode, denoted by Ψ^{co}.

We are interested in mode shifts that occur when evaluating conditionals of the form $(\Delta, \mathbb{D}\phi) \Rightarrow \mathbb{O}\psi$. The four interpretations of (near-)anankastic conditionals (Section. 3) are in fact *modes* for interpreting "if you want ϕ, you must ψ": i.e., anankastics Ψ^{tac}, near-anankastics Ψ^{tnc}, deontic anankastics Ψ^{dac}, and deontic near-anankastics Ψ^{dnc}. For instance, when evaluating in *anankastic* mode Ψ^{tac}, $\mathbb{D}\phi$ is interpreted as an action-relevant D^2, and $\mathbb{O}\psi$ via a teleological optimization O^2. Let us make these modes formally precise:

Definition 4.1 A *hyper* La-frame is a tuple $F = \langle W, \mathcal{N}_{D^1}, \mathcal{N}_{D^2}, \mathcal{N}_{O^1}, \mathcal{N}_{O^2} \mathcal{N}_{\Rightarrow}, \mathit{Modes}\rangle$, where $W \neq \emptyset$ is a non-empty set of worlds $w, v, u, ..$ and $\mathcal{N}_i$ ($i \in \{D^2, D^1, O^1, O^2, \Rightarrow\}$) are neighbourhood functions as defined in Def.3.2. F satisfies the constraints from Def.3.2. Let the set of modes be $\mathit{Modes} = \{\Psi^{co}, \Psi^{tac}, \Psi^{tnc}, \Psi^{dac}, \Psi^{dnc}\}$ (defined in Def.4.2, Def.4.4, Def.4.3, resp). The assignment $\underline{f}^i$ of modes ($i \in \{1,2,3\}$) is defined as:

- $\underline{f}^1 : \mathit{Modes} \mapsto \mathcal{P}(\mathit{Modes})$ (for monadic $\mathbb{D}$)
- $\underline{f}^2 : \mathit{Modes} \mapsto \mathcal{P}(\mathit{Modes} \times \mathit{Modes})$ (for dyadic $\mathbb{O}$)
- $\underline{f}^3 : \mathit{Modes} \mapsto \mathcal{P}(\mathit{Modes} \times \mathit{Modes} \times \mathit{Modes})$ (for triadic $\Rightarrow$)

A *hyper* La-model M consists of a *hyper*-frame F with a valuation V.

The function $\underline{f}^i$ in Def.4.1 determines, at a given mode, the possible modes available for evaluating a given modal formula. By default, we take as the *starting mode* for evaluating formulae the mode Ψ^{co}.

Definition 4.2 Given $\underline{f}^1$ of Def.4.1, we specify the following modes for $\mathbb{D}$:

(i) for $i \in \{tac, dac\}$, $\Psi^i_\phi(w, \mathbb{D})$ is $\|\phi\|_{\Psi^j} \in \mathcal{N}_{D^2}(w)$ with $\underline{f}^1(\Psi^i) = \Psi^j$

(ii) for $i \in \{tnc, dnc\}$, $\Psi^i_\phi(w, \mathbb{D})$ is $\|\phi\|_{\Psi^j} \in \mathcal{N}_{D^1}(w)$ with $\underline{f}^1(\Psi^i) = \Psi^j$

(iii) for $i = co$, $\Psi^i_\phi(w, \mathbb{D})$ is $\|\phi\|_{\Psi^j} \in \mathcal{N}_{D^2}(w)$ or $\|\phi\|_{\Psi^j} \in \mathcal{N}_{D^1}(w)$ with $\underline{f}^1(\Psi^i) = \Psi^j$

With $\underline{f}^1(\Psi^i) = \{\Psi^{co}\}$ for each $i \in \{co, tac, tnc, dac, dnc\}$.

To illustrate, consider (i) of Def.4.2: when evaluating in the teleological anankastic mode Ψ^{tac}, we interpret 'desire' $\mathbb{D}\phi$ as an action-relevant desire D^2

and interpret the internal goal ϕ in common mode. Condition (iii) states that, in the common mode, an ambiguous $\mathbb{D}\phi$ is satisfiable whenever it is satisfiable as a mere-desire or an action-relevant desire (possibly both).

For conditionals of the form $(\Delta, \mathbb{D}\phi) \Rightarrow \mathbb{O}\psi$ we find four possible (near-)anankastic interpretations of $\Rightarrow$. When a conditional is of the form $(\Delta, \theta) \Rightarrow \chi$ such that $\theta \neq \mathbb{D}\phi$ or $\chi \neq \mathbb{O}\psi$, we evaluate $\Rightarrow$ as a regular conditional.

Definition 4.3 Given $\underline{f}^3$ of Def.4.1, we specify the following modes for $\Rightarrow$.

(i) If $\phi = \mathbb{D}\phi'$ and $\psi = \mathbb{O}\psi'$, then $\Psi^{co}_{\Delta,\phi,\psi}(w, \Rightarrow)$ is $(\|\Delta\|_{\Psi^j}, \|\phi\|_{\Psi^k}, \|\psi\|_{\Psi^l}) \in \mathcal{N}_{\Rightarrow}(w)$, for some $(\Psi^j, \Psi^k, \Psi^l) \in \underline{f}^3(\Psi^{co}) \setminus \{(\Psi^{co}, \Psi^{co}, \Psi^{co})\}$

(ii) If $\phi \neq \mathbb{D}\theta$ or $\psi \neq \mathbb{O}\chi$, then $\Psi^{co}_{\Delta,\phi,\psi}(w, \Rightarrow)$ is $(\|\Delta\|_{\Psi^{co}}, \|\phi\|_{\Psi^{co}}, \|\psi\|_{\Psi^{co}}) \in \mathcal{N}_{\Rightarrow}(w)$, for $(\Psi^{co}, \Psi^{co}, \Psi^{co}) \in \underline{f}^3(\Psi^{co})$

With $\underline{f}^3(\Psi^i) = \{(\Psi^{co}, \Psi^{co}, \Psi^{co}), (\Psi^{co}, \Psi^{tac}, \Psi^{tac}), (\Psi^{co}, \Psi^{tnc}, \Psi^{tnc}), (\Psi^{co}, \Psi^{dac}, \Psi^{dac}), (\Psi^{co}, \Psi^{dnc}, \Psi^{dnc})\}$ for $i \in \{co, tac, tnc, dac, dnc\}$.

Def.4.3 ensures that conditionals are only evaluated in Ψ^{co} mode, namely, (near-)anankastic modes are reserved for the hyper modalities $\mathbb{D}$ and $\mathbb{O}$ occurring *within* such a conditional. Only $\mathbb{D}$ and $\mathbb{O}$ can be evaluated in (near-)anankastic modes, which are modes that arise by identifying a hyperconditional of the form $(\Delta, \mathbb{D}\phi) \Rightarrow \mathbb{O}\psi$. Hence, when evaluating $\mathbb{O}$ in a (near-)anankastic mode, we *come from* a mode that interprets a conditional: consequently, we have additional information (an antecedent and a context) at our disposal that facilitates interpreting $\mathbb{O}$. This is reflected in Def.4.4.

Definition 4.4 Given $\underline{f}^2$ of Def.4.1, we specify the following modes for $\mathbb{O}$.

(i) $\Psi^{tac}_{\Delta,\mathbb{D}\phi,\mathbb{O}\psi}(w, \mathbb{O})$ is $(\|\psi\|_{\Psi^j}, \|\Sigma^{D^2}_\Delta \wedge \phi\|_{\Psi^k}) \in \mathcal{N}_{O^1}(w)$ with $(\Psi^j, \Psi^k) \in \underline{f}(\Psi^{tac})$

(ii) $\Psi^{tnc}_{\Delta,\mathbb{D}\phi,\mathbb{O}\psi}(w, \mathbb{O})$ is $(\|\psi\|_{\Psi^j}, \|\Sigma^{D^2}_\Delta\|_{\Psi^k} \cap \|\mathbb{D}\phi\|_{\Psi^{tnc}}) \in \mathcal{N}_{O^1}(w)$ with $(\Psi^j, \Psi^k) \in \underline{f}(\Psi^{tnc})$

(iii) $\Psi^{dac}_{\Delta,\mathbb{D}\phi,\mathbb{O}\psi}(w, \mathbb{O})$ is $(\|\psi\|_{\Psi^j}, \|\phi\|_{\Psi^k}) \in \mathcal{N}_{O^2}(w)$ with $(\Psi^j, \Psi^k) \in \underline{f}(\Psi^{dac})$

(iv) $\Psi^{dnc}_{\Delta,\mathbb{D}\phi,\mathbb{O}\psi}(w, \mathbb{O})$ is $(\|\psi\|_{\Psi^j}, \|\mathbb{D}\phi\|_{\Psi^k}) \in \mathcal{N}_{O^2}(w)$ with $(\Psi^j, \Psi^k) \in \underline{f}(\Psi^{dnc})$

(v) $\Psi^{co}_{\top,\mathbb{O}\psi}(w, \mathbb{O})$ is $(\|\psi\|_{\Psi^j}, \|\top\|_{\Psi^k}) \in \mathcal{N}_{O^1}(w)$ or $(\|\psi\|_{\Psi^j}, \|\top\|_{\Psi^k},) \in \mathcal{N}_{O^2}(w)$ with $(\Psi^j, \Psi^k) \in \underline{f}(\Psi^{co})$

And $\underline{f}^2(\Psi^i) = \{(\Psi^{co}, \Psi^{co}\}$ for $i \in \{co, tac, tnc, dac\}$ and $\underline{f}^2(\Psi^{dnc}) = \{(\Psi^{dnc}, \Psi^{co})\}$. (Note that for (tnc) and (dnc), we require that $\mathbb{D}\phi$ in the second argument of $\mathbb{O}$ is interpreted as D^1.)

Consider (i) in Def.4.4, when evaluating $\mathbb{O}$ in anankastic mode Ψ^{tac}, we check whether in those cases where the agent's known action-relevant desires $D^2\theta \in \Delta$ have been optimally realized, together with the realization of ϕ, we find that ψ is the case. Hence, in anankastic mode Ψ^{tac}, we treat the antecedent $\mathbb{D}\phi$ as if the agent has an action-relevant desire $D^2\phi$, and evaluate $\mathbb{O}\psi$ teleologically as $O^1\psi$, while conditioning it explicitly on the context Δ.

Last, we define the semantics of hyper modalities $\mathbb{D}$, $\mathbb{O}$, and $\Rightarrow$. Note that modes are *only activated* whenever we encounter a hyper modality in a formula.

Definition 4.5 Let M be a hyper La-model of Def.4.1. For every $w \in W$ we have the regular clauses for non-hyper modalities of Def.3.3, extended with:

- $w \models_{\Psi^i} \mathbb{D}\phi$ iff $\mathbb{D}\phi$ is satisfied at w for some $\Psi^j \in \underline{f}^1(\Psi^i)$ of Def.4.2.
- $w \models_{\Psi^i} \mathbb{O}(\phi, \psi)$ iff $\mathbb{O}(\phi, \psi)$ is satisf. at w for some $\Psi^j \in \underline{f}^2(\Psi^i)$ of Def.4.4.
- $w \models_{\Psi^i} (\theta, \phi) \Rightarrow \psi$ iff $(\theta, \phi) \Rightarrow \psi$ is satisf. at w for some $\Psi^j \in \underline{f}^3(\Psi^i)$ of Def.4.3.

A formula is La-*satisfiable* if there is an La-model M with $w \in W$ of M and there is a mode $\Psi^i \in \mathcal{M}odes$ s.t. $M, w \models_{\Psi^i} \phi$.

For any ambiguous conditional $\theta = (\Delta, \mathbb{D}\phi) \Rightarrow \mathbb{O}\psi$ the hyper setting gives us the following: If *only* Ψ^{ac} is satisfiable we say θ is an *anankastic conditional*. If *only* Ψ^{nc} is satisfiable θ is a *teleological near-anankastic*. If *only* Ψ^{da} is satisfiable, we say θ is a action-relevant *deontic near-anankastic conditional*. If *only* Ψ^{dn} is satisfied θ is called a mere-desire *deontic near-anankastic conditional*. If several of (i)-(iv) are satisfied, the resulting interpretation is a disjunction reflecting the possible readings of $(\Delta, \mathbb{D}\phi) \Rightarrow \mathbb{O}\psi$ given Δ. If neither is satisfiable, we say the $(\Delta, \mathbb{D}\phi) \Rightarrow \mathbb{O}\psi$ has no interpretation given Δ, and hence is false. In interpreting hyper formulae of the form $(\Delta, \mathbb{D}\phi) \Rightarrow \mathbb{O}\psi$ in modes $\Psi^{ac}, \Psi^{nc}, \Psi^{da}$, and Ψ^{dn}, we employ the same semantic interpretations as used for the four formally defined conditionals (4), (6), (8), (9), of Section. 3.

The main difference between Section. 3 and the hyper-approach presented here, is that (a) we internalise the interpretation procedure (i.e., part of the pragmatics) through using hyper modalities and corresponding modes, and (b) we leave open the possibility that a conditional remains ambiguous (i.e., several modes may be satisfiable given a context Δ). As a consequence of (a) and (b), we can logically reason with ambiguous conditionals, such as P1 and P2 of the Tobacco shop scenario, without assuming a definite linguistic interpretation. We can use logic to determine, given a certain context, which interpretations of ambiguous (near-)anankastic conditionals are excluded, and which are jointly satisfiable. Let us look at the Tobacco shop scenario again.

5 Disambiguation and the Tobacco shop scenario

For sentences such as "if you want ϕ, you must ψ" the hyper-setting can help reducing ambiguity by determining which interpretations (i.e., modes) are excluded given certain contexts. To illustrate this, we have another look at the Tobacco shop scenario and consider two possible contexts. First, we recall the remarks by Prof. Pragmatics and Prof. Restraint, respectively:

If S wants to smoke, then S must buy cigarettes. (P1)

If S wants to smoke, then S must not buy cigarettes. (P2)

Using the hyper-modalities for 'desire' $\mathbb{D}$, 'must' $\mathbb{O}$, and the conditional '$\Rightarrow$', we obtain the following hyper-modal readings (adding $\top$ for an empty context):

$$(\top, \mathbb{D}\text{smoke}) \Rightarrow \mathbb{O}\text{buy} \tag{11}$$

$$(\top, \mathbb{D}\text{smoke}) \Rightarrow \mathbb{O}\neg\text{buy} \tag{12}$$

Suppose that at this point we do not yet know which readings, or contexts, Pragmatics and Restraint assign to their utterances. Can we already derive something from the joint utterance of (11) and (12)? The answer is yes. The hyper-modal setting tells us that (11) and (12) cannot be jointly satisfied under the same mode. For instance, if we interpret both formulae as anankastic conditionals (*tac*) the conjunction is not satisfiable (models are assumed to be hyper models from Def.4.1):

$$\text{For any } w \in W, \, w \not\models_{\Psi^{co}} (\top, \mathbb{D}\text{smoke}) \Rightarrow \mathbb{O}\neg\text{buy or } w \not\models_{\Psi^{co}} \\ (\top, \mathbb{D}\text{smoke}) \Rightarrow \mathbb{O}\neg\text{buy for } \Psi^{tac} \in f(\Psi^{co}) \tag{13}$$

In short, (13) depends on the consistency requirement on 'must' O^1 under consistent conditions, together with the exclusion of impossible conditionals (d_2, f_1, and f_2 of Def.4.1). Similar reasoning excludes identical interpretations of (11) and (12) for any of the four modes Ψ^i, $i \in \{tac, tns, dac, dnc\}$. (For space reasons, all semantic proofs will be omitted.) Hence, the formalism allows us to conclude that P1 and P2 must have *distinct* (near-)anankastic readings if they are to be jointly satisfiable.

Recall that P1 is commonly taken as an anankastic conditional, that is, 'buying' proves teleologically optimal for realizing the goal of 'smoking'. If we take (14) as given, what additional conclusions can we draw concerning P2?

$$(\top, D^2\text{smoke}) \Rightarrow O^1\text{buy} \tag{14}$$

If Pragmatics and Restraint agree on the fact that Smoke has an action-relevant desire to smoke, the only mode in which (12) may be satisfied is Ψ^{dac}. So far, we were able to draw some minimal conclusions about P1 and P2 without assuming any additional context, that is, these conclusions were drawn from the logical behaviour for the different modes of hyper-modalities $\mathbb{D}$ and $\mathbb{O}$.

Now, suppose Prof. Pragmatics asks Restraint to to explain herself. The latter recalls that Dr. Smoke has an action-relevant desire to stay healthy pointing out that smoking will obstruct that goal. We obtain the context $\Delta = \{D^2\text{health}, \Box(\text{smoke} \rightarrow \neg\text{health})\}$ and update the formalisation of P2:

$$(\Delta, \mathbb{D}\text{smoke}) \Rightarrow \mathbb{O}\neg\text{buy} \tag{15}$$

Independent of how we interpret P1, the additional context for P2 excludes the interpretation that (15) is an anankastic conditional: The action-relevant desire to be healthy $D^2\text{health}$ (with $\text{health} \in \Sigma_\Delta^{D^2}$), cannot be realised together with an action-relevant interpretation of $\mathbb{D}\text{smoke}$, namely, $D^2\text{smoke}$. In brief, goals expressed by an agent's action-relevant desires should be jointly realisable (cf. f_1, f_2 of Def. 4.1). The result is expressed in (16).

$$\text{For any } w \in W, w \not\models_{\Psi^{co}} (\Delta, \mathbf{D}\text{smoke}) \Rightarrow \bigcirc \neg \text{buy, for } \Psi^{tac} \in f(\Psi^{co}) \quad (16)$$

The above does not imply that an action-relevant reading of $\mathbf{D}$smoke is impossible for (12): a *deontic* reading of the consequent $\bigcirc \neg$buy interacts differently with the context and thus allows for other desire statement in the antecedent (cf. the discussion of (8) in Section. 3). Furthermore, we find that given Δ the realism requirement imposed on action-relevant (near-)anankastics in general is still satisfiable, i.e., if you smoke, you can still have an action-relevant desire to be healthy, but the latter goal cannot be attained.

Suppose another context Δ' in which Prof. Restraint recalls Dr. Smoke's promise to buy some cigarettes for a friend. Moreover, suppose she points out that keeping the promise prom is Smoke's duty, *irrespective* of whether he desires to smoke D^1smoke or actually does so, smoke. In other words, if Smoke keeps his promise, he will buy cigarettes $\Box$(prom $\to$ buy). We obtain the new context $\Delta' = \{O^2(\text{prom}, \text{smoke}), O^2(\text{prom}, D^1\text{smoke}), \Box(\text{prom} \to \text{buy})\}$.

$$\text{For any } w \in W, w \not\models_{\Psi^{co}} (\Delta', \mathbf{D}\text{smoke}) \Rightarrow \bigcirc \neg \text{buy, for } \Psi^{dac}, \Psi^{dnc} \in f(\Psi^{co}) \quad (17)$$

We find that the conditional expressed in (17) excludes any reading of P2 as a *deontic* near-anankastic conditional, i.e., under either desire reading. Namely, given Δ', the obligation to keep one's promise will conflict with the readings $O^2(\neg\text{buy}, \text{smoke})$ and $O^2(\neg\text{buy}, D^1\text{smoke})$ since not buying implies breaking one's promise (cf. f_1, f_2, and d_2 of Def. 4.1). Given Δ' and the anankastic reading of P1 (14), the only reading of $(\Delta', \mathbf{D}\text{smoke}) \Rightarrow \bigcirc \neg$buy which is not necessarily excluded is the teleological near-anankastic reading.

The analysis shows that, through interaction between contexts (such as Δ and Δ') and different interpretations of the antecedent and consequent (D^1 and D^2, respectively, O^1 and O^2) we may formally exclude certain interpretations of ambiguous linguistic expressions such as P1 and P2 of the Tobacco shop scenario. The example shows that certain restrictions on different interpretations of (near-)anankastics serve to reduce ambiguity, e.g., consistency of goals for teleological optimality in (16).

The hyper-modal setting enables us to represent ambiguity, and use formal machinery to (partially) resolve it, thus internalising some of the pragmatics of linguistic interpretations. Some of the benefits of this approach are that (i) we do not need to assume prior to formalisation that all ambiguity is resolved, (ii) we can formalise ambiguous sentences that will receive their interpretation at a later stage, and (iii) we can study those criteria that function as identifiers for rightly interpreting hyper-modalities. Still, future work should be devoted to identifying other conditions that enable us to draw conclusions from hyper-modal formulae concerning 'must', 'desire', and (near-)anankastic conditionals. Another point left unaddressed here is whether the logic of (near-)anankastic conditionals allows for (certain forms of) detachment (cf. [10]).

6 Conclusion and future work

In this work, we related semantics of deontic modality and deontic logic. We discussed the Tobacco shop scenario, highlighting the interaction between consequent, antecedent, and context in interpreting (near-)anankastics. We presented a logic inspired by [4], capturing four (near-)anankastic conditionals, while assuming linguistic interpretation of concrete conditionals prior to formalisation. We extended the formalism to the hyper setting, where hypermodalities bring ambiguity within the reach of logical analysis: i.e., internalising parts of the interpretation process of modalities, such as 'must' and 'desire'.

Perhaps the most unusual aspect of our approach is that we treat 'context' as part of the syntax of a formula. This means that one and the same natural language conditional must be translated differently in different contexts. This is unusual, since most approaches aim for a systematic analysis that accounts for the way in which the content of the sentence depends on context. Moreover, in Section. 5 we did not fully exploit the additional expressive power that comes with having the context in the language. We plan to do this in future research. For example, under suitable conditions, instead of assuming that we have a prohibition to smoke, we would be able to derive it.

This paper touched on several other points requiring future work: (i) Further the analysis of the Tobacco shop scenario, e.g., by relating it to existing approaches handling contrary-to-duty reasoning, nonmonotonic reasoning, and dynamic logic. (ii) Extend the analysis of hyper-modality (e.g., in the context of NLP). (iii) Investigate other aspects of pragmatics that can be studied in logic (e.g., detachment using nonmonotonic logic).

References

[1] Arlo-Costa, H., *The Logic of Conditionals*, in: E. N. Zalta, editor, *The Stanford Encyclopedia of Philosophy*, Metaphysics Research Lab, Stanford University, 2019, summer 2019 edition .

[2] Carmo, J. and A. J. Jones, *Deontic logic and contrary-to-duties*, in: *Handbook of philosophical logic*, Springer, 2002 pp. 265–343.

[3] Chellas, B. F., "Modal logic: an introduction," Cambridge university press, 1980.

[4] Condoravdi, C. and S. Lauer, *Anankastic conditionals are just conditionals*, Semantics & Pragmatics **9** (2016).

[5] Condoravdi, C. and L. van der Torre, *Anankastic conditionals deontic modality: linguistic and logical perspectives on oughts and ends* (2016).

[6] Gabbay, D. M., *A general theory of the conditional in terms of a ternary operator*, Theoria **38** (1972), pp. 97–104.

[7] Gabbay, D. M., *A theory of hypermodal logics: Mode shifting in modal logic*, Journal of Philosophical Logic **31** (2002), pp. 211–243.

[8] Hilpinen, R. and P. McNamara, *Deontic logic: A historical survey and introduction*, Handbook of deontic logic and normative systems **1** (2013), pp. 3–136.

[9] Kratzer, A., H.-J. Eikmeyer and H. Rieser, *The notional category of modality*, Formal semantics: The essential readings (1981), pp. 289–323.

[10] Lauer, S. and C. Condoravdi, *Preference-conditioned necessities: Detachment and practical reasoning*, Pacific Philosophical Quarterly **95** (2014), pp. 584–621.

[11] Parent, X. and L. van der Torre, *Detachment in normative systems: Examples, inference patterns, properties*, IfCoLog Journal of Logics and Their Applications **4** (2017), pp. 2295–3039.

[12] Sæbø, K. J., *Necessary conditions in a natural language*, Audiatur vox sapientiae: A Festschrift for Arnim von Stechow (2001), pp. 427–449.

[13] von Fintel, K. . S. I., *What to do if you want to go to harlem: Anankastic conditionals and related matters* .

[14] von Stechow, A., S. Krasikova and D. Penka, *Anankastic conditionals again*, A Festschrift for Kjell Johan Sæbø: In partial fulfilment of the requirements for the celebration of his 50th birthday (2006), pp. 151–171.

[15] Von Wright, G. H., "Norm and action," London:Routledge & Kegan Paul., 1963.

The Varieties of Ought-Implies-Can and Deontic STIT Logic

Kees van Berkel [1]

Institute of Logic and Computation,
Technische Universität Wien, 1040 Wien, Austria

Tim Lyon

Institute of Artificial Intelligence,
Technische Universität Dresden, 01069 Dresden, Germany

Abstract

STIT logic is a prominent framework for the analysis of multi-agent choice-making. In the available deontic extensions of STIT, the principle of Ought-implies-Can (OiC) fulfills a central role. However, in the philosophical literature a variety of alternative OiC interpretations have been proposed and discussed. This paper provides a modular framework for deontic STIT that accounts for a multitude of OiC readings. In particular, we discuss, compare, and formalize ten such readings. We provide sound and complete sequent-style calculi for all of the various STIT logics accommodating these OiC principles. We formally analyze the resulting logics and discuss how the different OiC principles are logically related. In particular, we propose an endorsement principle describing which OiC readings logically commit one to other OiC readings.

Keywords: Deontic logic, STIT logic, Ought implies can, Labelled sequent calculus

1 Introduction

From its earliest days, the development of deontic logic has been accompanied by the observation that reasoning about duties is essentially connected to *praxeology*, that is, the theory of agency (e.g. [13,31,44]). A prominent modal framework developed for the analysis of multi-agent interaction and choice is the logic of 'Seeing To It That' [7] (henceforth, STIT), and its potential for *deontic reasoning* was recognized from the outset [6]. Despite several philosophical investigations of the subject [5,24], concern for its formal specification lay dormant until the beginning of this century when a thorough investigation of deontic STIT logic was finally conducted [23,32]. Up to the present

---

[1] We would like to thank the reviewers of DEON2020 for their useful comments. This work is funded by the projects WWTF MA16-028, FWF I2982 and FWF W1255-N23. For questions and comments please contact kees@logic.at.

day, deontic STIT continues to receive considerable attention, being applied to epistemic [11], temporal [9], and juridical contexts [28].

The traditional deontic STIT setting [23] is rooted in a utilitarian approach to agential choice, which enforces certain minimal properties on its agent-dependent obligation operators. In particular, it implies a version of the eminent *Ought-implies-Can* principle (henceforth, OiC), a metaethical principle postulating that 'what an agent ought to do, the agent can do'. OiC has a long history within moral philosophy and can be traced back to, for example, Aristotle [2, VII-3], or the "Roman legal maxim *impossibilium nulla obligatio est*" [40]. Still, it is often accredited to the renowned philosopher Immanuel Kant [25, A548/B576]. Aside from debates on whether OiC should be adopted at all [19,35], most discussions revolve around which *version* of the principle should be endorsed. Notable positions have been taken up by Hintikka [22], Lemmon [27], Stocker [36], Von Wright [43], and, more recently, Vranas [40]. However, most of these authors advocate readings that are either weaker or stronger than the minimally implied OiC principle of traditional deontic STIT. In order to formally investigate these different readings, it is necessary to modify and fine-tune the traditional framework.

The contributions of this work are as follows: First, we discuss, compare, and formalize ten OiC principles occurring in the philosophical literature (Sect. 2). To the best of our knowledge, such a taxonomy of principles has not yet been undertaken (cf. [40] for an extensive bibliography). The intrinsically agentive setting provided within the STIT paradigm will enable us to conduct a fine-grained analysis of the various renditions of OiC. Still, the available utilitarian characterization of deontic STIT makes it cumbersome to accommodate this multiplicity of principles. For that reason, the present endeavour will take a more modular approach to STIT, adopting relational semantics [14] through which the use of utilities may be omitted [9] (Sect. 3).

Second, we provide sound and complete sequent-style calculi for all classes of deontic STIT logics accommodating the various kinds of formalized OiC principles (Sect. 4). In particular, we adopt labelled sequent calculi which explicitly incorporate useful semantic information into their rules [34,39]. A general benefit of using sequent-style calculi [37], in contrast to axiomatic systems, is that the former are suitable for applications (e.g. proof-search and counter-model extraction) [29]. Although this work is not the first to address STIT through alternative proof-systems [4,29,41], it is the first to address both the traditional deontic setting [23] and a large class of novel deontic STIT logics.

Last, we will use the resulting deontic STIT calculi to obtain a formal taxonomy of the OiC readings discussed. The benefit of employing proof theory is twofold: First, we classify the ten OiC principles according to the respective strength of the underlying STIT logics in which they are embedded (Sect. 5). The calculi can be used to determine which logics subsume each other, giving rise to what we call an *endorsement principle*; it demonstrates which endorsement of which OiC readings logically commits one to endorsing other OiC readings (from the vantage of STIT). Second, the calculi can be applied to show the mutual independence of certain OiC readings through the construction of counter-models from failed proof-search. This work will lay the foundations for an extensive investigation of OiC within the realm of agential choice, and future research directions will be addressed in Sect. 6.

2 A Variety of Ought-implies-Can Principles

The fields of moral philosophy and deontic logic have given rise to a variety of metaethical principles, such as "no vacuous obligations" [42], "deontic contingency" [3], "deontic consistency" [21], and the principle of "alternate possibilities" [15]. One of the most prevalent is perhaps the principle of *"Ought-implies-Can"*. In fact, we will see that each of the former metaethical canons is significant relative to different interpretations of OiC. In this section, we introduce and discuss ten such interpretations of OiC and indicate their relation to the aforementioned metaethical principles. Many philosophers have addressed OiC, and while earlier thinkers (e.g. Aristotle and Kant) only discussed it implicitly, it was made an explicit subject of investigation in the past century. We will focus solely on frequently recurring readings from authors that are—in our opinion—central to the debate. Despite the apparent relationships between some of the considered OiC readings, a precise taxonomy of their logical interdependencies can only be achieved through a formal investigation of their corresponding logics. We will provide such a taxonomy in Sect. 5.

One of the allures of OiC is that it releases agents from alleged duties which are impossible, strenuous, or over-demanding [16,30]. Namely, in its basic formulation—'what an agent ought to do, the agent can do'—the principle ensures that an agent can only be normatively bound by what it can do, i.e., 'what the agent can't do, the agent is not obliged to do'. Most disagreement concerning OiC can be understood in terms of the degree to which an agent must be burdened or relieved. In essence, such discussions revolve around the appropriate interpretation of the terms 'ought', 'implies', and predominantly, 'can'. In what follows, we take 'ought' to represent agent-dependent obligations and take 'implies' to stand for material implication (for a discussion see [1,40]). With respect to the term 'can', we roughly identify four readings: (i) possibility, (ii) ability, (iii) violability, and (iv) control. These four concepts give rise to eight OiC principles. We close the section with a discussion of two additional OiC principles which adopt a normative reading of the term 'can'.

Throughout our discussion we introduce logical formalizations of the proposed OiC readings that will be made formally precise in subsequent sections. Therefore, it will be useful at this stage to introduce some notation employed in our formal language: we let ϕ stand for an arbitrary STIT formula. The connectives $\neg, \wedge$, and $\rightarrow$ are respectively interpreted as 'not', 'and', and 'implies'. Let $[i]$ be the basic STIT operator such that, in the spirit of [7], we interpret $[i]\phi$ both as 'agent i sees to it that ϕ' and 'agent i chooses to ensure ϕ'. We use the operator $\Box$ to refer to what is 'settled true', such that $\Box\phi$ can be read as 'currently, ϕ is settled true'. The main use of $\Box$ is to discern between those state-of-affairs that can become true—i.e. actual—through an agent's choice and those state-of-affairs that are true—i.e. actual—independent of the agent's choice. For this reason we will also interchangeably employ the term 'actual' in referring to $\Box$ (for an extensive discussion see [7]). We take $\Diamond$ to be the dual of $\Box$, denoting that some state of affairs is actualizable, i.e., can become actual. Last, we read $\otimes_i$ as 'it ought to be the case for agent i that'.[2]

1. *Ought implies Logical Possibility*: $\otimes_i \phi \to \neg \otimes_i \neg \phi$ (OiLP). What is obligatory for an agent, should be consistent from an ideal point of view.

The first principle, which is one of the weakest interpretations of OiC, requires the content of an agent's obligations to be non-contradictory. Within the philosophical literature this interpretation has been referred to as "ought implies logical possibility" [40] and the principle has been generally equated with the metaethical principle of "deontic consistency" (e.g. [17,27]).[3] As a minimal constraint on deontic reasoning, the principle is a cornerstone of (standard) Deontic Logic [3,21,42], though it has been repudiated by some [27].

2. *Ought implies Actually Possible*: $\otimes_i \phi \to \Diamond \phi$ (OiAP). What is obligatory for an agent, should be actualizable.

The above principle is slightly stronger than the previous one: it rules out those conceptual consistencies that might not be realizable at the current moment.[4] That is, the principle requires that norm systems can only demand what can presently become *actual*. For example, 'although it is logically possible to open the window, it is currently not actualizable, since I am tied to the chair'.

However, both OiLP and OiAP are arguably too weak, and do not involve the concerned agent whilst interpreting 'can'. For instance, although 'a moon eclipse' is both logically and actually possible, it should not be considered as something an agent ought to bring about. For this reason, most renditions of OiC involve the agent explicitly:

3. *Ought implies Ability*: $\otimes_i \phi \to \Diamond [i] \phi$ (OiA). What is obligatory for an agent, the agent must have the ability to see to, i.e. the choice to realize.

The above reading enforces an explicitly agentive precondition on obligations: it requires ability as the agent's capacity to guarantee the realization of that which is prescribed.[5] The concept of ability has many formulations (cf. [11,12,18,43]); for example, it may denote general ability, present ability, potential ability, learnability, know-how, and even technical skill (also, see [30,36,40]

---

[2] We stress that OiC is essentially agentive, but not necessarily referring to choice in particular. For this reason, we distinguish 'it ought to be the case for agent i that' from the stronger 'agent i ought to see to it that'. The latter reading corresponds to the notion of '*dominance ought*' advocated by Horty [23]. Initially, the distinction will be observed for OiC. In Sect. 5 we show how the logics can be expanded to obtain the stronger reading proposed in [23].
[3] In [45], Von Wright baptizes OiLP 'Bentham's Law' and points out that the canon was already adopted by Mally in what is known as the first attempt to construct a deontic logic.
[4] In [21], OiC is named 'Kant's law' and OiLP and OiAP are classified as weak versions of the law. However, it is open to debate which reading of OiC Kant would admit (e.g. [26,38]).
[5] Similarly, Von Wright distinguishes between human and physical possibility (cf. OiA and OiAP, resp.), both implying logical possibility (cf. OiLP) as a necessary condition [44, p.50].

on the corresponding notion of 'inability'). In what follows, we take 'ability' to mean a *moment-dependent* possibility for an agent to guarantee that which is commanded through an available *choice*.

Observe that OiA is the principle implied by the traditional, utilitarian based deontic STIT logic [23,32]. However, this OiC reading does not completely capture the notion of 'ability' as generally encountered in the philosophical literature. That is, OiA merely requires that what is prescribed for the agent can be guaranteed through one of the agent's choices, but does not exclude what is called vacuously satisfied obligations. Agents could still have obligations (and corresponding 'abilities') to bring about inevitable states-of-affairs, such as the obligation to realize a tautology (cf. [9]). Philosophical notions of ability regularly ban such consequences by strengthening the concept of ability with either (i) the *possibility* that the obligation may be *violated*, (ii) the agent's *ability to violate* what is demanded (i.e. an agent may refrain from fulfilling a duty), (iii) the right *opportunity* for the agent to exercise its ability, or (iv) the agent's *control* over the situation (i.e. the agent's power to decide over the fate of what is prescribed). All of the above conceptions of agency are contingent in nature, that is, they range over state-of-affairs which are capable of being otherwise [24]. Each notion will be addressed in turn.

4. *Ought implies Violability*: $\otimes_i \phi \to \Diamond \neg \phi$ (OiV). An agent's obligation must be violable, that is, the opposite of what is prescribed must be possible.

The above principle corresponds to the metaethical principle of "no vacuous obligations", which ensures that neither tautologies are obligatory nor contradictions are prohibited [3,21,43]. However, in OiV a violation might still arise through causes external to the agent concerned; e.g. 'the prescribed opening of a window, might be closed through a strong gust of wind'.[6] The following principle strengthens this notion by making violability an agentive matter:

5. *Ought implies Refrainability*: $\otimes_i \phi \to \Diamond [i] \neg [i] \phi$ (OiR). An agent's obligation must be deliberately violable by the agent, that is, the agent must be able to refrain from satisfying its obligation.

In the jargon of STIT, we say that *refraining* from fulfilling one's duty requires "an embedding of a non-acting within an acting" [7, Ch.2]. That is, it requires the possibility to 'see to it that one does not see to it that'. However, the two violation principles above are insubstantial when that which is obliged is not possible in the first place.[7] For instance, it is not difficult for an agent to violate an obligation to 'create a moon eclipse' (it could not be done otherwise).[8] To avoid such cases, we often find that the ideas from **1–5** are combined:

---

[6] Already in [42] Von Wright posed the 'no vacuous obligations' principle as a central principle of deontic logic. There, he referred to it as "the principle of contingency", however, contingency requires that an obligation is not only violable, but also satisfiable (cf. OiO).
[7] We conjecture that this is why Vranas states that OiR is strictly not an OiC principle [40].
[8] Observe that violability relates strongly to the metaethical principle of "alternate possibility", stating that an agent is morally culpable if it could have done otherwise (e.g. [15,47]).

6. *Ought implies Opportunity* (OiO): $\otimes_i \phi \to (\Diamond \phi \wedge \Diamond \neg \phi)$. What is obligatory for an agent, must be a contingent state-of-affairs.

The above uses the terms 'opportunity' and 'contingency' intentionally in an interchangeable manner. Like previous terms, these terms know a variety of readings in the literature (cf. [15,16,40,42]). Nevertheless, what these readings share in relation to OiC is that they refer to the propriety of the circumstances in which the agent is required to fulfill its duty. Minimally, opportunity and contingency both require that a state-of-affairs within the scope of an active norm must be presently manipulable; i.e. the state-of-affairs can still become true or false.[9] This interpretation of OiO is related to what Von Wright has in mind when he talks about the opportunity to interfere with the course of nature [43], and to Anderson and Moore's claim that sanctions (i.e. violations) must be both provokable and avoidable, viz. contingent [3].

Taking the above a step further, agency can be more precisely described as the agent's *ability* together with the right *opportunity*. Following Vranas [40], the latter component specifies "the situation hosting the event in which the agent has to exercise her ability". The following principle merges these ideas:

7. *Ought implies Ability and Opportunity*: $\otimes_i \phi \to (\Diamond[i]\phi \wedge \Diamond \phi \wedge \Diamond \neg \phi)$ (OiA + O). What is obligatory for an agent, must be a contingent state-of-affairs whose truth the agent has the ability to secure.[10]

The above is the first completely agentive OiC principle, making that which is obligatory fall, in all its facets, within the reach of the agent. Such a reading of OiC can be said to be truly deliberative and both Vranas [40] and Von Wright [43] appear to endorse a principle similar to OiA+O. However, there is an even stronger reading which restricts norms to those state-of-affairs within the agent's complete *control*:

8. *Ought implies Control*: $\otimes_i \phi \to (\Diamond[i]\phi \wedge \Diamond[i]\neg \phi)$ (OiCtrl). What is obligatory for an agent, the agent must have the ability to see to and the agent must have the ability to see to it that the obligation is violated.

This reading, arguably advocated by Stocker [36], requires that an agent can act *freely*: "it has often been maintained that we act freely in doing or not doing an act only if we both can do it and are able not to do it" [36].[11] This last, perhaps too strong, instance of OiC implies that an agent is only subject to norms whose subject matter is within the *power* of the agent.

In all its readings, OiC has still been regarded as too strong. For example,

---

[9] A more fine-grained distinction can be made: in temporal settings a state-of-affairs can be occasionally true and false (i.e. contingent), despite the fact that at the present moment it is settled true and thus beyond the scope of the agent's influence (i.e. there is no opportunity). In the current atemporal STIT setting, this will not be explored.

[10] In basic atemporal STIT the occurrence of $\Diamond \phi$ in the consequent of OiA+O can be omitted since it is strictly implied by $\Diamond[i]\phi$; that is, if ϕ can be the result of an agent's choice, then by definition it can be actualized. For the sake of completion we leave $\Diamond \phi$ present in OiA+O.

[11] In the above quote, 'able not to do $[\phi]$' can also be interpreted as $\Diamond[i]\neg[i]\neg\phi$, instead of $\Diamond[i]\neg\phi$. The resulting principle would then equate with the weaker OiA+O in basic STIT.

Label	Ought implies...	Formalized	References
OiLP	Logical Possibility	$\otimes_i \phi \to \neg \otimes_i \neg \phi$	[3], [17], [42], [45]
OiAP	Actually Possible	$\otimes_i \phi \to \Diamond \phi$	[17], [23, Ch.3]
OiA	Ability	$\otimes_i \phi \to \Diamond [i] \phi$	[23, Ch.4], [43, Ch.7]
OiV	Violability	$\otimes_i \phi \to \Diamond \neg \phi$	[3], [16], [18], [43, Ch.8]
OiR	Refrainability	$\otimes_i \phi \to \Diamond [i] \neg [i] \phi$	[18]
OiO	Opportunity	$\otimes_i \phi \to (\Diamond \phi \wedge \Diamond \neg \phi)$	[3], [15], [16], [42], [44]
OiA+O	Ability and Opp.	$\otimes_i \phi \to (\Diamond [i] \phi \wedge \Diamond \phi \wedge \Diamond \neg \phi)$	[1], [26], [40], [43]
OiCtrl	Control	$\otimes_i \phi \to (\Diamond [i] \phi \wedge \Diamond [i] \neg \phi)$	[16], [36], [30]
OiNC	Normatively Can	$\otimes_i \phi \to \otimes_i \Diamond \phi$	[1], [22]
OiNA	Normatively Able	$\otimes_i \phi \to \otimes_i \Diamond [i] \phi$	[1], [22]

Fig. 1. List of the ten OiC principles together with their treatment in the literature.

Lemmon challenged the legitimacy of OiLP in light of the existence of moral dilemmas [27]. Other philosophers, like Hintikka [22], adopted more modest standpoints toward OiC, suggesting weaker, normative versions of the principle. In light of the latter, it has been argued that OiC is dispositional, merely capturing a normative attitude towards OiC [1]. Two approaches present themselves: (i) 'it *ought to be* the case that what morality prescribes is possible' or (ii) 'it *ought to be possible* for an agent to fulfill its obligations'.[12] The former does not correspond to an OiC principle, but only expresses that OiC *should* hold as a metaethical principle (we return to this in Sect. 5). The latter approach does provide OiC principles—we consider two possible readings:

9. *Ought implies Normatively Can*: $\otimes_i \phi \to \otimes_i \Diamond \phi$ (OiNC). What is obligatory for an agent, ought to be actually possible (for the agent).

10. *Ought implies Normatively Able*: $\otimes_i \phi \to \otimes_i \Diamond [i] \phi$ (OiNA). What is obligatory for an agent, ought to be actualizable through the agent's behavior.

Hence, both OiNC and OiNA require that, 'if ϕ ought to be the case for agent i, it ought to be the case for agent i that ϕ is actually possible (as a result of the agent's choice-making)'. In Fig. 1, the ten principles are collected and associated with references to the various authors that treat such principles.

It is not our aim to decide which OiC principle should be adopted, as good cases have been made for each. Instead, our present aim is as follows: first, we appropriate the framework of STIT such that all ten principles can be explicitly formulated (Sect. 4). Second, we use the resulting logics to formally determine the logical relations between the ten principles (Sect. 5). The final result will be a logical hierarchy of OiC principles, identifying which principles subsume others and which are mutually independent within the setting of STIT.

3 Deontic STIT Logic for Ought-implies-Can

In this section, we will introduce a general deontic STIT language and semantics whose modularity enables us to define a collection of deontic STIT logics that

---

[12] Hintikka advocates the first possibility; i.e. "$\mathcal{O}(\mathcal{O}\phi \to \Diamond \phi)$" [22]. However, one could argue that the first occurrence of $\mathcal{O}$ is actually agent-*in*dependent, and the latter agent-dependent.

will accommodate the variety of OiC principles discussed previously. It will suffice to consider a multi-agent modal language containing the basic STIT operator (i.e. the Chellas STIT) and the 'settled true' operator, extended with agent-dependent deontic operators.

Definition 3.1 (The Language $\mathcal{L}_n$) Let $Ag = \{1, 2, ..., n\}$ be a finite set of agent labels and let $Atm = \{p_1, p_2, p_3...\}$ be a denumerable set of propositional atoms. The language $\mathcal{L}_n$ is defined via the following BNF grammar:

$$\phi ::= p \mid \neg p \mid \phi \vee \phi \mid \phi \wedge \phi \mid \Box \phi \mid \Diamond \phi \mid [i]\phi \mid \langle i \rangle \phi \mid \otimes_i \phi \mid \ominus_i \phi$$

where $i \in Ag$ and $p \in Atm$.

We note that the formulae of $\mathcal{L}_n$ are defined in negation normal form. In line with [8,29], we opt for this notation because it will enhance the readability of the technical part of this paper. Namely, negation normal form will reduce the number of logical rules needed in our sequent-style calculi (see Sect. 4), and will simplify the structure of sequents used in derivations (see Sect. 5). Briefly, the negation of a formula $\phi \in \mathcal{L}_n$, denoted by $\neg \phi$, can be obtained by replacing each positive propositional atom p with its negation $\neg p$ (and vice versa), each $\wedge$ with $\vee$ (and vice versa), and each modal operator with its corresponding dual (and vice versa).

The logical connectives $\vee$ and $\wedge$ stand for 'or' and 'and', respectively. Other connectives and abbreviations are defined accordingly: $\phi \to \psi$ iff $\neg \phi \vee \psi$, $\phi \equiv \psi$ iff $(\phi \to \psi) \wedge (\psi \to \phi)$, $\top$ iff $p \vee \neg p$, and $\bot$ iff $p \wedge \neg p$. The modal operators $\Box$, $[i]$, and $\otimes_i$ express, respectively, 'currently, it is settled true that', 'agent i sees to it that', and 'it ought to be the case for agent i that'. We take $\Diamond$, $\langle i \rangle$, and $\ominus_i$ as their respective duals. Last, we interpret $\ominus_i$ as 'it is not obligatory for agent i that not' (a similar interpretation is applied to $\Diamond$ and $\langle i \rangle$). (NB. negation normal form requires us to take diamond-modalities as primitive.) [13]

3.1 Minimal Deontic STIT Frames

Since we are dealing with an atemporal STIT language, we can forgo the traditional semantics of branching time frames with agential choice functions [7]. Instead, we adopt a more modular approach using relational semantics [14]. As shown in [20], it suffices to semantically characterize basic STIT using frames that only model moments partitioned into equivalence classes, with the latter representing the choices available to the agents at the respective moment. As our starting point, we propose the following minimal deontic STIT models:

Definition 3.2 (Frames and Models for DS_n) A DS_n-*frame* is defined to be a tuple $F = \langle W, R_\Box, \{R_{[i]} \mid i \in Ag\}, \{R_{\otimes_i} \mid i \in Ag\} \rangle$ with $n = |Ag|$. Let $R_\alpha \subseteq W \times W$ and $R_\alpha(w) := \{v \in W \mid (w, v) \in R_\alpha\}$ for $\alpha \in \{\Box\} \cup \{[i], \otimes_i \mid i \in Ag\}$. Let W be a non-empty set of worlds $w, v, u...$ where:

---

[13] In line with [32], we take the concatenation $\otimes_i[i]$ to stand for 'agent i ought to see to it that', thus expressing the stronger agentive reading of obligation defended by [23] (also, see footnote 2). However, whether $\otimes_i[i]$ will capture the intended logical behavior of this reading will depend on the adopted class of STIT-frames. We will discuss this in Sect. 5.

C1 $R_\Box$ is an equivalence relation.
C2   For all $i \in Ag$, $R_{[i]} \subseteq R_\Box$ is an equivalence relation.
C3   For all $w \in W$ and all $u_1, ..., u_n \in R_\Box(w)$, $\bigcap_{i \in Ag} R_{[i]}(u_i) \neq \emptyset$.
D1   For all $w, v, u \in W$, if $R_\Box wv$ and $R_{\otimes_i} wu$, then $R_{\otimes_i} vu$.

A DS_n-*model* is a tuple $M = (F, V)$ where F is a DS_n-frame and V is a valuation function mapping propositional atoms to subsets of W, i.e. $V: Atm \mapsto \mathcal{P}(W)$.

In Def. 3.2, property **C1** stipulates that DS_n-frames are partitioned into $R_\Box$-equivalence classes, which we will refer to as *moments*. Intuitively, a moment is a collection of worlds that can become actual. For every agent in the language, **C2** partitions moments into equivalence classes, representing the agent's *choices* at such moments. The elements of a choice represent those worlds that can become actual through exercising that choice. **C3** captures the pivotal STIT principle called 'independence of agents', ensuring that all agents can jointly perform their available choices; i.e. simultaneous choices are consistent (cf. [7]). **D1** enforces that ideal worlds do not vary from different perspectives within a single moment; i.e. an ideal world is ideal from the perspective of the entire moment. In addition, **D1** states that obligations are moment-dependent; i.e. obligations might vary from moment to moment. We emphasize that the class of DS_n-frames does not require that worlds ideal at a certain moment lie within that very moment. Hence, what is ideal might not be realizable by any of the agents' (combined) choices, and so, might be beyond the grasp of agency.[14]

Definition 3.3 (Semantics for $\mathcal{L}_n$) Let M be a DS_n-model and let $w \in W$ of M. The *satisfaction* of a formula $\phi \in \mathcal{L}_n$ in M at w is defined accordingly:

1. $w \Vdash p$ iff $w \in V(p)$
2. $w \Vdash \neg p$ iff $w \notin V(p)$
3. $w \Vdash \phi \wedge \psi$ iff $w \Vdash \phi$ and $w \Vdash \psi$
4. $w \Vdash \phi \vee \psi$ iff $w \Vdash \phi$ or $w \Vdash \psi$
5. $w \Vdash \Box \phi$ iff $\forall u \in R_\Box(w)$, $u \Vdash \phi$
6. $w \Vdash \Diamond \phi$ iff $\exists u \in R_\Box(w)$, $u \Vdash \phi$
7. $w \Vdash [i]\phi$ iff $\forall u \in R_{[i]}(w)$, $u \Vdash \phi$
8. $w \Vdash \langle i \rangle \phi$ iff $\exists u \in R_{[i]}(w)$, $u \Vdash \phi$
9. $w \Vdash \otimes_i \phi$ iff $\forall u \in R_{\otimes_i}(w)$, $u \Vdash \phi$
10. $w \Vdash \ominus_i \phi$ iff $\exists u \in R_{\otimes_i}(w)$, $u \Vdash \phi$

Global truth, validity, and semantic entailment are defined as usual (see [10]). We define the *logic* DS_n as the set of $\mathcal{L}_n$ formulae valid on all DS_n-frames.

3.2 Expanded Deontic STIT Frames

In order to obtain an assortment of deontic STIT characterizations accommodating the different OiC principles, we proceed in two ways: first, we define more fine-grained deontic STIT operators capturing deliberative aspects of obligation, and second, we introduce a class of frame properties that change the behavior of the $\otimes_i$ operator when imposed on DS_n-frames.

---

[14] Traditional deontic STIT confines ideal worlds to moments since it restricts the evaluation of utilities to moments [23]. Consequently, $(\otimes_i \phi \rightarrow \ominus_i \phi) \equiv (\otimes_i \phi \rightarrow \Diamond \phi)$ is valid for the traditional approach, and thus, logical and actual possibility coincide. Our alternative semantics enables us to differentiate between OiLP, OiAP and a variety of other OiC principles.

Observe that in basic STIT the choice-operator $[i]$ is a normal modal operator, which implies that $[i]\top$ is one of its validities. In contrast, the more refined *deliberative* STIT operator—i.e. $[i]^d\phi$ iff $[i]\phi \wedge \Diamond\neg\phi$—is non-normal and, for this reason, has been taken as defined [24] (with the exception of [46]). (NB. For deliberative STIT, choices thus range over contingent state of affairs.) For the same reason that $\otimes_i\top$ is a validity of basic DS_n, we will similarly introduce two defined modalities for *deliberative obligations*. Namely, we take

$$\otimes_i^d \phi \text{ iff } \otimes_i \phi \wedge \Diamond\neg\phi$$

to define a *weak deliberative* obligation, expressing that an agent's obligations can be violated (cf. [32,9]). Furthermore, we introduce

$$\otimes_i^c \phi \text{ iff } \otimes_i \phi \wedge \Diamond[i]\neg\phi$$

as defining a *strong deliberative* obligation, asserting that the obligation is violable through the agent's behavior. These operators will be necessary to formally capture the deliberative versions of OiC in the present STIT setting.

Additionally, we provide four properties that may be imposed on DS_n-frames to change the logical behavior of the $\otimes_i$ operator:

D2 For all $w \in W$ there exists $v \in W$ s.t. $R_{\otimes_i}wv$.
D3 For all $w,v \in W$, if $R_{\otimes_i}wv$ then $R_\Box wv$.
D4 For all $w,v,u \in W$, if $R_{\otimes_i}wv$ and $R_{[i]}vu$, then $R_{\otimes_i}wu$.
D5 For all $w \in W$, there exists a $v \in W$, such that $R_{\otimes_i}wv$ and for all $u \in W$, if $R_{[i]}vu$, then $R_{\otimes_i}wu$.

Property **D2** requires that obligations are consistent; i.e. at every moment and for every agent, there exists an ideal situation for which the agent should strive (cf. seriality in Standard Deontic Logic [21]). **D3** enforces that ideal worlds are confined to moments (implying that every ideal world is realizable at its corresponding moment; cf. footnote 14). Subsequently, **D4** expresses that agent-dependent obligations are about choices, thus enforcing that every ideal world coincides with an ideal choice (cf. footnote 13): i.e. when 'it ought to be the case for agent i that' then 'agent i ought to see to it that' (the other direction follows from **C2** Def. 3.2). Lastly, **D5** states that for every agent i there always exists at least one ideal choice (depending on whether **D3** is adopted, this ideal choice will be guaranteed accessible by an agent or not). It must be noted that, as shown in [9], all four properties hold for the traditional approach to deontic STIT [32]. We return to this in Sect. 5.

We define the entire class of STIT logics considered in this paper as follows:

Definition 3.4 (The logics $\mathsf{DS}_n\mathsf{X}$) Let $\mathcal{D} = \{\mathbf{D2}, \mathbf{D3}, \mathbf{D4}, \mathbf{D5}\}$, $n = |Ag|$ and $\mathsf{X} \subseteq \mathcal{D}$. A $\mathsf{DS}_n\mathsf{X}$-*frame* is a tuple $F = \langle W, R_\Box, \{R_{[i]} \mid i \in Ag\}, \{R_{\otimes_i} \mid i \in Ag\}\rangle$ such that F satisfies all properties of a DS_n-frame (Def. 3.2) expanded with the frame properties X. A $\mathsf{DS}_n\mathsf{X}$-*model* is a tuple (F, V) where F is a $\mathsf{DS}_n\mathsf{X}$-*frame* and V is a valuation function as in Def. 3.2. We define the *logic* $\mathsf{DS}_n\mathsf{X}$ to be the set of formulae from $\mathcal{L}_n$ valid on all $\mathsf{DS}_n\mathsf{X}$-frames.

In the following section we provide sound and complete sequent-style calculi for all logics $\mathsf{DS}_n\mathsf{X}$ obtainable through Def. 3.4. Together with the defined deliberative obligation modalities $\otimes_i^d$ and $\otimes_i^c$, the resulting class of calculi will suffice to capture all the deontic STIT logics accommodating the different OiC principles of Sect. 2. This will be demonstrated in Sect. 5.

4 Deontic STIT Calculi for Ought-implies-Can

This section comprises the technical part of the paper: we introduce sound and complete sequent-style calculi $\mathsf{G3DS}_n\mathsf{X}$ for the multi-agent logics $\mathsf{DS}_n\mathsf{X}$ defined in Def. 3.4. In what follows, we build on a simplified version of the refined labelled calculi for basic STIT proposed in [29]. In the present work, we modify this framework to include the deontic setting. Due to space constraints, we refer to [29] for an extensive discussion on refined labelled calculi. For an introduction to sequent-style calculi in general see [37], and for labelled calculi in particular, see [34,39]. Labelled calculi offer a procedural, computational approach to making explicit semantic arguments. This approach not only allows for a precise understanding of the logical relationships between the different OiC readings and corresponding logics, but can additionally be harnessed to construct counter-models confirming the independence of certain OiC principles. We will demonstrate this in Sect. 5.

Definition 4.1 Let $Lab := \{x, y, z, ...\}$ be a denumerable set of *labels*. The language of our calculi consists of *sequents* Λ, which are syntactic objects of the form $\mathcal{R} \vdash \Gamma$. $\mathcal{R}$ and Γ are defined via the following BNF grammars:

$$\mathcal{R} ::= \varepsilon \mid R_\Box xy \mid R_{[i]}xy \mid R_{\otimes_i}xy \mid \mathcal{R},\mathcal{R} \qquad \Gamma ::= \varepsilon \mid x:\phi \mid \Gamma,\Gamma$$

with $i \in Ag$, $\phi \in \mathcal{L}_n$, and $x, y \in Lab$.

We refer to $\mathcal{R}$ as the *antecedent* of Λ and to Γ as the *consequent* of Λ. We use $\mathcal{R}, \mathcal{R}', \ldots$ to denote strings generated by the top left grammar and refer to formulae (e.g. $R_{[i]}xy$ and $R_{\otimes_i}xy$) occurring in such strings as *relational atoms*. We use $\Gamma, \Gamma', \ldots$ to denote strings generated by the top right grammar and refer to formulae (e.g. $x:\phi$) occurring in such strings as *labelled formulae*. We take the comma operator to commute and associate in $\mathcal{R}$ and Γ (i.e. $\mathcal{R}$ and Γ are multisets) and read its presence in $\mathcal{R}$ and Γ, respectively, as a conjunction and a disjunction (cf. Def. 4.5). We let ε represent the *empty string*.[15] Last, we use $Lab(\mathcal{R} \vdash \Gamma)$ to represent the set of labels contained in $\mathcal{R} \vdash \Gamma$.

The calculus $\mathsf{G3DS}_n$ for the minimal deontic STIT logic DS_n (with $n \in \mathbb{N}$) is shown in Fig. 2. Intuitively, $\mathsf{G3DS}_n$ can be seen as a transformation of the semantic clauses of Def. 3.3 and DS_n-frame properties of Def. 3.2 into inference rules. For example, the (id) rule encodes the fact that either a propositional atom p holds at a world in a DS_n-model, or it does not (recall that a comma

---

[15] The empty string ε serves as an identity element for comma (e.g. $R_\Box xy, \varepsilon \vdash x:p, \varepsilon, y:q$ identifies with $R_\Box xy \vdash x:p, y:q$). If ε is the entire antecedent or consequent, it is left empty by convention (e.g. $\varepsilon \vdash \Gamma$ identifies with $\vdash \Gamma$). In what follows, it suffices to leave ε implicit.

$$\frac{}{\mathcal{R} \vdash x : p, x : \neg p, \Gamma} \text{ (id)} \qquad \frac{\mathcal{R} \vdash x : \phi, w : \psi, \Gamma}{\mathcal{R} \vdash x : \phi \vee \psi, \Gamma} \text{ (}\vee\text{)}$$

$$\frac{\mathcal{R} \vdash x : \phi, \Gamma \quad \mathcal{R} \vdash x : \psi, \Gamma}{\mathcal{R} \vdash x : \phi \wedge \psi, \Gamma} \text{ (}\wedge\text{)} \qquad \frac{\mathcal{R}, R_{[1]}x_1y, ..., R_{[n]}x_ny \vdash \Gamma}{\mathcal{R} \vdash \Gamma} \text{ (IOA)}^{\dagger_2}$$

$$\frac{\mathcal{R}, R_\Box xy \vdash y : \phi, \Gamma}{\mathcal{R} \vdash x : \Box \phi, \Gamma} \text{ (}\Box\text{)}^{\dagger_1} \qquad \frac{\mathcal{R} \vdash x : \Diamond \phi, y : \phi, \Gamma}{\mathcal{R} \vdash x : \Diamond \phi, \Gamma} \text{ (}\Diamond\text{)}^{\dagger_3}$$

$$\frac{\mathcal{R}, R_{[i]}xy \vdash y : \phi, \Gamma}{\mathcal{R} \vdash x : [i]\phi, \Gamma} \text{ (}[i]\text{)}^{\dagger_1} \qquad \frac{\mathcal{R} \vdash x : \langle i \rangle \phi, y : \phi, \Gamma}{\mathcal{R} \vdash x : \langle i \rangle \phi, \Gamma} \text{ (}\langle i \rangle\text{)}^{\dagger_4}$$

$$\frac{\mathcal{R}, R_{\otimes_i}xy \vdash y : \phi, \Gamma}{\mathcal{R} \vdash x : \otimes_i \phi, \Gamma} \text{ (}\otimes_i\text{)}^{\dagger_1} \qquad \frac{\mathcal{R}, R_{\otimes_i}xy \vdash x : \ominus_i \phi, y : \phi, \Gamma}{\mathcal{R}, R_{\otimes_i}xy \vdash x : \ominus_i \phi, \Gamma} \text{ (}\ominus_i\text{)}$$

$$\frac{\mathcal{R}, R_{\otimes_i}xz, R_{\otimes_i}yz \vdash \Gamma}{\mathcal{R}, R_{\otimes_i}xz \vdash \Gamma} \text{ (D1}_i\text{)}^{\dagger_3}$$

Fig. 2. The calculi G3DS$_n$ (with $n = |Ag|$). $\dagger_1$ on ($\Box$), ($[i]$), and ($\otimes_i$) indicates that y is an eigenvariable, i.e. y does not occur in the rule's conclusion. $\dagger_2$ on (IOA) states that y is an eigenvariable and for all $i \in \{1, \ldots, n\}$, $x_i \sim_\Diamond^\mathcal{R} x_{i+1}$ (see Def. 4.3). $\dagger_3$ on ($\Diamond$) and (D1$_i$) and $\dagger_4$ on ($\langle i \rangle$) state, respectively, that $x \sim_\Diamond^\mathcal{R} y$ and $x \sim_i^\mathcal{R} y$ (see Def. 4.3 and Def. 4.2). We have ($[i]$), ($\langle i \rangle$), ($\otimes_i$), ($\ominus_i$), and (D1$_i$) rules for each $i \in Ag$.

in the consequent reads disjunctively). The rules (IOA) and (D1$_i$) encode, respectively, condition **C3** (i.e. independence of agents) and condition **D1** of Def. 3.2. A particular feature of refinement, is that we can incorporate the semantic behavior of modalities into their corresponding rules. For instance, the side condition $\dagger_4$ of the ($\langle i \rangle$) rule integrates the fact that $\langle i \rangle$ is semantically characterized as an equivalence relation. These side conditions—including those for the rules ($\Diamond$), ($\langle i \rangle$) and (D1$_i$)—rely on the notion of a $\Diamond$- and $\langle i \rangle$-path.

Definition 4.2 ($\langle i \rangle$-**path**) Let $x \sim_i y \in \{R_{[i]}xy, R_{[i]}yx\}$ and $\Lambda = \mathcal{R} \vdash \Gamma$. An $\langle i \rangle$-path of relational atoms from a label x to y occurs in Λ (written as $x \sim_i^\mathcal{R} y$) iff $x = y$, $x \sim_i y$, or there exist labels z_j ($j \in \{1, \ldots, k\}$) such that $x \sim_i z_1, \ldots, z_k \sim_i y$ occurs in $\mathcal{R}$.

Definition 4.3 ($\Diamond$-**path**) Let $x \sim_\Diamond y \in \{R_\Box xy, R_\Box yx\} \cup \{R_{[i]}xy, R_{[i]}yx \mid i \in Ag\}$, and $\Lambda = \mathcal{R} \vdash \Gamma$. An $\Diamond$-path of relational atoms from a label x to y occurs in Λ (written as $x \sim_\Diamond^\mathcal{R} y$) iff $x = y$, $x \sim_\Diamond y$, or there exist labels z_j ($j \in \{1, \ldots, k\}$) such that $x \sim_\Diamond z_1, \ldots, z_k \sim_\Diamond y$ occurs in $\mathcal{R}$.

The definition of an $\langle i \rangle$- and $\Diamond$-path captures a notion of reachability that simulates the fact that $R_{[i]}$ and $R_\Box$ are equivalence relations. Moreover, $\Diamond$-paths also incorporate the fact that choices are subsumed under moments (cf. **C2** of Def. 3.2). Observe that the $\Diamond$-path condition on (IOA) indicates that 'independence of agents' can only be applied to choices that occur at the same moment. One of the advantages of using such paths as side conditions is that it allows us to reduce the number of rules in our calculi [29].

Fig. 3 contains four additional structural rules with which the base calculi G3DS$_n$ can be extended. As their names suggest, these rules simulate their

$$\frac{\mathcal{R}, R_{\otimes_i}xy \vdash \Gamma}{\mathcal{R} \vdash \Gamma} \ (\mathsf{D2}_i)^{\dagger_1} \qquad \frac{\mathcal{R}, R_{\otimes_i}xy, R_\square xy \vdash \Gamma}{\mathcal{R}, R_{\otimes_i}xy \vdash \Gamma} \ (\mathsf{D3}_i) \qquad \frac{\mathcal{R}', R_{\otimes_i}xz \vdash \Gamma'}{\mathcal{R}' \vdash \Gamma'} \ (\mathsf{D5}_i^2)^{\dagger_2}$$

$$\vdots$$

$$\frac{\mathcal{R}, R_{\otimes_i}xy, R_{\otimes_i}xz \vdash \Gamma}{\mathcal{R}, R_{\otimes_i}xy \vdash \Gamma} \ (\mathsf{D4}_i)^{\dagger_2} \qquad \frac{\mathcal{R}, R_{\otimes_i}xy \vdash \Gamma}{\mathcal{R} \vdash \Gamma} \ (\mathsf{D5}_i^1)^{\dagger_1}$$

Fig. 3. The Deontic Structural Rules. Condition $\dagger_1$ on ($\mathsf{D2}_i$) and ($\mathsf{D5}_i^1$) states that y is a eigenvariable. Condition $\dagger_2$ on ($\mathsf{D4}_i$) and ($\mathsf{D5}_i^2$) indicates that $y \sim_i^{\mathcal{R}} z$ (Def. 4.2). Last, we let ($\mathsf{D5}_i)^\ddagger$ be $\langle(\mathsf{D5}_i^1), (\mathsf{D5}_i^2)\rangle$ with $\ddagger$ the global restriction (mentioned below), and have ($\mathsf{D2}_i$), ($\mathsf{D3}_i$), ($\mathsf{D4}_i$), ($\mathsf{D5}_i$) rules for each $i \in Ag$.

respective frame properties (cf. Def. 3.4). In doing so, we obtain calculi for the logics $\mathsf{DS}_n\mathsf{X}$. As an example, the logic $\mathsf{DS}_n\{\mathbf{D2},\mathbf{D4}\}$ corresponds to the calculus $\mathsf{G3DS}_n\{(\mathsf{D2}_i),(\mathsf{D4}_i) \mid i \in Ag\}$ (henceforth, we write $\mathsf{G3DS}_n\{\mathsf{D2}_i,\mathsf{D3}_i\}$).

Definition 4.4 (The calculi $\mathsf{G3DS}_n\mathsf{X}$) Let $\mathsf{DS}_n\mathsf{X}$ be a logic from Def. 3.4. Let $n = |Ag| \in \mathbb{N}$ and $\mathsf{X} \subseteq \{\mathbf{D2},\mathbf{D3},\mathbf{D4},\mathbf{D5}\}$. We define $\mathsf{G3DS}_n\mathsf{X}$ to consist of $\mathsf{G3DS}_n$ extended with (DK_i), if $\mathbf{DK} \in \mathsf{X}$ (with $K \in \{2,3,4,5\}$) for all $i \in Ag$.

We point out that the first-order condition $\mathbf{D5}$ (Def. 3.2) is a *generalized geometric axiom*. In [34], it was shown that properties of this form require *system of rules* in their corresponding calculi. We adopt this approach in our calculi as well and use ($\mathsf{D5}_i$) to denote the system of rules $\langle(\mathsf{D5}_i^1), (\mathsf{D5}_i^2)\rangle$ (see Fig. 3). The global restriction $\ddagger$ imposed on applying ($\mathsf{D5}_i$) is that, although we may use ($\mathsf{D5}_i^1$) wherever, if we use ($\mathsf{D5}_i^2$) we must also use ($\mathsf{D5}_i^1$) further down in the derivation. In Sect. 5, Ex. 5.1 demonstrates an application of ($\mathsf{D5}_i$).

To confirm soundness and completeness for our calculi—thus demonstrating an equivalence between the semantics ($\mathsf{DS}_n\mathsf{X}$) and proof-theory ($\mathsf{G3DS}_n\mathsf{X}$) of our logics—we need to provide a semantic interpretations of sequents:

Definition 4.5 (Sequent Semantics) Let M be a $\mathsf{DS}_n\mathsf{X}$-model with domain W and I an *interpretation function* mapping labels to worlds; i.e. $I\colon Lab \mapsto W$. A sequent $\Lambda = \mathcal{R} \vdash \Gamma$ is *satisfied* in M with I (written, $M, I \models \Lambda$) iff for all relational atoms $R_\alpha xy \in \mathcal{R}$ (where $\alpha \in \{\square\} \cup \{[i], \otimes_i \mid i \in Ag\}$), if $R_\alpha I(x)I(y)$ holds in M, then there exists a $z : \phi \in \Gamma$ such that $M, I(z) \Vdash \phi$. Λ is *valid relative to* $\mathsf{DS}_n\mathsf{X}$ iff it is satisfiable in any $\mathsf{DS}_n\mathsf{X}$-model M with any I.

Theorem 4.6 (Soundness and Completeness of $\mathsf{G3DS}_n\mathsf{X}$) *A sequent Λ is derivable in $\mathsf{G3DS}_n\mathsf{X}$ iff it is valid relative to $\mathsf{DS}_n\mathsf{X}$.*

Proof. Follows from Thm. A.1 and A.3. See the Appendix A for details. □

5 A formal analysis of Deontic STIT and OiC

In this section, we put our $\mathsf{G3DS}_n\mathsf{X}$ calculi to work. First, we make use of our calculi to organize our logics in terms of their strength—observing which are equivalent, distinct, or subsumed by another. Second, we discuss the logical (in)dependencies between our various OiC principles by confirming the minimal logic in which each principle is validated.

5.1 A Taxonomy of Deontic STIT Logics

In Fig. 4, a lattice is provided ordering the sixteen deontic STIT calculi of Def. 4.4 on the basis of their respective strength (reflexive and transitive edges are left implicit). We consider a calculus G3DS$_n$X stronger than another calculus G3DS$_n$Y whenever the former generates at least the same set of theorems as the latter. Consequently, the lattice simultaneously orders the deontic STIT logics of Def. 3.4, generated by these calculi, on the basis of their expressivity. In Fig. 4, the calculi are ordered bottom-up: G3DS$_n$ is the weakest system, generating the smallest logic subsumed by all others, whereas G3DS$_n${D2$_i$, D3$_i$, D4$_i$} is the strongest calculus with its logic subsuming all others. Notice that the latter calculus generates the traditional deontic STIT logic of [23,32]. To determine the existence of a directed edge from one calculus G3DS$_n$X to another G3DS$_n$Y in the lattice, we need to show that every derivation in the former can be transformed into a derivation in the latter. As an example of this procedure, we consider the edge from G3DS$_n${D3$_i$, D5$_i$} to G3DS$_n${D2$_i$, D3$_i$, D4$_i$}.

Example 5.1 To transform a G3DS$_n${D3$_i$, D5$_i$}-derivation into a derivation of G3DS$_n${D2$_i$, D3$_i$, D4$_i$}, it suffices to show that each instance of (D5$_i^1$) and (D5$_i^2$) can be replaced, respectively, by instances of (D2$_i$) and (D4$_i$). For example:

$$\dfrac{\dfrac{\dfrac{\dfrac{\dfrac{\dfrac{R_\Box xy, R_{\otimes_i} xy, R_{[i]} yz, R_{\otimes_i} xz \vdash z : \neg \phi, ..., z : \phi}{R_\Box xy, R_{\otimes_i} xy, R_{[i]} yz, R_{\otimes_i} xz \vdash x : \ominus_i \neg \phi, ..., z : \phi} (\ominus_i)}{R_\Box xy, R_{\otimes_i} xy, R_{[i]} yz \vdash x : \ominus_i \neg \phi, ..., z : \phi} (D5_i^2)}{R_\Box xy, R_{\otimes_i} xy \vdash x : \ominus_i \neg \phi, ..., y : [i]\phi} ([i])}{R_\Box xy, R_{\otimes_i} xy \vdash x : \ominus_i \neg \phi, x : \Diamond [i]\phi} (\Diamond)}{R_{\otimes_i} xy \vdash x : \ominus_i \neg \phi, x : \Diamond [i]\phi} (D3_i)}{\vdash x : \ominus_i \neg \phi, x : \Diamond [i]\phi} (D5_i^1)$$
$$\dfrac{}{\vdash x : \ominus_i \neg \phi \lor \Diamond [i]\phi} (\lor)$$
$$= \vdash x : \otimes_i \phi \to \Diamond [i]\phi$$

$\rightsquigarrow$

$$\dfrac{\dfrac{\dfrac{\dfrac{\dfrac{\dfrac{R_\Box xy, R_{\otimes_i} xy, R_{[i]} yz, R_{\otimes_i} xz \vdash z : \neg \phi, ..., z : \phi}{R_\Box xy, R_{\otimes_i} xy, R_{[i]} yz, R_{\otimes_i} xz \vdash x : \ominus_i \neg \phi, ..., z : \phi} (\ominus_i)}{R_\Box xy, R_{\otimes_i} xy, R_{[i]} yz \vdash x : \ominus_i \neg \phi, ..., z : \phi} (D4_i)}{R_\Box xy, R_{\otimes_i} xy \vdash x : \ominus_i \neg \phi, ..., y : [i]\phi} ([i])}{R_\Box xy, R_{\otimes_i} xy \vdash x : \ominus_i \neg \phi, x : \Diamond [i]\phi} (\Diamond)}{R_{\otimes_i} xy \vdash x : \ominus_i \neg \phi, x : \Diamond [i]\phi} (D3_i)}{\vdash x : \ominus_i \neg \phi, x : \Diamond [i]\phi} (D2_i)$$
$$\dfrac{}{\vdash x : \ominus_i \neg \phi \lor \Diamond [i]\phi} (\lor)$$
$$= \vdash x : \otimes_i \phi \to \Diamond [i]\phi$$

The non-existence of a directed edge in the opposite direction is implied by the fact that G3DS$_n${D2$_i$, D3$_i$, D4$_i$} $\vdash \otimes_i \phi \to \otimes_i [i]\phi$ and G3DS$_n${D3$_i$, D5$_i$} $\nvdash \otimes_i \phi \to \otimes_i [i]\phi$. The latter is shown through failed proof search (See Ex. 5.2 for an illustration of how failed proof-search can be used to determine underivability.)

To determine that two calculi G3DS$_n$X and G3DS$_n$Y are equivalent (i.e. G3DS$_n$X $\equiv$ G3DS$_n$Y), thus implying that the associated logics are identical, one shows that every derivation in the former can be transformed into a derivation in the latter, and vice-versa. Last, to prove that two calculi G3DS$_n$X and G3DS$_n$Y are independent—yielding incomparable logics—it is sufficient to show that there exist formulae ϕ and ψ such that G3DS$_n$X $\vdash \phi$, G3DS$_n$Y $\nvdash \phi$, G3DS$_n$Y $\vdash \psi$, and G3DS$_n$X $\nvdash \psi$. We come back to this in the following subsection when we consider an example of an underivable OiC formula.

5.2 Logical (In)Dependencies of OiC Principles

Fig. 4 also represents which deontic STIT calculi should at least be adopted to make certain OiC principles theorems of the corresponding logics. These principles were initially formalized in Sect. 2. However, as discussed in Sect. 3, in order to formally represent *deliberative* readings of OiC in a normal modal

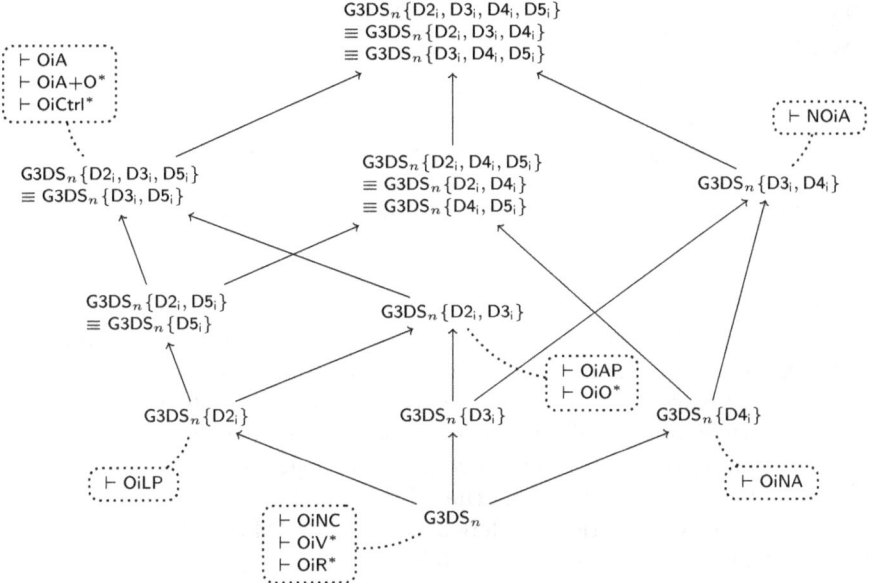

Fig. 4. The lattice of deontic STIT calculi. Directed edges point from weaker calculi to stronger calculi, consequently ordering the corresponding logics w.r.t. their expressivity (reflexive and transitive edges are left implicit). We use ≡ to denote equivalent calculi. Dotted nodes show which calculi should at least be adopted to make the indicated OiC principles theorems (for the final OiC formalizations see Fig. 5).

setting, we must replace the initial antecedent $\otimes_i \phi$ with its deliberative correspondent $\otimes_i^d \phi$ in OiV, OiR, OiO, OiA+O and with $\otimes_i^c \phi$ in OiCtrl. The final list of OiC formalizations is presented in Fig. 5. Although for now the above suffices—i.e. the approach being in line with the traditional treatment of deliberative agency [7,23,24]—the solution may be considered *ad hoc*. We note that these deliberative canons may alternatively be captured as follows: (i) through characterizing deliberation directly in the logic, taking $\otimes_i^d$ and $\otimes_i^c$ as primitive operators (cf. [46]), or (ii) through characterizing contingency via the use of sanction constants (cf. [3]). We leave this to future work.

In Ex. 5.1, we saw that OiA is derivable in both $G3DS_n\{D2_i, D3_i, D4_i\}$ and $G3DS_n\{D3_i, D5_i\}$. What is more, since $\otimes_i[i]\phi \to \otimes_i\phi$ is already a theorem of $G3DS_n$, we find that the weaker logic generated by $G3DS_n\{D3_i, D5_i\}$ already suffices to accommodate OiC of the traditional deontic STIT setting [23], that is, $G3DS_n\{D3_i, D5_i\} \vdash \otimes_i[i]\phi \to \Diamond[i]\phi$. We emphasize that only through the addition of $D4_i$ do we restore the position advocated by Horty in [23] (cf. footnote 2). Namely, by adding $D4_i$ to a calculus, the distinction between $\otimes_i$ and $\otimes_i[i]$ collapses—i.e. $G3DS_n\{D4_i\} \vdash \otimes_i\phi \equiv \otimes_i[i]\phi$—and the agent-dependent obligation operator will demonstrate the same logical behavior as the interpretation of obligation restricted to complete choices; i.e. the 'dominance ought'. (NB. In [9] it was shown that the relational characterization of $\otimes_i$ in $DS_n\{\mathbf{D2}, \mathbf{D3}, \mathbf{D4}\}$ is equivalent to the logic of 'dominance ought' [23,32].)

$\mathsf{G3DS}_n\{\mathsf{D2}_i\} \vdash \otimes_i \phi \to \neg \otimes_i \neg \phi$	OiLP	$\mathsf{G3DS}_n\{\mathsf{D2}_i, \mathsf{D3}_i\} \vdash \otimes_i^d \phi \to (\Diamond \phi \wedge \Diamond \neg \phi)$	OiO*
$\mathsf{G3DS}_n\{\mathsf{D2}_i, \mathsf{D3}_i\} \vdash \otimes_i \phi \to \Diamond \phi$	OiAP	$\mathsf{G3DS}_n\{\mathsf{D3}_i, \mathsf{D5}_i\} \vdash \otimes_i^d \phi \to (\Diamond [i]\phi \wedge \Diamond \phi \wedge \Diamond \neg \phi)$	OiA+O*
$\mathsf{G3DS}_n\{\mathsf{D3}_i, \mathsf{D5}_i\} \vdash \otimes_i \phi \to \Diamond[i]\phi$	OiA	$\mathsf{G3DS}_n\{\mathsf{D3}_i, \mathsf{D5}_i\} \vdash \otimes_i^c \phi \to (\Diamond[i]\phi \wedge \Diamond[i]\neg \phi)$	OiCtrl*
$\mathsf{G3DS}_n \vdash \otimes_i^d \phi \to \Diamond \neg \phi$	OiV*	$\mathsf{G3DS}_n \vdash \otimes_i \phi \to \otimes_i \Diamond \phi$	OiNC
$\mathsf{G3DS}_n \vdash \otimes_i^d \phi \to \Diamond[i]\neg[i]\phi$	OiR*	$\mathsf{G3DS}_n\{\mathsf{D4}_i\} \vdash \otimes_i \phi \to \otimes_i \Diamond[i]\phi$	OiNA

Fig. 5. STIT formalizations of OiC, with the minimal $\mathsf{G3DS}_n\mathsf{X}$ calculi entailing them.

From a philosophical perspective, Fig. 4 gives rise to what we will call the *endorsement principle* of the philosophy of OiC. Namely, the ordering of calculi tells us which endorsements of which OiC readings will logically commit us to endorsing other OiC readings (within the logical realm of agential choice). For instance, endorsing OiA tells us that we must also endorse the weaker OiLP and OiAP since they are logically entailed in the minimal calculus for OiA.

Furthermore, the taxonomy of deontic STIT logics shows which readings of OiC are independent from one another. In particular, we note that the normative principle OiNA is strictly independent of OiA, OiLP, OiAP. An advantage of the present proof theoretic approach is that we can constructively prove why certain readings of OiC fail to entail one another (relative to their calculi):

Example 5.2 To show that OiNA is not entailed by OiLP in $\mathsf{G3DS}_1\{\mathsf{D2}_1\}$ one attempts to prove an instance of OiNA via bottom-up proof-search (left):

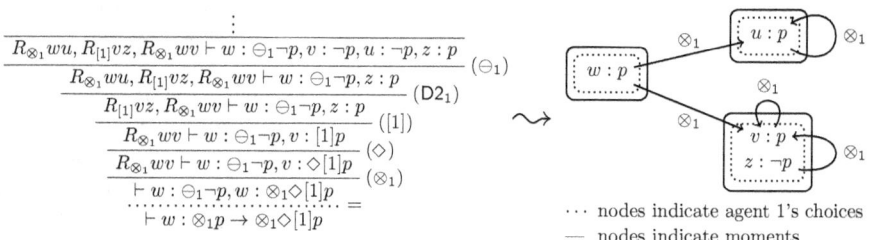

··· nodes indicate agent 1's choices
— nodes indicate moments

In theory, the left derivation will be infinite, but a quick inspection of the rules of $\mathsf{G3DS}_1\{\mathsf{D2}_1\}$ (with $Ag = \{1\}$) ensures us that no additional rule application will cause the proof to successfully terminate: $\neg p$ will never be propagated to z. The topsequent (left) will give the $\mathsf{DS}_1\{\mathbf{D2}\}$-counterexample for OiNA (right), provided that the model is appropriately closed under **D1** and **D2**: i.e. $M, w \not\models$ OiNA with $W = \{w, v, u, z\}$, $R_{[1]} = \{(v, z), (z, v)\}$, $R_\square = \{(v, z), (z, v)\}$, $R_{\otimes_1} = \{(w, u), (w, v), (u, u), (v, v), (z, v)\}$ and $V(p) = \{w, v, u\}$ (reflexivity is omitted for $R_{[1]}$ and $R_\square$). We leave development of terminating proof-search procedures with automated countermodel extraction to future work (cf. [29]).

We close with two remarks: First, recall Hintikka's position that OiC merely captures the normative disposition that 'it *ought* to be that OiC'. An agent-dependent variation of this principle (referred to as NOiA in Fig. 4) turns out to be a theorem of $\mathsf{G3DS}_n\{\mathsf{D3}_i, \mathsf{D4}_i\}$; i.e. $\mathsf{G3DS}_n\{\mathsf{D3}_i, \mathsf{D4}_i\} \vdash \otimes_i(\otimes_i \phi \to \Diamond[i]\phi)$. Second, we observe that the calculus $\mathsf{G3DS}_n\{\mathsf{D5}_i\}$ gives rise to an interesting, yet unaddressed, OiC principle which combines the ideas behind OiLP and

OiNA, namely, G3DS$_n${D5$_i$} ⊢ ⊗$_i\phi$ → ⊖$_i$◇[i]ϕ. Loosely, this principle expresses that 'ought implies that it is ideally consistent that the agent has the ability to fulfil its duties'. Future research will be directed toward further investigation of the philosophical consequences of our logical taxonomy of deontic STIT logics.

6 Conclusion

In this work, we analyzed, formalized, and compared ten distinct readings of Ought-implies-Can as taken from the philosophical literature. We modified the deontic STIT setting to accommodate this variety of OiC principles. Sound and complete deontic STIT calculi were provided of which the aforementioned OiC principles were shown to be theorems. We used these calculi to determine the logical interdependencies between these principles, resulting in a logical taxonomy of Ought-implies-Can according to each principle's respective strength. In particular, we proposed an endorsement principle describing which OiC readings commit one to other readings logically entailed by the former.

Future work will be twofold: First, from a technical perspective, we aim to provide decision algorithms based on the deontic STIT calculi G3DS$_n$X, following the work in [29]. Thus, we will leverage our calculi for the desired automation of normative reasoning within STIT. Furthermore, we aim to logically capture the deliberative OiC principles, bypassing the use of defined deliberative operators. Second, from a more philosophical perspective, future work will be directed toward the identification and analysis of further OiC principles derived from our logical taxonomy of deontic STIT logics.

Appendix
A Soundness and Completeness Proofs

Theorem A.1 (Soundness) *If a sequent Λ is derivable in G3DS$_n$X, then it is valid relative to DS$_n$X.*

Proof. It suffices to show that (id) is valid and each rule of G3DS$_n$X preserves validity relative to DS$_n$X. With the exception of (D5$_i$) = ⟨(D5$_i^1$), (D5$_i^2$)⟩, all cases are relatively straightforward (cf. [8,29]). The (D5$_i$) case follows from the general soundness result for systems of rules presented in [34]. □

Lemma A.2 *For any sequent Λ, either Λ is provable in G3DS$_n$X, or there exists a DS$_n$X-model M with I such that $M, I \not\models \Lambda$.*

Proof. For the proof we expand on the methods employed in [33]. In brief, we first (1) define a reduction-tree **RT** for an arbitrary sequent $\Lambda = \mathcal{R} \vdash \Gamma$. Either **RT** terminates and represents a proof in G3DS$_n$X, implying the provability of Λ, or it does not terminate. In the latter case the tree will be infinite and, using König's Lemma, we therefore know that (at least) one of **RT**'s branches is infinite. We use this infinite branch to show that (2) a DS$_n$X-model M can be constructed with an interpretation I such that $M, I \not\models \Lambda$.

(1) The inductive construction of **RT** consists of phases, each phase having two cases: (i) if every topmost sequent of every branch of **RT** is an initial se-

quent (id) the construction terminates. (ii) If not, then for those open branches, the construction proceeds and we continue applying—when possible—the rules of the calculus in a roundabout fashion. (NB. If no rule can be applied to a top sequent, yet it is not an initial sequent, then we copy the top sequent indefinitely.) We show how the ($\langle i \rangle$) and (D5$_i$) rules are applied (bottom-up) below; all remaining cases are similar or simple (cf. [8,33]).

We first consider the ($\langle i \rangle$) case, and suppose that m top sequents $\Lambda_j = \mathcal{R}_j \vdash \Gamma_j$ (with $1 \leq j \leq m$) are open in **RT** (i.e. no Λ_j is an instance of the (id) rule). Let $x_1 : \langle i \rangle \phi_1, ..., x_{k_j} : \langle i \rangle \phi_{k_j}$ be all labelled formulae in Λ_j prefixed with a $\langle i \rangle$ modality. Moreover, let $y_{l,1}, \ldots, y_{l,r_l} \in Lab(\Lambda_j)$ s.t. $x_l \sim_i^{\mathcal{R}_j} y_{l,s}$ (for $1 \leq l \leq k_j$ and $1 \leq s \leq r_l$). We add $\Lambda_{j+1} = \mathcal{R}_j \vdash y_{1,1} : \phi_1, \ldots, y_{1,r_1} : \phi_1, \ldots, y_{k_j,1} : \phi_{k_j}, \ldots, y_{k_j,r_{k_j}} : \phi_{k_j}, \Gamma_j$ on top of Λ_j. We apply this procedure for all $i \in Ag$.

For the (D5$_i$) case, assume that m top sequents $\Lambda_j = \mathcal{R}_j \vdash \Gamma_j$ (with $1 \leq j \leq m$) are still open in **RT**. First, for all $x_1, ..., x_{k_j} \in Lab(\Lambda_j)$, we set $\mathcal{R}_{j+1} := R_{\otimes_i} x_1 y_1, ..., R_{\otimes_i} x_{k_j} y_{k_j}, \mathcal{R}_j$, set $\Gamma_{j+1} := \Gamma_j$, and add $\Lambda_{j+1} = \mathcal{R}_{j+1} \vdash \Gamma_{j+1}$ on top of Λ_j, where $y_1, ..., y_{k_j}$ are fresh. (NB. This corresponds to applications of (D5$_i^1$).) Second, for all $z_1', \ldots, z_{l_r}' \in Lab(\Lambda_{j+1})$ such that $z_r \sim_i^{\mathcal{R}_{j+1}} z_1', \ldots, z_r \sim_i^{\mathcal{R}_{j+1}} z_{l_r}'$ and $R_{\otimes_i} x_r' z_r$ was introduced by an application of (D5$_i^1$) at any stage $s \leq j$ (with $1 \leq r \leq h$), we add $\Lambda_{j+2} = R_{\otimes_i} x_1' z_1', ..., R_{\otimes_i} x_1', z_{l_1}', \ldots, R_{\otimes_i} x_h' z_1', ..., R_{\otimes_i} x_h', z_{l_h}', \mathcal{R}_{j+1} \vdash \Gamma_{j+1}$ on top of Λ_{j+1}. We apply this procedure for all agents $i \in Ag$.

(2) If the construction of the **RT** for Λ terminates, we know that the topmost sequents of all branches are initial sequents and hence **RT** corresponds to a proof. If **RT** does not terminate, the tree is infinite and, with König's Lemma, we obtain an infinite branch from which we can construct a DS$_n$X countermodel for Λ. Let $\mathcal{R}_0 \vdash \Gamma_0, ..., \mathcal{R}_j \vdash \Gamma_j, ...$ be the sequence of sequents from the infinite branch, such that, (i) $\Lambda = \mathcal{R}_0 \vdash \Gamma_0$ and (ii) $\Lambda^+ = \mathcal{R}^+ \vdash \Gamma^+$, where $\mathcal{R}^+ = \bigcup_{j \geq 0} \mathcal{R}_j$ and $\Gamma^+ = \bigcup_{j \geq 0} \Gamma_j$.

We construct a model $M^+ = \langle W, R_\square, \{R_{[i]} | i \in Ag\}, \{R_{\otimes_i} | i \in Ag\}, V \rangle$ as follows: $W := Lab(\Lambda^+)$; $R_\square := \{(x,y) \mid x \sim_\Diamond^{\mathcal{R}^+} y\}$; $R_{[i]} := \{(x,y) \mid x \sim_i^{\mathcal{R}^+} y\}$ (for all $i \in Ag$); $R_{\otimes_i} := \{(x,y) \mid R_{\otimes_i} xy \in \mathcal{R}^+\}$ (for all $i \in Ag$); last, $x \in V(p)$ iff $x : \overline{p} \in \Gamma^+$. It is straightforward to show that M^+ is a DS$_n$X-model. We show that M^+ satisfies **C2** and **D5** (assuming that **D5** $\in$ X). The cases for all other conditions **C1**, **C3**, **D1**, and those in X are similar or simple.

To show that M^+ satisfies **C2** we need to show (i) $R_{[i]} \subseteq R_\square$, and (ii) $R_{[i]}$ is an equivalence relation. To show (i), assume that $(x,y) \in R_{[i]}$. This implies that $x \sim_i^{\mathcal{R}^+} y$ holds, which further implies that $x \sim_\Diamond^{\mathcal{R}^+} y$ holds by Def. 4.2 and 4.3. Therefore, by the definition of $R_\square$ in M^+ above, $(x,y) \in R_\square$. To see that $R_{[i]}$ is an equivalence relation, it suffices to observe that the relation is defined relative to $\sim_i^{\mathcal{R}^+}$, which is an equivalence relation.

To prove that M^+ satisfies **D5**, we assume $x \in W$. By the definition of **RT**, we know that there exists a Λ_j in the infinite branch such that $x \in Lab(\Lambda_j)$. Since the branch is infinite and rules are applied in a roundabout fashion we know that at some point $k > j$ the ($\mathsf{D5}_i$) step of the **RT** procedure must have been applied (and so, ($\mathsf{D5}_i^1$) must have been applied). Hence, $R_{\otimes_i} xy \in \mathcal{R}_{k+1}$ for $\Lambda_{k+1} = \mathcal{R}_{k+1} \vdash \Gamma_{k+1}$ with y fresh, implying that $(x,y) \in R_{\otimes_i}$. We aim to show that for all $z \in W$, if $(y,z) \in R_{[i]}$, then $(x,z) \in R_{\otimes_i}$. Take an arbitrary $z \in W$ for which $(y,z) \in R_{[i]}$. By the assumption that $(y,z) \in R_{[i]}$ and by the definition of **RT**, we know that at some point $m \geq k+1$ that the ($\mathsf{D5}_i$) step of the **RT** procedure must have been applied (and so, ($\mathsf{D5}_i^2$) must have been applied) with $y \sim_i^{\mathcal{R}_m} z$ for $\Lambda_m = \mathcal{R}_m \vdash \Gamma_m$. Hence, $R_{\otimes_i} xz \in \mathcal{R}_{m+1}$ in $\Lambda_{m+1} = \mathcal{R}_{m+1} \vdash \Gamma_{m+1}$, implying that $(x,z) \in R_{\otimes_i}$.

Let $I : Lab \mapsto W$ be the identity function (we may assume w.l.o.g. that $Lab = W$). By construction, M^+ satisfies each relational atom occurring in $\mathcal{R}^+$ with I, meaning that M^+ satisfies each relational atom in $\mathcal{R}$ with I (recall $\Lambda = \mathcal{R} \vdash \Gamma$). It can be shown by induction on the complexity of ϕ that for any $x : \phi \in \Gamma^+$, $M^+, I(x) \not\models \phi$. Consequently, since $\Gamma \subseteq \Gamma^+$, $M^+, I \not\models \Lambda$. □

Theorem A.3 (Completeness) *If a sequent Λ is valid relative to $\mathsf{DS}_n\mathsf{X}$, then it is derivable in $\mathsf{G3DS}_n\mathsf{X}$.*

Proof. Follows directly from A.2. □

References

[1] van Ackeren, M. and M. Kühler, *Ethics on (the) edge? introduction to moral demandingness and 'ought implies can'*, in: *The Limits of Moral Obligation*, Routledge, 2015 pp. 1–18.

[2] Ameriks, K. and D. M. Clarke, "Aristotle: Nicomachean Ethics," Cambridge University Press, 2000.

[3] Anderson, A. R. and O. K. Moore, *The formal analysis of normative concepts*, American Sociological Review **22** (1957), pp. 9–17.

[4] Arkoudas, K., S. Bringsjord and P. Bello, *Toward ethical robots via mechanized deontic logic*, in: *AAAI fall symposium on machine ethics*, The AAAI Press Menlo Park, CA, 2005, pp. 17–23.

[5] Bartha, P., *Conditional obligation, deontic paradoxes, and the logic of agency*, Annals of Mathematics and Artificial Intelligence **9** (1993), pp. 1–23.

[6] Belnap, N. and M. Perloff, *Seeing to it that: a canonical form for agentives*, Theoria **54** (1988), pp. 175–199.

[7] Belnap, N. D., M. Perloff and M. Xu, "Facing the future: agents and choices in our indeterminist world," Oxford University Press, Oxford, 2001.

[8] van Berkel, K. and T. Lyon, *Cut-free calculi and relational semantics for temporal stit logics*, in: *European Conference on Logics in Artificial Intelligence*, Springer, 2019, pp. 803–819.

[9] van Berkel, K. and T. Lyon, *A neutral temporal deontic stit logic*, in: *International Workshop on Logic, Rationality and Interaction*, Springer, 2019, pp. 340–354.

[10] Blackburn, P., M. de Rijke and Y. Venema, "Modal Logic," Camebridge University Press, 2001.

[11] Broersen, J., *Deontic epistemic stit logic distinguishing modes of mens rea*, Journal of Applied Logic **9** (2011), pp. 137–152.

[12] Brown, M. A., *On the logic of ability*, Journal of philosophical logic (1988), pp. 1–26.

[13] Castañeda, H.-N., *On the semantics of the ought-to-do*, in: *Semantics of natural language*, Springer, 1972 pp. 675–694.

[14] Ciuni, R. and E. Lorini, *Comparing semantics for temporal stit logic*, Logique et Analyse **61** (2017), pp. 299–339.

[15] Copp, D., *'ought' implies 'can', blameworthiness, and the principle of alternate possibilities*, in: *Moral responsibility and alternative possibilities*, Routledge, 2017 pp. 265–299.
[16] Dahl, N. O., *Ought implies can and deontic logic*, Philosophia **4** (1974), pp. 485–511.
[17] van Eck, J. A., *A system of temporally relative modal and deontic predicate logic and its philosophical applications 2*, Logique et analyse **25** (1982), pp. 339–381.
[18] Goldman, A. I., "Theory of human action," Princeton University Press, 1970.
[19] Graham, P. A., *'ought'and ability*, Philosophical Review **120** (2011), pp. 337–382.
[20] Herzig, A. and F. Schwarzentruber, *Properties of logics of individual and group agency.*, Advances in modal logic **7** (2008), pp. 133–149.
[21] Hilpinen, R. and P. McNamara, *Deontic logic: A historical survey and introduction*, Handbook of deontic logic and normative systems **1** (2013), pp. 3–136.
[22] Hintikka, J., *Some main problems of deontic logic*, in: *Deontic logic: Introductory and systematic readings*, Springer, 1970 pp. 59–104.
[23] Horty, J. F., "Agency and deontic logic," Oxford University Press, 2001.
[24] Horty, J. F. and N. Belnap, *The deliberative stit: A study of action, omission, ability, and obligation*, Journal of philosophical logic **24** (1995), pp. 583–644.
[25] Kant, I., "Critique of Pure Reason," Cambridge University Press, 2000.
[26] Kohl, M., *Kant and 'ought implies can'*, The Philosophical Quarterly **65** (2015), pp. 690–710.
[27] Lemmon, E. J., *Moral dilemmas*, The philosophical review **71** (1962), pp. 139–158.
[28] Lorini, E. and G. Sartor, *Influence and responsibility: A logical analysis.*, in: *JURIX*, 2015, pp. 51–60.
[29] Lyon, T. and K. van Berkel, *Automating agential reasoning: Proof-calculi and syntactic decidability for stit logics*, in: *International Conference on Principles and Practice of Multi-Agent Systems*, Springer, 2019, pp. 202–218.
[30] McConnell, T. C., *'ought'implies 'can' and the scope of moral requirements*, Philosophia **19** (1989), pp. 437–454.
[31] Meyer, J.-J. C. et al., *A different approach to deontic logic: deontic logic viewed as a variant of dynamic logic.*, Notre dame journal of formal logic **29** (1988), pp. 109–136.
[32] Murakami, Y., *Utilitarian deontic logic*, AiML-2004: Advances in Modal Logic **287** (2004), pp. 287–302.
[33] Negri, S., *Kripke completeness revisited*, Acts of Knowledge-History, Philosophy and Logic (2009), pp. 247–282.
[34] Negri, S., *Proof analysis beyond geometric theories: from rule systems to systems of rules*, Journal of Logic and Computation **26** (2014), pp. 513–537.
[35] Saka, P., *Ought does not imply can*, American Philosophical Quarterly **37** (2000), pp. 93–105.
[36] Stocker, M., *'ought'and 'can*, Australasian journal of philosophy **49** (1971), pp. 303–316.
[37] Takeuti, G., "Proof theory," Courier Corporation, 2013.
[38] Timmermann, J., *Kantian dilemmas? moral conflict in kant's ethical theory* (2013).
[39] Vigano, L., "Labelled non-classical logics," Kluwer Academic Publishers, 2000.
[40] Vranas, P. B., *I ought, therefore i can*, Philosophical studies **136** (2007), pp. 167–216.
[41] Wansing, H., *Tableaux for multi-agent deliberative-stit logic.*, Advances in modal logic **6** (2006), pp. 503–520.
[42] von Wright, G. H., *Deontic logic*, Mind **60** (1951), pp. 1–15.
[43] von Wright, G. H., "Norm and action," Routledge, 1963.
[44] von Wright, G. H., "An essay in deontic logic and the general theory of action," North-Holland Publishing Company, Amsterdam, 1968.
[45] von Wright, G. H., *On the logic of norms and actions*, in: R. Hilpinen, editor, *New studies in deontic logic*, Springer, 1981 pp. 3–35.
[46] Xu, M., *Axioms for deliberative stit*, Journal of Philosophical Logic **27** (1998), pp. 505–552.
[47] Yaffe, G., *'ought' implies 'can' and the principle of alternate possibilities*, Analysis **59** (1999), pp. 218–222.

Deontic Action Logics via Algebra

Pablo F. Castro

Universidad Nacional de Rio Cuarto and CONICET, Argentina

Valentin Cassano

Universidad Nacional de Rio Cuarto and CONICET, Argentina

Raul Fervari

Universidad Nacional de Cordoba and CONICET, Argentina

Carlos Areces

Universidad Nacional de Cordoba and CONICET, Argentina

Abstract

Deontic logics are dubbed the logics of normative or prescriptive reasoning. These logics can roughly be categorized into *ought-to-be*, dealing with the prescription of state of affairs, or *ought-to-do*, dealing with the prescription of actions. An important family of *ought-to-do* deontic logics have their origin in Segerberg's *Deontic Action Logic* (DAL, see [23]). In this work, we provide an algebraic characterization of DAL and some known variants. In brief, we capture actions and formulas as elements of different base algebras, and deontic operators as algebraic operations; different algebras capture the different variants. This algebraization enables us to obtain completeness results via standard algebraic means. Moreover, we argue that this algebraic framework offers a natural way of (re-)thinking many deontic logical issues at large.

Keywords: Deontic Action Logic, Algebraic Logic, Normative Reasoning.

1 Introduction

Deontic Logic (DL) is devoted to the study of norms and their logical foundations. The beginnings of DL can be traced back to the pioneer works of G. von Wright [28], J. Kalinowski [13], and O. Becker [5]. Since then, most deontic logics have been defined as particular classes of modal logics (see [7,6]). The most famous among these formal systems is *Standard Deontic Logic*, SDL for short. SDL extends the normal modal system K with the extra axiom D for *seriality*. An in-depth introduction to diverse formal systems of deontic logic is provided in [4], together with a historical presentation.

Deontic logics built on SDL are known as *ought-to-be*, as they deal with the prescription of states of affairs, i.e., propositions. However, G. von Wright pointed out that deontic logics are closely related to the concept of action (see [28,29]), and furthermore, they should be constructed upon a theory of actions (see [28,29]). These observations, also shared by other authors (e.g., [14,23,19,9,8,26,21]), have led to the development of deontic logics where prescriptions apply to actions instead of propositions. Deontic logics of this kind are called *ought-to-do*.

One of the first *ought-to-do* deontic logics was presented by K. Segerberg in [23]. Segerberg's logic formally distinguishes between actions and formulas. In this formalism, actions are built up from basic action names using action combinators. Then, deontic connectives apply to actions to yield formulas, and formulas are obtained from formulas using logical connectives. We illustrate this by means of a simple example. Let driving and drinking be basic action names; the formula ¬P(driving ⊓ drinking) states that drinking while driving is not permitted. In this formula, ⊓ is an action operator that can be understood as the parallel execution of actions; P is the deontic connective of permission, and ¬ is logical negation. The obtained logic is extremely simple and admits a sound and complete proof system. An interesting feature of Segerberg's logic is its two tier interpretation structure, i.e., actions are interpreted resorting to an algebra of events, whereas formulas are interpreted using truth values. Segerberg's initial formalism was revisited by other authors, for instance: [9] introduces action prescriptions and combines them with modal operators, and [25] investigates several fragments of Segerberg's logic. We follow the terminology from [25] and call these formalisms *deontic action logics*.

In this paper we provide an algebraic formulation of deontic action logics. More precisely, we develop an abstract view of deontic action logics in terms of algebraic structures. To this end, we follow some of the main ideas introduced by Halmos in [11], where Boolean algebras serve as an abstraction of propositions; Venema in [27], who introduced Boolean algebras with operators as an algebraic counterpart of modal logics; and Pratt in [20], who introduced dynamic algebras to investigate the theoretical properties of dynamic logics via many-sorted algebras. Intuitively, in our framework, formulas are captured as elements of a Boolean algebra, while actions are formalized by means of another (Boolean) algebra. In this setting, deontic operators are modeled as functions connecting both algebras. We put forth that the benefits of this algebraic version of deontic action logics are twofold. Firstly, algebraic logic has been shown useful when analyzing theoretical properties of logics and investigating the relations between different formalisms. Secondly, extensions to a deontic action logic can be obtained by considering different action and predicate algebras. We explore these ideas in Sec. 4.

Structure. In Sec. 2, we introduce some of the basic definitions of Segerberg's deontic action logic, called DAL. In Sec. 3, we present the basic algebraic framework, and prove an algebraic version of soundness and completeness for DAL using standard algebraic tools. Preliminary definitions about algebra used

in that section can be found in Appendix A. In Sec. 4, we discuss variants of
deontic action logics using particular classes of algebras. Lastly, in Sec. 5, we
offer some final remarks and discuss future work.

2 Segerberg's Deontic Action Logic

We cover the syntax and semantics of the deontic action logic originally introduced by Segerberg in [23]. We refer to this logic as DAL.

Syntax of DAL. The language of DAL is comprised of a set Act of *actions* and a set Form of *formulas* defined on a countable set $\mathsf{Act}_0 = \{\, \mathsf{a}_i \mid i \in \mathbb{N} \,\}$ of basic action symbols. The sets Act and Form are given by the grammars in Eq. (1) and Eq. (2), respectively:

$$\alpha ::= \mathsf{a}_i \mid \alpha \sqcup \alpha \mid \alpha \sqcap \alpha \mid \bar{\alpha} \mid 0 \mid 1 \tag{1}$$

$$\varphi ::= \mathsf{P}\alpha \mid \mathsf{F}\alpha \mid \alpha = \beta \mid \varphi \to \varphi \mid \neg \varphi. \tag{2}$$

Intuitively, any $\mathsf{a}_i \in \mathsf{Act}_0$ is a *basic* action; $\alpha \sqcup \beta$ is the *free-choice* between α and β; $\alpha \sqcap \beta$ is the *parallel* execution of α and β; $\bar{\alpha}$ is the *complement* of α, i.e., any action other than α; and 0 and 1 are the *impossible* and the *universal* actions, respectively. Turning to formulas, the connective = indicates *equality* of actions. The logical connectives $\to$ and $\neg$ stand for *material implication* and *negation*, respectively. We also consider the derived logical connectives: $\vee$ for *disjunction*, $\wedge$ for *conjunction*, $\top$ for *verum*, $\bot$ for *falsum*, and $\leftrightarrow$ for *material bi-implication*. The derived logical connectives are defined from $\to$ and $\neg$ in the usual way. The connectives P and F have a deontic reading: (a) P stands for *permitted*, i.e., α is allowed to be executed; (b) F stands for *forbidden*, i.e., the execution of α forbidden.

The axioms for DAL are listed in Fig. 1. A Hilbert-style notion of provability based on these axioms is defined in the usual way using the rule of *modus ponens*. More precisely, a proof of φ is a finite sequence $\psi_1, \ldots, \psi_n$ of formulas s.t. $\psi_n = \varphi$, and for each $k \leq n$, ψ_k is either: (i) an axiom; or (ii) obtained from two earlier formulas using *modus ponens*, i.e., there are $i, j < k$ s.t. $\psi_j = \psi_i \to \psi_k$. We say that φ is a theorem of DAL, written $\vdash \varphi$, if there is a proof of φ. The set of theorems of DAL is the set: $\{\, \varphi \mid\, \vdash \varphi \,\}$.

Semantics of DAL. A deontic action model is a tuple $\mathfrak{M} = \langle E, P, F \rangle$ where: (a) E is a set of elements; and (b) P and F are subsets of E satisfying $P \cap F = \emptyset$. Intuitively, in a deontic action model $\mathfrak{M}$, we can think of the set E the set of possible outcomes of actions, and of the sets P and F as sets of permitted and forbidden events. The condition $P \cap F = \emptyset$ in (b) can be understood as an indication that: *only the impossible is both permitted and forbidden*.

A valuation on a deontic model $\mathfrak{M}$ is a function $v : \mathsf{Act}_0 \to 2^E$. Every

> 1. The following set of axioms for actions α, β, and γ (see [10]):
> (3) $\quad \alpha \sqcup 0 = \alpha$ $\qquad$ (4) $\quad \alpha \sqcap 1 = \alpha$
> (5) $\quad \alpha \sqcup 1 = 1$ $\qquad$ (6) $\quad \alpha \sqcap 0 = 0$
> (7) $\quad \alpha \sqcup \beta = \beta \sqcup \alpha$ $\qquad$ (8) $\quad \alpha \sqcap \beta = \beta \sqcap \alpha$
> (9) $\quad \alpha \sqcup (\beta \sqcap \gamma) = (\alpha \sqcup \beta) \sqcap (\alpha \sqcup \gamma)$ $\quad$ (10) $\quad \alpha \sqcap (\beta \sqcup \gamma) = (\alpha \sqcap \beta) \sqcup (\alpha \sqcap \gamma)$
>
> 2. The following set of axioms for formulas φ, ψ, χ (see [18]):
> $\qquad$ (11) $\varphi \to (\psi \to \varphi)$
> $\qquad$ (12) $(\neg \varphi \to \neg \psi) \to ((\neg \varphi \to \psi) \to \varphi)$
> $\qquad$ (13) $(\varphi \to (\psi \to \chi)) \to ((\varphi \to \psi) \to (\varphi \to \chi))$
>
> 3. The following set of axioms for (=):
> (14) $\alpha = \alpha$ $\quad$ (15) $(\alpha = \beta) \to (\beta = \alpha)$ $\quad$ (16) $(\alpha = \beta) \wedge (\beta = \gamma) \to (\alpha = \gamma)$
>
> 4. The substitution axiom:
> $\qquad$ (17) $(\alpha = \beta) \to (\varphi \to \varphi_\alpha^\beta)$
> where φ_α^β is the formula obtained from replacing some ocurrences of α with β.
>
> 5. The deontic axioms:
> (18) $P(\alpha \sqcup \beta) \leftrightarrow (P\alpha \wedge P\beta)$ $\qquad$ (19) $F(\alpha \sqcup \beta) \leftrightarrow (F\alpha \wedge F\beta)$
> (20) $(P\alpha \wedge F\alpha) \leftrightarrow (\alpha = 0)$

Figure 1. Axioms for DAL

valuation v extends uniquely to a function $v^* : \mathsf{Act} \to 2^E$ defined as
$$v^*(\alpha \sqcup \beta) = v^*(\alpha) \cup v^*(\beta)$$
$$v^*(\alpha \sqcap \beta) = v^*(\alpha) \cap v^*(\beta)$$
$$v^*(\bar\alpha) = E \setminus v^*(\alpha)$$
$$v^*(0) = \emptyset$$
$$v^*(1) = E.$$

The notion of satisfiability in a deontic action model under a valuation v, written $\mathfrak{M}, v \vDash \varphi$, is inductively defined as:
$$\mathfrak{M}, v \vDash \alpha = \beta \iff v^*(\alpha) = v^*(\beta)$$
$$\mathfrak{M}, v \vDash P\alpha \iff v^*(\alpha) \subseteq P$$
$$\mathfrak{M}, v \vDash F\alpha \iff v^*(\alpha) \subseteq F$$
$$\mathfrak{M}, v \vDash \varphi \to \psi \iff \mathfrak{M}, v \nvDash \varphi \text{ or } \mathfrak{M}, v \vDash \psi$$
$$\mathfrak{M}, v \vDash \neg \varphi \iff \mathfrak{M}, v \nvDash \varphi.$$

A formula φ is universally valid, written $\vDash \varphi$, iff for any deontic action model $\mathfrak{M}$ and valuation v on $\mathfrak{M}$, it follows that $\mathfrak{M}, v \vDash \varphi$.

3 DAL via Algebra

The logical formalism introduced by Segerberg in [23] enjoys some interesting characteristics. In particular, it is a simple modal logic that provides a well-executed characterization of deontic operators. Moreover, it enjoys an elegant semantics via ideals and Boolean algebras, or dually via sets and collections of sets. Furthermore, Segerberg's formalism further accommodates for additional deontic operators to be added sistematically. More importantly, the formalism is sound and complete (Theorem 3.1 in [23]).

In this section, we revise Segerberg's formalism from an algebraic perspective. More precisely, we provide an algebraic generalization of DAL. This generalization preserves the aforementioned properties of the original system. In particular, the algebraic theory is simple and uses standard tools of algebras (Boolean algebras, homomorphisms, free generated algebras, etc). It is modular in the sense that the algebras described below can be straightforwardly extended to support other deontic operators. And it also addresses the soundness and completeness of DAL using standard algebraic tools. It is worth remarking that the framework described below is, arguably, mathematically more abstract that the original DAL. This is one of the characteristics of algebraic logics which can be exploited to discuss some deontic logical issues at large. We retake this point later on.

3.1 Algebraic Background

In what follows, we assume that the reader is familiar with the following algebraic concepts. A (many-sorted) *signature* $\Sigma = \langle S, \Omega \rangle$ is a pair of a set S of sort names, or sorts, and a set Ω of function names. Each $f \in \Omega$ is assigned a non-empty sequence of elements of S indicating its type; formally: $\text{type}(f) = s_0 \ldots s_n \to s$. A Σ-Algebra is a structure $\mathbf{A} = \langle \{A_s\}_{s \in S}, \{f_\mathbf{A}\}_{f \in F} \rangle$ where $f_\mathbf{A} : A_{s_0} \times \cdots \times A_{s_n} \to A_s$ iff $\text{type}(f) = s_0 \ldots s_n \to s$. Given a family of (mutually disjoint) sets of variables $X = \{X_s\}_{s \in S}$ and a signature Σ, $\mathbf{T}_\Sigma(X)$ denotes the term algebra constructed from Σ and X. An interpretation is a homomorphism $h : \mathbf{T}_\Sigma(X) \to \mathbf{A}$ which assigns meaning to the elements of $\mathbf{T}_\Sigma(X)$. A Σ-equation is a pair (t_1, t_2) of terms of $\mathbf{T}_\Sigma(X)$ written as $t_1 \approx t_2$. Given an algebra $\mathbf{A}$ and an interpretation h, we write $\mathbf{A}, i \vDash t_1 \approx t_2$ iff $h(t_1) = h(t_2)$. Moreover, we write $\mathbf{A} \vDash t_1 \approx t_2$ iff $\mathbf{A}, h \vDash t_1 \approx t_2$ holds for every interpretation h. We also assume some basics notions of Boolean algebras. Given a Boolean algebra $\mathbf{A}$, $\preccurlyeq_\mathbf{A}$ denotes its underlying partial order. An ideal is a lower subset of $\mathbf{A}$ w.r.t. $\preccurlyeq_\mathbf{A}$ closed under finite joins, and a filter is an upper subset of $\preccurlyeq_\mathbf{A}$ closed under finite meets. **2** is the Boolean algebra containing exactly two elements. A Boolean algebra is called *concrete* if it is a field of sets. We use Stone's representation theorem. In particular, for any Boolean algebra $\mathbf{A}$, $\mathsf{s}(\mathbf{A})$ denotes its isomorphic Stone space, and $\varphi_\mathbf{A} : \mathbf{A} \to \mathsf{s}(\mathbf{A})$ is the corresponding isomorphism. These and other useful notions are introduced in more detail in Appendix A.

3.2 Algebraizing DAL

One of the first steps in algebraizing a logic is to view formulas of a logical language as terms of an algebraic language. We begin by being clear about the algebraic language that we will use in the rest of this section.

Definition 3.1 The similarity type of DAL is a pair $\Sigma = (S, \Omega)$ where: (a) $S = \{a, f\}$ is a set of sort names and (b) $\Omega = \{\sqcup, \sqcap, ^-, 0, 1, \vee, \wedge, \neg, \bot, \top, =, \mathsf{P}, \mathsf{F}\}$ is a set of operation names s.t.:

(21) $\sqcup : a \times a \to a$ (22) $\sqcap : a \times a \to a$ (23) $^- : a \to a$ (24) $0 : a$
(25) $1 : a$ (26) $\vee : f \times f \to f$ (27) $\wedge : f \times f \to f$ (28) $\neg : f \to f$
(29) $\bot : f$ (30) $\top : f$ (31) $= : a \times a \to f$ (32) $\mathsf{P} : a \to f$
(33) $\mathsf{F} : a \to f$

Intuitively, we think of the elements a and f of S in the signature Σ as sort names for actions and formulas, respectively. In turn, we think of the operation names in Ω as names for operators on actions, operators on formulas, or heterogeneous operators. The algebraic language we will use in the rest of this section is the freely generated algebra over the similarity type Σ of DAL w.r.t. the set Act_0 of basic action symbols. We refer to this algebra, written $\mathbf{T}$, as the *deontic action term algebra*.

Having defined the algebraic language, we turn our attention to the way in which this language is interpreted in an algebra. In this regard, just as Boolean algebras are fundamental for the algebraization of Classical Propositional Logic, what we call *deontic action algebras* are fundamental for the algebraization of DAL. We introduce deontic action algebras in Def. 3.2 and discuss the technical details and the intuitions leading to this definition shortly after. (This notion borrows ideas and terminology from Pratt's *dynamic algebras* [20].)

Definition 3.2 A deontic action algebra is a tuple $\mathbf{D} = \langle \mathbf{A}, \mathbf{F}, \mathcal{E}, \mathcal{P}, \mathcal{F} \rangle$ where: (a) $\mathbf{A} = \langle A, +_\mathbf{A}, *_\mathbf{A}, -_\mathbf{A}, 0_\mathbf{A}, 1_\mathbf{A} \rangle$ and $\mathbf{F} = \langle F, +_\mathbf{F}, *_\mathbf{F}, -_\mathbf{F}, 0_\mathbf{F}, 1_\mathbf{F} \rangle$ are Boolean algebras; and (b) $\mathcal{E} : A \times A \to F$, $\mathcal{P} : A \to F$, and $\mathcal{F} : A \to F$, are total functions satisfying:

(34) $\mathcal{P}(a +_\mathbf{A} b) =_\mathbf{F} \mathcal{P}(a) *_\mathbf{F} \mathcal{P}(b)$ (35) $\mathcal{P}(a) *_\mathbf{F} \mathcal{F}(a) =_\mathbf{F} \mathcal{E}(a, 0_\mathbf{A})$
(36) $\mathcal{F}(a +_\mathbf{A} b) =_\mathbf{F} \mathcal{F}(a) *_\mathbf{F} \mathcal{F}(b)$ (37) $\mathcal{E}(a, b) *_\mathbf{F} \mathcal{P}(a) \preccurlyeq_\mathbf{F} \mathcal{P}(b)$
(38) $\mathcal{E}(a, b) *_\mathbf{F} \mathcal{F}(a) \preccurlyeq_\mathbf{F} \mathcal{F}(b)$ (39) $a =_\mathbf{A} b$ iff $\mathcal{E}(a, b) =_\mathbf{F} 1_\mathbf{F}$.

From an intuitive point of view, the elements in a deontic action algebra $\mathbf{D}$ may be understood as: (a) $\mathbf{A}$ and $\mathbf{F}$ correspond to an algebra of actions and an algebra of formulas, respectively; (b) $\mathcal{P}$ and $\mathcal{F}$ are abstract versions of the operations of an action being permitted and an action being forbidden, respectively; (c) $\mathcal{E}$ is an abstract version of the equality on actions at the level of formulas. From a technical point of view, Eq. (39) occupies a special place. This equation, in contrast to the others, is not expressed by an identity. Instead, it is expressed as a pair of conditional identities, or quasi-identities. This renders the class of deontic action algebras a quasi-variety (see [22]).

Definition 3.3 The quasi-variety of deontic action algebras is denoted by $\mathcal{D}_0$.

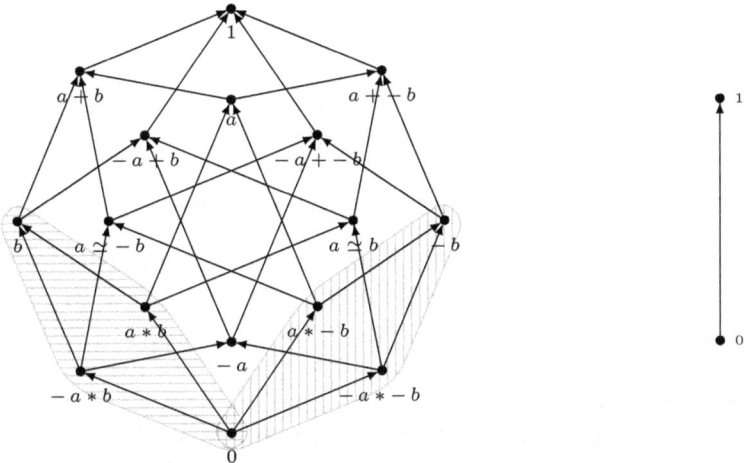

Figure 2. A Deontic Action Algebra

We give an example of a deontic action algebra $\mathbf{D} = \langle \mathbf{A}, \mathbf{F}, \mathcal{E}, \mathcal{P}, \mathcal{F} \rangle$ in Fig. 2. In this figure, the graph on the left illustrates the Boolean algebra $\mathbf{A}$ of actions. This algebra is the free Boolean algebra on the set of generators $\{a, b\}$. We use $x \simeq y$ as syntax sugar for $(x * y) + (-x * -y)$. The graph on the right illustrates the Boolean algebra $\mathbf{F}$ of formulas. This algebra is the Boolean algebra $\mathbf{2}$. We omitted subscripts on the operations of the Boolean algebras to improve legibility. The functions $\mathcal{P}$ and $\mathcal{F}$ are defined in Eqs. (40) and (41). The area shaded with horizontal lines illustrates the elements of $|\mathbf{A}|$ that $\mathcal{P}$ maps to 1, i.e., the elements of $|\mathbf{A}|$ that are permitted. Notice that these elements form an ideal in $\mathbf{A}$. In turn, the area shaded with vertical lines illustrates the elements of $|\mathbf{A}|$ that $\mathcal{F}$ maps to 1, i.e., the elements of $|\mathbf{A}|$ that are forbidden. Again, notice that these elements also form an ideal in $\mathbf{A}$. It can easily be seen in this example that: if $\mathcal{P}(x) = 1$ for all $x \in |\mathbf{A}|$, then, $\mathcal{F}(0) = 1$ and $\mathcal{F}(x) = 0$ for all $0 \neq x \in |\mathbf{A}|$. Similarly, if $\mathcal{F}(x) = 1$ for all $x \in |\mathbf{A}|$, then, $\mathcal{P}(0) = 1$ and $\mathcal{P}(x) = 0$ for all $0 \neq x \in |\mathbf{A}|$. These cases are known as *deontic heaven* and *deontic hell*, respectively. We will briefly discuss them later on.

$$(40) \quad \mathcal{P}(x) = \begin{cases} 1 & \text{if } x \preccurlyeq b \\ 0 & \text{otherwise} \end{cases} \qquad (41) \quad \mathcal{F}(x) = \begin{cases} 1 & \text{if } x \preccurlyeq -b \\ 0 & \text{otherwise} \end{cases}$$

We are now in a position to establish the connection between deontic action algebras and DAL.

Definition 3.4 Let $\mathbf{D}$ be a deontic algebra; an assignment on $\mathbf{D}$ is a function

$f : \mathsf{Act}_0 \to |\mathbf{A}|$. An interpretation on $\mathbf{D}$ is a homomorphism $h : \mathbf{T} \to \mathbf{D}$ s.t.:

$h(\alpha \sqcup \beta) = h(\alpha) +_{\mathbf{A}} h(\beta)$ $\quad h(\phi \vee \psi) = h(\phi) +_{\mathbf{F}} h(\psi)$ $\quad h(\mathsf{P}\alpha) = \mathscr{P}(h(\alpha))$

$h(\alpha \sqcap \beta) = h(\alpha) *_{\mathbf{A}} h(\beta)$ $\quad h(\phi \wedge \psi) = h(\phi) *_{\mathbf{F}} h(\psi)$ $\quad h(\mathsf{F}\alpha) = \mathscr{F}(h(\alpha))$

$h(\bar{\alpha}) = -_{\mathbf{A}} h(\alpha)$ $\qquad\qquad h(\neg\varphi) = -_{\mathbf{F}} h(\varphi)$ $\qquad\qquad h(\top) = 1_{\mathbf{F}}$

$h(0) = 0_{\mathbf{A}}$ $\qquad\qquad\qquad h(\bot) = 0_{\mathbf{F}}$

$h(1) = 1_{\mathbf{A}}$ $\qquad\qquad\qquad h(\alpha = \beta) = \mathscr{E}(h(\alpha), h(\beta))$

Fact 3.5 *Assignments extend uniquely to interpretations. Given an assignment f, f^* denotes its unique extension.*

Definition 3.6 An equation is a pair (τ_1, τ_2), written $\tau_1 \approx \tau_2$, where either $\tau_1, \tau_2 \in \mathsf{Act}$ or $\tau_1, \tau_2 \in \mathsf{Form}$. An equation $\tau_1 \approx \tau_2$ is valid under an interpretation h on a deontic algebra $\mathbf{D}$, written $\mathbf{D}, h \models \tau_1 \approx \tau_2$, iff $h(\tau_1) = h(\tau_2)$. An equation $\tau_1 \approx \tau_2$ is universally valid, written $\models \tau_1 \approx \tau_2$, iff for all deontic algebras $\mathbf{D}$ and interpretations h on $\mathbf{D}$, it follows that $\mathbf{D}, h \models \tau_1 \approx \tau_2$.

Theorem 3.7 (Soundness) *If $\vdash \varphi$, then, $\models \varphi \approx \top$.*

Proof [Sketch] By induction on the length of a proof of φ. We restrict our attention to some interesting cases. In particular, to the axioms displayed in Eqs. (17), (18) and (20). Let $\mathbf{D}$ be any deontic algebra and h be any homomorphism on $\mathbf{D}$:

Eq. (17): We need to show that $h((\alpha = \beta) \to (\varphi \to \varphi_\beta^\alpha)) = 1_{\mathbf{F}}$. The simple cases in which $\varphi = \mathsf{P}\alpha$ or $\varphi = \mathsf{F}\alpha$ entail all others. Then,

$$h((\alpha = \beta) \to (\mathsf{P}\alpha \to \mathsf{P}\beta)) = h(\neg(\alpha = \beta) \vee (\neg \mathsf{P}\alpha \vee \mathsf{P}\beta))$$
$$= -_{\mathbf{F}} h(\alpha = \beta) +_{\mathbf{F}} h(\neg \mathsf{P}\alpha) +_{\mathbf{F}} h(\mathsf{P}\beta)$$
$$= -_{\mathbf{F}} \mathscr{E}(h(\alpha), h(\beta)) +_{\mathbf{F}} -_{\mathbf{F}} h(\mathsf{P}\alpha) +_{\mathbf{F}} \mathscr{P}(h(\beta))$$
$$= -_{\mathbf{F}} \mathscr{E}(h(\alpha), h(\beta)) +_{\mathbf{F}} -_{\mathbf{F}} \mathscr{P}(h(\alpha)) +_{\mathbf{F}} \mathscr{P}(h(\beta))$$
$$= -_{\mathbf{F}}(\mathscr{E}(h(\alpha), h(\beta)) *_{\mathbf{F}} \mathscr{P}(h(\alpha))) +_{\mathbf{F}} \mathscr{P}(h(\beta))$$

From Eq. (37), $\mathscr{P}(h(\beta)) = \mathscr{E}(h(\alpha), h(\beta)) *_{\mathbf{F}} \mathscr{P}(h(\alpha)) +_{\mathbf{F}} \mathscr{P}(h(\beta))$. From this fact, $-_{\mathbf{F}}(\mathscr{E}(h(\alpha), h(\beta)) *_{\mathbf{F}} \mathscr{P}(h(\alpha))) +_{\mathbf{F}} \mathscr{P}(h(\beta)) = 1_{\mathbf{F}}$.

Eq. (18): We need to show that $h(\mathsf{P}(\alpha \sqcup \beta) \leftrightarrow (\mathsf{P}\alpha \wedge \mathsf{P}\beta)) = 1_{\mathbf{F}}$. Then,

$$h(\mathsf{P}(\alpha \sqcup \beta) \leftrightarrow (\mathsf{P}\alpha \wedge \mathsf{P}\beta)) = h((\neg \mathsf{P}(\alpha \sqcup \beta) \vee (\mathsf{P}\alpha \wedge \mathsf{P}\beta))$$
$$\wedge (\neg(\mathsf{P}\alpha \wedge \mathsf{P}\beta) \vee \mathsf{P}(\alpha \sqcup \beta)))$$
$$= h(\neg \mathsf{P}(\alpha \sqcup \beta) \vee (\mathsf{P}\alpha \wedge \mathsf{P}\beta))$$
$$*_{\mathbf{F}} h(\neg(\mathsf{P}\alpha \wedge \mathsf{P}\beta) \vee \mathsf{P}(\alpha \sqcup \beta))$$

We continue by cases. Consider first:

$$h(\neg \mathsf{P}(\alpha \sqcup \beta) \vee (\mathsf{P}\alpha \wedge \mathsf{P}\beta)) = h(\neg \mathsf{P}(\alpha \sqcup \beta)) +_{\mathbf{F}} h(\mathsf{P}\alpha \wedge \mathsf{P}\beta)$$
$$= -_{\mathbf{F}} h(\mathsf{P}(\alpha \sqcup \beta)) +_{\mathbf{F}} (h(\mathsf{P}\alpha) *_{\mathbf{F}} h(\mathsf{P}\beta))$$
$$= -_{\mathbf{F}}(h(\mathsf{P}\alpha) *_{\mathbf{F}} h(\mathsf{P}\beta)) +_{\mathbf{F}} (h(\mathsf{P}\alpha) *_{\mathbf{F}} h(\mathsf{P}\beta))$$
$$= 1_{\mathbf{F}}$$

Similarly, $h(\neg(\mathsf{P}\alpha \wedge \mathsf{P}\beta) \vee \mathsf{P}(\alpha \sqcup \beta)) = 1_{\mathbf{F}}$.

Eq. (20) We need to show that $h((\mathsf{P}\alpha \wedge \mathsf{F}\alpha) \to (\alpha = 0)) = 1_\mathbf{F}$. Then,
$$\begin{aligned} h((\mathsf{P}\alpha \wedge \mathsf{F}\alpha) \to (\alpha = 0)) &= h(\neg(\mathsf{P}\alpha \wedge \mathsf{F}\alpha) \vee (\alpha = 0)) \\ &= h(\neg(\mathsf{P}\alpha \wedge \mathsf{F}\alpha)) +_\mathbf{F} h(\alpha = 0) \\ &= -_\mathbf{F} h(\mathsf{P}\alpha \wedge \mathsf{F}\alpha) +_\mathbf{F} \mathcal{E}(h(\alpha), h(0)) \\ &= -_\mathbf{F} (\mathcal{P}(h(\alpha)) *_\mathbf{F} \mathcal{F}(h(\alpha))) +_\mathbf{F} \mathcal{E}(h(\alpha), 0_\mathbf{A}) \\ &= -_\mathbf{F} \mathcal{E}(h(\alpha), 0_\mathbf{A}) +_\mathbf{F} \mathcal{E}(h(\alpha), 0_\mathbf{A}) \\ &= 1_\mathbf{F} \end{aligned}$$
□

It is important to notice that, as expected, not every sentence is provable in DAL. In particular, if φ is a theorem, i.e., $\vdash \varphi$, then, $\neg\varphi$ is not provable, i.e., $\nvdash \neg\varphi$. This claim is substantiated as follows. Let $\mathbf{D} = \langle \mathbf{A}, \mathbf{F}, \mathcal{E}, \mathcal{P}, \mathcal{F} \rangle$ be the deontic action algebra in Fig. 2, and let h be any interpretation on $\mathbf{D}$; if φ is a theorem, then, $h(\varphi) = 1_\mathbf{F}$. Since h is a homomorphism, $h(\neg\varphi) = 0_\mathbf{F}$. Therefore, from Thm. 3.7, $\nvdash \neg\varphi$.

To prove the converse of Thm. 3.7, our sought after algebraic completeness result, we need to show that every non-theorem of DAL can be falsified on some deontic action algebra $\mathbf{D}$ (in the sense that there is some homorphism on $\mathbf{D}$ under which the non-theorem does not evaluate to $1_\mathbf{F}$). To this end, we introduce the notion of a Lindenbaum-Tarski deontic action algebra.

Fact 3.8 *Let $\cong_a \subseteq \mathsf{Act} \times \mathsf{Act}$ and $\cong_f \subseteq \mathsf{Form} \times \mathsf{Form}$ be defined as:*
$$\alpha \cong_a \beta \text{ iff } \vdash \alpha = \beta \qquad \varphi \cong_f \psi \text{ iff } \vdash \varphi \leftrightarrow \psi,$$
then the relations $\{\cong_a, \cong_f\}$ form a congruence on the deontic action term algebra $\mathbf{T}$. This congruence is denoted with the symbol $\cong$.

Definition 3.9 The Lindenbaum-Tarski deontic action algebra is the structure $\mathbf{L} = \langle \mathbf{A}, \mathbf{F}, \mathcal{E}, \mathcal{P}, \mathcal{F} \rangle$ where:

$\mathbf{A} = \langle \mathsf{Act}/_{\cong_a}, \sqcup_{\cong_a}, \sqcap_{\cong_a}, {}^{-\cong_a}, [0]_{\cong_a}, [1]_{\cong_a} \rangle \qquad \mathcal{E}([\alpha]_{\cong_a}, [\beta]_{\cong_a}) = [\alpha = \beta]_{\cong_a}$

$\mathbf{F} = \langle \mathsf{Form}/_{\cong_f}, \vee_{\cong_f}, \wedge_{\cong_f}, \neg_{\cong_f}, [\bot]_{\cong_f}, [\top]_{\cong_f} \rangle \qquad \mathcal{P}([\alpha]_{\cong_a}) = [\mathsf{P}\alpha]_{\cong_f}$

$\qquad\qquad\qquad\qquad\qquad\qquad\qquad\qquad\qquad\quad \mathcal{F}([\alpha]_{\cong_a}) = [\mathsf{F}\alpha]_{\cong_f}.$

Proposition 3.10 *The Lindenbaum-Tarski deontic action algebra $\mathbf{L}$ is a deontic action algebra.*

Proof [Sketch] That $\mathbf{A}$ and $\mathbf{F}$ are Boolean algebras is more or less immediate. We show that the functions $\mathcal{E}$, $\mathcal{P}$, and $\mathcal{F}$ satisfy axioms from Eqs. (34), (35) and (39). The proof for axioms from Eqs. (36) to (38) are similar.

Eq. (34) We need to show that $\mathcal{P}([\alpha \sqcup \beta]_{\cong_a}) = \mathcal{P}([\alpha]_{\cong_a}) \wedge_{\cong_f} \mathcal{P}([\beta]_{\cong_a})$. Then,
$$\begin{aligned} \mathcal{P}([\alpha \sqcup \beta]_{\cong_a}) &= [\mathsf{P}(\alpha \sqcup \beta)]_{\cong_f} \\ &= [\mathsf{P}\alpha \wedge \mathsf{P}\beta]_{\cong_f} &&\text{see Eq. (18)} \\ &= [\mathsf{P}\alpha]_{\cong_f} \wedge_{\cong_f} [\mathsf{P}\beta]_{\cong_f} \\ &= \mathcal{P}([\alpha]_{\cong_a}) \wedge_{\cong_f} \mathcal{P}([\beta]_{\cong_a}) \end{aligned}$$

Eq. (35) We need to show that $\mathscr{P}([\alpha]_{\cong_a}) \wedge_{\cong_f} \mathscr{F}([\alpha]_{\cong_a}) = \mathcal{E}([\alpha]_{\cong_a}, [0]_{\cong_a})$. Then,

$$\begin{aligned}
\mathscr{P}([\alpha]_{\cong_a}) \wedge_{\cong_f} \mathscr{F}([\alpha]_{\cong_a}) &= [\mathsf{P}\alpha]_{\cong_f} \wedge_{\cong_f} [\mathsf{F}(\alpha \sqcup \beta)]_{\cong_f} \\
&= [\mathsf{P}\alpha \wedge \mathsf{F}\alpha]_{\cong_f} \\
&= [\alpha = 0]_{\cong_f} \quad\quad\quad \text{see Eq. (20)} \\
&= \mathcal{E}([\alpha]_{\cong_a}, [0]_{\cong_a})
\end{aligned}$$

Eq. (39) We need to show that $[\alpha]_{\cong_a} = [\beta]_{\cong_b}$ iff $\mathcal{E}([\alpha]_{\cong_a}, [\beta]_{\cong_a}) = [\top]_{\cong_f}$. Suppose that $[\alpha]_{\cong_a} = [\beta]_{\cong_b}$; it follows that $\vdash \alpha = \beta$; and so $\vdash (\alpha = \beta) \leftrightarrow \top$. Then, $\mathcal{E}([\alpha]_{\cong_a}, [\beta]_{\cong_a}) = [\alpha = \beta]_{\cong_f} = [\top]_{\cong_f}$. Similarly, if $\mathcal{E}([\alpha]_{\cong_a}, [\beta]_{\cong_a}) = [\top]_{\cong_f}$, then, $[\alpha]_{\cong_a} = [\beta]_{\cong_b}$.

□

The following result connects logical deduction in DAL with the Lindenbaum-Tarski Algebra. Roughly speaking, it says that the Lindenbaum-Tarski algebra captures DAL theoremhood.

Theorem 3.11 (Completeness) $\vdash \varphi$ iff $\mathbf{L} \vDash \varphi \approx \top$.

Proof The left to right direction is immediate from Thm. 3.7. For the right to left direction we show that if $\nvdash \varphi$, then $\mathbf{L} \nvDash \varphi \approx \top$. Suppose that $\nvdash \varphi$; then $\nvdash \varphi \leftrightarrow \top$. This means that $[\varphi]_{\cong_f} \neq [\top]_{\cong_f}$. Construct an assignment $f : \mathsf{Act}_0 \to |\mathbf{L}|$ that sends each $\mathsf{a}_i \in \mathsf{Act}_0$ to the equivalence class $[\mathsf{a}_i]_{\cong_a}$. Using induction, we construct a homomorphism f^* which agrees on f that is such that $f^*(\varphi) = [\varphi]_{\cong_f}$. Then, from our assumption, we have $f^*(\varphi) = [\varphi]_{\cong_f} \neq [\top]_{\cong_f} = f^*(\top)$. Therefore, $\mathbf{L} \nvDash \varphi \approx \top$.

□

The following corollary can be obtained using a standard argument in algebraic logic.

Corollary 3.12 If $\vDash \varphi \approx \top$, then $\vdash \varphi$.

Proof Assume $\nvdash \varphi$, then by Theorem 3.11, we have that $\mathbf{L} \nvDash \varphi \approx \top$ and therefore $\nvDash \varphi \approx \top$.

□

In other words, the Lindenbaum-Tsarski algebra can be thought as a canonical (algebraic) model which provides counterexamples of non-valid formulas.

3.3 Deontic Action Algebras and Deontic Action Models

We connect deontic action algebras and deontic action models via a Stone's representation. This gives us another proof of the completeness of Segerberg's deduction system w.r.t. the original semantics. Recall that the Stone's representation theorem [24] establishes that every Boolean algebra is isomorphic to a certain field of sets. We will prove a similar result for deontic action algebras.

We begin by introducing some additional concepts. First, just as Boolean algebras made of sets (i.e., fields of sets) are sometimes named *concrete Boolean algebras* in Algebraic Logic, we define concrete deontic action algebras as deontic action algebras whose action and formula algebras are fields of sets. Concrete deontic algebras allow us to establish the connection with Segerberg's original semantics for DAL.

Definition 3.13 A deontic action algebra $\mathbf{D} = \langle \mathbf{F}, \mathbf{A}, \mathcal{E}, \mathcal{P}, \mathcal{F} \rangle$ is called concrete iff $\mathbf{F}$ and $\mathbf{A}$ are fields of sets. The class of concrete deontic algebras is denoted by $\mathcal{C}_0$.

Using Stone duality we can prove that algebraic validity can be reduced to validity in concrete deontic algebras.

Theorem 3.14 *For any* DAL *formula φ, we have:* $\vDash \varphi \approx \top$ *iff* $\mathcal{C}_0 \vDash \varphi \approx \top$.

Proof The left to right direction is straightforward. For the other direction, assume that $\mathcal{C}_0 \vDash \varphi \approx \top$ and $\nvDash \varphi \approx \top$. This means that we have a deontic action algebra $\mathbf{D} = \langle \mathbf{F}, \mathbf{A}, \mathcal{E}, \mathcal{P}, \mathcal{F} \rangle$ and a valuation v s.t. $\mathbf{D}, v \nvDash \varphi \approx \top$. Applying Stone duality we have a concrete deontic action algebra $\mathbf{D}' = \langle \mathbf{F}', \mathbf{A}', \mathcal{E}', \mathcal{P}', \mathcal{F}' \rangle$ that is isomorphic to $\mathbf{D}$. On this concrete deontic algebra, we can define valuation $v'(a_i) = \varphi_{\mathbf{A}'}(v(a_i))$ (being $\varphi_{\mathbf{A}'}$ the Stone isomorphism for $\mathbf{A}'$). Then, we have $\mathbf{D}', v' \nvDash \varphi \approx \top$. From this fact, we obtain a contradiction. □

We relate Segerberg's models to concrete deontic action algebras as follows.

Definition 3.15 Let $\mathfrak{M} = \langle E, P, F \rangle$ and $v : \mathsf{Act}_0 \to E$ be a deontic action model and a valuation, resp.; we associate with $\mathfrak{M}$ and v the deontic action algebra $\mathrm{alg}(\mathfrak{M}, v) = \langle \mathbf{F}_\mathfrak{M}^v, \mathbf{A}_\mathfrak{M}^v, \mathcal{E}_\mathfrak{M}^v, \mathcal{P}_\mathfrak{M}^v, \mathcal{F}_\mathfrak{M}^v \rangle$ where:

(a) $\mathbf{F}_\mathfrak{M} = \mathbf{2}$;

(b) $\mathbf{A}_\mathfrak{M}$ is the field of sets generated from $\{ v(\mathsf{a}_i) \mid \mathsf{a}_i \in \mathsf{Act}_0 \}$.

(c) $\mathcal{E}_\mathfrak{M}^v(x, y) = \begin{cases} 1 & \text{if } x = y \\ 0 & \text{otherwise} \end{cases}$

(d) $\mathcal{P}_\mathfrak{M}^v(x) = \begin{cases} 1 & \text{if } x \subseteq P \\ 0 & \text{otherwise} \end{cases}$

(e) $\mathcal{F}_\mathfrak{M}^v(x) = \begin{cases} 1 & \text{if } x \subseteq F \\ 0 & \text{otherwise} \end{cases}$

Similarly, deontic action models form concrete deontic action algebras.

Definition 3.16 Let $\mathbf{D} = \langle \mathbf{F}, \mathbf{A}, \mathcal{E}, \mathcal{P}, \mathcal{F} \rangle$ be a concrete deontic action algebra and $f : \mathsf{Act}_0 \to A$ an assignment in $\mathbf{D}$; we associate with $\mathbf{D}$ and f a deontic action model $\mathrm{mod}(\mathbf{D}) = \langle E_\mathbf{D}, P_\mathbf{D}, F_\mathbf{D} \rangle$ and a valuation $v_f : \mathsf{Act}_0 \to E_\mathbf{D}$ where:

(a) $E_\mathbf{D} = |\mathbf{A}|$

(b) $P = \bigcup \{ x \mid \mathbf{D}, f \vDash \mathcal{P}(x) \approx \top \}$

(c) $F = \bigcup \{ x \mid \mathbf{D}, f \vDash \mathcal{F}(x) \approx \top \}$

The following are important properties of alg and mod.

Theorem 3.17 $\mathbf{D}, f \vDash \varphi \approx \top$ *iff* $\mathrm{mod}(\mathbf{D}), v_f \Vdash \varphi$.

Theorem 3.18 $\mathfrak{M}, v \vDash \varphi$ *iff* $\mathrm{alg}(\mathfrak{M}), f_v \vDash \varphi \approx 1$.

Interestingly, when seen as operators, mod and alg are inverses of each other and therefore the two are isomorphisms.

Theorem 3.19 *For all deontic action algebra $\mathbf{D}$ and deontic action model $\mathfrak{M}$:*

$$\mathrm{alg}(\mathrm{mod}(\mathbf{D})) = \mathbf{D} \quad \text{and} \quad \mathrm{mod}(\mathrm{alg}(\mathfrak{M})) = \mathfrak{M}$$

Then, we can prove the completeness of the Segerberg's deductive system w.r.t. deontic models in an algebraic way.

Theorem 3.20 $\vdash \varphi$ iff $\vDash \varphi$.

Proof Suppose $\vdash \varphi$. By algebraic completeness, this is equivalent to $\vDash\!\!\!\approx \varphi \approx \top$ and, by Thm. 3.14, also to $C_0 \vDash\!\!\!\approx \varphi \approx \top$; by Thm. 3.19 and Thm. 3.17, this is equivalent to $\vDash \varphi$. □

4 Algebraizing Other Deontic Action Logics

The work of Segerberg in [23] gave rise to a family of closely related deontic logics. The logics DAL^i for $1 \leq i \leq 5$ reported in [26] are particularly interesting. Each DAL^i deals with a particular deontic issue, and is obtained from DAL^j (with $j < i$) by adding additional axioms to those in Fig. 1. Here, we show how to extend the algebraic framework in Sec. 3 to each of these variants.

The first of these extensions, DAL^1, is obtained from DAL by adding, for each $\mathsf{a}_i \in \mathsf{Act}_0$, $\mathsf{Fa}_i \vee \mathsf{Pa}_i$ to the set of axioms in Fig. 1. Intuitively, these axioms intend to capture what is called the *Principle of Deontic Closure* in deontic logics: what is not forbidden is permitted (alt., every action is either permitted or forbidden). As noted in [26], these axioms capture closeness only at the level of action generators, and they are not able to capture closeness for other (perhaps more fine-grained) actions. The algebraic counterpart of DAL^1 is determined by the class of deontic action algebras: (i) whose algebra of actions is generated by a set G of generators; and (ii) that satisfy Eq. (42) below.

$$\mathcal{F}(x) +_\mathbf{F} \mathcal{P}(x) = 1_\mathbf{F} \qquad \text{for every generator } x \in |\mathbf{A}| \qquad (42)$$

In turn, the extension DAL^2 is obtained from DAL^1 by: (i) requiring the set Act_0 of basic actions to be a finite, i.e., $\mathsf{Act}_0 = \{\, \mathsf{a}_i \mid 0 \leq i \leq n \,\}$; and (ii) adding the axioms $\mathsf{P}(\overline{\mathsf{a}_0} \sqcap \cdots \sqcap \overline{\mathsf{a}_n}) \vee \mathsf{F}(\overline{\mathsf{a}_0} \sqcap \cdots \sqcap \overline{\mathsf{a}_n})$. Intuitively, the additional axiom states that not performing any of the basic actions is permitted or forbidden. On the algebraic side, by considering a finite set Act_0 of basic actions, we obtain that the algebra $\mathbf{A}$ of actions is an atomic Boolean algebra. The atoms in this algebra allow us to focus on the most basic actions being considered. Then, the algebraic counterpart of DAL^2 is determined by the class of deontic action algebras: (i) that are finitely generated by a set $G = \{\, \mathsf{a}_i \mid 0 \leq i \leq n \,\}$; and that satisfy Eq. (43) below.

$$\mathcal{P}((-_\mathbf{A}\, a_1) +_\mathbf{A} \cdots +_\mathbf{A} (-_\mathbf{A}\, a_n)) +_\mathbf{F} \mathcal{F}((-_\mathbf{A}\, a_1) +_\mathbf{A} \cdots +_\mathbf{A} (-_\mathbf{A}\, a_n)) = 1_\mathbf{F} \quad (43)$$

The extension DAL^3 is obtained from DAL^2 by adding the following axiom: $(\mathsf{a}_1 \sqcup \cdots \sqcup \mathsf{a}_n) = 1_\mathbf{A}$. Intuitively, this axiom can be read as stating that the actions $\mathsf{a}_1, \ldots, \mathsf{a}_n$ are the sole actions that the agent can perform. The algebraic counterpart of DAL^3 is determined by the subclass of deontic action algebras of DAL^2 that further satisfy Eq. (44) below.

$$a_0 +_A \cdots +_A a_n = 1_A \qquad (44)$$

The extension DAL^4 is obtained from DAL by requiring closedness at the level of "atomic" actions. Formally, DAL^4 considers a finite number of actions $\mathsf{Act}_0 = \{\, \mathsf{a}_i \mid 0 \leq i \leq n \,\}$ and a collection $\{\, \alpha_i \mid 0 \leq i \leq 2^n \,\}$ of action terms s.t.: each α_i is of the form $*\mathsf{a}_0 \sqcap \cdots \sqcap *\mathsf{a}_n$, where $*\mathsf{a}_i \in \{\mathsf{a}_i, \overline{\mathsf{a}_i}\}$. Syntactically, each α_i represents a possible atomic action. Closedness is then obtained by adding the following set of axioms to those of DAL: $\mathsf{P}\alpha_i \vee \mathsf{F}\alpha_i$, for all $0 \leq i \leq 2^n$. The algebraic counterpart of DAL^4 is determined by the class of deontic action algebras: (i) that are finitely generated; and (ii) that satisfy Eq. (45) below.

$$\mathscr{P}(a) +_{\mathbf{F}} \mathscr{F}(a) = 1_{\mathbf{F}} \qquad \text{for all atoms } a \in |\mathbf{A}| \tag{45}$$

Finally, the extension DAL^5 is obtained by putting together DAL^3 and DAL^4. The algebraic counterpart of DAL^5 is obtained from the deontic action algebras that are deontic action algebras of DAL^3 and DAL^4.

Following from the above, we obtain for each DAL^i an associated class $\mathcal{D}_i$ of deontic action algebras. Each of these classes accommodates for a corresponding soundness and completeness result. This is made precise in Thm. 4.1. (The proof of Thm. 4.1 is a routine extension of the proof of Thm. 3.11.)

Theorem 4.1 *For every $0 \leq i \leq 5$, let $\vdash_{\mathsf{DAL}^i}$ be theoremhood relation of DAL^i and $\vDash_{\mathcal{D}_i}$ equational validity in the class $\mathcal{D}_i$; then, $\vdash_{\mathsf{DAL}^i} \varphi$ iff $\vDash_{\mathcal{D}_i} \varphi \approx \top$.*

5 Final Remarks

We presented an algebraic treatment of Sergerberg's deontic action logic and some of known extensions via deontic action algebras. As is commonly done in the algebraization of a logic, along the way we discussed concepts such as: actions and formulas algebras, operators of permission and prohibition, and Lindenbaum-Tarski algebras. Moreover, we established that the algebraic characterization is correct by proving soundness and completeness theorems. In our opinion the overall picture is just as important. Our algebraic treatment can be thought of as an abstract version of deontic action logics which can be used to establish connections between deontic action logics and mathematical areas such as topology, category theory, probability, etc.

In addition to the obvious mathematical benefits of having an algebraization of deontic action logics, we believe that the algebraic framework introduced above paves the way for interesting future work. First, deontic action algebras are modular in their formulation; i.e., action and formula algebras can be replaced to obtain new systems. For instance, by changing the algebra of actions we can obtain systems where it is possible to reason about other action combinators. Interesting cases are those of: action composition (denoted by ;), and action iteration (denoted by *). In this line, the work of Meyer in [12] was one of the first in considering a deontic logic containing action composition. Meyer named the system Dynamic Deontic Logic (DDL). This system is not without challenges. As observed in [3], one of the main problems of DDL is that action composition (and so action iteration) makes it possible to formulate some paradoxes. Regarding action iteration (*), in [8], Broersen pointed out that dynamic deontic logics can be divided into: (i) *goal norms*, where prescriptions over a sequence of actions only take into account the last action performed; or

(ii) *process norms*, where a sequence of actions is permitted/forbidden iff every action in the sequence is permittedd/forbidden. It is a matter of discussion which one of these approaches is better, but both have cons and pros. The interested reader is referred to [8] for an in-depth discussion on this issue. To the best of our knowledge, we are not aware of any extension of Segerberg's logic that provides action composition or action iteration. This said, notice that deontic action algebras can be straightforwardly modified to admit these operators. More precisely, we may consider deontic action algebras $\langle \mathbf{F}, \mathbf{A}, \mathscr{P}, \mathscr{F} \rangle$ where $\mathbf{F}$ is a Boolean algebra; $\mathbf{A} = \langle A, +, ;, * \rangle$ is a Kleene algebra (see [15]); and $\mathscr{P}$ and $\mathscr{F}$ are deontic operators of permitted and forbidden on these algebras. Intuitively, in $\mathbf{A}$, ; captures action composition, + captures action choice, and $*$ captures the iteration of actions. Kleene algebras enjoy some nice properties. They are quasi-varieties, and they are complete w.r.t. equality of regular expressions (see [16]). In this respect, Kleene algebras provide a robust framework for reasoning about action composition and iteration. Similarly, one can extend deontic action algebras with other interesting algebras; e.g., relation algebras (see [17]). Relation algebras would provide other action operators, most notably, action converse. We leave it as further work investigating the properties the operators $\mathscr{P}$ and $\mathscr{F}$ in these new algebraic settings.

In turn, another interesting line of research consists in investigating other algebras for formulas. In this paper, we have used Boolean algebras as an abstraction of formulas, but there are different kinds of algebras that may provide alternative ways for reasoning about norms. Some immediate examples are: Heyting Algebras, semi-lattices, metric spaces, etc. We draw attention to the fact that changing the algebra of formulas in deontic action algebras may bridge the way for designing deontic logics that are not logics of normative propositions. More precisely, von Wright in [29], and Alchourron in [1,2], both noted the distinction between logics of normative propositions and logics of norms. The former are Boolean logics where their formulas express assertions about the existence of norms; i.e., a formula s.t. Pφ states that *there is a norm allowing the occurrence of* φ – SDL and DAL fall into this category. In contrast, logics of norms allow to express prescriptions that, as observed by von Wright, are not necessarily evaluated to a Boolean value (i.e., true or false). To deal with logics of norms, we can use other algebras to generalize formulas. For instance, by taking a meet semi-lattice as the algebra of formulas we can capture a theory of norms where norms can be put together, and where some norms are in contradiction with each other (but not necessarily where norms are true or false). Of course, there are several other appealing algebras that could play this role as well: metric spaces, rings, etc. We leave all this as a further work.

Appendix
A Many Sorted Algebras in a Nutshell

In this section we introduce some basic concepts used in the paper. These serve as a way to fix terminology and notation. The interested reader is referred to [10,22] for an in-depth introduction to this topic.

Definition A.1 A many-sorted signature is a pair $\Sigma = \langle S, \Omega \rangle$ where: (a) S is a set of sort names; and (b) $\Omega = \{ f : s_1 \ldots s_n \to s \mid s_i, s \in S \}$ is a set of operation names. A Σ-algebra $\mathbf{A}$ consists of: (c) an S-indexed family of sets, written $|\mathbf{A}| = \{ A_s \mid s \in S \}$; and (d) for each $f : s_1 \ldots s_n \to s \in \Omega$ a function $f_\mathbf{A} : A_{s_1} \ldots A_{s_n} \to A_s$.

Note that standard algebras can be seen as many-sorted algebras with only one sort. A special kind of Σ-algebras are the so-called Σ-term algebras.

Definition A.2 Let $\Sigma = \langle S, \Omega \rangle$ be a signature and $X = \{ X_s \mid s \in S \}$ be an S-indexed family of sets; a Σ-term algebra with variables in X is a Σ-algebra $\mathbf{T}$ in which:

(a) $|\mathbf{T}| = \{ T_s \mid s \in S \}$ is the $\subseteq$-smallest S-indexed family of sets s.t. for all $x \in X_s$, the string 'x' $\in T_s$; and for all $f : s_1 \ldots s_n \to s \in \Omega$ and strings $t_i \in T_{s_i}$, the string '$f(t_1 \ldots t_n)$' $\in T_s$;

(b) for each $f : s_1 \ldots s_n \to s \in \Omega$, there is a function $f_\mathbf{T} : T_{s_1} \ldots T_{s_n} \to T_s$ s.t. for all strings $t_i \in T_{s_i}$, $f_{\mathbf{T}(X)}(t_1 \ldots t_n)$ equals the string '$f(t_1 \ldots t_n)$'.

Definition A.3 Let $\mathbf{A}$ and $\mathbf{B}$ be Σ-algebras; a Σ-homomorphism $h : \mathbf{A} \to \mathbf{B}$ is an S-indexed family of functions $h = \{ h_s : A_s \to B_s \mid s \in S \}$ such that: for all $f : s_1 \ldots s_n \to s \in \Omega$ and $a_i \in A_{s_i}$, it follows that $h_s(f_\mathbf{A}(a_1 \ldots a_n)) = f_\mathbf{B}(h_{s_1}(a_1) \ldots h_{s_n}(a_n))$.

Definition A.4 Let $\mathbf{A}$ be a Σ-algebra; a Σ-congruence $\cong$ on $\mathbf{A}$ is an S-sorted family of relations $\cong \, = \{ \cong_s \, \subseteq A_s^2 \mid s \in S \}$ such that: each $\cong_s$ is an equivalence relation on A_s^2; and for all $f : s_1 \ldots s_n \to s \in \Omega$ and $a_i, a'_i \in A_{s_i}$, if $a_i \cong_{s_i} a'_i$, then $f_\mathbf{A}(a_1 \ldots a_n) \cong_s f_\mathbf{A}(a'_1 \ldots a'_n)$.

Definition A.5 Let $\mathbf{A}$ be a Σ-algebra and $\cong$ be a Σ-congruence on $\mathbf{A}$; the quotient Σ-algebra of $\mathbf{A}$ under $\cong$, written $\mathbf{A}/{\cong}$, has: (a) $|\mathbf{A}/{\cong}| = \{ A_s/{\cong_s} \mid s \in S \}$; and (b) for all $f : s_1 \ldots s_n \to s \in \Omega$ and $a_i \in A_{s_i}$, $f_{\mathbf{A}/{\cong}}([a_1]_{\cong_{s_1}} \ldots [a_n]_{\cong_{s_n}}) = [f_\mathbf{A}(a_1 \ldots a_n)]_{\cong_s}$.

We omit making sorts and indices from signatures explicit when they can easily be understood from the context. We also omit making an explicit distinction between signatures and algebras. Moreover, making an abuse of notation, we indicate a Σ-algebras by its signature Σ. By this, we mean a Σ-algebra which has no other function than those named in Σ. We conclude this section by recalling some basics definitions of Boolean algebras.

Definition A.6 A Boolean algebra is an algebra $\mathbf{A} = \langle A, +, *, -, 0, 1 \rangle$ where: (a) $A = |\mathbf{A}|$ is a non-empty set of elements; and (b) $+, * : A^2 \to A$ are commutative and associative; $- : A \to A$ is idempotent; and $0, 1 : A$, called top and bottom, are neutral elements for $+$ and $*$, resp., further satisfying for all $a \in A$, $a + -a = 1$ and $a * -a = 0$.

Definition A.7 Every Boolean algebra $\mathbf{A}$ is equipped with a partial order defined as $x \preceq_\mathbf{A} y$ iff $x = x * y$. An ideal is a non-empty subset $I \subseteq |\mathbf{A}|$ s.t.: (a) for all $x, y \in I$, there is $z \in I$ s.t. $z \preceq_\mathbf{A} x * y$; and (b) for all $x \in I$ and $a \in |\mathbf{A}|$, if $a \preceq_\mathbf{A} x$, then, $a \in I$. An ideal I is proper if $I \neq |\mathbf{A}|$; otherwise it is trivial. An ideal I is maximal if there is no other ideal J s.t. $I \subset J$. The smallest ideal containing an element $a \in |\mathbf{A}|$, called a principal ideal, is the set $\downarrow a = \{ x \mid x \preceq_\mathbf{A} a \}$. The dual notion of an ideal is called a filter and is obtained by reversing $\preceq_\mathbf{A}$ and exchanging $*$ with $+$.

Two other notions that are important in our constructions are: freely generated and finitely generated algebras.

Definition A.8 Let $\mathbf{A} = \langle A, +, *, -, 0, 1 \rangle$ be a Boolean algebra; a subset $E \subseteq A$ is called a set of generators for $\mathbf{A}$ iff the following facts hold: (a) the intersection of all subalgebras of $\mathbf{A}$ including E is a subalgebra; (b) that intersection is the smallest subalgebra of $\mathbf{A}$ including E. Such algebra is called the generated algebra. It is called finitely generated, if the set of generators E is finite.

Definition A.9 A set E of generators of a Boolean algebra $\mathbf{B}$ is called free if every mapping from E to an arbitrary Boolean algebra $\mathbf{A}$ can be uniquely extended to an homomorphism $h : \mathbf{B} \to \mathbf{A}$. An algebra is called freely generated (or free) if it has a free set of generators.

References

[1] Alchourrón, C. E., *Logic of norms and logic of normative propositions*, Logique et Analyse **12** (1969).
[2] Alchourrón, C. E. and E. Bulygin, "Normative Systems," Springer-Verlag, 1971.
[3] Anglberger, A. J. J., *Dynamic deontic logic and its paradoxes*, Studia Logica **89** (2008).
[4] Åqvist, L., *Deontic logic*, in: D. M. Gabbay and F. Guenthner, editors, Handbook of Philosophical Logic: Volume 8, Springer Netherlands, Dordrecht, 2002 pp. 147–264.
[5] Becker, O., "Untersuchungen Über den Modalkalkül," A. Hain, 1952.
[6] Blackburn, P., M. de Rijke and Y. Venema, "Modal Logic," Cambridge University Press, 2001.
[7] Blackburn, P., J. van Benthem and F. Wolter, editors, "Handbook of Modal Logic," Elsevier, 2007.
[8] Broersen, J., "Modal Action Logic for Reasoning about Reactive Systems," Ph.D. thesis, Vrije Universiteit (2003).
[9] Castro, P. F. and T. S. E. Maibaum, *Deontic action logic, atomic boolean algebras and fault-tolerance.*, Journal of Applied Logic **7** (2009).
[10] Givant, S. and P. Halmos, "Introduction to Boolean Algebras," Undergraduate Texts in Mathematics, Springer, 2009.
[11] Halmos, P. and S. Givant, "Logic as Algebra," The Dociani Mathematical Expositions **21**, The Mathematical Association of America, 1998.
[12] jules Meyer, J., *A different approach to deontic logic: Deontic logic viewed as variant of dynamic logic*, Notre Dame Journal of Formal Logic **29** (1988).
[13] Kalinowski, J., *Theorie des propositions normatives*, Studia Logica **1** (1953), pp. 147–182.

[14] Khosla, S. and T. Maibaum, *The prescription and description of state based systems*, in: B. Banieqbal, H. Barringer and A. Pnueli, editors, *Proceedings of Temporal Logic in Specification*, LNCS **398** (1987), pp. 243–294.
[15] Kozen, D., *On kleene algebras and closed semirings*, in: *Mathematical Foundations of Computer Science 1990, MFCS'90, Banská Bystrica, Czechoslovakia, August 27-31, 1990, Proceedings*, 1990, pp. 26–47.
[16] Kozen, D., *A completeness theorem for kleene algebras and the algebra of regular events*, in: *Logic in Computer Science, 1991. LICS '91., Proceedings of Sixth Annual IEEE Symposium on*, 1991.
[17] Maddux, R., "Relation Algebras," Elsevier, 2006.
[18] Mendelson, E., Textbooks in Mathematics, CRC Press, 2015, 6 edition.
[19] Meyer, J.-J. C., F. P. M. Dignum and R. J. Wieringa, *The paradoxes of deontic logic revisited: a computer science perspective*, Technical report, University of Utrecht (1994).
[20] Pratt, V., *Dynamic algebras: Examples, constructions, applications*, Studia Logica **50** (1991), pp. 571–605.
[21] Prisacariu, C. and G. Schneider, *A dynamic deontic logic for complex contracts*, J. Log. Algebr. Program. **8** (2012).
[22] Sannella, D. and A. Tarlecki, "Foundations of Algebraic Specification and Formal Software Development," Monographs in Theoretical Computer Science. An EATCS Series, Springer, 2012.
[23] Segerberg, K., *A deontic logic of action*, Studia Logica **41** (1982), pp. 269–282.
[24] Stone, M. H., *The theory of representation for boolean algebras*, Transactions of the American Mathematical Society **40** (1936), pp. 37–111.
[25] Trypuz, R. and P. Kulicki, *Towards metalogical systematisation of deontic action logics based on boolean algebra*, in: *Deontic Logic in Computer Science, 10th International Conference, DEON 2010, Fiesole, Italy, July 7-9, 2010. Proceedings*, 2010, pp. 132–147.
[26] Trypuz, R. and P. Kulicki, *On deontic action logics based on boolean algebra*, Journal of Logic and Computation **25** (2015), pp. 1241–1260.
[27] Venema, Y., *Algebras and general frames*, [6] pp. 263–333.
[28] von Wright, G., *Deontic logic*, Mind **60** (1951).
[29] von Wright, G., *Deontic logic: A personal view*, Ratio Juris **12** (1999), pp. 26–38.

Sequent Rules for Reasoning and Conflict Resolution in Conditional Norms

Agata Ciabattoni   Björn Lellmann

TU Wien, Vienna, Austria
{*agata, lellmann*}@*logic.at*

Abstract

We introduce a sequent based method for reasoning with deontic assumptions using specificity and superiority for conflict resolution. Starting from a base logic, we apply strengthening of the antecedent to the assumptions wherever possible unless this would yield an inconsistency. The method applies to logics with an arbitrary finite number of dyadic deontic operators of type MP or MD (in the sense of Chellas) with inclusions among the operators. We illustrate the method using various examples. An implementation is also available.

1 Introduction

Legal, ethical, religious and behavioral norms often have a conditional form. A common way for formalising such conditional norms is via dyadic deontic operators, historically introduced to represent Contrary-To-Duty (CTD) obligations, i.e. obligations which are applicable only if another norm is violated. Although the dyadic representation can solve notorious CTD paradoxes, it also introduces new difficulties; in particular, how to reason on the conditions (i.e., the second argument of dyadic operators) without reintroducing possible deontic conflicts. Roughly speaking, a deontic conflict occurs when two or more obligations/prohibitions cannot be mutually realized.

Various general conflict resolution principles are considered in the literature. Here we focus on two major ones, widely used in law and AI: specificity and superiority. The former, known in law as *lex specialis derogat legi generali*, states that specific obligations/prohibitions override more general ones, while the latter refers to prioritized obligations/prohibitions coming from normative authorities of different strength (*lex superior*) or, e.g., in a different chronological order (*lex posterior*).

In this article we extend the most basic dyadic deontic logics with a general and purely syntactic mechanism for reasoning on the conditions of deontic

---

[*] This work was partly supported by WWTF Project MA16-28 and by BRISE-Vienna (UIA04-081), a European Union Urban Innovative Actions project.

assumptions, resolving conflicts using specificity on conditions and superiority between assumptions. The mechanism generalizes and extends to superiority the calculus introduced in [8] for a particular logic (see also Sec. 6.1).

Our starting point are logics based on finite combinations of operators $\heartsuit$, which are dyadic versions of non-normal (upwards or downwards) monotone modal logic M, extended with the (dyadic versions of) the axioms P $\neg(\heartsuit(\bot/B) \wedge \heartsuit(\top/B))$ or D $\neg(\heartsuit(A/B) \wedge \heartsuit(\neg A/B))$. For an upwards monotone, i.e., *obligation type*, operator this yields, e.g., the dyadic version of *minimal deontic logic* MP from [6]. Although well behaved, these logics are not useful for reasoning on the conditions of deontic formulae. E.g, for a downwards monotone, i.e., *prohibition type*, operator $\mathcal{F}$ we can derive $\mathcal{F}(\mathsf{park}/\top) \to \mathcal{F}(\mathsf{park} \wedge \mathsf{ride}/\top)$, but not $\mathcal{F}(\mathsf{park}/\top) \to \mathcal{F}(\mathsf{park}/\neg\mathsf{permit})$. The naive solution of adding unrestricted strengthening of the antecedent, i.e., an unrestricted downwards monotonicity rule for the second argument, quickly leads to conflicting norms, and in presence of axiom D to a contradiction. To avoid this, we consider sequent rules incorporating a limited form of strengthening of the antecedent / downwards monotonicity "up to conflicting assumptions". Starting from prima-facie deontic assumptions and propositional background facts, our sequent rules intuitively permit to derive every formula resulting from strengthening the antecedent, unless this would lead to an inconsistency over the base logic. Deontic conflicts are resolved using specificity and superiority. The resulting system satisfies the *disjunctive response* of [10], see Ex. 4.3, and can be used to model permissions as exceptions as well as some forms of CTD reasoning, see Ex. 4.2, Sect. 6.2 and Rem. 6.3.

As in sequent calculi for non-monotonic logics [3,25], our rules use statements expressing that certain sequents are not derivable. In contrast with other calculi for non-monotonicity in normative reasoning like [13,29], our calculi enjoy cut-elimination, which yields decidability and complexity results. A further corollary is that we can define the set of consequences of deontic assumptions iteratively, thus avoiding fixed-point constructions like those in [17].

The generality of our system is demonstrated with case studies including the logic simulating the reasoning of the Mīmāṃsā school from [7,8], a modelling of permissions as exceptions, and the operators of sanction and violation.

The system is implemented in the Prolog system deonticProver2.0 (http://subsell.logic.at/bprover/deonticProver/version2.0/). For any finite set of dyadic operators of type M, MP, or MD, with (possible) inclusions, the system constructs sequent rules to deal with specificity and superiority, and uses them to answer the question: Given an input of deontic assumptions and background facts, which conditional norms are in force, i.e., which formulae are derivable? In addition to a web interface for the prover, the website contains a number of examples and illustrates the behaviour of the system with respect to some standard deontic puzzles and paradoxa.

2 Restricted strengthening of the antecedent

Before delving into the technicalities, we briefly illustrate the intuitions behind our approach. As mentioned above, given deontic assumptions such as

(i) You ought not to eat with your fingers,

(ii) You ought to put your napkin on your lap,

(iii) If you are served asparagus, you ought to eat it with your fingers,

from the *asparagus example* (e.g. [31,16]) we would like to be able to apply strengthening of the antecedent to (ii) to derive "If you are served asparagus, you ought to put your napkin on your lap". However, as is well-known, adopting an unrestricted form of strengthening of the antecedent would also yield "If you are served asparagus, you ought not to eat with your fingers". Together with (iii) this yields a pair of conflicting obligations, and hence an inconsistency in any logic satisfying the D-axiom for obligations.

Our proposal for dealing with this situation is based on two main aspects: First, it is parametric in the *base logic*, and second it follows what could be called a *generous approach* towards applying strengthening of the antecedent. The latter means that given a set of deontic assumptions we apply strengthening of the antecedent whenever this is possible without resulting in inconsistencies over the base logic. In particular, this aims at keeping in force as many prima-facie norms as possible. Conflicts between norms are resolved following the *specificity principle*, i.e., assuming that conditional norms with more specific conditions like (iii) above overrule those with more general conditions like (i), and an (optional) *superiority relation* on the deontic assumptions. Note that inconsistencies are always evaluated with respect to the base logic. Hence for logics containing no principles ruling out conflicting or impossible norms there are no conflicts to avoid, and we obtain unrestricted strengthening of the antecedent/downwards monotonicity in the second argument.

In the asparagus example above, given a base logic ruling out conflicting obligations, we thus should derive "If you are served asparagus, you ought to put your napkin on your lap" as well as, e.g., "If you are served asparagus at your grandparents', you ought to eat it with your fingers": For the former, there are no assumptions which could yield a conflict; for the latter, the assumption (i) could be used to derive a conflicting obligation, but this assumption is overruled by the more specific assumption (iii).

The situation becomes more interesting if we consider the following additional deontic assumption: [1]

(iv) If you are at your grandparents', you ought not to eat with your fingers.

Now neither of the two assumptions (iii) and (iv) is more specific than the other. Hence, in order to keep the derived obligations consistent over the base logic we cannot derive the obligation "If you are served asparagus at your

---

[1] We are grateful to the anonymous reviewer for bringing this and the following examples to our attention.

grandparents', you ought to eat it with your fingers" anymore, because then by symmetry we should also be able to derive the conflicting obligation "If you are served asparagus at your grandparents', you ought not to eat it with your fingers" from assumption (iv). Note that this shows the difference to credulous approaches, where both the above statements would be derivable. The situation changes again, however, if we add the permission (see Sect. 6.2):

(v) If you are served asparagus at your grandparents', you may eat it with your fingers.

Assuming that the base logic contains the principle that there are no conflicting pairs of obligations and permissions, this assumption would prevent the derivation of the obligation "If you are served asparagus at your grandparents', you ought not to eat it with your fingers" from assumption (iv), since it is more specific. But then we cannot derive any obligation which would conflict with "If you are served asparagus at your grandparents', you ought to eat it with your fingers". Hence, following the generous approach towards applying strengthening of the antecedent, we don't have any reason to refrain from deriving this obligation from assumption (iii). While it has been argued, e.g., in [29] that it might be undesired if more specific permissions reinstate less specific obligations, this is in line with the idea of preventing the derivation of only those obligations which would result in inconsistencies over the base logic.

The generous approach of deriving every obligation which would not result in an inconsistency over the base logic further motivates the idea that the notion of a conflicting assumption is evaluated with respect to the obligation we want to derive, and not the assumption we want to derive it from. As an example, consider the additional assumption:

(vi) You ought not to eat with your fingers and not to pick your nose.

While the obligation "If you are served asparagus, you ought not to eat with your fingers and not to pick your nose" is in conflict with the more specific assumption (iii) and hence should not be derivable, the obligation "If you are served asparagus, you ought not to pick your nose" is not. Thus, we don't refrain from deriving the latter, even though the content of the assumption we derived it from is inconsistent with the content of the more specific applicable assumption (iii).

This focus on what we want to derive instead of the assumptions we derive it from has the additional benefit that we do not need to worry about chains of more and more specific assumptions, each in conflict with the previous one: Given that the set of deontic assumptions is finite, such a chain will contain a most specific applicable assumption. To see whether we should refrain from deriving an obligation which would follow from one of the more general ones, we thus only need to check the most specific assumption which is in conflict with what we want to derive. If this one is overruled by an even more specific assumption, then we can use the latter to derive the obligation in question; otherwise we refrain from doing so.

We would like to stress again that the approach is parametric in the base logic. Hence the resulting systems inherit some of the limitations imposed by the latter, both in terms of what is removed as inconsistent and of what can be derived from the assumptions. In this work we consider only relatively weak base logics. In particular, they neither permit to aggregate obligations, nor rule out conflicts between more than two obligations, where each pair of these is nonconflicting. We are, however, confident that the general method can be extended to stronger base logics as well (see Sec. 7). Note also that since we only aim to consistently close a set of conditional deontic assumptions under strengthening of the antecedent with respect to a base logic, the derivable formulae are still conditional statements, and hence we do not incorporate factual detachment principles.

3 The base system

Formally, the basic logical systems we consider are *propositional deontic logics*. Our logics extend the language of classical propositional logic consisting of *variables* $(p, q, \ldots)$, *falsum* ($\bot$) and *implication* ($\rightarrow$), with *dyadic deontic operators* $\heartsuit(./.)$ where the first argument represents the *content* of a conditional norm, while the second argument represents its *condition*. We distinguish two kinds of operators, depending on what it takes to comply with the norm:

- An operator $\heartsuit$ is of *obligation-type* if the norm $\heartsuit(A/B)$ is complied with whenever A is true;
- An operator $\heartsuit$ is of *prohibition-type* if the norm $\heartsuit(A/B)$ is complied with whenever A is false.

Note that this makes our operators upwards monotone in the first argument for obligation type operators, and downwards monotone for prohibition type ones. To capture relations between operators and their properties, given a set Op of deontic operators with associated types, we assume a reflexive and transitive *inclusion relation* $\rightarrow$, a symmetric *conflict relation* $\sharp$, and a unary *nontriviality predicate* nt with the following intended meaning:

- If $\heartsuit \rightarrow \spadesuit$ for two operators $\heartsuit, \spadesuit \in$ Op of the same type, then complying with $\heartsuit(A/B)$ implies complying with $\spadesuit(A/B)$.
- If $\heartsuit \sharp \spadesuit$ for two operators of the same type $\heartsuit, \spadesuit \in$ Op, then complying with one of $\heartsuit(A/B), \spadesuit(\neg A/B)$ entails violating the other.
- If nt($\heartsuit$) for an operator $\heartsuit \in$ Op, then $\heartsuit$ is *non-trivial*, in that it is logically possible to comply with it.

For operators $\heartsuit, \spadesuit$ of different type we flip the polarity of A in one of the assumptions, i.e., we replace $\spadesuit(A/B)$ with $\spadesuit(\neg A/B)$ and vice versa. We assume that the relations $\sharp$ and nt are closed under preimages of the implication relation, i.e., if $\heartsuit \sharp \spadesuit$ and $\Diamond \rightarrow \heartsuit$, then also $\Diamond \sharp \spadesuit$. In the following, an *operator characterisation* is a tuple $\mathfrak{D} = (\mathsf{Op}, \rightarrow, \sharp, \mathsf{nt})$ consisting of a set Op of operators with types together with inclusion, conflict, and non-triviality relations.

The base logic we will consider then contains the Hilbert-style rules and axioms of propositional classical logic together with the rules and axioms in

$$\left\{\frac{A \to C}{\heartsuit(A/B) \to \heartsuit(C/B)} : \heartsuit \text{ obl type}\right\} \quad \left\{\frac{C \to A}{\heartsuit(A/B) \to \heartsuit(C/B)} : \heartsuit \text{ proh type}\right\}$$

$$\{\heartsuit(A/B) \to \spadesuit(A/B) \;:\; \heartsuit \to \spadesuit, \text{ same type }\}$$
$$\cup \;\; \{\heartsuit(A/B) \to \spadesuit(\neg A/B) \;:\; \heartsuit \to \spadesuit, \text{ different type }\}$$
$$\cup \;\; \{\neg(\heartsuit(A/B) \wedge \spadesuit(\neg A/B)) \;:\; \heartsuit \mathbin{\not\hspace{-1pt}\mathrel{}} \spadesuit, \text{ same type }\}$$
$$\cup \;\; \{\neg(\heartsuit(A/B) \wedge \spadesuit(A/B)) \;:\; \heartsuit \mathbin{\not\hspace{-1pt}\mathrel{}} \spadesuit, \text{ different type }\}$$
$$\cup \;\; \{\neg(\heartsuit(\bot/B) \wedge \heartsuit(\top/B)) \;:\; \mathsf{nt}(\heartsuit)\}$$

Fig. 1. The deontic axioms and rules for $\mathfrak{D} = (\mathsf{Op}, \to, \mathbin{\not\hspace{-1pt}\mathrel{}}, \mathsf{nt})$.

Fig. 1. Note that due to upwards and downwards monotonicity respectively, for operators $\heartsuit$ with $\mathsf{nt}(\heartsuit)$ the axiom $\neg(\heartsuit(\bot/B) \wedge \heartsuit(\top/B))$ is equivalent to $\neg\heartsuit(\bot/B)$ for $\heartsuit$ of obligation type and to $\neg\heartsuit(\top/B)$ for $\heartsuit$ of prohibition type.

Example 3.1 (i) Setting $\mathsf{Op} = \{\mathcal{O}\}$ with $\mathcal{O}$ of obligation type and $\mathsf{nt}(\mathcal{O})$ yields the dyadic version of *minimal deontic logic* MP from [6].

(ii) Replacing $\mathsf{nt}(\mathcal{O})$ with $\mathcal{O} \mathbin{\not\hspace{-1pt}\mathrel{}} \mathcal{O}$ in (i) yields the dyadic version of monotone modal logic M extended with the D axiom $\neg(\mathcal{O}(A/B) \wedge \mathcal{O}(\neg A/B))$.

(iii) Setting $\mathsf{Op} = \{\mathcal{O}, \mathcal{F}\}$ with $\mathcal{O}$ of obligation type, $\mathcal{F}$ of prohibition type, and $\mathcal{O} \mathbin{\not\hspace{-1pt}\mathrel{}} \mathcal{F}$ yields a logic with upwards monotone obligations $\mathcal{O}$, downwards monotone prohibitions $\mathcal{F}$, and no conflicts between obligations and prohibitions, i.e., the axiom $\neg(\mathcal{O}(A/B) \wedge \mathcal{F}(A/B))$. Note that this does not rule out conflicts between obligations or between prohibitions. This could be added by stipulating $\mathcal{O} \mathbin{\not\hspace{-1pt}\mathrel{}} \mathcal{O}$ and $\mathcal{F} \mathbin{\not\hspace{-1pt}\mathrel{}} \mathcal{F}$, respectively.

(iv) Let $\mathsf{Op} = \{\mathsf{must}, \mathsf{ought}, \mathsf{should}\}$ with all operators of obligation type. Setting $\mathsf{must} \to \mathsf{ought}, \mathsf{must} \to \mathsf{should}, \mathsf{ought} \to \mathsf{should}$ with $\mathsf{must} \mathbin{\not\hspace{-1pt}\mathrel{}} \mathsf{must}$ and $\mathsf{nt}(\mathsf{ought})$ illustrates the possibility of using different operators for analysing different strengths of obligations. The intuition is that must behaves like an obligation, while the weaker ought behaves more like a recommendation, hence satisfies only the P axiom instead of D. See, e.g., [1].

To facilitate automated reasoning and prove useful meta-logical properties, we switch from Hilbert-style calculi to *sequent calculi*. As usual, a *sequent* is a tuple $\Gamma \Rightarrow \Delta$ of multisets of formulae, with formula interpretation $\bigwedge \Gamma \to \bigvee \Delta$, see, e.g., [30]. To write the rules with a coincise notation we introduce the following two abbreviations:

$$\mathsf{Impl}_{\heartsuit, \spadesuit}(A, B) := \begin{cases} A \Rightarrow B & \heartsuit, \spadesuit \text{ obligation type} \\ A, B \Rightarrow & \heartsuit \text{ obligation type}, \spadesuit \text{ prohibition type} \\ B \Rightarrow A & \heartsuit, \spadesuit \text{ prohibition type} \\ \Rightarrow A, B & \heartsuit \text{ prohibition type}, \spadesuit \text{ obligation type} \end{cases}$$

$$\left\{ \frac{\mathsf{Impl}_{\heartsuit,\spadesuit}(A,C) \quad B \Rightarrow D \quad D \Rightarrow B}{\Gamma, \heartsuit(A/B) \Rightarrow \spadesuit(C/D), \Delta} \; \mathsf{Mon}_{\heartsuit,\spadesuit} \;:\; \heartsuit, \spadesuit \in \mathsf{Op}, \; \heartsuit \to \spadesuit \right\}$$

$$\left\{ \frac{\mathsf{Confl}_{\heartsuit,\spadesuit}(A,C) \quad B \Rightarrow D \quad D \Rightarrow B}{\Gamma, \heartsuit(A/B), \spadesuit(C/D) \Rightarrow \Delta} \; \mathsf{D}_{\heartsuit,\spadesuit} \;:\; \heartsuit, \spadesuit \in \mathsf{Op}, \; \heartsuit \nleftrightarrow \spadesuit \right\}$$

$$\left\{ \frac{\mathsf{Confl}_{\heartsuit,\heartsuit}(A,A)}{\Gamma, \heartsuit(A/B) \Rightarrow \Delta} \; \mathsf{P}_\heartsuit \;:\; \heartsuit \in \mathsf{Op}, \; \heartsuit \nleftrightarrow \heartsuit \text{ or } \mathsf{nt}(\heartsuit) \right\}$$

$$\overline{\Gamma, p \Rightarrow p, \Delta} \; \mathsf{init} \qquad \overline{\Gamma, \bot \Rightarrow \Delta} \; \bot_L \qquad \frac{\Gamma, B \Rightarrow \Delta \quad \Gamma \Rightarrow A, \Delta}{\Gamma, A \to B \Rightarrow \Delta} \to_L \qquad \frac{\Gamma, A \Rightarrow B, \Delta}{\Gamma \Rightarrow A \to B, \Delta} \to_R$$

Fig. 2. The base calculus for a given operator characterisation $\mathfrak{D} = (\mathsf{Op}, \to, \nleftrightarrow, \mathsf{nt})$

$$\mathsf{Confl}_{\heartsuit,\spadesuit}(A,B) := \begin{cases} A, B \Rightarrow & \heartsuit, \spadesuit \text{ obligation type} \\ A \Rightarrow B & \heartsuit \text{ obligation type}, \spadesuit \text{ prohibition type} \\ \Rightarrow A, B & \heartsuit, \spadesuit \text{ prohibition type} \\ B \Rightarrow A & \heartsuit \text{ prohibition type}, \spadesuit \text{ obligation type} \end{cases}$$

The intuition is that, e.g., for two operators $\heartsuit, \spadesuit$ of obligation type, complying with $\heartsuit(A/C)$ means that A is true, whereas violating $\spadesuit(B/C)$ means that B is false. Hence complying with $\heartsuit(A/C)$ implies violating $\spadesuit(B/C)$ if A implies $\neg B$. This is captured in $\mathsf{Confl}_{\heartsuit,\spadesuit}(A, B)$, i.e., the sequent $A, B \Rightarrow$. Using these abbreviations, converting the Hilbert-style axioms into sequent rules using the general method from [19] then gives the deontic rules $\mathsf{Mon}_{\heartsuit,\spadesuit}, \mathsf{D}_{\heartsuit,\spadesuit}, \mathsf{P}_\heartsuit$ of the base calculus in Fig. 2. Note that since the relation $\to$ is reflexive, we have for every operator $\heartsuit$ either the upwards or downwards monotonicity rule:

$$\frac{A \Rightarrow C \quad B \Rightarrow D \quad D \Rightarrow B}{\Gamma, \heartsuit(A/B) \Rightarrow \heartsuit(C/D), \Delta} \; \mathsf{Mon} \uparrow \qquad \frac{C \Rightarrow A \quad B \Rightarrow D \quad D \Rightarrow B}{\Gamma, \heartsuit(A/B) \Rightarrow \heartsuit(C/D), \Delta} \; \mathsf{Mon} \downarrow$$

The resulting calculi are equivalent and admit cut-elimination, see [19].

4 Reasoning from assumptions

To reason on the conditions of norms in the above systems we will introduce special sequent rules. These allow us to reason from *deontic assumptions*, i.e., a finite set $\mathfrak{L}$ of deontic formulae $\heartsuit(A/B)$ with $\heartsuit \in \mathsf{Op}$. To ensure well-definedness and termination of proof search we require that these assumptions are *non-nested*, i.e., that A, B are purely propositional. The main idea is to make the second argument downwards monotone "up to conflicting assumptions". Since a formula A with $A \to B$ can be seen as "more specific" than B, this captures the *specificity principle*, that more specific conflicting deontic assumptions overrule less specific ones. Before considering the rules in detail we mention two more features of the system.

To be able to reason with non-deontic propositions as well, we also consider *propositional facts* as assumptions. W.l.o.g. we assume that these are given in the form of a finite set $\mathfrak{F}$ of atomic sequents, i.e., sequents of the form

$p_1, \ldots, p_n \Rightarrow q_1, \ldots, q_m$ with the p_i, q_j propositional variables. Since every purely propositional formula is equivalent to a formula in conjunctive normal form, this is equivalent to permitting arbitrary purely propositional formulae as assumptions. The sequent rules for these assumptions then are given by

$$\left\{ \overline{\Sigma, \Gamma \Rightarrow \Delta, \Pi} \ \mathfrak{F} : \Gamma \Rightarrow \Delta \in \mathfrak{F} \right\}.$$

Often obligations and prohibitions further come with a priority order. To capture this, we follow the standard approach and say that a *superiority relation* is a binary relation $\succ$ on the set of deontic assumptions. The intuition is that for two deontic assumptions A, B with $A \succ B$, the former is superior, or has higher authority, than the latter, and hence A cannot be overruled by B, even if the latter is more specific. For technical reasons we impose that for every two assumptions A, B we have $A \not\succ B$ or $B \not\succ A$. Note that this rules out cycles of length one or two, but due to lack of transitivity not those of greater length.

Sequent calculus rules: We then extend the base calculus with sequent rules for capturing the specificity principle in presence of prioritized deontic assumptions. The idea is that we can use downwards monotonicity in the second argument to derive, e.g., $\heartsuit(A/B)$ from a deontic assumption $\heartsuit(A/B \vee C)$ unless the latter is overruled by a more specific conflicting deontic assumption or in conflict with the P axiom. In addition, we also need to rule out that there is another conflicting assumption which is not overruled by a more specific one. The crucial feature needed for this is the addition of *underivability statements* in the premises of the rules. These are used for stating, e.g., that we cannot derive a conflict between two formulae. The general conditions for deriving an obligation or a prohibition from a deontic assumption then are as follows:

Given a list $\mathfrak{L}$ of deontic assumptions, we can derive $\heartsuit(A/B)$ from the assumption $\spadesuit(C/D) \in \mathfrak{L}$ with $\spadesuit \to \heartsuit$ if:

- the assumption $\spadesuit(C/D)$ is *applicable*, i.e., if we can derive that the condition B implies the condition D; AND
- complying with the assumption $\spadesuit(C/D)$ implies complying with $\heartsuit(A/B)$, i.e., if we can derive $\mathsf{Impl}_{\spadesuit, \heartsuit}(C, A)$; AND
- there is no conflict with the non-triviality axiom P for $\heartsuit$, i.e., we cannot derive $\mathsf{Confl}_{\heartsuit, \heartsuit}(A, A)$ provided that $\mathsf{nt}(\heartsuit)$; AND
- the assumption $\spadesuit(C/D)$ is neither overruled by a more specific one, nor in conflict with another assumption which is not overruled. I.e., for every assumption $\clubsuit(E/F) \in \mathfrak{L}$ with $\clubsuit \not\to \spadesuit$ and $\spadesuit(C/D) \not\succ \clubsuit(E/F)$ we have:
  · the assumption $\clubsuit(E/F)$ is not applicable, i.e., we cannot derive that the condition B implies F; OR
  · the assumption $\clubsuit(E/F)$ is not in conflict with what we want to derive, i.e., we cannot derive $\mathsf{Confl}_{\clubsuit, \heartsuit}(E, A)$; OR
  · the assumption $\clubsuit(E/F)$ is not more specific than $\spadesuit(C/D)$ and it is not overruled by another more specific one, i.e.:
    * the assumption $\clubsuit(E/F)$ is not more specific than $\spadesuit(C/D)$, i.e., we

$$\cfrac{\begin{array}{l}B \Rightarrow D\\ \mathsf{Impl}_{\spadesuit,\heartsuit}(C, A)\\ \{\nvdash \mathsf{Confl}_{\clubsuit,\clubsuit}(A, A) : \mathsf{nt}(\clubsuit), \clubsuit = \heartsuit\}\\ \left\{\left[\begin{array}{l}\nvdash B \Rightarrow F\\ \nvdash \mathsf{Confl}_{\clubsuit,\heartsuit}(E, A)\\ \left[\left[\begin{array}{l}\nvdash F \Rightarrow D,\\ B \Rightarrow Y,\\ Y \Rightarrow F,\\ \mathsf{Impl}_{\diamond,\heartsuit}(X, A)\end{array}\right] : \begin{array}{l}\diamond(X/Y) \in \mathfrak{L},\\ \diamond \nmid \clubsuit\\ \clubsuit(E/F) \not\succ \diamond(X/Y)\end{array}\right]\end{array}\right] : \begin{array}{l}\clubsuit(E/F) \in \mathfrak{L},\\ \clubsuit \nmid \spadesuit\\ \spadesuit(C/D) \not\succ \clubsuit(E/F)\end{array}\right\}\\ \Gamma \Rightarrow \heartsuit(A/B), \Delta\end{array}}{\Gamma \Rightarrow \heartsuit(A/B), \Delta} \heartsuit_R^{\spadesuit(C/D)}$$

$$\cfrac{\begin{array}{l}B \Rightarrow D\\ \mathsf{Confl}_{\spadesuit,\heartsuit}(C, A)\\ \left\{\left[\begin{array}{l}\nvdash B \Rightarrow F\\ \nvdash \mathsf{Impl}_{\clubsuit,\heartsuit}(E, A)\\ \left[\left[\begin{array}{l}\nvdash F \Rightarrow D,\\ B \Rightarrow Y,\\ Y \Rightarrow F,\\ \mathsf{Confl}_{\diamond,\heartsuit}(X, A)\end{array}\right] : \begin{array}{l}\diamond(X/Y) \in \mathfrak{L},\\ \diamond \nmid \clubsuit\\ \clubsuit(E/F) \not\succ \diamond(X/Y)\end{array}\right]\end{array}\right] : \begin{array}{l}\clubsuit(E/F) \in \mathfrak{L},\\ \clubsuit \nmid \spadesuit\\ \spadesuit(C/D) \not\succ \clubsuit(E/F)\end{array}\right\}\\ \Gamma, \heartsuit(A/B) \Rightarrow \Delta\end{array}}{\Gamma, \heartsuit(A/B) \Rightarrow \Delta} \heartsuit_L^{\spadesuit(C/D)}$$

Fig. 3. The deontic assumption rules.

cannot derive that the condition F implies D; AND
* there is another more specific applicable assumption $\diamond(X/Y)$, complying with which implies complying with $\heartsuit(A/B)$, i.e., for one of $\diamond(X/Y) \in \mathfrak{L}$ with $\diamond \nmid \clubsuit$ and $\clubsuit(E/F) \not\succ \diamond(X/Y)$ we have:
  - the assumption $\diamond(X/Y)$ applies, i.e., we can derive that the condition B implies Y; AND
  - the condition Y is more specific than the condition of $\clubsuit(E/F)$, i.e., we can derive that the condition Y implies F; AND
  - complying with the assumption $\diamond(X/Y)$ implies complying with $\heartsuit(A/B)$, i.e., we can derive $\mathsf{Impl}_{\diamond,\heartsuit}(X, A)$.

In order to formalise this as sequent rules we use the following abbreviation. Let $\mathfrak{S} = \{\mathcal{S}_1, \ldots, \mathcal{S}_n\}$ be a finite set of sets of premises. Then we write

$$\frac{\mathcal{P} \cup [\mathfrak{S}]}{\mathcal{C}} \quad \text{for the set of rules} \quad \left\{\frac{\mathcal{P} \cup \mathcal{S}_1}{\mathcal{C}}, \ldots, \frac{\mathcal{P} \cup \mathcal{S}_n}{\mathcal{C}}\right\}.$$

The general assumption right rules $\heartsuit_R^{\spadesuit(C/D)}$ are given in Fig. 3, where we write $\nvdash \Gamma \Rightarrow \Delta$ for an underivability statement. Note that in this notation sets essentially correspond to conjunctive conditions on the premises and capture the "AND" and "for all" above, while the choice notation [.] essentially corresponds to disjunctive conditions and captures the "OR" and "there is". In particular, the notation $[\mathcal{S}_{\diamond(X/Y)} : \diamond(X/Y) \in \mathfrak{L}]$ corresponds to the big disjunction over the $\diamond(X/Y) \in \mathfrak{L}$ of the $\mathcal{S}_{\diamond(X/Y)}$ and hence the existential quantification over the finite set $\mathfrak{L}$. To abbreviate the notation we equivalently incorporated the premiss $\nvdash F \Rightarrow D$ into the following choice block.

Remark 4.1 The D axiom is equivalent to $\heartsuit(A/B) \to \neg\heartsuit(\neg A/B)$, and hence from an assumption $\heartsuit(A/B)$ we should be able to derive $\neg\heartsuit(\neg A/B)$. The assumption right rules allow us to do that *only if* we use the cut rule, see Sec. 5. As the presence of this rule destroys useful properties of the calculus, we introduce in the system the corresponding left rules $\heartsuit_L^{\spadesuit(C/D)}$ in Fig. 3, obtained by absorbing cuts between the assumption right rules $\heartsuit_R^{\spadesuit(C/D)}$ and the D-rules $D_{\heartsuit,\clubsuit}$. As usual, introducing a formula $\heartsuit(A/B)$ on the left hand side of the sequent, amounts to deriving $\neg\heartsuit(A/B)$.

Note that the nonderivability premiss for removing conflicts with the P axiom is no longer present – if $\mathsf{nt}(\heartsuit)$ and $\mathsf{Confl}_{\heartsuit,\heartsuit}(A,A)$ is derivable, then we immediately obtain the conclusion using the rule $\mathsf{P}_\heartsuit$. The full calculus then contains the base rules of Fig. 2 together with the rules:

$$\left\{\heartsuit_R^{\spadesuit(C/D)} \; : \; \spadesuit \to \heartsuit, \spadesuit(C/D) \in \mathfrak{L}\right\} \cup \left\{\heartsuit_L^{\spadesuit(C/D)} \; : \; \spadesuit \not\downarrow \heartsuit, \spadesuit(C/D) \in \mathfrak{L}\right\} .$$

4.1 Examples

The examples below can be checked at http://subsell.logic.at/bprover/deonticProver/version2.0/, where also more examples are available.

Example 4.2 Continuing Ex. 3.1.(ii), consider $\mathcal{O}$ of obligation type with $\mathcal{O} \not\downarrow \mathcal{O}$ and the deontic assumptions corresponding to the *asparagus example* [31,16] (see also Sec. 2) given by $\mathfrak{L} = \{\mathcal{O}(\neg\mathsf{fingers}/\top), \mathcal{O}(\mathsf{fingers}/\mathsf{asparagus}), \mathcal{O}(\neg\mathsf{asparagus}/\top)\}$. Since $\mathsf{asparagus} \to \top$ and there is no conflicting assumption, we can derive $\mathcal{O}(\neg\mathsf{asparagus}/\mathsf{asparagus})$, hence the contrary-to-duty obligation $\mathcal{O}(\mathsf{fingers}/\mathsf{asparagus})$ does not override the primary obligation $\mathcal{O}(\neg\mathsf{asparagus}/\top)$. However, the more specific obligation (or exception) $\mathcal{O}(\mathsf{fingers}/\mathsf{asparagus})$ overrides $\mathcal{O}(\neg\mathsf{fingers}/\top)$. Moreover, exemplifying Rem. 4.1, since we can derive $\mathcal{O}(\mathsf{fingers}/\mathsf{asparagus})$, due to $\mathcal{O} \not\downarrow \mathcal{O}$ and the assumption left rule $\mathcal{O}_L^{\mathcal{O}(\mathsf{fingers}/\mathsf{asparagus})}$ we also derive $\neg\mathcal{O}(\neg\mathsf{fingers}/\mathsf{asparagus})$.

Example 4.3 Consider the classical *drowning twins example*, for the same operator $\mathcal{O}$ as in the previous example, deontic assumptions $\mathfrak{L} = \{\mathcal{O}(\mathsf{save\_twin\_1}/\top), \mathcal{O}(\mathsf{save\_twin\_2}/\top)\}$ and the propositional fact $\mathsf{save\_twin\_1}, \mathsf{save\_twin\_2} \Rightarrow \bot$ which stipulates that saving both twins is impossible. Neither of the two assumptions is derivable because it is in conflict with the other one. However, the formula $\mathsf{save\_twin\_1} \lor \mathsf{save\_twin\_2}$ is noncontradictory, hence we can derive $\mathcal{O}(\mathsf{save\_twin\_1} \lor \mathsf{save\_twin\_2}/\top)$. This shows that norms which are nonderivable can still serve to derive other norms, and in particular that our system satisfies the *disjunctive response* of [10] for two conflicting deontic assumptions. Adding superiority between the two assumptions, e.g., stipulating $\mathcal{O}(\mathsf{save\_twin\_1}/\top) \succ \mathcal{O}(\mathsf{save\_twin\_2}/\top)$, would break the tie and make the $\mathcal{O}(\mathsf{save\_twin\_1}/\top)$ derivable.

$$\dfrac{\Gamma, A, A \Rightarrow \Delta}{\Gamma, A \Rightarrow \Delta}\ \text{Con}_L \qquad \dfrac{\Gamma \Rightarrow A, A, \Delta}{\Gamma \Rightarrow A, \Delta}\ \text{Con}_R \qquad \dfrac{\Gamma \Rightarrow \Delta}{\Sigma, \Gamma \Rightarrow \Delta, \Pi}\ W$$

Fig. 4. The structural rules.

Example 4.4 Continuing Ex. 3.1.(iv), with the operator characterisation given there for the operators must, ought, should and the assumptions {must(¬murder/⊤), ought(help_friend/⊤)} as well as the unfortunate factual assumption help_friend ⇒ murder we can derive must(¬murder/⊤), ought(¬murder/⊤), should(¬murder/⊤). We also derive ought(murder/⊤) using ought$_R^{\text{ought(help\_friend}/\top)}$, but since ought behaves like a recommendation and hence doesn't satisfy the D axiom, these two are not in conflict.

Example 4.5 Consider the *order puzzle* from, e.g., [14], with the operator $\mathcal{O}$ as in Ex. 4.2 and deontic assumptions given by the ordered list $\mathcal{O}(\neg\text{open\_window}/\text{heating}) \succ \mathcal{O}(\text{open\_window}/\top) \succ \mathcal{O}(\text{heating}/\top)$. For the situation where the window is open and the heating is off we can derive $\mathcal{O}(\text{open\_window}/\text{open\_window} \wedge \neg\text{heating})$ as well as $\mathcal{O}(\text{heating}/\text{open\_window} \wedge \neg\text{heating})$, but not $\mathcal{O}(\neg\text{open\_window}/\text{open\_window} \wedge \neg\text{heating})$, since the assumption $\mathcal{O}(\neg\text{open\_window}/\text{heating})$ does not apply. This illustrates that deontic detachment/transitivity does not hold (since these principles are not present in the base logic). In particular, there also is no aggregation of priorities along chains of obligations which could make the assumption $\mathcal{O}(\text{open\_window}/\top)$ overrule the inferior $\mathcal{O}(\text{heating}/\top)$. A similar effect could be achieved, however, by adding the assumption $\mathcal{O}(\neg\text{heating}/\text{open\_window})$, since by specificity this would block the derivation of $\mathcal{O}(\text{heating}/\text{open\_window} \wedge \neg\text{heating})$.

5 Cut-elimination and Consequences

We now consider the formal details of the introduced calculi. Due to the underivability statements in the rules we proceed in two stages.

Definition 5.1 We call *deontic assumptions* a finite set $\mathfrak{L}$ of non-nested deontic formulae. We further call *propositional facts* a finite set $\mathfrak{F}$ of atomic sequents closed under applications of the *cut rule* below and the *contraction rules* Con$_L$, Con$_R$ of Fig. 4.

$$\dfrac{\Gamma \Rightarrow \Delta, A \quad A, \Sigma \Rightarrow \Pi}{\Gamma, \Sigma \Rightarrow \Delta, \Pi}\ \text{cut}$$

A *normative basis* is a triple $\mathfrak{N} = (\mathfrak{O}, \mathfrak{L}, \succ, \mathfrak{F})$ consisting of an operator characterisation $\mathfrak{O}$, deontic assumptions $\mathfrak{L}$ with a superiority relation $\succ$, and propositional facts $\mathfrak{F}$. Given a normative basis, the rules of the system G$_\mathfrak{N}$ are those of the base calculus for $\mathfrak{O}$ from Fig. 2, the factual assumption rules $\mathfrak{F}$, the deontic assumption rules of Fig. 3 and the structural rules of Fig. 4. The system G$_\mathfrak{N}$cut extends G$_\mathfrak{N}$ with the rule cut.

Definition 5.2 Given a normative basis $\mathfrak{N} = (\text{Op}, \mathfrak{L}, \succ, \mathfrak{F})$, a *proto-derivation* in G$_\mathfrak{N}$ (or G$_\mathfrak{N}$cut) is a finite labelled tree, with every internal node labelled with a sequent which is obtained from the labels of the node's children using a rule of $G_\mathfrak{N}$ (or $G_\mathfrak{N}$ plus cut, respectively), and every leaf node labelled with the conclusion of a zero-premiss rule in $G_\mathfrak{N}$ or an *underivability statement* $\nvdash \Gamma \Rightarrow \Delta$. The *conclusion* of a proto-derivation is the label of its root. A *proto-derivation of rank n* is a proto-derivation where the nesting depth of operators from Op in every formula occurring in it is at most n. A proto-derivation (of rank n) is a *derivation* (of rank n), if none of the underivability statements occurring in it have a derivation in $G_\mathfrak{N}$cut (of rank $n-1$). We write $\vdash_{G_\mathfrak{N}} \Gamma \Rightarrow \Delta$ if there is a derivation of $\Gamma \Rightarrow \Delta$ and $\vdash_{G_\mathfrak{N}}^n \Gamma \Rightarrow \Delta$ if there is a derivation of rank n.

Note that underivability statements always range over $G_\mathfrak{N}$cut, i.e., the system with the cut rule. Since the definition of a derivation refers to itself, we need to show that it is well-defined. This follows from the observation that the modal nesting depth of the underivability statements occurring in the premisses of the assumption rules is strictly smaller than that of the conclusion, together with the main result of this section, stating that cut is admissible.

Before proving this theorem (in its stronger version, namely that the cut rule is eliminable) we show some preliminary results:

Proposition 5.3 *The following rules are derivable in $G_\mathfrak{N}$cut:*

$$\frac{\mathsf{Impl}_{\heartsuit,\spadesuit}(A,B) \quad \mathsf{Impl}_{\spadesuit,\clubsuit}(B,C)}{\mathsf{Impl}_{\heartsuit,\clubsuit}(A,C)} \text{ cut} \qquad \frac{\mathsf{Impl}_{\heartsuit,\spadesuit}(A,B) \quad \mathsf{Confl}_{\spadesuit,\clubsuit}(B,C)}{\mathsf{Confl}_{\heartsuit,\clubsuit}(A,C)} \text{ cut}$$

Proof. By applying cut and spelling out the cases for Impl and Confl. □

Lemma 5.4 *The generalised initial sequents $\Gamma, A \Rightarrow A, \Delta$ are derivable.*

Proof. By induction on the depth of the derivation, using $\mathsf{Mon}_{\heartsuit,\heartsuit}$. □

The proof of the cut-elimination theorem generalizes the one in [8], which was tailored to the particular rules of the modalities for the dyadic version of the non-normal deontic logic MD [6] (see Section 6.1).

Theorem 5.5 (Cut elimination) *If $\vdash_{G_\mathfrak{N}\text{cut}} \Gamma \Rightarrow \Delta$, then $\vdash_{G_\mathfrak{N}} \Gamma \Rightarrow \Delta$.*

Proof. By eliminating topmost applications of *multicut*, i.e., the rule

$$\frac{\Gamma \Rightarrow \Delta, A^n \quad A^m, \Sigma \Rightarrow \Pi}{\Gamma, \Sigma \Rightarrow \Delta, \Pi} \text{ mcut}$$

using a double induction on the complexity of the *cut formula A* and the sum of the depths of the two premisses of the application of multicut. The interesting case is for A being a deontic formula, the propositional cases are standard.

The case of the last applied rules being modal is straightforward, e.g., for

$$\frac{\mathsf{Impl}_{\heartsuit,\spadesuit}(A,C) \quad B \Rightarrow D \quad D \Rightarrow B}{\Gamma, \heartsuit(A/B) \Rightarrow \spadesuit(C/D), \Delta} \mathsf{Mon}_{\heartsuit,\spadesuit} \qquad \frac{\mathsf{Confl}_{\spadesuit,\clubsuit}(C,E) \quad D \Rightarrow F \quad F \Rightarrow D}{\Sigma, \spadesuit(C/D), \clubsuit(E/F) \Rightarrow \Pi} \mathsf{D}_{\spadesuit,\clubsuit}$$

we replace the cut on $\spadesuit(C/D)$ by cuts on the premisses and an application of $\mathsf{D}_{\heartsuit,\clubsuit}$ using Prop.5.3 and the fact that since $\heartsuit \to \spadesuit$ and $\spadesuit \nleftrightarrow \clubsuit$, also $\heartsuit \nleftrightarrow \clubsuit$. The case of both rules being Mon is similar. A multicut between the conclusions of

$$\frac{\mathsf{Impl}_{\heartsuit,\spadesuit}(A,C) \quad B \Rightarrow D \quad D \Rightarrow B}{\Gamma, \heartsuit(A/B) \Rightarrow \spadesuit(C/D), \Delta} \mathsf{Mon}_{\heartsuit,\spadesuit} \qquad \frac{\mathsf{Confl}_{\spadesuit,\spadesuit}(C,C) \quad D \Rightarrow D \quad D \Rightarrow D}{\Sigma, \spadesuit(C/D), \spadesuit(C/D) \Rightarrow \Pi} \mathsf{D}_{\spadesuit,\spadesuit}$$

is replaced by two cuts of smaller complexity, obtaining first $\mathsf{Confl}_{\heartsuit,\spadesuit}(A,C)$ and then $\mathsf{Confl}_{\heartsuit,\heartsuit}(A,A)$ using Prop.5.3. Then we apply $\mathsf{D}_{\heartsuit,\heartsuit}$. The case of a cut between the conclusions of the rules $\mathsf{Mon}_{\heartsuit,\spadesuit}$ and $\mathsf{P}_{\spadesuit}$ is analogous.

The case involving the right assumption rule and the monotonicity rule is as follows (strictly speaking, the first denotes a set of rules). Suppose we have

$$\frac{\begin{array}{l} B \Rightarrow D \\ \mathsf{Impl}_{\spadesuit,\heartsuit}(C,A) \\ \{\nvdash \mathsf{Confl}_{\clubsuit,\clubsuit}(A,A) : \mathsf{nt}(\clubsuit), \clubsuit = \heartsuit\} \\ \left\{ \left[\begin{array}{l} \nvdash B \Rightarrow F \\ \nvdash \mathsf{Confl}_{\clubsuit,\heartsuit}(E,A) \\ \cup \left\{ \begin{array}{l} \nvdash F \Rightarrow D \\ B \Rightarrow Y, \\ Y \Rightarrow F, \\ \mathsf{Impl}_{\diamond,\heartsuit}(X,A) \end{array} \right\} : \begin{array}{c} \diamond(X/Y) \in \mathfrak{L}, \\ \diamond \downarrow \clubsuit \\ \clubsuit(E/F) \nsucc \diamond(X/Y) \end{array} \right] : \begin{array}{c} \clubsuit(E/F) \in \mathfrak{L}, \\ \clubsuit \downarrow \spadesuit \\ \spadesuit(C/D) \nsucc \clubsuit(E/F) \end{array} \right\} \\ \Gamma \Rightarrow \heartsuit(A/B), \Delta \end{array}}{} \heartsuit_R^{\spadesuit(C/D)}$$

(1)

and

$$\frac{\mathsf{Impl}_{\heartsuit,\heartsuit'}(A,G) \quad B \Rightarrow H \quad H \Rightarrow B}{\Sigma, \heartsuit(A/B) \Rightarrow \heartsuit'(G/H), \Pi} \mathsf{Mon}_{\heartsuit,\heartsuit'}$$

By induction hypothesis on the cut complexity we obtain the premisses of

$$\frac{\begin{array}{l} H \Rightarrow D \\ \mathsf{Impl}_{\spadesuit,\heartsuit'}(C,G) \\ \{\nvdash \mathsf{Confl}_{\clubsuit,\clubsuit}(G,G) : \mathsf{nt}(\clubsuit), \clubsuit = \heartsuit'\} \\ \left\{ \left[\begin{array}{l} \nvdash H \Rightarrow F \\ \nvdash \mathsf{Confl}_{\clubsuit,\heartsuit'}(E,G) \\ \cup \left\{ \begin{array}{l} \nvdash F \Rightarrow D \\ H \Rightarrow Y, \\ Y \Rightarrow F, \\ \mathsf{Impl}_{\diamond,\heartsuit'}(X,G) \end{array} \right\} : \begin{array}{c} \diamond(X/Y) \in \mathfrak{L}, \\ \diamond \downarrow \clubsuit \\ \clubsuit(E/F) \nsucc \diamond(X/Y) \end{array} \right] : \begin{array}{c} \clubsuit(E/F) \in \mathfrak{L}, \\ \clubsuit \downarrow \spadesuit \\ \spadesuit(C/D) \nsucc \clubsuit(E/F) \end{array} \right\} \\ \Gamma, \Sigma, \Rightarrow \heartsuit'(G/H), \Delta, \Pi \end{array}}{} \heartsuit_R'^{\spadesuit(C/D)}$$

This uses Prop. 5.3 for obtaining $\mathsf{Impl}_{\spadesuit,\heartsuit'}(C,G)$ and $\mathsf{Impl}_{\diamond,\heartsuit'}(X,G)$, as well as obtaining $\nvdash \mathsf{Confl}_{\clubsuit,\heartsuit'}(E,G)$ from $\nvdash \mathsf{Confl}_{\clubsuit,\heartsuit}(E,A)$ and $\mathsf{Impl}_{\heartsuit,\heartsuit'}(A,G)$. Finally, Prop. 5.3 also yields $\nvdash \mathsf{Confl}_{\heartsuit',\heartsuit'}(G,G)$ from $\nvdash \mathsf{Confl}_{\heartsuit,\heartsuit}(A,A)$ and $\mathsf{Impl}_{\heartsuit,\heartsuit'}(A,A)$ in case we have $\mathsf{nt}(\heartsuit')$ and the premiss needs to be present – in that case we also have $\mathsf{nt}(\heartsuit)$ and the corresponding premiss is in (1) as well. The cases involving $\heartsuit_L^{\spadesuit(C/D)}$ and $\mathsf{Mon}_{\heartsuit,\heartsuit'}$ or $\heartsuit_R^{\spadesuit(C/D)}$ and $\mathsf{D}_{\heartsuit,\heartsuit'}$ are similar.

For the case of a multicut between $\heartsuit_R^{\spadesuit(C/D)}$ and both principal formulae of the D rule, we claim that this cannot happen. For suppose we had (1) and

$$\frac{\mathsf{Confl}_{\heartsuit,\heartsuit}(A,A) \quad B \Rightarrow B \quad B \Rightarrow B}{\Sigma, \heartsuit(A,B), \heartsuit(A/B) \Rightarrow \Pi} \mathsf{D}_{\heartsuit,\heartsuit}$$

If $\mathsf{nt}(\heartsuit)$ we immediately obtain a contradiction since $\mathsf{Confl}_{\heartsuit,\heartsuit}(A,A)$ is both derivable and not derivable. Otherwise, since $\mathsf{D}_{\heartsuit,\heartsuit}$ is in the system, we have $\heartsuit \not\downarrow \heartsuit$, and since the rule $\heartsuit_R^{\spadesuit(C/D)}$ was used, we have $\spadesuit \to \heartsuit$. Hence we also have $\spadesuit \not\downarrow \spadesuit$. Thus one instance of $\clubsuit(E/F)$ in the set of premises of $\heartsuit_R^{\spadesuit(C/D)}$ is the assumption $\spadesuit(C/D)$. But for this formula the first premiss gives us $B \Rightarrow D$, hence in the choice block the instantiation $\nvdash B \Rightarrow D$ of the first underivability statement $\nvdash B \Rightarrow F$ does not hold. Further, from Prop. 5.3 with the premises $\mathsf{Impl}_{\spadesuit,\heartsuit}(C,A)$ and $\mathsf{Confl}_{\heartsuit,\heartsuit}(A,A)$ we get $\mathsf{Confl}_{\spadesuit,\heartsuit}(C,A)$, hence this instantiation of the second underivability statement $\nvdash \mathsf{Confl}_{\spadesuit,\heartsuit}(E,A)$ of the choice block also does not hold. Finally, the instantiation $\nvdash D \Rightarrow D$ of the third underivability statement $\nvdash F \Rightarrow D$ also does not hold due to Lem. 5.4, and hence the proto-derivation ending in (1) cannot have been a derivation. The case involving $\heartsuit_R^{\spadesuit(C/D)}$ and $\mathsf{P}_\heartsuit$ is completely analogous.

Also in the case of the assumption right rule versus the assumption left rule we claim that this cannot happen. Suppose we would have (1) and

$$\frac{\begin{array}{c} B \Rightarrow D' \\ \mathsf{Confl}_{\spadesuit',\heartsuit}(C',A) \\ \left\{\left[\begin{array}{c} \nvdash B \Rightarrow F' \\ \nvdash \mathsf{Impl}_{\spadesuit',\heartsuit}(E',A) \\ \{\nvdash F' \Rightarrow D'\} \end{array}\right] \cup \left\{\left[\begin{array}{c} B \Rightarrow Y', \\ Y' \Rightarrow F', \\ \mathsf{Confl}_{\Diamond',\heartsuit}(X',A) \end{array}\right] : \begin{array}{c} \Diamond'(X'/Y') \in \mathfrak{L}, \\ \Diamond' \not\downarrow \clubsuit', \\ \clubsuit'(E'/F') \not\succ \Diamond'(X'/Y') \end{array}\right] : \begin{array}{c} \clubsuit'(E'/F') \in \mathfrak{L}, \\ \clubsuit' \not\downarrow \spadesuit', \\ \spadesuit'(C'/D') \not\succ \clubsuit'(E'/F') \end{array}\right\}}{\Gamma, \heartsuit(A/B) \Rightarrow \Delta} \heartsuit_L^{\spadesuit'(C'/D')}$$

Since both rules are in the system, we have $\spadesuit \to \heartsuit$ and $\spadesuit' \not\downarrow \heartsuit$, and hence also $\spadesuit' \not\downarrow \spadesuit$. Further, since the superiority relation is acyclic, we have either $\spadesuit(C/D) \not\succ \spadesuit'(C'/D')$ or $\spadesuit'(C'/D') \not\succ \spadesuit(C/D)$. Suppose $\spadesuit(C/D) \not\succ \spadesuit'(C'/D')$. Then instantiating $\clubsuit(E/F)$ in the premises of $\heartsuit_R^{\spadesuit(C/D)}$ with $\spadesuit'(C'/D')$ we have either $\nvdash B \Rightarrow D'$, or $\nvdash \mathsf{Confl}_{\spadesuit',\heartsuit}(C',A)$ or $\nvdash D' \Rightarrow D$ together with the choice. The first of these cannot be the case, because from $\heartsuit_L^{\spadesuit'(C'/D')}$ we have $B \Rightarrow D'$. The second also cannot be the case because again from $\heartsuit_L^{\spadesuit'(C'/D')}$ we get $\mathsf{Confl}_{\spadesuit',\heartsuit}(C',A)$. So assume that $\nvdash D' \Rightarrow D$ and for some $\Diamond(X/Y) \in \mathfrak{L}$ with $\Diamond \not\downarrow \spadesuit'$ and $\spadesuit'(C'/D') \not\succ \Diamond(X/Y)$ we have all three of

$$B \Rightarrow Y \quad Y \Rightarrow D' \quad \mathsf{Impl}_{\Diamond,\heartsuit}(X,A) \qquad (2)$$

But then instantiating this assumption $\Diamond(X/Y)$ for $\clubsuit'(E'/F')$ in the premises of $\heartsuit_L^{\spadesuit'(C'/D')}$ yields that one of $\nvdash B \Rightarrow Y$ or $\nvdash \mathsf{Impl}_{\Diamond,\heartsuit}(X,A)$ or $\nvdash Y \Rightarrow D'$ holds. This is clearly in contradiction to (2). Hence every possibility yields a contradiction, and thus one of the two proto-derivations was not a derivation. The case of $\spadesuit'(C'/D') \not\succ \spadesuit(C/D)$ is analogous, starting with instantiating the formula $\clubsuit'(E'/F')$ in the premises of the rule $\heartsuit_L^{\spadesuit'(C'/D')}$ with the assumption $\spadesuit(C/D)$ and then reasoning as in the first case. $\square$

An important corollary of this result is that we can reduce derivability to derivability of bounded rank, and hence obtain well-definedness of the former notion:

Theorem 5.6 (Derivability is well-defined) *Let the maximal nesting depth of operators in $\Gamma \Rightarrow \Delta$ be n. Then we have $\vdash_{\mathsf{G}_{\mathfrak{N}}\mathrm{cut}} \Gamma \Rightarrow \Delta$ iff $\vdash_{\mathsf{G}_{\mathfrak{N}}} \Gamma \Rightarrow \Delta$ iff $\vdash^n_{\mathsf{G}_{\mathfrak{N}}} \Gamma \Rightarrow \Delta$. Hence derivability in $\mathsf{G}_{\mathfrak{N}}$ is well-defined.*

Proof. The first equivalence follows straightforwardly from cut elimination (Thm. 5.5). The proof for the second equivalence is by induction on n. For $n = 0$ the sequent is purely propositional. Hence the derivation cannot contain underivability statements, and the statement is straightforward. Suppose the statement holds for all $m < n$. Due to the shape of the rules, every sequent in a derivation of $\Gamma \Rightarrow \Delta$ has nesting depth $\leq n$, and the underivability statements mention sequents of depth $\leq n - 1$. Thus by induction hypothesis on the underivability statements the derivation is of rank n and we have $\vdash^n_{\mathsf{G}_{\mathfrak{N}}} \Gamma \Rightarrow \Delta$. Similarly, if $\vdash^n_{\mathsf{G}_{\mathfrak{N}}} \Gamma \Rightarrow \Delta$, then by induction hypothesis on the underivability statements occurring in the derivation we obtain $\vdash_{\mathsf{G}_{\mathfrak{N}}} \Gamma \Rightarrow \Delta$. □

As a further corollary we obtain decidability of the system and complexity results. Notably, the complexity of reasoning from assumptions is the same as that of reasoning without assumptions in Standard Deontic Logic [18]:

Theorem 5.7 *Given $\mathfrak{N}$, the problem of deciding whether $\vdash_{\mathsf{G}_{\mathfrak{N}}} \Gamma \Rightarrow \Delta$ is decidable in space polynomial in the size of $\Gamma \Rightarrow \Delta$.*

Proof. (Sketch) The idea is to perform backwards proof search to find a proto derivation. For each underivability statement we then recursively call the algorithm and flip the answer. To prevent loops caused by contraction, we copy the principal formula of the implication rules into the premises and omit the weakening and contraction rules. Standard inductions on the depth of the proto derivation then show admissibility of the contraction and weakening rules. The proof search procedure existentially guesses the last applied rule, checks that its application is non-redundant, i.e., introduces at least one new formula, then universally chooses its premises and checks derivability. Since each backwards application of a rule adds at least one new subformula of the conclusion or reduces the maximal nesting depth of the sequent, the depth of the search tree is polynomial in the size of the conclusion. Since moreover its branching factor only depends on the number of rules, i.e., deontic assumptions, it is independent of the size of the input. Hence the procedure runs in alternating polynomial time, which is equivalent to polynomial space [5]. □

6 Applications

We apply our methodology to the case studies of Mīmāṃsā-inspired logic, permissions as exception, and a logic of sanction and violation, showing how contrary-to-duties can be modeled as instance of defeasible reasoning [26].

6.1 Mīmāṃsā-inspired logic

The specificity rules in [8] for the Mīmāṃsā-inspired logic are a *particular* case of our general rule schemas. Before showing how to model these rules, and how to extend them with prioritized obligations, we briefly recall the logic in question, introduced to formalize and provide a better understanding of the deontic reasoning of Mīmāṃsā authors. Mīmāṃsā is an ancient influential school of Indian philosophy mainly focusing on the exegesis of the prescriptive portions of the *Vedas* – the Sacred Texts of Hinduism. In order to explain the deontic content of the Vedas and interpret them in a noncontradictory way, Mīmāṃsā authors proposed a rich body of deontic, hermeneutical and linguistic principles called *nyāyas*. In [7] some of the deontic *nyāyas* were transformed into Hilbert axioms for a non-normal dyadic deontic logic, which yielded a formal analysis of a famous deontic controversy contained in the Vedas. Interestingly, this solution coincided with that of Prabhākara, one of the chief Mīmāṃsā authors, which previous approaches failed to make sense of. As shown in [9] the □-free fragment of this logic is the dyadic version of the non-normal deontic logic MD [6].

Not all *nyāyas* can be converted into Hilbert axioms. These include more general interpretative principles to resolve apparent contradictions in the Vedas like the specificity principle, discussed already by Mīmāṃsā author Śabara (3rd-5th c. CE) under the name *guṇapradhāna*. Hence the dyadic version of MD was extended in [8] with sequent rules for specificity. These rules can be seen as a particular case of the general scheme described here by considering an operator characterisation with only one obligation type operator $\mathcal{O}$ with $\mathcal{O} \notin \mathcal{O}$ and no superiority relation. Going beyond [8], the superiority relation in the rules of Fig. 3 lets us deal with the Mīmāṃsā interpretative principle called *hierarchy of sources* (*śrutismṛtyādibādha*). This principle states that out of two apparently clashing commands, the one issued by a less authoritative source is to be suspended. Indeed, Mīmāṃsā author Kumārila describes four sources of duty, in decreasing order of authority: *śruti* (the Vedas), *smṛti* (the 'recollected texts', based on the Vedas), *sadācāra* (the behaviour of good people, who are learned in the Vedas) and *ātmatuṣṭi* (the inner feeling of approval by people who are learned in the Vedas). Hence, the considered norms can be formalized by four obligation type operators $\mathcal{O}_V, \mathcal{O}_{rt}, \mathcal{O}_{gp}, \mathcal{O}_{if}$ with $\heartsuit \notin \spadesuit$ for each $\heartsuit, \spadesuit \in \{\mathcal{O}_V, \mathcal{O}_{rt}, \mathcal{O}_{gp}, \mathcal{O}_{if}\}$, with the transitive closure of the priorities $\mathcal{O}_V(A/B) \succ \mathcal{O}_{rt}(C/D)$, $\mathcal{O}_{rt}(C/D) \succ \mathcal{O}_{gp}(E/F)$ and $\mathcal{O}_{gp}(E/F) \succ \mathcal{O}_{if}(G/H)$ between any assumptions using these operators.

6.2 Permissions as exceptions

Considered often as the dual of obligation, permission has been treated as primitive operator as well [22,11]. Here we model the notion of permissions as *exceptions* to other deontic operators (compare [2] for an analogous treatment in the context of input-output logics). Intuitively, a permission $\mathcal{P}^\heartsuit(A/B)$ acts as an exception to deontic assumptions in $\heartsuit$, in that it blocks the derivation of a formula $\heartsuit(C/D)$ whenever A and C are in conflict. To define what "in conflict" means, we assume that what is permitted is not forbidden, i.e., that given $\mathcal{P}^\heartsuit(A/B)$ we have not $\heartsuit(A/B)$ if $\heartsuit$ is of prohibition type and not $\heartsuit(\neg A/B)$ for $\heartsuit$ of obligation type. This suggests that permission operators are of obligation type, i.e., upwards monotone in the first argument, in line with the standard notion that if something is permitted, everything which follows from this is also permitted. Thus, to model permissions for an operator $\heartsuit$, we add an obligation type operator $\mathcal{P}^\heartsuit$ with $\heartsuit \nmid \mathcal{P}^\heartsuit$. Note that $\heartsuit$ could be of obligation or prohibition type, and it can but does not need to satisfy $\heartsuit \nmid \heartsuit$ and $\mathrm{nt}(\heartsuit)$.

Example 6.1 To model the sentence "Parking is forbidden, unless one has a permit" we use a prohibition-type operator $\mathcal{F}$ with $\mathcal{F} \nmid \mathcal{F}$ and the corresponding (obligation type) permission operator $\mathcal{P}^\mathcal{F}$ with $\mathcal{F} \nmid \mathcal{P}^\mathcal{F}$. The deontic assumptions are $\{\mathcal{F}(\mathsf{parking}/\top), \mathcal{P}^\mathcal{F}(\mathsf{parking}/\mathsf{permit})\}$. We can then derive, e.g., $\mathcal{F}(\mathsf{parking}/\top)$ and $\mathcal{F}(\mathsf{parking}/\mathsf{lazy})$, but neither $\mathcal{F}(\mathsf{parking}/\mathsf{permit})$ nor $\mathcal{F}(\mathsf{parking}/\mathsf{permit} \wedge \mathsf{lazy})$. Hence the permission $\mathcal{P}^\mathcal{F}(\mathsf{parking}/\mathsf{permit})$ acts as an explicit exception to the more general prohibition $\mathcal{F}(\mathsf{parking}/\top)$.

Note that adding permission operators also makes permission formulae derivable, e.g., $\mathcal{P}^\mathcal{F}(\mathsf{parking}/\mathsf{permit} \wedge \mathsf{lazy})$ in Ex. 6.1. These could be read as "explicit" or "strong" permissions in that they are derived from permissions explicitly mentioned in the assumptions. To keep them implicit, we can consider permissions in the assumptions, but not as derived formulae. Note also that to introduce a more general permission operator $\mathcal{P}$ which acts as exception to several other operators $\heartsuit_1, \ldots, \heartsuit_n$, it is enough to add $\heartsuit_i \nmid \mathcal{P}$ for every $i \le n$.

6.3 Sanctions and violations

We can also use our approach to differentiate between *exceptions* to a primary norm (as above), and *secondary* norms, which come into effect after a primary one has been violated. The crucial difference is that for exceptions to a more general norm there is no violation, whereas for secondary norms the primary one stays in force, and hence can be violated. This is similar to the distinction between *violations* of norms and *sanctions* as a result of violations. We model this using two prohibition type operators $\mathcal{S}$ and $\mathcal{V}$ with corresponding permission operators $\mathcal{P}^\mathcal{S}$ and $\mathcal{P}^\mathcal{V}$ as in Sec. 6.2. The intuitive reading of $\mathcal{S}(A/B)$ is that A is forbidden given B, and doing A results in a sanction. For $\mathcal{V}(A/B)$ we read that A is forbidden given B, and doing A results in a violation but not necessarily a sanction. Here we assume that there is no sanction without violation, $\mathcal{S} \to \mathcal{V}$, and that $\mathcal{V} \nmid \mathcal{V}$, $\mathcal{V} \nmid \mathcal{P}^\mathcal{V}$, $\mathcal{S} \nmid \mathcal{P}^\mathcal{S}$. Closure under $\to$ then yields $\mathcal{S} \nmid \mathcal{V}$, $\mathcal{S} \nmid \mathcal{S}$, $\mathcal{S} \nmid \mathcal{P}^\mathcal{V}$. The latter means that exceptions to violations can overrule sanctions, but in absence of $\mathcal{V} \nmid \mathcal{P}^\mathcal{S}$ exceptions to sanctions cannot overrule violations. Hence there might be a violation, even though there is no sanction.

Example 6.2 Consider the assumptions $\{\mathcal{S}(\mathsf{parking}/\top), \mathcal{V}(\mathsf{parking}/\top), \mathcal{P}^\mathcal{V}(\mathsf{parking}/\mathsf{permit}), \mathcal{P}^\mathcal{S}(\mathsf{parking}/\mathsf{fine\_paid})\}$, modelling the fact that once a fine for illegal parking has been paid, there is no further sanction. We derive all three of $\mathcal{S}(\mathsf{parking}/\top)$, $\mathcal{V}(\mathsf{parking}/\top)$, $\mathcal{V}(\mathsf{parking}/\mathsf{fine\_paid})$. However, we cannot derive either of $\mathcal{S}(\mathsf{parking}/\mathsf{fine\_paid})$, $\mathcal{S}(\mathsf{parking}/\mathsf{permit})$, $\mathcal{V}(\mathsf{parking}/\mathsf{permit})$. The first of these is overruled by $\mathcal{P}^\mathcal{S}(\mathsf{fine\_paid})$, the second and third ones by $\mathcal{P}^\mathcal{V}(\mathsf{parking}/\mathsf{permit})$. So if there is no permit, but the fine has been paid, there is no further sanction but still a violation of the prohibition to park.

Remark 6.3 Similarly, we can model contrary-to-duty (CTD) obligations while maintaining the distinction between defeasibility and violation of primary obligations. Indeed, borrowing the example from [26], we can model "There must be no fence", as $\mathcal{S}(\text{fence}/\top) \wedge \mathcal{V}(\text{fence}/\top)$, and "If there is a fence, it must be a white fence" as $\mathcal{P}^{\mathcal{S}}(\text{white\_fence}/\text{fence})$. Then we derive that the primary obligation is in force ($\mathcal{S}(\text{fence}/\top) \wedge \mathcal{V}(\text{fence}/\top)$) and having a white fence results in a violation of the primary norm ($\mathcal{V}(\text{white\_fence}/\text{fence})$), but does not violate the secondary norm ($\not\vdash \mathcal{S}(\text{white\_fence}/\text{fence})$). This distinction between violations of primary and secondary norms is somewhat similar to the distinction between *instrumental/actual* and *proper/ideal* obligations in [28] and [4] respectively: Roughly speaking, proper or ideal obligations, i.e., all obligations that apply to a context, including violated primary ones, correspond to the violation operator, while instrumental or actual ones, i.e., those detailing what to do in a particular situation, correspond to the sanction operator.

In general, CTDs of other CTDs are modeled by as many different operators as nested CTDs +1. A similar approach is in [13], that employs the (n-ary) substructural connective $\otimes$ where $A \otimes B$ stands for "the violation of A can be repaired by B" to reduce CTD to a special kind of normative exception.

7 Conclusions and Related Work

We introduced sequent rules for reasoning with deontic assumptions using specificity in presence of prioritized deontic operators. The method, which relies on cut elimination in presence of underivability premises, captures systems with an arbitrary finite number of dyadic deontic operators based on M possibly extended with axioms P or D and inclusions among the operators. The method is applied to various case studes and implemented in deonticProver2.0.

Related work. The approaches closest to ours are those in the framework of *dyadic deontic logic*, e.g., [33,6,32,20,26]. The main difference is that we consider reasoning from deontic assumptions to be inherently nonmonotonic, and hence do not attempt to capture it purely axiomatically. Indeed, while from the assumption $\mathcal{O}(A/\top)$ we derive $\mathcal{O}(A/\top)$, this no longer holds if we add the conflicting assumption $\mathcal{O}(\neg A/\top)$. This aspect cannot be captured in a purely axiomatic setting, since propositional logic already gives $\mathcal{O}(A/\top) \wedge \mathcal{O}(\neg A/\top) \to \mathcal{O}(A/\top)$. Additionally, unlike our system, most dyadic deontic logics derive $\mathcal{O}(A/A)$, which rules out, e.g., the derivation of a formula like $\mathcal{O}(\neg\text{asparagus}/\text{asparagus})$ in Ex. 4.2.

In the nonmonotonic setting, different methods have been introduced to deal with conflicts using specificity and/or superiority; these are either logic-tailored, e.g. [29,27], or are handled within general frameworks like the following.

Deontic default logic [15,16] uses semantical extensions to provide a credulous or skeptical approach (an obligation is derivable if it belongs to at least one or all extensions, respectively). While our system is heavily influenced by the notions of specificity and overriding in [15,16], it avoids the fixpoint construction necessary there, accounts for explicit exceptions, and permits nested obligations on the logic level.

Defeasible deontic logic (DDL), introduced in [12], uses facts, strict and defeasible rules, undercutting rules, and a binary superiority relation on the rules to solve conflicts between defeasible rules. The main differences with our approach are that in DDL propositional reasoning is defeasible, tractable complexity is paid for by the omission of binary connectives, specificity is handled "manually" by adding the superiority relation to all rules where it should apply.

A very influential logic expressing conditional norms is *Input-Output Logic* [21,23,24]. The main difference w.r.t. our approach is that their base logic is based fundamentally on (deontic or factual) detachment principles. Perhaps more in line with the notion of *contextual obligations* [26], neither of these holds in our system, nor, e.g., in the Mīmāṃsā-inspired logic (see Section 6.1).

Limitations and future work. An obvious limitation of our proposal is that the underlying non-normal deontic logics are rather weak. In particular, it would be interesting to extend the logic with an aggregation principle $\heartsuit(A/C) \wedge \heartsuit(B/C) \rightarrow \heartsuit(A \wedge B/C)$. We anticipate that this is possible by suitably adjusting the assumption rules, albeit at a severe cost to the complexity. The more interesting question is how to extend the assumption rules to additional axioms in a general way. We'd also like to solve the limitation mentioned in [16] and rule out conflicts between more than two deontic assumptions, i.e., to incorporate the rules $\vdash \neg(A_1 \wedge \cdots \wedge A_n/B) / \vdash \neg(\heartsuit(A_1/B) \wedge \cdots \wedge \heartsuit(A_n/B))$ in the base logic. This should be possible using methods similar to those for aggregation. A perhaps more challenging extension would be to incorporate principles like deontic detachment / transitivity. It is not entirely clear whether it is possible to avoid a fixpoint construction in this case. Finally, while neighbourhood semantics for the base logics as in [6] are reasonably straightforward, the big challenge is to find a suitable semantic characterisation for the assumption rules. These topics are left for future work.

References

[1] Björnsson, G. and R. Shanklin, *'must', 'ought' and the structure of standards*, in: F. Cariani, D. Grossi, J. Meheus and X. Parent, editors, *DEON 2014*, LNAI **8554**, Springer, 2014 pp. 33–48.

[2] Boella, G. and L. van der Torre, *Institutions with a hierarchy of authorities in distributed dynamic environments*, Artif. Intell. Law **16** (2008), pp. 53–71.

[3] Bonatti, P. A. and N. Olivetti, *Sequent calculi for propositional nonmonotonic logics*, ACM Trans. Comput. Log. **3** (2002), pp. 226–278.

[4] Carmo, J. and A. J. Jones, *Deontic logic and contrary-to-duties*, in: D. Gabbay and F. Guenthner, editors, *Handbook of Philosophical Logic, Volume 8*, Kluwer Academic Publishers, 2002, 2 edition pp. 265–343.

[5] Chandra, A. K., D. C. Kozen and L. J. Stockmeyer, *Alternation*, J. Assoc. Comput. Mach. **28** (1981), pp. 114–133.

[6] Chellas, B. F., "Modal Logic," Cambridge University Press, 1980.

[7] Ciabattoni, A., E. Freschi, F. Genco and B. Lellmann, *Mīmāṃsā deontic logic: proof theory and applications*, in: H. De Nivelle, editor, *TABLEAUX 2015*, LNCS **9323**, Springer, 2015 pp. 323–338.

[8] Ciabattoni, A., F. Gulisano and B. Lellmann, *Resolving conflicting obligations in Mīmāṃsā: A sequent-based approach*, in: J. Broersen, C. Condoravdi, S. Nair and G. Pigozzi, editors, *DEON 2018 proceedings*, College Publications, 2018 pp. 91–109.
[9] Freschi, E., A. Ollett and M. Pascucci, *Duty and sacrifice. a logical analysis of the Mīmāṃsā theory of Vedic injunctions*, History and Philosophy of Logic **40(4)** (2019), pp. 323–354.
[10] Goble, L., *Prima facie norms, normative conflicts, and dilemmas*, in: *Handbook of Deontic Logic and Normative Systems*, College Publications, 2013 pp. 241–351.
[11] Governatori, G., F. Olivieri, A. Rotolo and S. Scannapieco, *Computing strong and weak permissions in defeasible logic*, J. Philos. Logic **42** (2013), pp. 799–829.
[12] Governatori, G. and A. Rotolo, *Defeasible logic: Agency, intention and obligation*, in: *DEON 2004 proceedings*, LNAI **3065**, Springer, 2004 pp. 114–128.
[13] Governatori, G. and A. Rotolo, *Logic of violations: A Gentzen system for reasoning with contrary-to-duty obligations*, The Australasian Journal of Logic **4** (2006), pp. 193–215.
[14] Horty, J., *Defaults with priorities*, J. Philos. Logic **36** (2007), pp. 367–413.
[15] Horty, J. F., *Deontic logic as founded on nonmonotonic logic*, Ann. Math. Artif. Intell. **9** (1993), pp. 69–91.
[16] Horty, J. F., *Nonmonotonic foundations for deontic logic*, in: D. Nute, editor, *Defeasible Deontic Logic*, Kluwer, 1997 pp. 17–44.
[17] Horty, J. F., "Reasons as Defaults," Oxford University Press, 2012.
[18] Ladner, R. E., *The computational complexity of provability in systems of modal logic*, SIAM J. Comput. **6** (1977), pp. 467–480.
[19] Lellmann, B. and D. Pattinson, *Constructing cut free sequent systems with context restrictions based on classical or intuitionistic logic*, in: K. Lodaya, editor, *ICLA 2013*, LNCS **7750**, Springer, 2013 pp. 148–160.
[20] Lewis, D., *Semantic analyses for dyadic deontic logics*, in: S. Stendlund, editor, *Logical Theory and Semantic Analysis*, Reidel, 1974 pp. 1–14.
[21] Makinson, D. and L. W. N. van der Torre, *Input/output logics*, J. Philos. Logic **29** (2000), pp. 383–408.
[22] Makinson, D. and L. W. N. van der Torre, *Permission from an input/output perspective*, J. Philos. Logic **32** (2003), pp. 391–416.
[23] Parent, X. and L. van der Torre, *Input/output logic*, in: D. Gabbay, J. Horty, X. Parent, R. van der Meyden and L. van der Torre, editors, *Handbook of Deontic Logic and Normative Systems*, College Publications, 2013 pp. 495–544.
[24] Parent, X. and L. van der Torre, "Introduction to Deontic Logic and Normative Systems," College Publications, 2018.
[25] Piazza, M. and G. Pulcini, *Unifying logics via context-sensitiveness*, Journal of Logic and Computation **1** (2017), pp. 21–40.
[26] Prakken, H. and M. J. Sergot, *Contrary-to-duty obligations*, Studia Logica **57** (1996), pp. 91–115.
[27] Putte, F. V. D. and C. Strasser, *A logic for prioritized normative reasoning*, J. Log. Comput. **23** (2013), pp. 563—583.
[28] Straßer, C., *A deontic logic framework allowing for factual detachment*, J. Appl. Log. **9** (2011), pp. 61–80.
[29] Straßer, C. and O. Arieli, *Normative reasoning by sequent-based argumentation*, J. Log. Comput. **29** (2019), pp. 381–415.
[30] Troelstra, A. S. and H. Schwichtenberg, "Basic Proof Theory," Cambridge Tracts In Theoretical Computer Science **43**, Cambridge University Press, 2000, 2 edition.
[31] van der Torre, L., *Violated obligations in a defeasible deontic logic*, in: A. Cohn, editor, *ECAI 94*, Wiley, 1994 pp. 371–375.
[32] van Fraassen, B. C., *The logic of conditional obligation*, J. Philos. Logic **1** (1972), pp. 417–438.
[33] von Wright, G., *A note on deontic logic and derived obligation*, Mind **65** (1956), pp. 507–509.

Proof Systems for the Logics of Bringing-It-About

Tiziano Dalmonte, Charles Grellois, and Nicola Olivetti [1]

Aix Marseille Univ, Université de Toulon, CNRS, LIS, Marseille, France

Abstract

The logic of Bringing-it-About was introduced by Elgesem to formalise the notions of agency and capability. It contains two families of modalities indexed by agents, the first one expressing what an agent brings about (does), and the second expressing what she *can* bring about (can do). We first introduce a new neighbourhood semantics, defined in terms of bi-neighbourhood models for this logic, which is more suited for countermodel construction than the semantics defined in the literature. We then introduce a hypersequent calculus for this logic, which leads to a decision procedure allowing for a practical countermodel extraction. We finally extend both the semantics and the calculus to a coalitional version of Elgesem logic proposed by Troquard.

Keywords: Logic of agency, logic of ability, coalition logic, sequent calculus, countermodel extraction, decision procedure.

1 Introduction

The logic of Bringing-It-About was originally proposed by Elgesem [5], and provides one possible formalisation of agents' actions in terms of their results: that an agent "does something" is interpreted as the fact that the agent brings about something, for instance "John does a bank transfer" is interpreted as "John does that the bank transfer is done". The logical system proposed by Elgesem contains two modalities indexed by agents $\mathbb{E}_i$ and $\mathbb{C}_i$ (this is not his original notation), the former expressing the agentive modality of bringing-it-about, and the latter expressing capability, roughly speaking $\mathbb{E}_{lucy}$ *BankTransfer* means that Lucy makes a bank transfer, whereas $\mathbb{C}_{lucy}$ *BankTransfer* means that Lucy *can* make a bank transfer. Elgesem's logic is then intended to capture the effect of the action "what is brought about" and the agency relation, abstracting away from any temporal and game-theoretic aspect. In this way it provides a terse formalism, that has become a standard, quite simpler than other formalisms such as STIT-logic [2,8]. Elgesem's logic is well-suited for expressing notions of *responsibility* and formalising notions of control, power, and delegation, for instance: "Sara prevents Lucy from making a bank transfer" will be captured just

---

[1] tiziano.dalmonte@gmail.com, {charles.grellois,nicola.olivetti}@lis-lab.fr. This work has been partially supported by the ANR project TICAMORE ANR-16-CE91-0002-01.

by $\mathbb{E}_{sara} \neg \mathbb{E}_{lucy}$ *BankTransfer*; moreover it can be easily combined with deontic modalities in order to express e.g. that an agent is *obliged* to do something and so on.

Elgesem proposed an axiomatisation of his logic and a (almost matching) semantics based on selection function models. Notice that the intended notion of capability is rather weak, the only characterising axioms are that (i) agency implies capability $\mathbb{E}_i A \to \mathbb{C}_i A$ and (ii) $\neg \mathbb{C}_i \top$, the latter expressing that an agent i is not capable of doing anything that is always true, whence also $\neg \mathbb{E}_i \top$: an agent cannot do anything that will happen anyway, no matter her own involvement and responsibility.

Elgesem's logic was further studied by Governatori and Rotolo [7], who proposed an alternative semantics in terms of neighbourhood models. In their semantics, models contain two neighbourhood functions corresponding to the two operators $\mathbb{E}_i$ and $\mathbb{C}_i$ assigning for each agent i the propositions (identified with their truth sets) that the agent i brings/can bring about. They also proved that Elgesem's semantics entails the validity of the further axiom $\neg \mathbb{C}_i \bot$ meaning that an agent cannot bring about something which is contradictory.

Elgesem's logic deals with actions of a single agent, who might be either a human individual, or an institution, or a group conceived as an indivisible entity. A natural extension of this logic is to handle groups or coalitions that act *jointly* to bring about an action. This has been proposed by Troquard [11] who has developed an extension of Elgesem logic to handle "coalitions": individuals may gather in coalitions to bring about a joint action. In a joint action, each participant must be involved, so that the logic rejects coalition monotonicity: $\mathbb{E}_g A \to \mathbb{E}_{g'} A$ whenever $g \subseteq g'$ is *not* considered as valid. Troquard provided a computational analysis of his logic and determined its complexity by providing a decision procedure for his logic, whence for Elgesem's.

While the semantics of Elgesem logic, as well as its coalitional extension are well-understood, its proof-theory is mainly unexplored: the only known proof system for this logic was proposed by Lellmann [9]. In particular, no proof system connecting the syntax and the semantics is known. By this we mean that there is no proof system so far that permits the construction of countermodels of non-valid formulas. Moreover, no proof system is known at all for the coalitional extension. In particular, the decision procedure developed by Troquard [11] computes a reduction of a question about validity in his coalition logic to a set of SAT problems. This is in the spirit of the approach of Vardi [12] and Giunchiglia et al. [6] for non-normal modal logics. But this algorithms based on SAT-reduction does not provide neither *derivations*, nor *countermodels*.

This is precisely the purpose of this work. We take our move by redefining the semantics of Elgesem logic: we consider bi-neighbourhood models, a variant of neighbourhood models defined in [7]. Like the models in [7], our models contain, for each agent i, two neighbourhood functions corresponding to the two operators $\mathbb{E}_i$ and $\mathbb{C}_i$. But contrary to the neighbourhood models of [7], these functions assign to each world a set of *pairs* of neighbourhoods (α, β).

Although it would be pretentious to suggest here a new semantics of actions, we can suggest some intuitive interpretations of the pairs of neighbourhoods (α, β): given a proposition A representing the result of an action of an agent i, the two components (α, β) of the pairs can be understood respectively as specifying independently a set of situations α enabling i to bring about A and β preventing i from doing A. An alternative interpretation is as follows: since A must be true in all worlds (situations) in α and false in all worlds in β, the former can also be thought as a set of *possible outcomes* of A and the latter as a set of *impossible outcomes* of A.[2] In this second interpretation, each pair (α, β) can also be thought of as expressing a *lower* and an *upper* approximation of propositions that the agent brings/can bring about given a proposition.

Note that a bi-neighbourhood model can be transformed into a standard neighbourhood model of [7], and conversely.

No matter its intuitive interpretation, the bi-neighbourhood semantics has a clear technical advantage as it makes easier to compute countermodels of non-valid formulas than the standard neighbourhood semantics, by avoiding the exact determination of the truth sets of formulas.

We next move to proof theory by proposing a hypersequent calculus. A hypersequent can be thought of as a disjunction of ordinary sequents. While the hypersequent structure is not needed to obtain a complete calculus (as witnessed by [9] itself), the use of hypersequents allows us to define a calculus with invertible rules, as a difference with the one in [9]. The main advantage is that from *one* failed hypersequent occurring as a leaf of *one* derivation tree, a countermodel can directly be extracted in the bi-neighbourhood semantic of the formula under verification. In this sense, our calculus provides not only a decision procedure for this logic, but also the first practical procedure to compute countermodels. Observe that it is not possible to compute directly countermodels by ordinary sequent calculi: because the rules are not invertible, the fact that *one* specific derivation fails, does not mean that the sequent is unprovable, so that in order to build a countermodel (for a non-valid formula), all possible derivations must be attempted and inspected. Another syntactic feature of our calculi is that hypersequents contain additional structural constructs, the blocks, which are necessary for countermodel construction, but also to capture the logic in a clean and modular way, reflecting its axiomatisation.

The hypersequent calculus has nonetheless good proof-theoretic properties, as it enjoys *a syntactic proof of cut elimination*, from which also follows its completeness with respect to the axiomatisation. We then turn to the coalitional version of Elgesem's logic proposed by Troquard [11]: we are able to extend both the bi-neighbourhood semantics and the calculus to this setting, needing only to add the rules for handling the empty coalition and coalition fusion. Our calculus then provides a decision procedure for Troquard coalitional logic, with derivations and countermodels.

---

[2] We are grateful to one reviewer for suggesting this latter interpretation.

RE$_\mathbb{E}$	$\dfrac{A \leftrightarrow B}{\mathbb{E}_i A \leftrightarrow \mathbb{E}_i B}$	RE$_\mathbb{C}$	$\dfrac{A \leftrightarrow B}{\mathbb{C}_i A \leftrightarrow \mathbb{C}_i B}$
C$_\mathbb{E}$	$\mathbb{E}_i A \wedge \mathbb{E}_i B \to \mathbb{E}_i(A \wedge B)$	Q$_\mathbb{C}$	$\neg \mathbb{C}_i \top$
T$_\mathbb{E}$	$\mathbb{E}_i A \to A$	P$_\mathbb{C}$	$\neg \mathbb{C}_i \bot$
Int$_\mathbb{EC}$	$\mathbb{E}_i A \to \mathbb{C}_i A$		

Fig. 1. Modal axioms and rules of Elgesem's logic **ELG**.

2 Elgesem's logic and bi-neighbourhood semantics

In this section, we present Elgesem's agency and ability logic, which we denote by **ELG**. Then we define the bi-neighbourhood models for this logic.

Let $\mathcal{A} = \{a, b, c, ...\}$ be a set of agents. The logic **ELG** is defined on a propositional language $\mathcal{L}_{Elg}$ containing, for every $i \in \mathcal{A}$, two unary modalities $\mathbb{E}_i$ and $\mathbb{C}_i$, respectively of "agency" and "ability". The formulas of $\mathcal{L}_{Elg}$ are defined by the following grammar:

$$A ::= p \mid \bot \mid \top \mid \neg A \mid A \wedge B \mid A \vee B \mid A \to B \mid \mathbb{E}_i A \mid \mathbb{C}_i A,$$

where $\mathbb{E}_i A$ and $\mathbb{C}_i A$ are respectively read as "the agent i brings it about that A", and "the agent i is capable of realising A". The logic **ELG** is defined by extending classical propositional logic (formulated in language $\mathcal{L}_{Elg}$) with the modal axioms and rules in Fig. 1.[3]

Notice that $\neg \mathbb{E}_i \bot$ and $\neg \mathbb{E}_i \top$ are derivable in **ELG**. By contrast, the axioms C and T hold only for the modality $\mathbb{E}$, meaning respectively that if an agent realises two things, then she realises both, and that if A is brought about by some agent, then it is actually the case that A.

Semantic characterisations of the logic **ELG** are provided by Elgesem [5] in terms of selection function models and by Governatori and Rotolo [7] in terms of neighbourhood models, the latter having separate neighbourhood functions for the modalities $\mathbb{E}$ and $\mathbb{C}$. Here we propose an alternative semantics based on bi-neighbourhood models [4]. We explain the advantages of this alternative semantics just after its definition.

Definition 2.1 A *bi-neighbourhood model* for **ELG** is a tuple $\mathcal{M} = \langle \mathcal{W}, \mathcal{N}_i^\mathbb{E}, \mathcal{N}_i^\mathbb{C}, \mathcal{V} \rangle$, where $\mathcal{W}$ is a non-empty set, $\mathcal{V}$ is a valuation function, and for each agent i, $\mathcal{N}_i^\mathbb{E}$ and $\mathcal{N}_i^\mathbb{C}$ are two bi-neighbourhood functions $\mathcal{W} \longrightarrow \mathcal{P}(\mathcal{P}(\mathcal{W}) \times \mathcal{P}(\mathcal{W}))$ satisfying the following conditions:

(C$_\mathbb{E}$) If $(\alpha, \beta), (\gamma, \delta) \in \mathcal{N}_i^\mathbb{E}(w)$, then $(\alpha \cap \gamma, \beta \cup \delta) \in \mathcal{N}_i^\mathbb{E}(w)$.
(T$_\mathbb{E}$) If $(\alpha, \beta) \in \mathcal{N}_i^\mathbb{E}(w)$, then $w \in \alpha$.
(Q$_\mathbb{C}$) If $(\alpha, \beta) \in \mathcal{N}_i^\mathbb{C}(w)$, then $\beta \neq \emptyset$.
(P$_\mathbb{C}$) If $(\alpha, \beta) \in \mathcal{N}_i^\mathbb{C}(w)$, then $\alpha \neq \emptyset$.
(Int$_\mathbb{EC}$) $\mathcal{N}_i^\mathbb{E}(w) \subseteq \mathcal{N}_i^\mathbb{C}(w)$.

---

[3] A variant of Elgesem's logic not containing axiom P$_\mathbb{C}$ is considered in [7,9]. All results presented in this work can be extended to this variant just by dropping the corresponding condition in the bi-neighbourhood semantics and the corresponding rule in the calculus.

The forcing relation ⊩ is defined as usual for atomic formulas and boolean connectives, whereas for $\mathbb{E}$- and $\mathbb{C}$-formulas it is defined as follows:

$\mathcal{M}, w \Vdash \mathbb{E}_i A$ iff   there is $(\alpha, \beta) \in \mathcal{N}_i^{\mathbb{E}}(w)$ s.t.
                    for all $v \in \alpha$, $\mathcal{M}, v \Vdash A$, and for all $u \in \beta$, $\mathcal{M}, u \nVdash A$.

$\mathcal{M}, w \Vdash \mathbb{C}_i A$ iff   there is $(\alpha, \beta) \in \mathcal{N}_i^{\mathbb{C}}(w)$ s.t.
                    for all $v \in \alpha$, $\mathcal{M}, v \Vdash A$, and for all $u \in \beta$, $\mathcal{M}, u \nVdash A$.

Notice that if we denote by $[\![A]\!]$ the set $\{v \mid \mathcal{M}, v \Vdash A\}$, i.e., the *truth set* of A, the above clauses can be rewritten as $\mathcal{M}, w \Vdash \mathbb{E}_i A$ if and only if there is $(\alpha, \beta) \in \mathcal{N}_i^{\mathbb{E}}(w)$ s.t. $\alpha \subseteq [\![A]\!]$ and $\beta \subseteq [\![\neg A]\!]$, and similarly for $\mathbb{C}$-formulas. As usual, we omit to specify the model $\mathcal{M}$ when it is clear from context, and then we simply write $w \Vdash A$.

The main reason for considering bi-neighbourhood semantics is that is offers a much easier and natural way to extract countermodels from failed proofs. To see this, in the standard neighbourhood semantics, to make w satisfy $\mathbb{E}_i A$, *exactly* the truth set of A must belong to $\mathcal{N}_i^{\mathbb{E}}(w)$, whereas in the bi-neighbourhood semantics it is sufficient to find a pair (α, β) such that $\alpha \subseteq [\![A]\!]$ and $\beta \subseteq [\![\neg A]\!]$. Observe that this condition can be rewritten as $\alpha \subseteq [\![A]\!] \subseteq \mathcal{W} \setminus \beta$: in this way the pair (α, β) can be thought of as specifying a lower and upper approximation of the truth set of A. The fact that the exact determination of truth sets is not needed in the bi-neighbourhood semantics makes countermodels extraction from failed proofs substantially easier than in the standard semantics: a failed proof only specifies "partial" information, from which one can directly compute bi-neighbourhood pairs, but not exact truth-sets. For this reason bi-neighbourhood semantics is more natural for direct countermodel extraction than the standard one.

As mentioned in the introduction, bi-neighbourhood semantics can also have some intuitive meaning in terms of agency, we have suggested two possible interpretations: a bi-neighbourhood pair can be interpreted as a specification of enabling and preventing conditions for the realisation of actions, or as a set of possible/impossible outcomes of an action. In both interpretations, the conditions ($P_{\mathbb{C}}$) and ($Q_{\mathbb{C}}$), i.e., $\alpha \neq \emptyset$ and $\beta \neq \emptyset$ have a natural meaning: the former imposes that an action must be enabled or possible (non-empty possible outcomes), so that a contradiction cannot be realised; the latter imposes that an action must be preventable (non-empty impossible outcomes), so that a tautology cannot be realised.

Notice also that, because of the validity of $\neg \mathbb{E}_i \top$ and of the axiom $T_\mathbb{E}$, formulas of the form $\mathbb{E}_i A$ are never valid in models for **ELG**, this is the semantic counterpart of the idea that actions can be always prevented.

Theorem 2.2 (Characterisation) *A is derivable in **ELG** if and only if it is valid in all bi-neighbourhood models for **ELG**.*

Proof. The proof of soundness is easy and amounts to showing that all axioms are valid and all rules are validity-preserving. Completeness can be proved by the canonical model construction as it is done in [4] for classical non-normal modal logics. Let us call **ELG**-maximal any set Φ of formulas of $\mathcal{L}_{Elg}$ such that $\Phi \nvdash_{\mathbf{ELG}} \bot$ and if $A \notin \Phi$, then $\Phi \cup \{A\} \vdash_{\mathbf{ELG}} \bot$. We denote by $\uparrow A$ the class of **ELG**-maximal sets containing A, and we define the canonical model for **ELG** as the tuple $\langle \mathcal{W}, \mathcal{N}_i^{\mathbb{E}}, \mathcal{N}_i^{\mathbb{C}}, \mathcal{V} \rangle$, where $\mathcal{W}$ is the class of maximal sets, $\mathcal{V}(p) = \{\Phi \in \mathcal{W} \mid p \in \Phi\}$, and for every $i \in \mathcal{A}$ and $\mathbb{X} \in \{\mathbb{E}, \mathbb{C}\}$, $\mathcal{N}_i^{\mathbb{X}}(\Phi) = \{(\uparrow A, \mathcal{W} \setminus \uparrow A) \mid \mathbb{X}_i A \in \Phi\}$. We can prove that $\Phi \Vdash A$ if and only if $A \in \Phi$, (truth lemma, cf. [4]) and that the canonical model is a bi-neighbourhood model for **ELG**. We show as an example that it satisfies the conditions ($Q_\mathbb{C}$) and (Int$_{\mathbb{EC}}$): ($Q_\mathbb{C}$) Assume $(\uparrow A, \mathcal{W} \setminus \uparrow A) \in \mathcal{N}_i^{\mathbb{C}}(\Phi)$. Then there is $\mathbb{C}_i B \in \Phi$ such that $\uparrow B = \uparrow A$, whence $\vdash B \leftrightarrow A$. If $\uparrow A = \mathcal{W}$, then $\vdash A \leftrightarrow \top$. Thus by RE$_\mathbb{C}$, $\vdash \mathbb{C}_i B \leftrightarrow \mathbb{C}_i \top$, and since Φ is closed under derivation, $\mathbb{C}_i \top \in \Phi$, against the fact that $\neg \mathbb{C}_i \top \in \Phi$ and Φ is **ELG**-consistent. Therefore $\uparrow A \neq \mathcal{W}$, that is $\mathcal{W} \setminus \uparrow A \neq \emptyset$. (Int$_{\mathbb{EC}}$) Assume $(\alpha, \beta) \in \mathcal{N}_i^{\mathbb{E}}(\Phi)$. Then there is $\mathbb{E}_i A \in \Phi$ such that $\alpha = \uparrow A$ and $\beta = \mathcal{W} \setminus \uparrow A$. Since $\mathbb{E}_i A \to \mathbb{C}_i A \in \Phi$ and Φ is closed under derivation, $\mathbb{C}_i A \in \Phi$. Thus $(\uparrow A, \mathcal{W} \setminus \uparrow A) = (\alpha, \beta) \in \mathcal{N}_i^{\mathbb{C}}(\Phi)$. □

Similarly to the transformation described in [3,4], a bi-neighbourhood model for **ELG** can be transformed into a neighbourhood model for it as follows (the proof is easy by induction on A):

Proposition 2.3 (Model transformation) *Let $\mathcal{M}_{bi} = \langle \mathcal{W}, \mathcal{N}_{bi}, \mathcal{V} \rangle$ be a bi-neighbourhood model for **ELG**, and $\mathcal{M}_n = \langle \mathcal{W}, \mathcal{N}_n, \mathcal{V} \rangle$ be the neighbourhood model defined by taking the same $\mathcal{W}$ and $\mathcal{V}$ and, for all $w \in \mathcal{W}$,*

$$\mathcal{N}_n(w) = \{\gamma \subseteq \mathcal{W} \mid \text{there is } (\alpha, \beta) \in \mathcal{N}_b(w) \text{ such that } \alpha \subseteq \gamma \subseteq \mathcal{W} \setminus \beta\}.$$

Then, for every $A \in \mathcal{L}_{Elg}$ and every $w \in \mathcal{W}$, $\mathcal{M}_n, w \Vdash A$ if and only if $\mathcal{M}_{bi}, w \Vdash A$.

As the above transformation shows, bi-neighbourhood models have in general smaller functions than their equivalent neighbourhood models. The reason is that every bi-neighbourhood pair (α, β) – whose elements can be thought of as lower and upper bounds of neighbourhoods – might validate more than one modal formula.

3 Hypersequent calculus

We now focus on proof theory. To our knowledge, the only proof-theoretic investigation of Elgesem's logic is carried on in [9], where a cut-free sequent calculus is defined. That calculus provides a decision procedure for Elgesem's logic, but has no link with the semantics.

We propose here a hypersequent calculus (see [1]) for Elgesem's logic, in the same style of calculi for basic non-normal modal logics presented in [3]. A hypersequent can be loosely interpreted as a disjunction of sequents. The hypersequents considered in this article rely on an additional structure, called *blocks*. A block is used to collect $\mathbb{E}$- and $\mathbb{C}$-formulas: more precisely it represents a conjunction of formulas under the scope of the *same* $\mathbb{E}$ or $\mathbb{C}$. Since neither $\mathbb{E}$, nor $\mathbb{C}$ distribute over conjunction, blocks are not an abbreviation, they are a proper structural construct, and specific structural rules of the calculus handle them. Blocks within hypersequents are primarily needed for building countermodels of non-derivable formulas: as we will see, they are used to define bi-neighbourhood pairs. Blocks also have two other advantages: by using blocks we can encode in a clean (close to the axiomatisation) and analytic way the relation between the modalities $\mathbb{E}$ and $\mathbb{C}$; in addition the rules governing the modalities $\mathbb{E}$ and $\mathbb{C}$ are independent one of the other, so that the two $\mathbb{E}$ and $\mathbb{C}$-fragments are *separated*, and the interaction between the two modalities is captured just by a structural rule on blocks. We consider the following definitions:

Definition 3.1 (Block, sequent, hypersequent) A *block* is a structure $\langle \Sigma \rangle_i^{\mathbb{E}}$ or $\langle \Sigma \rangle_i^{\mathbb{C}}$, where i is an agent, and Σ is a multiset of formulas of $\mathcal{L}_{Elg}$. A *sequent* is a pair $\Gamma \Rightarrow \Delta$, where Γ is a multiset of formulas and blocks, and Δ is a multiset of formulas. We sometimes consider $\mathsf{set}(\Gamma)$, the *support* of a multiset Γ, i.e., the set of its elements disregarding multiplicities. A *hypersequent* is a multiset $S_1 \mid ... \mid S_n$, where $S_1, ..., S_n$ are sequents. $S_1, ..., S_n$ are called the *components* of the hypersequent.

Definition 3.2 (Formula interpretation) Single sequents are interpreted as formulas of the logic as follows:

$$i(A_1, ..., A_n, \langle \Sigma_1 \rangle_{a_1}^{\mathbb{E}}, ..., \langle \Sigma_m \rangle_{a_m}^{\mathbb{E}}, \langle \Pi_1 \rangle_{b_1}^{\mathbb{C}}, ..., \langle \Pi_k \rangle_{b_k}^{\mathbb{C}} \Rightarrow B_1, ..., B_\ell)$$
$$=$$
$$\bigwedge_{i \leq n} A_i \wedge \bigwedge_{j \leq m} \mathbb{E} a_j \bigwedge \Sigma_j \wedge \bigwedge_{s \leq k} \mathbb{C} a_s \bigwedge \Pi_s \to \bigvee_{t \leq \ell} B_t.$$

Definition 3.3 (Semantic interpretation) We say that a sequent S is *valid* in a bi-neighbourhood model $\mathcal{M}$, denoted $\mathcal{M} \models S$, if for all $w \in \mathcal{M}$, $\mathcal{M}, w \Vdash i(S)$. We say that a hypersequent H is *valid* in $\mathcal{M}$, denoted $\mathcal{M} \models H$, if $\mathcal{M} \models S$ for some $S \in H$.

The rules of the hypersequent calculus $\mathbf{HS_{ELG}}$ are presented in Fig. 2. They are expressed in the cumulative version: the principal formulas or blocks are copied into the premiss(es). This allows us to extract a countermodel from a single saturated hypersequent. The propositional rules are just the hypersequent versions of the ordinary corresponding sequent rules (we omit the rules for $\neg, \vee, \to$, which are standard). As usual, initial sequents init are restricted to propositional variables, but it is easy to see that $G \mid A, \Gamma \Rightarrow \Delta, A$ is derivable for every A. Similarly to propositional connectives, $\mathbb{E}$- and $\mathbb{C}$-formulas are handled by separate left and right rules. The rules $R_{\mathbb{E}}$ and $R_{\mathbb{C}}$ have multiple premisses, but their number is fixed by the cardinality of the principal blocks $\langle \Sigma \rangle_i^{\mathbb{E}}$ and $\langle \Sigma \rangle_i^{\mathbb{C}}$. For every axiom of $\mathbf{ELG}$ there is a corresponding rule in the calculus. Blocks have a central role in all modal rules. Observe in particular that $\mathbb{E}$-blocks can be merged by means of the rule $C_{\mathbb{E}}$, but there is no analogous rule for $\mathbb{C}$-blocks. However, once complex $\mathbb{E}$-blocks are created, they can be converted into $\mathbb{C}$-blocks by means of the rule $\mathsf{Int}_{\mathbb{EC}}$. In Fig. 3 we show two examples of derivation in $\mathbf{HS_{ELG}}$.

Proposition 3.4 (Soundness) *If H is derivable in $\mathbf{HS_{ELG}}$, then it is valid in all bi-neighbourhood models for $\mathbf{ELG}$.*

$$\text{init} \frac{}{G \mid \Gamma, p \Rightarrow p, \Delta} \qquad \text{L}\bot \frac{}{G \mid \Gamma, \bot \Rightarrow \Delta} \qquad \text{R}\top \frac{}{G \mid \Gamma \Rightarrow \top, \Delta}$$

$$\text{L}\wedge \frac{G \mid \Gamma, A \wedge B, A, B \Rightarrow \Delta}{G \mid \Gamma, A \wedge B \Rightarrow \Delta} \qquad \text{R}\wedge \frac{G \mid \Gamma \Rightarrow A, A \wedge B, \Delta \qquad G \mid \Gamma \Rightarrow B, A \wedge B, \Delta}{G \mid \Gamma \Rightarrow A \wedge B, \Delta}$$

$$\text{L}_{\mathbb{E}} \frac{G \mid \Gamma, \mathbb{E}_i A, \langle A\rangle_i^{\mathbb{E}} \Rightarrow \Delta}{G \mid \Gamma, \mathbb{E}_i A \Rightarrow \Delta} \qquad \text{L}_{\mathbb{C}} \frac{G \mid \Gamma, \mathbb{C}_i A, \langle A\rangle_i^{\mathbb{C}} \Rightarrow \Delta}{G \mid \Gamma, \mathbb{C}_i A \Rightarrow \Delta}$$

$$\text{Int}_{\mathsf{EC}} \frac{G \mid \Gamma, \langle\Sigma\rangle_i^{\mathbb{E}}, \langle\Sigma\rangle_i^{\mathbb{C}} \Rightarrow \Delta}{G \mid \Gamma, \langle\Sigma\rangle_i^{\mathbb{E}} \Rightarrow \Delta}$$

$$\text{R}_{\mathbb{E}} \frac{G \mid \Gamma, \langle\Sigma\rangle_i^{\mathbb{E}} \Rightarrow \mathbb{E}_i A, \Delta \mid \Sigma \Rightarrow A \qquad \{G \mid \Gamma, \langle\Sigma\rangle_i^{\mathbb{E}} \Rightarrow \mathbb{E}_i A, \Delta \mid A \Rightarrow B\}_{B \in \Sigma}}{G \mid \Gamma, \langle\Sigma\rangle_i^{\mathbb{E}} \Rightarrow \mathbb{E}_i A, \Delta}$$

$$\text{R}_{\mathbb{C}} \frac{G \mid \Gamma, \langle\Sigma\rangle_i^{\mathbb{C}} \Rightarrow \mathbb{C}_i A, \Delta \mid \Sigma \Rightarrow A \qquad \{G \mid \Gamma, \langle\Sigma\rangle_i^{\mathbb{C}} \Rightarrow \mathbb{C}_i A, \Delta \mid A \Rightarrow B\}_{B \in \Sigma}}{G \mid \Gamma, \langle\Sigma\rangle_i^{\mathbb{C}} \Rightarrow \mathbb{C}_i A, \Delta}$$

$$\text{C}_{\mathbb{E}} \frac{G \mid \Gamma, \langle\Sigma\rangle_i^{\mathbb{E}}, \langle\Pi\rangle_i^{\mathbb{E}}, \langle\Sigma, \Pi\rangle_i^{\mathbb{E}} \Rightarrow \Delta}{G \mid \Gamma, \langle\Sigma\rangle_i^{\mathbb{E}}, \langle\Pi\rangle_i^{\mathbb{E}} \Rightarrow \Delta} \qquad \text{T}_{\mathbb{E}} \frac{G \mid \Gamma, \langle\Sigma\rangle_i^{\mathbb{E}}, \Sigma \Rightarrow \Delta}{G \mid \Gamma, \langle\Sigma\rangle_i^{\mathbb{E}} \Rightarrow \Delta}$$

$$\text{Q}_{\mathbb{C}} \frac{\{G \mid \Gamma, \langle\Sigma\rangle_i^{\mathbb{C}} \Rightarrow \Delta \mid \Rightarrow B\}_{B \in \Sigma}}{G \mid \Gamma, \langle\Sigma\rangle_i^{\mathbb{C}} \Rightarrow \Delta} \qquad \text{P}_{\mathbb{C}} \frac{G \mid \Gamma, \langle\Sigma\rangle_i^{\mathbb{C}} \Rightarrow \Delta \mid \Sigma \Rightarrow}{G \mid \Gamma, \langle\Sigma\rangle_i^{\mathbb{C}} \Rightarrow \Delta}$$

Fig. 2. The calculus **HS**$_{\mathbf{ELG}}$.

$$\cfrac{\cfrac{\cfrac{\mathbb{E}_i A, \langle A\rangle_i^{\mathbb{E}}, \langle A\rangle_i^{\mathbb{C}} \Rightarrow \mathbb{C}_i A \mid A \Rightarrow A \qquad \mathbb{E}_i A, \langle A\rangle_i^{\mathbb{E}}, \langle A\rangle_i^{\mathbb{C}} \Rightarrow \mathbb{C}_i A \mid A \Rightarrow A}{\mathbb{E}_i A, \langle A\rangle_i^{\mathbb{E}}, \langle A\rangle_i^{\mathbb{C}} \Rightarrow \mathbb{C}_i A} \text{R}_{\mathbb{C}}}{\mathbb{E}_i A, \langle A\rangle_i^{\mathbb{E}} \Rightarrow \mathbb{C}_i A} \text{Int}_{\mathsf{EC}}}{\mathbb{E}_i A \Rightarrow \mathbb{C}_i A} \text{L}_{\mathbb{E}}$$

$$\cfrac{\cfrac{\cfrac{\cfrac{..., \langle A, B\rangle_i^{\mathbb{E}} \Rightarrow \mathbb{E}_i(A \wedge B) \mid A, B \Rightarrow A \wedge B \qquad ... \mid A \wedge B \Rightarrow A \qquad ... \mid A \wedge B \Rightarrow B}{\mathbb{E}_i A \wedge \mathbb{E}_i B, \mathbb{E}_i A, \mathbb{E}_i B, \langle A\rangle_i^{\mathbb{E}}, \langle B\rangle_i^{\mathbb{E}}, \langle A, B\rangle_i^{\mathbb{E}} \Rightarrow \mathbb{E}_i(A \wedge B)} \text{R}_{\mathbb{E}}}{\mathbb{E}_i A \wedge \mathbb{E}_i B, \mathbb{E}_i A, \mathbb{E}_i B, \langle A\rangle_i^{\mathbb{E}}, \langle B\rangle_i^{\mathbb{E}} \Rightarrow \mathbb{E}_i(A \wedge B)} \text{C}_{\mathbb{E}}}{\mathbb{E}_i A \wedge \mathbb{E}_i B, \mathbb{E}_i A, \mathbb{E}_i B, \langle A\rangle_i^{\mathbb{E}} \Rightarrow \mathbb{E}_i(A \wedge B)} \text{L}_{\mathbb{E}}}{\cfrac{\mathbb{E}_i A \wedge \mathbb{E}_i B, \mathbb{E}_i A, \mathbb{E}_i B \Rightarrow \mathbb{E}_i(A \wedge B)}{\mathbb{E}_i A \wedge \mathbb{E}_i B \Rightarrow \mathbb{E}_i(A \wedge B)} \text{L}\wedge} \text{L}_{\mathbb{E}}$$

Fig. 3. Derivations of axioms Int$_{\mathbb{E}\mathbb{C}}$ and C$_{\mathbb{E}}$ in **HS**$_{\mathbf{ELG}}$.

Proof. As usual, we have to show that the initial sequents are valid, and that whenever the premiss(es) of a rule are valid, so is the conclusion. We show the following illustrative cases.

(R$_{\mathbb{E}}$) Assume $\mathcal{M} \models G \mid \Gamma, \langle\Sigma\rangle_i^{\mathbb{E}} \Rightarrow \mathbb{E}_i A, \Delta \mid \Sigma \Rightarrow A$ and $\mathcal{M} \models G \mid \Gamma, \langle\Sigma\rangle_i^{\mathbb{E}} \Rightarrow \mathbb{E}_i A, \Delta \mid A \Rightarrow B$ for all $B \in \Sigma$. Then (i) $\mathcal{M} \models G$, or (ii) $\mathcal{M} \models \Gamma, \langle\Sigma\rangle_i^{\mathbb{E}} \Rightarrow \mathbb{E}_i A, \Delta$, or (iii) $\mathcal{M} \models \Sigma \Rightarrow A$ and $\mathcal{M} \models A \Rightarrow B$ for all $B \in \Sigma$. If (i) or (ii) we are done. If (iii), then $\mathcal{M} \models \bigwedge \Sigma \to A$ and $\mathcal{M} \models A \to B$ for all $B \in \Sigma$, that is $\mathcal{M} \models \bigwedge \Sigma \leftrightarrow A$. Since $RE_\mathbb{E}$ is valid, $\mathcal{M} \models \mathbb{E}_i \bigwedge \Sigma \to \mathbb{E}_i A = i(\langle \Sigma \rangle_i^\mathbb{E} \Rightarrow \mathbb{E}_i A)$. Thus $\mathcal{M} \models \Gamma, \langle \Sigma \rangle_i^\mathbb{E} \Rightarrow \mathbb{E}_i A, \Delta$.

(Int$_{\mathbb{EC}}$) Assume $\mathcal{M} \models G \mid \Gamma, \langle \Sigma \rangle_i^\mathbb{E}, \langle \Sigma \rangle_i^\mathbb{C} \Rightarrow \Delta$. Then $\mathcal{M} \models G$ or $\mathcal{M} \models \Gamma, \langle \Sigma \rangle_i^\mathbb{E}, \langle \Sigma \rangle_i^\mathbb{C} \Rightarrow \Delta$. In the first case we are done. In the second case, $\mathcal{M} \models i(\Gamma, \langle \Sigma \rangle_i^\mathbb{E}, \langle \Sigma \rangle_i^\mathbb{C} \Rightarrow \Delta)$, which is equivalent to $\mathbb{E}_i \bigwedge \Sigma \wedge \mathbb{C}_i \bigwedge \Sigma \to i(\Gamma \Rightarrow \Delta)$. By the validity of axiom Int$_{\mathbb{EC}}$, this is in turn equivalent to $\mathbb{E}_i \bigwedge \Sigma \to i(\Gamma \Rightarrow \Delta)$. Therefore $\mathcal{M} \models i(\Gamma, \langle \Sigma \rangle_i^\mathbb{E} \Rightarrow \Delta)$. □

We now investigate the structural properties of our calculus, and show that it is complete with respect to the axiomatisation. A purely syntactic completeness proof is significant because it is independent from the choice of any specific semantics. As usual, this proof requires to show the admissibility of the cut rule, that we formulate as follows:

$$\text{cut} \frac{G \mid \Gamma \Rightarrow \Delta, A \qquad G \mid A, \Gamma \Rightarrow \Delta}{G \mid \Gamma \Rightarrow \Delta}$$

This means that whenever the premises of cut are derivable, the conclusion is also derivable. In turn, admissibility of cut depends upon the admissibility of the structural rules of weakening and contraction, that in the hypersequent framework must be formulated both in their internal and in their external variants as follows:

Proposition 3.5 (Admissibility of structural rules) *The following rules are admissible in* **HS$_{\text{ELG}}$**, *where ϕ is any formula A or block $\langle \Sigma \rangle_i^\mathbb{E}$ or $\langle \Sigma \rangle_i^\mathbb{C}$:*

$$\text{Lwk} \frac{G \mid \Gamma \Rightarrow \Delta}{G \mid \phi, \Gamma \Rightarrow \Delta} \qquad \text{Lctr} \frac{G \mid \phi, \phi, \Gamma \Rightarrow \Delta}{G \mid \phi, \Gamma \Rightarrow \Delta} \qquad \text{Bctr} \frac{G \mid \langle A, A, \Sigma \rangle, \Gamma \Rightarrow \Delta}{G \mid \langle A, \Sigma \rangle, \Gamma \Rightarrow \Delta}$$

$$\text{Rwk} \frac{G \mid \Gamma \Rightarrow \Delta}{G \mid \Gamma \Rightarrow \Delta, A} \qquad \text{Rctr} \frac{G \mid \Gamma \Rightarrow \Delta, A, A}{G \mid \Gamma \Rightarrow \Delta, A}$$

$$\text{Ewk} \frac{G}{G \mid \Gamma \Rightarrow \Delta} \qquad \text{Ectr} \frac{G \mid \Gamma \Rightarrow \Delta \mid \Gamma \Rightarrow \Delta}{G \mid \Gamma \Rightarrow \Delta}$$

The proof of admissibility of weakening and contraction is standard by induction on the derivation of the premises. Observe that as an immediate consequence of the admissibility of weakening all rules are invertible, which means that whenever the conclusion of a rule is derivable, so are the premises. This is important because if a formula is derivable we get a derivation no matter the order in which the rules are applied (see Sec. 4).

By contrast, the proof of admissibility of cut is a bit more intricate and deserves more attention. We shall prove simultaneously the admissibility of cut and of the following rule sub, which states that a formula A inside one or more blocks can be replaced by any equivalent set of formulas Σ:

$$\text{sub} \frac{G \mid \Sigma \Rightarrow A \qquad \{G \mid A \Rightarrow B\}_{B \in \Sigma} \qquad G \mid \overrightarrow{\langle A^n, \Pi \rangle_i^\mathbb{E}}, \overrightarrow{\langle A^m, \Omega \rangle_j^\mathbb{C}}, \Gamma \Rightarrow \Delta}{G \mid \overrightarrow{\langle \Sigma^n, \Pi \rangle_i^\mathbb{E}}, \overrightarrow{\langle \Sigma^m, \Omega \rangle_j^\mathbb{C}}, \Gamma \Rightarrow \Delta}$$

where for instance $\overrightarrow{\langle A^n, \Pi \rangle_i^\mathbb{E}}$ stays for $\langle A^{n_1}, \Pi_1 \rangle_{i_1}^\mathbb{E}, ..., \langle A^{n_k}, \Pi_k \rangle_{i_k}^\mathbb{E}$, and A^{n_ℓ} is a compact way to denote n_ℓ occurrences of A. In the proof we use the following definition of weight of formulas and blocks.

Definition 3.6 (Weight of formulas and blocks) The weight of formulas and blocks is recursively defined as follows: $\mathsf{w}(\bot) = \mathsf{w}(\top) = \mathsf{w}(p) = 0$; $\mathsf{w}(A \wedge B) = \mathsf{w}(A \vee B) = \mathsf{w}(A \to B) = \mathsf{w}(A) + \mathsf{w}(B) + 1$; $\mathsf{w}(\langle A_1, ..., A_k \rangle_i^\mathbb{E}) = \mathsf{w}(\langle A_1, ..., A_k \rangle_j^\mathbb{C}) = max_{1 \leq n \leq k}\{\mathsf{w}(A_n)\} + 1$, $\mathsf{w}(\mathbb{E}_i A) = \mathsf{w}(\mathbb{C}_i A) = \mathsf{w}(A) + 2$.

Theorem 3.7 (Cut elimination) *The rules* cut *and* sub *are admissible in* $\mathbf{HS_{ELG}}$.

Sketch of Proof. Let $Cut(c, h)$ mean that all applications of cut of height h on a cut formula of weight c are admissible, and $Sub(c)$ mean that all applications of sub where A has weight c are admissible. Then the theorem is a consequence of the following claims: **(A)** $\forall c.Cut(c, 0)$; **(B)** $\forall h.Cut(0, h)$; **(C)** $\forall c.(\forall h.Cut(c, h) \to Sub(c))$; **(D)** $\forall c.\forall h. ((\forall c' < c.(Sub(c') \wedge \forall h'.Cut(c', h')) \wedge \forall h'' < h.Cut(c, h'')) \to Cut(c, h))$. The proof is in the Appendix. □

As a consequence of admissibility of cut we can prove the following completeness theorem.

Theorem 3.8 (Axiomatic completeness) *If A is derivable in* **ELG**, *then* $\Rightarrow A$ *is derivable in* $\mathbf{HS_{ELG}}$.

Proof. All modal axioms and rules of **ELG** are derivable in $\mathbf{HS_{ELG}}$. As examples, in Fig. 3 we have shown the derivations of axioms $\text{Int}_{\mathbb{E}\mathbb{C}}$ and $\text{C}_\mathbb{E}$. Moreover, the rule $\text{RE}_\mathbb{E}$ (and analogously the rule $\text{RE}_\mathbb{C}$) is derived as follows:

$$\dfrac{\dfrac{\dfrac{A \Rightarrow B}{\mathbb{E}_i A, \langle A \rangle_i^\mathbb{E} \Rightarrow \mathbb{E}_i B \mid A \Rightarrow B}\;\text{Ewk} \quad \dfrac{B \Rightarrow A}{\mathbb{E}_i A, \langle A \rangle_i^\mathbb{E} \Rightarrow \mathbb{E}_i B \mid B \Rightarrow A}\;\text{Ewk}}{\dfrac{\mathbb{E}_i A, \langle A \rangle_i^\mathbb{E} \Rightarrow \mathbb{E}_i B}{\mathbb{E}_i A \Rightarrow \mathbb{E}_i B}\;\text{L}_\mathbb{E}}\;\text{R}_\mathbb{E}}$$

The derivation contains applications of Ewk, which has been proved admissible. Finally, Modus Ponens is simulated by cut, which has been proved admissible, in the usual way. □

As mentioned, hypersequents are not strictly necessary for making derivations, and in particular one can show that a hypersequent is derivable in $\mathbf{HS_{ELG}}$ if and only if one of its components is derivable. However, the use of hypersequents allows us to obtain a calculus where all rules are invertible, which entails that the order of rule applications does not matter: essentially, modulo the order of rule applications, every formula has a *single* derivation, or a *single* failed proof, whence in particular proof search does not require backtracking. Moreover, hypersequents are crucial for a direct computation of countermodels from every single unprovable hypersequent occurring as a leaf of a failed derivation. We shall see all this in the next section.

4 Proof search and countermodel extraction

In this section, we define a procedure for checking the validity/derivability of formulas in Elgesem's logic by means of our hypersequent calculus. The procedure is based on a simple root-first proof search strategy. We show that the strategy always terminates and constructs a derivation for every valid formula. Moreover, we show that whenever the proof fails it possible to directly extract a countermodel of the non-valid formula. The strategy is based on the notion of saturation. Intuitively, a saturated hypersequent is such that the backward application of any rule to it cannot add any information, in the sense that one of the premisses of the rule is already included in the hypersequent.

Definition 4.1 (Saturated hypersequent) Let $H = \Gamma_1 \Rightarrow \Delta_1 \mid ... \mid \Gamma_k \Rightarrow \Delta_k$ be a hypersequent occurring in a proof for H'. The saturation conditions associated to each application of a rule of **HS$_{\mathbf{ELG}}$** are as follows:

- Unprovability: (init) $\Gamma_n \cap \Delta_n = \emptyset$. ($\perp_\mathsf{L}$) $\perp \notin \Gamma_n$. ($\top_\mathsf{R}$) $\top \notin \Delta_n$.
- Propositional rules: ($\wedge_\mathsf{L}$) If $A \wedge B \in \Gamma_n$, then $A \in \Gamma_n$ and $B \in \Gamma_n$. ($\wedge_\mathsf{R}$) If $A \wedge B \in \Delta_n$, then $A \in \Delta_n$ or $B \in \Delta_n$. Analogous for the rules for $\neg, \vee, \rightarrow$.
- Modal rules: ($\mathsf{L}_\mathbb{E}$) If $\mathbb{E}_i A \in \Gamma_n$, then $\langle A \rangle_i^\mathbb{E} \in \Gamma_n$. ($\mathsf{R}_\mathbb{E}$) If $\Gamma, \langle \Sigma \rangle_i^\mathbb{E} \Rightarrow \mathbb{E}_i B, \Delta$ is in H, then there is $\Gamma', \Sigma \Rightarrow B, \Delta'$ in H or there is $\Gamma', B \Rightarrow A, \Delta'$ in H for some $A \in \Sigma$. ($\mathsf{L}_\mathbb{C}$) and ($\mathsf{R}_\mathbb{C}$) are analogous. ($\mathsf{C}_\mathbb{E}$) If $\langle \Sigma \rangle_i^\mathbb{E}, \langle \Pi \rangle_i^\mathbb{E} \in \Gamma_n$, then there is $\langle \Omega \rangle_i^\mathbb{E} \in \Gamma_n$ such that $\mathsf{set}(\Sigma, \Pi) = \mathsf{set}(\Omega)$. ($\mathsf{T}_\mathbb{E}$) If $\langle \Sigma \rangle_i^\mathbb{E} \in \Gamma_n$, then $\mathsf{set}(\Sigma) \subseteq \Gamma_n$. ($\mathsf{Q}_\mathbb{C}$) If $\Gamma, \langle \Sigma \rangle_i^\mathbb{C} \Rightarrow \Delta$ is in H, then there is $\Gamma' \Rightarrow B, \Delta'$ in H for some $B \in \Sigma$. ($\mathsf{P}_\mathbb{C}$) If $\Gamma, \langle \Sigma \rangle_i^\mathbb{C} \Rightarrow \Delta$ is in H, then there is $\Gamma' \Rightarrow \Delta'$ in H such that $\mathsf{set}(\Sigma) \subseteq \Gamma'$. ($\mathsf{Int}_{\mathbb{E}\mathbb{C}}$) If $\langle \Sigma \rangle_i^\mathbb{E} \in \Gamma_n$, then there is $\langle \Omega \rangle_i^\mathbb{C} \in \Gamma_n$ such that $\mathsf{set}(\Sigma) = \mathsf{set}(\Omega)$.

We say that H is saturated with respect to an application of a rule R if it satisfies the corresponding saturation condition (R) for that particular rule application, and that it is saturated with respect to **HS$_{\mathbf{ELG}}$** if it is saturated with respect to every possible application of any rule of **HS$_{\mathbf{ELG}}$**.

The *proof search strategy* is simple: (i) do not apply any rule to initial sequents, and (ii) do not apply a rule to a hypersequent which is already saturated with respect to that particular application of that rule.

The strategy essentially amounts to avoiding applications of rules that do not add any additional information to the hypersequents. We can prove that this strategy leads to a terminating proof search algorithm.

Proposition 4.2 (Termination of proof search) *Every branch of a proof of a hypersequent H built in accordance with the strategy is finite. Thus, the proof search procedure for H always terminates. Moreover, every branch ends either with an initial hypersequent or a saturated one.*

Proof. Let $\mathfrak{P}$ be a proof of H. Then all formulas occurring in $\mathfrak{P}$ (both inside and outside blocks) are subformulas of formulas of H, so they are finitely many. Moreover, saturation conditions prevent duplications of the same formulas (both inside and outside blocks) and same blocks. Therefore every branch of $\mathfrak{P}$ can contain only finitely many hypersequents. $\square$

Hypersequents occurring in a proof of H can be exponentially large with respect to the size of H. This is due to the presence of the rule $\mathsf{C}_\mathbb{E}$ that, given n formulas $\mathbb{E}_i A_1, \ldots, \mathbb{E}_i A_n$, allows one to build a block for every subset of $\{A_1, ..., A_n\}$. In this respect, our decision procedure does not match the PSPACE complexity upper bound established for Elgesem's logic by Schröder and Pattinson [10] and Troquard [11].

An optimal calculus could be obtained either by considering the sequent calculus in [9], or (similarly to [3]) by reformulating the rules in Fig. 2 in such a way that the principal formulas are *not copied* into the premises. However, in this way we would lose the invertibility of the rules, whence the possibility to directly extract countermodels from single failed proofs. The situation is analogous to the one of modal logic **K**: while a PSPACE complexity upper bound can be obtain with the sequent calculus, the same is not possible with a calculus with only invertible rules allowing for direct countermodel extraction of non-valid formulas. This essentially shows the existence of a necessary trade-off in the logic **ELG** between the optimal complexity of the calculus and the possibility to directly extract countermodels from failed proofs.

We now show how to directly build a countermodel in the bi-neighbourhood semantics from a saturated hypersequent.

Definition 4.3 (Countermodel construction) Let H be a saturated hypersequent occurring in a proof for H'. Moreover, let $e : \mathbb{N} \longrightarrow H$ be an enumeration of the components of H. Given e, we can write H as $\Gamma_1 \Rightarrow \Delta_1 \mid \ldots \mid \Gamma_k \Rightarrow \Delta_k$. The model $\mathcal{M} = \langle \mathcal{W}, \mathcal{N}, \mathcal{V} \rangle$ is defined as follows:

- $\mathcal{W} = \{n \mid \Gamma_n \Rightarrow \Delta_n \in H\}$.
- $\mathcal{V}(p) = \{n \mid p \in \Gamma_n\}$.
- For every block $\langle \Sigma \rangle_i^{\mathbb{E}}$ or $\langle \Sigma \rangle_i^{\mathbb{C}}$ occurring in a component $\Gamma_m \Rightarrow \Delta_m$ of H,
$$\Sigma^+ = \{n \in \mathcal{W} \mid \mathsf{set}(\Sigma) \subseteq \Gamma_n\} \text{ and } \Sigma^- = \{n \in \mathcal{W} \mid \Sigma \cap \Delta_n \neq \emptyset\}.$$
- For every $i \in \mathcal{A}$ and every $n \in \mathcal{W}$,
$$\mathcal{N}_i^{\mathbb{E}}(n) = \{(\Sigma^+, \Sigma^-) \mid \langle \Sigma \rangle_i^{\mathbb{E}} \in \Gamma_n\} \text{ and } \mathcal{N}_i^{\mathbb{C}}(n) = \{(\Sigma^+, \Sigma^-) \mid \langle \Sigma \rangle_i^{\mathbb{C}} \in \Gamma_n\}.$$

Lemma 4.4 *Let $\mathcal{M}$ be defined as in Def. 4.3. Then for every A, $\langle \Sigma \rangle_i^{\mathbb{E}}$, $\langle \Pi \rangle_j^{\mathbb{C}}$ and every $n \in \mathcal{W}$, we have: If $A \in \Gamma_n$, then $n \Vdash A$; if $\langle \Sigma \rangle_i^{\mathbb{E}} \in \Gamma_n$, then $n \Vdash \mathbb{E}_i \bigwedge \Sigma$; if $\langle \Pi \rangle_j^{\mathbb{C}} \in \Gamma_n$, then $n \Vdash \mathbb{C}_j \bigwedge \Pi$; and if $A \in \Delta_n$, then $n \not\Vdash A$. Moreover, $\mathcal{M}$ is a bi-neighbourhood model for **ELG**.*

Proof. The first claim is proved by mutual induction on A and $\langle \Sigma \rangle_i^{\mathbb{E}}, \langle \Sigma \rangle_i^{\mathbb{C}}$. We only consider the inductive cases of modal formulas and blocks.

($\langle \Sigma \rangle_i^{\mathbb{E}} \in \Gamma_n$) By definition, $(\Sigma^+, \Sigma^-) \in \mathcal{N}_i^{\mathbb{E}}(n)$. We show that $\Sigma^+ \subseteq [\![\bigwedge \Sigma]\!]$ and $\Sigma^- \subseteq [\![\neg \bigwedge \Sigma]\!]$, which implies $n \Vdash \mathbb{E}_i \bigwedge \Sigma$. If $m \in \Sigma^+$, then $\mathsf{set}(\Sigma) \subseteq \Gamma_m$. By i.h. $m \Vdash A$ for all $A \in \Sigma$, then $m \Vdash \bigwedge \Sigma$. If $m \in \Sigma^-$, then there is $B \in \Sigma \cap \Delta_m$. By i.h. $m \not\Vdash B$, then $m \not\Vdash \bigwedge \Sigma$.

($\mathbb{E}_i B \in \Gamma_n$) By saturation of rule $\mathsf{L}_\mathbb{E}$, $\langle B \rangle_i^{\mathbb{E}} \in \Gamma_n$. Then by i.h. $n \Vdash \mathbb{E}_i B$.

($\mathbb{E}_i B \in \Delta_n$) Assume $(\alpha, \beta) \in \mathcal{N}_i^{\mathbb{E}}(n)$. Then there is $\langle \Sigma \rangle_i^{\mathbb{E}} \in \Gamma_n$ such that $\Sigma^+ = \alpha$ and $\Sigma^- = \beta$. By saturation of rule $\mathsf{R}_\mathbb{E}$, there is $m \in \mathcal{W}$ such that $\Sigma \subseteq \Gamma_m$ and $B \in \Delta_m$, or there is $m \in \mathcal{W}$ such that $\Sigma \cap \Delta_m \neq \emptyset$ and $B \in \Gamma_m$. In the first case, $m \in \Sigma^+ = \alpha$ and by i.h. $m \not\Vdash B$, thus $\alpha \not\subseteq [\![B]\!]$. In the second case, $m \in \Sigma^- = \beta$ and by i.h. $m \Vdash B$, thus $\beta \not\subseteq [\![\neg B]\!]$. Therefore $n \not\Vdash \mathbb{E}_i B$.

For blocks $\langle \Sigma \rangle_i^{\mathbb{C}}$ and formulas $\mathbb{C}_i B$ the proof is analogous. Now we prove that $\mathcal{M}$ is a model for **ELG**.

($\mathsf{C}_\mathbb{E}$) Assume that $(\alpha, \beta), (\gamma, \delta) \in \mathcal{N}_i^{\mathbb{E}}(n)$. Then there are $\langle \Sigma \rangle_i^{\mathbb{E}}, \langle \Pi \rangle_i^{\mathbb{E}} \in \Gamma_n$ such that $\Sigma^+ = \alpha$, $\Sigma^- = \beta$, $\Pi^+ = \gamma$ and $\Pi^- = \delta$. By saturation of rule $\mathsf{C}_\mathbb{E}$, there is $\langle \Omega \rangle \in \Gamma_n$ such that $\mathsf{set}(\Omega) = \mathsf{set}(\Sigma, \Pi)$, thus $(\Omega^+, \Omega^-) \in \mathcal{N}_i^{\mathbb{E}}(n)$. We show that (i) $\Omega^+ = \alpha \cap \gamma$ and (ii) $\Omega^- = \beta \cup \delta$. (i) $m \in \Omega^+$ iff $\mathsf{set}(\Omega) = \mathsf{set}(\Sigma, \Pi) \subseteq \Gamma_m$ iff $\mathsf{set}(\Sigma) \subseteq \Gamma_m$ and $\mathsf{set}(\Pi) \subseteq \Gamma_m$ iff $m \in \Sigma^+ = \alpha$ and $m \in \Pi^+ = \gamma$ iff $m \in \alpha \cap \gamma$. (ii) $m \in \Omega^-$ iff $\Omega \cap \Delta_m \neq \emptyset$ iff $\Sigma, \Pi \cap \Delta_m \neq \emptyset$ iff $\Sigma \cap \Delta_m \neq \emptyset$ or $\Pi \cap \Delta_m \neq \emptyset$ iff $m \in \Sigma^- = \beta$ or $m \in \Pi^- = \delta$ iff $m \in \beta \cup \delta$.

($\mathsf{Int}_{\mathbb{E}\mathbb{C}}$, $\mathsf{T}_\mathbb{E}$) If $(\alpha, \beta) \in \mathcal{N}_i^{\mathbb{E}}(n)$, then there is $\langle \Sigma \rangle_i^{\mathbb{E}} \in \Gamma_n$ such that $\Sigma^+ = \alpha$ and $\Sigma^- = \beta$. By saturation of rule $\mathsf{T}_\mathbb{E}$, $\mathsf{set}(\Sigma) \subseteq \Gamma_n$, then $n \in \Sigma^+ = \alpha$. Moreover, by saturation of rule $\mathsf{Int}_{\mathbb{E}\mathbb{C}}$, then there is $\langle \Omega \rangle_i^{\mathbb{C}} \in \Gamma_n$ such that $\mathsf{set}(\Sigma) = \mathsf{set}(\Omega)$. Then $(\Omega^+, \Omega^-) = (\Sigma^+, \Sigma^-) = (\alpha, \beta) \in \mathcal{N}_i^{\mathbb{C}}(n)$.

($\mathsf{P}_\mathbb{C}$, $\mathsf{Q}_\mathbb{C}$) If $(\alpha, \beta) \in \mathcal{N}_i^{\mathbb{C}}(n)$, then there is $\langle \Sigma \rangle_i^{\mathbb{C}} \in \Gamma_n$ such that $\Sigma^+ = \alpha$ and $\Sigma^- = \beta$. By saturation of rule $\mathsf{P}_\mathbb{C}$, there is $m \in \mathcal{W}$ such that $\mathsf{set}(\Sigma) \subseteq \Gamma_m$. Then $m \in \Sigma^+ = \alpha$, that is $\alpha \neq \emptyset$. Moreover, by saturation of rule $\mathsf{Q}_\mathbb{C}$, there is $\ell \in \mathcal{W}$ such that $\Sigma \cap \Delta_\ell \neq \emptyset$. Then $\ell \in \Sigma^- = \beta$, that is $\beta \neq \emptyset$. □

Observe that since all rules are cumulative, the countermodel $\mathcal{M}$ of H is also a countermodel of the root hypersequent H'. Then for every hypersequent we either get a derivation (if the hypersequent is valid) or obtain a countermodel. This entails the following theorem.

Theorem 4.5 (Semantic completeness) *If H is valid in all bi-neighbourhood models for **ELG**, then it is derivable in $\mathbf{HS_{ELG}}$.*

The proof search procedure for the calculus $\mathbf{HS_{ELG}}$ can be used to automatically and constructively check the validity/derivability of formulas in Elgesem logic. For every formula, the proof search procedure either provides a derivation if the formula is valid, or returns a countermodel if it is not.

Example 4.6 (Failure of delegation) The treatment of delegation represents a main difference between Elgesem's account of agency and other accounts, such as for instance the one formalised by STIT logics. It is explicitly rejected by Elgesem [5]: "a person is normally not considered the agent of some consequence of his action if another agent interferes in the causal chain." For instance, we can say that having the car repaired is not the same as repairing the car by yourself. Let us represent Anna by a, Beatrice by b, and "repairing the car" by p. Then $\mathbb{E}_a \mathbb{E}_b p \to \mathbb{E}_a p$ expresses the sentence "If Anna gets Beatrice to repair her car, then Anna repairs her car". By using our calculus we can automatically obtain a countermodel of $\mathbb{E}_a \mathbb{E}_b p \to \mathbb{E}_a p$. First, in Fig. 4 we find a failed proof of $\mathbb{E}_a \mathbb{E}_b p \to \mathbb{E}_a p$ in $\mathbf{HS_{ELG}}$. Then we consider the following enumeration of the components of the saturated hypersequent: $1 \mapsto \langle \mathbb{E}_b p \rangle_a^{\mathbb{E}}, \langle p \rangle_b^{\mathbb{E}}, \langle \mathbb{E}_b p \rangle_a^{\mathbb{C}}, \langle p \rangle_b^{\mathbb{C}}, p, \mathbb{E}_b p, \mathbb{E}_a \mathbb{E}_b p \Rightarrow \mathbb{E}_a p$; $2 \mapsto p \Rightarrow \mathbb{E}_b p$; and $3 \mapsto \Rightarrow p$. We obtain the following countermodels:

<u>Bi-neighbourhood countermodel:</u> By applying the construction in Def. 4.3 we

$$\cfrac{\cfrac{\cfrac{\cfrac{\cfrac{\cfrac{\cfrac{\cfrac{\cfrac{\overset{\text{saturated}}{\langle \mathbb{E}_b p \rangle_a^{\mathbb{E}}, \langle p \rangle_b^{\mathbb{E}}, \langle \mathbb{E}_b p \rangle_a^{\mathbb{C}}, \langle p \rangle_b^{\mathbb{C}}, p, \mathbb{E}_b p, \mathbb{E}_a \mathbb{E}_b p \Rightarrow \mathbb{E}_a p \mid p \Rightarrow \mathbb{E}_b p \mid \Rightarrow p}}{\langle \mathbb{E}_b p \rangle_a^{\mathbb{E}}, \langle p \rangle_b^{\mathbb{E}}, \langle \mathbb{E}_b p \rangle_a^{\mathbb{C}}, \langle p \rangle_b^{\mathbb{C}}, p, \mathbb{E}_b p, \mathbb{E}_a \mathbb{E}_b p \Rightarrow \mathbb{E}_a p \mid p \Rightarrow \mathbb{E}_b p} \mathsf{Q}_{\mathsf{C}}}{\langle \mathbb{E}_b p \rangle_a^{\mathbb{E}}, \langle p \rangle_b^{\mathbb{E}}, \langle \mathbb{E}_b p \rangle_a^{\mathbb{C}}, p, \mathbb{E}_b p, \mathbb{E}_a \mathbb{E}_b p \Rightarrow \mathbb{E}_a p \mid p \Rightarrow \mathbb{E}_b p} \mathsf{Int}_{\mathsf{EC}}}{\langle \mathbb{E}_b p \rangle_a^{\mathbb{E}}, \langle p \rangle_b^{\mathbb{E}}, p, \mathbb{E}_b p, \mathbb{E}_a \mathbb{E}_b p \Rightarrow \mathbb{E}_a p \mid p \Rightarrow \mathbb{E}_b p} \mathsf{Int}_{\mathsf{EC}}}{\langle \mathbb{E}_b p \rangle_a^{\mathbb{E}}, \langle p \rangle_b^{\mathbb{E}}, \mathbb{E}_b p, \mathbb{E}_a \mathbb{E}_b p \Rightarrow \mathbb{E}_a p \mid p \Rightarrow \mathbb{E}_b p} \mathsf{T}_{\mathsf{E}}}{\langle \mathbb{E}_b p \rangle_a^{\mathbb{E}}, \mathbb{E}_b p, \mathbb{E}_a \mathbb{E}_b p \Rightarrow \mathbb{E}_a p \mid p \Rightarrow \mathbb{E}_b p} \mathsf{L}_{\mathsf{E}}}{\langle \mathbb{E}_b p \rangle_a^{\mathbb{E}}, \mathbb{E}_a \mathbb{E}_b p \Rightarrow \mathbb{E}_a p \mid p \Rightarrow \mathbb{E}_b p} \mathsf{T}_{\mathsf{E}}}{\langle \mathbb{E}_b p \rangle_a^{\mathbb{E}}, \mathbb{E}_a \mathbb{E}_b p \Rightarrow \mathbb{E}_a p} \mathsf{R}_{\mathsf{E}}}{\mathbb{E}_a \mathbb{E}_b p \Rightarrow \mathbb{E}_a p} \mathsf{L}_{\mathsf{E}}$$

$$\text{init} \cfrac{}{\ldots \mid \langle p \rangle_b^{\mathbb{E}}, p, \mathbb{E}_b p \Rightarrow p} \mathsf{T}_{\mathsf{E}} \qquad \cfrac{}{\ldots \mid \langle p \rangle_b^{\mathbb{E}}, \mathbb{E}_b p \Rightarrow p} \mathsf{L}_{\mathsf{E}} \qquad \cfrac{}{\ldots \mid \mathbb{E}_b p \Rightarrow p}$$

Fig. 4. Failed proof in $\mathbf{HS_{ELG}}$.

$\mathsf{RE}_{\mathbb{E}}$	$\cfrac{A \leftrightarrow B}{\mathbb{E}_g A \leftrightarrow \mathbb{E}_g B}$	$\mathsf{RE}_{\mathbb{C}}$	$\cfrac{A \leftrightarrow B}{\mathbb{C}_g A \leftrightarrow \mathbb{C}_g B}$
$\mathsf{C}_{\mathbb{E}}$	$\mathbb{E}_g A \wedge \mathbb{E}_g B \to \mathbb{E}_g (A \wedge B)$	$\mathsf{Q}_{\mathbb{C}}$	$\neg \mathbb{C}_g \top$
$\mathsf{T}_{\mathbb{E}}$	$\mathbb{E}_g A \to A$	$\mathsf{P}_{\mathbb{C}}$	$\neg \mathbb{C}_g \bot$
$\mathsf{Int}_{\mathbb{EC}}^1$	$\mathbb{E}_g A \to \mathbb{C}_g A$	$\mathsf{F}_{\mathbb{C}}$	$\neg \mathbb{C}_\emptyset A$
$\mathsf{Int}_{\mathbb{EC}}^2$	$\mathbb{E}_{g_1} A \wedge \mathbb{E}_{g_2} B \to \mathbb{C}_{g_1 \cup g_2}(A \wedge B)$		

Fig. 5. Modal axioms and rules of Troquard's logic **COAL**.

obtain the bi-neighbourhood countermodel $\mathcal{M} = \langle \mathcal{W}, \mathcal{N}_i^{\mathbb{E}}, \mathcal{N}_i^{\mathbb{C}}, \mathcal{V} \rangle$, where $\mathcal{W} = \{1, 2, 3\}$; $\mathcal{V}(p) = \{1, 2\}$; $\mathcal{N}_a^{\mathbb{E}}(1) = \mathcal{N}_a^{\mathbb{C}}(1) = \{(\{1\}, \{2\})\}$ – since $\mathcal{N}_a^{\mathbb{E}}(1) = \mathcal{N}_a^{\mathbb{C}}(1) = \{(\mathbb{E}_b p^+, \mathbb{E}_b p^-)\}$, $\mathbb{E}_b p^+ = \{1\}$, and $\mathbb{E}_b p^- = \{2\}$ –; $\mathcal{N}_b^{\mathbb{E}}(1) = \mathcal{N}_b^{\mathbb{C}}(1) = \{(\{1, 2\}, \{3\})\}$ – since $\mathcal{N}_b^{\mathbb{E}}(1) = \mathcal{N}_b^{\mathbb{C}}(1) = \{(p^+, p^-)\}$, $p^+ = \{1, 2\}$, and $p^- = \{3\}$ –; $\mathcal{N}_i^{\mathbb{E}}(n) = \mathcal{N}_i^{\mathbb{C}}(n) = \emptyset$ for $i = a, b$ and $n = 2, 3$.

<u>Neighbourhood countermodel:</u> By applying the transformation in Prop. 2.3 we obtain the neighbourhood countermodel $\mathcal{M} = \langle \mathcal{W}, \mathcal{N}_i^{\mathbb{E}}, \mathcal{N}_i^{\mathbb{C}}, \mathcal{V} \rangle$, where $\mathcal{W} = \{1, 2, 3\}$; $\mathcal{V}(p) = \{1, 2\}$; $\mathcal{N}_a^{\mathbb{E}}(1) = \mathcal{N}_a^{\mathbb{C}}(1) = \{\{1\}, \{1, 3\}\}$; $\mathcal{N}_b^{\mathbb{E}}(1) = \mathcal{N}_b^{\mathbb{C}}(1) = \{\{1, 2\}\}$; and $\mathcal{N}_i^{\mathbb{E}}(n) = \mathcal{N}_i^{\mathbb{C}}(n) = \emptyset$ for $i = a, b$ and $n = 2, 3$.

5 Extension to Troquard's coalition logic

A coalition version of Elgesem's logic is proposed by Troquard [11]. In Troquard's logic, called **COAL**, single agents are replaced by *groups* of agents. The aim is to represent what agents do and can do when acting in coalitions. The logic **COAL** is defined by extending classical propositional logic with the modal axioms and rules in Fig. 5.

Apart from $\mathsf{F}_{\mathbb{C}}$ and $\mathsf{Int}_{\mathbb{EC}}^2$, the axioms and rules of **COAL** are just the coalition versions of the corresponding ones in **ELG**, with agents i replaced by groups g. The peculiar aspects of group agency are represented in **COAL** by the axioms $\mathsf{F}_{\mathbb{C}}$ and $\mathsf{Int}_{\mathbb{EC}}^2$. In particular, the axiom $\mathsf{F}_{\mathbb{C}}$ expresses that the empty group cannot realise anything, whereas the axiom $\mathsf{Int}_{\mathbb{EC}}^2$ says that if a group realises A and another group realises B, then by joining their forces they *could* realise both A and B. Observe that the axiom $\mathsf{Int}_{\mathbb{EC}}^1$ is derivable from $\text{Int}^2_{\mathbb{EC}}$. Nevertheless we keep it in the axiomatisation, as in [11], to keep the correspondence with the calculus where a specific rule for $\text{Int}^1_{\mathbb{EC}}$ is needed to ensure the admissibility of contraction. As for **ELG**, we can define bi-neighbourhood models for **COAL**.

Definition 5.1 A bi-neighbourhood model for **COAL** is a tuple $\mathcal{M} = \langle \mathcal{W}, \mathcal{N}^{\mathbb{E}}_g, \mathcal{N}^{\mathbb{C}}_g, \mathcal{V} \rangle$, where in particular for every group of agents g, $\mathcal{N}^{\mathbb{E}}_g$ and $\mathcal{N}^{\mathbb{C}}_g$ are two bi-neighbourhood functions satisfying the conditions ($C_{\mathbb{E}}$), ($T_{\mathbb{E}}$), ($Q_{\mathbb{C}}$), and ($P_{\mathbb{C}}$) of Def. 2.1 (but with $\mathcal{N}^{\mathbb{E}}$ and $\mathcal{N}^{\mathbb{C}}$ indexed by g instead of i), and also the following additional conditions:

($F_{\mathbb{C}}$) $\quad \mathcal{N}^{\mathbb{C}}_{\emptyset}(w) = \emptyset$.
($\text{Int}^2_{\mathbb{EC}}$) $\quad$ If $(\alpha, \beta) \in \mathcal{N}^{\mathbb{E}}_{g_1}(w)$ and $(\gamma, \delta) \in \mathcal{N}^{\mathbb{E}}_{g_2}(w)$, then $(\alpha \cap \gamma, \beta \cup \delta) \in \mathcal{N}^{\mathbb{C}}_{g_1 \cup g_2}(w)$.

The forcing relation $\Vdash$ is defined as in Def. 2.1, in particular:

$\mathcal{M}, w \Vdash \mathbb{E}_g A$ iff there is $(\alpha, \beta) \in \mathcal{N}^{\mathbb{E}}_g(w)$ s.t.
for all $v \in \alpha$, $\mathcal{M}, v \Vdash A$, and for all $u \in \beta$, $\mathcal{M}, u \not\Vdash A$.
$\mathcal{M}, w \Vdash \mathbb{C}_g A$ iff there is $(\alpha, \beta) \in \mathcal{N}^{\mathbb{C}}_g(w)$ s.t.
for all $v \in \alpha$, $\mathcal{M}, v \Vdash A$, and for all $u \in \beta$, $\mathcal{M}, u \not\Vdash A$.

Similarly to logic **ELG** we can prove the following completeness theorem.

Theorem 5.2 *A is derivable in **COAL** if and only if it is valid in all bi-neighbourhood models for **COAL**.*

Moreover, by a transformation analogous to the one in Prop. 2.3 we can convert the bi-neighbourhood models for **COAL** into equivalent neighbourhood models for it, as they are defined in [11]: it suffices to assign to the $\mathbb{E}_g$-neighbourhood (resp. the $\mathbb{C}_g$-neighbourhood) of each world w, the subsets γ such that $\alpha \subseteq \gamma \subseteq \mathcal{W} \setminus \beta$ and $(\alpha, \beta) \in \mathcal{N}^{\mathbb{E}}_g(w)$ (resp. $(\alpha, \beta) \in \mathcal{N}^{\mathbb{C}}_g(w)$).

The hypersequent calculus HS_{COAL} is defined by the propositional rules in Fig. 2 and the modal rules in Fig. 6. As before, each axiom has a corresponding rule in the calculus. An example of derivation is the following.

$$\cfrac{\cfrac{\cfrac{\cfrac{..., \langle A, B \rangle^{\mathbb{C}}_{g_1 \cup g_2} \Rightarrow \mathbb{C}_{g_1 \cup g_2}(A \wedge B) \mid A, B \Rightarrow A \wedge B \quad ... \mid A \wedge B \Rightarrow A \quad ... \mid A \wedge B \Rightarrow B}{\mathbb{E}_{g_1} A \wedge \mathbb{E}_{g_2} B, \mathbb{E}_{g_1} A, \mathbb{E}_{g_2} B, \langle A \rangle^{\mathbb{E}}_{g_1}, \langle B \rangle^{\mathbb{E}}_{g_2}, \langle A, B \rangle^{\mathbb{C}}_{g_1 \cup g_2} \Rightarrow \mathbb{C}_{g_1 \cup g_2}(A \wedge B)} R_{\mathbb{C}}}{\mathbb{E}_{g_1} A \wedge \mathbb{E}_{g_2} B, \mathbb{E}_{g_1} A, \mathbb{E}_{g_2} B, \langle A \rangle^{\mathbb{E}}_{g_1}, \langle B \rangle^{\mathbb{E}}_{g_2} \Rightarrow \mathbb{C}_{g_1 \cup g_2}(A \wedge B)} \text{Int}^2_{\mathbb{EC}}}{\mathbb{E}_{g_1} A \wedge \mathbb{E}_{g_2} B, \mathbb{E}_{g_1} A, \mathbb{E}_{g_2} B \Rightarrow \mathbb{C}_{g_1 \cup g_2}(A \wedge B)} L_{\mathbb{E}} \times 2}{\mathbb{E}_{g_1} A \wedge \mathbb{E}_{g_2} B \Rightarrow \mathbb{C}_{g_1 \cup g_2}(A \wedge B)} L\wedge$$

By extending the proofs for HS_{ELG} we can obtain the following theorem.

Theorem 5.3 *All structural rules including* cut *are admissible in* HS_{COAL}. *Moreover,* HS_{COAL} *is axiomatically complete with respect to* **COAL**, *that is, if A is derivable in* **COAL**, *then* $\Rightarrow A$ *is derivable in* HS_{COAL}.

Termination of proof search can be obtained by considering a proof search strategy analogous to the one in HS_{ELG}. We only need to consider the following two additional saturation conditions: ($F_{\mathbb{C}}$) $\langle \Sigma \rangle^{\mathbb{C}}_{\emptyset} \notin \Gamma_n$, and ($\text{Int}^2_{\mathbb{EC}}$) if

$$L_\mathbb{E} \frac{G \mid \Gamma, \mathbb{E}_g A, \langle A \rangle_g^\mathbb{E} \Rightarrow \Delta}{G \mid \Gamma, \mathbb{E}_g A \Rightarrow \Delta} \qquad L_\mathbb{C} \frac{G \mid \Gamma, \mathbb{C}_g A, \langle A \rangle_g^\mathbb{C} \Rightarrow \Delta}{G \mid \Gamma, \mathbb{C}_g A \Rightarrow \Delta}$$

$$\mathsf{Int}_{\mathbb{EC}}^1 \frac{G \mid \Gamma, \langle \Sigma \rangle_g^\mathbb{E}, \langle \Sigma \rangle_g^\mathbb{C} \Rightarrow \Delta}{G \mid \Gamma, \langle \Sigma \rangle_g^\mathbb{E} \Rightarrow \Delta}$$

$$R_\mathbb{E} \frac{G \mid \Gamma, \langle \Sigma \rangle_g^\mathbb{E} \Rightarrow \mathbb{E}_g A, \Delta \mid \Sigma \Rightarrow A \quad \{ G \mid \Gamma, \langle \Sigma \rangle_g^\mathbb{E} \Rightarrow \mathbb{E}_g A, \Delta \mid A \Rightarrow B \}_{B \in \Sigma}}{G \mid \Gamma, \langle \Sigma \rangle_g^\mathbb{E} \Rightarrow \mathbb{E}_g A, \Delta}$$

$$R_\mathbb{C} \frac{G \mid \Gamma, \langle \Sigma \rangle_g^\mathbb{C} \Rightarrow \mathbb{C}_g A, \Delta \mid \Sigma \Rightarrow A \quad \{ G \mid \Gamma, \langle \Sigma \rangle_g^\mathbb{C} \Rightarrow \mathbb{C}_g A, \Delta \mid A \Rightarrow B \}_{B \in \Sigma}}{G \mid \Gamma, \langle \Sigma \rangle_g^\mathbb{C} \Rightarrow \mathbb{C}_g A, \Delta}$$

$$C_\mathbb{E} \frac{G \mid \Gamma, \langle \Sigma \rangle_g^\mathbb{E}, \langle \Pi \rangle_g^\mathbb{E}, \langle \Sigma, \Pi \rangle_g^\mathbb{E} \Rightarrow \Delta}{G \mid \Gamma, \langle \Sigma \rangle_g^\mathbb{E}, \langle \Pi \rangle_g^\mathbb{E} \Rightarrow \Delta} \qquad T_\mathbb{E} \frac{G \mid \Gamma, \langle \Sigma \rangle_g^\mathbb{E}, \Sigma \Rightarrow \Delta}{G \mid \Gamma, \langle \Sigma \rangle_g^\mathbb{E} \Rightarrow \Delta}$$

$$Q_\mathbb{C} \frac{\{ G \mid \Gamma, \langle \Sigma \rangle_g^\mathbb{C} \Rightarrow \Delta \mid \Rightarrow B \}_{B \in \Sigma}}{G \mid \Gamma, \langle \Sigma \rangle_g^\mathbb{C} \Rightarrow \Delta} \qquad P_\mathbb{C} \frac{G \mid \Gamma, \langle \Sigma \rangle_g^\mathbb{C} \Rightarrow \Delta \mid \Sigma \Rightarrow}{G \mid \Gamma, \langle \Sigma \rangle_g^\mathbb{C} \Rightarrow \Delta}$$

$$F_\mathbb{C} \frac{}{G \mid \Gamma, \langle \Sigma \rangle_\emptyset^\mathbb{C} \Rightarrow \Delta} \qquad \mathsf{Int}_{\mathbb{EC}}^2 \frac{G \mid \Gamma, \langle \Sigma \rangle_{g_1}^\mathbb{E}, \langle \Pi \rangle_{g_2}^\mathbb{E}, \langle \Sigma, \Pi \rangle_{g_1 \cup g_2}^\mathbb{C} \Rightarrow \Delta}{G \mid \Gamma, \langle \Sigma \rangle_{g_1}^\mathbb{E}, \langle \Pi \rangle_{g_2}^\mathbb{E} \Rightarrow \Delta}$$

Fig. 6. Modal rules of **HS**$_{\mathbf{COAL}}$.

$\langle \Sigma \rangle_{g_1}^\mathbb{E}, \langle \Pi \rangle_{g_2}^\mathbb{E} \in \Gamma_n$, then $\langle \Omega \rangle_{g_1 \cup g_2}^\mathbb{C} \in \Gamma_n$ such that $\mathsf{set}(\Omega) = \mathsf{set}(\Sigma, \Pi)$. As for **HS**$_{\mathbf{ELG}}$ we can prove that proof search always terminates, whence we obtain a decision procedure for the logic **COAL**. Again, proof search is not optimal since derivation can have an exponential size whereas the logic is in PSPACE, as proved by Troquard [11].

We can also prove that the calculus is semantically complete. As before, the proof consists in showing how to extract a countermodel of a non-derivable hypersequent using the information provided by the failed proof.

Theorem 5.4 *If H is valid in all bi-neighbourhood models for* **COAL**, *then it is derivable in* **HS**$_{\mathbf{COAL}}$.

Proof. Given a saturated hypersequent H we define a model $\mathcal{M}$ as in Def. 4.3 (replacing agents i with groups g). We can prove that formulas and blocks in the left-hand side of the components are satisfied in the corresponding worlds, and that formulas in the right-hand side are falsified, whence $\mathcal{M}$ is a countermodel of H. Moreover, we can prove that $\mathcal{M}$ is a bi-neighbourhood model for **COAL**. The proofs are as in Lemma 4.4. We only consider the following two conditions.

($\mathsf{Int}_{\mathbb{EC}}^2$) Assume that $(\alpha, \beta) \in \mathcal{N}_{g_1}^\mathbb{E}(n)$ and $(\gamma, \delta) \in \mathcal{N}_{g_2}^\mathbb{E}(n)$. If $(\alpha, \beta) \neq (\gamma, \delta)$ or $g_1 \neq g_2$, then there are $\langle \Sigma \rangle_{g_1}^\mathbb{E}, \langle \Pi \rangle_{g_2}^\mathbb{E} \in \Gamma_n$ such that $\Sigma^+ = \alpha$, $\Sigma^- = \beta$, $\Pi^+ = \gamma$ and $\Pi^- = \delta$. By saturation or rule $\mathsf{Int}_{\mathbb{EC}}^2$, there is $\langle \Omega \rangle_{g_1 \cup g_2}^\mathbb{C} \in \Gamma_n$ such that $\mathsf{set}(\Omega) = \mathsf{set}(\Sigma, \Pi)$, thus $(\Omega^+, \Omega^-) \in \mathcal{N}_{g_1 \cup g_2}^\mathbb{C}(n)$, where, as shown in the proof of Lemma 4.4 case ($C_\mathbb{E}$), $\Omega^+ = \alpha \cap \gamma$ and $\Omega^- = \beta \cup \delta$. If instead $(\alpha, \beta) = (\gamma, \delta)$ and $g_1 = g_2$, then there is $\langle \Sigma \rangle_{g_1}^\mathbb{E} \in \Gamma_n$ such that $\Sigma^+ = \alpha$

and $\Sigma^- = \beta$. Then by saturation of rule $\mathsf{Int}_{\mathsf{EC}}$ there is $\langle\Omega\rangle_i^{\mathsf{C}} \in \Gamma_n$ such that $\mathsf{set}(\Sigma) = \mathsf{set}(\Omega)$. Then $(\Omega^+, \Omega^-) = (\Sigma^+, \Sigma^-) = (\alpha, \beta) \in \mathcal{N}_i^{\mathsf{C}}(n)$.

($\mathsf{F_C}$) By saturation of $\mathsf{F_C}$ there is no block $\langle\Sigma\rangle_\emptyset^{\mathsf{C}} \in \Gamma_n$, then $\mathcal{N}_\emptyset^{\mathsf{C}}(n) = \emptyset$. □

We conclude this section by showing that *coalition monotonicity* is not valid in **COAL**. We present the countermodel directly extracted from a failed proof.

Example 5.5 (No coalition monotonicity) We show that the formula $\mathbb{E}_{\{a\}}p \to \mathbb{E}_{\{a,b\}}p$ is not valid in **COAL**. A failed proof is as follows:

$$\dfrac{\dfrac{\dfrac{\dfrac{\dfrac{\overset{\text{saturated}}{\langle p\rangle_{\{a\}}^{\mathbb{E}}, \langle p\rangle_{\{a\}}^{\mathsf{C}}, p, \mathbb{E}_{\{a\}}p \Rightarrow \mathbb{E}_{\{a,b\}}p \mid \Rightarrow p}}{\langle p\rangle_{\{a\}}^{\mathbb{E}}, \langle p\rangle_{\{a\}}^{\mathsf{C}}, p, \mathbb{E}_{\{a\}}p \Rightarrow \mathbb{E}_{\{a,b\}}p} Q_{\mathsf{C}}}{\langle p\rangle_{\{a\}}^{\mathbb{E}}, p, \mathbb{E}_{\{a\}}p \Rightarrow \mathbb{E}_{\{a,b\}}p} \mathsf{Int}_{\mathsf{EC}}}{\langle p\rangle_{\{a\}}^{\mathbb{E}}, \mathbb{E}_{\{a\}}p \Rightarrow \mathbb{E}_{\{a,b\}}p} \mathsf{T_E}}{\mathbb{E}_{\{a\}}p \Rightarrow \mathbb{E}_{\{a,b\}}p} \mathsf{L_E}$$

Let $1 \mapsto \langle p\rangle_a^{\mathbb{E}}, \langle p\rangle_a^{\mathsf{C}}, p, \mathbb{E}_{\{a\}}p \Rightarrow \mathbb{E}_{\{a,b\}}p$, and $2 \mapsto \Rightarrow p$. We obtain the following countermodels:

Bi-neighbourhood countermodel: $\mathcal{M} = \langle \mathcal{W}, \mathcal{N}_g^{\mathbb{E}}, \mathcal{N}_g^{\mathsf{C}}, \mathcal{V}\rangle$, where $\mathcal{W} = \{1, 2\}$; $\mathcal{V}(p) = \{1\}$; $\mathcal{N}_{\{a\}}^{\mathbb{E}}(1) = \mathcal{N}_{\{a\}}^{\mathsf{C}}(1) = \{(p^+, p^-)\} = \{(\{1\}, \{2\})\}$; and $\mathcal{N}_g^{\mathbb{E}}(k) = \mathcal{N}_g^{\mathsf{C}}(k) = \emptyset$ for $g \neq \{a\}$ or $k \neq 1$.

Neighbourhood countermodel: $\mathcal{M} = \langle \mathcal{W}, \mathcal{N}_g^{\mathbb{E}}, \mathcal{N}_g^{\mathsf{C}}, \mathcal{V}\rangle$, where $\mathcal{W} = \{1, 2\}$; $\mathcal{V}(p) = \{1\}$; $\mathcal{N}_{\{a\}}^{\mathbb{E}}(1) = \mathcal{N}_{\{a\}}^{\mathsf{C}}(1) = \{\{1\}\}$; and $\mathcal{N}_g^{\mathbb{E}}(k) = \mathcal{N}_g^{\mathsf{C}}(k) = \emptyset$ for $g \neq \{a\}$ or $k \neq 1$.

6 Conclusion

We have presented hypersequent calculi for Elgesem's logic of agency and ability and its coalition extension proposed by Troquard. The calculi have good structural properties, including the syntactical admissibility of cut. Furthermore, we have defined a terminating proof search strategy which ensures that a derivation or a countermodel will be found for every formula. In particular, in case of a failed proof it is possible to directly extract a countermodel of the non-valid formula in the bi-neighbourhood semantics, whence by an easy transformation in the standard neighbourhood semantics. All in all, the calculi provide constructive decision procedures for the two logics.

Troquard has proposed several extensions of his coalition logic with further principles for group agency, such as delegation and strict-joint agency, the latter stating that if a group brings about that A, then any strict subgroup of it cannot bring about that A. We plan to extend our calculi to cover also these extensions, and possibly others. Moreover, our calculi are well-suited for automatisation. We plan to implement them in order to realise the first theorem provers for the logics of agency and ability.

Appendix

Proof of Theorem 3.7. Recall that, for an application of cut, the *cut formula* is the formula which is deleted by that application, while the *cut height* is the sum of the heights of the derivations of the premises of cut. We prove that:
(A) $\forall c.Cut(c,0)$. **(B)** $\forall h.Cut(0,h)$. **(C)** $\forall c.(\forall h.Cut(c,h) \to Sub(c))$. **(D)** $\forall c. \forall h.((\forall c' < c.(Sub(c') \wedge \forall h'.Cut(c',h')) \wedge \forall h'' < h.Cut(c,h'')) \to Cut(c,h))$.

(A) and **(B)** are trivial. **(C)** Assume $\forall h Cut(c,h)$. The proof is by induction on the height m of the derivation of $G \mid \overrightarrow{\langle A^n, \Pi \rangle_i^{\mathbb{E}}}, \overrightarrow{\langle A^m, \Omega \rangle_j^{\mathbb{C}}}, \Gamma \Rightarrow \Delta$. We only consider the case where $m > 0$ and at least one block among $\overrightarrow{\langle A^n, \Pi \rangle_i^{\mathbb{E}}}, \overrightarrow{\langle A^m, \Omega \rangle_j^{\mathbb{C}}}$ is principal in the last rule application. We consider as an example the case where the last rule applied is $\mathsf{Int}_{\mathbb{EC}}$:

$$\dfrac{G \mid \langle A^{n_k}, \Pi_k \rangle_i^{\mathbb{E}}, \langle A^{n_k}, \Pi_k \rangle_i^{\mathbb{C}}, \Gamma \Rightarrow \Delta}{G \mid \langle A^{n_k}, \Pi_k \rangle_i^{\mathbb{E}}, \Gamma \Rightarrow \Delta} \; \mathsf{Int}_{\mathbb{EC}}$$

By applying the inductive hypothesis to the premiss we obtain $G \mid \langle \Sigma^{n_k}, \Pi_k \rangle_i^{\mathbb{E}}, \langle \Sigma^{n_k}, \Pi_k \rangle_i^{\mathbb{C}}, \Gamma \Rightarrow \Delta$, then by $\mathsf{Int}_{\mathbb{EC}}$ we derive $G \mid \langle \Sigma^{n_k}, \Pi_k \rangle_i^{\mathbb{C}}, \Gamma \Rightarrow \Delta$.

(D) Assume $\forall c' < c.(Sub(c') \wedge \forall h'.Cut(c',h'))$ and $\forall h'' < h.Cut(c,h'')$. We show that all applications of cut of height h on a cut formula of weight c can be replaced by different applications of cut, either of smaller height or on a cut formula of smaller weight. We only consider the cases where $h, c > 0$ and the cut formula is $\mathbb{E}_i B$, principal in the derivation of both premises of cut:

$$\mathsf{R}_\mathbb{E} \dfrac{G \mid \langle \Sigma \rangle_i^{\mathbb{E}}, \Gamma \Rightarrow \Delta, \mathbb{E}_i B \mid \Sigma \Rightarrow B \quad \{G \mid \langle \Sigma \rangle_i^{\mathbb{E}}, \Gamma \Rightarrow \Delta, \mathbb{E}_i B \mid B \Rightarrow C\}_{C \in \Sigma}}{G \mid \langle \Sigma \rangle_i^{\mathbb{E}}, \Gamma \Rightarrow \Delta, \mathbb{E}_i B} \quad \mathsf{L}_\mathbb{E} \dfrac{G \mid \langle B \rangle, \mathbb{E}_i B, \langle \Sigma \rangle_i^{\mathbb{E}}, \Gamma \Rightarrow \Delta}{G \mid \mathbb{E}_i B, \langle \Sigma \rangle_i^{\mathbb{E}}, \Gamma \Rightarrow \Delta}$$
$$\mathsf{cut}\ \dfrac{}{G \mid \langle \Sigma \rangle_i^{\mathbb{E}}, \Gamma \Rightarrow \Delta}$$

The derivation is converted as follows, with several applications of cut of smaller height and an admissible application of sub.

$$\textcircled{1}\; \dfrac{G \mid \langle \Sigma \rangle_i^{\mathbb{E}}, \Gamma \Rightarrow \Delta, \mathbb{E}_i B \mid \Sigma \Rightarrow B \quad \dfrac{G \mid \mathbb{E}_i B, \langle \Sigma \rangle_i^{\mathbb{E}}, \Gamma \Rightarrow \Delta}{G \mid \mathbb{E}_i B, \langle \Sigma \rangle_i^{\mathbb{E}}, \Gamma \Rightarrow \Delta \mid \Sigma \Rightarrow B} \; \mathsf{Ewk}}{G \mid \langle \Sigma \rangle_i^{\mathbb{E}}, \Gamma \Rightarrow \Delta \mid \Sigma \Rightarrow B} \; \mathsf{cut}$$

$$\mathsf{Lwk} \; \textcircled{2}\; \dfrac{\dfrac{\dfrac{G \mid \langle \Sigma \rangle_i^{\mathbb{E}}, \Gamma \Rightarrow \Delta, \mathbb{E}_i B}{G \mid \langle B \rangle_i^{\mathbb{E}}, \langle \Sigma \rangle_i^{\mathbb{E}}, \Gamma \Rightarrow \Delta, \mathbb{E}_i B} \quad G \mid \langle B \rangle, \mathbb{E}_i B, \langle \Sigma \rangle_i^{\mathbb{E}}, \Gamma \Rightarrow \Delta}{G \mid \langle B \rangle_i^{\mathbb{E}}, \langle \Sigma \rangle_i^{\mathbb{E}}, \Gamma \Rightarrow \Delta} \; \mathsf{cut}}{G \mid \langle \Sigma \rangle_i^{\mathbb{E}}, \Gamma \Rightarrow \Delta \mid \langle B \rangle_i^{\mathbb{E}}, \langle \Sigma \rangle_i^{\mathbb{E}}, \Gamma \Rightarrow \Delta} \; \mathsf{Ewk}$$

$$\dfrac{\textcircled{1} \quad \left\{ \dfrac{G \mid \langle \Sigma \rangle_i^{\mathbb{E}}, \Gamma \Rightarrow \Delta, \mathbb{E}_i B \mid B \Rightarrow C \quad \dfrac{G \mid \mathbb{E}_i B, \langle \Sigma \rangle_i^{\mathbb{E}}, \Gamma \Rightarrow \Delta}{G \mid \mathbb{E}_i B, \langle \Sigma \rangle_i^{\mathbb{E}}, \Gamma \Rightarrow \Delta \mid B \Rightarrow C} \; \mathsf{Ewk}}{G \mid \langle \Sigma \rangle_i^{\mathbb{E}}, \Gamma \Rightarrow \Delta \mid B \Rightarrow C} \; \mathsf{cut} \right\}_{C \in \Sigma} \quad \textcircled{2}}{\dfrac{\dfrac{G \mid \langle \Sigma \rangle_i^{\mathbb{E}}, \Gamma \Rightarrow \Delta \mid \langle \Sigma \rangle_i^{\mathbb{E}}, \langle \Sigma \rangle_i^{\mathbb{E}}, \Gamma \Rightarrow \Delta}{G \mid \langle \Sigma \rangle_i^{\mathbb{E}}, \Gamma \Rightarrow \Delta \mid \langle \Sigma \rangle_i^{\mathbb{E}}, \Gamma \Rightarrow \Delta} \; \mathsf{Lctr}}{G \mid \langle \Sigma \rangle_i^{\mathbb{E}}, \Gamma \Rightarrow \Delta} \; \mathsf{Ectr}} \; \mathsf{sub}$$

□

References

[1] Avron, A., "The Method of Hypersequents in the Proof Theory of Propositional Non-Classical Logics," Clarendon Press, USA, 1996 p. 1–32.
[2] Belnap, N. D., M. Perloff and M. Xu, "Facing the future: agents and choices in our indeterminist world," Oxford University Press on Demand, 2001.
[3] Dalmonte, T., B. Lellmann, N. Olivetti and E. Pimentel, *Countermodel construction via optimal hypersequent calculi for non-normal modal logics*, in: S. N. Artëmov and A. Nerode, editors, *Logical Foundations of Computer Science - International Symposium, LFCS 2020*, Lecture Notes in Computer Science **11972** (2020), pp. 27–46.
[4] Dalmonte, T., N. Olivetti and S. Negri, *Non-normal modal logics: Bi-neighbourhood semantics and its labelled calculi*, in: *Advances in Modal Logic 12, proceedings of the 12th conference on "Advances in Modal Logic"* (2018), pp. 159–178.
[5] Elgesem, D., *The modal logic of agency*, Nordic Journal of Philosophical Logic **2** (1997), pp. 1–46.
[6] Giunchiglia, E., A. Tacchella and F. Giunchiglia, *Sat-based decision procedures for classical modal logics*, Journal of Automated Reasoning **28** (2002), pp. 143–171.
[7] Governatori, G. and A. Rotolo, *On the axiomatisation of elgesem's logic of agency and ability*, J. Philosophical Logic **34** (2005), pp. 403–431.
[8] Horty, J. F., "Agency and deontic logic," Oxford University Press, 2001.
[9] Lellmann, B., "Sequent Calculi with Context Restrictions and Applications to Conditional Logic," Ph.D. thesis, Imperial College London (2013).
[10] Schröder, L. and D. Pattinson, *Shallow models for non-iterative modal logics*, in: A. Dengel, K. Berns, T. M. Breuel, F. Bomarius and T. Roth-Berghofer, editors, *KI 2008: Advances in Artificial Intelligence, 31st Annual German Conference on AI, KI 2008*, Lecture Notes in Computer Science **5243** (2008), pp. 324–331.
[11] Troquard, N., *Reasoning about coalitional agency and ability in the logics of "bringing-it-about"*, Autonomous Agents and Multi-Agent Systems **28** (2014), pp. 381–407.
URL https://doi.org/10.1007/s10458-013-9229-x
[12] Vardi, M. Y., *On the complexity of epistemic reasoning*, in: *Proceedings of the Fourth Annual Symposium on Logic in Computer Science (LICS '89), Pacific Grove, California, USA, June 5-8, 1989* (1989), pp. 243–252.

The Original Position: A Logical Analysis

Thijs De Coninck [1]

Ghent University

Frederik Van De Putte [2]

University of Bayreuth
Erasmus University of Rotterdam

Abstract

Rawls famously claimed that choices based on the Difference Principle coincide with the choices of any rational individual in the Original Position. In this paper, we develop a logic in which we can express and prove Rawls' thesis in its object language. Starting from a standard semantics of choice under uncertainty, we enrich our models in order to represent uncertainty about one's position. We then introduce a sound and strongly complete logic that allows us to speak about agents' positions and their derived utilities, and that can express changes in the uncertainty about those positions using dynamic operators. Finally, we show how this logic allows us to define various types of obligation based on a Rawlsian notion of procedural fairness.

Keywords: The Original Position, Choice under uncertainty, Deontic logic, Fairness.

1 Introduction

In his *Theory of Justice*, John Rawls puts forward principles of justice that he argues should be used to determine the basic structure of society [16]. What made Rawls innovative, however, were not the principles themselves, but the way in which he argued for them [15]. Famously, he makes use of a methodological device known as the *Original Position*, which he describes as

> [...] a purely hypothetical situation characterized so as to lead to a certain conception of justice. Among the essential features of this situation is that no one knows his place in society, his class position or social status, nor does any one know his fortune in the distribution of natural assets and abilities, his intelligence, strength, and the like. [16, p. 11]

---

[1] Thijs De Coninck is a PhD fellow of the Research Foundation – Flanders supported by a fundamental research grant (1167619N).

[2] Frederik Van De Putte's research was funded by a grant from the Research Foundation – Flanders (FWO-Vlaanderen), by a Marie Skłodowska-Curie Fellowship (grant agreement ID: 795329), and by a grant from the Dutch Research Council (NWO), no. VI.Vidi.191.105.

So Rawls' basic idea is to conceive of a situation in which a person is deprived of morally irrelevant knowledge and to ask: what would such a person choose?

One of the principles that would characterize the resulting choices, according to Rawls, is the *Difference Principle*, which states that "social and economic inequalities are to be arranged so that they are to the greatest benefit of the least advantaged." [16, p. 266]. Rawls claims that what one ought to choose according to the Difference Principle coincides with what a rational individual would choose, if it were fully uncertain about the position it occupies in society. Henceforth we refer to this claim as *Rawls' thesis*.[3]

In this paper, we provide a logical analysis of Rawls' thesis. We first set out general models of choice under uncertainty and define notions of individually rational choices and fair choices (Section 2). Next, we refine these models in such a way that we can verify Rawls' thesis (Section 3). In Section 4 we introduce a logic that can express key features of those models and has Rawls' thesis as a validity. Finally, we show how our logic can handle various deontic operators based on Rawls' notion of procedural fairness (Section 5).

Existing Formal Models Rawls' publication of *A Theory of Justice* has spawned research both of informal and formal nature. Most of the formal literature is focused on the Rawls/Harsanyi dispute over how exactly to characterize the Original Position and how agents would choose, once placed in such a situation. John Harsanyi [9] conceives of the situation as one of choice under risk where, for any given outcome, the agent can reason based on some probability estimate of how likely it is that that outcome occurs. With this in place, Harsanyi argues that a rational individual would choose according to the principles of expected utility theory. In contrast, Rawls thinks of the situation as one of choice under uncertainty, where no such probabilities are given [16, p. 134].

Given a great deal of uncertainty and the risks associated with choosing suboptimal options, Rawls argues that individuals would choose according to the *Maximin rule* (cf. Section 2.2). In contrast, most of the formal work on the Original Position follows Harsanyi's characterization by relying on a uniform probability distribution that assigns a chance of $\frac{1}{n}$ to an individual ending up in one of the n possible positions (see e.g. [6,8,13,17]).

In the present paper, we bracket the Rawls/Harsanyi dispute and stay as close as possible to Rawls' conception of the Original Position as a situation of non-strategic choice under uncertainty (cf. [7]).

---

[3] Rawls states: "To say that a certain conception of justice would be chosen in the original position is equivalent to saying that rational deliberation satisfying certain conditions and restrictions would reach a certain conclusion. If necessary, the argument to this result could be set out more formally." [16, p. 120]. What we call Rawls' thesis is thus the more concrete version of this claim where the Difference Principle is put forward as the conception of justice in question.

2 Choice Under Uncertainty

In this section we present general models of choice under uncertainty and introduce a formal language that can express some of Rawls's fundamental concepts.

2.1 Models of Choice Under Uncertainty

Our models are inspired by the tradition of STIT logics, i.e. logics that feature modal operators of the type "agent i sees to it that", which are interpreted in terms of the states of affairs that are guaranteed by the (past or current) choice(s) of i. The classic exposition of STIT logic is Belnap et al. [4]. In [10], Horty shows that this framework can be combined with utilitarian ideas, in order to interpret various deontic notions such as individual and group oughts. Kooi and Tamminga [14,18] use STIT models without a temporal component, but including agent-relative utilities. Here, we further simplify the models of [14] by working with a single set of choices and a finite[4] set of utility values.

Fix a finite set Agt of agents, a finite set $N = \{1, 2, \ldots\} \subset \mathbb{N}$ of utilities, and a countable set $Q = \{q, q', \ldots\}$ of propositional variables. We use i, j and n, m as metavariables for agents and values respectively.

Definition 2.1 A *model of choice under uncertainty* is a tuple $\mathfrak{M} = \langle S, U, C, V \rangle$, where $S \neq \emptyset$ is a set of *states*, $U : S \times Agt \to N$ is a *utility function*, C is a partition of S into *choices*, and $V : Q \to \wp(S)$ is a *valuation function*.

Each state $s \in S$ can be seen as a possible outcome of the choice situation. The utility function U specifies, for each state s and agent i, the utility $U(s, i)$ that i receives at s. Note that choices are *sets* of states $X \in C$. This means that, as in the traditional STIT-based accounts, we identify choices with the set of states they leave open. Unlike in STIT, we do not attribute choices to (a) particular (group of) agents. The focus is rather on how choices *affect* agents, not on who is choosing or acting. Depending on the particular perspective we take, e.g. that of an individual or that of society at large, some of the choices will be better or worse than others. Correspondingly, one may interpret the choices as those of a social planner or policy-maker, even if that person is herself a member of Agt.

If $s \in X$, then s is a possible outcome of choosing X. We write $C(s)$ for the unique choice $X \in C$ such that $s \in X$. Figure 1 represents a simple model of choice under uncertainty for two agents i and j, with two choices $X = \{s_1, s_2\}$ and $Y = \{s_3, s_4\}$. Here, the couples (n, m) represent the utility function, where $n = U(s, i)$ and $m = U(s, j)$. For instance, at state s_1, agent i receives a utility of 3 whereas agent j receives a utility of 1.

2.2 Two Standards of Admissibility

Given a model of choice under uncertainty $\mathfrak{M}$, the utility function U induces agent-relative preferences over outcomes: i weakly prefers s over s' iff $U(s, i) \geq U(s', i)$. For example, in Figure 1 agent i weakly prefers s_1 over s_2

---
[4] The generalization to an infinite set of utility values is left for future work.

X		Y	
s_1	s_2	s_3	s_4
(3,1)	(2,4)	(2,2)	(1,4)

Fig. 1. A model of choice under uncertainty.

since $U(s_1, i) \geq U(s_2, i)$. However, since we are considering what choices a rational agent should make, we should specify how preferences over states induce preferences over choices. In other words, we need to specify a *lifting criterion*. Four such lifting criteria are given in Table 1, which is based on [19]. Each of these lifting criteria give us a weak preference relation over the set of choices in a model.

Where $l \in \{\forall\forall, \forall\exists, \exists\forall, \exists\exists\}$ the strict preference relation $\sqsupset_i^l$ is defined as: $X \sqsupset_i^l Y$ iff $X \sqsupseteq_i^l Y$ and $Y \not\sqsupseteq_i^l X$. In words, X being strictly preferred to Y means that X is preferred to Y while Y is not preferred to X. Following common practice, we assume that it is rational to choose X for an agent i iff there is no other choice Y such that i strictly prefers Y to X. We call such rational choices *admissible* for the agent in question, and treat rationality and admissibility as interchangeable notions.

Definition 2.2 Where $\mathfrak{M} = \langle S, U, C, V \rangle$ is a model of choice under uncertainty, $i \in Agt$, and $l \in \{\forall\forall, \forall\exists, \exists\forall, \exists\exists\}$: the set of i-admissible^l choices in $\mathfrak{M}$ is

$$Adm_i^l(\mathfrak{M}) =_{df} \{X \in C \mid \text{for no } Y \in C : Y \sqsupset_i^l X\}.$$

$l =$	Preference Relation		
$\forall\forall$	$X \sqsupseteq_i^{\forall\forall} Y$	$=_{df}$	$\forall s \in X, \forall s' \in Y : U(s,i) \geq U(s',i)$
$\forall\exists$	$X \sqsupseteq_i^{\forall\exists} Y$	$=_{df}$	$\forall s \in X, \exists s' \in Y : U(s,i) \geq U(s',i)$
$\exists\forall$	$X \sqsupseteq_i^{\exists\forall} Y$	$=_{df}$	$\exists s \in X, \forall s' \in Y : U(s,i) \geq U(s',i)$
$\exists\exists$	$X \sqsupseteq_i^{\exists\exists} Y$	$=_{df}$	$\exists s \in X, \exists s' \in Y : U(s,i) \geq U(s',i)$

Table 1
Lifting criteria. Here, X and Y are sets of states.

Maximin In what follows, we focus on the *Maximin* criterion, i.e. the lifting criterion denoted by $\forall\exists$. We hence take $Adm_i^{\forall\exists}$ as defining rational choice under uncertainty. We return to the other lifting criteria in Section 5. Until then, we omit the superscript l in notation.

The Maximin principle is usually considered typical for risk-averse agents. Rawls states that "the maximin rule is not, in general, a suitable guide for choices under uncertainty" while he does defend Maximin in situations "marked by certain special features" [16, pp. 133]. These features are:

- knowledge of likelihoods is impossible, or at best extremely insecure;
- the person choosing has a conception of the good such that he cares very little, if anything, for what he might gain above the minimum stipend that he can, in fact, be sure of by following the maximin rule;
- the rejected alternatives have outcomes that one can hardly accept.

Rawls concludes that because the Original Position has these three features, the Maximin criterion is the most appropriate one in this context.

The Difference Principle The principle of justice that we focus on in this paper is the Difference Principle. Informally, it states that we should maximize the gains of the least well-off. Rawls warns us that the Difference Principle should not be mistaken for the Maximin rule [16, p. 72]:

> The maximin criterion is generally understood as a rule for choice under great uncertainty, whereas the difference principle is a principle of justice. It is undesirable to use the same name for two things that are so distinct.

To define the Difference Principle in exact terms, we need some more notation. For any state s in a given model, let $U(s,*)$ denote the smallest $n \in N$ such that, for some $i \in Agt$, $U(s,i) = n$. Intuitively, $U(s,*)$ is the utility of the agent that is the least well-off at state s. One may say that, according to the Difference Principle, a state s is at least as good as a state s' if and only if $U(s,*) \geq U(s',*)$, i.e. whenever the least well-off at state s is at least as well-off as the least well-off at state s'.

Just as before, we need to lift this preference relation on states in order to obtain preferences over choices. In line with the preceding, we use the Maximin criterion.[5] This gives us the following definitions:

Definition 2.3 Where $\mathfrak{M} = \langle S, U, C, V \rangle$ is a model of choice under uncertainty and $X, Y \in C$: $X \sqsupseteq_*^{\forall\exists} Y$ iff $\forall s \in X, \exists s' \in Y$: $U(s,*) \geq U(s',*)$.

Definition 2.4 Where $\mathfrak{M} = \langle S, U, C, V \rangle$ is a model of choice under uncertainty, the set of *Difference admissible choices in* $\mathfrak{M}$ is

$$Adm_*(\mathfrak{M}) =_{df} \{X \in C \mid \text{For no } Y \in C : Y \sqsupset_*^{\forall\exists} X\}.$$

In our example from Figure 1, it can be easily verified that $X \sqsupseteq_i^{\forall\exists} Y$, $Y \sqsupseteq_j^{\forall\exists} X$, $X \sqsupseteq_*^{\forall\exists} Y$, and $Y \sqsupseteq_*^{\forall\exists} X$. Hence, $Adm_i(\mathfrak{M}) = \{X\}, Adm_j = \{Y\}$, and $Adm_* = \{X, Y\}$. In other words, both X and Y are difference admissible in this model, while only X is admissible for i and only Y is admissible for j.

In what follows, we will sometimes use "$*$" to denote "the least well-off" (at a given state in a given model). This convention allows us to present our results in a compact way.

---

[5] One can define alternative "fairness rankings", using the other lifting criteria from Table 1. We leave the study of such rankings for future work.

2.3 Expressing Admissibility in a Formal Language

Here, we introduce a formal language that allows us to express i.a. that the current choice is i-admissible and/or difference admissible. Let $\mathcal{L}$ be defined by the following Backus-Naur Form (BNF):

$$\varphi := q \mid \mathsf{u}_i^n \mid \neg\varphi \mid \varphi \vee \varphi \mid \Box\varphi \mid \boxed{c}\varphi$$

where q ranges over Q, i over Agt, and n over N. The constant u_i^n expresses that agent i receives a utility of n. $\Box$ is a universal modality: $\Box\varphi$ means that φ is true at all states in the model; $\Diamond$ denotes its dual. $\boxed{c}\varphi$ expresses that the current choice guarantees that φ is the case; the dual of $\boxed{c}$ is denoted by $\diamondsuit$. $\boxed{c}$ is a normal modal operator, similar in spirit to the "Chellas STIT" [5,11]. Both $\Box$ and $\boxed{c}$ are modal operators of type S5.

Definition 2.5 Where $\mathfrak{M} = \langle S, C, U, V \rangle$ is a model of choice under uncertainty and $s \in S$:

(SC1) $\mathfrak{M}, s \models q$ iff $s \in V(q)$

(SC2) $\mathfrak{M}, s \models \neg\varphi$ iff $\mathfrak{M}, s \not\models \varphi$

(SC3) $\mathfrak{M}, s \models \varphi \vee \psi$ iff $\mathfrak{M}, s \models \varphi$ or $\mathfrak{M}, s \models \psi$

(SC4) $\mathfrak{M}, s \models \boxed{c}\varphi$ iff for all $s' \in C(s)$, $\mathfrak{M}, s' \models \varphi$

(SC5) $\mathfrak{M}, s \models \Box\varphi$ iff for all $s' \in S$, $\mathfrak{M}, s' \models \varphi$

(SC6) $\mathfrak{M}, s \models \mathsf{u}_i^n$ iff $U(s, i) = n$.

Let us use the example from Figure 1 to illustrate some of these semantic clauses. Let $\mathfrak{M}$ correspond to the model in Figure 1 with $V(q) = \{s_1, s_3, s_4\}$. Since $U(s_1, i) = 3$ and by applying (SC6), we obtain that $\mathfrak{M}, s_1 \models \mathsf{u}_i^3$. In view of (SC4) and since $U(s_2, i) = 2$, $\mathfrak{M}, s_1 \not\models \boxed{c}\mathsf{u}_i^3$. Likewise, since q is false at s_2, $\mathfrak{M}, s_1 \not\models \boxed{c}q$. However, by (SC6) and since q is true at both s_3 and s_4, we have $\mathfrak{M}, s_1 \models \Diamond\boxed{c}q$.

With the language $\mathcal{L}$, we can express the notions of individual admissibility and difference admissibility that were introduced in Section 2.2. In order to explain this, we need some preparatory work. Where $\dagger \in Agt \cup \{*\}$ and where s is a state in some model $\mathfrak{M}$, let $G^{\mathfrak{M}}(s, \dagger)$ be the set of all $n \in N$ such that for all $s' \in C(s)$, $U(s', \dagger) \geq n$. When $n \in G^{\mathfrak{M}}(s, \dagger)$, we say that utility n is *guaranteed* for $\dagger$ at s. A little reflection on the Maximin criterion and our definitions of admissibility gives us:

Lemma 2.6 $C(s) \in Adm_\dagger(\mathfrak{M})$ iff for all $s' \in S$: $G^{\mathfrak{M}}(s', \dagger) \subseteq G^{\mathfrak{M}}(s, \dagger)$.

Let s be a state in some model of choice under uncertainty, and let $X = C(s)$. Using the formal language $\mathcal{L}$, we can express that, for any utility $n \in N$ and for any other choice Y in the model, if Y guarantees n, then so does X – see Table 2. Relying on Lemma 2.6, we immediately obtain:

Theorem 2.7 *Where $\mathfrak{M} = \langle S, U, C, V \rangle$ is a model of choice under uncertainty, $s \in S$, and $\dagger \in Agt \cup \{*\}$: $C(s) \in Adm_\dagger(\mathfrak{M})$ iff $\mathfrak{M}, s \models \mathsf{Adm}_\dagger$.*

Abbr.	Definition	Interpretation
$\mathsf{u}_i^{\geq n}$	$\bigvee_{m \geq n} \mathsf{u}_i^m$	The utility of i is at least n.
u_*^n	$(\bigvee_{i \in Agt} \mathsf{u}_i^n) \wedge (\bigwedge_{j \in Agt} \mathsf{u}_j^{\geq n})$	The utility of the least well-off is n.
$\mathsf{u}_*^{\geq n}$	$\bigvee_{m \geq n} \mathsf{u}_*^m$	The utility of the least well-off is at least n.
g_i^n	$\boxdot \mathsf{u}_i^{\geq n}$	A utility of n is guaranteed for i.
g_*^n	$\boxdot \mathsf{u}_*^{\geq n}$	A utility of n is guaranteed for the least well-off.
Adm_i	$\bigwedge_{n \in N}(\Diamond \mathsf{g}_i^n \to \mathsf{g}_i^n)$	The given choice is i-admissible.
Adm_*	$\bigwedge_{n \in N}(\Diamond \mathsf{g}_*^n \to \mathsf{g}_*^n)$	The given choice is difference admissible.

Table 2
Some useful abbreviations. Here, † ranges over $Agt \cup \{*\}$.

Proof. $C(s) \in Adm_\dagger(\mathfrak{M})$ iff [by Lemma 2.6] for all $s' \in S$, $G^{\mathfrak{M}}(s', \dagger) \subseteq G^{\mathfrak{M}}(s, \dagger)$ iff for all $s' \in S$, for all $n \in N$, if $n \in G^{\mathfrak{M}}(s', \dagger)$ then $n \in G^{\mathfrak{M}}(s, \dagger)$ iff [in view of Table 2] for all $n \in N$, for all $s' \in S$, if $\mathfrak{M}, s' \models \mathsf{g}_\dagger^n$, then $\mathfrak{M}, s \models \mathsf{g}_\dagger^n$ iff for all $n \in N$, if there is an $s' \in S$ such that $\mathfrak{M}, s' \models \mathsf{g}_\dagger^n$, then $\mathfrak{M}, s \models \mathsf{g}_\dagger^n$ iff [by the semantic clauses] for all $n \in N$, $\mathfrak{M}, s \models \Diamond \mathsf{g}_\dagger^n \to \mathsf{g}_\dagger^n$ iff $\mathfrak{M}, s \models \mathsf{Adm}_\dagger$. □

Theorem 2.7 tells us that we can express that a given option is individually admissible (according to the Maximin criterion) or difference admissible in $\mathfrak{L}$. Recall however that, according to Rawls' thesis, these two notions are only related given a specific type of uncertainty, viz. uncertainty about the position one occupies in society. In what follows, we show how our semantics and formal language can be refined in order to represent such uncertainty.

3 A Semantics for Rawls' Thesis

The kind of uncertainty we are dealing with in the Original Position is, at bottom, uncertainty about who gets which position; from that, one then derives uncertainty about the agent's utilities. To make this idea precise, we introduce a more specific class of models of choice under uncertainty in Section 3.1. Next, we define a type of updates on those models, which capture changes in position uncertainty (Section 3.2). Finally, we show how, with the formal instrumentarium thus introduced, we can make Rawls' thesis exact (Section 3.3).

3.1 Models of Choice Under Position Uncertainty

Fix a finite, non-empty set of positions $P = \{p, p', \ldots\}$, with $|P| \leq |Agt|$.[6] Here, one should think of a position in rather abstract terms: a position is

---

[6] We require that the number of positions does not exceed that of agents because we will need the presupposition that every position is occupied by at least one agent for Rawls' thesis to hold – see also footnote 8.

simply that which determines the utility of the agent at a given state.

Definition 3.1 A *model of choice under position uncertainty* is a tuple $\mathfrak{M}^0 = \langle W, \Pi, C^0, U^0, V^0 \rangle$ where $W \neq \emptyset$ is the set of *worlds*, Π is a non-empty set of *position assignments* $\pi : Agt \to P$ that are surjective, C^0 is a partition of W, $U^0 : W \times P \to N$ is a *position-utility function*, and $V^0 : Q \to \wp(W)$ is a *valuation function*.

Definition 3.2 Where $\mathfrak{M}^0 = \langle W, \Pi, C^0, U^0, V^0 \rangle$ is a model of choice under position uncertainty, the corresponding model of choice under uncertainty is $\mathfrak{M} = \langle S, C, U, V \rangle$, where:

- $S =_{df} W \times \Pi$
- for all $(w, \pi) \in W \times \Pi : U((w, \pi), i) =_{df} U^0(w, \pi(i))$
- $C =_{df} \{\{(w, \pi) \mid w \in X, \pi \in \Pi\} \mid X \in C^0\}$
- $V(q) =_{df} \{(w, \pi) \mid w \in V^0(q), \pi \in \Pi\}$

In a model of choice under position uncertainty, states are made up of two components: a world w that determines what factual states of affairs obtain and what utilities each position gets, and a position assignment π that determines the position of each agent.[7] Note that we require the position assignment functions to be surjective. This means that every position in society is occupied by at least one agent.[8]

This in turn allows us to decompose the utility function U from Section 2 into two parts. First, U^0 specifies the utilities of every position, for every way the world may end up being. So $U^0(w, p) = n$ means that at world w, any agent with position p receives a utility of n. Second, the position assignment π specifies the position an agent gets in society. The *agent-utility of i at a state* $s = (w, \pi)$ is then defined as $U^0(w, \pi(i))$: it is the position-utility at w of the position to which i is assigned at s.

In view of Definition 3.2, each model of choice under position uncertainty corresponds to a model of choice under uncertainty. Given this, we can apply our earlier definitions of individual and difference admissibility to models of choice under position uncertainty.

Figure 2 represents two models of choice under position uncertainty. In $\mathfrak{M}_1^0$, Π is a singleton $\{\pi_1\}$. In $\mathfrak{M}_2^0$, Π consists of two position assignments. Note that this difference affects which choices are admissible for each of the agents, though it does not affect which choices are difference admissible. In

---

[7] Here, a warning is in place: since π determines which agent gets which utility, the "factual states of affairs" are limited to those statements that do not depend, logically speaking, on who gets what. For instance, "agent 2 gets a utility of 5" is not a "factual state of affairs" on this reading. In principle, we could also make the truth of propositional variables dependent on both w and π. This would not affect our main results in this paper.

[8] This presupposition is necessary for Rawls' thesis. Indeed, otherwise the "worst-off" agent given the current position assignment may be guaranteed to get a higher utility than what some agents could have in positions that are currently not occupied.

	X		Y	
	w_1	w_2	w_3	w_4
U^0	$(3,1)$	$(2,4)$	$(2,2)$	$(1,4)$
π_1	$(3,1)$	$(2,4)$	$(2,2)$	$(1,4)$

(a) The model $\mathfrak{M}_1^0$ with $\Pi = \{\pi_1\}$.

	X		Y	
	w_1	w_2	w_3	w_4
U^0	$(3,1)$	$(2,4)$	$(2,2)$	$(1,4)$
π_1	$(3,1)$	$(2,4)$	$(2,2)$	$(1,4)$
π_2	$(1,3)$	$(4,2)$	$(2,2)$	$(4,1)$

(b) The model $\mathfrak{M}_2^0$ with $\Pi = \{\pi_1, \pi_2\}$.

Fig. 2. Two models of choice under position uncertainty. Where $(\mathbf{n}, \mathbf{m}) \in N \times N$, $\mathbf{n}$ denotes the utility of p_1, and $\mathbf{m}$ denotes the utility of p_2 at the given world. The two position assignments are: $\pi_1(i) = p_1, \pi_1(j) = p_2$ and $\pi_2(i) = p_2, \pi_2(j) = p_1$.

particular, $Adm_i(\mathfrak{M}_1^0) = \{X\}, Adm_j(\mathfrak{M}_1^0) = \{Y\}$, and $Adm_*(\mathfrak{M}_1^0) = \{X,Y\}$, while $Adm_i(\mathfrak{M}_2^0) = Adm_j(\mathfrak{M}_2^0) = Adm_*(\mathfrak{M}_2^0) = \{X,Y\}$.

3.2 Updates of Position Uncertainty

Given a model $\mathfrak{M}^0 = \langle W, \Pi, C^0, U^0, V^0 \rangle$, the parameter Π specifies our uncertainty about who gets what position in society. Importantly, and in line with our agent-independent notion of choice, this uncertainty is agent-independent. For instance, if there are $\pi, \pi' \in \Pi$ and $p, p' \in P$ such that $\pi(i) = p$ and $\pi'(i) = p'$ (with $p \neq p'$), then this means that whoever is choosing does not know whether i occupies position p, or rather position p'.

Consequently, a change in position uncertainty amounts to an update of the parameter Π. We will define such updates in general, after which we apply them to Rawls' thesis. In what follows, let Π_* denote the set of *all* position assignments, i.e. all surjective functions $\pi : Agt \to P$.

Definition 3.3 Where $\mathfrak{M}^0 = \langle W, \Pi, C^0, U^0, V^0 \rangle$ is a model of choice under position uncertainty and where $\emptyset \neq \Pi' \subseteq \Pi_*$, $\mathfrak{M}_{\Pi'}^0 = \langle W, \Pi', C^0, U^0, V^0 \rangle$.

In other words, all that is changed by an update (if anything) is the set of position assignments that are considered possible. With this general type of update, we can model both increasing and decreasing uncertainty about position assignments. At one end of the spectrum, updates with a singleton $\{\pi\}$ amount to restricting the model to a single position assignment. At the other end, updates with Π_* amount to making every position assignment possible.

Returning to our example in Figure 2, it can be easily observed that the model on the right hand side is obtained by updating the model on the left hand side with $\{\pi_1, \pi_2\}$, and conversely, the model on the left hand side is obtained by updating the model on the right hand side with $\{\pi_1\}$.

3.3 Rawls' Thesis

Recall that in the Original Position, we do not know anything about our position in society. So if, for a given model $\mathfrak{M}^0$ of choice under position uncertainty, we ask what an agent i would choose in the Original Position, we are in fact asking what i would choose in the updated model $\mathfrak{M}^0_{\Pi_*}$. On this analysis, Rawls' thesis says that a given choice is difference admissible in $\mathfrak{M}^0$ if and only if the "corresponding" choice in $\mathfrak{M}^0_{\Pi_*}$ is i-admissible in $\mathfrak{M}^0_{\Pi_*}$.

In order to make this notion of correspondence precise we need some extra notation. Given any model $\mathfrak{M}^0 = \langle W, \Pi, C^0, U^0, V^0 \rangle$, we let $\mathfrak{M}^0_* = \mathfrak{M}^0_{\Pi_*} = \langle W, \Pi_*, C^0, U^0, V^0 \rangle$, and we use C_*, U_*, and V_* to refer to the set of choices, the agent-utility function, and the valuation function of the model $\mathfrak{M}_*$ of choice under uncertainty that corresponds to $\mathfrak{M}^0_*$ (cf. Definition 3.2).

Our proof of Rawls' thesis crucially relies on the observation that the set of guaranteed utilities for the least well-off at a given state in the original model equals the set of guaranteed utilities for any individual i in the corresponding state in the Original Position. Formally:

Lemma 3.4 *Where $\mathfrak{M}^0 = \langle W, \Pi, C^0, U^0, V^0 \rangle$ is a model of choice under position uncertainty, $s \in W \times \Pi$, and $i \in Agt$: $G^{\mathfrak{M}^0}(s, *) = G^{\mathfrak{M}^0_*}(s, i)$.*

Proof. Let $i \in Agt$ and $n \in N$. We have: $n \in G^{\mathfrak{M}^0}(s, *)$ iff [by the definition of $G^{\mathfrak{M}^0}(s, *)$] for all $s' \in C(s)$, $U(s', *) \geq n$ iff [by the definition of $U(*, s)$] for all $i \in Agt$ and $s' \in C(s)$, $U(s', i) \geq n$ iff [since every position assignment is surjective] for all $p \in P$ and all $w' \in C^0(w)$, $U^0(w', p) \geq n$ iff [by the definition of Π_*] for all $s' \in C_*(s)$, $U_*(s, i) \geq n$ iff [by the definition of $G^{\mathfrak{M}^0_*}(s, i)$] $n \in G^{\mathfrak{M}^0_*}(s, i)$. □

Note also that, whatever utility is guaranteed for i at a state (w, π) in a model $\mathfrak{M}^0$, is also guaranteed for i at every state (w, π') in $\mathfrak{M}^0$. Formally:

Fact 3.5 *Where $\mathfrak{M}^0 = \langle W, \Pi, C^0, U^0, V^0 \rangle$ is a model of choice under position uncertainty, $w \in W$, $\pi, \pi' \in \Pi$, and $i \in Agt \cup \{*\}$: $G^{\mathfrak{M}^0}((w, \pi), i) = G^{\mathfrak{M}^0}((w, \pi'), i)$.*

Theorem 3.6 *Where $\mathfrak{M}^0 = \langle W, \Pi, C^0, U^0, V^0 \rangle$ is a model of choice under position uncertainty, $(w, \pi) \in W \times \Pi$, and $i \in Agt$: $C(w, \pi) \in Adm_*(\mathfrak{M}^0)$ iff $C_*(w, \pi) \in Adm_i(\mathfrak{M}^0_*)$. (Rawls' Thesis)*

Proof. $C(w, \pi) \in Adm_*(\mathfrak{M}^0)$ iff [by Lemma 2.6] for all $(w', \pi') \in W \times \Pi$, $G^{\mathfrak{M}^0}((w', \pi'), *) \subseteq G^{\mathfrak{M}^0}((w, \pi), *)$ iff [by Lemma 3.4] for all $(w', \pi') \in W \times \Pi$, $G^{\mathfrak{M}^0_*}((w', \pi'), i) \subseteq G^{\mathfrak{M}^0_*}((w, \pi), i)$ iff [by Fact 3.5] for all $(w', \pi') \in W \times \Pi_*$, $G^{\mathfrak{M}^0_*}((w', \pi'), i) \subseteq G^{\mathfrak{M}^0_*}((w, \pi), i)$ iff [by Lemma 2.6] $C_*(w, \pi) \in Adm_i(\mathfrak{M}^0_*)$. □

4 A Logic of Choice Under Position Uncertainty

In order to express Rawls' thesis syntactically, we enrich the formal language $\mathcal{L}$ from Section 2. First, we define a static modal language in which we can express position utilities and position assignments, and provide an axiomatization for the resulting logic (Section 4.1). Next, we add dynamic operators that can express changes in position uncertainty and give reduction axioms for them (Section 4.2). After this preparatory work, we show that Rawls' thesis corresponds to a validity of the resulting logic (Section 4.3).

4.1 Static Part

Formal Language Let $\mathcal{L}^+$ be defined by the BNF:

$$\varphi := q \mid \mathsf{a}_{ip} \mid \mathsf{u}_p^n \mid \neg\varphi \mid \varphi \vee \varphi \mid \Box\varphi \mid \boxdot\varphi \mid \boxplus\varphi$$

where q ranges over Q, i over N, p over P, and n over $\mathbb{N}$. The constant a_{ip} expresses that agent i occupies position p, while u_p^n expresses that any agent with position p gets utility n. The only new modality is $\boxplus$. This operator allows us to talk about all states that have the same world component (see (SC9) below). In other words, $\boxplus\varphi$ expresses that "φ is the case, no matter which position the agents occupy".

The following definition gives the semantic clauses for a_{ip}, u_p^n, and $\boxplus$; note that the clauses for the variables, connectives, and other operators are exactly as in Definition 2.5, relying on the fact that every model of choice under position uncertainty is also a model of choice under uncertainty (cf. Definition 3.2).

Definition 4.1 Where $\mathfrak{M}^0 = \langle W, \Pi, C^0, U^0, V^0 \rangle$ is a model of choice under position uncertainty and $s = (w, \pi) \in W \times \Pi$,

(SC7) $\mathfrak{M}^0, s \models \mathsf{a}_{ip}$ iff $\pi(i) = p$

(SC8) $\mathfrak{M}^0, s \models \mathsf{u}_p^n$ iff $U^0(w, p) = n$

(SC9) $\mathfrak{M}^0, s \models \boxplus\varphi$ iff for all $\pi \in \Pi$, $\mathfrak{M}^0, (w, \pi) \models \varphi$

The formal language introduced above is an extension of $\mathcal{L}$. The constants that expressed agent-utilities in $\mathcal{L}$ can now be defined:

$$\mathsf{u}_i^n =_{\mathsf{df}} \bigvee_{p \in P} (\mathsf{a}_{ip} \wedge \mathsf{u}_p^n)$$

Consequently, we can reuse all the definitions from Table 2 to express that a given choice is i-admissible or difference admissible in $\mathcal{L}^+$. However, we now also have the additional expressive power that allows us to talk about position uncertainty, which is crucial for Rawls' thesis.

Axiomatization The set of validities in $\mathcal{L}^+$ is axiomatized by the axioms and rules in Table 3. Axiom QW (resp. PW) expresses that the truth of a propositional variable (resp. the utility of a position) depends only on the world-component of a state. I1-I3 capture interactions between the various modalities. I1 is an immediate result of the fact that $\Box$ is a universal modality. I2 follows from the fact that choices are defined in terms of the world-components of states, and hence one cannot choose between two states with the same world-component. I3 captures the property that, if a certain position assignment π is possible in the model at hand, then there is some state with the same world component as the current state and the possition assignment π. Finally, PA1 and PA2 (PU1 and PU2) express that every π (U) is a function; PA3 expresses that every π is surjective.

Theorem 4.2 $\vdash \varphi$ iff $\models \varphi$. *(Soundness and Completeness)*

Proof. Soundness is a matter of routine. For completeness, observe that every

CL	Classical Logic	
S5	S5 for $\blacksquare \in \{\Box, \boxdot, \boxplus\}$	
QW	$\boxplus q \vee \boxplus \neg q$	$(q \in Q)$
PW	$\boxplus \mathsf{u}_p^n \vee \boxplus \neg \mathsf{u}_p^n$	$(p \in P, n \in N)$
I1	$\Box \varphi \rightarrow \boxdot \varphi$	
I2	$\boxdot \varphi \rightarrow \boxplus \varphi$	
I3	$\boxplus \bigwedge_{i \in Agt, \pi(i)=p} \mathsf{a}_{ip} \rightarrow \Box \bigwedge_{i \in Agt, \pi(i)=p} \mathsf{a}_{ip}$	$(\pi \in \Pi_*)$
PA1	$\bigvee_{p \in P} \mathsf{a}_{ip}$	$(i \in Agt)$
PA2	$\mathsf{a}_{ip} \rightarrow \neg \mathsf{a}_{ip'}$	$(i \in Agt, p, p' \in P, p \neq p')$
PA3	$\bigvee_{i \in Agt} \mathsf{a}_{ip}$	$(p \in P)$
PU1	$\bigvee_{n \in N} \mathsf{u}_p^n$	$(p \in P)$
PU2	$\mathsf{u}_p^n \rightarrow \neg \mathsf{u}_p^m$	$(p \in P, n, m \in N, n \neq m)$
MP	if $\vdash \varphi \rightarrow \psi$ and $\vdash \varphi$ then $\vdash \psi$	
NEC	if $\vdash \varphi$ then $\vdash \Box \varphi$	

Table 3

model $\mathfrak{M}^0$ of choice under position uncertainty can be rewritten as a Kripke-model of the type $\mathfrak{M}^K = \langle S, \sim^{\boxdot}, \sim^{\boxplus}, V \rangle$, where $S \neq \emptyset$ is a set of states, $\sim^{\boxdot}$ is the equivalence relation that corresponds to the choices in $\mathfrak{M}^0$, $\sim^{\boxplus}$ is the equivalence relation that corresponds to the worlds in $\mathfrak{M}^0$, and $V : Q \cup \{\mathsf{a}_{ip} \mid i \in Agt, p \in P\} \cup \{u_p = n \mid p \in P, n \in N\} \rightarrow \wp(S)$ is a valuation function. Conversely, given suitable conditions on such Kripke-models, we can rewrite them as models of position uncertainty — cf. Table 4. Proving that, taken jointly, these conditions ensure translatability to a model of choice under position uncertainty is a tedious but routine job, which we omit for reasons of space.

Let MCS be the set of all maximal consistent subsets of $\mathfrak{L}^+$. Where $\blacksquare \in \{\Box, \boxdot, \boxplus\}$ and $\Delta \in \mathsf{MCS}$, let $\Delta^{\blacksquare} = \{\blacksquare \varphi \in \mathfrak{L}^+ \mid \blacksquare \varphi \in \Delta\}$. Fix a $\Gamma \in \mathsf{MCS}$. Let $\mathfrak{M}_\Gamma^K = \langle S_\Gamma, \sim_\Gamma^{\boxdot}, \sim_\Gamma^{\boxplus}, V_\Gamma \rangle$, where
1. S_Γ is the set of all maximal consistent sets Δ such that $\Delta^\Box = \Gamma^\Box$.
2. Where $\Delta, \Theta \in S_\Gamma$, $\Delta \sim_\Gamma^{\boxdot} \Theta$ iff $\Delta^{\boxdot} = \Theta^{\boxdot}$
3. Where $\Delta, \Theta \in S_\Gamma$, $\Delta \sim_\Gamma^{\boxplus} \Theta$ iff $\Delta^{\boxplus} = \Theta^{\boxplus}$
4. $V_\Gamma(\varphi) = \{\Delta \in S_\Gamma \mid \varphi \in \Delta\}$ for all $\varphi \in Q \cup \{\mathsf{a}_{ip} \mid i \in Agt, p \in P\} \cup \{u_p = n \mid p \in P, n \in N\}$

The truth lemma is proven for $\mathfrak{M}_\Gamma^K$ in the standard way. By an induction on the complexity of formulas, we can moreover prove that for any $s, s' \in S_\Gamma$, condition (C) is satisfied. For the other conditions, we can rely on the corresponding axioms to prove they hold for $\mathfrak{M}_\Gamma^K$. In sum, $\mathfrak{M}_\Gamma^K$ satisfies all the conditions from Table 4. □

(C)	if $s \sim^{\boxplus} s'$ and $s, s' \in \bigcap_{i \in Agt, \pi(i)=p} V(\mathsf{a}_{ip})$, then $s = s'$
(CQW)	if $s \sim^{\boxplus} s'$, then $s \in V(q)$ iff $s' \in V(q)$
(CPW)	if $s \sim^{\boxplus} s'$, then $s \in V(\mathsf{u}_p^n)$ iff $s' \in V(\mathsf{u}_p^n)$
(CI2)	$\sim^{\boxplus} \subseteq \sim^{\boxdot}$
(CI3)	if $s \in \bigcap_{i \in Agt, \pi(i)=p} V(\mathsf{a}_{ip})$, then $\forall s' \in S, \exists s'' \in S$: $(\pi \in \Pi_*)$ $s' \sim^{\boxplus} s''$ and $s'' \in \bigcap_{i \in Agt, \pi(i)=p} V(\mathsf{a}_{ip})$
(CPA1)	$\forall i \in Agt, \exists p \in P: s \in V(\mathsf{a}_{ip})$
(CPA2)	if $s \in V(\mathsf{a}_{ip})$, then $s \notin V(\mathsf{a}_{ip'})$ $(p \neq p')$
(CPA3)	$\forall p \in P, \exists i \in Agt: s \in V(\mathsf{a}_{ip})$
(CPU1)	$\forall p \in P, \exists n \in N: s \in V(\mathsf{u}_p^n)$
(CPU2)	if $s \in V(\mathsf{u}_p^n)$, then $s \notin V(\mathsf{u}_p^m)$ $(n \neq m)$

Table 4

4.2 Dynamic Operators

In order to express what holds given an update of the set of position assignments, we rely on well-known ideas from dynamic epistemic logic [3,20]. In particular, we consider *pointed* updates of *pointed* models. As we will show in Section 4.3, we can use the resulting dynamic operators to express what holds in the Original Position.

Henceforth, an *update model* is a couple (Π, π), where $\Pi \subseteq \Pi_*$ and $\pi \in \Pi$. Intuitively, the update model expresses the new set of position assignments that become possible, and the specific position assignment that becomes actual. Update models are used to change a given pointed model of position uncertainty, i.e. a model together with a given state (w, π) in that model:

Definition 4.3 Where $\mathfrak{M}^0 = \langle W, \Pi, C^0, U^0, V^0 \rangle$, $(w, \pi) \in W \times \Pi$, and (Π', π') is an update model: the *update of* $(\mathfrak{M}^0, (w, \pi))$ *with* (Π', π') is $(\mathfrak{M}^0, (w, \pi)) \circ (\Pi', \pi') =_{\mathsf{df}} (\mathfrak{M}^0_{\Pi'}, (w, \pi'))$.

Given these conventions, we can introduce dynamic operators $[\Pi, \pi]$ for every update model (Π, π), and interpret them using the following standard clause:

(SC10) $\mathfrak{M}^0, s \models [\Pi, \pi]\varphi$ iff $(\mathfrak{M}^0, s) \circ (\Pi, \pi) \models \varphi$

In dynamic epistemic logic terminology, our updates are a specific type of (finitary) ontic updates with an empty precondition. Relying on this observation, we can easily find reduction axioms for the dynamic operators. These are listed in Table 5. Given these reduction axioms and Theorem 4.2, we obtain a sound and strongly complete axiomatization for the extension of $\mathfrak{L}^+$ with all dynamic operators of the type $[\Pi, \pi]$.

RA1	$[\Pi, \pi]q \leftrightarrow q$ (for all $q \in Q$)
RA2	$[\Pi, \pi]\mathsf{u}_p^n \leftrightarrow \mathsf{u}_p^n$ (for all $p \in P$ and $n \in N$)
RA3	$[\Pi, \pi]\mathsf{a}_{ip} \leftrightarrow \top$ if $\pi(i) = p$
RA4	$[\Pi, \pi]\mathsf{a}_{ip} \leftrightarrow \bot$ if $\pi(i) \neq p$
RA5	$[\Pi, \pi]\neg\varphi \leftrightarrow \neg[\Pi, \pi]\varphi$
RA6	$[\Pi, \pi](\varphi \vee \psi) \leftrightarrow ([\Pi, \pi]\varphi \vee [\Pi, \pi]\psi)$
RA7	$[\Pi, \pi]\blacksquare\varphi \leftrightarrow \bigwedge_{\pi' \in \Pi} \blacksquare[\Pi, \pi']\varphi$ (for $\blacksquare \in \{\Box, \boxdot, \boxplus\}$)

Table 5
Reduction axioms for the dynamic operators.

4.3 Rawls' Thesis in $\mathfrak{L}^+$

Recall that Π_* denotes the set of all position assignments. By means of the dynamic operators $[\Pi_*, \pi]$ we can define an operator that expresses what holds in the Original Position:

$$\boxtimes\varphi =_{\mathsf{df}} \bigwedge_{\pi \in \Pi_*} \Big(\bigwedge_{i \in Agt, \pi(i)=p} \mathsf{a}_{ip} \to [\Pi_*, \pi]\varphi \Big)$$

Theorem 4.4 *Where $\mathfrak{M}^0$ and $\mathfrak{M}_*^0$ are models of choice under position uncertainty, we have: $\mathfrak{M}^0, (w, \pi) \models \boxtimes\varphi$ iff $\mathfrak{M}_*^0, (w, \pi) \models \varphi$.*

Proof. For all $\pi \in \Pi$, let $\mathsf{a}_\pi = \bigwedge_{i \in Agt, \pi(i)=p} \mathsf{a}_{ip}$. We have: $\mathfrak{M}^0, (w, \pi) \models \boxtimes\varphi$ iff [by the definition of $\boxtimes$] for all $\pi' \in \Pi_*$, $\mathfrak{M}^0, (w, \pi) \models \mathsf{a}_{\pi'} \to [\Pi_*, \pi']\varphi$ iff [since only a_π is true at $\mathfrak{M}^0, (w, \pi)$] $\mathfrak{M}^0, (w, \pi) \models [\Pi_*, \pi]\varphi$ iff [by the semantic clause for $[\Pi, \pi]$] $(\mathfrak{M}^0, (w, \pi)) \circ (\Pi_*, \pi) \models \varphi$ iff [by the definition of pointed updates and since $\mathfrak{M}_*^0 = \mathfrak{M}_{\Pi_*}^0$] $\mathfrak{M}_*^0, (w, \pi) \models \varphi$. $\square$

Theorem 3.6 is now expressible as a formula in the object-language:

Theorem 4.5 $\models \mathsf{Adm}_* \leftrightarrow \boxtimes\mathsf{Adm}_i$. *(Rawls' Thesis in $\mathfrak{L}^+$)*

Proof. Let $\mathfrak{M}^0 = \langle W, \Pi, C^0, U^0, V^0 \rangle$ and $s \in W \times \Pi$. We have: $\mathfrak{M}^0, s \models \mathsf{Adm}_*$ iff [by Theorem 2.7] $C(s) \in Adm_*(\mathfrak{M}^0)$ iff [by Theorem 3.6] $C_*(s) \in Adm_i(\mathfrak{M}_*^0)$ iff [by Theorem 2.7] $\mathfrak{M}_*^0, s \models \mathsf{Adm}_i$ iff [by Theorem 4.4] $\mathfrak{M}^0, s \models \boxtimes\mathsf{Adm}_i$. $\square$

5 Deontic Logics Based on Fairness

In this last, somewhat more programmatic section, we show the potential of our models and logic from the viewpoint of deontic logic. We first show how admissibility based on the other lifting criteria can be formalized in $\mathfrak{L}$ (Section 5.1). This in turn gives us a general recipe for expressing various other notions of fairness (Section 5.2), and deontic operators based on them (Section 5.3).

5.1 Other Lifting Criteria

In Section 2 we introduced four lifting criteria that can be used to determine which choices are admissible in a given choice situation. Moreover, we demonstrated that the current choice being admissible for i according to the Maximin lifting ($\forall\exists$) can be expressed in the language using object-level formulas. In Table 6, we give an overview of how admissibility of the current choice for i can be expressed for the other three lifting criteria from Section 2. The logical relations between these notions are depicted in Figure 3, where the arrows stand for logical consequence.

Abbreviation	Definition
$\mathsf{u}_i^{\leq n}$	$\bigvee_{m \leq n} \mathsf{u}_i^m$
$\mathsf{Adm}_i^{\forall\forall}$	$\bigwedge_{n \in N}(\boxdot \mathsf{u}_i^{\leq n} \to ((\Diamond \mathsf{g}_i^n \to \mathsf{g}_i^n) \land \Box(\mathsf{g}_i^n \to \boxdot \mathsf{u}_i^n)))$
$\mathsf{Adm}_i^{\exists\forall}$	$\bigwedge_{n \in N}(\Diamond\mbox{\large$\diamond$} \mathsf{u}_i^{\geq n} \to \mbox{\large$\diamond$} \mathsf{u}_i^{\geq n})$
$\mathsf{Adm}_i^{\exists\exists}$	$\bigwedge_{n \in N}(\Diamond \mathsf{g}_i^n \to \mbox{\large$\diamond$} \mathsf{u}_i^{\geq n})$

Table 6

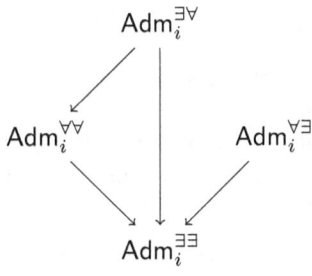

Fig. 3.

5.2 Other Notions of Fairness

Recall that the defined operator ⊠ talks about what holds in the Original Position (cf. Theorem 4.4). We can use this operator and our l-admissibility formulas to define three additional, distinct notions of fairness admissibility. That is, the formula $\boxtimes\mathsf{Adm}_i^l$ expresses that in the Original Position, if our standard of rational choice under uncertainty is determined by lifting criterion l, then the given choice is l-admissible for i. So, if one agrees with Rawls that fair choices are the choices a rational agent would make in the Original Position, then $\boxtimes\mathsf{Adm}_i^l$ expresses that the given choice is fair (modulo l).

The logical relations depicted in Figure 3 immediately transfer to the corresponding notions of fairness admissibility, in view of the following:

Theorem 5.1 *Where $i \in Agt$ and $l, l' \in \{\forall\forall, \forall\exists, \exists\forall, \exists\exists\}$:* $\vdash \mathsf{Adm}_i^l \to \mathsf{Adm}_i^{l'}$ *iff* $\vdash \boxtimes\mathsf{Adm}_i^l \to \boxtimes\mathsf{Adm}_i^{l'}$.

Proof. For left to right, one should show that ⊠ is a normal modal operator. For the other direction, suppose that ⊬ $\mathsf{Adm}_i^l \to \mathsf{Adm}_i^{l'}$. So there is a model $\mathfrak{M}^0$ and state s such that $\mathfrak{M}^0, s \models \mathsf{Adm}_i^l$ and $\mathfrak{M}^0, s \not\models \mathsf{Adm}_i^{l'}$. Consider the model $\mathfrak{M}_e^0$ that differs only from $\mathfrak{M}^0$ in that, at every state, *all* the agents receive the utility that i receives in the corresponding state in $\mathfrak{M}^0$. In this model, individual admissibility and fairness admissibility coincide, and hence $\mathfrak{M}_e^0, s \models \boxtimes \mathsf{Adm}_i^l$, $\mathfrak{M}_e^0, s \not\models \boxtimes \mathsf{Adm}_i^{l'}$. □

Thus, e.g. fairness admissibility using the Maximin criterion is strictly stronger than fairness admissibility using criterion ∀∀, which in turn implies fairness admissibility using ∃∃. In contrast, fairness using ∃∀ is logically incomparable to fairness admissibility with Maximin or with ∀∀.

5.3 Deontic Operators

By employing the familiar Kangerian reduction [1,2,12] we can use our admissibility formulas to define two types of deontic operators:

$$\mathsf{O}_i^l \varphi =_{\mathsf{df}} \Box(\mathsf{Adm}_i^l \to \varphi)$$

$$\mathsf{O}_*^l \varphi =_{\mathsf{df}} \Box(\boxtimes\mathsf{Adm}_i^l \to \varphi)$$

The formula $\mathsf{O}_i^l \varphi$ can be read as "it ought to be that φ for i" (where l determines a particular standard of rational choice under uncertainty). This contrasts with the formula $\mathsf{O}_*^l \varphi$ which can be read as "from the viewpoint of fairness, it ought to be that φ". Both O_i^l and O_*^l are normal modal operators in virtue of their definition.

Because the admissibility formulas stand in logical relations with each other, we can expect there to be logical relations between obligation statements as well. For example, we have:

Theorem 5.2 ⊢ $\mathsf{Adm}_i^l \to \mathsf{Adm}_i^{l'}$ *iff* ⊢ $\mathsf{O}_i^{l'} \varphi \to \mathsf{O}_i^l \varphi$.

Proof. Left to right of the equivalence is safely left to the reader. For the other direction, let $\varphi = \mathsf{Adm}_i^{l'}$. Then, the right hand side implies that ⊢ $\Box(\mathsf{Adm}_i^l \to \mathsf{Adm}_i^{l'})$ and hence, by the T-axiom for $\Box$, ⊢ $\mathsf{Adm}_i^l \to \mathsf{Adm}_i^{l'}$. □

Consequently, if Adm_i^l and $\mathsf{Adm}_i^{l'}$ are incomparable, then $\mathsf{O}_i^l \varphi$ and $\mathsf{O}_i^{l'} \varphi$ are incomparable as well. We can also expect there to be logical relations between the individual oughts and fairness oughts, in line with Rawls' thesis. For example, what ought to be for agent i (given the Maximin criterion) and what ought to be from the viewpoint of fairness coincide in the Original Position:

Theorem 5.3 ⊢ $\bigwedge_{\pi \in \Pi_*} \Diamond \bigwedge_{i \in Agt, \pi(i)=p} \mathsf{a}_{ip} \to (\mathsf{O}_i^{\forall\exists} \varphi \leftrightarrow \mathsf{O}_*^{\forall\exists} \varphi)$.

Proof. Note that, if the left hand side of the implication is true in a model $\mathfrak{M}^0 = \langle W, \Pi, C^0, U^0, V^0 \rangle$, then $\Pi = \Pi^*$. By our earlier results, individual admissibility and fairness admissibility coincide in such models, and hence so do the corresponding ought-operators. □

To summarize, by using a Kangerian reduction, we obtain various kinds of deontic logics, based on individual standards of rationality and Rawls' procedural account of fairness. All these logics are fragments of the logic presented in Section 4. Here we merely sketched the various possibilities this generates; we leave a full investigation for future work.

6 Conclusion

We have given a logical analysis of Rawls' thesis that choices motivated by the Difference Principle coincide with the choices of any rational individual in the Original Position. In particular, we presented models of choice under position uncertainty, inspired by simple models for STIT logic. With the help of these models and a suitable formal language, we showed how to capture Rawls' thesis both in semantic and in syntactic terms. Finally, we demonstrated the potential of our logical analysis for the study of deontic notions related to fairness.

Future Work We chose to work with a finite set of utility values as this removes some complexities. However, one may ask to which extent our results still go through when working with infinite sets of utility values such as $\mathbb{N}$ or $\mathbb{R}$. While the semantic results (e.g. Theorem 3.3) seem easy to generalize to such richer settings, this is far less obvious on the syntactic side. In particular, can the language be modified in order to cope with infinite sets of values, while keeping the logic well-behaved meta-theoretically (e.g. axiomatizable and compact)?

We focused on the four lifting criteria from Table 1. An open question is whether it is possible to express more complex lifting criteria, such as e.g. lexicographic preferences. Finally, both the notion of choice and the notion of uncertainty are agent-independent in our models. A natural generalization would be to have models where the choices and/or uncertainty are agent-dependent. Here again, semantics seem relatively easy to obtain, but complexity grows rapidly at the syntactic level.

References

[1] Anderson, A. R., *Some nasty problems in the formal logic of ethics*, Noûs (1967), pp. 345–360.
[2] Åqvist, L., *Deontic logic*, in: *Handbook of philosophical logic*, Springer, 2002 pp. 147–264.
[3] Baltag, A., L. S. Moss and S. Solecki, *The logic of public announcements, common knowledge, and private suspicions*, in: *Proceedings of the 7th Conference on Theoretical Aspects of Rationality and Knowledge*, TARK 98 (1998), pp. 43–56.
[4] Belnap, N. D., M. Perloff, M. Xu et al., "Facing the future: agents and choices in our indeterminist world," Oxford University Press on Demand, 2001.
[5] Chellas, B. F., "The Logical Form of Imperatives," Ph.D. thesis, Stanford University (1969).
[6] Chung, H., *Rawls's self-defeat: A formal analysis*, Erkenntnis (2018), pp. 1–29.
[7] Gaus, G. and J. Thrasher, *Rational choice and the original position: the (many) models of Rawls and Harsanyi* (2015).
[8] Giraud, G. and C. Renouard, *Is the veil of ignorance transparent?* (2010).
[9] Harsanyi, J. C., *Can the maximin principle serve as a basis for morality? a critique of john rawls's theory*, American political science review **69** (1975), pp. 594–606.
[10] Horty, J. F., "Agency and deontic logic," Oxford University Press, 2001.

[11] Horty, J. F. and N. Belnap, *The deliberative stit: A study of action, omission, ability, and obligation*, Journal of Philosophical Logic **24** (1995), pp. 583–644.
[12] Kanger, S., *New foundations for ethical theory*, in: *Deontic logic: Introductory and systematic readings*, Springer, 1970 pp. 36–58.
[13] Kariv, S. and W. R. Zame, *Piercing the veil of ignorance* (2008).
[14] Kooi, B. and A. Tamminga, *Moral conflicts between groups of agents*, Journal of Philosophical Logic **37** (2008), pp. 1–21.
[15] Kukathas, C. and P. Pettit, "Rawls: A Theory of Justice and Its Critics," Key contemporary thinkers, Polity, 1990.
[16] Rawls, J., "A Theory of Justice," Belknap Press, 1999, rev sub edition.
[17] Svensson, L.-G., *Fairness, the veil of ignorance and social choice*, Social Choice and Welfare **6** (1989), pp. 1–17.
[18] Tamminga, A., *Deontic logic for strategic games*, Erkenntnis **78** (2013), pp. 183–200.
[19] Van Benthem, J., P. Girard and O. Roy, *Everything else being equal: A modal logic for ceteris paribus preferences*, Journal of philosophical logic **38** (2009), pp. 83–125.
[20] Van Benthem, J., J. van Eijck and B. Kooi, *Logics of communication and change*, Information and Computation **204** (2006), pp. 1620 – 1662.

Impossible and Conflicting Obligations in Justification Logic

Federico L. G. Faroldi [1]

Centre for Logic and Philosophy of Science
Ghent University, Belgium

Meghdad Ghari [2]

Department of Philosophy
University of Isfahan, Iran
and
School of Mathematics
Institute for Research in Fundamental Sciences (IPM), Iran

Eveline Lehmann [3]

Institute of Computer Science
University of Bern, Switzerland

Thomas Studer [4]

Institute of Computer Science
University of Bern, Switzerland

Abstract

Different notions of the consistency of obligations collapse in standard deontic logic. In justification logics, which feature explicit reasons for obligations, the situation is different. Their strength depends on a constant specification and on the available set of operations for combining different reasons. We present different consistency principles in justification logic and compare their logical strength. Further, we propose a novel semantics for which justification logics with the explicit version of axiom D, **jd**, are complete for arbitrary constant specifications. We then discuss the philosophical implications with regard to some deontic paradoxes.

Keywords: Justification logic, consistency of obligations, completeness.

---

[1] Supported by FWO and FWF Lise Meitner grant M 25-27 G32.
[2] This research was in part supported by a grant from IPM (No. 99030420).
[3] Supported by the Swiss National Science Foundation grant 200020_184625.
[4] Supported by the Swiss National Science Foundation grant 200020_184625.

1 Introduction

Deontic logic is the logic of obligations, permissions, and sometimes other (primitive or derived) normative notions. What has emerged as the benchmark version, a system called Standard Deontic Logic (SDL), is nothing more than **KD**, the smallest normal modal logic with the D axiom schema added. For an introduction and historical overview, see [10].

The D axiom is in place to ensure the consistency of obligations, but can take different formulations, for instance $\neg\mathcal{O}\bot$, or $\mathcal{O}A \to \neg\mathcal{O}\neg A$, or $\neg(\mathcal{O}A \wedge \mathcal{O}\neg A)$. In a normal modal logic, all these formulations are provably equivalent, and therefore it does not matter much which one is chosen. In a non-normal (but still classical) setting, for instance when an aggregation principle is missing, different versions of D are not interderivable, and it therefore matters which one is chosen, both for philosophical and for technical reasons (for deontic logic in a paraconsistent setting, see e.g. [7]). One might want to distinguish, conceptually, between an obligation for an impossible or logically contradictory state of affairs ($\mathcal{O}\bot$) on one hand, and multiple obligations for jointly inconsistent states of affairs ($\mathcal{O}A \wedge \mathcal{O}\neg A$) on the other, because the former might thought to be self-defeating or conceptually impossible, whereas the latter can derive from different background or contingent normative systems (e.g. ethics and the law) and are only practically unenforceable, but logically possible. Moreover, SDL and its variants in the standard modal language lack the power to distinguish the source of one situation (an obligation for the impossible) from the source of the other (inconsistent obligations), or to exclude one situation for logical reasons and admit the other for contingent reasons.

Justification logic [2,17] is an explicit version of modal logic originally developed to provide a logic of proofs [1,16]. Instead of formulas such as $\Box A$, the language of justification logic includes formulas such as $t : A$ saying, for instance, that *t justifies knowledge of A* or *A is obligatory because of reason t*, where t is a term representing the reason. Systems of justification logic are parameterized by a so-called *constant specification* that states which logical axioms do have a justification. Hence the constant specification can be used to calibrate the strength of a justification logic. Of particular interest are axiomatically appropriate constant specifications where every axiom has a justification. In that case the justification logic enjoys a constructive analogue of the modal necessitation rule. (See Sect. 3 for a formal definition of constant specification).

The explicit counterpart in justification logic of one version of D (in standard modal logic) was first formulated by Brezhnev [3] as axiom **jd**, i.e. $\neg(t : \bot)$. This axiom turned out to be rather notorious. Usually one can establish completeness of a justification logic for an arbitrary constant specification. However, in the presence of **jd** this is not the case. Systems that include **jd** usually need an axiomatically appropriate constant specification in order to be complete. Kuznets [12] defined M-models for justification logics with **jd** and Pacuit [24] presented F-models for **jd**. Modular models for **jd** have been studied in [15] and subset models for **jd** are introduced in [18]. For all these different semantics, an axiomatically appropriate constant specification is required in order to obtain a completeness result. Notable exceptions to this phenomenon are M-models (defined in [12]) and Fk-models (defined in [13]) for which completeness holds for arbitrary constant specifications.

The requirement of an axiomatically appropriate constant specification is often overlooked. In particular, this requirement is omitted in the completeness theorems given in [9] and [18] (although for the latter paper it seems that it has been corrected later [19]). In the case of [9] the appropriateness requirement is important since the solution to avoid some of the known paradoxes, such as Ross', is to restrict the constant specification. But then the resulting justification logic is not complete anymore.

In this paper we study in detail the **jd** axiom and related principles. We compare their logical strength and we investigate the role of the constant specification. After an informal discussion in Section 2, in Sections 3 and 4 we present the basic syntax and semantics of system JD (with axiom **jd**). In Section 5 we propose a novel semantics for which justification logics with **jd** are complete for arbitrary constant specifications. In Section 6, we consider system JNoC, which has a different version of the consistency axiom, **noc**. In Section 7, we establish that various formulations of consistency are equivalent only with an axiomatically appropriate constant specification.

Acknowledgements. We are grateful to the anonymous reviewers of DEON 2020 for many helpful comments.

2 Impossible vs Inconsistent Obligations: An Overview

Standard (implicit) systems of deontic logic conflate impossible and conflicting obligations. One thing is to say that nothing logically impossible can be obligatory, i.e. $\neg\mathcal{O}\bot$, another to say that there are not (or there should not be) conflicting provisions that are obligatory, i.e. $\neg(\mathcal{O}A \wedge \mathcal{O}\neg A)$. Standard systems can derive $\mathcal{O}\bot$ from $\mathcal{O}A \wedge \mathcal{O}\neg A$ and vice versa, thus suffering a collapse. One way to see the difference is that the former might be argued to be unacceptable for conceptual or logical reasons (e.g. that such an obligation would be conceptually self-defeating), whereas the latter might be argued to be unacceptable for contingent reasons (e.g. that such obligations cannot be fulfilled in reality, although can potentially still arise in real-life situations). [5,6] use minimal models, [25] uses multiple accessibility relations in the disjunctive truth condition of the ought operator: in such ways the authors avoid aggregation and therefore the collapse of impossible to inconsistent obligations (multi-relational semantics has also been used more recently, cf. e.g. [4]).

In justification logic we have the explicit counterparts **jd**: $\neg(t : \bot)$ and **noc**: $\neg(t : A \wedge t : \neg A)$, respectively, of the above implicit principles, giving rise to systems we call JD and JNoC (respectively). Corollary 7.2 establishes that the former implies the latter. Lemma 7.5 shows that the converse direction holds in the presence of an axiomatically appropriate constant specification. In this situation we have the same collapse as in the standard implicit systems. There are two options to avoid this consequence:

- In justification logic we can use the constant specification to adjust the power of the logical systems and thus avoid the collapse. Lemma 7.6 shows that $\neg(t : A \wedge t : \neg A)$ does not imply $\neg(t : \bot)$ if the constant specification is not axiomatically appropriate. Theorems 5.3 and 6.3 prove that JD and JNoC with an arbitrary constant specification are complete with regard to a novel semantics we develop.
- As explained in Remark 7.7, we can avoid the collapse even in the presence of an axiomatically appropriate CS. It suffices to consider a language without the + operation. We denote this system JNoC$^-$.

Avoiding this collapse is important in situations with conflicting obligations. Let us look at Sartre's Dilemma [22] as presented in [23]:

(i) It is obligatory that I now meet Jones (say, as promised to Jones, my friend).

(ii) It is obligatory that I now do not meet Jones (say, as promised to Smith, another friend).

In implicit standard deontic logic featuring the principle $\neg(\mathcal{O}A \wedge \mathcal{O}\neg A)$, we immediately get a contradiction if we represent (1) and (2) as $\mathcal{O}A$ and $\mathcal{O}\neg A$, respectively. However, in a system such as JNoC$^-$, there is no conflict as there are two different reasons in (1) and (2). Hence (1) and (2) are represented by $s : A$ and $t : \neg A$ for two different terms s and t, which is consistent with axiom **noc**.

Moreover, in normal deontic logic, one can pass from *two* inconsistent obligations to *one* impossible obligation. This is dubious on philosophical grounds: we have pointed out that one may argue that one impossible obligation is conceptually self-defeating, whereas two inconsistent obligations may be in place for contingent reasons (e.g. different promises).

Justification logic gives us the means to not conflate the two, without loosing too much reasoning power. Even more, one can do justice to the background philosophical intuitions to exclude impossible obligations for logical reasons, for instance by focusing on the system JNoC and calibrating the constant specification. Keeping track of the source of obligations, for instance through reasons, opens up the possibility to solve conflicts if one has a priority ordering on reasons (see for instance [11], and [8] for an implementation in justification logic).

In the rest of the paper we present the formal results starting from the basic syntax and semantics of system JD (with axiom **jd**).

3 Syntax

Justification terms are built from countably many constants c_i and variables x_i according to the following grammar:

$$t ::= c_i \mid x_i \mid t \cdot t \mid (t + t) \mid \,!t$$

The set of terms is denoted by Tm.

Formulas are built from countably many atomic propositions P_i and the symbol $\bot$ according to the following grammar:

$$F ::= P_i \mid \bot \mid F \to F \mid t : F$$

The set of atomic propositions is denoted by Prop and the set of all formulas is denoted by $\mathcal{L}_J$. The other classical Boolean connectives $\neg, \top, \land, \lor, \leftrightarrow$ are defined as usual, in particular we have $\neg A := A \to \bot$ and $\top := \neg \bot$. Informally, $+$ mimics the aggregation of reasons, $\cdot$ embodies modus ponens reasoning, and ! is positive introspection. We keep ! for ease of exposition, but it can be dispensed with. For a discussion on the interpretation of the operations in a deontic context, see [9].

The axioms of JD are the following:

cl all axioms of classical propositional logic;
j+ $s : A \lor t : A \to (s+t) : A$;
j $s : (A \to B) \to (t : A \to s \cdot t : B)$;
jd $\neg(t : \bot)$.

Note that since $\neg$ is a defined notion, **jd** actually stands for $t : \bot \to \bot$.

Justification logics are parameterized by a so-called constant specification, which is a set

$$\mathsf{CS} \subseteq \{(c, A) \mid c \text{ is a constant and } A \text{ is an axiom of JD}\}.$$

Our logic $\mathsf{JD}_{\mathsf{CS}}$ is now given by the axioms of JD and the rules modus ponens:

$$\frac{A \quad A \to B}{B} \text{ (MP)}$$

and axiom necessitation

$$\frac{}{\underbrace{!...!}_{n} c : \underbrace{!...!}_{n-1} c : \, ... \, : !!c : !c : c : A} \text{ (AN!)} \quad \forall n \in \mathbb{N}, \text{ where } (c, A) \in \mathsf{CS}$$

Definition 3.1 [Axiomatically appropriate CS] A constant specification CS is called *axiomatically appropriate* if for each axiom A, there is a constant c with $(c, A) \in \mathsf{CS}$.

Axiomatically appropriate constant specifications are important as they provide a form of necessitation [1,2,17].

Lemma 3.2 *Let CS be an axiomatically appropriate constant specification. For each formula A with*

$$\mathsf{JD}_{\mathsf{CS}} \vdash A,$$

there exists a term t such that

$$\mathsf{JD}_{\mathsf{CS}} \vdash t : A.$$

4 Semantics

We recall the basic definitions and results about subset models for justification logic [18,20,21].

Definition 4.1 [General subset model] Given some constant specification CS, then a general CS-subset model $\mathcal{M} = (W, W_0, V, E)$ is defined by:

- W is a set of objects called worlds.
- $W_0 \subseteq W$ and $W_0 \neq \emptyset$.
- $V : W \times \mathcal{L}_J \to \{0, 1\}$ such that for all $\omega \in W_0$, $t \in \mathsf{Tm}$, $F, G \in \mathcal{L}_J$:
  · $V(\omega, \bot) = 0$;
  · $V(\omega, F \to G) = 1$ iff $V(\omega, F) = 0$ or $V(\omega, G) = 1$;
  · $V(\omega, t : F) = 1$ iff $E(\omega, t) \subseteq \{ v \in W \mid V(v, F) = 1 \}$.
- $E : W \times \mathsf{Tm} \to \mathcal{P}(W)$ that meets the following conditions where we use

$$[A] := \{\omega \in W \mid V(\omega, A) = 1\}. \tag{1}$$

For all $\omega \in W_0$, and for all $s, t \in \mathsf{Tm}$:
 · $E(\omega, s + t) \subseteq E(\omega, s) \cap E(\omega, t)$;
 · $E(\omega, s \cdot t) \subseteq \{v \in W \mid \forall F \in \mathsf{APP}_\omega(s,t)(v \in [F])\}$ where APP contains all formulas that can be justified by an application of s to t, see below;
 · $\exists v \in W_0$ with $v \in E(\omega, t)$;
 · for all $n \in \mathbb{N}$ and for all $(c, A) \in \mathsf{CS} : E(\omega, c) \subseteq [A]$ and

$$E(\omega, \underbrace{!...!}_{n} c) \subseteq [\underbrace{!...!}_{n-1} c : ....!c : c : A].$$

The set APP is formally defined as follows:

$$\mathsf{APP}_\omega(s,t) := \{F \in \mathcal{L}_J \mid \exists H \in \mathcal{L}_J \text{ s.t.}$$
$$E(\omega, s) \subseteq [H \to F] \text{ and } E(\omega, t) \subseteq [H]\};$$

W_0 is the set of *normal* worlds. The set $W \setminus W_0$ consists of the *non-normal* worlds. Moreover, using the notation introduced by (1), we can read the condition on V for justification formulas $t : F$ as:

$$V(\omega, t : F) = 1 \quad \text{iff} \quad E(\omega, t) \subseteq [F]$$

In subset semantics terms are not treated only syntactically (as in most other semantics for justification logics), but they get assigned a set of worlds.

$E(\omega, t)$ tells us the states that are ideal according to t from ω's perspective. Then $t : F$ at ω is true just in case F is true at those ideal states. We have seen that a formula of the form $t : F$ is true at a world w just in case the interpretation of t at w (a set of worlds) is a subset of the truth set of F (the set of worlds where F is true). However, take two axioms A and B. They are true in all possible worlds. Therefore, every term that is a reason for the former will also be a reason for the latter (if terms get assigned sets of *possible* worlds). But in this way, there is no control on the constant specification. Using *impossible* worlds, however, lets us solve this problem, because at impossible worlds classically logically equivalent propositions can differ in truth value, and a justification for one may not be a justification for the other. This makes the semantics able to capture hyperintensionality.

Since the valuation function V is defined on worlds and formulas, the definition of truth is standard.

Definition 4.2 [Truth] Given a subset model

$$\mathcal{M} = (W, W_0, V, E)$$

and a world $\omega \in W$ and a formula F we define the relation $\Vdash$ as follows:

$$\mathcal{M}, \omega \Vdash F \quad \text{iff} \quad V(\omega, F) = 1.$$

Validity is defined with respect to the normal worlds.

Definition 4.3 [Validity] Let CS be a constant specification. We say that a formula F is *general* CS-*valid* if for each general CS-subset model

$$\mathcal{M} = (W, W_0, V, E)$$

and each $\omega \in W_0$, we have $\mathcal{M}, \omega \Vdash F$.

As expected, we have soundness [18].

Theorem 4.4 (Soundness) *Let* CS *be an arbitrary constant specification. For each formula F we have that if* $\mathsf{JD}_{\mathsf{CS}} \vdash F$, *then F is general* CS-*valid.*

However, completeness only holds if the constant specification is axiomatically appropriate [19].

Theorem 4.5 (Completeness) *Let* CS *be an axiomatically appropriate constant specification. For each formula F we have that if F is general* CS-*valid, then* $\mathsf{JD}_{\mathsf{CS}} \vdash F$.

One might need more control on the constant specification, e.g. by relinquishing the requirement that each axiom be justified. For instance, [9] argued that restricting the constant specification is one way to avoid certain deontic paradoxes, such as Ross'. In the next section, we prove soundness and completeness with regard to an arbitrary constant specification.

5 D-arbitrary subset models

We present a novel class of subset models for JD and establish soundness and completeness.

Definition 5.1 [D-arbitrary subset model] A *D-arbitrary* CS-*subset model* $\mathcal{M} = (W, W_0, V, E)$ is defined like a general CS-subset model with the condition

$$\exists v \in W_0 \text{ with } v \in E(\omega, t)$$

being replaced with

$$\exists v \in W_{\not\perp} \text{ with } v \in E(\omega, t)$$

where $W_{\not\perp} := \{\omega \in W \mid V(\omega, \perp) = 0\}$.

The notion of D-arbitrary CS-validity is now as expected.

Definition 5.2 [D-arbitrary validity] Let CS be a constant specification. We say that a formula F is *D-arbitrary CS-valid* if for each D-arbitrary CS-subset model $\mathcal{M} = (W, W_0, V, E)$ and each $\omega \in W_0$, we have $\mathcal{M}, \omega \Vdash F$.

We have soundness and completeness with respect to arbitrary constant specifications.

Theorem 5.3 (Soundness and Completeness) *Let* CS *be an arbitrary constant specification. For each formula F we have*

$$\mathsf{JD}_{\mathsf{CS}} \vdash F \quad \textit{iff} \quad F \textit{ is D-arbitrary CS-valid.}$$

The completeness proof is by a canonical model construction as in the case of general subset models [18]. We will only sketch main steps here. The canonical model is given as follows.

Definition 5.4 [Canonical Model] Let CS be an arbitrary constant specification. We define the canonical model $\mathcal{M}^C = (W^C, W_0^C, V^C, E^C)$ by:

- $W^C = \mathcal{P}(\mathcal{L}_J)$.
- $W_0^C = \{\, \Gamma \in W^C \mid \Gamma \text{ is maximal } \mathsf{JD}_{\mathsf{CS}}\text{-consistent set of formulas}\,\}$.
- $V^C(\Gamma, F) = 1 \quad \text{iff} \quad F \in \Gamma$;
- $E^C(\Gamma, t) = \{\, \Delta \in W^C \mid \Delta \supseteq \Gamma/t \,\}$ where

$$\Gamma/t := \{F \in \mathcal{L}_J \mid t : F \in \Gamma\}.$$

The essential part of the completeness proof is to show that the canonical model is a D-arbitrary CS-subset model.

Lemma 5.5 *Let* CS *be an arbitrary constant specification. The canonical model $\mathcal{M}^C$ is a D-arbitrary CS-subset model.*

Proof. Let us only show the condition

$$\exists v \in W_{\not\perp}^C \text{ with } v \in E(\omega, t) \tag{2}$$

for all $\omega \in W_0$ and all terms t.

So let t be an arbitrary term and $\Gamma \in W_0^C$. Since Γ is a maximal $\mathsf{JD}_{\mathsf{CS}}$-consistent set of formulas, we find $\neg(t : \perp) \in \Gamma$ and thus $t : \perp \notin \Gamma$. Let $\Delta := \Gamma/t$. We find that $\perp \notin \Delta$ and by definition $V^C(\Delta, \perp) = 0$. Thus $\Delta \in W_{\not\perp}^C$. Moreover, again by definition, $\Delta \in E^C(\Gamma, t)$. Thus (2) is established. □

Now the Truth lemma and the completeness theorem follow easily as in [18].

Remark 5.6 In subset models, it is possible to reduce application to sum by introducing a new term $\mathsf{c}^\star$, see [18]. Our completeness result also holds in the setting with $\mathsf{c}^\star$. However, the proof that the canonical model is well-defined is a bit more complicated as one has to consider the case of $\mathsf{c}^\star$ separately.

6 No conflicts

So far, we have considered the explicit version of $\neg \mathcal{O}\bot$. In normal modal logic, this is provably equivalent to $\neg(\mathcal{O}A \wedge \mathcal{O}\neg A)$. In this section we study the explicit version of this principle, which we call NoC (*No Conflicts*), saying that reasons are self-consistent. That is A and $\neg A$ cannot be obligatory for one and the same reason. The axioms of JNoC are the axioms of JD where **jd** is replaced with:

noc $\neg(t : A \wedge t : \neg A)$.

Accordingly, a constant specification for JNoC is defined like a constant specification for JD except that the constants justify axioms of JNoC.

Given a constant specification CS for JNoC, the logic $\mathsf{JNoC_{CS}}$ is defined by the axioms of JNoC and the rules modus ponens and axiom necessitation.

Definition 6.1 [NoC subset model] A *NoC CS-subset model*

$$\mathcal{M} = (W, W_0, V, E)$$

is defined like a general CS-subset model with the condition

$$\exists v \in W_0 \text{ with } v \in E(\omega, t)$$

being replaced with

$$\exists v \in W_{\mathsf{nc}} \text{ with } v \in E(\omega, t)$$

where $W_{\mathsf{nc}} := \{\omega \in W \mid \text{for all formulas A } (V(\omega, A) = 0 \text{ or } V(\omega, \neg A) = 0)\}$.

The notion of NoC CS-validity is now as expected.

Definition 6.2 [NoC validity] Let CS be a constant specification. We say that a formula F is *NoC CS-valid* if for each NoC CS-subset model $\mathcal{M} = (W, W_0, V, E)$ and each $\omega \in W_0$, we have $\mathcal{M}, \omega \Vdash F$.

Theorem 6.3 (Soundness and Completeness) *Let CS be an arbitrary constant specification. For each formula F we have*

$$\mathsf{JNoC_{CS}} \vdash F \quad \text{iff} \quad F \text{ is NoC CS-valid.}$$

Again the completeness proof uses the canonical model construction from Definition 5.4 except that we set

- $W_0^C = \{\, \Gamma \in W^C \mid \Gamma \text{ is maximal } \mathsf{JNoC_{CS}}\text{-consistent set of formulas}\,\}$.

Now we have to show that the defined structure is an NoC CS-subset model.

Lemma 6.4 *Let CS be an arbitrary constant specification. The canonical model $\mathcal{M}^C$ is an NoC CS-subset model.*

Proof. As before, we only show the condition

$$\exists v \in W_{\mathsf{nc}}^C \text{ with } v \in E(\omega, t) \tag{3}$$

for all $\omega \in W_0$ and all terms t.

So let t be an arbitrary term and $\Gamma \in W_0^C$. Let A be an arbitrary formula. Since Γ is a is maximal $\mathsf{JNoC}_{\mathsf{CS}}$-consistent set of formulas, we find

$$\neg(t : A \wedge t : \neg A) \in \Gamma$$

and thus $t : A \wedge t : \neg A \notin \Gamma$. Thus, again by maximal consistency,

$$t : A \notin \Gamma \text{ or } t : \neg A \notin \Gamma.$$

Let $\Delta := \Gamma/t$. We find that

$$A \notin \Delta \text{ or } \neg A \notin \Delta$$

and hence, by definition,

$$V^C(\Delta, A) = 0 \text{ or } V^C(\Delta, \neg A) = 0.$$

Thus $\Delta \in W_{\mathsf{nc}}^C$. Moreover, again by definition, $\Delta \in E^C(\Gamma, t)$. Thus (3) is established. □

Again the Truth lemma and the completeness theorem follow easily.

7 Formal comparison

The following lemmas establish the exact relationship between JD and JNoC. First we show that $\mathsf{JD}_{\mathsf{CS}}$ proves that reasons are consistent among them, i.e. that $\neg(s : A \wedge t : \neg A)$ holds for arbitrary terms s and t, which is the consistency principle used in [9].

Lemma 7.1 *Let CS be an arbitrary constant specification. Then $\mathsf{JD}_{\mathsf{CS}}$ proves $\neg(s : A \wedge t : \neg A)$ for all terms s, t and all formulas A.*

Proof. Suppose towards a contradiction that $s : A \wedge t : \neg A$. Thus we have $s : A$ and $t : \neg A$ where the latter is an abbreviation for $t : (A \to \bot)$ (by the definition of the symbol $\neg$). Thus using axiom **j**, we get $t \cdot s : \bot$ and by axiom **jd** we conclude $\bot$. □

Corollary 7.2 *For any constant specification CS, $\mathsf{JD}_{\mathsf{CS}}$ proves every instance of noc.*

Remark 7.3 It is only by coincidence that Lemma 7.1, and thus also Corollary 7.2, hold for arbitrary constant specifications. If we base our propositional language on different connectives (say $\wedge$ and $\neg$ instead of $\to$ and $\bot$), then Lemma 7.1 and Corollary 7.2 only hold for axiomatically appropriate constant specifications.

The proof of Lemma 7.1 is as follows. Since CS is axiomatically appropriate, there exists a term r such that

$$r : (\neg A \to (A \to \bot)) \qquad (4)$$

is provable where $\bot$ is defined as $P \land \neg P$ (for some fixed P) and $F \to G$ is defined as $\neg(F \land \neg G)$. From (4) and axiom **j** we get

$$t : \neg A \to r \cdot t : (A \to \bot).$$

Thus from $s : A \land t : \neg A$, we obtain $(r \cdot t) \cdot s : \bot$, which contradicts axiom **jd** as before.

Next we show that also JNoC$_{\mathsf{CS}}$ proves that reasons are consistent among them.

Lemma 7.4 *Let* CS *be an arbitrary constant specification. Then* JNoC$_{\mathsf{CS}}$ *proves* $\neg(s : A \land t : \neg A)$ *for all terms* s, t *and all formulas* A.

Proof. Suppose towards a contradiction that $s : A \land t : \neg A$ holds. Using axiom **j+** we immediately obtain $s + t : A \land s + t : \neg A$. By axiom **noc** we conclude $\bot$, which establishes $\neg(s : A \land t : \neg A)$. □

Next we show that JNoC$_{\mathsf{CS}}$ with an axiomatically appropriate constant specification proves $\neg(t : \bot)$.

Lemma 7.5 *Let* CS *be an axiomatically appropriate constant specification. Then* JNoC$_{\mathsf{CS}}$ *proves* $\neg(t : \bot)$ *for each term* t.

Proof. Because CS is axiomatically appropriate, there are terms r and s such that

$$r : (\bot \to P) \quad \text{and} \quad s : (\bot \to \neg P).$$

Therefore, we get

$$t : \bot \to r \cdot t : P \quad \text{and} \quad t : \bot \to s \cdot t : \neg P.$$

Thus we have $t : \bot \to (r \cdot t : P \land s \cdot t : \neg P)$. Together with the previous lemma, this yields $t : \bot \to \bot$, which is $\neg(t : \bot)$. □

Here the requirement of an axiomatically appropriate constant specification is necessary.

Lemma 7.6 *There exists a NoC* CS*-subset model* $\mathcal{M} = (W, W_0, V, E)$ *with some* $\omega \in W_0$ *such that*

$$\mathcal{M}, \omega \Vdash t : \bot$$

for some term t.

Proof. Consider the empty CS and the following model:

(i) $W = \{\omega, \nu\}$ and $W_0 = \{\omega\}$

(ii) $V(\nu, \bot) = 1$ and $V(\nu, F) = 0$ for all other formulas F

(iii) $E(\omega, t) = \{\nu\}$ for all terms t.

We observe that $\nu \in W_{\mathsf{nc}}$. So the model is well-defined. Further, we find $E(\omega, t) \subseteq [\bot]$. Since $\omega \in W_0$, we get $V(\omega, t : \bot) = 1$. We conclude

$$\mathcal{M}, \omega \Vdash t : \bot.$$

□

Remark 7.7 For Lemmas 7.4 and 7.5, the presence of the $+$ operation is essential. Consider a term language without $+$ and the logic JNoC^- being JNoC without $\mathbf{j+}$. Let CS be an axiomatically appropriate CS for JNoC^-. There is a NoC CS-subset model $\mathcal{M}$ for $\mathsf{JNoC}^-_{\mathsf{CS}}$ with a normal world ω such that
$$\mathcal{M}, \omega \Vdash s : P \wedge t : \neg P$$
for some terms s and t and some proposition P.

Hence if we drop the $+$ operation, we can have self consistent reasons without getting reasons that are consistent among them even in the presence of an axiomatically appropriate constant specification.

Instead of using an axiomatically appropriate constant specification, we could also add the schema $s : \top$ to $\mathsf{JNoC}_{\mathsf{CS}}$ in order to derive **jd**.

Lemma 7.8 *Let CS be an arbitrary constant specification. Let $\mathsf{JNoC}^+_{\mathsf{CS}}$ be $\mathsf{JNoC}_{\mathsf{CS}}$ extended by the schema $s : \top$ for all terms s. We find that*
$$\mathsf{JNoC}^+_{\mathsf{CS}} \vdash \neg(t : \bot) \quad \text{for each term } t.$$

Proof. The following is an instance of axiom **noc**
$$\neg(t : \bot \wedge t : \neg\bot).$$
Using the definition $\top := \neg\bot$ and propositional reasoning, we obtain
$$t : \top \to \neg(t : \bot).$$
Using $t : \top$ and modus ponens, we conclude $\neg(t : \bot)$. □

8 Remarks

There are two main advantages in using the justification logic framework to deal with deontic matters. First, one can explicitly track which reasons are reasons for what and perform operation on them, thus having a higher degree of accuracy in formal representations of normative reasoning: every obligation has a source. Puzzles and paradoxes such as Ross' are very easy to identify and, under a plausible set-up, disappear. In the present paper we have seen how justification logic provides a means to keep track of the source of impossible and inconsistent obligations, thus helping not to conflate the two.

Second, the framework allows for the hyperintensionality of obligation, namely that logically equivalent contents may not be normatively equivalent. In general it is not the case that if $t : F$ and $F \equiv G$, then $t : G$. This also ensures a finer-grained formal approach to everyday normative reasoning that is currently unavailable in more standard approaches.

When we come to the specific topic of the present paper, however, we have to remark that it is possible to distinguish between $\neg\mathcal{O}\bot$ and $\neg(\mathcal{O}A \wedge \mathcal{O}\neg A)$ also in some non-normal implicit modal systems, as we noted in Sect. 2, and in particular in Chellas' system **D** (cf. [5,6]), which dispenses with axiom schema M: $\mathcal{O}(A \wedge B) \to \mathcal{O}A \wedge \mathcal{O}B$.

Chellas minimal monadic deontic logic **D** builds as usual on PC, adds $\neg\mathcal{O}\bot$ as an axiom, and has rule ROM: $A \to B/\mathcal{O}A \to \mathcal{O}B$. In Chellas' logic the collapse is indeed avoided, because $\neg(\mathcal{O}A \wedge \mathcal{O}\neg A)$ is not derivable from $\neg\mathcal{O}\bot$.

How does Chellas' approach compare to the one developed in the present paper? Given the apparent similarities, let's focus on the differences, both technical and philosophical. Rule ROM could be questioned in a deontic context: however, this rule is fundamental in Chellas' system, therefore one cannot ignore it (selectively or not); whereas in a justification logic context we can have a finer-grained control on which axioms get an "automatic", as it were, normative justification, by fine-tuning the constant specification.

Philosophically, we can start from the semantic interpretation of the obligation operator. For Chellas, "OA is true at a possible world just in case the world has a non-empty class of deontic alternatives throughout which A is true. The picture is one of possibly empty collections of non-empty classes of worlds functioning as moral standards: what ought to be true is what is entailed by one of these moral standards [5, p.24]". Chellas uses a neighborhood semantics. A standard, for him, is a collection of propositions. A term, in the context of the present paper, is instead interpreted as a set of worlds.

Moreover, Chellas' system is still an implicit modal logic, so it cannot keep track and reason with the sources of obligations. And indeed this reading is consistent with Chellas' intended interpretation of the obligation operator: What ought to be true is what is entailed by one of these moral standards. But which? In a justification logic context, for instance, if one wants to retain Chellas' ideas to interpret terms as moral standards, one can keep track of which moral standard requires what.

9 Conclusion

We provided a novel semantics for justification logics with axiom D that does *not* require an axiomatically appropriate constant specification, i.e. not every axiom needs to be justified by a constant. This can be crucial to have more control on the logic and solve some traditional puzzles such as Ross'. Axiom D can be formulated in at least two equivalent ways in normal modal logic, either with inconsistent obligations ($\neg(\mathcal{O}A \wedge \mathcal{O}\neg A)$) or with one impossible obligation ($\neg\mathcal{O}\bot$). We proved that their explicit versions are interderivable in JD and JNoC only when the constant specification is axiomatically appropriate. In particular, our technical results are:

(i) JD$_{CS}$ proves **noc** for axiomatically appropriate CS and vice versa

(ii) JNoC$_{CS}$ proves **jd** for axiomatically appropriate CS.

(iii) JD$_{CS}$ proves **noc** for arbitrary CS only if the language is based on the Boolean connectives $\to$ and $\bot$.

(iv) JNoC$_{CS}$ does not prove **jd** for arbitrary CS.

(v) JNoC$_{CS}^-$ does not prove **jd** for axiomatically appropriate CS.

Having more control not only on how to formulate D, but also on how to specify the constant specification is philosophically perspicuous: it avoids conflating impossible and conflicting obligations and can encode why this is the case, e.g. for conceptual (logical) or contingent reasons.

Recently, it was shown that principle **noc** is also very useful for analyzing epistemic situations in the context of quantum physics [26].

References

[1] Artemov, S., *Explicit provability and constructive semantics*, Bulletin of Symbolic Logic **7** (2001), pp. 1–36.

[2] Artemov, S. and M. Fitting, "Justification Logic: Reasoning with Reasons," Cambridge University Press, 2019.

[3] Brezhnev, V. N., *On explicit counterparts of modal logics*, Technical Report CFIS 2000-05, Cornell University (2000).

[4] Calardo, E. and A. Rotolo, *Quantification in some non-normal modal logics*, Journal of Philosophical Logic **46** (2017), pp. 541–576.

[5] Chellas, B. F., *Conditional obligation*, in: S. Stenlund, editor, *Logical Theory and Semantic Analysis*, Reidel, 1974 pp. 23–33.

[6] Chellas, B. F., "Modal Logic. An Introduction," Cambridge University Press, Cambridge, 1980.

[7] Da Costa, N. and W. A. Carnielli, *On paraconsistent deontic logic*, Philosophia **16** (1986), pp. 293–305.

[8] Faroldi, F. L. G. and T. Protopopescu, *All-things-considered oughts* (2018), ms.

[9] Faroldi, F. L. G. and T. Protopopescu, *A hyperintensional logical framework for deontic reasons*, Logic Journal of the IGPL **27** (2019), pp. 411–433.

[10] Hilpinen, R. and P. McNamara, *Deontic logic: A historical survey and introduction*, in: D. Gabbay, J. Horty, X. Parent, R. van der Meyden and L. van der Torre, editors, *Handbook of Deontic Logic and Normative Systems*, College Publications, 2013 pp. 3–136.

[11] Horty, J. F., "Reasons as Defaults," Oxford University Press, 2012.

[12] Kuznets, R., *On the complexity of explicit modal logics*, in: P. G. Clote and H. Schwichtenberg, editors, *Computer Science Logic, CSL 2000, Proceedings*, LNCS **1862**, Springer, 2000 pp. 371–383, errata concerning the explicit counterparts of $\mathcal{D}$ and $\mathcal{D}4$ are published as [14].

[13] Kuznets, R., "Complexity Issues in Justification Logic," Ph.D. thesis, City University of New York (2008).
URL http://gradworks.umi.com/33/10/3310747.html

[14] Kuznets, R., *Complexity through tableaux in justification logic*, in: *Logic Colloquium 2008*, Bulletin of Symbolic Logic **15(1)**, Association for Symbolic Logic, 2009 p. 121, abstract.

[15] Kuznets, R. and T. Studer, *Justifications, ontology, and conservativity*, in: T. Bolander, T. Braüner, S. Ghilardi and L. Moss, editors, *Advances in Modal Logic, Volume 9*, College Publications, 2012 pp. 437–458.

[16] Kuznets, R. and T. Studer, *Weak Arithmetical Interpretations for the Logic of Proofs*, Logic Journal of the IGPL **24** (2016), pp. 424–440.

[17] Kuznets, R. and T. Studer, "Logics of Proofs and Justifications," College Publications, 2019.

[18] Lehmann, E. and T. Studer, *Subset models for justification logic*, in: R. Iemhoff, M. Moortgat and R. de Queiroz, editors, *Logic, Language, Information, and Computation - WoLLIC 2019* (2019), pp. 433–449.

[19] Lehmann, E. and T. Studer, *Subset models for justification logic*, E-print 1902.02707, arXiv.org (2019).

[20] Lehmann, E. and T. Studer, *Belief expansion in subset models*, in: S. Artemov and A. Nerode, editors, *Proceedings of Logical Foundations of Computer Science LFCS'20* (2020), pp. 85–97.
[21] Lehmann, E. and T. Studer, *Exploring subset models for justification logic* (in print).
[22] Lemmon, E. J., *Moral dilemmas*, Philosophical Review **71** (1962), pp. 139–158.
[23] McNamara, P., *Deontic logic*, in: E. N. Zalta, editor, *The Stanford Encyclopedia of Philosophy*, Stanford University, 2019, summer 2019 edition .
URL https://plato.stanford.edu/archives/sum2019/entries/logic-deontic
[24] Pacuit, E., *A note on some explicit modal logics*, in: *Proceedings of the 5th Panhellenic Logic Symposium* (2005), pp. 117–125.
[25] Schotch, P. K. and R. E. Jennings, *Non-Kripkean deontic logic*, in: R. Hilpinen, editor, *New Studies in Deontic Logic: Norms, Actions, and the Foundations of Ethics*, Springer, 1981 pp. 149–162.
[26] Studer, T., *A conflict tolerant logic of explicit evidence*, Logical Inverstigations (in print).

The Manchester Twins: Conflicts Between Directed Obligations

Stef Frijters [1]

Ghent University

Thijs De Coninck [2]

Ghent university

Abstract

Term-modal logic uses modal operators that are indexed with terms of the language, which allows for quantification over these operators. Term-modal *deontic* logics (TMDL) can capture reasoning with rules, directed, and undirected obligations. Using the rich language of TMDL, we identify different types of deontic conflicts between directed obligations and describe reasoning in the face of these conflicts. We develop several monotonic logics in the TMDL family and show that none is capable of capturing all plausible deontic principles, while also being conflict-tolerant. To remedy this we develop several non-monotonic extensions in the format of adaptive logics. We end by isolating one of these, **TMDL^m**, and commenting on it.

Keywords: Conflict-tolerant deontic logic, term-modal logic, first-order, undirected obligations, directed obligations.

1 Introduction

In deontic reasoning, one often encounters conflicting obligations. These conflicting obligations do not always result from conflicting moral theories or legal systems. Take, for example, the commonly accepted general rule: 'Doctors have an obligation to their patients to benefit the health of these patients'.[3] Taken on its own, this rule is perfectly consistent. However, in certain specific

---

[1] stef.frijters@kuleuven.be This paper was written while Stef Frijters held a PhD grant of the Research Foundation - Flanders on the research project "Towards a more integrated formal account of actual ethical reasoning, with applications in medical ethics." (G0D2716N).

[2] thijs.deconinck@ugent.be Thijs De Coninck holds a PhD grant fundamental research of the Research Foundation - Flanders (1167619N).

[3] We have a distributive reading of this rule, instead of a collective one. Thus we interpret it as "Every individual doctor has an obligation to each of their patients to benefit the health of that patient." and not as "The group of all doctors have an obligation ..." or "Each doctor has an obligation to the group of all of his/her patients ...". This sentence is also not meant to be interpreted as a generic sentence.

situations it can lead to deontic conflicts. Let us illustrate this with an example loosely based on the Manchester twins case [14,11], summarised by Kaveny:[4]

A pair of conjoined twins, known by the pseudonyms of "Jodie" and "Mary," were born in Manchester, England, hospital in August 2000. Mary's heart and lungs were essentially non-functioning; she was entirely dependent upon her connection with her stronger sister for survival. But Jodie's cardiovascular system could not continue to do the work necessary to support both babies indefinitely. Physicians predicted that without an operation to separate the twins, both babies soon would die, probably before their first birthday. Unfortunately, however, the surgical separation would be able to save only Jodie. Although likely to need several reconstructive operations, she was predicted to live a long and virtually normal life once her body was liberated from the burden of providing life support to her sister. Mary's fate would be very different; she was predicted to die in the course of the procedure. [11, p. 115]

In this specific situation, benefitting Jodie's health implies performing the operation, while benefitting Mary's health implies refraining from it. Both Jodie and Mary are patients of the same physician.[5] Thus, this physician has an obligation to Jody to perform the operation, and an obligation to Mary not to do so: a genuine *deontic conflict* [7].

We define a deontic conflict as a situation in which multiple obligations hold that are individually, but not jointly fulfillable. In our example, the physician can perform the surgery, or she can refrain from it, but she cannot do both. Thus, these two obligations are individually fulfillable, but not jointly. This differs from a situation in which one is faced with multiple obligations none of which is fulfillable. These are excluded by our definition of a deontic conflict.

We can be more precise about the kind of deontic conflict with which the physician is faced. This is a conflict between *directed obligations*. A directed obligation is characterized by the fact that it has both a *bearer* and a *counterparty*. The bearer of an obligation is the person who is (in principle) blamed if the obligation is not fulfilled. In the Manchester Twins case, the physician is the bearer of both conflicting obligations. A counterparty is the person *to whom* the bearer has the obligation [10,5]. In the Manchester twins case, Jodie is the counterparty to the directed obligation that the physician has to operate. Mary is the counterparty to the directed obligation that the physician has to not operate.

Under normal circumstances, i.e. at least when there are no conflicts, it is plausible that directed obligations imply *undirected obligations*. With undirected obligations, we mean obligations that are only tied to a bearer and not to a counterparty [10,5]. In this paper we consider undirected obligations to be

---

[4] We say that this example is 'loosely based on' the case, as the actual case was much more complicated than this summary suggests [14,11].

[5] In reality there was a team of physicians, all responsible for both Jodie and Mary, but we make abstraction of this.

action guiding in the sense that they should not offer contradictory demands [20]. Suppose that a has an obligation toward b to tutor b's daughter c (as a has promised b to do so). This directed obligation normally implies the undirected obligation on the part of a to tutor c. Such an implication is, however, not so straightforward in cases with a deontic conflict between directed obligations.

In this paper we develop several logics with the aim of capturing reasoning with possibly conflicting directed obligations. The logics should enable us to derive conflicts from general premise sets, while at the same time being weak enough not to trivialize these conflicts. Specifically, we will develop term-modal deontic logics (TMDL) in the vein of [5], based on the more general framework of term-modal logics [4].

Term-modal logics are first-order modal logics with modal operators that are indexed by terms of the language (variables and constants). This allows one to quantify over (the indexes of) modal operators. In [5], these term-modal operators are given a deontic interpretation, to allow for the formalisation of general deontic rules, directed, and undirected obligations. However, the logic presented in [5] is not conflict-tolerant. To develop conflict-tolerant TMDL, we will use the neighborhood semantics for term-modal logics developed in [6], instead of the relational semantics of [4] and [5].

The paper is organised as follows. We begin in Section 2 by setting out **DE**, a very weak term-modal deontic logic. In the same section, we also discuss a number of monotonic extensions of **DE**. These logics all allow us to derive directed obligations from more general premises and to capture different principles of reasoning with both directed and undirected obligations. The next section is devoted to deontic conflicts. We distinguish two kinds of conflicts between directed obligations and then describe reasoning in the face of these conflicts. We show that the monotonic logics of Section 2 cannot at the same time capture all plausible principles, while also tolerating conflicts. To remedy this, Section 4 is devoted to defeasible versions of two principles of deontic logic. We show how we can use these to extend the monotonic logics to non-monotonic adaptive logics [1,2,3,19]. We end the paper by presenting some avenues of future research (Section 5).

2 A family of monotonic term-modal deontic logics

This section is divided into four subsections. The first of these presents the formal language that will be used in all of the logics in this article. Section 2.2 is dedicated to a semantic characterization of the weakest logic that we present: **DE**. A sound and complete axiomatisation of **DE** is given in Section 2.3. After this we discuss some other plausible principles of deontic logic and the ways in which we can extend **DE** to obtain these.

2.1 The formal language and its interpretation

Let $C = \{a, b, \ldots\}$ be a countable set of constants and $V = \{x, y, \ldots\}$ a countable set of variables. We let $\alpha, \beta, \alpha_1, \ldots$ range over C and $\nu, \xi, \nu_1, \ldots$ over V. Let $T = C \cup V$ be the set of terms and let $\theta, \kappa, \theta_1, \ldots$ be the metavariables ranging over it. For each $n \in \mathbb{N}$, let $\mathcal{P}^n$ be a countable set of n-ary predicate symbols and let $\mathcal{P}$ denote the union of all $\mathcal{P}^n$. Note that our language includes propositional variables, i.c. the 0-ary predicate symbols.

The formal language $\mathcal{L}$ is defined by the following Backus-Naur form, where $\Pi \in \mathcal{P}^n$, $\theta, \kappa \in T$ and $\nu \in V$:

$$\varphi ::= \Pi(\theta_1, \ldots, \theta_n) \mid \theta = \kappa \mid \neg\varphi \mid \varphi \vee \varphi \mid \mathsf{O}_\theta\varphi \mid \mathsf{O}^\theta_\kappa\varphi \mid (\forall\nu)\varphi \mid [\mathsf{U}]\varphi$$

The other Boolean connectives are defined in the standard way. Additionally, $(\exists\nu)\varphi =_{df} \neg(\forall\nu)\neg\varphi$, $\mathsf{P}_\theta\varphi =_{df} \neg\mathsf{O}_\theta\neg\varphi$, $\mathsf{P}^\theta_\kappa\varphi =_{df} \neg\mathsf{O}^\theta_\kappa\neg\varphi$ and $\langle\mathsf{U}\rangle\varphi =_{df} \neg[\mathsf{U}]\neg\varphi$. We will write $\theta \neq \kappa$ instead of $\neg(\theta = \kappa)$.[6]

The notions of free and bound variables are as usual, with two additions (cf. Fitting et al. [4]): (1) The free occurrences of variables in $\mathsf{O}_\theta\varphi$ are all free occurrences of variables in φ and in addition θ if θ is a variable, and (2) the free occurrences of variables in $\mathsf{O}^\kappa_\theta\varphi$ are θ, if θ is a variable, κ, if κ is a variable, and all free occurrences of variables in φ. A wff φ is a sentence iff all the variables in φ are bound. Let $\mathcal{S}$ be the set of sentences of $\mathcal{L}$.

We interpret $\mathsf{O}^b_a\varphi$ as the directed obligation 'a has an obligation towards b that φ' and $\mathsf{O}_a\varphi$ as the undirected obligation 'a has an obligation that φ'. We will only use terms to refer to agents, and not to other objects, such as apples. In this way we can avoid being able to express sentences such as 'this apple has an obligation'.

$[\mathsf{U}]$ is a universal modal operator and we interpret $[\mathsf{U}]\varphi$ as 'φ is settled true'. This operator allows us to express more conflicts. As an example, we can look back at the tutoring case. Here, a had promised b to tutor c, say at three in the afternoon. As a result, a has an obligation towards b that a tutors c at three in the afternoon. Suppose that a has also promised their friend d to meet for an afternoon of playing computer games. The resulting (directed) obligation conflicts with the obligation that a has towards c, but only because it is impossible to fulfill both obligations. This is not a logical impossibility, but for all intents and purposes it is *settled true* that b does not both tutor c at three and also meets d for an afternoon of playing computer games. We can express this with the $[\mathsf{U}]$-operator.

$\mathcal{L}$ allows for a great deal of precision. Let Sx be interpreted as 'x performs the surgery'. In $\mathcal{L}$ we can express that it is obligatory for our physician (a), to perform the surgery, $\mathsf{O}_a Sa$, or that she has this obligation towards Jodie (j), $\mathsf{O}^j_a Sa$. $\mathcal{L}$ also has the expressive power to formalise sentences where the agent of the obligatory action is not the bearer of the obligation, such as in 'it is obligatory for the head of the hospital, b, that someone else performs the surgery': $\mathsf{O}_b(\exists x)(x \neq b \wedge Sx)$.[7] It is also possible to distinguish 'there is someone for whom it is obligatory to perform the surgery', $(\exists x)\mathsf{O}_x Sx$, from 'it

---

[6] Note that the brackets around $(\theta = \kappa)$ are strictly speaking unnecessary.

[7] The sentence 'it is obligatory for the head of the hospital, b, that someone else performs the surgery' should not be confused with 'it is obligatory for the head of the hospital, b, that b *brings it about that* someone else performs the surgery'. In the second sentence the agent of the obligatory action is also the bearer of the obligation, whereas that is not the

is obligatory for someone that someone performs the surgery', $(\exists x)O_x(\exists y)Sy$. Finally, we can express general rules such as the one from the introduction, i.e. that if x is a patient of y (Pxy), then y has an obligation towards x to benefit the health of x (Byx): $(\forall x)(\forall y)(Pxy \to O_y^x Byx)$.

2.2 DE, the weakest logic

We now present a semantic characterization of **DE**, the weakest logic in the TMDL-family. These semantics are based on the neighborhood semantics for term-modal logics in [6]. A **DE**-model is a tuple $M = \langle W, \mathcal{A}, N^P, N^D, I, w_a \rangle$. W is a state domain, consisting of possible worlds $w, w_1, \ldots$ and $\mathcal{A}$ is an agent-domain, consisting of agents $p, p_1, p_2, \ldots$. Both are non-empty and are allowed to be at most countably infinite. I is an interpretation function. The actual world w_a is used to determine validity in the model (Definition 2.6, this becomes important in Section 4).

Definition 2.1 A **DE**-model is a tuple $M = \langle W, \mathcal{A}, N^P, N^D, I, w_a \rangle$, where:
1. $W \neq \emptyset$
2. $\mathcal{A} \neq \emptyset$
3. $N^P : W \times \mathcal{A} \to \wp(\wp(W))$ is a *neighborhood function* of M
3.1 for all $w \in W$ and $p \in \mathcal{A}$: if $X \in N^P(w,p)$ and $X \subseteq Y \subseteq W$, then $Y \in N^P(w,p)$
3.2 for all $w \in W$ and $p \in \mathcal{A}$: $W \in N^P(w,p)$
3.3 for all $w \in W$ and $p \in \mathcal{A}$: $\emptyset \notin N^P(w,p)$
3.4 for all $w \in W$ and $p \in \mathcal{A}$: if $X, Y \in N^P(w,p)$, then $X \cap Y \in N^P(w,p)$
4. $N^D : W \times \mathcal{A} \times \mathcal{A} \to \wp(\wp(W))$ is a *neighborhood function* of M
4.1 For all $w \in W$ and $p_1, p_2 \in \mathcal{A}$: $\emptyset \notin N^D(w, p_1, p_2)$
5. I is an *interpretation* function such that:
5.1 $I : T \to \mathcal{A}$
5.2 $I : \mathcal{P}^n \times W \to \wp(\mathcal{A}^n)$ for every natural number $n \in \mathbb{N}$ such that $1 \leq n$
5.3 $I : \mathcal{P}^0 \to \wp(W)$
6. $w_a \in W$.

The neighborhood function N^P assigns to each world-agent pair a set of propositions that are obligatory for this agent (each proposition being a set of worlds). This will be used to interpret the undirected obligation operator. N^P has a number of conditions. The first of these ensures inheritance: if a proposition is obligatory, then what necessarily follows from this proposition will also be obligatory. The second condition ensures that what is necessary is obligatory, and the third ensures that what is impossible cannot be obligatory. The final condition corresponds to aggregation: if two propositions are obligatory, then their conjunction is obligatory as well. Taken together, this means

---

case in the first sentence. That obligations exist where the bearer is not the agent of the obligatory action has been argued in [5,12,10]. To properly express the second sentence, we could extend our language with a term-modal 'bring it about'-operator. The technical results in [6] combined with the neighborhood semantics of [9] allow one to give a sound and complete logic for this extended language. However, since this extension is not essential for what follows, we leave a development of this approach for future work.

that the undirected obligation operator behaves in much the same way as the obligation operator of standard deontic logic.

The neighborhood function N^D assigns to every triple consisting of a world and two agents a set of propositions that are obligatory for the first agent towards the second agent. Condition 4.1. ensures that what is obligatory, is also possible. The reason for this condition is that we do not want the logic to model unfulfillable directed obligations. We defined a conflict as a situation in which multiple (directed) obligations hold that can each be individually fulfilled, but which are not jointly fulfillable. The ought-implies-can principle for directed obligations that is expressed by condition 4.1. ensures that all directed obligations can indeed be individually fulfilled.

To interpret quantifiers, we define ν-alternatives, before we give the semantic clauses. As usual, for any $\varphi \in \mathcal{L}$ and **DE**-model $M = \langle W, \mathcal{A}, N^P, N^D, I, w_a \rangle$, $[\![\varphi]\!]_M =_{df} \{w \in W \mid M, w \vDash \varphi\}$.

Definition 2.2 [ν-alternative] For any $\nu \in V$, $M' = \langle W, \mathcal{A}, N^P, N^D, I', w_a \rangle$ is a ν-alternative to $M = \langle W, \mathcal{A}, N^P, N^D, I, w_a \rangle$ iff I' differs at most from I in the member of $\mathcal{A}$ that I' assigns to ν.

Definition 2.3 [Semantic Clauses] For any **DE**-model $M = \langle W, \mathcal{A}, N^P, N^D, I, w_a \rangle$:
SC1 $M, w \vDash P(\theta_1, \ldots, \theta_n)$ iff $\langle I(\theta_1), \ldots, I(\theta_n) \rangle \in I(P, w)$
SC1' $M, w \vDash P$ iff $w \in I(P)$
SC2 $M, w \vDash \neg \varphi$ iff $M, w \nvDash \varphi$
SC3 $M, w \vDash \varphi \vee \psi$ iff $M, w \vDash \varphi$ or $M, w \vDash \psi$
SC4 $M, w \vDash \theta = \kappa$ iff $I(\theta) = I(\kappa)$
SC5 $M, w \vDash O_\theta \varphi$ iff $[\![\varphi]\!]_M \in N^P(w, I(\theta))$
SC6 $M, w \vDash O_\theta^\kappa \varphi$ iff $[\![\varphi]\!]_M \in N^D(w, I(\theta), I(\kappa))$
SC7 $M, w \vDash (\forall \nu) \varphi$ iff for every ν-alternative M': $M', w \vDash \varphi$
SC8 $M, w \vDash [U] \varphi$ iff $M, w' \vDash \varphi$ for all $w' \in W$.

In the following three definitions we define semantic consequence, validity and validity in a model. In this last definition, we use the actual world.

Definition 2.4 Where $\Gamma \subseteq \mathcal{S}$ and $\varphi \in \mathcal{S}$, φ is a semantic consequence of Γ, $\Gamma \Vdash \varphi$, iff for every **DE**-model $M = \langle W, \mathcal{A}, N^P, N^D, I \rangle$ and $w \in W$: if $M, w \vDash \psi$ for all $\psi \in \Gamma$, then $M, w \vDash \varphi$.

Definition 2.5 Where $\Gamma \subseteq \mathcal{S}$ and $\varphi \in \mathcal{S}$, **DE** validates φ iff for every **DE**-model $M = \langle W, \mathcal{A}, N^P, N^D, I, w_a \rangle$ and $w \in W$: $M, w \vDash \varphi$.

Definition 2.6 Where $\varphi \in \mathcal{S}$, φ is valid in a model M, $M \vDash \varphi$, iff $M, w_a \vDash \varphi$

2.3 Axiomatisation of DE

A sound and strongly complete axiomatisation of **DE** is obtained by closing a complete axiomatisation of classical propositional logic (**CL**) with all instances of the axiom schemata in Table 1 under the rules of Table 2.[8] $\varphi(\theta/\kappa)$ is

---

[8] Soundness and completeness follow from previous results in [6]. See also [4,17,5].

the result of replacing all free occurrences of κ in φ by θ, relettering bound variables if necessary to avoid rendering new occurrences of θ bound in $\varphi(\theta/\kappa)$. $\varphi(\theta//\kappa)$ is the result of replacing various (not necessarily all or even any) free occurrences of θ in φ by occurrences of κ, again relettering if necessary [18, p. 57].

(UK)	$[U](\varphi \to \psi) \to ([U]\varphi \to [U]\psi)$	(UI)	$(\forall \nu)\varphi \to \varphi(\alpha/\nu)$
(UT)	$[U]\varphi \to \varphi$	(REF)	$\alpha = \alpha$
(U5)	$\langle U \rangle \varphi \to [U]\langle U \rangle \varphi$	(SUB)	$(\alpha = \beta) \to (\varphi \to \varphi(\alpha//\beta))$
(UBF)	$(\forall \nu)[U]\varphi \to [U](\forall \nu)\varphi$	(ND)	$\alpha \neq \beta \to [U]\alpha \neq \beta$
(DREU)	$(O_\alpha^\beta \varphi \wedge [U](\varphi \leftrightarrow \psi)) \to O_\alpha^\beta \psi$		
(DIC)	$O_\alpha^\beta \varphi \to \langle U \rangle \varphi$		
(PK)	$O_\alpha(\varphi \to \psi) \to (O_\alpha \varphi \to O_\alpha \psi)$		
(PIC)	$O_\alpha \varphi \to \langle U \rangle \varphi$		
(PN)	$[U]\varphi \to O_\alpha \varphi$		

Table 1
Axiom schemata

(MP)	if $\varphi \to \psi$ and φ, then ψ
(UG)	if $\vdash \varphi \to \psi(\alpha/\nu)$ and α not in φ or ψ, then $\vdash \varphi \to (\forall \nu)\psi$.
(UNEC)	if $\vdash \varphi$, then $\vdash [U]\varphi$

Table 2
Rules

There is little that is surprising in this axiomatisation. $[U]$ is an **S5**-operator, O_α is a normal modal operator satisfying the ought-implies-can principle and O_α^β is a classical modal operator with the ought-implies-can principle. The other schemes are familiar from first-order modal logic. What might be surprising is that we do not have the Barcan formula for the obligation-operators even though we work with a constant domain semantics. This is a result of using neighborhood semantics instead of relational semantics [6].

2.4 Some further principles for the directed obligation operator

In this section we discuss four more logical principles for the directed obligation operator: (DP), necessitation, inheritance and aggregation. Standard deontic logic (**SDL**) satisfies the last three, but each of these can also be given up (see for example [19]). By adding any combination of the four conditions to Definition 2.1, we can define extensions of **DE**. We do so in Table 3 on page 175. The first column gives the name of the logic, the next four columns the conditions that it satisfies.

As we stated in the introduction, under normal circumstances directed and undirected obligations are related to each other in a natural way. If a has towards b a directed obligation to tutor c, then a has an undirected obligation to tutor c. This principle, stating that directed obligations imply undirected obligations, will be called (DP): $O_\alpha^\beta \varphi \to O_\alpha \varphi$. We can validate it easily by adding the following condition (that we call (dp)): for all $w \in W$, $p_1, p_2 \in \mathcal{A}$ and $X \subseteq W$: if $X \in N^D(w, p_1, p_2)$ then $X \in N^P(w, p_1)$.

Necessitation is the principle that anything that is settled true, is also obligatory: $[\mathsf{U}]\varphi \to \mathsf{O}_\alpha^\beta \varphi$. We can validate it by adding the condition (n) to our models: for all $p_1, p_2 \in \mathcal{A}$ and $w \in W$: $W \in N^D(w, p_1, p_2)$.

Inheritance is the principle: $\mathsf{O}_\alpha^\beta(\varphi \wedge \psi) \to (\mathsf{O}_\alpha^\beta \varphi \wedge \mathsf{O}_\alpha^\beta \psi)$. It is validated by models satisfying the condition (m): for all $w \in W$ and $p_1, p_2 \in \mathcal{A}$, if $X \in N^D(w, p_1, p_2)$ and $X \subseteq Y \subseteq W$, then $Y \in N^D(w, p_1, p_2)$. Note that any models satisfying condition (m) also validates the principle that we will call inheritance*: $(\mathsf{O}_\alpha^\beta \varphi \wedge [\mathsf{U}](\varphi \to \psi)) \to \mathsf{O}_\alpha^\beta \psi$.

Finally, aggregation (between directed obligations with the same bearer) says that if φ and ψ are obligatory, then their conjunction is also obligatory: $(\mathsf{O}_\alpha^\beta \varphi \wedge \mathsf{O}_\alpha^\beta \psi) \to \mathsf{O}_\alpha^\beta(\varphi \wedge \psi)$. It corresponds to the condition (c): for all $w \in W$, $p_1, p_2 \in \mathcal{A}$ and $X, Y \in N^D(w, p_1, p_2)$, $X \cap Y \in N^D(w, p_1, p_2)$.

3 Deontic conflicts

We distinguish two different types of conflicts between directed obligations, before discussing the kind of reasoning that is employed when encountering such conflicts.

3.1 Types of deontic conflict

In the introduction we distinguished deontic conflicts from situations in which an impossible proposition is obligatory. We see a *deontic conflict* as a situation in which two or more obligations hold that are not jointly fulfillable, but neither of which is impossible to fulfill on its own. In the Manchester twins case, the doctor has an obligation towards Jodie to perform the surgery, and another obligation towards Mary not to perform the surgery. In this article, we focus on such conflicts between directed obligations with the same bearer.

We distinguish two kinds of deontic conflicts between directed obligations with the same bearer: bilateral and multilateral conflicts. Multilateral conflicts are conflicts between directed obligations with distinct counterparties (for example $\{\mathsf{O}_a^b Qa, \mathsf{O}_a^c \neg Qa\}$ or $\{\mathsf{O}_a^b Pa, \mathsf{O}_a^b Qa, \mathsf{O}_a^c \neg (Qa \wedge Pa)\}$ in a context where $b \neq c$). In the Manchester twins case there is such a multilateral conflict: Mary is the counterparty of one obligation, and Jodie of the other obligation. Bilateral conflicts are conflicts where all the obligations involved are directed and where the counterparty is the same for all those obligations (for example $\{\mathsf{O}_a^b Qa, \mathsf{O}_a^b \neg Qa\}$).[9]

Consider the following case of a bilateral conflict: A patient w with cystic fibrosis is in need of a life-saving blood transfusion by doctor b. However, w is a Jehovah's witness, and refuses the transfusion on religious grounds [15, pp. 34-35]. The same general rule holds as in the Manchester twins case: 'Doctors have an obligation to their patients to benefit the health of these

---

[9] In this paper we will not consider conflicts between directed obligations with different bearers, but it is possible to make analogous constructions for these.

patients'. From this rule and the information at hand it follows that 'Doctor b has an obligation towards w to administer a blood transfusion to w'. However, this time there is also a second rule in play: 'Doctors have an obligation to their patients to respect the autonomy of these patients'. Since patient w refuses a blood transfusion, respecting the autonomy of w necessarily implies not administering a blood transfusion to w. Hence, b is faced with a bilateral conflict: b has an obligation towards w to administer the blood transfusion, and b has another obligation towards w not to administer the blood transfusion.

Not every conflict is a conflict between the obligatoriness of a proposition and its negation. Sometimes, as in the tutoring and gaming example above, the incompatibility of obligatory propositions is not due to logical impossibility, but due to contingent circumstances. We can use the [U]-operator to express that two propositions are mutually incompatible. At other times, we will have conflicts between three or more obligations, e.g. $\{O_d^a(P \vee Q), O_d^b \neg P, O_d^c \neg Q\}$. Finally, it is also possible to have existentially quantified formulas as part of a conflict. Thus, we consider $(\exists x)(O_x^a Px \wedge O_x^b \neg Px)$ to be a multilateral conflict as well.

Premise sets will usually not explicitly contain formulas that fit neatly into the definition of deontic conflicts above. Instead, we have to deduce these by means of deontic reasoning. In the Manchester twins case, the premises are: (1) all doctors have an obligation to their patients to benefit the health of these patients, (2) Jodie and Mary are patients of physician a, (3) it is necessary that if physician a acts to benefit Jodie, then she does perform the surgery and (4) if physician a acts to benefit Marie, then she does not perform the surgery. We can express these premises in the language as follows:

(i) $(\forall x)(\forall y)(Pxy \rightarrow O_y^x Byx)$

(ii) $Pja \wedge Pma$

(iii) $[U](Baj \rightarrow Sa)$

(iv) $[U](Bam \rightarrow \neg Sa)$

No combination of these formulas fits the definition of a deontic conflict. We need a logic that is strong enough to derive a conflict from such a premise set, but does not lead to triviality once it does so. **DE** allows us to derive the conflict $\{O_a^j Baj, O_a^m \neg Bam, \neg \langle U \rangle (Baj \wedge Bam)\}$ (by (UI), (MP), (UK) and (UNEC)).[10]

3.2 Reasoning in the face of deontic conflicts

When we are faced with a deontic conflict, we do not throw our hands up in the air and forego any further reasoning. We also do not conclude that everything is suddenly obligatory. This leads us to our first desideratum: deontic conflicts should not be trivialized. This means that if we have a deontic conflict in our premises, we should not be able to derive $\bot$.

---

[10] With any of the logics in Table 3 that validate inheritance, we can derive $\{O_a^j Sa, O_a^m \neg Sa\}$ as well.

Any extension of **DE** that validates aggregation trivialises bilateral conflicts. Consider, for example, the premise set $\{O_a^b Qa, O_a^b \neg Qa\}$. By aggregation, we can derive $O_a^b(Qa \wedge \neg Qa)$. By the ought-implies-can principle, we can derive $\langle U \rangle (Qa \wedge \neg Qa)$. By **CL** and the S5 properties of $[U]$, we derive $\bot$.

When an extension of **DE** validates (DP), then it tolerates none of the conflicts identified above. From a conflict between directed obligations, we can derive a conflict between undirected obligations. Since O_α is a normal modal operator, **DE** trivialises such conflicts.

Name	(m)	(c)	(n)	(dp)	bilateral	multilateral
DE					✓	✓
DM	x				✓	✓
DC		x				✓
DR	x	x				✓
DN			x		✓	✓
DMN	x		x		✓	✓
DCN		x	x			✓
DK	x	x	x			✓
DE + DP				x		
DM + DP	x			x		
DC + DP		x		x		
DR + DP	x	x		x		
DN + DP			x	x		
DMN + DP	x		x	x		
DCN + DP		x	x	x		
DK + DP	x	x	x	x		

Table 3
The different monotonic logics

However, there are other possible desiderata than conflict-tolerance that we must take into account. For obligations that are not tainted by conflicts we want to be able to apply all the principles from Section 2.4 that we deem to be plausible. However, since (DP) and aggregation are incompatible with a logic that is conflict-tolerant, this means that we need defeasible versions of these principles.[11] If we find the principle (DP) plausible, then we should, for example, be able to derive $O_a Qa$ from $\{O_a^b Qa\}$ or from $\{O_a^b Qa, O_a^c Pa, O_a^d \neg Pa\}$, but not from $\{O_a^b Qa, O_a^b \neg Qa\}$. Similarly, if one finds aggregation of directed obligations plausible, but also wants the logic to be conflict-tolerant, then one would want to be able to derive $O_a^b(Qa \wedge Pa)$ from $\{O_a^b Qa, O_a^b Pa\}$, but not $O_a^b(Qa \wedge \neg Qa)$ from $\{O_a^b Qa, O_a^b \neg Qa\}$. This is the second desideratum.

---

[11] This kind of problem is typical for the type of solution to normative conflicts that we propose here. Goble notes the same problem for conflict-tolerant variants of **SDL** [7, p. 297]: if the logic is weak enough to be conflict-tolerant, then it does not validate all principles of **SDL**, and if the logic does validate all principles of **SDL**, then it is not conflict-tolerant.

Finally, we should note that we do not want conflicts between undirected obligations to be derivable from conflicting directed obligations. In this paper we are only concerned with conflicts between directed obligations. Our undirected obligations should be action guiding in the sense that they should not offer contradictory demands [20]. All undirected obligations of an agent should be jointly fulfillable. Therefore we are only concerned with logics that satisfy ought-implies-can and aggregation for undirected obligations.

4 Adaptive extensions

The principles (DP) and aggregation seem incompatible with tolerating the conflicts between directed obligations. Nevertheless, one can argue that these principles are plausible for obligations not involved in a conflict (cf. the second desideratum). In this section we develop adaptive logics that take this idea into account. These logics allow us to apply (DP) or aggregation in unproblematic cases, but block the application of the same principles for obligations that are in conflict.[12]

In Section 4.1 we explain the basic ideas of adaptive logics, using a toy logic. Section 4.2 sets out two problems with this toy logic. The last two sections, 4.3 and 4.4, develop adaptive logics based on the conflict-tolerant logics that were presented in Section 3, while taking the two problems of the toy logic into account.

4.1 Adaptive logic, a toy example

To explain adaptive logics, we will use a running toy example of such a logic. The motivating idea behind this toy logic is that we would like to have a logic where the principle (DP) is blocked only in case the obligations involved are in conflict with other obligations. For example, we should be able to derive $O_a \varphi$ from $\{O_a^b \varphi\}$, but not from $\{O_a^b \varphi, O_a^c \neg \varphi\}$. In this last premise set, $O_a^b \varphi$ is part of a conflict and so applying (DP) would lead to triviality.

Our toy logic uses the standard format of adaptive logics [19]. Every logic in the standard format is defined by a lower limit logic (LLL), a set of abnormalities and an adaptive strategy. For our present purposes the LLL can be any of the logics in Table 3 that does not validate (DP). The adaptive logic validates all of the valid formulas of the LLL, so taking a logic that validates (DP) as LLL will result in an adaptive logic that does not block (DP) in any circumstance. For this toy example, we will use **DK** as the LLL.

Abnormalities are those formulas that we want the logic to falsify as often as possible. How this 'as often as possible' is interpreted exactly is determined by the adaptive strategy. In our case we want all negations of instances of (DP) to be falsified as long as this does not lead to triviality. So we use $\Omega = \{O_\alpha^\beta \varphi \land \neg O_\alpha \varphi \mid \alpha, \beta \in C \text{ and } \varphi \in \mathcal{L}\}$ as the set of abnormalities.

---

[12] The main advantage of adaptive logic over other non-monotonic formalisms is that adaptive logics (in the standard format) give us a dynamic proof theory [19, p. 528]. This dynamic proof theory "explicates the dynamics of defeasible reasoning" [16, p. 9]. (Another advantage is the transparent handling of premise sets [16, pp. 87-88].)

For all logics in this section, including our toy logic, we will use the adaptive strategy known as 'minimal abnormality' [19]. To explain this strategy, we first need some preliminary definitions. We say that a model M is a model of Γ iff for all $\varphi \in \Gamma$, $M \models \varphi$. For any model M, $\text{Ab}(M) =_{\text{df}} \{\varphi \mid \varphi \in \Omega \text{ and } M \models \varphi\}$.

Definition 4.1 An LLL-model M of Γ is *minimally abnormal* iff there is no LLL-model M' of Γ such that $\text{Ab}(M') \subset \text{Ab}(M)$.

The models of our adaptive logic are those models of the LLL that are minimally abnormal. Take our toy logic and the premise set $\{O_a^b Q\}$. There are **DK** models M of this premise set such that $\text{Ab}(M) = \emptyset$. In all of these models $M \models O_a Q$ (otherwise $O_a^b Q \wedge \neg O_a Q \in \text{Ab}(M)$ and thus $\text{Ab}(M) \neq \emptyset$). Hence, $O_a Q$ is a semantic consequence of $\{O_a^b Q\}$ in our toy logic. However, if we take the premise set $\{O_a^b Q, O_a^b \neg Q\}$, then for all LLL-models M: $M \models O_a^b Q \wedge \neg O_a Q$ or $M \models O_a^b \neg Q \wedge \neg O_a \neg Q$. So $O_a Q$ is not a consequence of this premise set.

The standard format of adaptive logic also gives us a proof theory, and soundness and completeness for our adaptive logic. Due to space constraints we will not elaborate on this here. Instead, we refer interested readers to [19].

4.2 Two problems with the toy logic

In the previous section, we presented a toy logic that gives us an adaptive version of (DP). In this section, we present two problems of this toy logic. Then, in the next section, we will present an adaptive logic that solves these problems.

The first problem with the toy logic is that it is what adaptive logicians call a *flip-flop*. An adaptive logic is a flip-flop iff, for all premise sets from which an abnormality is derivable, the formulas that are derivable with the adaptive logic are the same as those derivable with the LLL [19]. In other words, in the presence of an abnormality, the adaptive logic collapses into the LLL.

To illustrate this, let us take the premise set $\{O_a^b Pa, O_a^c \neg Pa, O_a^d Qa\}$ as an example. Here there is a conflict between the obligations that a has towards b and c. So we would want to block the derivation of $O_a Pa$ and $O_a \neg Pa$. However, $O_a^d Qa$ is unproblematic and so we do want $O_a Qa$ to be derivable.

Sadly, this is impossible in the toy logic. To see this, consider the following three abnormalities: $O_a^b(Pa \vee \neg Qa) \wedge \neg O_a(Pa \vee \neg Qa)$, $O_a^c(\neg Pa \vee \neg Qa) \wedge \neg O_a(\neg Pa \vee \neg Qa)$ and $O_a^d Qa \wedge \neg O_a Qa$. Each minimally abnormal model validates at least one of these abnormalities and every one of these formulas is validated in at least one of the minimally abnormal models. Since the last formula is validated in some minimally abnormal models, one cannot derive $O_a Qa$ in the toy logic.

A second problem with the abnormalities of the toy logic can be illustrated by taking as a premise set $\{(\exists x) O_x^a Pb\}$. From this, we would want to be able to derive $(\exists x) O_x Pb$. However, there are minimally abnormal models of $\{(\exists x) O_x^a Pb\}$ that do not validate $\{(\exists x) O_x Pb\}$.

Take, for instance, a **DK**-model $M = \langle W, \mathcal{A}, N^P, N^D, I, w_a \rangle$ where $W = \{w_a, w_b, w_c\}$ and $\mathcal{A} = \{p_1, p_2\}$. For every $w \in W$ and $p \in \mathcal{A}$, let $N^P(w, p) = \{w_c\}$. Let $N^D(w_a, p_1, p_2) = \{w_b\}$ and for every $\langle w, p, p' \rangle \in \{\langle w, p, p' \rangle \mid w \in$

W and $p, p' \in \mathcal{A}\} \setminus \langle w_a, p_1, p_2 \rangle$, let $N^D(w, p, p') = \{w_c\}$. Let I be such that for all $\theta \in T$, $I(\theta) = p_2$, $I(P, w_b) = \{p_2\}$ and $I(P, w_1) = I(P, w_2) = \emptyset$. M validates $(\exists x)O_x^b Pb \wedge \neg(\exists x)O_x Pb$, but this formula is not an abnormality. M is a minimally abnormal model of $\{(\exists x)O_x^a Pb\}$ and does not validate $(\exists x)O_x Pb$, thus we cannot derive this in our toy logic.

4.3 Adaptive (DP)

The problem of flip-flops is well-known in the study of adaptive logics. We can use the following solution, taken from [13]. [13]

Let $\mathcal{L}^a$ be the literals in $\mathcal{L}$. Where $\Theta \subseteq \mathcal{L}^a$ is finite and non-empty, we define $\sigma_\theta^\kappa(\Theta)$ as follows:

$$\sigma_\theta^\kappa(\Theta) =_{df} \{O_\theta^\kappa(\bigvee \Theta') \wedge \neg O_\theta(\bigvee \Theta') \mid \Theta' \subseteq \Theta \text{ and } \Theta' \neq \emptyset\}$$

We define the set of abnormalities, Ω^I, as follows:

$$\Omega^I =_{df} \{\bigvee(\sigma_\alpha^\beta(\Theta)) \mid \Theta \subseteq \mathcal{L}^a, \Theta \neq \emptyset, \Theta \text{ is finite and } \alpha, \beta \in C\}$$

This approach gets rid of our flip-flop problem. Recall our example premise set from above: $\{O_a^b Pa, O_a^c \neg Pa, O_a^d Qa\}$. We could not derive $O_a Qa$, since there were minimally abnormal models validating $O_a^b Qa \wedge \neg O_a Qa$, as every model validated at least one of $O_a^b(Pa \vee \neg Qa) \wedge \neg O_a(Pa \vee \neg Qa)$, $O_a^c(\neg Pa \vee \neg Qa) \wedge \neg O_a(\neg Pa \vee \neg Qa)$ and $O_a^d Qa \wedge \neg O_a Qa$. However, with the new definition of abnormalities, the first two of these three formulas are no longer abnormalities, while the last still is. Thus, models that validate $O_a^b Qa \wedge \neg O_a Qa$ are no longer minimally abnormal. Hence, we can derive $O_a Qa$.

The second problem with the abnormalities of the toy logic (that is not solved by taking Ω^I as abnormalities) was illustrated by the premise set $\{(\exists x)O_x^a Pb\}$. From this, we would want to be able to derive $(\exists x)O_x Pb$. However, there are minimally abnormal models of $\{(\exists x)O_x^a Pb\}$ that do not validate $\{(\exists x)O_x Pb\}$. To solve both the first and second problem of the toy logic, we define the following two sets of abnormalities:

$$\Omega^1 =_{df} \{(\exists \nu) \bigvee(\sigma_\alpha^\nu(\Theta)) \mid \Theta \subseteq \mathcal{L}^a, \Theta \neq \emptyset, \Theta \text{ is finite, } \alpha \in C \text{ and } \nu \in V\}$$

$$\Omega^2 =_{df} \{(\exists \nu)(\exists \xi) \bigvee(\sigma_\nu^\xi(\Theta)) \mid \Theta \subseteq \mathcal{L}^a, \Theta \neq \emptyset, \Theta \text{ is finite and } \nu, \xi \in V\}$$

Let $\Omega^{DP} = \Omega^1 \cup \Omega^2$. Models that validate $\{(\exists x)O_x^a Pb\}$, but not $\{(\exists x)O_x Pb\}$ are no longer minimally abnormal with these new abnormalities. Hence, $\{(\exists x)O_x Pb\}$ is derivable.

If we had taken only Ω^2 as our set of abnormalities, then we could not derive $O_a \varphi$ from $\{O_a^b \varphi\}$, but only $(\exists x)O_x \varphi$. By taking only Ω^1, we run into the opposite problem: not being able to derive $(\exists x)O_x \varphi$ from $\{(\exists x)O_x^a \varphi\}$. Thus, we need the union of both.

---

[13] See also [8,19].

Taking the set Ω^{DP} as the abnormalities solves both problems. We can use this set to get a defeasible form of (DP) for any of the logics in Table 3 that do not validate (DP). We simply take the selected monotonic logic from Table 3 as LLL, Ω^{DP} as the set of abnormalities and minimal abnormality as the strategy. Each of these logics is tolerant to the same conflicts as its LLL, but is stronger than its LLL. In particular, it satisfies the second desideratum for (DP): when there are no conflicts, then we are able to apply (DP).

4.4 Adaptive aggregation

To satisfy desideratum 2 for aggregation of directed obligations, we need an adaptive form of this aggregation. A first suggestion might be to use as abnormalities the set of all formulas of the form $O_\alpha^\beta \varphi \wedge O_\alpha^\beta \psi \wedge \neg O_\alpha^\beta (\varphi \wedge \psi)$. However, this leads to similar problems as the two we identified in Section 4.2. Luckily, the solution of these problems is analogous to those presented in Section 4.3 for the adaptive form of (DP).

Again let $\mathcal{L}^a$ be the literals in $\mathcal{L}$ and let $\Theta \subseteq \mathcal{L}^a$ be finite and non-empty.

$$\tau_\theta^\kappa(\Theta, K) =_{df} \{O_\theta^\kappa(\bigvee \Theta') \wedge O_\theta^\kappa(\bigvee K') \wedge \neg O_\theta^\kappa(\bigvee \Theta' \wedge \bigvee K') \mid \Theta' \subseteq \Theta, K' \subseteq K$$

$$\text{and } \Theta', K' \neq \emptyset\}$$

$$\Omega_C^1 =_{df} \{\bigvee(\tau_\alpha^\beta(\Theta)) \mid \Theta \subseteq \mathcal{L}^a, \Theta \neq \emptyset, \Theta \text{ is finite and } \alpha, \beta \in C\}$$

$$\Omega_C^2 =_{df} \{(\exists \nu)(\exists \xi) \bigvee(\tau_\nu^\xi(\Theta)) \mid \Theta \subseteq \mathcal{L}^a, \Theta \neq \emptyset, \Theta \text{ is finite and } \nu, \xi \in V\}$$

$$\Omega_C^3 =_{df} \{(\exists \nu) \bigvee(\tau_\nu^\beta(\Theta)) \mid \Theta \subseteq \mathcal{L}^a, \Theta \neq \emptyset, \Theta \text{ is finite}, \nu \in V \text{ and } \beta \in C\}$$

$$\Omega_C^4 =_{df} \{(\exists \nu) \bigvee(\tau_\alpha^\nu(\Theta)) \mid \Theta \subseteq \mathcal{L}^a, \Theta \neq \emptyset, \Theta \text{ is finite}, \alpha \in C \text{ and } \nu \in V\}$$

$$\Omega^C =_{df} \Omega_C^1 \cup \Omega_C^2 \cup \Omega_C^3 \cup \Omega_C^4$$

The flip-flop problem would already have been solved by only taking Ω_C^1 as our set of abnormalities. This solution is analogous to the one in [13, p. 10] and can be seen as an adaptation to aggregation of the solution to the flip-flop problem in Section 4.3.

To solve the second problem we need all four of Ω_C^1-Ω_C^4. Without Ω_C^4 we would not be able to derive $(\exists x)O_a^x(Pa \wedge Qa)$ from $\{(\exists x)(O_a^x Pa \wedge O_a^x Qa\}$, i.e. we would not be able to apply aggregation to formulas with existential quantification over the counterparty. Similarly, without Ω_C^3 we would not be able to apply aggregation to formulas with existential quantification over the bearer, without Ω_C^2 we would have trouble with existential quantification over both the bearer and the counterparty, and without Ω_C^1 we would have problems with formulas without existential quantification.

Now we can take any of the monotonic logics from Table 3 that do not validate aggregation for directed obligations and use this logic as the LLL of an adaptive logic. We take Ω^C as the set of abnormalities and minimal abnormality as the strategy. The resulting adaptive logic satisfies desideratum

2 for aggregation and does not suffer from the two problems from Section 4.2. In addition, it is tolerant to the same conflicts as its LLL.

Let us illustrate that desideratum 2 is satisfied by taking **DMN** as the LLL. If we take as premises the set $\{O_a^b Qa, O_a^b Pa\}$, then we can derive $O_a^b(Qa \wedge Pa)$. Any models that do not validate $O_a^b(Qa \wedge Pa)$ are not minimally abnormal, as they validate the abnormality $O_a^b Qa, \wedge O_a^b Pa \wedge \neg O_a^b(Qa \wedge Pa)$. Similarly, from $\{O_a^b Qa, O_a^b Pa, O_a^c \neg Pa\}$ we can derive $O_a^b(Qa \wedge Pa)$. All models of the premise set that do not validate $O_a^b(Qa \wedge Pa)$, do validate the abnormality $O_a^b Qa \wedge O_a^b Pa \wedge \neg O_a^b(Qa \wedge Pa)$ and are therefore not minimally abnormal.[14]

4.5 Combining adaptive (DP) and aggregation

We can also combine adaptive aggregation and adaptive (DP). Take as LLL any logic from Table 3 that does not validate (DP) nor aggregation for directed obligations, take as abnormalities $\Omega^C \cup \Omega^{DP}$ and as a strategy minimal abnormality. The resulting logic is tolerant to both kinds of conflicts (as its LLL is tolerant to both) and satisfies desideratum 2 for both aggregation and (DP). We consider for a moment the strongest of these logics, the one with **DMN** as its LLL. For ease of reference, we will call it **TMDL**^{m}.

Imagine an extension of the Manchester Twins case where it is necessary for performing the surgery to prepare surgical equipment, $[U](Sa \to Ea)$. With **TMDL**^{m} we can derive from this and the premises (i)-(iv) from Section 3.1 that a has an obligation towards Jodie to prepare the surgical equipment: By (UI) and (MP), we can derive $O_a^j Baj$ from (i). By two applications of inheritance* (see Section 2.4), we first derive $O_a^j Sa$ and then $O_a^j Ea$. This seems appropriate for cases of multilateral conflicts where it is not clear which obligation (if any) should be given up. When there is only a multilateral conflict, then we can still derive the obligations that the bearer has towards each counterparty. However, when we decide for some extra-logical reason that the obligation not to perform the surgery prevails, then we might no longer be willing to make this derivation.

5 Conclusion

In this article we distinguished bilateral and multilateral conflicts. We developed a number of monotonic extensions of the term-modal deontic logic **DE**, and showed which of these tolerate what kinds of conflicts. We then noted that none of the conflict-tolerant extensions validate aggregation of directed obligations with the same bearer, or the derivation of undirected obligations from directed obligations. They did not even validate these for obligations that were not involved in any conflict. Since these principles are arguably plausible, we developed defeasible versions of (DP) and aggregation. This allows us to construct non-monotonic logics that validate these principles as much as possible.

All of this gives us a broad range of logics that tolerate bilateral and multilateral conflicts. Whatever combination of the principles discussed in Section

---

[14] Naturally, any logics whose LLL validates (DP) will trivialise this last premise set.

2.4 one finds plausible can be used to construct a conflict-tolerant logic. If one finds inheritance, necessitation, both or neither plausible, then one can use the conflict-tolerant logic in Table 3 that validates exactly these principles. If, in addition, one finds aggregation, (DP) or both plausible, then one can add these as defeasible principles, as was described in Section 4. We see this as the main result of the present paper.

This opens the door to different avenues of future research. One can also consider conflicts between directed or undirected obligations with different bearers. Another option is to involve impersonal obligations, i.e. obligations not tied to any bearer or counterparty. One could ask whether, in a conflict between an impersonal obligation and an undirected obligation, one of the kinds prevails over the other.

It is also possible to consider more involved formalisations of general rules, based on a more in depth account of conditional obligations. For this article we have used the material implication to interpret both general rules and conditional obligations. However, this does not take into account the possibility of exceptions to general rules, nor the problem of contrary-to-duty obligations. Integrating a richer account of conditional obligations with term-modal deontic logics might open the way to different conflict-tolerant deontic logics. This is especially interesting in conjunction with a point made in Section 4.5, that in the case of a multilateral conflict, **TMDL**^{m} allows one to derive the obligations of every bearer-counterparty pair as if there was no conflict.[15]

References

[1] Batens, D., *Dialectical dynamics within formal logics*, Logique et Analyse **29** (1986), pp. 161–173.
[2] Batens, D., *Dynamic dialectical logics*, Paraconsistent Logic. Essays on the Inconsistent (1989), pp. 187–217.
[3] Batens, D., *A universal logic approach to adaptive logics*, Logica Universalis **1** (2007), pp. 221–242.
[4] Fitting, M., L. Thalmann and A. Voronkov, *Term-modal logics*, Studia Logica **69** (2001), pp. 133–169.
    URL https://doi.org/10.1023/A:1013842612702
[5] Frijters, S., J. Meheus and F. Van De Putte, *Representing reasoning with rules and rights: Term-modal deontic logic*, in: S. Rahman, M. Armgardt and H. C. Nordtveit Kvernenes, editors, *New Systematic and Historic Studies in Legal Reasoning and Logic*, In Press .
[6] Frijters, S. and F. Van De Putte, *Classical term-modal logics* (Submitted).
[7] Goble, L., *Prima facie norms, normative conflicts, and dilemmas*, Handbook of deontic logic and normative systems **1** (2013), pp. 241–351.
[8] Goble, L., *Deontic logic (adapted) for normative conflicts*, Logic Journal of the IGPL **22** (2014), pp. 206–235.
[9] Governatori, G. and A. Rotolo, *On the axiomatisation of Elgesem's logic of agency and ability*, Journal of Philosophical Logic (2005), pp. 403–431.
[10] Herrestad, H. and C. Krogh, *Deontic logic relativised to bearers and counterparties*, Anniversary Anthology in Computers and Law (1995), pp. 453–522.

---

[15] The authors would like to thank Frederik Van De Putte, Joke Meheus, Federico Faroldi, Nathan Wood and two anonymous reviewers for their valuable feedback and suggestions.

[11] Kaveny, M. C., *Conjoined twins and catholic moral analysis: Extraordinary means and casuistical consistency*, Kennedy Institute of Ethics Journal **12** (2002), pp. 115–140.
[12] McNamara, P., *Agential obligation as non-agential personal obligation plus agency*, Journal of Applied Logic **2** (2004), pp. 117–152.
[13] Meheus, J., M. Beirlaen, F. Van De Putte and C. Straßer, *Non-adjunctive deontic logics that validate aggregation as much as possible* (2010), unpublished.
[14] Paris, J. J. and A. C. Elias-Jones, *"Do we murder Mary to save Jodie?" an ethical analysis of the separation of the Manchester conjoined twins*, Postgraduate Medical Journal **77** (2001), pp. 593–598.
URL http://pmj.bmj.com/content/77/911/593
[15] Parker, M. and D. Dickenson, "The Cambridge medical ethics workbook: Case studies, commentaries and activities," Cambridge University Press, 2001.
[16] Straßer, C., "Adaptive Logics for Defeasible Reasoning," Springer, 2014.
[17] Thalmann, L., "Term-modal logic and quantifier-free dynamic assignment logic," Ph.D. thesis, Acta Universitatis Upsaliensis (2000).
[18] Thomason, R. H., *Some completeness results for modal predicate calculi*, in: K. Lambert, editor, *Philosophical Problems in Logic*, Synthese Library **29**, Springer Netherlands, 1970 pp. 56 – 76.
[19] Van De Putte, F., M. Beirlaen and J. Meheus, *Adaptive deontic logics: a survey*, Journal of Applied Logics - IfCoLog Journal of Logics and their Applications (2019).
[20] Van De Putte, F. and C. Straßer, *A Logic for prioritized normative reasoning*, Journal of Logic and Computation **23** (2012), pp. 563–583.
URL https://doi.org/10.1093/logcom/exs008

Reason-Based Deontic Logic

Alessandro Giordani

Catholic University of Milan
Largo A. Gemelli, 1 - 20123 Milan, Italy
alessandro.giordani@unicatt.it

Abstract

This paper introduces a system of deontic logic based on the idea that obligations are grounded on reasons. A reason-based deontic system is worth considering for at least three reasons: it may shed light on the way in which obligations are generated; it allows us to cope with conflicts between reasons while avoiding conflicts between obligations; finally, it may help us to assess the question as to whether standard deontic logic is appropriate to model basic deontic reasoning. The system I propose is developed in a framework that combines standard and neighborhood semantics and it is proved to be sufficiently powerful to represent ordinary deontic reasoning and to successfully address some significant problems in deontic logic.

Keywords: practical reasons, *pro tanto* obligations, all things considered obligations.

1 Introduction

The aim of this paper is to develop a modal system of deontic logic based on the idea that consistent obligations are grounded on possibly inconsistent reasons.[1] This project, whose significance is due to the prominent role currently attributed to reasons in the study of normative concepts and normative systems,[2] has to address two general issues: (*i*) from the philosophical side, to devise what basic principles about reasons are to be assumed to deduce consistent oughts without incurring in counter-intuitive consequences; (*ii*) from the logical side, to construct a system of deontic logic characterized by those principles in a suitable semantic framework. These issues are taken into account in the following three sections. In section 2 the basic principles underlying the system are proposed, as emerging from the recent debate on the connections between reasons and obligations. In section 3, after having characterized the system both from a semantic and from an axiomatic point of view, it is shown that it can be exploited to solve some interesting deontic problems and that it

---

[1] The notions of reason and obligation we will cope with are the notions of *pro tanto* objective normative reason and *pro toto* objective normative obligation [7, ch.1 and ch.4].
[2] See [7], [19, part I], [20, ch.1], [21, ch.4], and [22, ch.4] for extensive and insightful presentations of the topic. See [25, especially part III and V] for an up-to-date discussion of the structure and the role played by reasons in practical argumentation and deliberation.

enables us to vindicate standard deontic logic as the logic of basic deontic reasoning. In the final section two recent accounts, similar in scope, are considered and compared with the present one, and it is shown that the basic intuitions on which those accounts rely can be appropriately interpreted in it.

2 Intuitive principles

We want to be able to capture basic forms of deontic reasoning involving connections between reasons and obligations. In this respect, two inference schemata are considered as highly desirable in the literature [11,16,17]: a schema for implicative reasoning (*IR*) and a schema for disjunctive reasoning (*DR*). These schemata can be shown to be valid if we assume two intuitive principles concerning why certain reasons follow from the presence of other reasons: *Consistent Closure* and *Consistent Conjunction*. However, the assumption of these principles generates a problem of deontic explosion. What we want is then a system of deontic logic enabling us both to derive *IR* and *DR*, given an appropriate interpretation of *Consistent Closure* and *Consistent Conjunction*, and to avoid undesired consequences.[3] The idea underlying the system here developed is that obligations are based on reasons, which are distinguished into three kinds: basic reasons, which are the central elements of deontic reasoning; combined reasons, which allow for a principle of *Consistent Conjunction*, but are not closed under consistent closure; and derivative reasons, which allow for a principle of *Consistent Closure*, but are not closed under consistent conjunction. As we will see, in such system, besides solving paradigmatic cases of deontic dilemmas, we can derive specific versions of *IR* and *DR* (theorem 3.15 below), while avoiding explosions (theorem 3.13 below).

Let us begin with presenting *IR* and *DR*. Let a be a generic agent.

(i) **Implicative reasoning** (*IR*):
  (a) it is obligatory for a to do ϕ;
  (b) ϕ entails ψ;
  (c) therefore, a has a reason to do ψ.

(ii) **Disjunctive reasoning** (*DR*):
  (a) it is obligatory for a to do $\phi \vee \psi$;
  (b) a has a reason to do $\neg \phi$;
  (c) therefore, a has a reason to do ψ.

Both kinds of reasoning are acceptable, provided we assume that obligations are based on reasons and that some intuitive closure principles concerning reasons are logically valid. Specifically, as to *IR*, suppose that obligations are based on reasons and that it is obligatory for a to do ϕ; then a has a reason to do ϕ; thus, *if* having a reason to do ϕ implies having a reason to do all that ϕ entails, *then* a has a reason to do ψ. Similarly, as to *DR*, suppose that obligations are based on reasons and that it is obligatory for a to do $\phi \vee \psi$; then a has a reason

---

[3] See [16] for an exposition of the current debate and an analysis of the analogies between cases of conflicting reasons and cases of conflicting obligations.

to do $\phi \vee \psi$; thus, *if a has a reason to do $\neg\phi$ and having reasons to do two conjuncts implies having a reason to do the conjunction, then a has a reason to do* $(\phi \vee \psi) \wedge \neg\phi$; hence, if having a reason to do something entails having a reason to do all that is entailed by that thing, a has a reason to do ψ.

In sum, if we allow for principles like:

(*Closure*) if ψ is a necessary condition of ϕ, then having a reason to do ϕ implies having a reason to do ψ;

(*Conjunction*) having a reason to do ϕ and having a reason to do ψ implies having a reason to do $\phi \wedge \psi$;

then we are able to account for the validity of IR and DR. Similarly, conditioned versions of these schemata, involving the possibility of ϕ and $\phi \wedge \psi$, are derivable from the following conditioned versions of *Closure* and *Conjunction*:

(*Consistent Closure*) if ϕ is possible and ψ is a necessary condition of ϕ, then having a reason to do ϕ implies having a reason to do ψ;

(*Consistent Conjunction*) having a reason to do ϕ and having a reason to do ψ implies having a reason to do $\phi \wedge \psi$, if it is possible to do $\phi \wedge \psi$.

Hence, conditioned versions of IR and DR turn out to hold under very mild assumptions.

2.1 Problems

When considering the consequences of adopting IR and DR, we encounter two basic problems [11]. The first and lighter one is that, under the intuitive assumption that reasons can conflict and that there is no reason for doing something impossible, *Conjunction* is untenable. To be sure, it is impossible to allow for conflicts of reasons, since having a reason for ϕ and a reason for $\neg\phi$ would immediately entail having a reason for $\phi \wedge \neg\phi$. Thus, *Conjunction* is to be abandoned. The second problem is more pressing. Suppose that we have a reason to do ϕ and a reason to do $\neg\phi$, and that both ϕ and $\neg\phi$ are possible. Suppose also that something, say doing ψ, does not entail doing ϕ, so that we can do ψ without doing ϕ. Since ϕ entails $\phi \vee \psi$, we have a reason to do $\phi \vee \psi$, by *Consistent Closure*. Since it is possible to do $\neg\phi \wedge \psi$, it is also possible to do $\neg\phi \wedge (\phi \vee \psi)$, by standard modal logic. Hence, by *Consistent Conjunction*, we have a reason to do $\neg\phi \wedge (\phi \vee \psi)$. Still, $\neg\phi \wedge (\phi \vee \psi)$ entails ψ, and so, by *Consistent Closure* we have a reason to do ψ. Thus, assuming that reasons can conflict, these two principles allow us to derive the following

Principle of Explosion. If we have conflicting reasons, then we have reasons to do anything independent of the content of the conflict.

The key problem to be addressed, in a framework allowing for rules corresponding to IR and DR, is then the following

Problem of Explosion. How to avoid that conflicting reasons generate explosions.

2.2 Strategies of solution

In light of the current debate on conflicts in deontic logic and the logic of reasons, two main strategies can be pursued in order to solve this problem.[4] According to the first one, we can put into question the validity of *Consistent Closure* and adopt a more limited principle to the effect that, if ψ is a necessary condition of ϕ, then having a reason to do ϕ entails having a reason to do ψ provided that we have no reason to avoid to do ϕ. It is not difficult to see that limiting the application of *Consistent Closure* this way blocks the possibility of inferring that we have a reason to do $\phi \vee \psi$ if we have a reason to do ϕ since, in the case we have considered, we also have a reason to do $\neg \phi$. According to the second strategy, we can put into question the validity of *Consistent Conjunction* and limit the application of the principle to a certain class of reasons, typically basic reasons, thus blocking the possibility of inferring that we have a reason to do $\neg \phi \wedge (\phi \vee \psi)$ if we have a reason to do $\neg \phi$ and a reason to do $\phi \vee \psi$, given that having a reason to do $\phi \vee \psi$ derives from having a reason to do ϕ. While both strategies are effective in preventing the derivation of explosions, the first one can do that only at a high cost, due to the fact that it prevents us from deriving that we have a reason to do $\phi \vee \psi$ if we have a reason to do ϕ and a reason to do ψ. In fact, when we have a reason to do ϕ and a reason to do ψ, we would like to accept that we also have a reason to do one of ϕ and ψ, even though ϕ and ψ cannot be done together.[5]

The system we are going to introduce is designed, among other things, to allow for rules like IR and DR and to provide a solution to the problem of explosion along the lines of the second strategy.

3 A system of reason-based deontic logic

In this section system RDL of reason-based deontic logic is introduced. Its language should be rich enough to describe different ways of operating with reasons. In particular, when arguing about what to do, we typically combine reasons and infer the existence of reasons from the presence of other reasons. Then, if we become aware that some reasons generate conflicts, we select the strongest ones, or the ones that seem to be the strongest in the circumstances, and combine them to identify a definite course of action. In addition, when assessing our actions, we discern things which are done for a reason and things done without reason. To take into account these distinctions, modal operators are introduced for saying that a reason to do something is basic (R_B), obtained by aggregation (R_C), by derivation (R_D), or by aggregating selected reasons (S_C). Furthermore, I introduce two modal operators for saying that something which is the case is supported by a reason (R) or by a selected reason (S). Finally, two deontic operators are considered, for obligations based on generic reasons (O_R) and obligations based on selected reasons (O_S).

---

[4] See [17] and [10,11,12] for comprehensive presentations.
[5] This point is cogently defended in [6,13,14,16,17].

Definition 3.1 The language $\mathcal{L}_{RDL}$ of RDL is based on a set $\{p_i\}_{i \in \mathbb{N}}$ of propositional variables and is defined according to the following grammar.

$$\phi ::= p_i \mid \neg\phi \mid \phi \wedge \phi \mid \Box\phi \mid \mathsf{R_B}\phi \mid \mathsf{R_C}\phi \mid \mathsf{R_D}\phi \mid \mathsf{R}\phi \mid \mathsf{O_R}\phi \mid \mathsf{S}\phi \mid \mathsf{S_C}\phi \mid \mathsf{O_S}\phi$$

$\mathcal{L}_{RDL}$ is a powerful language. This notwithstanding it can be interpreted in a very intuitive fashion on the basis of appropriate modal frames. Let us first present the intended meaning of the modal formulas. $\Box\phi$ states that ϕ is an unavoidable state of affairs under the circumstances. $\mathsf{R_B}\phi$ states that there is a *basic reason* to do ϕ. The notion of basic reason is here introduced as a primitive notion.[6] $\mathsf{R_C}\phi$ states that there is a *combined reason* to do ϕ, i.e., that ϕ is supported by a set of basic reasons opportunely combined, while $\mathsf{R_D}\phi$ states that there is a *derivative reason* to do ϕ, i.e., that ϕ is a consequence of something that is supported by a set of basic reasons opportunely combined. $\mathsf{R}\phi$ states that ϕ is the case in accordance with a reason, i.e., in typical cases, that the agent has seen to it that ϕ based on a certain reason. $\mathsf{S_C}\phi$ states that there is a *strong combined reason* to do ϕ, viewed as a reason that has passed a process of deliberation and selection run by the agent under specific circumstances, and $\mathsf{S}\phi$ states that ϕ is the case in accordance with a *strong reason*. Finally, a formula like $\mathsf{O_R}\phi$ states that ϕ is obligatory given the set of available reasons, while $\mathsf{O_S}\phi$ states that ϕ is obligatory given the set of available strong reasons. I will refer to $\mathsf{O_R}\phi$ and $\mathsf{O_S}\phi$ as reason-based obligations.

Remark 3.2 *Intuitively, we can use $\mathsf{R_B}$ $\mathsf{R_C}$, $\mathsf{R_D}$ to model different kinds of* pro tanto *practical reason, $\mathsf{O_R}$ to model the notion of* pro toto *practical reason and $\mathsf{O_S}$ to model the notion of* pro toto or all things considered *obligation.*

3.1 Semantics

The semantics for RDL builds on suitable combinations of neighborhood and standard semantics recently proposed in epistemic logic [7] and incorporates both a distinction between non-derivative and derivative reasons and a distinction between reasons and strong reasons.[8]

---

[6] This notion is widely used in epistemology, where a basic reason is associated with a basic source of justification, as acknowledged by standard foundationalist accounts. See [9] for an introduction and [1, ch.1 and ch.3] for further discussion. It is also becoming popular in ethics, where it parallels the notion of basic obligation: in some approaches, basic reasons are assumed to be primitive [7,16,21,22]; in others they are identified either with basic intrinsic desires and values [2] or with propositions constituting the antecedents of basic rules of action. In particular, the last interpretation is consistent both with the approach proposed in [21] and with the one developed in [13,14].

[7] Specifically, these semantics are used in evidence-based epistemic logic [26,27,28] and topological epistemic logic [3,4,5]. See [8,18] for an introduction to neighborhood semantics.

[8] The distinction between non-derivative and derivative reasons is the key element that will allow us to separate *Closure*, which is valid with respect to derivative reasons, from *Consistent Conjunction*, which is valid with respect to non-derivative combined reasons. A strategy based on this distinction is pursued in [16] to address the problem of explosion relative to reasons. The distinction between reasons and strong reasons is a version of the distinction between defeasible and defeasible but undefeated reasons [12,13,14].

Definition 3.3 A frame for $\mathcal{L}_{RDL}$ is a tuple $(W, \mathcal{R}, \mathcal{R}^+, \mathcal{S}^+)$, where $\mathcal{R}, \mathcal{R}^+, \mathcal{S}^+ \subseteq \wp(W)$ and $W, \mathcal{R}, \mathcal{R}^+, \mathcal{S}^+$ satisfy the following conditions

1. $\varnothing \neq W$;
2. $W \in \mathcal{R}$;
3. $\mathcal{R} \subseteq \mathcal{R}^+$;
4. if $X \in \mathcal{R}^+$, then $X \neq \varnothing$;
5. if $X \in \mathcal{R}^+$ and $Y \in \mathcal{R}^+$ and $X \cap Y \neq \varnothing$, then $X \cap Y \in \mathcal{R}^+$;
6. if $r(P) \neq \varnothing$, then $r(P) \in \mathcal{R}^+$, where $r(P) = \bigcup \{X \in \mathcal{R}^+ : X \subseteq P\}$;
7. if $X \in \mathcal{S}^+$ and $Y \in \mathcal{S}^+$ and $X \cap Y \neq \varnothing$, then $X \cap Y \in \mathcal{S}^+$;
8. if $s(P) \neq \varnothing$, then $s(P) \in \mathcal{S}^+$, where $s(P) = \bigcup \{X \in \mathcal{S}^+ : X \subseteq P\}$;
9. $W \in \mathcal{S}^+ \subseteq \mathcal{R}^+$.

In light of conditions 5, 6, we say that $\mathcal{R}^+$ is closed under *consistent aggregation* and *conditioned addition*. Let us comment on these elements in turn.

W is a set of states, viewed as the set of scenarios that are consistent with the background situation in which an agent is located, that is the set of scenarios that are possible given what is settled in the background. Condition 1 ensures that the background itself is consistent, so that there are indeed possible states.

$\mathcal{R}$ is a set of elements related to the basic reasons of an agent. In this framework $\mathcal{R}$ is sufficiently abstract to allow for different interpretations. In more detail, $\mathcal{R}$ can be interpreted in at least two different ways.

(i) As *a set of values identified with the agent's basic reasons*. The intuitive sense of $X \in \mathcal{R}$ is then that X is a basic reason viewed as an intrinsic value to be realized, so that X is the set of states where that value is actually realized. Accordingly, the interpretation of $X \subseteq P$ is that P is implied by one of the agent's basic reasons.

(ii) As *a set of propositions supported by the agent's basic reasons*. The intuitive sense of $X \in \mathcal{R}$ is then that there is a basic reason to do X, so that X is a proposition supported by the agent's basic reasons. Accordingly, the interpretation of $X \subseteq P$ is that P is *indirectly* supported by the agent's basic reasons, being entailed by X, which is *directly* supported by them.

Here I assume the second interpretation, under the general proviso that having a reason to do something, say P, is to be understood as having a reason (i) to do P, if P is not settled given the background and not realized, (ii) to preserve P, if P is not settled but realized, or (iii) to take P into account, if P is settled given the background. Hence, in light of (iii), condition 2, stating that W is in $\mathcal{R}$, captures the intuitive principle that an agent has always to take into account what is settled given the background.

$\mathcal{R}^+$ is the set of propositions supported by the *combined reasons* available to an agent, that is the set containing the propositions that an agent can support by combining basic reasons. Condition 3 states that $\mathcal{R}$ is a subset of $\mathcal{R}^+$, which corresponds to the requirement that taking a reason as it stands is a way of combining reasons. Condition 4 states that propositions supported by combined reasons, and so also by basic reasons, are consistent. The underlying idea is that an agent is able to combine reasons in a consistent way, and so that no combination of basic reasons supports a contradiction. Crucially, the fact that combined reasons are consistent in themselves does not exclude the possibility of conflicting reasons, that is of reasons supporting inconsistent propositions. To be sure, the fact that $X \in \mathcal{R}^+$ implies $X \neq \varnothing$ does not exclude the possibility that, for some $X, Y \in \mathcal{R}^+$, $X \cap Y = \varnothing$; what is excluded is only that $X \cap Y$ can be supported by a combined reasons, i.e., that $X \cap Y \in \mathcal{R}^+$. Finally, conditions 5 and 6 specify what kinds of operation of combination are available to an agent. Condition 5 underpins a principle that allows for operations of consistent aggregation, on the basis of which reasons that support two mutually consistent propositions are aggregated into a reason supporting their conjunction. Condition 6 allows for operations of conditioned addition, on the basis of which all the reasons supporting propositions that entail a proposition P can be added to obtain a new reason, which is in fact *the most stable reason* that supports P. This follows from the fact that $r(P)$ is the union of all propositions that entail P and that are supported by some available reasons. Therefore, all the reasons supporting $r(P)$ provide support for propositions that are stronger than P, and so are less stable, being reasons that can be attacked with less difficulty.

Remark 3.4 *I will refer to $\mathcal{R}^+$ as the set of all reasons and to $\mathcal{S}^+$ as the set of all strong reasons, thus identifying the concepts of reason and strong reason with the concept of combined reason and combined strong reason.*

Finally, $\mathcal{S}^+$ is the set of propositions supported by the *strong combined reasons* available to an agent. Conditions 7 and 8 are analogous to the corresponding conditions on $\mathcal{R}^+$ and ensure that the operations of composition available to the agent are operative with respect to the reasons in $\mathcal{S}^+$ as well. Condition 9 states that $\mathcal{S}^+$ is a subset of $\mathcal{R}^+$, which follows from the definition of $\mathcal{S}^+$, and that W is in $\mathcal{S}^+$, which is intuitive given the characterization of the notion of reason. The idea behind the introduction of $\mathcal{S}^+$ is that, given a certain background and a certain set of initial reasons, and given the possibility of conflicts, an agent has to weigh up the reasons that are stronger under the circumstances and arrive at a decision based on them. In this respect, let us note that the present framework is not committed to a specific procedure for weighing reasons, since what is important for our purposes is just the outcome of the process, that is the set of reasons that are eventually selected.

Definition 3.5 A model for $\mathcal{L}_{RDL}$ is a tuple $M = (W, \mathcal{R}, \mathcal{R}^+, \mathcal{S}^+, V)$, where $(W, \mathcal{R}, \mathcal{R}^+, \mathcal{S}^+)$ is a frame for $\mathcal{L}_{RDL}$ and $V : \{p_i\}_{i \in \mathbb{N}} \to \wp(W)$ is a modal valuation assigning propositions to propositional variables.

The notion of truth is defined as follows.

Definition 3.6 Let $M = (W, \mathcal{R}, \mathcal{R}^+, \mathcal{S}^+, V)$ be a model for $\mathcal{L}_{RDL}$. The truth of ϕ at a world $w \in W$ in M is defined through the following conditions, where $[\phi]^M = \{w : M, w \models \phi\}$.

$M, w \models p_i \Leftrightarrow w \in V(p_i)$
$M, w \models \neg \phi \Leftrightarrow M, w \not\models \phi$
$M, w \models \phi \wedge \psi \Leftrightarrow M, w \models \phi$ and $M, w \models \psi$
$M, w \models \Box \phi \Leftrightarrow [\phi]^M = W$
$M, w \models \mathsf{R_B} \phi \Leftrightarrow [\phi]^M \in \mathcal{R}$
$M, w \models \mathsf{R_C} \phi \Leftrightarrow [\phi]^M \in \mathcal{R}^+$
$M, w \models \mathsf{R_D} \phi \Leftrightarrow \exists X \in \mathcal{R}^+ (X \subseteq [\phi]^M)$
$M, w \models \mathsf{R} \phi \Leftrightarrow w \in r([\phi]^M)$
$M, w \models \mathsf{O_R} \phi \Leftrightarrow \forall X \in \mathcal{R}^+ \exists Y \in \mathcal{R}^+ (Y \subseteq X \cap [\phi]^M)$
$M, w \models \mathsf{S_C} \phi \Leftrightarrow [\phi]^M \in \mathcal{S}^+$
$M, w \models \mathsf{S} \phi \Leftrightarrow w \in s([\phi]^M)$
$M, w \models \mathsf{O_S} \phi \Leftrightarrow \forall X \in \mathcal{S}^+ \exists Y \in \mathcal{S}^+ (Y \subseteq X \cap [\phi]^M)$

The notion of logical consequence is defined as usual. So, $\Delta \Vdash_{RDL} \phi$ iff $M, w \models \Delta$ entails $M, w \models \phi$ for all $w \in W$ and models M for $\mathcal{L}_{RDL}$.

Definition 3.7 *RDL* is the logic of the class of models for $\mathcal{L}_{RDL}$.

The truth conditions reflect the intended meaning of the modal formulas. As expected, $\Box \phi$ is true just in case ϕ is true at all the states in W, $\mathsf{R_B}\phi$ is true just in case ϕ is supported by a basic reason in $\mathcal{R}$, and $\mathsf{R_C}\phi$ is true just in case ϕ is supported by a combined reason in $\mathcal{R}^+$. As to $\mathsf{R_D}$, the definition clarifies the distinction between non-derivative and derivative reasons in terms of the distinction between directly and indirectly supported propositions.[9] Thus, $\mathsf{R_D}\phi$ is true iff ϕ is supported by a derivative reason, i.e., iff ϕ is entailed by a proposition supported by a non-derivative reason in $\mathcal{R}^+$, i.e., iff ϕ is indirectly supported by a non-derivative reason in $\mathcal{R}^+$. $\mathsf{R}\phi$ is true just in case ϕ is true in a state in which what is supported by the most stable reason for ϕ is realized: ϕ is the case and that ϕ is the case is in accordance with a reason, since ϕ is supported by its most stable reason. Similarly, $\mathsf{S}\phi$ is true just in case ϕ is true in a state in which what is supported by the most stable strong reason for ϕ is realized, while $\mathsf{S_C}\phi$ is true just in case ϕ is supported by a strong combined reason in $\mathcal{S}^+$. Lastly, $\mathsf{O_R}\phi$ is true just in case every reason in $\mathcal{R}^+$ can be strengthened to a reason for ϕ. Hence, being obligatory given the whole set of available reasons is interpreted as being supported in a set of reasons that do not conflict on what is obligatory.[10] Similarly, $\mathsf{O_S}\phi$ is true just in case every reason in $\mathcal{S}^+$ can be strengthened to a strong reason for ϕ.

It is worth noting that *RDL* models are generalizations of uniform models in standard deontic logic, which are models of the form $(W, Ideal)$. Indeed, let $M = (W, \mathcal{R}, \mathcal{R}^+, \mathcal{S}^+, V)$ be such that $\mathcal{R} = \mathcal{R}^+ = \mathcal{S}^+ = \{W, Ideal\}$, where $\emptyset \neq Ideal \subseteq W$. Then $Ideal \subseteq [\phi]^M$ iff $\forall X \in \mathcal{S}^+ \exists Y \in \mathcal{S}^+(Y \subseteq X \cap [\phi]^M)$.

---

[9] In accordance with this definition, both basic and combined reasons count as non-derivative reasons, since both provide direct support to a proposition.

[10] It is not difficult to see that $\forall X \in \mathcal{R}^+ \exists Y \in \mathcal{R}^+(Y \subseteq X \cap [\phi]^M)$ if and only if $\forall X \in \mathcal{R}^+ \exists Y \in \mathcal{R}^+(Y \cap X \neq \emptyset$ and $Y \subseteq [\phi]^M)$. Thus, every reason in $\mathcal{R}$ is consistent with a reason for ϕ, and so ϕ is supported by reasons that are not in conflict relative to ϕ itself.

Thus, uniform standard deontic logic can be viewed as the logic determined by the class of models like M, where the standard notion of obligation is captured by O_S or, equivalently, by O_R. The underlying assumption in this case is that there is only one normative reason to be considered, namely the reason that is encoded in a consistent deontic code.

3.2 All things considered obligations

A standard approach for generating all things considered obligations from a set of available reasons has it that ϕ is obligatory when there is a *good* reason to perform ϕ, where the notion of good reason is defined in terms of an ordering relation on the set of reasons.[11] As an alternative, in line with the approach developed in [3,4,5,26,27,28], we may assume that it is obligatory to do ϕ when every reason can be strengthened to a reason for ϕ, that is when every reason is part of a set of reasons supporting ϕ. If every available reason can be strengthened to a reason for ϕ, then no available reason for ϕ can be outweighed by a stronger reason, and so the second approach is stricter than the first and generates less obligations. The approach I propose here is a combination of the ones just sketched and can be divided into two ideal stages. Suppose an agent is confronted with a deontic problem, e.g. whether she should do p. In the first stage, the agent implements the first approach by selecting within the set $\mathcal{R}^+$ of available reasons the set $\mathcal{S}^+$ of strong reasons, to be identified with the good reasons obtained after a process of deliberation. In the second stage, she checks whether every reason in $\mathcal{S}^+$ can be strengthened to a reason for p. If so, then she concludes that p is all things considered obligatory. Here, $\mathcal{S}^+$ can be thought of as generated from $\mathcal{R}^+$ by virtue of a suitable choice function, along the lines originally proposed in [23,24]. The reason why this procedure is adopted, instead of introducing an ordering relation on $\mathcal{R}^+$, is that it provides us with a more flexible device for modeling the outcome of a process of deliberation. In a more general setting, this approach can be developed in such a way that a set of triggered reasons $\mathcal{T}_w^+ = \tau(w) \subseteq \mathcal{R}^+$ is assigned to each state w by a specific function τ, and then a set of strong reasons $\mathcal{S}_w^+ = \sigma(w) \subseteq \mathcal{T}_w^+$ is selected by a choice function σ. In such a setting, various σ can be defined based on different properties of a choice function and the connections between the corresponding notions of all things considered obligation can be explored.

3.3 Axiomatization

Let us consider the following groups of axioms and rules.

Group 1: $KT5$ axioms and rules for $\Box$.

Group 2: KT axioms and rules for R and S.

---

[11] This approach can be implemented in different ways. A common option is to assume that there is a good reason to perform ϕ iff there is an *undefeated* reason for ϕ, i.e., iff there is a reason for ϕ and there is no stronger reason, given the ordering, against the performance of ϕ. A more sophisticated option, which incorporates the notion of undefeated reasons, is put forward in [13,14]. Section 4.2 gives a hint of how this option can be handled in RDL.

Group 3: minimal axioms for R_B, R_C, S_C, R_D.

BN: $\Box(\phi \leftrightarrow \psi) \land R_B\phi \to R_B\psi$
RN: $\Box(\phi \leftrightarrow \psi) \land R_C\phi \to R_C\psi$
SN: $\Box(\phi \leftrightarrow \psi) \land S_C\phi \to S_C\psi$
DN: $\Box(\phi \to \psi) \land R_D\phi \to R_D\psi$

Consistent conjunction for R_C and S_C:

RC: $R_C\phi \land R_C\psi \land \Diamond(\phi \land \psi) \to R_C(\phi \land \psi)$
SC: $S_C\phi \land S_C\psi \land \Diamond(\phi \land \psi) \to S_C(\phi \land \psi)$

Group 4: inclusions between modalities.

B1: $R_B\phi \to \Box R_B\phi$
B2: $\Box\phi \to R_B\phi$

R1: $R_C\phi \to \Box R_C\phi$
R2: $\Box\phi \to R_C\phi$
R3: $R_C\phi \land \phi \to R\phi$
R4: $\Diamond R\phi \to R_C R\phi$
R5: $O_R\phi \leftrightarrow \Box\neg R\neg\phi$

S1: $S_C\phi \to \Box S_C\phi$
S2: $\Box\phi \to S_C\phi$
S3: $S_C\phi \land \phi \to S\phi$
S4: $\Diamond S\phi \to S_C S\phi$
S5: $O_S\phi \leftrightarrow \Box\neg S\neg\phi$

I1: $S\phi \to R\phi$
I2: $S_C\phi \to R_C\phi$
I3: $R_B\phi \to R_C\phi$
I4: $R_C\phi \to R_D\phi$
I5: $R_D\phi \leftrightarrow \Diamond R\phi$

Theorem 3.8 *Axioms and rules in groups 1 – 5 are sound and complete with respect to the class of all models for $\mathcal{L}_{RDL}$.*

The proof is rather long and is presented in the extended version of the paper.

Fact 3.9 $O_R\phi \to R_D\phi$ and $O_S\phi \to R_D\phi$.

It follows from R5 and S5 by factivity of R and S, the logic of $\Box$, I5 and I1.

Fact 3.10 R_C *is not closed under necessary implication.*

To provide a counter-model for $\Box(\phi \to \psi) \land R_C\phi \to R_C\psi$ let M be such that $W = \{w_1, w_2, w_3\}$, $\mathcal{R} = \mathcal{R}^+ = \mathcal{S}^+ = \{W, \{w_1\}\}$, $V(p_1) = \{w_1\}$ and $V(p_2) = \{w_2\}$. Then, for all $w \in W$, $M, w \models \Box(p_1 \to p_1 \lor p_2)$, by the truth conditions of $\Box$, and $M, w \models R_C p_1$, since $[p_1]^M = \{w_1\} \in \mathcal{R}^+$. Still, $M, w \models R_C(p_1 \lor p_2)$ for no $w \in W$, since $\{w_1, w_2\} \notin \mathcal{R}^+$.

Fact 3.11 R_D *is not closed under consistent conjunction.*

To provide a counter-model for $R_D\phi \land R_D\psi \land \Diamond(\phi \land \psi) \to R_D(\phi \land \psi)$ let M be such that $W = \{w_1, w_2, w_3\}$, $\mathcal{R} = \mathcal{R}^+ = \mathcal{S}^+ = \{W, \{w_1\}, \{w_2, w_3\}\}$, $V(p_1) = \{w_1\}$ and $V(p_2) = \{w_2\}$. Then, for all $w \in W$, $M, w \models R_C(p_1 \lor p_2)$, since $[p_1]^M = \{w_1\} \in \mathcal{R}^+$ and $[p_1]^M \subseteq [p_1 \lor p_2]^M$, and $M, w \models R_D \neg p_1$, since $[\neg p_1]^M = \{w_2, w_3\} \in \mathcal{R}^+$ and $M, w \models \Diamond((p_1 \lor p_2) \land \neg p_1)$, since $[(p_1 \lor p_2) \land \neg p_1]^M = \{w_2\}$. Still, $M, w \models R_D((p_1 \lor p_2) \land \neg p_1)$, for no $w \in W$: $X \subseteq \{w_2\}$ for no $X \in \mathcal{R}^+$.

Fact 3.12 *Basic reasons do not entail obligations.*

To provide a counter-model for $R_B\phi \to O_R\phi$ let M be such that $W = \{w_1, w_2\}$, $\mathcal{R} = \mathcal{R}^+ = \mathcal{S}^+ = \{W, \{w_1\}, \{w_2\}\}$, $V(p_1) = \{w_1\}$. Then, for all $w \in W$, $M, w \models R_B p_1$, since $[p_1]^M = \{w_1\} \in \mathcal{R}$. Still, $M, w \models O_R p_1$ for no $w \in W$, since there is no $Y \in \mathcal{R}^+$ such that $Y \subseteq \{w_2\} \cap [\phi]$, given that $\{w_2\} \cap [\phi] = \varnothing$.

By I3 and I4 we obtain that neither derivative nor non-derivative reasons entail obligations. Furthermore, in accordance with the intuitive interpretation given in remark 3.2, we conclude that *pro tanto* obligations do not entail *pro toto* obligations, as it should be.

3.4 Solving a paradigmatic dilemma

In order to show how deontic reasoning is modeled in the present framework let us provide a solution to two versions of a paradigmatic dilemma.

> Alice promised both Bob and Carl that she would dine with them. The promises are equally important, but she prefer not to dine with them together, given that Bob doesn't like Carl.

In the first version Alice has no particular preference, while in the second version she would prefer to dine with Bob, since she is interested in him. Intuitively, we would like to derive that, in the first version, Alice ought to dine with one of them and that, in the second version, she ought to dine with Bob.

Model 1 $W = \{w_1, w_2, w_3\}$, $\mathcal{R} = \{W, \{w_1, w_2\}, \{w_1, w_3\}, \{w_2, w_3\}\}$, $\mathcal{R}^+ = \mathcal{S}^+ = \wp(W) - \{\varnothing\}$. Set $V(p_1) = \{w_1, w_2\}$ and $V(p_2) = \{w_2, w_3\}$, where p_1 stands for dining with Bob and p_2 stands for dining with Carl.

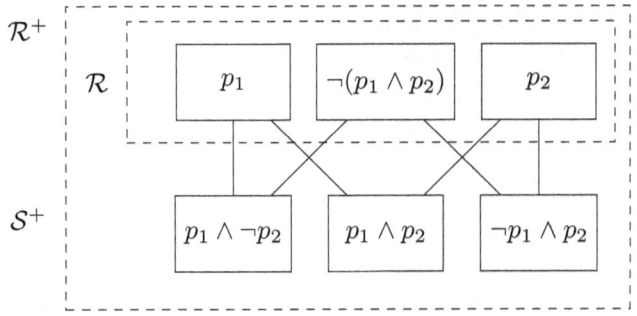

Model 1: only reasons different from W are represented

There is a reason for dining with Bob and a reason for dining with Carl, but there is no obligation for dining with one of them in particular, or with both, since the reasons in $\mathcal{R}^+$ cannot be strengthened to a reason for one of $p_1 \wedge \neg p_2$, $p_1 \wedge p_2$, $\neg p_1 \wedge p_2$. Still, there is an obligation to do $p_1 \vee p_2$, since any reason in $\mathcal{S}^+ = \mathcal{R}^+$ can be strengthened to a reason for doing $p_1 \vee p_2$.

Model 2 Let M be as before except that $\mathcal{S}^+ = \{W, \{w_1, w_2\}, \{w_1, w_3\}, \{w_1\}\}$.

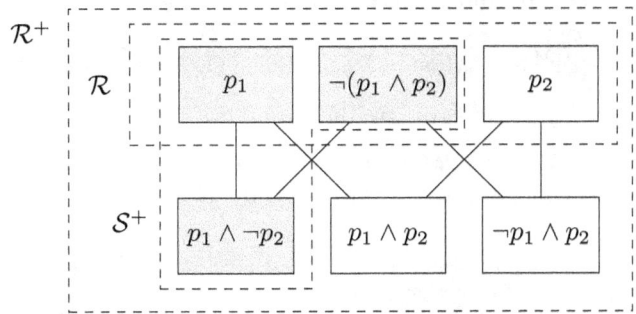

Model 2: only reasons different from W are represented

Again, there is a reason for dining with Bob and a reason for dining with Carl, but now there is an obligation to dine with Bob, since any reason in $\mathcal{S}^+$ can be strengthened to a reason for doing $p_1 \land \neg p_2$, which is $\{w_1\}$.

3.5 Solving the problem of explosion

The present framework allows for a solution of the problem of explosion based on the distinction between non-derivative and derivative reasons.[12] In particular, in our system we can prove that there is no valid principle of explosion of the following form, where $C(\phi_1, \phi_2)$ stand for $\mathsf{R}_\mathsf{B}\phi_1 \land \mathsf{R}_\mathsf{B}\phi_2 \land \neg \Diamond(\phi_1 \land \phi_2)$ [13]

E1: $\mathsf{R}_\mathsf{B}\phi \land \mathsf{R}_\mathsf{B}\neg\phi \to \mathsf{R}_\mathsf{D}\psi$
E2: $\mathsf{R}_\mathsf{B}\phi \land \mathsf{R}_\mathsf{B}\neg\phi \land \Diamond\psi \to \mathsf{R}_\mathsf{D}\psi$
E3: $\mathsf{R}_\mathsf{B}\phi \land \mathsf{R}_\mathsf{B}\neg\phi \land \neg\mathsf{R}_\mathsf{D}\neg\psi \to \mathsf{R}_\mathsf{D}\psi$

GE1: $C(\phi_1, \phi_2) \to \mathsf{R}_\mathsf{D}\psi$
GE2: $C(\phi_1, \phi_2) \land \Diamond\psi \to \mathsf{R}_\mathsf{D}\psi$
GE3: $C(\phi_1, \phi_2) \land \neg\mathsf{R}_\mathsf{D}\neg\psi \to \mathsf{R}_\mathsf{D}\psi$

E2 entails E1, by propositional logic, and E3 entails E2, since $\Diamond\psi$ follows from $\neg\mathsf{R}_\mathsf{D}\neg\psi$, by B2, I3, I4. In addition, the invalidity of the basic principles Ei entails the invalidity of the corresponding generalized principles GEi, and therefore it is sufficient to prove the following

Theorem 3.13 $\mathsf{R}_\mathsf{D}\psi$ *is not a logical consequence of* $\{\mathsf{R}_\mathsf{B}\phi, \mathsf{R}_\mathsf{B}\neg\phi, \neg\mathsf{R}_\mathsf{D}\neg\psi\}$. *Hence, basic reasons can conflict without implying that anything independent of the content of the conflict be supported by a reason.*

Let $W = \{w_1, w_2, w_3, w_4\}$, $\mathcal{R} = \mathcal{R}^+ = \mathcal{S}^+ = \{W, \{w_1, w_2\}, \{w_3, w_4\}\}$. Set $V(p_1) = \{w_1, w_2\}$ and $V(p_2) = \{w_1, w_3\}$, so that p_2 is independent of the content of p_1. Then, for all $w \in W$, $M, w \models \mathsf{R}_\mathsf{B}p_1 \land \mathsf{R}_\mathsf{B}\neg p_1$, by the definition of truth, and $M, w \models \neg\mathsf{R}_\mathsf{D}\neg p_2$, since $X \cap [p_2]^M \neq \varnothing$ for all $X \subseteq \mathcal{R}^+$. Still, $M, w \models \mathsf{R}_\mathsf{D}p_2$ for no $w \in W$, since $X \subseteq [p_2]^M$ for no $X \subseteq \mathcal{R}^+$.

Corollary 3.14 *None of* $\mathsf{R}_\mathsf{B}\psi, \mathsf{R}_\mathsf{C}\psi, \mathsf{R}_\mathsf{D}\psi$ *follows from one of the sets obtained by substituting one of* $\{\mathsf{R}_\mathsf{C}\psi, \mathsf{R}_\mathsf{D}\psi\}$ *for* $\mathsf{R}_\mathsf{B}\psi$ *in* $\{\mathsf{R}_\mathsf{B}\phi, \mathsf{R}_\mathsf{B}\neg\phi, \neg\mathsf{R}_\mathsf{D}\neg\psi\}$ *or from one of the antecedents of GE1, GE2, GE3.*

Next, we show that *RDL* is powerful enough to derive *IR* and *DR*.

Theorem 3.15 *In RDL the rules corresponding to schemata IR and DR are derivable, when the reasons are interpreted as derivative reasons.*

Suppose $M, w \models \mathsf{O}_\mathsf{R}\phi$ and $M, w \models \Box(\phi \to \psi)$. Then $\forall X \in \mathcal{R}^+ \exists Y \in \mathcal{R}^+(Y \subseteq X \cap [\phi]^M)$ and $[\phi]^M \subseteq [\psi]^M$, so that $\forall X \in \mathcal{R}^+ \exists Y \in \mathcal{R}^+(Y \subseteq X \cap [\psi]^M)$. Thus $\exists Y \in \mathcal{R}^+(Y \subseteq [\psi]^M)$; so $M, w \models \mathsf{R}_\mathsf{D}\phi$. Hence $\mathsf{O}_\mathsf{R}\phi, \Box(\phi \to \psi) \Vdash_{RDL} \mathsf{R}_\mathsf{D}\psi$, and

$\mathsf{O}_\mathsf{R}\phi, \Box(\phi \to \psi) \vdash_{RDL} \mathsf{R}_\mathsf{D}\psi$, by theorem 3.8.

Suppose now $M, w \models \mathsf{O}_\mathsf{R}(\phi \lor \psi)$ and $M, w \models \mathsf{R}_\mathsf{C}\neg\phi$. Then $\forall X \in \mathcal{R}^+ \exists Y \in \mathcal{R}^+(Y \subseteq X \cap [\phi \lor \psi]^M)$ and $R \subseteq [\neg\phi]^M$ for some $R \in \mathcal{R}^+$. Thus $Y \subseteq R \cap [\phi \lor \psi]^M$

---

[12] See [16]. Since the present problem (involving conflicting reasons) has the same structure as the deontic problem of explosion (involving conflicting obligations), the solution in [16] is based on solutions for the deontic problem put forward e.g. in [14,15,29,30]. All these solutions share the idea of distinguishing two kinds of obligations.

[13] See [11] and [12, sec.5] for an in-depth discussion of the deontic versions of these principles.

for some $Y \in \mathcal{R}^+$; $Y \subseteq [\neg\phi]^M \cap [\phi \vee \psi]^M$ for some $Y \in \mathcal{R}^+$; $Y \subseteq [\psi]^M$ for some $Y \in \mathcal{R}^+$, and so $M, w \models \mathsf{R_D}\phi$. Hence $\mathsf{O_R}(\phi \vee \psi), \mathsf{R_B}\neg\phi \Vdash_{RDL} \mathsf{R_D}\psi$, so that

$\mathsf{O_R}(\phi \vee \psi), \mathsf{R_B}\neg\phi \vdash_{RDL} \mathsf{R_D}\psi$, by theorem 3.8.

A similar theorem can be proved when $\mathsf{O_S}$ is substituted for $\mathsf{O_R}$.

3.6 Logic of obligation

Finally, it can be proved that $\mathsf{O_R}$ and $\mathsf{O_S}$ are $KD45$ modalities, so that the logic of obligation is the system SDL of standard deontic logic.

Theorem 3.16 *The logic of* $\mathsf{O_R}$, *respectively* $\mathsf{O_S}$, *is* $KD45$.

It is sufficient to show that, for all sets $\Delta \cup \{\phi\}$ of formulas in the sublanguage of $\mathcal{L}_{RDL}$ containing $\mathsf{O_R}$, respectively $\mathsf{O_S}$, as the only modality, $\Delta \vdash_{KD45} \phi \Rightarrow \Delta \Vdash_{RDL} \phi \Rightarrow \Delta \Vdash_{KD45} \phi$, where $\Vdash_{KD45}$ is the relation of logical consequence based on the class of models $M = (W, R, V)$ in which $R : W \to \wp(W)$ is such that $v \in R(w) \Rightarrow R(v) = R(w)$. The proof is based on the fact that, as said above, uniform models for standard deontic logic can be viewed as specific RDL models, together with the fact that RDL models validates all $KD45$ axioms and rules. The full proof is included in the extended version of the paper.

As $\mathsf{O_R}$ and $\mathsf{O_S}$ are $KD45$ modality, obligations cannot conflict, while conflicts between reasons are allowed (theorem 3.13). This result is of interest inasmuch as it allows us to interpret SDL plus axioms 4 and 5 as the logic concerning consistent obligations selected on the basis of deliberation, and so to vindicate this extension of SDL as apt to model deontic reasoning about this kind of obligations. Note that, in the present context, axioms 4 and 5 are not problematic once properly understood. In fact, as per axioms $R2$ and $S2$, we have reasons to take into account what is settled given the background; but what is supported by a reason, and therefore also what is all things considered obligatory and permitted, is settled given the background. As a consequence, we have reasons to take into account our all things considered obligations and permissions, and this is what axioms 4 and 5 state.

4 Comparison with two related accounts

In this section I consider two versions of the systems put forward by McNamara and Horty to deal with conflicting obligations and reasons and show how they can be interpreted in the present framework. The choice of these systems is due to the fact that they inspired me in the construction of RDL.

4.1 McNamara's two-level system

This system accounts for the possibility of aggregating obligations in conflict-tolerant contexts.[14] The key idea is to introduce a distinction between *basic*, *derived* and *unproblematic derived* obligations, with corresponding operators

---

[14] See [15]. I will be only interested in the final part, which presents a distinction between different kinds of obligation in a minimal setting.

O_0, O_1, O_U. In this framework a two-level model can be defined as a triple (W, Φ, V) where $W \neq \varnothing$, Φ is a finite set of formulas, and V is a modal valuation. To simplify the comparison, the truth conditions for modal formulas are given as follows, where $\subseteq_{Fin}$ is set-theoretical inclusion of a finite set.[15]

$M, w \models \Box \phi$ iff $[\phi]^M = W$;
$M, w \models O_0 \phi$ iff $\exists \phi_i \in \Phi ([\phi]^M = [\phi_i]^M)$;
$M, w \models O_1 \phi$ iff $\exists \Delta \subseteq_{Fin} \Phi (\varnothing \neq [\wedge \Delta]^M \subseteq [\phi]^M)$;
$M, w \models O_U \phi$ iff $\exists \Delta \subseteq_{Fin} \Phi (\varnothing \neq [\wedge \Delta]^M \subseteq [\phi]^M)$ and $M, w \models \neg O_1 \neg (\wedge \Delta))$.

As we can see, being unproblematically obligatory entails being obligatory.

Interpreting the two-level framework The connection between the two-level framework and the present one is the following. Let $M = (W, \Phi, V)$ be a two-level model and define $M^* = (W^*, \mathcal{R}, \mathcal{R}^+, \mathcal{S}^+, V^*)$ so that $W^* = W$, $\mathcal{R} = \{[\phi]^M : \phi \in \Phi\}$, $\mathcal{S}^+ = \mathcal{R}^+$ is the closure of $\mathcal{R}$ under consistent aggregation and conditioned addition; $V^* = V$.

Proposition 4.1 $M, w \models O_0 \phi$ iff $M^*, w \models R_B \phi$.

Proof. $M, w \models O_0 \phi$ iff $\exists \phi_i \in \Phi ([\phi]^M = [\phi_i]^M)$; iff $[\phi_i]^{M^*} \in \mathcal{R}$. □

Proposition 4.2 $M, w \models O_1 \phi$ iff $M^*, w \models R_D \phi$.

Proof. $M, w \models O_1 \phi$
iff $\exists \Delta \subseteq_{Fin} \Phi ([\wedge \Delta]^M \neq \varnothing$ and $[\wedge \Delta]^M \subseteq [\phi]^M)$
iff $\exists \mathcal{X} \subseteq \mathcal{R} (\bigcap \mathcal{X} \neq \varnothing$ and $\bigcap \mathcal{X} \subseteq [\phi]^{M^*})$, by def. $\mathcal{R}$
iff $\exists X \in \mathcal{R}^+ (X \subseteq [\phi]^{M^*})$, by def. $\mathcal{R}^+$, since $\mathcal{X}$ is finite □

As a corollary we get that $M, w \models \neg O_1 \neg \phi$ iff $\forall X \subseteq \mathcal{R}^+ (X \cap [\phi]^{M^*} \neq \varnothing)$.

Proposition 4.3 $M, w \models O_U \phi$ iff $M^*, w \models O_R \phi$.

Proof. $M, w \models O_U \phi$
iff $\exists \Delta \subseteq_{Fin} \Phi ([\wedge \Delta]^M \neq \varnothing, [\wedge \Delta]^M \subseteq [\phi]^M$ and $M, w \models \neg O_1 \neg (\wedge \Delta))$
iff $\exists \Delta \subseteq_{Fin} \Phi ([\wedge \Delta]^M \neq \varnothing, [\wedge \Delta]^M \subseteq [\phi]^M, \forall X \subseteq \mathcal{R}^+ (X \cap [\wedge \Delta]^{M^*} \neq \varnothing))$
iff $\exists \mathcal{Y} \subseteq \mathcal{R} (\bigcap \mathcal{Y} \neq \varnothing$ and $\bigcap \mathcal{Y} \subseteq [\phi]^M$ and $\forall X \subseteq \mathcal{R}^+ (X \cap \bigcap \mathcal{Y} \neq \varnothing))$
iff $\exists Y \in \mathcal{R}^+ (Y \subseteq [\phi]^M$ and $\forall X \subseteq \mathcal{R}^+ (X \cap Y \neq \varnothing))$, since $\mathcal{X}$ is finite
iff $\exists Y \in \mathcal{R}^+ \forall X \in \mathcal{R}^+ (X \cap Y \neq \varnothing$ and $Y \subseteq [\phi]^M)$, by logic
iff $\forall X \in \mathcal{R}^+ (X \cap r([\phi]^M))$ by the definition of r □

As a consequence, *obligations turn out to coincide with derivative reasons, while unproblematic obligations coincide with reason-based obligations.* Hence, McNamara's two-level framework can be interpreted as the fragment of *RDL* dealing with derivative reasons and reason-based obligations.

---

[15] This is a semantic version of the truth conditions proposed in [15, 148-150]. See [15, sec.2] for a detailed presentation of the system and a justification of the truth conditions.

4.2 Horty's default system

Let $(\mathcal{L}, \vdash)$ be a system of classical propositional logic.[16] A *default rule* r is a pair $(\mathbf{a}[r], \mathbf{c}[r])$, where $\mathbf{a}[r], \mathbf{c}[r] \in \mathcal{L}$ are the antecedent and the consequent of r. In terms of reasons, r states that $\mathbf{a}[r]$ is a reason to do $\mathbf{c}[r]$. A *default theory* is a triple $(\mathcal{W}, \mathcal{D}, <)$ where $\varnothing \neq \mathcal{W} \subseteq \mathcal{L}$ is a consistent set of background information, $\mathcal{D} \neq \varnothing$ a set of default rules and $<$ an irreflexive and transitive relation on $\mathcal{D}$. A *scenario* is a set of rules. If S is a scenario, then $\mathbf{a}[S] = \{\mathbf{a}[r] : r \in S\}$ and $\mathbf{c}[S] = \{\mathbf{c}[r] : r \in S\}$. If S is a scenario and $r \in \mathcal{D}$, we say that

(i) $Tr[S] = \{r \in \mathcal{D} : \mathcal{W}, \mathbf{c}[S] \vdash \mathbf{a}[r]\}$
   is the set of rules that are *triggered* in S.

(ii) $Cr[S] = \{r \in \mathcal{D} : \mathcal{W}, \mathbf{c}[S] \vdash \neg \mathbf{c}[r]\}$
   is the set of rules that are *conflicted* in S.

(iii) $Dr[S] = \{r \in \mathcal{D} : \exists d \in Tr[S] (r < d \text{ and } r \in Cr[d])\}$
   is the set of rules that are *defeated* in S.

If S is a scenario, then S is *consistent* iff $S = S - Cr[S]$ and S is *proper* iff $S = Tr[S] - Cr[S] - Dr[S]$. Thus, a scenario is consistent provided that it is conflict free and it is proper provided that it contains all and only the triggered rules that are not conflicted or defeated in it.

Remark 4.4 Say that a default theory is basic when $\forall \phi (\mathcal{W}, \mathbf{c}[\mathcal{D}] \Vdash \phi \Leftrightarrow \mathcal{W} \Vdash \phi)$. Then, $Tr[S] = Tr[\mathcal{D}]$ and $Dr[S] = Dr[\mathcal{D}]$ for all $S \subseteq \mathcal{D}$. Thus, there is a unique set $Tr[\mathcal{D}] - Dr[\mathcal{D}]$ of undefeated rules and S is proper iff S is consistent, that is if and only if $S = S - Cr[S] \subseteq Tr[\mathcal{D}] - Dr[\mathcal{D}]$.[17]

Let $\mathcal{D}^*$ be the union of the set of proper scenarios, $\mathbf{c}[\mathcal{D}^*]$ be the set of consequents of rules in $\mathcal{D}^*$, and $\mathcal{E}[\mathbf{c}[\mathcal{D}^*]]$ be the set of maximal consistent subsets of $\mathbf{c}[\mathcal{D}^*]$. The elements of $\mathcal{E}[\mathbf{c}[\mathcal{D}^*]]$ are then the most inclusive *objectives* available to an agent given the reasons in $\mathcal{D}$ and the background information $\mathcal{W}$.[18]

Definition 4.5 Let $(\mathcal{W}, \mathcal{D}, <)$ be a default theory. Then, we can define two operators, S_h and O_h, corresponding to derivative reason and obligation a là Horty, by introducing the following truth conditions.

1. $(\mathcal{W}, \mathcal{D}, <) \models \mathsf{S}_h \phi$ iff $\Delta \Vdash \phi$ for some $\Delta \in \mathcal{E}[\mathbf{c}[\mathcal{D}^*]]$.
2. $(\mathcal{W}, \mathcal{D}, <) \models \mathsf{O}_h \phi$ iff $\Delta \Vdash \phi$ for every $\Delta \in \mathcal{E}[\mathbf{c}[\mathcal{D}^*]]$.

Hence, there is a derivative reason to do ϕ iff ϕ is entailed by some objective and there is an obligation to do ϕ iff ϕ is entailed by every objective.

Interpreting the default framework To provide a representation of a default theory in the framework of *RDL*, I assume that the theory we want to represent contains a finite number of rules and a rule to the effect that agents have to take into account the background information, so that what is set-

---

[16] See [13,14]. In [16] a version of this system is used to model the connection between reasons and obligations. Here I will only consider fixed priority default versions.

[17] The system in [16] is essentially a basic default theory.

[18] Here it is assumed that $\mathcal{D}^* \neq \varnothing$. This entails that there is a proper scenario in $\mathcal{D}$.

tled given $\mathcal{W}$ is something that cannot be contrasted. A rule of this kind is $(\top, \wedge\{\mathcal{W}\})$ and requires that $\mathcal{W}$ be finite. So, let us say that a default theory $(\mathcal{W}, \mathcal{D}, <)$ is *suitable* when $\mathcal{W}$ and $\mathcal{D}$ are *finite* and $(\top, \wedge\{\mathcal{W}\})$ is contained in every proper scenario. Let $[\phi]$ be the set of maximal RDL-consistent sets containing ϕ and $M = (W, \mathcal{R}, \mathcal{R}^+, \mathcal{S}^+, V)$ be such that

W is the set of maximal RDL-consistent sets including $\mathcal{W}$;
$\mathcal{R} = \{[\phi] : \phi \in \mathbf{c}[\mathcal{D}]\}$;
$\mathcal{R}^+$ is the closure of $\mathcal{R}$ under combinations of reasons;
$\mathcal{S}^+$ is the closure of $\{[\phi] : \phi \in \mathbf{c}[\mathcal{D}^*]\}$ under combinations of reasons.

$W \neq \varnothing$, since $\mathcal{W}$ is consistent, and is to be identified with the set of states that are possible in light of what is settled given the background information. It is evident that $M = (W, \mathcal{R}, \mathcal{R}^+, \mathcal{S}^+, V)$ is a model for RDL. Now, let $(\mathcal{W}, \mathcal{D}, <)$ be *suitable*.

Proposition 4.6 *If no modality is in ϕ, then $(\mathcal{W}, \mathcal{D}, <) \models \mathsf{S}_h\phi$ iff $M \models \Diamond\mathsf{S}\phi$.*

Proof. $M, w \models \Diamond\mathsf{S}\phi$ iff $\exists X \in \mathcal{S}^+(X \subseteq [\phi]^M)$
iff $\exists \theta_1, ..., \theta_N \in \mathbf{c}[\mathcal{D}^*](\varnothing \neq [\theta_1]^M \cap ... \cap [\theta_N]^M \subseteq [\phi]^M)$
iff $[\Delta]^M \subseteq [\phi]^M$ for some $\Delta \in \mathcal{E}[\mathbf{c}[\mathcal{D}^*]]$, since $\mathcal{D}$ is finite
iff $\mathcal{W}, \Delta \vdash_{RDL} \phi$ for some $\Delta \in \mathcal{E}[\mathbf{c}[\mathcal{D}^*]]$, by the definition of W
iff $\Delta \vdash_{RDL} \phi$ for some $\Delta \in \mathcal{E}[\mathbf{c}[\mathcal{D}^*]]$, since $\wedge\{\mathcal{W}\}$ is in every Δ
iff $\Delta \vdash \phi$ for some $\Delta \in \mathcal{E}[\mathbf{c}[\mathcal{D}^*]]$, since $\phi \in \mathcal{L}$
iff $(\mathcal{W}, \mathcal{D}, <) \models \mathsf{S}_h\phi$ □

Proposition 4.7 *If no modality is in ϕ, then $(\mathcal{W}, \mathcal{D}, <) \models \mathsf{O}_h\phi$ iff $M \models \mathsf{O}_\mathsf{S}\phi$.*

Proof. $M, w \models \mathsf{O}_\mathsf{S}\phi$ iff $\forall X \in \mathcal{S}^+ \exists Y \in \mathcal{S}^+(Y \subseteq X \cap [\phi]^M)$; by the definition of $\mathcal{S}^+$ and the finiteness of $\mathcal{D}$, this is equivalent to $[\Delta]^M \subseteq [\phi]^M$ for all $\Delta \in \mathcal{E}[\mathbf{c}[\mathcal{D}^*]]$; the rest of the proof is then similar to the previous one. □

Models for RDL can be regarded as semantic generalizations of default theories.[19] To be sure, the notions of being conflicted and being defeated are definable in terms of the consequents of the rules in $\mathcal{D}$: two rules conflict when their consequents cannot be realized together, while the ordering on $\mathcal{D}$ can be based on an ordering on consequents, which are the items that are assessed as to their practical weight. In RDL we abstract both from the structure of the rules and from the specific procedure used to identify what scenarios are proper. This is a benefit in terms of flexibility, but a cost in terms of transparency, since the work done by the ordering relation is completely incorporated in the implicit choice function that allows us to pick out $\mathcal{S}^+$ from $\mathcal{R}^+$.

5 Conclusion

Consider the following core principles of standard deontic logic.

---

[19] Since reasons derive from triggered rules, we are also able to interpret conditional rules like $(\mathbf{a}[r], \mathbf{c}[r])$ in terms of conjunctive reasons $W - (X \cap -Y)$, where X is the set of states where $\mathbf{a}[r]$ is realized and Y is the set of states where $\mathbf{c}[r]$ is realized.

1. $\Box(\phi \leftrightarrow \psi) \wedge O\phi \to O\psi$;
2. $\Box\phi \to O\phi$;
3. $O\phi \to \Diamond\phi$;
4. $O\phi \to \neg O\neg\phi$;
5. $O\phi \wedge O\psi \to O(\phi \wedge \psi)$;
6. $\Box(\phi \to \psi) \wedge O\phi \to O\psi$;
7. $O\phi \wedge O\psi \wedge \Diamond(\phi \wedge \psi) \to O(\phi \wedge \psi)$.

Systems including $\{3,5\}$ or $\{4,6\}$ do not allow for conflicts, while systems including $\{5,6\}$ or $\{6,7\}$ do not avoid explosions. So, in order to allow for conflicts and avoid explosions one principle in each of $\{3,5\}$, $\{4,6\}$, $\{5,6\}$, $\{6,7\}$ is to be discarded. In reason-based deontic logic, the picture is as follows.

	1.	2.	3.	4.	5.	6.	7.
R_B	✓	✓	✓				
R_C	✓	✓	✓			✓	
R_D	✓	✓	✓		✓		
O_R and O_S	✓	✓	✓	✓	✓	✓	✓

O_R and O_S are conflict-free operators, given that they model kinds of all things considered obligation resulting from agent deliberation. In addition, since principles $4, 5, 6$ are invalid relative to R_B and R_C and $4, 5, 7$ are invalid relative to R_D, all of R_B, R_C, R_D are conflict tolerant operators that enable us to avoid explosions (theorem 3.13) and to construct arguments based on IR and DR (theorem 3.15). Thus, RDL provides us with an intuitive way to integrate reasons and obligations in a coherent system. This integration is based on the notion of combined reason, which is worth considering both from a philosophical point of view (it allows us to focus on two basic operations characterizing practical reasoning, namely consistent aggregation and conditioned addition) and from a logical point of view (it allows us to connect the logic of reasons with the logic of obligation and to develop a complete system of reason-based deontic logic). Also, RDL gives us a principled framework for addressing issues concerning deontic principles. Indeed, RDL was not obtained by constraining some deontic principles in order to avoid counter-intuitive conclusions.[20] To the contrary, we have first introduced elementary principles on how to combine reasons, and then demonstrated how solutions to pressing deontic problems follow from these principles, given suitable definitions of the deontic operators. Finally, RDL is connected with evidence-based systems of epistemic logic, thus providing us with a helpful basis for developing a unified account of reasons in deontic and epistemic contexts.

---

[20] This complaint is expressed in [12, p. 311]: "The kind of neighborhood semantics described above, while valuable for establishing results about the logics, such as determining what is derivable from what within the systems, do not yield much illumination into the concepts being formalized. The conditions on the neighborhoods that validate the various principles merely mimic, at the level of propositions, the principles being validated".

References

[1] Audi, R., *The Architecture of Reason: The Structure and Substance of Rationality*, Oxford University Press, Oxford, 2001.
[2] Audi, R., *Practical Reasoning and Ethical Decision*, Routledge, London, 2006.
[3] Baltag, A. and S. Smets, *A qualitative theory of dynamic interactive belief revision*, in: van der Hoek, W. and M. Wooldridge, editors, *Texts in Logic and Games 3*, Amsterdam University Press, Amsterdam, 2008, pp. 11–58.
[4] Baltag, A., N. Bezhanishvili, A. Özgün and S. Smets, *Justified belief and the topology of evidence*, WoLLIC, Springer, Berlin, 2016, pp. 83–103.
[5] Baltag, A., N. Bezhanishvili, A. Özgün and S. Smets, *A topological approach to full belief*, Journal of Philosophical Logic, 48 (2019), pp. 205–244.
[6] Brink, D., *Moral conflict and its structure*, Philosophical Review, 103 (1994), pp. 215–247.
[7] Broome, J., *Rationality Through Reasoning*, Blackwell, Oxford, 2013.
[8] Chellas, B., *Modal logic. An Introduction*, Cambridge University Press, Cambridge, 1980.
[9] Fumerton, R., *Epistemology*, Blackwell, Oxford, 2006.
[10] Goble, L., *A Logic for Deontic Dilemmas*, Journal of Applied Logic, 3 (2005), pp. 461–483.
[11] Goble, L. *Normative conflicts and the logic of 'ought'*, Nous, 43 (2009), pp. 450–489.
[12] Goble, L., *Prima facie norms, normative conflicts, and dilemmas*, in: D. Gabbay, J. Horty, X. Parent, R. van der Meyden and L. van der Torre, editors, *Handbook of deontic logic and normative systems*, College Publications, Milton Keynes, 2013, pp. 241–352.
[13] Horty, J., *Reasoning with moral conflicts*, Nous, 37 (2003), pp. 557–605.
[14] Horty, J., *Reasons as defaults*, Oxford University Press, Oxford, 2012.
[15] McNamara, P., *Agential obligation as non-agential personal obligation plus agency*, Journal of Applied Logic, 2 (2004), pp. 117–152.
[16] Nair, S., *Conflicting reasons, unconflicting 'ought's*, Philosophical Studies, 173 (2016), pp. 629–663.
[17] Nair, S. and J. Horty, *The logic of reasons*, in: D. Star, editor, *The Oxford Handbook of Reasons and Normativity*, Oxford University Press, Oxford, 2018, pp. 67–85.
[18] Pacuit, E., *Neighborhood Semantics for Modal Logic*. Springer, Berlin, 2017.
[19] Parfit, D., *On what matters*, Volume 1, Oxford University Press, Oxford, 2011.
[20] Raz, J., *Practical reasoning and norms*, Oxford University Press, Oxford, 1999.
[21] Raz, J., *Engaging Reason: On the Theory of Value and Action*, Oxford University Press, Oxford, 2002.
[22] Scanlon, T. M., *What we owe to each other*, Harvard University Press, Cambridge, 1998.
[23] Sen, A., *Choice Functions and Revealed Preference*, The Review of Economic Studies, 38 (1971), pp. 307–317.
[24] Sen, A., *Collective Choice and Social Welfare. An Expanded Edition*, Harvard University Press, Cambridge, 2017.
[25] Star, D., editor, *The Oxford Handbook of Reasons and Normativity*, Oxford University Press, Oxford, 2016.
[26] van Benthem, J. and E. Pacuit, *Dynamic logics of evidence-based beliefs*, Studia Logica, 99 (2011), pp. 61–92.
[27] van Benthem, J., D. Fernández-Duque and E. Pacuit, *Evidence Logic: A New Look at Neighborhood Structures*, Advances in modal logic, 9 (2012), pp. 97–118.
[28] van Benthem, J., D. Fernandez-Duque and E. Pacuit, *Evidence and plausibility in neighborhood structures*, Annals of Pure and Applied Logic, 165 (2014), pp. 106–133.
[29] van der Torre, L., and Y. H. Tan, *Two phase deontic logic*, Logique et Analyse, 171–172 (2000), pp. 411–456.
[30] van Fraassen, B., *Values and the heart's command*, Journal of Philosophy, 70 (1973), pp. 5–19.

Axiomatizing Norms Across Time and the 'Paradox of the Court'

Daniela Glavaničová [1]

Department of Analytic Philosophy
Slovak Academy of Sciences

Matteo Pascucci [2]

Department of Analytic Philosophy
Slovak Academy of Sciences

Abstract

In normative reasoning one typically refers to intervals of time across which norms are intended to hold, as well as to alternative possibilities representing hypothetical developments of a given scenario. Thus, deontic modalities are naturally intertwined with temporal and metaphysical ones. Furthermore, contemporary debates in philosophy suggest that a proper understanding of fundamental ethical principles, such as the Ought-Implies-Can thesis, requires a simultaneous analysis of these three families of concepts. In the present article we propose a general formal framework which allows for fine-grained multimodal reasoning in the normative domain. We provide an axiomatization for a novel system of propositional logic encoding the way in which possibilities and norms arising from different sources change over intervals of time. The usefulness of our framework is illustrated by analysing an ancient and particularly challenging 'cold case', the Paradox of the Court.

Keywords: Multimodal Reasoning, Norms Across Time, Ought-Implies-Can, Paradox of the Court, Temporal Opportunities.

1 Introduction

One of the most frequently debated principles in moral philosophy, the *Ought-Implies-Can* thesis (OIC), suggests that deontic modalities are essentially connected with other families of modalities. The naive formulation of OIC is the following: if an agent A is obliged to bring about ϕ, then it is possible that

---

[1] daniela.glavanicova@gmail.com
[2] matteopascucci.academia@gmail.com
This work was supported by the grant VEGA No. 2/0117/19. The authors are grateful to Vladimír Marko for navigating them through the literature on the Paradox of the Court and to the anonymous referees for their valuable comments. The two authors equally contributed to the contents of the article.

A brings about ϕ. But what kind of possibility is here involved? Logical or metaphysical possibility is too broad for the message that one wants to convey in terms of OIC. Indeed, what really matters is whether A has a certain *ability* and an *opportunity* of acting in appropriate circumstances to bring about ϕ. While there is a considerable amount of work in the literature focusing on the role played by agents' abilities in normative reasoning (see, for instance, [6], [7], [15] and [13]), much less has been said on the role played by agents' opportunities (see, for instance, [28], [21] and the analysis of spatial opportunities in [5]). It seems that both ability and opportunity involve many conceptual dimensions which would be very hard to represent in a single formal framework. In the present article we deal with opportunities from the perspective of *time*: norms are usually expected to apply to specific *temporal intervals* and the possibility of acting in appropriate circumstances to bring about something can be gained or lost during an interval.

For instance, consider the following scenario, adapted from [29]: at 9:00 a student received an order to write a five-page paper by 17:00 as part of an exam. Time passed by and it is now 16:57. The student has not started writing the paper yet. Is there an obligation which applies to the interval between 16:57 and 17:00? At first glance, one would be tempted to give a positive answer, since the student should not be able to justify the outcome of her behaviour by just relying on the flow of time. However, after a closer look at the problem one could argue that the student has actually no obligation to write a five-page paper between 16:57 and 17:00, namely that her original obligation expired. The reason is that the student does not have an opportunity of exercising her ability to write such a long paper in such a short amount of time.

This example shows that an obligation applying to a certain interval of time I is not automatically inherited by all subintervals of I; not even by those subintervals having the same final point as I (e.g., 17:00), namely those subintervals by which I *is finished*, according to the terminology in [1]. Therefore, the fact that at 16:57 it is no longer possible to fulfil the obligation is not a counterexample to OIC; as it is claimed in [30], "obligations that become infeasible at a given time are lost at that time". This does not mean that no trace of an obligation is left once new conditions make its fulfilment impossible. After 17:00 the student will be blamed and she will not pass the exam, since an obligation applying to a past interval of time, the one between 9:00 and 17:00, will have been violated due to her behaviour.

Furthermore, losing the opportunity to behave in a certain way is often not due to the flow of time *per se*; it is rather due to the fact that new norms become effective over time. A norm applying to an interval I may be overridden by a new norm that is introduced within I when it is not possible to comply with both. Therefore, *normative sources* can play a relevant role in determining when a norm expires. For instance, we can imagine a variation of the scenario above in which the student received a phone call at 9:10 and the person on the phone urgently asked her to go home and assist a family member, thus making impossible for her to write the paper by 17:00. The new obligation clearly takes priority over the old one.

The conclusion we draw from such discussion is that a fine-grained analysis of OIC and other philosophical problems requires making reference not only to deontic and metaphysical modalities, but also —and at least— to temporal ones. In the first part of the present article we will develop a very general formal framework to represent the way in which the three families of modalities at issue are intertwined in normative reasoning. More precisely, we provide an axiomatization for a new logic over a multimodal language making reference to temporal intervals across which norms arising from different sources are expected to hold, as well as to alternative possibilities. Our contribution can be located within the rich and long-lasting tradition of studies on the foundations of *multimodal reasoning with deontic modalities* (some examples are [26], [4], [3] and [24]).

In the second part of the article we will put our framework at work by discussing the *Paradox of the Court*, which is arguably the oldest puzzle for normative reasoning. Such puzzle is described, for instance, in [12]. In Ancient Greece a wealthy young man, Euathlus, became a student of Protagoras, paying him a half of the cost of teaching, and promising to pay him the remaining half on the day he would win his first case. After the end of his education, however, Euathlus changed his mind and decided not to undertake the career of a lawyer. What remained then of the original agreement? Was a payment of the education fee still due? Clever Protagoras thought that there was a way to make sure that the payment would take place. He decided to sue Euathlus, arguing in the following manner ([12], 407):

> Let me tell you, most foolish of youths, that in either event you will have to pay what I am demanding, whether judgment be pronounced for or against you. For if the case goes against you, the money will be due me in accordance with the verdict, because I have won; but if the decision be in your favour, the money will be due me according to our contract, since you will have won a case.

Euathlus, being a clever pupil himself, was not willing to let Protagoras win the argument. He rather saw this as an occasion to make sure that *no* payment would take place. He argued for the opposite conclusion as follows ([12], 409):

> I shall not have to pay what you demand, whether judgment be pronounced for or against me. For if the jurors decide in my favour, according to their verdict nothing will be due you, because I have won; but if they give judgment against me, by the terms of our contract I shall owe you nothing, because I have not won a case.

Prima facie, it seems that both Protagoras and Euathlus are right; despite this, their arguments lead to mutually contradicting conclusions: if Protagoras is right, Euathlus should pay the promised amount of money whether he wins or loses. If Euathlus is right, he is not obliged to pay the promised amount of money whether he wins or loses. This short presentation reveals that the Paradox of the Court is a paradigmatic problem of normative reasoning rooted in the connection between deontic, temporal and metaphysical modalities.

The structure of our article is as follows. In Section 2 we introduce the formal language and the axiomatic basis of our logic **DTM** for reasoning with Deontic, Temporal and Metaphysical modalities; in Section 3 we provide a semantic analysis of **DTM** and a characterization result in terms of a class of intended models, discussing also how the principle OIC can be represented within it. In Section 4, we review some accounts of the Paradox of the Court proposed in the literature. Subsequently, in Section 5, we present our analysis of the paradox in terms of the new logical framework introduced. The article is concluded with an overview of possible applications of **DTM**.

2 Syntax

In this section we describe the multimodal logic **DTM** for reasoning with Deontic, Temporal and Metaphysical modalities. We start by introducing the formal language $\mathcal{L}$ on which the logic is based.

Definition 2.1 (Primitive symbols) *The language $\mathcal{L}$ contains the following primitive symbols:*

- *a countable set of propositional variables VAR, denoted by p, q, r, etc.;*
- *a countable set of normative sources SOU, denoted by s_1, s_2, s_3, etc.;*
- *a countable set of temporal indices IND, denoted by i, j, k, etc.;*
- *the monadic modal operators $\Box_\infty$, $\Box_{[i,j]}$, $\Box_{\Leftarrow i}$ and $\Box_{i\Rightarrow}$, for $i,j \in IND$;*
- *the monadic modal operator L;*
- *the monadic modal operator O^s, for $s \in SOU$;*
- *the binary predicate E taking temporal indices as arguments;*
- *the boolean connectives $\neg$ (negation) and $\to$ (material implication);*
- *round brackets.*

A temporal index can be conceived of as a *non-indexical temporal reference*, namely a particular date or time. For instance, "11 January 2020" or "three days after 5 February 2020" or "Christmas 2020 at 3pm". The intended reading of the primitive symbols in $\mathcal{L}$ will be clarified below, after having specified the set of well-formed formulas.

Definition 2.2 (Well-formed formulas) *The set WFF of well-formed formulas of $\mathcal{L}$ is defined by the grammar below (where $p \in VAR$, $i,j \in IND$ and $s \in SOU$), provided that the following two restrictions apply:*

- *in formulas of kind $O^s\phi$, ϕ neither contains occurrences of the predicate E nor of any modal operator different from $\Box_\infty$, $\Box_{[i,j]}$, $\Box_{\Leftarrow i}$ and $\Box_{i\Rightarrow}$;*
- *formulas of kind $\Box_\infty\phi$, $\Box_{[i,j]}\phi$, $\Box_{\Leftarrow i}\phi$ and $\Box_{i\Rightarrow}\phi$ do not include occurrences of operators of kind O^s.*

$$\phi ::= p \mid E(i,j) \mid \neg\phi \mid \phi \to \phi \mid \Box_\infty\phi \mid \Box_{[i,j]}\phi \mid \Box_{\Leftarrow i}\phi \mid \Box_{i\Rightarrow}\phi \mid L\phi \mid O^s\phi$$

Let $ATO = VAR \cup \{E(i,j) : i,j \in IND\}$ be the set of propositional atoms in WFF; elements of ATO will be denoted by a, a', a'', etc. Furthermore, we will denote by WFF^O the subset of WFF including only formulas in which the predicate E never occurs and where the only modal operators (if any) are $\Box_\infty$, $\Box_{[i,j]}$, $\Box_{\Leftarrow i}$ and $\Box_{\Rightarrow i}$ (this set includes precisely the formulas that can be in the scope of an operator O^s, according to the first restriction in Definition 2.2).

A formula of kind $E(i,j)$ means "temporal index i is earlier than temporal index j"; for instance, 11 January 2020 is earlier than 12 January 2020. A formula of kind $\Box_\infty \phi$ means "it is always the case that ϕ"; $\Box_{[i,j]}\phi$ means "throughout the interval between i and j it is always the case that ϕ"; $\Box_{\Leftarrow i}\phi$ means "ϕ is always the case until i"; $\Box_{i \Rightarrow}\phi$ means "ϕ is always the case starting from i"; $O^s\phi$ means "according to normative source s it is obligatory that ϕ"; finally, $L\phi$ means "it is necessarily the case that ϕ".[3] The following formulas can be used as abbreviations, according to usual definitions of boolean and modal operators: $\phi \land \psi$, $\phi \lor \psi$, $\phi \equiv \psi$, $\Diamond_\infty \phi$ ("it is sometimes the case that ϕ"), $\Diamond_{[i,j]}\phi$ ("throughout the interval between i and j it is sometimes the case that ϕ"), $\Diamond_{\Leftarrow i}\phi$ ("it is sometimes the case that ϕ until i"), $\Diamond_{i \Rightarrow}\phi$ ("it is sometimes the case that ϕ starting from i"), $P^s\phi$ ("according to normative source s it is permitted that ϕ") and $M\phi$ ("it is possibly the case that ϕ"). For instance, $\Diamond_{[i,j]}\phi := \neg\Box_{[i,j]}\neg\phi$ and $P^s\phi := \neg O^s\neg\phi$. The fact that i is the left index and j the right index in $\Box_{[i,j]}\phi$ does not bear any consequence on whether i is earlier than j; indeed, $\Box_{[j,i]}\phi$ is a well-formed formula as well. The relation earlier/later is rather associated with the predicate E. We will use the expressions $\Box_i$ and $\Diamond_i$ as abbreviations for $\Box_{[i,i]}$ and $\Diamond_{[i,i]}$.

We will now provide a step by step presentation of the axiomatic basis for **DTM**, assuming some familiarity with correspondence theory for modal logic (see, e.g., [27]). First of all, **DTM** is an extension of the classical propositional calculus (**PC**); therefore, we can start developing the axiomatic basis with the following set of principles:[4]

A0+RX   All WFF-substitution instances of axioms and rules of **PC**.

Then, we add the following two axioms for the predicate E, which will make the relation of temporal precedence a strict partial order:

---

[3] The expression "until" in the reading of $\Box_{\Leftarrow i}\phi$ has an inclusive sense: ϕ is expected to hold also at instant i. For such reason, this operator is more closely related to the "release" operator than to the "until" operator in temporal logics of computation (see, e.g., [9]). Analogously, in a formula of the form $\Box_{\Leftarrow i}\phi$ (respectively, $\Box_{[i,j]}\phi$) the interval considered is inclusive with respect to index i (and index j).

[4] For the sake of brevity, a label of kind An, where n is a natural number, may denote a set of distinct axioms (more precisely, axiom-schemata) and a label of kind Rλ, where λ is an upper case letter, may denote a set of distinct rules. A label of kind An+Rλ denotes the union of all axioms associated with An and all rules associated with Rλ.

A1 $E(i,j) \rightarrow (E(j,k) \rightarrow E(i,k))$;
A2 $\neg E(i,i)$.

A1 and A2 can be used to introduce functions specifying the first and the last temporal index in an interval (analogous functions can be found in a logic for characterizing deadlines in [14]). We use the expression $S(i,j)$ ('i and j are simultaneous') as an abbreviation for $\neg E(i,j) \wedge \neg E(j,i)$. Thus, the predicate S will satisfy the property: $S(i,j) \equiv S(j,i)$.

Definition 2.3 (First and last index in an interval) *Given $i,j \in IND$, let $\alpha[i,j]$ ("the first index in the interval $[i,j]$") be:*

- *i if $E(i,j)$ holds;*
- *j if $E(j,i)$ holds;*
- *both i and j otherwise (that is, if $S(i,j)$ holds).*

Furthermore, let $\omega[i,j]$ ("the last index in the interval $[i,j]$") be:

- *i if $E(j,i)$ holds;*
- *j if $E(i,j)$ holds;*
- *both i and j otherwise (that is, if $S(i,j)$ holds).*

We can now define additional *relations among intervals* in a very simple way, exploiting the functions α and ω, the predicates E and S, and boolean connectives. The labels for these relations are: Ide ("is identical with"), Bef ("is before than"), Mee ("meets"), Ove ("overlaps"), Fin ("is finished by"), Con ("contains") and Sta ("is started by").

Definition 2.4 (Allen-style interval algebra) *Given two intervals $[i,j]$ and $[k,l]$, we have the following fundamental relations among them:*

$$Ide([i,j],[k,l]) := S(\alpha[i,j],\alpha[k,l]) \wedge S(\omega[i,j],\omega[k,l])$$

$$Bef([i,j],[k,l]) := E(\omega[i,j],\alpha[k,l])$$

$$Mee([i,j],[k,l]) := E(\alpha[i,j],\omega[i,j]) \wedge E(\alpha[k,l],\omega[k,l]) \wedge S(\omega[i,j],\alpha[k,l])$$

$$Ove([i,j],[k,l]) := E(\alpha[i,j],\alpha[k,l]) \wedge E(\alpha[k,l],\omega[i,j]) \wedge E(\omega[i,j],\omega[k,l])$$

$$Fin([i,j],[k,l]) := E(\alpha[i,j],\alpha[k,l]) \wedge S(\omega[i,j],\omega[k,l])$$

$$Con([i,j],[k,l]) := E(\alpha[i,j],\alpha[k,l]) \wedge E(\omega[k,l],\omega[i,j])$$

$$Sta([i,j],[k,l]) := S(\alpha[i,j],\alpha[k,l]) \wedge E(\omega[k,l],\omega[i,j])$$

Furthermore, for any interval relation R defined above, one can denote by R^{-1} its converse relation. For instance, $Sta^{-1}([i,j],[k,l]) = Sta([k,l],[i,j])$. Thus, one can represent within **DTM** all thirteen relations described by Allen in [1]. Notice that the following property holds in **DTM**, due to A0+RX, A1 and A2: $Ide^{-1}([i,j],[k,l]) \equiv Ide([i,j],[k,l])$.

Then we move to the analysis of the deductive properties of the operators L and $\square_\infty$, which are intended to represent metaphysical necessity and temporal necessity (in the sense of truth over any interval of time), respectively. As it is argued in [22], logics of metaphysical necessity should be in the range between **KT** and **S5**; we opt for the strongest logic in this range, since it is a very common choice in approaches combining metaphysical and temporal modalities (see, for instance, the approaches to the Paradox of the Court discussed in Section 4).[5] We choose to adopt an **S5** basis for $\Box_\infty$ as well, in order to treat this operator as an interval-based analogue of the notion of Aristotelian necessity defined over linear and transitive temporal structures (see, e.g., [20]). The axiomatic basis is thus extended with the principles below:

A3+RY   All axioms and rules of **S5** for L and $\Box_\infty$.

Operators of kind $\Box_{\Leftarrow i}$, $\Box_{i \Rightarrow}$ and $\Box_{[i,j]}$, instead, do not satisfy the axiom T, since they may concern intervals of time to which the current time does not belong. Therefore, we add:

A4+RZ   All axioms and rules of **KD45** for operators of kind $\Box_{[i,j]}$, $\Box_{\Leftarrow i}$ and $\Box_{i \Rightarrow}$.

Now, we need to ensure that modal operators can capture all intended properties of temporal intervals. One of these properties is that there is only one point in an interval of kind $[i,i]$. Therefore, we add the following principle:

A5 $\Diamond_i \phi \to \Box_i \phi$.

Then, we need to ensure that $[i,j]$ and $[j,i]$ are two ways of looking at the same interval, and this is a consequence of adding the principle below:

A6 $\Box_{[i,j]} \phi \equiv \Box_{[j,i]} \phi$.

After this, we encode the temporal algebra over the set of intervals via the following axioms:

A7 $Ide([i,j],[k,l]) \to (\Box_{[i,j]}\phi \to \Box_{[k,l]}\phi)$;
A8 $Ove([i,j],[k,l]) \to (\Box_{[i,j]}\phi \to \Box_{\alpha[k,l]}\phi) \wedge (\Box_{[k,l]}\psi \to \Box_{\omega[i,j]}\psi)$;
A9 $Mee([i,j],[k,l]) \to (\Diamond_{\omega[i,j]}\phi \equiv \Diamond_{\alpha[k,l]}\phi)$;
A10 $Con([i,j],[k,l]) \to (\Box_{[i,j]}\phi \to \Box_{[k,l]}\phi)$;
A11 $Sta([i,j],[k,l]) \to ((\Box_{[i,j]}\phi \to \Box_{[k,l]}\phi) \wedge (\Box_{[k,l]}\psi \to \Diamond_{\alpha[i,j]}\psi))$;
A12 $Fin([i,j],[k,l]) \to ((\Box_{[i,j]}\phi \to \Box_{[k,l]}\phi) \wedge (\Box_{[k,l]}\psi \to \Diamond_{\omega[i,j]}\psi))$.

For instance, A7, together with the fact that **DTM** is closed under the schema

---

[5] Names of systems and axioms of modal logic adhere to the presentation in [8].

$Ide([i,j],[k,l]) \equiv Ide^{-1}([i,j],[k,l])$, says that two identical temporal intervals are indistinguishable with respect to the truth of formulas in states they contain: if something is always the case between February 14 (i) and December 25 (j) of a particular year, then it is always the case between Valentine's Day (k) and Christmas (l) of that year, and vice versa.

The next step is making sure that operators of kind $\Box_\infty$, $\Box_{[i,j]}$, $\Box_{\Leftarrow i}$ and $\Box_{i \Rightarrow}$ are related in an appropriate way. This requires also principles combining them with the predicate E. Thus, we add:[6]

A13 $\quad \Box_\infty \phi \equiv (\Box_{\Leftarrow i} \phi \wedge \Box_{i \Rightarrow} \phi)$;
A14 $\quad (\Box_{\Leftarrow i} \phi \wedge \neg E(i,j) \wedge \neg E(i,k)) \rightarrow \Box_{[j,k]} \phi$;
A15 $\quad (\Box_{i \Rightarrow} \phi \wedge \neg E(j,i) \wedge \neg E(k,i)) \rightarrow \Box_{[j,k]} \phi$.

Furthermore, let int and int' be arbitrary intervals, that is, strings of any of the following kinds: either ∞ or $[i,j]$ or $\Leftarrow i$ or $i \Rightarrow$; we add to the axiomatic basis the following bridge-axioms connecting different modalities:

A16 $\quad \Box_{int} \phi \equiv \Box_{int'} \Box_{int} \phi$;
A17 $\quad E(i,j) \rightarrow L \Box_{int} E(i,j)$;
A18 $\quad \Box_{int} L \phi \equiv L \Box_{int} \phi$.

Finally, given that **DTM** is intended to capture minimal relations among deontic, temporal and metaphysical modalities, and that there are several arguments in the literature supporting the idea that deontic modalities are hyperintensional (see, e.g., [10] and [11]), we do not impose any deductive property on operators of kind O^s, except for the following bridge-axiom:

A19 $\quad O^s \phi \rightarrow M \phi$.

In the end, we get the following definition:

Definition 2.5 (Axiomatic basis) *The axiomatic basis for the logic* **DTM** *corresponds with the list of axioms A0-A19 and the rules RX, RY and RZ.*

The principle A19 can be taken as the formal analogue of the naive formulation of the Ought-Implies-Can thesis. However, in the present framework we can provide a more-fine grained analysis of OIC, taking into account temporal intervals. For instance, as the example from [29] that we discussed in Section 1 shows, one could say that if it is obligatory that ϕ occurs within a certain interval $[i,j]$ and we are at a point k within $[i,j]$, then there is a possible development of the world in which ϕ occurs within the interval $[k,j]$. This is a way of explicitly taking into account the temporal opportunity of bringing about

---

[6] Axiom A13 can be also taken as a definition. This allows one to remove $\Box_\infty$ from the set of primitive symbols, provided that some changes in the axiomatic basis are made; for instance, A17 becomes $E(i,j) \rightarrow L(E(i,j) \wedge \Box_{int} E(i,j))$.

ϕ between k and the time in which the obligation was originally supposed to expire (j). Thus, we can formulate this version of OIC as follows:

$$(O^s \Diamond_{[i,j]}\phi \land E(i,k) \land \neg E(j,k)) \rightarrow (O^s \Diamond_{[k,j]}\phi \rightarrow M \Diamond_{[k,j]}\phi)$$

One can use this schema to distinguish between those obligations that are still in effect at a time and those that are not: even if an obligation concerning a temporal interval I is assumed for deductive reasoning, this does not entail that that obligation is effective when we reason about some point within I.

3 Semantics

In this section we describe the *intended* class of frames and models to interpret the logic **DTM**. Let $\mathbb{R}$ be the set of all relations R_{int} such that int is an interval. In analogy with what we did in the syntactic part, we will use R_i as a shorthand for $R_{[i,i]}$.

Definition 3.1 (Frames) *The language $\mathcal{L}$ is interpreted on relational frames of kind $\mathfrak{F} = \langle W, \mathbb{R}, A, < \rangle$ where:*

- W *is a set of* states *denoted by w, v, u, etc.;*
- *for any $R_{int} \in \mathbb{R}$, $R_{int} \subseteq W \times W$ is a "temporal inspection" relation;*
- $A \subseteq W \times W$ *is a "metaphysical inspection" relation;*
- $< \; \subseteq W \times IND \times IND$ *is a "temporal precedence" relation.*

For any $w \in W$, we have $R_{int}(w) = \{v : wR_{int}v\}$ and this can be called the R_{int}-sphere of w. An analogous notation can be used with reference to the other relations in a frame.

Definition 3.2 (Models) *A model over a frame $\mathfrak{F}$ is a structure of kind $\mathfrak{M} = \langle \mathfrak{F}, V, N \rangle$ such that:*

- $V : ATO \longrightarrow \wp(W)$ *is a valuation function;*
- *for any $s \in SOU$, $N^s : W \longrightarrow WFF^O$ is a norm assignment with respect to source s.*

For any $s \in SOU$ and $w \in W$, $N^s(w) \subseteq WFF^O$ is the N^s-sphere of w. We want to highlight the fact that the N^s-sphere of a state is model-dependent, whereas any R_{int}-sphere of a state is frame-dependent. Furthermore, in a frame the $<$-sphere of a state w can be different (in general) from the $<$-sphere of a state v; therefore, we will speak of the Allen relation between two intervals $[i,j]$ and $[k,l]$ *as seen from* a state w.

Definition 3.3 (Truth-conditions) *The truth of a formula with reference to a state w in a model $\mathfrak{M}$ is defined below, where $a \in ATO$ and $s \in SOU$:*

- $\mathfrak{M}, w \vDash a$ *iff $w \in V(a)$;*
- $\mathfrak{M}, w \vDash \neg \phi$ *iff $\mathfrak{M}, w \nvDash \phi$;*
- $\mathfrak{M}, w \vDash \phi \rightarrow \psi$ *iff either $\mathfrak{M}, w \nvDash \phi$ or $\mathfrak{M}, w \vDash \psi$;*
- $\mathfrak{M}, w \vDash \Box_{int}\phi$ *iff $\mathfrak{M}, v \vDash \phi$ for all $v \in R_{int}(w)$;*

- $\mathfrak{M}, w \vDash O^s \phi$ iff $\phi \in N^s(w)$;
- $\mathfrak{M}, w \vDash L\phi$ iff $\mathfrak{M}, v \vDash \phi$ for all $v \in A(w)$.

A formula ϕ is valid in a model $\mathfrak{M}$ iff ϕ is true at all states in the domain of $\mathfrak{M}$; ϕ is valid in a frame $\mathfrak{F}$ iff it is valid in all models over $\mathfrak{F}$. Validity in a class of frames/models is validity in all frames/models of the class.

We will denote by $R \circ R'$ the *composition* or two relations R and R'.

Definition 3.4 (Intended frames) *The class of intended frames for* **DTM**, *denoted by* C_f, *is the class of all frames such that:*

Πa for every $i, j \in IND$, $R_{[i,j]}$, $R_{\Leftarrow i}$ and $R_{i \Rightarrow}$ are serial, transitive and euclidean relations;

Πb A and R_∞ are equivalence relations;

Πc $<$ is a strict partial order;

Πd for any $i \in IND$, if $v, u \in R_i(w)$, then $u = v$;

Πe for any $i, j \in IND$, $R_{[i,j]} = R_{[j,i]}$;

Πf for any $i, j, k, l \in IND$, the following properties are associated with the Allen relation among $[i,j]$ and $[k,l]$ as seen from w:
- if $[i,j]$ and $[k,l]$ are identical, then $R_{[i,j]}(w) = R_{[k,l]}(w)$
- if $[i,j]$ overlaps with $[k,l]$, then $R_{\omega[i,j]}(w) \subseteq R_{[k,l]}(w)$ and $R_{\alpha[k,l]}(w) \subseteq R_{[i,j]}(w)$
- if $[i,j]$ meets $[k,l]$, then $R_{\omega[i,j]}(w) = R_{\alpha[k,l]}(w)$
- if $[i,j]$ is finished by $[k,l]$, then $R_{[k,l]}(w) \subseteq R_{[i,j]}(w)$ and $R_{\omega[i,j]}(w) \cap R_{[k,l]}(w) \neq \emptyset$
- if $[i,j]$ is started by $[k,l]$, then $R_{[k,l]}(w) \subseteq R_{[i,j]}(w)$ and $R_{\alpha[i,j]}(w) \cap R_{[k,l]}(w) \neq \emptyset$
- if $[i,j]$ contains $[k,l]$, then $R_{[k,l]}(w) \subseteq R_{[i,j]}(w)$;

Πg for any $i \in IND$, $R_\infty = R_{\Leftarrow i} \cup R_{i \Rightarrow}$;

Πh for any $i, j, k \in IND$, if $(i,j), (i,k) \notin <(w)$ then $R_{[j,k]}(w) \subseteq R_{\Leftarrow i}(w)$;

Πi for any $i, j, k \in IND$, if $(j,i), (k,i) \notin <(w)$ then $R_{[j,k]}(w) \subseteq R_{i \Rightarrow}(w)$;

Πj $R_{int} \circ R_{int'} = R_{int}$;

Πk $<(w) = <(v)$ whenever $v \in (A \circ R_{int})(w)$ (for some R_{int});

Πl $R_{int} \circ A = A \circ R_{int}$.

Definition 3.5 (Intended models) *The class of intended models for* **DTM**, *denoted by* C_m, *is the class of all models over frames in* C_f *such that:*[7]

Πx for any $i, j \in IND$, $w \in V(E(i,j))$ iff $(i,j) \in <(w)$;

Πy for any $s \in SOU$, $\phi \in N^s(w)$ only if there is $v \in A(w)$ s.t. $\mathfrak{M}, v \vDash \phi$.

---

[7] Due to the syntactic restrictions specified in Definition 2.2, the N^s-sphere of a state w (for every $s \in SOU$) includes only formulas where deontic operators never occur, and the truth of such formulas at a state v of a model $\mathfrak{M}$ can be established without any reference to the N^s-sphere of w. This ensures that property Πy is not defined in a circular way.

Before moving to the semantic characterization of **DTM**, we would like to briefly comment on some philosophical points. In the present logical framework the interaction between truth of formulas, states in a model and temporal indices captures the relation between the *flow of time* and *change* along the following lines. First, we can say that a maximal and **DTM**-consistent set of formulas in WFF constitutes a *configuration of the world*. Second, states in a model can be said to be *pictures of a configuration of the world*, since each of them is associated with a maximal and **DTM**-consistent set of formulas (though, this is not in general a one-to-one correspondence, since there are models in which two states are associated with the same configuration). Third, according to property Πd, exactly one picture of a configuration of the world is associated with each temporal index. However, the same picture can be associated with successive temporal indices, since this depends on the level of temporal granularity of a representation (for instance, one might have a new picture of a configuration of the world every second day, every second month, etc.).

Proposition 3.6 *The system* **DTM** *is sound w.r.t. the class* C_m.

Proof. An induction on the length of derivations. First we consider axioms: in the case of A0-A6, A13, A16, A18 and A19 the proof is a standard procedure in propositional (multimodal) reasoning; we here illustrate the other cases.

Consider A7. Assume that we have a model $\mathfrak{M}$ in C_m and a state w in its domain s.t. $\mathfrak{M}, w \vDash Ide([i,j],[k,l])$ and $\mathfrak{M}, w \vDash \Box_{[i,j]}\phi$ but $\mathfrak{M}, w \nvDash \Box_{[k,l]}\phi$. Thus, (I) for all states $v \in R_{[i,j]}(w)$, we have $\mathfrak{M}, v \vDash \phi$ and (II) there is a state $u \in R_{[k,l]}(w)$ s.t. $\mathfrak{M}, u \nvDash \phi$. However, the intervals $[i,j]$ and $[k,l]$ are identical as seen from w and so, due to property Πf, $R_{[i,j]}(w) = R_{[k,l]}(w)$ and a contradiction can be obtained.

Consider A8. Assume that $\mathfrak{M}, w \vDash Ove([i,j],[k,l])$ but $\mathfrak{M}, w \nvDash (\Box_{[k,l]}\phi \to \Box_{\omega[i,j]}\phi) \land (\Box_{[i,j]}\psi \to \Box_{\alpha[k,l]}\psi)$, for some $\phi, \psi \in WFF$. Let $\mathfrak{M}, w \nvDash \Box_{[k,l]}\phi \to \Box_{\omega[i,j]}\phi$. We know that (I) $[i,j]$ overlaps $[k,l]$ as seen from w, and (II) for all $v \in R_{[k,l]}(w)$ we have $\mathfrak{M}, v \vDash \phi$. Due to properties Πa and Πf, there is some state $u \in R_{\omega[i,j]}$, and $u \in R_{[k,l]}(w)$. Therefore, $\mathfrak{M}, u \vDash \phi$. Furthermore, due to Πd, u is the only state in $R_{\omega[i,j]}(w)$, so $\mathfrak{M}, w \vDash \Box_{\omega[i,j]}\phi$: contradiction. The argument for $\mathfrak{M} \nvDash \Box_{[i,j]}\psi \to \Box_{\alpha[k,l]}\psi$ is analogous.

Consider A9. Assume that $\mathfrak{M}, w \vDash Mee([i,j],[k,l])$ but $\mathfrak{M}, w \nvDash \Diamond_{\omega[i,j]}\phi \equiv \Diamond_{\alpha[k,l]}\phi$ for some $\phi \in WFF$. We can focus, without loss of generality, on the case in which $\mathfrak{M}, w \vDash \Diamond_{\omega[i,j]}\phi$ and $\mathfrak{M}, w \nvDash \Diamond_{\alpha[k,l]}\phi$. Thus, (I) there is $v \in R_{\omega[i,j]}(w)$ s.t. $\mathfrak{M}, v \vDash \phi$, and (II) for all $u \in R_{\alpha[k,l]}(w)$ we have $\mathfrak{M}, u \nvDash \phi$. However, since $[i,j]$ meets $[k,l]$ as seen from w, then $R_{\omega[i,j]}(w) = R_{\alpha[k,l]}(w)$ and we get a contradiction.

Consider A10. Assume $\mathfrak{M}, w \vDash Con([i,j],[k,l])$ and $\mathfrak{M}, w \nvDash \Box_{[i,j]}\phi \to \Box_{[k,l]}\phi$ for some $\phi \in WFF$. From this one can infer that (I) for all $v \in R_{[i,j]}(w)$, we have $\mathfrak{M}, v \vDash \phi$, and (II) for some $u \in R_{[k,l]}(w)$, we have $\mathfrak{M}, u \nvDash \phi$. However, since $[i,j]$ contains $[k,l]$ as seen from w, then $R_{[k,l]}(w) \subseteq R_{[i,j]}$ and we get a contradiction.

Consider A11. Assume that $\mathfrak{M}, w \vDash Sta([i,j],[k,l])$ and $\mathfrak{M}, w \nvDash (\Box_{[i,j]}\phi \to \Box_{[k,l]}\phi) \wedge (\Box_{[k,l]}\psi \to \Diamond_{\alpha[i,j]}\psi)$, for some $\phi, \psi \in WFF$. Let $\mathfrak{M}, w \nvDash \Box_{[i,j]}\phi \to \Box_{[k,l]}\phi$. Therefore, (I) for all $v \in R_{[i,j]}(w)$, we have $\mathfrak{M}, v \vDash \phi$, and (II) there is some $u \in R_{[k,l]}(w)$ s.t. $\mathfrak{M}, u \nvDash \phi$. However, since $[i,j]$ is started by $[k,l]$ as seen from w, then $R_{[k,l]}(w) \subseteq R_{[i,j]}(w)$ and we get a contradiction. Let $\mathfrak{M}, w \nvDash \Box_{[k,l]}\psi \to \Diamond_{\alpha[i,j]}\psi$. Therefore, (I) for all $v \in R_{[k,l]}(w)$ we have $\mathfrak{M}, v \vDash \psi$, and (II) there is no $u \in R_{\alpha[i,j]}(w)$ s.t. $\mathfrak{M}, u \vDash \psi$. However, since $[i,j]$ is started by $[k,l]$ as seen from w, then $R_{\alpha[i,j]}(w) \cap R_{[k,l]}(w) \neq \emptyset$ and this leads to a contradiction. The argument for A12 is analogous.

Consider A14. Assume $\mathfrak{M}, w \vDash (\Box_{\Leftarrow i}\phi \wedge \neg E(i,j) \wedge \neg E(i,k))$ and $\mathfrak{M}, w \nvDash \Box_{[j,k]}\phi$ for some $\phi \in WFF$. Then, for all $v \in R_{\Leftarrow i}(w)$ we have $\mathfrak{M}, v \vDash \phi$; furthermore, $(i,j),(i,k) \notin <(w)$. Due to property Πh we have that $R_{[j,k]}(w) \subseteq R_{\Leftarrow i}(w)$ and we get a contradiction. The argument for A15 is analogous.

Consider A17. Assume $\mathfrak{M}, w \vDash E(i,j)$ but $\mathfrak{M}, w \nvDash L\Box_{int}E(i,j)$ for some interval int. Then, $(i,j) \in <(w)$ and there is $v \in (A \circ R_{int})(w)$ s.t. $\mathfrak{M}, v \nvDash E(i,j)$. However, by property Πk, $<(w) = <(v)$, and we get a contradiction.

The fact that rules RX, RY and RZ preserve validity in every model in C_m is straightforward. $\square$

Proposition 3.7 *The system* **DTM** *is complete w.r.t. the class* C_m.

Proof. The canonical frame $\mathfrak{F}$ for **DTM** can be built following the usual steps for systems of modal logic, with the only difference that for every maximal consistent set of formulas w, every $i,j \in IND$ and every $s \in SOU$ we have:

- $<(w) = \{(i,j) : E(i,j) \in w\}$;
- $N^s(w) \subseteq WFF^O$.

The canonical model $\mathfrak{M}$ over $\mathfrak{F}$ is such that, for every maximal consistent set of formulas w, and every $a \in ATO$, we have:

- $V(a) = \{w : a \in w\}$.

We now illustrate that the canonical model belongs to the class C_m.

The proof that $\mathfrak{M}$ satisfies properties Πa–Πe, Πg and Πj–Πl relies on standard arguments in completeness results for modal propositional logic (in the case of Πc only basic propositional reasoning with A1 and A2 is needed). We will analyse how the remaining properties of models in C_m (and of the underlying frames) are satisfied.

In the case of Πf we illustrate one example. Assume that the interval $[i,j]$ contains the interval $[k,l]$ as seen from a state w but $R_{[k,l]}(w) \nsubseteq R_{[i,j]}(w)$. Then there is a state $v \in W$ s.t. $v \in R_{[k,l]}(w)$ and $v \notin R_{[i,j]}(w)$. From this one can infer that $\{\phi : \Box_{[k,l]}\phi \in w\} \subseteq v$ and that there is some $\psi \in WFF$ s.t. $\Box_{[i,j]}\psi \in w$ and $\psi \notin v$. However, since $[i,j]$ contains $[k,l]$ as seen from w, then $(\alpha(i,j), \alpha(k,l)), (\omega(k,l), \omega(i,j)) \in <(w)$ and this entails $Con([i,j],[k,l]) \in w$. Furthermore, since w is closed under A10, then $\Box_{[k,l]}\psi \in w$, whence $\psi \in v$: contradiction.

In the case of Πh, assume that $(i,j),(i,k) \notin <(w)$ but $R_{[j,k]}(w) \nsubseteq R_{\Leftarrow i}(w)$. Then there is some state $v \in R_{[j,k]}(w)$ such that $v \notin R_{\Leftarrow i}(w)$. From this one can infer that $\{\phi : \square_{[j,k]}\phi \in w\} \subseteq v$ and that there is ψ s.t. $\square_{\Leftarrow i}\psi \in w$ and $\psi \notin v$. However, since $\neg E(i,j) \wedge \neg E(i,k) \in w$ and w contains all instances of A14, then $\square_{[j,k]}\psi \in w$ and $\psi \in v$: contradiction.

Property Πx is satisfied due to the definition of $\mathfrak{F}$ and $\mathfrak{M}$. Finally, consider property Πy. Suppose there are $s \in SOU$ and $\phi \in WFF^O$ s.t. $\phi \in N^s(w)$, and that for no $v \in A(w)$ we have $\mathfrak{M}, v \vDash \phi$. Then, for any such v we have $\phi \notin v$, whence $M\phi \notin w$; however $\mathfrak{M}, w \vDash O^s\phi$, so $O^s\phi \in w$ and we get a contradiction in the light of A19.

□

4 The Paradox of the Court: Proposed Solutions

Many formal accounts of the Paradox of the Court, also known as *Protagoras v. Euathlus*, have been provided in the literature. In [16] Lenzen offers an analysis within a so-called *base logic*. As [18] neatly puts it, this logic "is defined by the axioms of the classical sentence calculus, axioms of necessity operator of the modal system **S5**, and axioms of identity predicate. The only inference rule is Modus Ponens." This approach is improved by Åqvist in [2], in terms of temporal deontic logic and the definition of several interesting notions, such as (in)validity as applied to an agreement, (in)correctness as applied to a verdict, or the import of an agreement (what follows from it). Both Lenzen and Åqvist aim at *solving the paradox* by deriving in their formal systems a way of getting paid the established fee.

Smullyan in [25] proposes an informal solution, which is suggested to him by "a lawyer" and goes as follows:

> The court should award the case to the student —the student shouldn't have to pay, since he hasn't yet won his first case. After the termination of the case, *then* the student owes money to Protagoras, so Protagoras should then turn around and sue the student a second time. This time, the court should award the case to Protagoras, since the student has now won his first case.

Rescher [23] agrees with Smullyan that the two-trials solution appears to have the strongest claim.

Łukowski summarizes and criticizes various solutions to the Paradox of the Court, proposing his own solution, in [17] and [18]. He objects that many accounts available in the literature, including the logical reconstructions by Lenzen and Åqvist, substitute a *legal pseudo-problem* (getting paid the established fee) for the ancient *logical dilemma*. The original logical problem consists in the contradiction resulting from putting the conclusions of two arguments that are *equally plausible* —one formulated by Protagoras, another one by Euathlus. The legal pseudo-problem can be managed only within the second case mentioned by Smullyan. But Łukowski notes that no second case is mentioned in the original paradox, and that the real problem clearly pertains to the first (and only) case.

We partly agree with Łukowski's criticism. However, we are also afraid that his own solution is open to similar objections as those he raises. He argues as follows:

> Let us use two expressions: 'pay the agreed *fee*' and 'pay the court ordered *damages*' rather than 'pay for the education'. It is easy to see that such a simple operation eliminates the unwanted contradiction. If Euathlus wins the court case, he must pay the agreed fee, even though he does not pay the damages. If Protagoras is the winner, the situation will be quite the opposite.

The conclusion of the argument is that Euathlus has to pay the fee in both cases, and the contradiction thus vanishes. The problem with this account is that, just as the *second case* has not been mentioned in the original paradox, *damages* have not been mentioned either. Because of this, while Łukowski's solution appears to be on the right track, it is not entirely satisfactory. Lenzen and Åqvist importantly show that the temporal aspect is important to understand the paradox; Łukowski shows that an ambiguity of a certain kind is lurking behind the paradox. We will combine these intuitions by proposing a novel account of the paradox within the formal framework introduced in this article.

5 Representing the Paradox of the Court in Our Framework

Where is the ambiguity at the basis of the Paradox of the Court to be located, exactly? In the original formulation, there is no ambiguity in what should be paid, that is, there is no distinction between the *fee* versus the court ordered *damages*. Damages are not mentioned at all, it is all about the money for education (the fee). The ambiguity is rather rooted in *different sources of norms*.[8] Recall this passage from [12] (italics added): "For if the case goes against you, *the money will be due me in accordance with the verdict*, because I have won; but if the decision be in your favour, *the money will be due me according to our contract*, since you will have won a case." We thereby suggest that the difference is between being obliged (or not being obliged) to pay the fee in accordance with the verdict versus to pay the fee in accordance with the agreement.

Euathlus might not be obliged to pay the fees on the basis of the court decision (if he wins the court case); but it does not follow that there is no other obligation - on the basis of the agreement - to pay Protagoras for his teaching. And the other way around, Euathlus might not be obliged to pay the fee on the basis of the agreement (if he loses the court case), but it does not follow that there is no other obligation to pay Protagoras for his teaching.

Let us now capture these intuitions in the proposed formal language. To begin with, we need to distinguish five different things surrounding the paradox:

- the agreement between Protagoras and Euathlus;

---

[8] Given that the promise or the agreement in question is sufficient for generating the obligation; cf. [19] for a more complex procedural treatment of promises.

- the argument by Protagoras leading to the conclusion that Euathlus is obliged to pay;
- the argument by Euathlus leading to the conclusion that Euathlus is not obliged to pay;
- the scenario that (apparently) took place;
- the scenario that should have taken place.

Let p be the proposition that Euathlus wins the first court case and let q be the proposition that Euathlus pays the fee. In addition, let O^a represent an obligation on the basis of the agreement between Protagoras and Euathlus and O^c represent an obligation on the basis of the court decision. Let i be the date when the education terminated.

The Agreement. In the proposed language, the agreement can be analysed by saying that in all possible courses of events, if there is a case that Euathlus wins at a time j after i, then Euathlus is obliged to pay the fee starting from j. Thus, we have the following schema, for all $j \in IND$:

$$E(i,j) \to L(\Diamond_j p \to O^a \Diamond_{j \Rightarrow} q)$$

The Argument Formulated by Protagoras. Protagoras breaks down the possible outcomes of his trial against the former scholar into two options:

- Euathlus wins the case.
- Euathlus does not win the case.

Let r stand for the proposition that Euathlus wins the case and let k be the date of the court decision. Thus, at k, either r holds or $\neg r$ holds. If r holds, Protagoras argues, $O^a \Diamond_{k \Rightarrow} q$ holds too, since the conditions of the agreement are satisfied. If, on the other hand, $\neg r$ holds, $O^c \Diamond_{k \Rightarrow} q$ holds too, because the court decided in favour of Protagoras. We can immediately identify one mistake in the argument formulated by Protagoras: If Euathlus does not win, it does not follow that the court ordered Euathlus to pay (i.e., that Protagoras wins), as it will be clear in the description of the scenario that (apparently) took place.

The Argument Formulated by Euathlus. Euathlus too, breaks the possible outcomes of the trial into two options. If r holds, Euathlus argues, the jurors decided in his favour, so he does not have to pay; i.e., $\neg O^c \Diamond_{k \Rightarrow} q$. If, on the other hand, $\neg r$ holds, then he has not won any case yet, so he is not obliged to pay according to the agreement; i.e., $\neg O^a \Diamond_{k \Rightarrow} q$. Here we can again spot a mistake: it does not follow from not being obliged to pay on the basis of one normative source that there is no other normative source that obliges one to pay. In other words, if r holds, then really $\neg O^c \Diamond_{k \Rightarrow} q$ holds, but also $O^a \Diamond_{k \Rightarrow} q$ holds. And if $\neg r$ holds, then really $\neg O^a \Diamond_{k \Rightarrow} q$ holds, but it can be that $O^c \Diamond_{k \Rightarrow} q$ holds too (i.e., that Protagoras wins).

The Actual Scenario. In the actual scenario (as presented by Gellius; [12], 409), "the jurors, thinking that the plea on both sides was uncertain and insoluble, for fear that their decision, for whichever side it was rendered, might annul itself, left the matter undecided and postponed the case to a distant day." At k, the conditions that there is some j such that $E(i,j)$ and $\Diamond_j p$ are not satisfied, because Euathlus has not won the given case (no one has), nor has he won any other case yet. He is thus not obliged to pay the fees —neither on the basis of the court decision nor on the basis of the agreement. We thus have $\neg O^c \Diamond_{k \Rightarrow} q$, but also $\neg O^a \Diamond_{k \Rightarrow} q$.

The (Legally) Ideal Scenario. In the legally ideal scenario described above by Smullyan, Euathlus wins the first case because, till the court decision, he has not won any case, and thus is not obliged to pay. However, after this victory he can be sued another time and be obliged to pay. In other words, until k, the conditions that there is some j such that $E(i,j)$ and $\Diamond_j p$ are not satisfied. However, after k, those conditions can be satisfied.

6 Final Remarks

This paper proposed a fine-grained formal framework for normative reasoning that combines deontic, temporal, and metaphysical modalities. The main motivation for this choice is the fact that in normative reasoning, as well as in fundamental ethical principles, such as the Ought-Implies-Can principle (OIC), modalities of these three kinds are intertwined. Admittedly, OIC is a rather complex principle; in the present paper, we have dealt with only one aspect of OIC – the agent's need of a temporal opportunity to fulfil an obligation.

Furthermore, we illustrated how the framework works in formalizing a troublesome ancient paradox, the Paradox of the Court. We proposed a new account of the paradox, which takes some inspiration from existing accounts and highlights the following aspects: the temporal dimension; the presence of ambiguity; the hidden mistakes of the two arguments leading to contradictory results.

There are several interesting directions for future research. One direction is exploring how the families of modalities in question are intertwined in other issues related to ethics and morality; for instance, in debates around moral or legal *responsibility*, where reference to alternative possibilities and temporal opportunities is fundamental to evaluating the behaviour of a normative party. Exploring these issues in a formal framework that is very simple (being based on a propositional multimodal language), but rich enough to specify how obligations and possibilities are lost or gained across intervals of time, could be very useful. Yet another direction would be developing axiomatic extensions of the minimal logic **DTM** presented here in order to encode further principles of normative reasoning. For instance, in order to account for an agent's freedom, one could extend the formal representation of OIC with a condition making reference to the metaphysical possibility of violating obligations.

References

[1] Allen, F., *Maintaining knowledge about temporal intervals*, Communications of the ACM **26** (1983), pp. 832–843.
[2] Åqvist, L., *The Protagoras case: an exercise in elementary logic for lawyers*, in: J. Bjarup and M. Blegvad, editors, *Time, Law, and Society*, Franz Steiner, 1981, pp. 73–84.
[3] Åqvist, L., *Combinations of tense and deontic modality*, in: A. Lomuscio and D. Nute, editors, *Proceedings of DEON 2004*, Springer, 2004, pp. 3–28.
[4] Bailhache, P., *Canonical models for temporal deontic logic*, Logique et Analyse **149** (1995), pp. 3–21.
[5] Broersen, J., *Agents necessitating effects in Newtonian time and space: from power and opportunity to effectivity*, Synthese **196** (2019), pp. 31–68.
[6] Brown, M.A., *On the logic of ability*, Journal of Philosophical Logic **17** (1988), pp. 1–26.
[7] Brown, M.A., *Normal bimodal logics of ability and action*, Studia Logica **51** (1992), pp. 519–532.
[8] Chellas, B.F., "Modal Logic", Cambridge University Press, 1980.
[9] Demri, S., V. Goranko and M. Lange, "Temporal Logics in Computer Science", Cambridge University Press, 2016.
[10] Faroldi, F.L.G., "Hyperintensionality and Normativity", Springer, 2019.
[11] Faroldi, F.L.G., *Deontic modals and hyperintensionality*, Logic Journal of the IGPL **27** (2019) 387–410.
[12] Gellius, A., "The Attic Nights of Aulus Gellius: with an English Translation by John C. Rolfe", Harvard University Press, 1927.
[13] Giordani, A., *Ability and responsibility in general action logic*, in: J.M. Broersen, C.Condoravdi, N. Shyam and G. Pigozzi, editors, *Proceedings of DEON 2018*, College Publications, 2018, pp. 121–138.
[14] Governatori, G., J. Hulstijn, R. Riveret and A. Rotolo, *Characterising deadlines in temporal modal defeasible logic*, in: M. A. Orgun and J. Thornton, editors, *Proceedings of AI 2007*, ACM, 2007, pp. 486–496.
[15] Horty, J. and N. Belnap, *The deliberative STIT: A study of action, omission, ability and obligation*, Journal of Philosophical Logic **24** (1995), pp. 583–644.
[16] Lenzen, W., *Protagoras versus Euathlus: Reflections on a so-called paradox*, Ratio **19** (1977), pp. 176–180.
[17] Łukowski, P., *O co chodzi w paradoksie Protagorasa?*, Acta Universitatis Lodziensis **17** (2006), pp. 17–38.
[18] Łukowski, P., *Paradoxes of Ambiguity*, in: P. Łukowski, editor, *Paradoxes*, Springer, 2010, 37–74.
[19] Marko, V., *Sľuby a procedúry*, Filozofia **74** (2019), pp. 735–753.
[20] McArthur, R.P., "Tense Logic", Kluwer, 2019.
[21] Meyer, J.-J.Ch., W. van der Hoek, and B. van Linder, *A logical approach to the dynamics of commitments*, Artificial Intelligence **113** (1999), pp. 1–40.
[22] Quinn, P.L., *Metaphyisical necessity and modal logics*, The Monist **65** (1982), pp. 444–455.
[23] Rescher, N., "Philosophical Clarifications: Studies Illustrating the Methodology of Philosophical Elucidation", Springer, 2019.
[24] Rönnedal, D., *Quantified temporal alethic-deontic logic*, Logic and Logical Philosophy **24** (2015), pp. 19–59.
[25] Smullyan, R.M., "What is the Name of This Book? The Riddle of Dracula and Other Logical Puzzles", Prentice-Hall, 1978.
[26] Thomason, R., *Deontic logic as founded on tense logic*, in: R. Hilpinen, editor, *New Studies in Deontic Logic*, Springer, 1981, pp. 165–176.
[27] Van Benthem, J., "The Logic of Time", 1982, Springer.
[28] van der Hoek, W., B. van Linder, and J.-J.Ch. Meyer, *An integrated modal approach to rational agents*, in: M. Wooldridge and A. Rao, editors, *Foundations of Rational Agency*, Springer, 1998, pp. 133–168.
[29] Vranas, P.B.M., *I ought, therefore I can*, Philosophical Studies **136** (2007), pp. 167–216.

[30] Vranas, P.B.M., *"Ought" implies "Can" but does not imply "Must": an asymmetry between becoming infeasible and becoming overridden*, The Philosophical Review **127** (2018), pp. 487–514.

How Deontic Logic Ought to Be: Towards a Many-Sorted Framework for Normative Reasoning

Valentin Goranko [1]

Department of Philosophy, Stockholm University
and School of Mathematics, University of the Witwatersrand (visiting professorship)

Abstract

Formalising adequately normative logical reasoning with deontic logic has been notoriously problematic. Here I argue that one of the major reasons is that a typical deontic inference combines different types of sentences, expressing (inter alia) propositions, norms, and actions. These have different logical properties and formally mixing them can leads to unnatural (or, plainly absurd) conclusions, of which deontic logic abounds. Thus, I argue that deontic logical reasoning is inherently many-sorted and that an adequate logical formalisation of such reasoning ought to involve separate, yet inter-related syntactic sorts, at least including *norms, actions, and propositions*. Here I propose such formal logical framework, illustrate its use for formalising commonsense normative reasoning, and provide formal semantics for a large fragment of it.

Keywords: logic-based normative reasoning, many-sorted deontic logic, actions, norms, propositions

1 Introduction

The questions of how logic-based normative reasoning should be formalised, and what it should apply to, have permeated the entire history of deontic logic and have been driving much of its agenda, ever since (and even before) G.H. von Wright's seminal 1951 paper [40]. Von Wright himself struggled with these questions for over 50 years and changed his views and opinions more than once meanwhile. His original system of deontic logic proposed in [40] was not a logic of propositions, but a logic of norms over actions ('acts') [2], to which the deontic operators apply, thus producing normative propositions. That led to various problems, both formal and conceptual, which von Wright tried hard to resolve over the following years, meanwhile gradually moving towards the

---

[1] Email: valentin.goranko@philosophy.su.se
[2] Von Wright wrote there *"First a preliminary question must be settled. What are the "things" which are pronounced obligatory, permitted, forbidden, etc.? We shall call these "things" acts."*

(so called) 'standard deontic logic' of propositions, cf. [42,43,44]. This was a shift, in von Wright's terms, from an *ought-to-do* ("Tun-sollen") approach to an *ought-to-be* ("Sein-sollen") approach to Deontic logic (cf. [47]). That shift was also motivated by the vigorous development of modal logic in the 1960s and von Wright (as he himself admits, e.g. in [48]) was strongly influenced by some leading logicians of the time and proponents of the possible worlds semantics, incl. Anderson, Prior, and others.

The so-called 'standard' propositional deontic logics emerging as a result of that shift not only did not resolve the fundamental problems of logical formalisation of normative reasoning, but actually aggravated some of them, by bringing to the surface numerous formalised versions of deontic paradoxes and puzzles, such as Ross' paradoxes, Prior's paradoxes of derived obligations, etc. Much of the mainstream research in deontic logic has been devoted to attempting to resolve these paradoxes, either one at a time, or *"all of them in one fell swoop"* [4]. Indeed, much progress has been made over the years, but also many problems arising in the area have not been resolved yet in a satisfactory way, and more have arisen meanwhile. In particular, the fundamental question *"(how) is formal logic applicable to normative reasoning?"* remains, I would argue, not definitively resolved yet. Just one very telling fact about the long-lasting drama around that question is that 40 years after his original 1951 paper in *Mind* von Wright published in 1991 the paper [46] titled *"Is there a logic of norms?."* In the abstract he wrote *"If norms are neither true nor false, can logical relations such as contradiction and entailment obtain between them? Earlier logical positivists [...] have answered the question with No. While appreciating the seriousness of the problem, the author of the present paper makes a fresh attempt to answer the question with Yes. [...]"*

Arguably, most of the problems with formalising normative reasoning are inherent in its very nature, which combines usual propositional reasoning with reasoning about norms, about agents' actions, and judgements about the agents' compliance with norms while performing these actions. It is also widely acknowledged that it is quite challenging (if possible at all) to capture all these in 'traditional' logical systems, involving a single sort of formal expressions, viz. formulae expressing propositions. Numerous attempts have been made to develop more elaborated such systems that would capture better the essence of normative reasoning. In particular, several systems of deontic logic have been proposed (see notes on related work in Section 6) putting together actions and norms, or actions and normative propositions. Still, it appears that these attempts have not yet led to a full and seamless integration of various deontic logics of norms, logics of propositions, and logics of actions, into a *deontic logic combining reasoning about norms, propositions and actions on a par*. This, in a nutshell, is the essence of this paper's proposal.

Here I argue that such more elaborated approach is not optional, but necessary for the design of adequate logical systems for normative reasoning. In particular, I claim that adequate logical formalisation of normative reasoning ought to be *many-sorted*, involving separate syntactic sorts, at least for *norms, actions, and propositions*. Then I propose a concrete, yet generic such new logical framework, for which I present here the basic building blocks of its language, provide formal semantics for a large fragment of it, and illustrate its use for formalising 'everyday' normative reasoning.

2 Some fundamental issues of deontic logic revisited

2.1 Jørgensen's dilemma: is there a deontic logical consequence?

As noted in the introduction, a fundamental question arising even before the birth of formal deontic logic is *"What do deontic sentences express: norms or propositions?"* It has been widely assumed that these two uses are mutually exclusive. The traditional logical positivism view was definite: deontic sentences expressing norms can be neither true nor false, so they are not propositions and cannot be treated with formal logic. One of the leading representatives of that view, the Danish philosopher J. Jørgensen, stated that issue – now known as *Jørgensen's dilemma* – in his 1937 paper [20] essentially as follows:

- *either* the notion of logical consequence is defined in terms of truth, in which case there can be no deontic logical consequence, hence no possibility for deontic logic;
- *or* a logic of norms is possible, but the notion of logical consequence should not be defined in terms of truth, which contradicts a fundamental assumption in logic (according to the logical positivism).

Von Wright's position (cf. e.g. [41]) was strongly in favour of the latter: *"Yes, logic of norms is possible, as logic has a wider reach than truth!"*.

Let me also note that the question whether there can be a coherent notion of respective logical consequence is not exclusive to normative reasoning. Such questions arise, on similar grounds, for instance regarding reasoning about imperatives, as well as about interrogatives [3].

2.2 Norms vs normative propositions

One of the fundamental issues arising in normative reasoning is the important distinction that is to be made between a *norm* and a *normative proposition*, perhaps first explicitly pointed out (according to von Wright) by I. Hedenius [17]. What makes this distinction so subtle is that it is often noted only on level of pragmatics. For instance, a normative sentence such as *"Parking here is forbidden"* may have a *prescriptive (norming) meaning*, if stated by an authority (parking attendant or traffic police), or a *descriptive (informative) meaning*, if uttered by a possibly informed passer-by. The former case yields a *norm*, whereas the latter – a *(normative) proposition*. That makes a major difference, as norms prescribe what should, or may (or not) be done, whereas normative propositions describe the normative status of actions according to the existing

---

[3] However, it should also be noted that there has been a significant recent progress on the latter issue, by means of the so called inquisitive semantics and logics, cf. [6]. Compared to that development, formalising logical normative reasoning is now lagging behind.

norms. Thus, norms can be obeyed, fulfilled or violated, but *cannot be true or false*, whereas normative propositions are naturally assigned truth values.

According to von Wright, the appearance of many paradoxes (such as Ross' letter paradox) arises from a confusion between norms and norm-propositions, as norms do not satisfy some basic logical laws, e.g.:
"*A norm to the effect of 'p' does not imply a norm to effect that 'p or q'*".

Several scholars on normative reasoning, incl. C. Alchourrón and E. Bulygin [2], kept raising the question *Is Deontic Logic a logic of norms or a logic of normative propositions?*. According to the early von Wright, the 'real deontic logic' is the former. But, as noted earlier, he changed his views more than once over time, cf. [45], and later he took the view that logic of norms is impossible and deontic logic can only be a logic of propositions about (the existence of) norms [4] – from which he backtracked again still later, cf. e.g. [48] [5].

The answer advocated in the present work is that Deontic Logic ought to be *both* a logic of norms *and* a logic of normative propositions. Also, in my view, Deontic Logic should combine the *Sein-sollen* and the *Tun-sollen* perspectives [6], rather than opposing them to each other.

2.3 Can there be formal logical deontic reasoning? Yes, there can, and there is!

Jørgensen's dilemma raises the existential for deontic logic question: *can there be deontic logical consequence, at all?*. As already noted, von Wright's position, albeit sometimes shrouded in doubts, was positive. This view is shared by most (if not all) formal deontic logicians, and even by those who view the so called 'standard systems of deontic logic' to be a failed attempt to adequately formalise deontic reasoning. For the present author, the question is but rhetoric, because deontic logical reasoning *does exist in real life* and we do make deontic logical inferences on a daily basis, even though usually without realising that. The real question is: *how to formalise properly deontic logical reasoning?* In several papers von Wright explored the notions of consistency and entailment between norms, but (for all I know) stopped short of putting them on a par with normative propositions. As I argued above, a major problem arising with normative logical reasoning is that it is intrinsically many-sorted, combining

---

[4] As he confessed in [46]: "*Over the years my view became more "radical", and I came to think that logical relations such as contradiction and entailment could not hold between (genuine) norms and that therefore, in a sense, there could be no such thing as a "logic of norms."* " [...] "*The notion of rationality came to my help and so I arrived at a position according to which deontic logic is neither a logic of norms nor a logic of norm-propositions but a study of conditions which must be satisfied in rational norm-giving activity*".

[5] von Wright adds there in a footnote: "*This was what I thought initially to be the lesson of the coming into existence of deontic logic. Later I thought differently. In the end, it seems, I have gone full circle back to my original position. But I still think the journey was worth making.*"

[6] This alternative is distinct, but in my view essentially related to the previous one, as deontic logic of norms links more naturally with the *ought-to-do* approach, whereas a logic of normative propositions is closer to the *ought-to-be* perspective.

propositions, norms, actions, and that only many-sorted formalisms for normative reasoning can captured it adequately.

Let us consider a few simple examples of what one can call 'everyday normative reasoning' to see the issue at hand.

(i) *Pippi is buying beer from the local liquor store.* (action; fact)

(ii) *Everyone*[7] *of age over 20 is allowed to buy alcohol from liquor stores.*
 (norm (permission))

(iii) *Pippi is 21 years old.* (fact)

(iv) *Therefore, Pippi is allowed to buy alcohol at liquor stores.*
 (conclusion: derived individual norm? normative proposition? both?)

(v) *Therefore, Pippi's buying beer from the local liquor store is legal.*
 (conclusion: a normative proposition, stating compliance with a norm)

Another example:

(i) *Smoking in the building is forbidden.* (common norm (prohibition))

(ii) *Chuck Norris is in the building.* (fact)

(iii) *Therefore, Chuck Norris is not allowed to smoke.*
 (derived individual norm (prohibition))

(iv) *Chuck Norris is smoking.* (action; fact)

(v) *Therefore, Chuck Norris is violating the non-smoking rule.*
 (fact; normative proposition, stating a norm violation)

(vi) *Chuck Norris may violate any rule.* (???... ok, a Chuck Norris joke)

Clearly, these are proper logical inferences (skipping a few trivial steps). Yet, they are *not* instances of traditional logical reasoning, because they involve *not only propositions* (facts), but also actions and (common and individual) norms, and these do not necessarily obey the rules of propositional reasoning, at least because they cannot be assigned truth values. Still, we should all agree that these inferences are – intuitively – *logically correct*, in a sense that ought to be made precise.

The logical inference problems get amplified when *conditional* norms are involved, often leading to non-monotonic reasoning, as in the following example.

(i) *To be allowed to drive a car, one must have a valid driving licence.*
 (norm (conditional obligation))

---

[7] This sentence formally requires universal quantification over agents. However, a language with full-fledged quantification over agents is hardly necessary for normative reasoning, as norms usually involve universal quantification over agents and very seldom existential quantification (e.g. as in "at least one author of each accepted paper must register for the conference"). The universal quantification over agents can thus be omitted and assumed implicitly. So, to avoid such quantification and keep the arguments propositional, I will use in the formal framework so called 'common norms', applying by default to all agents.

(ii) Noone under the age of 18 may have a driving licence.
(norm (prohibition))
(iii) Tommy is 16 years old. (fact)
(iv) Therefore, Tommy is not allowed to drive a car.
(conclusion 1: a norm and a normative proposition)
(v) A person aged between 16 and 18 may practice driving, but only if accompanied by a licensed driver. (norm (conditional permission))
(vi) So, Tommy may drive a car, if accompanied by a licensed driver.
(conclusion 2: a conditional norm, and a normative proposition)
(vii) Noone is allowed to drive under the influence of alcohol.
(norm (prohibition))
(viii) Tommy has just had two beers and is a little dizzy. (fact)
(ix) Therefore, Tommy is not allowed to drive a car (now)[8].
(conclusion 3: a norm and a normative proposition)
(x) Tommy is driving his dad's Volvo back home.
(action; fact; norm violation)

Note that some of the expressions in the inference above can be classified in more than one way, e.g. the last one is an action, but also a fact, and a proposition implying a violation of the prohibition norm stated in (g). In order to treat properly these type-sharing phenomena we need syntactic mechanisms that generically transform one type of statement into another.

3 MS-Deon: a multi-sorted logical language for normative reasoning

As discussed earlier, initially von Wright conceived deontic logic as a logical theory of *ought-to-do*. However, he and others gradually transformed it into a logical theory of *ought-to-be*. I argue that neither of these approaches suffices alone to capture the distinction and interaction between *actions, norms and normative propositions*, but they must be combined in a more sophisticated and versatile language, with different syntactic sorts of formal expressions, involving (at least) each of these. Furthermore, *agentivity* should be brought to the fore of normative reasoning, as norms should (primarily) apply to agents and their actions, not to propositions, i.e. von Wright's original idea was the right one! Importantly, *existence and validity of norms* should be separated already on syntactic level from statements about *agents' compliance with norms*, as truth values apply to the latter, but not the former. Eventually, both validity and compliance with norms should be postulated for the basic norms, and then *computed, or deductively derived* for complex norms.

Before introducing the formal language, here are some guiding principles

---

[8] This example raises more issues, such as the role of time and temporality in normative reasoning, which I will not discuss further here.

followed in its construction.

- *Norms* are constructs applying deontic operators D (*Obligatory, Permitted, Forbidden*) to actions, to produce *D-to-do* expressions. Norms can be applied to actions of specific agents, or universally, to all agents. The latter can be done generically ("smoking is forbidden") or by explicit quantification ("no passenger is allowed to smoke"). I will follow the former approach (cf. footnote 7). Furthermore, often norms are *conditional* (e.g. "persons of age below 18 are not allowed to buy alcohol").
- Norms can be combined by boolean and other operations to produce complex norms. More generally, a *norm-building sub-language* emerges.
- *Actions* can be atomic, primitive entities, or can be built in a compositional style of dynamic logic, from (names of) atomic actions, by using operations on actions. Thus, an *action-building sub-language* emerges, too.
- Actions also involve specific STIT-like constructions, where only the required effect of the action, and possibly the agent executing that action, are specified, thus relating *D-to-be* and *D-to-do* expressions.
- *Propositions* can be of different sub-sorts, too, including *factual, normative*, and *performative* (explained further). They are built separately, by imposing suitable restrictions on the language, but can be eventually combined by applying standard logical connectives.

3.1 The formal 3-sorted deontic language MS-Deon

Here I present a generic multi-sorted language of **Deontic Logic for Norms, Actions, and Propositions**, hereafter denoted MS-Deon, with the three main sorts defined by mutual recursion, along with the intuitive semantics of the respective language constructs. This language will generically involve various possible constructs and sub-sorts, but these need not all be included in any concrete instantiation of it, which would only select the constructs that are naturally needed for the concrete purpose of specific normative reasoning.

To define the language, I first fix sets of atomic propositions PROP, atomic actions ACT, and agents Agt.These will usually (though, not necessarily) be assumed finite and common for all formal models that will be defined further and can be thought as specifying the signature of a concrete instantiation of MS-Deon, analogously to the non-logical symbols in a concrete first-order language. I will use specific names for agents, for which I will use metavariables such as A, B, C. I also use *agent parameters*, i.e., free variables ranging over agents, typically denoted by a, b, c. These parameters will only play an auxiliary role, to avoid explicit universal quantification over agents (cf. footnote 7) which is typically needed for common or conditional norms.

- **Actions.** The sort for actions is built compositionally, generally following the style of Propositional Dynamic Logic PDL, by this inductive definition:

$$\alpha := \alpha_{at} \mid \bar{\alpha} \mid \alpha;\alpha \mid \alpha \cup \alpha \mid \alpha \cap \alpha \mid \cdots \mid \varphi? \mid [\text{stit}]\varphi$$

Here α_{at} are atomic actions and $\bar{}, ;, \cup, \cap$ are respectively the operations of 'negation', sequential composition, choice, and parallel execution of actions, whereas $\varphi?$ is the action 'test' applied to a formula φ. Intuitively, the action $\varphi?$ succeeds without changing the current state if φ is true in it, else it fails, leading to no outcome state ('crashes'). Lastly, $[\text{stit}]\varphi$ is an action [9] which is only described by its effect, viz. 'bringing about (the truth of) φ'. Note that actions are so far agent-less, but in the context of the full language they will be usually attributed to agents.

- **Norms.** I consider three sub-sorts of norms: *common* (applicable generically to any agent), *individual agentive* (applicable to an explicitly specified agent), and *conditional agentive* (applicable to all agents satisfying the given condition). In the definitions below: α is an action; $\varphi(\mathsf{a})$ is a proposition (possibly) containing an agent parameter a; A is (a generic name of) an agent; a is a parameter ranging over agents; $\mathsf{Ought}^{do}, \mathsf{Perm}^{do}, \mathsf{Proh}^{do}$ are the main deontic to-do operators, viz. 'ought-to-do', 'permitted-to-do', and 'prohibited-to-do'; and $\bar{}, \heartsuit, \diamondsuit$ are the respective boolean operations on norms, the precise meaning of which can vary (cf. e.g. [42,45,34,37,38] for possible interpretations) and will not be fixed here. Formally:

    Common norms:

    $$\mathsf{N} := \mathsf{Ought}^{do}\alpha \,|\, \mathsf{Perm}^{do}\alpha \,|\, \mathsf{Proh}^{do}\alpha \,|\, \bar{\mathsf{N}} \,|\, \mathsf{N} \heartsuit \mathsf{N} \,|\, \mathsf{N} \diamondsuit \mathsf{N}$$

    Individual agentive norms:

    $$\mathsf{N} := \mathsf{Ought}^{do}_{\mathrm{A}}\alpha \,|\, \mathsf{Perm}^{do}_{\mathrm{A}}\alpha \,|\, \mathsf{Proh}^{do}_{\mathrm{A}}\alpha \,|\, \bar{\mathsf{N}} \,|\, \mathsf{N} \heartsuit \mathsf{N} \,|\, \mathsf{N} \diamondsuit \mathsf{N}$$

    Conditional agentive norms:

    $$\mathsf{N} := \mathsf{Ought}^{do}(\varphi(\mathsf{a}), \alpha) \,|\, \mathsf{Perm}^{do}(\varphi(\mathsf{a}), \alpha) \,|\, \mathsf{Proh}^{do}(\varphi(\mathsf{a}), \alpha) \,|\, \bar{\mathsf{N}} \,|\, \mathsf{N} \heartsuit \mathsf{N} \,|\, \mathsf{N} \diamondsuit \mathsf{N}$$

    Intuitively, $\mathsf{Ought}^{do}(\varphi(\mathsf{a}), \alpha)$ says that doing α is obligatory for every agent a satisfying the condition $\varphi(\mathsf{a})$; likewise for Perm^{do} and Proh^{do}.

    These sub-sorts of norms can be combined further by using the boolean operations on norms.

- **Propositions.** I consider 3 most sub-sorts of propositions that naturally and most commonly occur in normative reasoning.

    Factual propositions:

    $$\phi := p \,|\, p_{\mathrm{A}} \,|\, p_{\mathsf{a}} \,|\, \neg\phi \,|\, \phi \wedge \phi \,|\, [\alpha]_{\mathrm{A}}\varphi \,|\, [\alpha]_{\mathsf{a}}\varphi \,|\, [\alpha]\varphi$$

    Factual propositions are built up from atomic propositions to describe facts of the world. They can be agent-less, or referring to a specific agent (p_{A},

---

[9] NB: even though the idea comes from STIT theories, there is a fundamental difference: here $[\text{stit}]\varphi$ is an action, not a formula.

meaning that the agent A satisfies a property p), or agent-parameterised (p_a, with likewise meaning). Factual propositions can also express statements about facts holding after executions of actions, such as $[\alpha]_A \varphi$, meaning to say that 'after the agent A executes the action α (the fact expressed by) φ will hold'; likewise for $[\alpha]_a \varphi$; respectively, $[\alpha]\varphi$ means 'after any execution of the action α, φ holds'.

Performative propositions:

$$\psi := \mathsf{Perform}_A(\alpha) \mid \mathsf{Perform}_a(\alpha) \mid \mathsf{Perform}(\alpha) \mid \neg \psi \mid \psi \wedge \psi$$

Performative propositions express claims about actions being performed, by specific agents, or in general. Optionally, they can also involve propositions expressing *abilities* of agents to perform actions, e.g. $\mathsf{Able}_A \alpha$, etc., as well as other attitudes towards actions (desires, intensions, etc.).

Normative propositions:

$$\theta := \mathtt{InForce}(\mathsf{N}) \mid \mathtt{Sat}_A(\mathsf{N}) \mid \mathtt{Sat}(\mathsf{N}) \mid \mathtt{Legal}_A(\alpha) \mid \neg \theta \mid \theta \wedge \theta$$

Normative propositions express claims about the validity of norms, where $\mathtt{InForce}(N)$ means to say that "*the norm* N *is (currently) 'in-force'*", as well as claims about the compliance or violation of norms, by specific agents or generally. For instance, $\mathtt{Sat}_A(\mathsf{N})$ means to say that '*the agent* A *is complying with the norm* N' and $\mathtt{Legal}_A(\alpha)$ says that '*the performance of action* α *by agent* A *legal (norm-compliant)*'.

Inter-sort propositions can be combined freely by applying boolean connectives to these sorted propositions.

3.2 Some notes on the expressiveness and use of MS-Deon

Inter-sort transitions. MS-Deon allows for seamless transitions from one sort to another, whenever that makes good sense. For instance, an action α transforms to norms $\mathsf{Ought}^{do}\alpha$, $\mathsf{Ought}^{do}_A \alpha$ and likewise to $\mathsf{Perm}^{do}\alpha$ and $\mathsf{Proh}^{do}\alpha$, as well as to propositions like $\mathsf{Perform}(\alpha)$. A norm N transforms to normative propositions $\mathtt{InForce}(\mathsf{N})$ and $\mathtt{Sat}(\mathsf{N})$, whereas a normative proposition $\mathtt{InForce}(\mathsf{N})$ can be tested with a test action $\mathtt{InForce}(\mathsf{N})?$. Further, $\mathtt{Sat}(\mathsf{N})$ can be transformed to an action $[\mathsf{stit}]\mathtt{Sat}(\mathsf{N})$, which can then produce a performative proposition $\mathsf{Perform}_A([\mathsf{stit}]\mathtt{Sat}(\mathsf{N}))$, etc. All that enables natural formalisation of informal normative inferences, like the examples in Section 2.3, into formal propositional logical inferences in a MS-Deon-based deductive system (see Section 4).

Expressing 'To-Be' norms. MS-Deon can express uniformly both agentive (indicated by the optional index A) and non-agentive 'To-Be' norms and normative propositions from the respective 'To-Do' norms and propositions:

	Norms	Norm-propositions
$\text{Ought}^{be}_{(A)}\varphi :=$	$\text{Ought}^{do}_{(A)}[\text{stit}]\varphi$	$\texttt{InForce}(\text{Ought}^{do}_{(A)}[\text{stit}]\varphi)$
$\text{Perm}^{be}_{(A)}\varphi :=$	$\text{Perm}^{do}_{(A)}[\text{stit}]\varphi$	$\texttt{InForce}(\text{Perm}^{do}_{(A)}[\text{stit}]\varphi)$
$\text{Proh}^{be}_{(A)}\varphi :=$	$\text{Proh}^{do}_{(A)}[\text{stit}]\varphi$	$\texttt{InForce}(\text{Proh}^{do}_{(A)}[\text{stit}]\varphi)$

These apply likewise to produce normative propositions like $\texttt{Sat}(\text{Ought}^{be}\varphi)$.

Expressing individual conditional norms. Individual conditional norms, e.g. of the type *"if agent A satisfies property ϕ then A ought to do α"* can be expressed as the agentive conditional norm $\text{Ought}^{do}(\phi(\text{A}) \wedge \psi_A(\text{a}), \alpha)$ if ψ_A is a characteristic property that is satisfied by A and by no other agent. Otherwise, expressing that is still possible, but in a somewhat cumbersome roundabout way: first, take the proposition $\psi = \phi(\text{A}) \rightarrow \texttt{InForce}(\text{Ought}^{do}_A \alpha)$ which can then be turned into a norm by using the construct $\text{Ought}^{be}(\psi)$.

Adding imperatives. MS-Deon can be extended with a sort for *imperatives*, which can be both individual (referring to a specific agent, indicated below by the optional index (A)) and common, involving for instance constructs like:

$$[\texttt{stit } (\text{A})]!\varphi \mid \text{Do}_{(A)}\alpha \mid \text{Don't}_{(A)}\alpha \mid \text{MayDo}_{(A)}\alpha$$

The construct $[\texttt{stit }]!\varphi$ intuitively means to express the common imperative *'See to it that ϕ!'*, whereas $[\texttt{stit A}]!\varphi$ means the same, but addressing only the agent A; the rest are self-explanatory. The idea of adding imperatives is that they can be applied by authorities or agents for producing new norms, i.e. bringing about obligation, permission, or prohibition requirements in force, thus enabling the expression of *norm creation*. This idea will be explored further in a follow-up work.

Expressing some deontic principles and relationships. MS-Deon enables expressing various *to-do* norms in a uniform way, by translating them to *seeing-to-it* norms, whenever appropriate, following the scheme:

'Agent A *ought-to-do X'* $\Rightarrow$
'Agent A *ought to see-to-it-that (or, ought to bring-it-about-that)* A *does X*.

Respectively,
'*X ought-to-be-done*' $\Rightarrow$
'It *ought-to-be-seen-to-it-that (ought-to-be-brought-about-that) X is done*.

On the other hand, MS-Deon also enables expressing the agentive *stit* action by the non-agentive one:

$$[\text{A stit}]\varphi := \text{Perform}_A([\texttt{stit}]\varphi)$$

Further, the agentive *to-do* norms can be reduced to non-agentive *to-be* ones by using the well-known *Meinong-Chisholm Reduction* principle:
'*Agent A ought to see to it that p holds iff
it ought to be that agent A sees to it that p holds*'.

or,

'Agent A *is obliged to bring it about that p holds iff it is obligatory that agent* A *brings it about that p holds'.*

The Meinong-Chisholm reduction principle is simply formalised in MS-Deon:

$$\text{Ought}_A^{do}[\text{stit}]\varphi \equiv \text{Ought}^{be}[\text{A stit}]\varphi$$

which, when translated back becomes:

$$\text{Ought}_A^{do}[\text{stit}]\varphi \equiv \text{Ought}^{do}[\text{stit}]\text{Perform}_A([\text{stit}]\varphi)$$

(Note the two different uses of 'stit' above.)

On the other hand, intuitively, the following equivalence should hold:

$$\text{Ought}_A^{do}\alpha \equiv \text{Ought}^{be}[\text{A stit}]\text{Perform}_A(\alpha).$$

This and other such intuitive validities impose natural requirements on the formal semantics.

4 Towards a system of deduction for MS-Deon

The main purpose of a logic-based framework is to provide a platform for logical deduction. This is especially the case for MS-Deon. Indeed, the full language seems too rich and generic to build a complete, yet feasible logical system for it. So, in reality only suitable fragments of MS-Deon would be formalised deductively and used for specific purposes.

The many-sortedness of MS-Deon suggests two different approaches to building systems of deduction for it:

(i) *many-sorted deduction*, involving inter-sort inference rules and allowing for the derivation of logical consequences not only in terms of propositions, but also in terms of norms. That, inter alia, resolves the dilemma of 'deontic logic of norms' vs 'deontic logic of normative propositions', by putting both together.

(ii) *'flat' single-sort deduction*, where reasoning about actions and norms is transformed to reasoning about performative and normative propositions.

The 'flat', proposition-based approach is more traditional and can be naturally based on any standard logical deductive system, e.g., a suitably enriched system of natural deduction, sequent calculus, or even a Hilbert-style system for axiomatic deduction [10], also involving intermediate steps of inter-sort transformations. A system of many-sorted deduction would necessarily be rule-based [11]

---

[10] Hilbert-style systems are of limited practical use. Still, a provably complete such system can be used as a basis for building a more efficiently structured deductive systems, involving mechanisms for goal-oriented proof search. Resolution based proof systems are case in point.
[11] The rules are needed to avoid the clash of sorts that would occur if these were replaced by implications within formulae.

and closer to a natural deduction style, which I find more practically useful, though also perhaps more challenging. Thus, both approaches have pros and cons, but, eventually, due to the inter-sort reductions, both should yield the same deductive power.

In either case, building a practically useful system of deduction which is sound, both with respect to the prevailing common sense and (when the common sense is inconclusive) with respect to the formal semantics provided further, and is also sufficiently rich to capture non-trivial normative inferences, is a big and long-term project which goes beyond the limitations of this paper. Still, here are some initial construction steps for a rule-based system of deduction for MS-Deon, with some relevant references:

- To begin with, such system should contain as a core a system of deduction (e.g., natural deduction or sequent calculus) [12] for the underlying classical logical reasoning. That could be a version or a fragment of some complete system of natural deduction for first-order logic, cf. e.g. [29,39], or [12].

- In addition, it should contain subsystems of rules capturing the logical properties of each of the other two main sorts, viz. actions and norms. To build the sub-system of reasoning about actions compositionally, in PDL style, possibly suitable deductive systems to base it on are [19], [18], [9] and particularly [11], which explicitly incorporates the many-sorted approach. As for logics of norms, I am aware of few works on structured, rule-based deduction systems, starting with the pioneering [10], the non-technical but conceptually very relevant to the present work [13], [15] and the more recent [30], [8].

- The most challenging task is to develop a sufficiently rich system of *inter-sort rules* that enable truly many-sorted reasoning. Some of the earlier mentioned works enable structured multi-sorted deductive reasoning, but very few works that I am aware of, incl. [7] and [11], focus on that issue, and none of them on many-sorted *normative* reasoning.

Here are a few samples of many-sorted rules needed for formalising deductive reasoning in MS-Deon:

- If any agent a satisfying the condition φ ought to perform the action α and the agent A satisfies the condition φ, then A ought to perform the action α:

$$\frac{\mathrm{InForce}(\mathrm{Ought}^{do}(\varphi(\mathsf{a}),\alpha)),\ \varphi(\mathrm{A})}{\mathrm{InForce}(\mathrm{Ought}^{do}_{\mathrm{A}}\alpha)}$$

Hereafter, to keep the rules simpler, I will omit the construct InForce(·) but will just write the norm N itself as a premise or conclusion of a rule, meaning that it represents the normative proposition InForce(N). An advantage of

---

[12] I do not mention here tableaux-based systems, as they are based on proofs by contradiction, which only applies when one already knows what conclusion one wants to prove, whereas practical normative reasoning is often open-ended. However, I should at least mention tableaux-based systems for deontic and other related logics by Rönnedal, cf. [31], [33], [32].

the many-sorted framework is that it allows such freedom of expression without formally producing the sort mismatch that would occur if N itself, rather than InForce(N), is used in a formula in a flat language of propositions.

- If any agent a is permitted to perform the action α and A performs α then A's performing that action is legal [13]:

$$\frac{\text{Perm}_\text{A}^{do}\alpha,\ \text{Perform}_\text{A}(\alpha)}{\text{Legal}_\text{A}(\alpha)}$$

- If an agent A is prohibited from (respectively, obliged to) performing the action α and A performs (respectively, does not perform) α then A is violating that prohibition (respectively, obligation) norm:

$$\frac{\text{Proh}_{(\text{a})}^{do}\alpha,\ \text{Perform}_\text{A}(\alpha)}{\neg\text{Sat}_\text{A}(\text{Proh}_{(\text{a})}^{do}\alpha)}, \quad \frac{\text{Ought}_\text{a}^{do}\alpha,\ \neg\text{Perform}_\text{a}(\alpha)}{\neg\text{Sat}_\text{A}(\text{Ought}_{(\text{a})}^{do}\alpha)}$$

- If any agent a is prohibited from seeng to it that φ, and if φ obtains for sure after a performs the action α, then a is prohibited from performing α:

$$\frac{\text{Proh}_{(\text{a})}^{do}[\text{stit}]\varphi,\ [\alpha]_\text{a}\varphi}{\text{Proh}_\text{a}^{do}\alpha}$$

A stronger rule is obtained by replacing the premise $[\alpha]_\text{a}\varphi$ with $\langle\alpha\rangle_\text{a}\varphi$.

To illustrate how the MS-Deon framework can be used to formalise common-sense normative reasoning, consider the first example from Section 2.3. Let

- β be the action *"buying beer from the liquor store."*
- $o(\text{a})$ be the agent-parameterised atomic proposition *"a is over 20 years old."*

Here is a formalised many-sorted derivation of that example, skipping a few trivial steps. Note the use of the (ad hoc introduced) inter-sort inference rules.

(i) $\text{Perform}_{\text{Pippi}}(\beta)$ \hfill (action performative proposition)

(ii) $\text{Perm}^{do}(o(\text{a}),\beta)$ \hfill (norm (permission))

(iii) $o(\text{Pippi})$ \hfill ((essentially given) fact)

(iv) *Inter-sort inference rule*:

$$\frac{o(\text{A}),\ \text{Perm}^{do}(o(\text{a}),\beta)}{\text{Perm}_\text{A}^{do}\beta}$$

(v) $\text{Perm}_{\text{Pippi}}^{do}\beta$ \hfill (individual norm, derived from i-iv)

(vi) *Inter-sort inference rule*:

$$\frac{\text{Perm}_\text{A}^{do}\beta}{\text{Legal}_\text{A}(\beta)}$$

---

[13] We implicitly assume here that the permission excludes the possibility of a conflicting prohibition. In general, that may not be the case, so such proviso is to be added to the rule.

(vii) $\text{Legal}_{\text{Pippi}}(\beta)$ (normative proposition, derived from v-vi)

The inference above can be transformed to a flat propositional inference, by using the $\texttt{InForce}(N)$ construct to convert norms to normative propositions. Then, the inter-sort inference rules can be transformed to axiom schemes.

I leave the systematic development of a suitable full-fledged system of deduction for MS-Deon as an open challenge.

Lastly, I just note here that many well-known deontic paradoxical inferences, such as Prior's paradoxes of derived obligations, are naturally blocked in a many-sorted inference system for MS-Deon because of involving disallowed inter-sort inferences; again, this is to be discussed in a future work.

5 Semantics for MS-Deon

Designing adequate formal semantics for such rich language as MS-Deon is not less challenging task than designing an adequate system of deduction for it. That semantics is still partly under construction, as some clauses depend on resolving quite non-trivial questions beyond the scope of this paper, e.g., of how norms extend from atomic over to composite actions. I will first present some underlying principles of the semantics, then will define the proposed formal models, and then will give the semantic clauses for a large fragment of the language, leaving some clauses subject to additionally specified conditions resolving the issues mentioned above.

5.1 Semantics intuitively: main features

The semantics proposed here is influenced by several related works mentioned earlier and blends features from several semantics of well-known logics, incl. PDL, STIT, and the Coalition Logic CL, with some key new ideas:

- Models consist of states, describing snapshots of the world. Each state has:
  – a *propositional label*, being the set of atomic propositions true there;
  – a *normative label*, describing the atomic actions that are obligatory, permitted, or forbidden (commonly, or for a given agent) to perform from that state.

- Agents act concurrently from states of the model, each choosing independently to perform *a set of atomic actions* available to her at that state. (Such multiple actions are often referred to by norms, e.g. 'do not drink while driving'.) The result is an *action choice set profile*. Then, every agent performs independently all actions in her choice set (and only them).

- Transitions between states occur in a discrete manner and are determined by all agents performing 'simultaneously' their choices of actions. Sequences of states and transitions between them form *histories*.

- The composite actions are computed over histories, like in PDL.

- Norms are evaluated for *validity* (being *in-force*) at states, and for *agents' compliance* at histories and current transitions, in terms of the choice profile labels, inductively on their structure.

- Factual propositions are evaluated at states. Normative and performative propositions are evaluated at histories and current transitions.

5.2 Multi-agent deontic models

Given fixed (and usually assumed finite) sets of atomic propositions PROP, atomic actions ACT, and agents Agt, a **multi-agent deontic (MAD) model** over these is a structure

$$\mathcal{M} = \langle St, \text{act}, \pi, \delta, \Delta, \tau \rangle, \text{ where:}$$

- St is a set of *states*.
- $\text{act} : St \times \text{Agt} \to \mathcal{P}(\text{ACT})$ is a mapping assigning to every state s and agent A a set $\text{act}(s, \text{A})$ of *actions available to* A *at* s.
- $\pi : St \to \mathcal{P}(\text{PROP})$ is a *state description function*, assigning to each state s its *propositional label* $\pi(s)$.
- $\delta : St \times \text{Agt} \to \mathcal{P}(\text{ACT}) \times \mathcal{P}(\text{ACT}) \times \mathcal{P}(\text{ACT})$ is a *normative function*, such that $\delta(s, \text{A}) = (\delta_o(s, \text{A}), \delta_p(s, \text{A}), \delta_f(s, \text{A}))$ is the *normative label* of s, consisting of the sets of obligatory, permitted and forbidden actions for each agent A at state s. These satisfy the following natural constraints [14]:
  $\delta_o(s, \text{A}) \subseteq \delta_p(s, \text{A})$,
  $\delta_p(s, \text{A}) \cap \delta_f(s, \text{A}) = \emptyset$, and
  $\delta_p(s, \text{A}) \cup \delta_f(s, \text{A}) \subseteq \text{act}(s, \text{A})$.
- a *general normative function* $\Delta : St \to \mathcal{P}(\text{ACT}) \times \mathcal{P}(\text{ACT}) \times \mathcal{P}(\text{ACT})$ is defined likewise, to specify the *general normative label* of s, applying to all agents [15].
- $\tau : St \times \mathcal{P}(\text{ACT})^{\text{Agt}} \to St$ is the *transition function*, which for every $s \in St$ and a *choice of actions profile* σ determines the *successor state* $\tau(s, \sigma)$ of s, where a *choice of actions profile* at a state s is a mapping $\sigma : \text{Agt} \to \mathcal{P}(\text{ACT})$ such that $\sigma(\text{A}) \subseteq \text{act}(s, \text{A})$ for each $\text{A} \in \text{Agt}$, representing the selection of available actions that the agent chooses to perform at the current state. Thus, the mapping σ is not part of the description of the model, but is a component of the context of evaluation of norms and propositions.

A *history* [16] in $\mathcal{M}$ is a finite sequence of states and the transitions between them. The last state of the history h will be denoted by $l(h)$, and $h \circ s$ will denote the history h extended with the state s.

---

[14] The semantics presented here only handles norms that are applied in a particular situation (state), but not norms that are *not* applied because of being in conflict with higher priority norms applied at that state. Thus, the issue of conflicting norms and mechanisms for their resolution arises here, which will not be addressed in the present work.

[15] Of course, Δ can be subsumed in all individual normative labels, but that would lead to an unnecessary repetition of all common norms for all existing and newly appearing agents.

[16] NB: this is distinct from the notion of 'history' in STIT models, where it is a primitive abstract entity, representing a potentially infinite possible course of events.

5.3 Truth, norm validity, norm compliance, performance values

Given a MAD model $\mathcal{M} = \langle St, \text{act}, \pi, \delta, \Delta, \tau \rangle$ we define *truth of propositions* $\models$, *validity of norms* $\models$, and then *compliance with norms*, with respect to pairs *(history, choice profile)* [17] in $\mathcal{M}$ by a mutual induction as follows (omitting the standard boolean cases, and given here only for atomic actions):

- $\mathcal{M}, h, \sigma \models p$ iff $p \in \pi(l(h))$.
  (Here $l(h)$ is the current state of the evolution of the system with history h.)

- $\mathcal{M}, h, \sigma \models \text{Ought}^{do}\alpha$ iff $\alpha \in \Delta_o(l(h), \text{A})$; likewise for $\text{Perm}^{do}\alpha$ and $\text{Proh}^{do}\alpha$.
  $\mathcal{M}, h, \sigma \models \text{Ought}^{do}_A\alpha$ iff $\alpha \in \delta_o(l(h), \text{A})$; likewise for $\text{Perm}^{do}_A\alpha$ and $\text{Proh}^{do}_A\alpha$.

  We read $\mathcal{M}, h, \sigma \models \text{N}$ as "The norm N is in force at $\mathcal{M}, h, \sigma$".

  Thus, the normative labels at the current state determine the norms regarding atomic actions that are in force at that state [18].

- $\mathcal{M}, h, \sigma \models \text{Ought}^{do}(\varphi(\text{a}), \alpha)$ iff $\alpha \in \delta_o(l(h), \text{A})$ holds for all agents A such that $\mathcal{M}, h, \sigma \models \varphi(\text{A})$ (assuming that each $\varphi(\text{A})$ is already evaluated).
  Likewise for $\text{Perm}^{do}(\varphi(\text{a}), \alpha)$ and $\text{Proh}^{do}(\varphi(\text{a}), \alpha)$.

- $\models$ is extended to all norms inductively on the construction of norms, following additionally specified semantics of the norm-building operations.

- Now, we define $\mathcal{M}, h, \sigma \models \text{InForce(N)}$ iff $\mathcal{M}, h, \sigma \models \text{N}$,

- $\mathcal{M}, h, \sigma \models \text{Sat}_A(\text{Ought}^{do}_A\alpha)$ iff $\alpha \in \sigma(\text{A})$ [19];
  $\mathcal{M}, h, \sigma \models \text{Sat}_A(\text{Perm}^{do}_A\alpha)$ iff $\alpha \in \delta_p(s, \text{A})$;
  $\mathcal{M}, h, \sigma \models \text{Sat}_A(\text{Proh}^{do}_A\alpha)$ iff $\alpha \notin \sigma(\text{A})$;
  Analogously for $\text{Sat}(\text{Ought}^{do}\alpha)$, $\text{Sat}(\text{Perm}^{do}\alpha)$, $\text{Sat}(\text{Proh}^{do}\alpha)$.
  The clauses for compliance with composite actions, providing semantics of $\text{Legal}_A(\alpha)$, are to be given inductively on the definition of actions, according to externally specified conditions. These are generally non-trivial and subject to ongoing research and debate, so they will not be provided here.

- $\mathcal{M}, h, \sigma \models \text{Perform}_A(\alpha_{at})$ iff $\alpha_{at} \in \sigma(\text{A})$;

- $\mathcal{M}, h, \sigma \models \text{Perform}_A([\text{stit}]\phi)$ iff $\mathcal{M}, h, \sigma' \models \phi$ for every σ' with $\sigma'(\text{A}) = \sigma(\text{A})$

- The clauses for performance of composite actions are given inductively on their structure, according to their operational semantics, as in PDL.

---

[17] The truth of some propositions, e.g. the factual ones, would only depend on the current state, but others – typically normative and performative – also depend on the history and the choice profile.

[18] An essential point, raised by Karl Nygren: there is a difference between "existence" of norms and norms being "in force". The former typically refers to laws or other agentless and timeless norms, whereas the latter apply locally, at "states" possibly involving both a time instant and a location, i.e. apply "here and now". Ideally, both types of norms should be included and treated on a par in the formal framework. However, to keep it simple, here I will assume that globally existing norms are included in all normative labels of states.

[19] This definition makes the provision that one can comply with a norm even if it is not "in force" (by being in the current normative label). Thus, we can keep the judgment of compliance/non-compliance separate from the validity of norms. That may be useful e.g. for counterfactual normative reasoning.

- $\mathcal{M}, h, \sigma \models [\alpha]_A(\phi)$ iff $\mathcal{M}, h \circ \tau(l(h), \sigma'), \sigma'' \models \phi$ for every σ' such that $\alpha \in \sigma'(A)$ and every choice of actions profile σ'' at $\tau(l(h), \sigma')$.
  Likewise for $\mathcal{M}, h, \sigma \models [\alpha](\phi)$.

- The cases of agent-parameterised propositions will not be treated in full generality here, to avoid having to deal with assignments over agents. So far, we will only be interested in formulae where all agent-parameterised propositions are in the scope of conditional agentive norms, where their use is reduced to checking their truth for each agent in **Agt**.

This completes the semantics, modulo the mentioned externally specified mechanisms, e.g., for the extension of deontic operators to composite actions.

6 Some related work

In addition to the earlier cited works by von Wright, here is a selective and inevitably incomplete list of earlier publications, some of which have inspired, and others just anticipated, some ideas in the present work.

- In [1] C. Alchourrón proposed a "normative logic", viz. a logic of normative propositions (rather than a deontic logic of norms). These are propositions stating that an agent has 'issued a norm'. The logic has two sorts of formulae, building on top of a modal deontic logic in early von Wright style, by applying a 'norm issuing' operator N to deontic formulae. A couple of years later, Alchourrón and Bulygin proposed in their 1971 book [2] another two sorted deontic language, involving a 'universe of actions' and 'universe of properties', with deontic operators applying to the former, to build normative systems, all in a semi-formal style.

- Castañeda proposes in [4] a 'calculus containing proposition-practition distinction', where he distinguishes ought-to-do norms applying to actions ('practitions'), from those applying to 'action propositions', but without providing formal semantics or a system of deduction for these.

- Several researchers have proposed (with different motivations) and developed essentially two sorted deontic logics of actions, by analogy with the (propositional) dynamic logic of programs PDL. In particular:
  · K. Segerberg [34,35,36] developed a 'dynamic deontic logic' of actions, involving formulae and event (action) terms, evaluated over histories of events. *Norms* are defined only in the semantics, as functions N such that for any history h, $N(h)$ selects a set of extensions of h considered normal (legal) according to that norm. Still, Segerberg's system is closer to a logic of normative propositions, than to a logic of norms.
  · J.-J. Meyer proposed in [27] a different approach to two-sorted deontic logic as a variant of dynamic logic. He revisited 'Segerberg style' of dynamic deontic logic in [28].
  · P. Kulicki and R. Trypuz have developed in a series of papers [22,37,38,23] 'deontic action logics', based on algebraic approach to actions. These are formally logics of normative propositions about actions, not logics of norms.

- Other two-sorted deontic action logics were proposed by J. Broersen [3] and P. Castro and T. Maibaum [5].
- J. Hage has proposed and developed rule-based methodology and systems for normative and legal reasonings in a series of non-technical but conceptually close to the present proposal works [20], including [14] (summarized in [13]), [15], and the more recent [16], amongst others.
- Motivated by the study of conditional norms, D. Makinson and L. van der Torre, proposed the *"Input-Output Logic"* framework [24,25,26]. Notably, they develop an inference mechanism not based on transfer of truth-values.
- M. Knobbout, M. Dastani and J.-J. Meyer introduce in [21] a deontic logic for distinct types of norms, viz. state-based and action-based, with semantics based on formally defined normative systems.

7 Concluding remarks

This paper stemmed from my (growing over the years) dissatisfaction with how modal logic – in the guise of 'standard deontic logic' – handles normative reasoning. I have argued here that a more elaborated formal framework for deontic logic is needed for adequately formalising (especially, multi-agent) normative reasoning. I have then proposed such a many-sorted [21] formal framework, have outlined its language and formal semantics, and indicated briefly how it could be used for formalising normative reasoning. This work is still in progress and some features are still under construction, in particular the formal semantics for the full language and the development of adequate deductive systems for it.

I emphasise again that the proposed framework is generic and includes many syntactic features and constructs on actions, norms and propositions, only some of which would be applicable to any concrete instantiation of the framework. On the other hand, I also admit that a truly adequate logical formalisation of normative reasoning would require much more than what the proposed framework offers. It should also involve, *inter alia*:

- a full-fledged *first-order (or higher-order) language* for the domain of normative discourse, including constant, function, and predicate symbols, as well as quantification over individuals.
- an elaborated *multi-agent framework*, involving individual, as well as group and collective actions and respective deontic operators for individual, group, and collective obligations, permissions, and prohibitions, in their interaction; possibly, also explicit quantification over agents.
- *multi-agent knowledge*: individual, common and distributed, as well as *beliefs*. These are crucial for realistic and meaningful normative reasoning.

---

[20] I have only discovered these publications during the last days of work on the final version of this paper, hence my comments are brief and more superficial than deserved.

[21] As noted by a reviewer, the virtues of using sorted/typed logical formalisms for knowledge representation and reasoning go well beyond normative reasoning and deontic logics.

Further, a more *refined theory of agency*, that relates the deontic attitudes with agents' knowledge, beliefs, and abilities to act is needed.

- *explicit temporality*, as norms exist and agents act over time; so, judging norm compliance and violation *must* take temporality and timing into account.

All these are inherent aspects of adequate normative reasoning and should all be brought together, under one (enormous, but necessary) formal umbrella. That is a long-term (and multi-agent) endeavour, left to future ought-to-do's.

Acknowledgments

The occasion that give the initial impetus for writing this paper was the Georg Henrik von Wright Centenary Symposium held in Helsinki in May 2016. I wish to thank Sara Negri for inviting me to give there the talk which started this work. I am particularly indebted to Patrick Blackburn for his enthusiastic response and strong encouragement for me to develop these ideas further and eventually to write this paper. I thank also the anonymous referees, as well as Mark Brown, Fengkui Ju, and Karl Nygren for stimulating criticism and for several corrections and numerous useful comments.

References

[1] Alchourrón, C., *Logic of norms and logic of normative propositions*, Logique et Analyse **12** (1969), pp. 242–268.

[2] Alchourrón, C. and E. Bulygin, "Normative Systems," Springer-Verlag, 1971.

[3] Broersen, J. M., *Action negation and alternative reductions for dynamic deontic logics*, J. Appl. Log. **2** (2004), pp. 153–168.

[4] Castañeda, H.-N., *The paradoxes of deontic logic: The simplest solution to all of them in one fell swoop*, in: R. Hilpinen, editor, *New Studies in Deontic Logic: Norms, Actions, and the Foundations of Ethics*, Springer Netherlands, Dordrecht, 1981 pp. 37–85.

[5] Castro, P. F. and T. S. E. Maibaum, *Deontic action logic, atomic boolean algebras and fault-tolerance*, J. Appl. Log. **7** (2009), pp. 441–466.

[6] Ciardelli, I., J. Groenendijk and F. Roelofsen, "Inquisitive semantics," Oxford University Press, 2018.

[7] Cimatti, A., F. Giunchiglia and R. W. Weyhrauch, *A many-sorted natural deduction*, Comput. Intell. **14** (1998), pp. 134–149.

[8] Dalmonte, T., B. Lellmann, N. Olivetti and E. Pimentel, *Hypersequent calculi for non-normal modal and deontic logics: Countermodels and optimal complexity*, CoRR **abs/2006.05436** (2020).

[9] Docherty, S. and R. N. S. Rowe, *A non-wellfounded, labelled proof system for propositional dynamic logic*, CoRR **abs/1905.06143** (2019).

[10] Fitch, F., *Natural deduction rules for obligation*, American Philosophical Quaterly **3** (1966), pp. 27–38.

[11] Frittella, S., G. Greco, A. Kurz and A. Palmigiano, *Multi-type display calculus for propositional dynamic logic*, J. Log. Comput. **26** (2016), pp. 2067–2104.

[12] Goranko, V., "Logic as a Tool - A Guide to Formal Logical Reasoning," Wiley, 2016.

[13] Hage, J., *A theory of legal reasoning and a logic to match*, Artif. Intell. Law **4** (1996), pp. 199–273.

[14] Hage, J., "Reasoning with Rules," Kluwer, 1997.

[15] Hage, J., *Formalizing legal coherence*, in: *Proceedings of ICAIL 2001*, ACM, 2001 pp. 22–31.

[16] Hage, J., *Of norms*, in: G. B. et al, editor, *Handbook of Legal Reasoning and Argumentation*, Springer Nature, Dordrecht, 2018 pp. 103–138.
[17] Hedenius, I., "Om Rätt och Moral ('On Law and Moral')," Tidens, Stockholm, 1941.
[18] Hill, B. and F. Poggiolesi, *A contraction-free and cut-free sequent calculus for propositional dynamic logic*, Studia Logica **94** (2010), pp. 47–72.
[19] Honsell, F. and M. Miculan, *A natural deduction approach to dynamic logic*, in: *Proceedings of TYPES'95*, 1995, pp. 165–182.
[20] Jörgensen, J., *Imperatives and logic*, Erkenntnis **7** (1937), pp. 288–296.
[21] Knobbout, M., M. Dastani and J. C. Meyer, *A dynamic logic of norm change*, in: *Proceedings of ECAI 2016*, 2016, pp. 886–894.
[22] Kulicki, P. and R. Trypuz, *A deontic action logic with sequential composition of actions*, in: *Proceedings of DEON 2012*, 2012, pp. 184–198.
[23] Kulicki, P. and R. Trypuz, *Connecting actions and states in deontic logic*, Studia Logica **105** (2017), pp. 915–942.
[24] Makinson, D. and L. W. N. van der Torre, *Input/output logics*, J. Philosophical Logic **29** (2000), pp. 383–408.
[25] Makinson, D. and L. W. N. van der Torre, *Constraints for input/output logics*, J. Philosophical Logic **30** (2001), pp. 155–185.
[26] Makinson, D. and L. W. N. van der Torre, *Permission from an input/output perspective*, J. Philosophical Logic **32** (2003), pp. 391–416.
[27] Meyer, J. C., *A different approach to deontic logic: deontic logic viewed as a variant of dynamic logic*, Notre Dame Journal of Formal Logic **29** (1988), pp. 109–136.
[28] Meyer, J.-J. C., *Dynamic deontic logic, Segerberg-style*, in: R. Trypuz, editor, *Krister Segerberg on Logic of Actions*, Springer Netherlands, Dordrecht, 2014 pp. 119–132.
[29] Negri, S. and J. von Plato, "Structural proof theory," Cambridge University Press, 2001.
[30] Orlandelli, E., *Proof analysis in deontic logics*, in: *Proceedings of DEON 2014*, 2014, pp. 139–148.
[31] Rönnedal, D., *Dyadic deontic logic and semantic tableaux*, Logic and Logical Philosophy **18** (2009), pp. 221–252.
[32] Rönnedal, D., "Extensions of Deontic Logic : An Investigation into some Multi-Modal Systems," Ph.D. thesis, Stockholm University, Department of Philosophy (2012).
[33] Rönnedal, D., *Temporal alethic–deontic logic and semantic tableaux*, Journal of Applied Logic **10** (2012), pp. 219–237.
[34] Segerberg, K., *A deontic logic of action*, Studia Logica **41** (1982), pp. 269–282.
[35] Segerberg, K., *Blueprint for a dynamic deontic logic*, J. Appl. Log. **7** (2009), pp. 388–402.
[36] Segerberg, K., *DΔl: a dynamic deontic logic*, Synthese **185** (2012), pp. 1–17.
[37] Trypuz, R. and P. Kulicki, *A deontic logic of actions and states*, in: *Proceedings of DEON 2014*, 2014, pp. 258–272.
[38] Trypuz, R. and P. Kulicki, *On deontic action logics based on boolean algebra*, J. Log. Comput. **25** (2015), pp. 1241–1260.
[39] von Plato, J., "Elements of Logical Reasoning," Cambridge University Press, 2014.
[40] von Wright, G., *Deontic logic*, Mind **60** (1951), pp. 1–15.
[41] von Wright, G., "Logical Studies," Routledge, London, 1957.
[42] von Wright, G., "Norm and Action: A Logical Enquiry," Humanities Press, New York, 1963.
[43] von Wright, G., *A new system of deontic logic*, Danish Yearbook of Philosophy **1** (1964), pp. 173–182.
[44] von Wright, G., *Deontic logic and the theory of conditions*, Critica **2** (1968), pp. 3–25.
[45] von Wright, G., *On the logic of norms and actions*, in: R. Hilpinen, editor, *New Studies in Deontic Logic*, D. Reidel, Dordrecht-Holland, 1981 .
[46] von Wright, G., *Is there a logic of norms?*, Ratio Juris **4** (1991), pp. 265–283.
[47] von Wright, G., *Ought to be – ought to do*, in: E. G. Valdés, W. Krawietz, G. von Wright and R. Zimmerling, editors, *Normative Systems in Legal and Moral Theory*, Duncker & Humblot, Berlin, 1997 .
[48] von Wright, G., *Deontic logic: A personal view*, Ratio Juris **12** (1999), pp. 26–39.

A Defeasible Deontic Logic for Pragmatic Oddity

Guido Governatori

Data61, CSIRO, Australia

Silvano Colombo Tosatto

Data61, CSIRO, Australia

Antonino Rotolo

Alma Human AI, University of Bologna, Italy

Abstract

We introduce a variant of Deontic Defeasible Logic to handle the issue of Pragmatic Oddity. The key idea is that a conjunctive obligation is allowed only when each individual obligation is independent from the violation of the other obligations. The solution makes essential use of the constructive proof theory of the logic while maintaining a feasible computational complexity.

Keywords: Pragmatic Oddity, Defeasible Deontic Logic

1 Introduction

A differentiator between norms and other constraints is that, typically, (legal) norms can be violated. Moreover, normative systems contain provisions about other norms that become effective when violations occur. Since the seminal work by Chisholm [3] the obligations in force triggered by violations have been dubbed contrary-to-duty obligations (CTDs). The treatment of CTDs has proven problematic for formal (logical) representations of normative systems. Accordingly, CTDs are the source for many paradoxes and the driver for the development of many formalisms and deontic logics. The contribution in this paper follows the tradition: we are going to propose an extension of a logic (Defeasible Deontic Logic) that addresses the Pragmatic Oddity CTD paradox.

The problem of Pragmatic Oddity, introduced by Prakken and Sergot [12], is illustrated by the scenario that when you make a promise, you have to keep it. But if you do not, then you have to apologise. The oddity is that when you fail to keep your promise, you have the obligation to keep the promise and the obligation to apologise. In our view, what is odd, is not that the two obligations are in force at the same time, but that if one admits for form a conjunctive obligation from the two individual obligations then we get an obligation that is impossible to comply with. In the scenario, when the promise is broken, we have the conjunctive obligation obligation to keep the promise and to apologise for not having kept the promise.

The Pragmatic Oddity arises when we have a conjunctive obligation, i.e., $O(a \wedge b)$, derived from the two individual obligations (Oa and Ob) where one of the conjuncts is a contrary-to-duty obligation triggered by the violation of the other individual obligation, for example when $\neg a$ entails that Ob is in force.

Most of the work on Pragmatic Oddity (e.g., [12,2]) focuses on the issue of how to distinguish the mechanisms leading to the derivation of the two individual obligations, and create different classes of obligations. Consequently, the solution to the Pragmatic Oddity problem is to prevent the conjunction when the obligations are from different classes. Accordingly, if the problem is to prevent that a conjunctive obligation is in force when the individual obligations are in force themselves, the simplest solution is to have a deontic logic that does not support the aggregation axiom [1]:

$$(Oa \wedge Ob) \to O(a \wedge b)$$

However, a less drastic solution, advocated by Parent and van der Torre [10,11], is to restrict the aggregation axiom to independent obligations (meaning that one obligation should not depend on the violation of the other obligation).

We are going to take Parent and van der Torre's suggestion and propose a simple mechanism in Defeasible Deontic Logic to guard the derivation of conjunctive obligations. The mechanism guarantees that the obligations of a conjunctive obligation are independent from the violations of the individual obligations. The mechanism is founded on the proof theory of the logic.

2 Defeasible Deontic Logic

Defeasible Deontic Logic [6] is a sceptical computationally oriented rule-based formalism designed for the representation of norms. The logic extends Defeasible Logic [1] with deontic operators to model obligations and (different types of) permissions and provides an integration with the logic of violation proposed in [8]. The resulting formalism offers features for the natural and efficient representation of exceptions, constitutive and prescriptive rules and of compensatory norms. The logic is based on a constructive proof theory that allows for full traceability of the conclusions, and flexibility to handle and combine different facets of non-monotonic reasoning. In the rest of this section we are going to show how the proof theory can be used to propose a simple and (arguably) elegant treatment of the issue of Pragmatic Oddity.

We restrict ourselves to the fragment of Defeasible Deontic Logic that excludes permission and permissive rules, since they do not affect the way we handle Pragmatic Oddity: Definitions 2.12 and 2.13, the definitions that describe the mechanisms we adopt for a solution to Pragmatic Oddity, are independent

---

[1] See, among others, [4].

from any issue related to permission. The definitions can be used directly in the full version of the logic. Accordingly, we consider a logic whose language is defined as follows.

Definition 2.1 Let PROP be a set of propositional atoms, O the modal operator for obligation.

- The set Lit = PROP $\cup$ $\{\neg p \,|\, p \in$ PROP$\}$ is the set of *literals*.
- The *complement* of a literal q is denoted by $\sim q$; if q is a positive literal p, then $\sim q$ is $\neg p$, and if q is a negative literal $\neg p$, then $\sim q$ is p.
- The set of *deontic literals* is DLit = $\{Ol, \neg Ol \,|\, l \in$ Lit$\}$.
- If $c_1, \ldots, c_n \in$ Lit, then $O(c_1 \wedge \cdots \wedge c_n)$ is a *conjunctive obligation*.

In the rest of the paper, when relevant to the discussion, we will refer to elements of Lit as plain literals, and often we will use the unmodified term 'literal' to indicate either a plain literal or a deontic literal.

We introduce the compensation operator $\otimes$. This operator is used to build chains of compensation called $\otimes$-expressions. The formation rules for well-formed $\otimes$-expressions are:

(i) every literal $l \in$ Lit is an $\otimes$-expression;

(ii) if $c_1, \ldots, c_k \in$ Lit, then $c_1 \otimes \cdots \otimes c_k$ is an $\otimes$-expression;

(iii) nothing else is an $\otimes$-expression.

In addition we stipulate that $\otimes$ obeys the following property (duplication and contraction on the right):

$$\bigotimes_{i=1}^{n} a_i = \left(\bigotimes_{i=1}^{k-1} a_i\right) \otimes \left(\bigotimes_{i=k+1}^{n} a_i\right)$$

where there exists j such that $a_j = a_k$ and $j < k$.

Given an $\otimes$-expression A, the *length* of A is the number of literals in it. Given an $\otimes$-expression $A \otimes b \otimes C$ (where A and C can be empty), the *index* of b is the length of $A \otimes b$. We also say that b appears at index n in $A \otimes b$ if the length of $A \otimes b$ is n.

The meaning of a compensation chain

$$c_1 \otimes c_2 \otimes \cdots \otimes c_n$$

as proposed by [8] and further discussed in [5], is that Oc_1 is the primary obligation, and when violated (i.e., $\neg c_1$ holds), then Oc_2 is in force and it compensates for the violation of the obligation of c_1. Moreover, when Oc_2 is violated, then Oc_3 is in force, and so on until we reach the end of the chain when a violation of the last element is a non-compensable violation where the norm corresponding to the rule in which the chain appears is not complied with.

We adopt the standard DL definitions of *strict rules*, *defeasible rules*, and *defeaters* [1]. However, for the sake of simplicity, and to better focus on the non-monotonic aspects that DL offers, in the remainder we use only defeasible rules and defeaters. Also, we have to take the obligation operator into account.

Definition 2.2 Let Lab be a set of arbitrary labels. Every rule is of the type

$$r\colon A(r) \hookrightarrow C(r)$$

where

(i) $r \in$ Lab is the name of the rule;

(ii) $A(r) = \{a_1, \ldots, a_n\}$, the *antecedent* (or *body*) of the rule, is the set of the premises of the rule (alternatively, it can be understood as the conjunction of all the elements in it). Each a_i is either a literal, a deontic literal or a conjunctive obligation;

(iii) $\hookrightarrow \in \{\Rightarrow, \Rightarrow_O, \leadsto, \leadsto_O\}$ denotes the type of the rule. If $\hookrightarrow$ is $\Rightarrow$, the rule is a *defeasible rule*, while if $\hookrightarrow$ is $\leadsto$, the rule is a *defeater*. Rules without the subscript O are constitutive rules, while rules with such a subscript are prescriptive rules.

(iv) $C(r)$ is the *consequent* (or *head*) of the rule. It is a single literal for defeaters and constitutive rules, and an $\otimes$-expressions for prescriptive defeasible rules.

As we will see, prescriptive rules are used to derive obligations.

Given a set of rules R, we use the following abbreviations for specific subsets of rules:

- R_d denotes the set of defeasible rules in the set R;
- $R[q,n]$ is the set of rules where q appears at index n in the consequent. The set of rules where q appears at any index n is denoted by $R[q]$;
- R^O denotes the set of prescriptive rules in R, i.e., the set of rules with O as their subscript;
- R^C denotes the set of constitutive rules in R, i.e., $R \setminus R^O$.

The above notations can be combined. Thus, for example, $R_d^O[q,n]$ stands for the set of defeasible prescriptive rules such that q appears at index n in the consequent of the rule.

Definition 2.3 A *Defeasible Theory* is a structure $D = (F, R, >)$, where F, the set of facts, is a set of literals and deontic literals, R is a set of rules and $>$, the superiority relation, is a binary relation over R.

A theory corresponds to a normative system, i.e., a set of norms, where every norm is modelled by some rules. The superiority relation is used for conflicting rules, i.e., rules whose conclusions are complementary literals, in case both rules fire. We do not impose any restriction on the superiority relation: it just determines the relative strength between two rules.

Definition 2.4 A *proof* (or *derivation*) P in a defeasible theory D is a linear sequence $P(1) \ldots P(z)$ of *tagged literals* in the form of $+\partial q$, $-\partial q$, $+\partial_O q$, $-\partial_O q$,

$+\partial_O c_1 \wedge \cdots \wedge c_m$ and $-\partial_O c_1 \wedge \cdots \wedge c_m$ where $P(1) \ldots P(z)$ satisfy the proof conditions given in Definitions 2.8–2.13.

The tagged literal $+\partial q$ means that q is *defeasibly provable* as an institutional statement, or in other terms, that q holds in the normative system encoded by the theory. The tagged literal $-\partial q$ means that q is *defeasibly refuted* by the normative system. Similarly, the tagged literal $+\partial_O q$ means that q is *defeasibly provable* in D as an obligation, while $-\partial_O q$ means that q is *defeasibly refuted* as an obligation. For $+\partial_O c_1 \wedge \cdots \wedge c_m$ the meaning is that the conjunctive obligation $O(c_1 \wedge \cdots \wedge c_m)$ is defeasibly derivable; and that a conjunctive obligation $O(c_1 \wedge \cdots \wedge c_m)$ is defeasibly refuted corresponds to $-\partial_O(c_1 \wedge \cdots \wedge c_m)$. The initial part of length i of a proof P is denoted by $P(1..i)$.

The first thing to do is to define when a rule is applicable or discarded. A rule is *applicable* for a literal q if q occurs in the head of the rule, all elements in the antecedent have been defeasibly proved (eventually with the appropriate modalities). On the other hand, a rule is *discarded* if at least one of the modal literals in the antecedent has not been proved. However, as literal q might not appear as the first element in an $\otimes$-expression in the head of the rule, some additional conditions on the consequent of rules must be satisfied. Defining when a rule is applicable or discarded is essential to characterise the notion of provability for constitutive rules and then for obligations ($\pm\partial_O$).

Definition 2.5 Given a proof P, a rule $r \in R$ is *body-applicable* at step $P(n+1)$ iff for all $a_i \in A(r)$:

(i) if $a_i = Ol$ then $+\partial_O l \in P(1..n)$;

(ii) if $a_i = \neg Ol$ then $-\partial_O l \in P(1..n)$;

(iii) if $a_i = O(c_1 \wedge \cdots \wedge c_m)$ then $+\partial_O c_1 \wedge \cdots \wedge c_m \in P(1..n)$;

(iv) if $a_i = l \in \text{Lit}$ then $+\partial l \in P(1..n)$.

A rule $r \in R[q, j]$ is *body-discarded* at step $P(n+1)$ iff $\exists a_i \in A(r)$ such that

(i) if $a_i = Ol$ then $-\partial_O l \in P(1..n)$;

(ii) if $a_i = \neg Ol$ then $+\partial_O l \in P(1..n)$;

(iii) if $a_i = O(c_1 \wedge \cdots \wedge c_m)$ then $-\partial_O c_1 \wedge \cdots \wedge c_m \in P(1..n)$;

(iv) if $a_i = l \in \text{Lit}$ then $-\partial l \in P(1..n)$.

Definition 2.6 Given a proof P, a rule $r \in R^O[q, j]$ such that $C(r) = c_1 \otimes \cdots \otimes c_m$ is *applicable* for literal q at index j at step $P(n+1)$ (or, simply, applicable for q), with $1 \leq j < m$, in the condition for $\pm\partial_O$ iff
(i) r is body-applicable at step $P(n+1)$; and
(ii) for all $c_k \in C(r)$, $1 \leq k < j$, $+\partial_O c_k \in P(1..n)$ and $+\partial \sim c_k \in P(1..n)$.

Conditions (i) represents the requirements on the antecedent stated in Definition 2.5; condition (ii) on the head of the rule states that each element c_k prior to q must be derived as an obligation, and a violation of such obligation has occurred.

Definition 2.7 Given a proof P, a rule $r \in R^O[q, j]$ such that $C(r) = c_1 \otimes \cdots \otimes c_m$ is *discarded* for literal q at index j at step $P(n+1)$ (or, simply, discarded for q), with $1 \leq j \leq m$, in the condition for $\pm \partial_O$ iff
  (i) r is body-discarded at step $P(n+1)$; or
  (ii) there exists $c_k \in C(r)$, $1 \leq k < l$, such that either $-\partial_O c_k \in P(1..n)$ or $+\partial c_k \in P(1..n)$.

In this case, condition (institutional) ensures that an obligation prior to q in the chain is not in force or has already been fulfilled (thus, no reparation is required).

We now introduce the proof conditions for $\pm \partial$ and $\pm \partial_O$:

Definition 2.8 The proof condition of *defeasible provability for an institutional statement* is
$+\partial$: If $P(n+1) = +\partial q$ then
(1) $q \in F$ or
  (2.1) $\sim q \notin F$ and
  (2.2) $\exists r \in R_d[q]$ such that r is applicable for q, and
  (2.3) $\forall s \in R[\sim q]$, either
    (2.3.1) s is discarded for $\sim q$, or
    (2.3.2) $\exists t \in R[q]$ such that t is applicable for q and $t > s$.

As usual, we use the strong negation to define the proof condition for $-\partial$

Definition 2.9 The proof condition of *defeasible refutability for an institutional statement* is
$-\partial$: If $P(n+1) = -\partial q$ then
(1) $q \notin F$ and
  (2.1) $\sim q \in F$ or
  (2.2) $\forall r \in R_d[q]$: either r is discarded for q, or
  (2.3) $\exists s \in R[\sim q]$, such that
    (2.3.1) s is applicable for $\sim q$, and
    (2.3.2) $\forall t \in R[q]$ either t is discarded for q or not $t > s$.

The proof conditions for $\pm \partial$ are the standard conditions in defeasible logic, see [1] for the full explanations.

Definition 2.10 The proof condition of *defeasible provability for obligation* is
$+\partial_O$: If $P(n+1) = +\partial_O q$ then
(1) $Oq \in F$ or
  (2.1) $O\sim q \notin F$ and $\neg Oq \notin F$ and
  (2.2) $\exists r \in R_d^O[q, i]$ such that r is applicable for q, and
  (2.3) $\forall s \in R^O[\sim q, j]$, either
    (2.3.1) s is discarded for $\sim q$, or
    (2.3.2) $\exists t \in R^O[q, k]$ such that t is applicable for q and $t > s$.

To show that q is defeasibly provable as an obligation, one must show either that: (1) the obligation of q is a fact, or (2) q must be derived by the rules of the theory. In the second case, three conditions must hold: (2.1) q does not appear as not obligatory as a fact, and $\sim q$ is not provable as an obligation using the set of deontic facts at hand; (2.2) there must be a rule introducing the obligation for q which can apply; (2.3) every rule s for $\sim q$ is either discarded or defeated by a stronger rule for q.

The strong negation of Definition 2.10 gives the negative proof condition for obligation.

Definition 2.11 The proof condition of *defeasible refutability for obligation* is
$-\partial_O$: If $P(n+1) = -\partial_O q$ then
(1) $Oq \notin F$ and either
   (2.1) $O\sim q \in F$ or $\neg Oq \in F$ or
   (2.2) $\forall r \in R_d^O[q,i]$ either r is discarded for q, or
   (2.3) $\exists s \in R^O[\sim q, j]$ such that
      (2.3.1) s is applicable for $\sim q$, and
      (2.3.2) $\forall t \in R^O[q,k]$, either t is discarded for q or $t \not> s$.

Notice that, given the intended correspondence between Ol and $+\partial_O l$, see Definition 2.5, we will refer to "the derivation of Ol" when, strictly speaking, we should use "the derivation of $+\partial_O l$".

We are now ready to provide the proof condition under which a conjunctive obligation can be derived. The condition essentially combines two requirements: the first that a conjunction holds only when all the conjuncts hold (individually). The second requirement is that the derivation of one of the individual obligations does not depend on the violation of the other conjunct. To achieve this, we determine the line of the proof when the obligation appears, and then we check that the negation of the other elements of the conjunction does not occur in the previous derivation steps.

Definition 2.12 The proof condition of *defeasible provability for a conjunctive obligation* is
If $P(n+1) = +\partial_O c_1 \wedge \cdots \wedge c_m$, then
$\forall c_i, 1 \leq i \leq m$,
(1) $+\partial_O c_i \in P(1..n)$ and
(2) if $P(k) = +\partial_O c_1 \wedge \cdots \wedge c_m$, $k \leq n$, then
   $\forall c_j, 1 \leq j \leq m$ and $c_j \neq c_i$, $+\partial\sim c_j \notin P(1..k)$.

Again, the proof condition to refute a conjunctive obligation is obtained by strong negation from the condition to defeasibly derive a conjunctive obligation.

Definition 2.13 The proof condition of *defeasible refutability for a conjunctive obligation* is
If $P(n+1) = -\partial_O c_1 \wedge \cdots \wedge c_m$, then
$\exists c_i, 1 \leq i \leq m$, such that either
(1) $-\partial_O c_i \in P(1..n)$ or
(2) if $P(k) = +\partial_O c_1 \wedge \cdots \wedge c_m$, $k \leq n$, then
   $\exists c_j, 1 \leq j \leq m$ such that $c_j \neq c_i$ and $+\partial\sim c_j \in P(1..k)$.

In case of a binary conjunctive obligation the positive proof condition boils down to

$+\partial_{O\wedge}$: If $P(n+1) = +\partial_O p \wedge q$ then
(1) $+\partial_O p \in P(1..n)$ and
(2) $+\partial_O q \in P(1..n)$ and
(3) if $P(k) = +\partial_O p$ ($k \leq n$), then $+\partial{\sim}q \notin P(1..k)$ and
(4) if $P(k) = +\partial_O q$ ($k \leq n$), then $+\partial{\sim}p \notin P(1..k)$.
Similarly, for the condition for $-\partial_{O\wedge}$.

Before moving on proving some theoretical results about the logic defined we give some examples that illustrate the behaviour of the logic. In what follows we use $\cdots \Rightarrow c$ to refer to an applicable rule for c where we assume that the elements are not related (directly or indirectly) to the other literals used in the examples.

Compensatory Obligations The first case we want to discuss is when the conjunctive obligation corresponding to the Pragmatic Oddity has as conjuncts an obligation and its compensation. This scenario is illustrated by the rule:

$$\cdots \Rightarrow_O a \otimes b$$

In this case, it is clear that we cannot derive the conjunctive obligation of a and b, since the proof condition that allows us to derive $+\partial_O b$ explicitly requires that $+\partial{\sim}a$ has been already derived (condition 2 of Definition 2.6). In this case, it is impossible to have the obligation of b without the violation of the obligation of a.

Contrary-to-duty The second case is when we have a CTD. The classical representation of a CTD is given by the following two rules:

$$\cdots \Rightarrow_O a \qquad \neg a \Rightarrow_O b$$

In this case, it is possible to have situations when the obligation of b is in force without having a violation of the obligation of a, namely, when a is not obligatory. However, as soon as we have Oa, we need to derive $\neg a$ to trigger the derivation of Ob (Definition 2.5).

Pragmatic Oddity via Intermediate Concepts The situations in the previous two cases can be easily detected by a simple inspection of the rules involved; there could be more complicated cases. Specifically, when the second conjunct does not immediately depends on the first conjunct, but it depends through a reasoning chain. The simplest structure for this case is illustrated by the following three rules:

$$\cdots \Rightarrow_O a$$
$$\neg a \Rightarrow b$$
$$b \Rightarrow_O c$$

Here to derive Oc, we need first to prove b. To prove b we require that $\neg a$ has already been proved.

Negative Support In the previous case the support was through an intermediate concepts. However, given the non-monotonic nature of Defeasible Deontic Logic, we can have cases where the support is not to directly derive the other obligation from the violation, but the violation prevents the derivation of the prohibition (or the permission of the opposite) of the other conjunct. This situation is illustrated by the following set of rules: [2]

$$\cdots \Rightarrow_O a$$
$$\cdots \Rightarrow_O b$$
$$c \Rightarrow_O \neg b$$
$$\cdots \Rightarrow c$$
$$\neg a \rightsquigarrow \neg c$$

To derive Ob, we have to ensure that the rule for $O\neg b$ is discarded. This means that c should be rejected (i.e., $-\partial c$). We have two options, either the rule for c is discarded, or the rule for $\neg c$ is applicable. This implies that to prove $+\partial_O b$ we have to prove first $+\partial \neg a$. Thus, one of the two elements of the conjunctive obligation $O(a \wedge b)$ depends on the violation of the other.

Pragmatic Un-pragmatic Oddity What about when there are multiple norms both prescribing the contrary-to-duty obligation, and at least one of the norms is not related to the violation of the primary norm?

$$r_1: \cdots \Rightarrow_O a \otimes b$$
$$r_2: \cdots \Rightarrow_O b$$
$$\neg a$$

In this situation you can have a derivation:

(1) $+\partial \neg a$ fact
(2) $+\partial_O a$ from r_1
(2) $+\partial_O b$ from r_1 and (1) and (2)

where the derivation of Ob ($+\partial_O b$) depends on the violation of the primary obligation of r_1. In this case, we cannot derive the conjunctive obligation of a and b. However, there is an alternative derivation, namely:

(1) $+\partial_O a$ from r_1
(2) $+\partial_O b$ from r_2
(3) $+\partial \neg a$ fact
(4) $+\partial_O a \wedge b$ from (1) and (2)

that demonstrates the independence of Ob from $\neg a$, given that the derivation of $\neg a$ occurs in a line after the line where $+\partial_O b$ is derived.

---

[2] It is worth noting that, in the theory below, the rules for $\neg b$ and $\neg c$ can be either defeasible rules or defeaters producing the same result as far as the derivation of $O(a \wedge b)$ is concerned.

3 Independence

As we have discussed the idea of the proof conditions above is to ensure that the individual obligations do not depend on the violations of the others. Accordingly, the question now is what does it mean that a formula is independent from another formula. In classical logic, given a theory T, a formula A depends on the formula B if $T \cup B \vdash A$, but $T \setminus B \nvdash A$. In Defeasible Deontic Logic, we have to remove all possible reasons to conclude the literal; this means we have to remove it from the facts and we have to remove the rules where it appears in the head of the rule. Since we are interested in removing only non deontic literals we can restrict the removal to the constitutive rules whose head is the literal to be removed. Accordingly, we can define the following transformation.

Definition 3.1 Given a defeasible theory $D = (F, R, >)$ and a literal l, the *Pragmatic Oddity Transformation* of D based on l, noted as $pot(D, l)$ is the defeasible theory $D' = (F', R', >')$ satisfying the following conditions:

(i) $F' = F \setminus \{l\}$;

(ii) $R' = R \setminus R[l]$;

(iii) $>' => \setminus \{(r, s) \colon r \notin R' \vee s \notin R'\}$.

The transformation is to create a theory similar to the original theory but, as we said, without l. The condition on F is obvious. The second condition ensures that the rules that can derive the literal are removed. Then the literal is no longer derivable, since the resulting theory does not contain rules for the literal anymore. Given that $R'[l] = \emptyset$, the following result is immediate.

Observation 1 *Given a Defeasible Theory D and a literal l, $-\partial l$ is not derivable in $pot(D, l)$.*

It worth noting that we do not have to remove rules where the literal appears in the antecedent of the rule. Such rules are immediately discarded. Similarly, for prescriptive rules where the complement of the removed literal appears in the head of the rules. Such rules are no longer applicable for any elements appearing after the complement of the removed literal. Thus if you have a rule with the $\otimes$-chain $c_1 \otimes \cdots \otimes c_n \otimes \neg l \otimes c_{n+1} \cdots$, the rules in $R^O[c, m]$ for any $m \geq n+1$ are not applicable. Remember, that to derive $+\partial_O c_{n+1}$ we have to prove both $+\partial_O \neg l$ and $+\partial l$. The transformation pot is then extended to the case of a (finite) set of literals $L = \{l_1, \ldots, l_n\}$ by applying the transformation to all the literals in L; thus $pot(D, L) = pot(\cdots (pot(D, l_1), \cdots l_n)$ for an arbitrary sequence of all the elements in L.

We can now specify when a (deontic) literal is independent from a set of plain literals in Defeasible Deontic Logic

Definition 3.2 Given a defeasible theory D, a set L of plain literals and a literal m, m is *independent from L* iff m is defeasibly provable in D and in $pot(D, L)$.

We can now show that the condition (2) in the proof conditions for a conjunctive obligation ensures the independence of the obligations from the violations. However, before proving this result we have to recall a general property about Defeasible (Deontic) Logic: First of all a defeasible theory is consistent if F does not contain a literal l and its complement $\neg l$. Second, given a logical formula expressing a proof condition the strong negation of the formula/conditions is obtained by replacing every occurrence of a positive proof tag with the corresponding negative proof tag, replacing conjunctions with disjunctions, disjunctions with conjunctions, existential with universal and universal with existential. It is immediate to observe that all negative proof conditions given in this section are the strong negation of the corresponding positive one (and the other way around). If corresponding proof conditions are defined using the principle of strong negation outlined above, then, given a derivation, it is not possible to have that the literal (conjunctive obligation) is both derivable and refutable in the same derivation.

Proposition 3.3 *[7] Given a consistent defeasible theory D, a derivation P, a literal l, and proof tag $\# \in \{\partial, \partial_O\}$ it is not possible that $+\#l, -\#l \in P$.*

Armed with this result we can prove the result linking independence and the proof conditions for conjunctive obligations.

Proposition 3.4 *Given a consistent defeasible theory D, a deontic literal m and a set L of plain literals. m is independent from L iff there is a derivation P in D such that*

- $P(n) = +\partial_O m$ and
- $\forall l \in L, +\partial l \notin P(1..n)$.

4 Complexity

In this section, we are going to study the computational complexity of the problem of computing whether a conjunctive obligation is derivable from a given defeasible theory. To this end, we adapt the algorithm proposed in [6] to compute the extension of a defeasible theory, where the computation of the extension is linear in the size of the theory. The algorithm is based on a series of transformations that reduce the complexity of the theory, by either removing elements from rules when some elements are provable, and removing rules when they become discarded (and so no longer able to produce positive conclusions). Using the idea in [6] the extension of a defeasible theory D is defined as follows:

Definition 4.1 Given a theory D, the *literal extension* of D is the tuple

$$\langle \partial^+(D), \partial^-(D), \partial_O^+(D), \partial_O^-(D) \rangle$$

where

- $\partial^+(D)$ is the set of literals appearing in D that are defeasibly provable as institutional statements;
- $\partial^-(D)$ is the set of literals appearing in D that are defeasibly refutable as institutional statements;

- $\partial_O^+(D)$ is the set of literals appearing in D that are defeasibly provable as obligations;
- $\partial_O^-(D)$ is the set of literals appearing in D that are defeasibly refutable as obligations;

The aim of the paper is to determine when conjunctive obligations are either provable or discarded. Accordingly, we have to extend the definition to account for conjunctive obligation. However, if we want to maintain a feasible computational complexity we have to limit the conjunctions we are going to consider: given a set of n literals the set of all possible non logically equivalent conjunctions that can be formed by the n literals contains 2^n conjunctions; hence, we cannot compute in polynomial time for such a set if any element is derivable or refuted by the theory. However, we are going to show that for each individual conjunction we can compute in polynomial time whether it is derivable or refuted.

Definition 4.2 Given a defeasible theory D the *conjunctive extension* of the theory is the tuple:

$$\langle \partial^+(D), \partial^-(D), \partial_O^+(D), \partial_O^-(D), \partial_\wedge^+(D), \partial_\wedge^-(D) \rangle$$

where $\partial^+(D)$, $\partial^-(D)$, $\partial_O^+(D)$ and $\partial_O^-(D)$ are as in Definition 4.1 and

- $\partial_\wedge^+(D)$ is the set of conjunctive obligations appearing in D (i.e., $c = O(c_1 \wedge \cdots \wedge c_n)$ and $\exists r \in R$ such that $c \in A(r)$) that are defeasibly provable in D (Definition 2.12);
- $\partial_\wedge^-(D)$ is the set of conjunctive obligations appearing in D that are defeasibly refutable in D (Definition 2.13).

The algorithm to determine the conjunctive extension of a theory is based on the following data structure (for the full details we refer the reader to [6]). We create a list of the atoms appearing in the theory. Every entry in the list of atoms has an array associated to it. The array has ten cells, where every cell contains pointers to rules depending on whether and how the atom appears in the rule. The first cell is where the atom appears in the head of a constitutive rule, the second where the negation of the atom appears in the head of a constitutive rule, the third where the atom appears in the head of a prescriptive rule, the fourth where the negation of atom appears in the head of a prescriptive rule, the fifth where the atom appears in the body of a rule, the sixth where the negation of the atom appears in the body of a rule, the seventh where the atom appears as an obligation in the body of a rule, the eighth where the negation of the atom appears as an obligation in the body of a rule, the ninth where the atom appears as a negative obligation in the body of a rule, and the tenth where the negation of the atom appears as a negative obligation in the body of a rule. In addition, we maintain a list of conjunctive obligations occurring in the theory, and for every conjunction we associate it to the rules where it appears in the body.

The algorithm works as follows: at every round we scan the list of atoms. For every atom (excluding the entries for the conjunctions) we look if the atom appears in the head of some rules. If it does not appears in any of the cells for the heads, we can set the corresponding literals as refuted; and we can remove rules, from corresponding cells. So, for example, given an atom p, if there are no prescriptive rules for $\neg p$, then, we can conclude that the theory proves $-\partial_O \neg p$; accordingly, all rules where $\neg O \neg p$ occurs in the body are (body)-discarded, and we can remove them from the data structure. Similarly, if there are no constitutive rules for $\neg p$, then we can prove $-\partial \neg p$, and, then (i) all the rules where it appears in the body are body-discarded, but also, for each rule r in whose head p appears as an obligation, no elements following p in r can any longer be derived using r and such elements are removed from the appropriate cells. If an atom appears in the head of a rule, we determine (i) if the body of the rule is empty, and (ii) for prescriptive rules, if the atom is the first element of the head. If this is the case, then, the rule is applicable, and we check if there are rules for the negation. If there are no rules for the negation, or the rules are weaker than applicable rules, then the atom/literal is provable with the suitable proof tag, and then we remove the atom/literal from the appropriate rules. We repeat the above steps until we are no longer able to obtain new conclusions. When, we are no able to derive new conclusion we turn our attention to the list of the conjunctive obligations, where we invoke the following (sub)algorithm for every conjunction $c = O(c_1 \wedge \cdots \wedge c_n)$ in the list (where $C = \{\sim c_i, 1 \leq i \leq n\}$)

Algorithm 1 Evaluate Conjunctive Obligation

1: **for** $i \in 1..n$ **do**
2:    **if** $c_i \in \partial_O^-(D)$ **then**
3:       $c \in \partial_\wedge^-(D)$ remove all rules r where $c \in A(R)$
4:       Exit
5:    **end if**
6:    **if** $c_i \in \partial_O^+(D)$ **then**
7:       **if** $\forall c_j \neg c_i, \sim c_j \in \partial^+(D)$ **then**
8:          **if** $c_i \in +\partial_O^+(pot(D, C \setminus \{\sim c_i\})$ **then**
9:             $i := i + 1$
10:          **else** $c \in \partial_\wedge^-(D)$ remove all rules r where $c \in A(R)$
11:             Exit
12:          **end if**
13:       **if** $\exists c_j \neq c_i, \sim c_i \in \partial^-(D)$ **then**
14:          $i := i + 1$
15:       **end if**
16:       **end if**
17:    **end if**
18:    Exit
19: **end for**
20: $c \in \partial_\wedge^+(D)$, remove c from all rules r where $c \in A(r)$

For every conjunction the algorithm iterates over the conjuncts. If a conjunct is not provable as an obligation the conjunction is not provable (line 2–4). If the conjunct is provable as an obligation, it checks whether the violations of the other obligations are provable; if so, it has to check whether the obligation of the conjunct is independent from the violations. To determine this, we can repeat the whole algorithm with the the sub-theory obtained by the transformation $pot(D, C \setminus \{c_i\})$. If it is independent we continue with the next element of the conjunction; otherwise, the conjunction is not derivable. Similarly, if some of violations are not derivable we continue with the iteration. The conjunction is provable when the iteration is successful for all the elements of the conjunction.

At the end of the sub-routine, we return to the main algorithm, if there are changes in the rules we repeat the process, otherwise the process terminates.

The algorithm outline above is sound and complete; hence, we can state the following proposition. Essentially, the correctness of the algorithm depends on Proposition 3.4.

Proposition 4.3 *Given a defeasible theory D*

- $+\partial l$ *is defeasibly provable in D iff $l \in \partial^+(D)$;*
- $-\partial l$ *is defeasibly provable in D iff $l \in \partial^-(D)$;*
- $+\partial_O l$ *is defeasibly provable in D iff $l \in \partial_O^+(D)$;*
- $-\partial_O l$ *is defeasibly provable in D iff $l \in \partial_O^-(D)$;*
- $+\partial_O c_1 \wedge \cdots \wedge c_n$ *is defeasibly provable in D iff $c_1 \wedge \cdots \wedge c_n \in \partial_\wedge^+(D)$;*
- $-\partial_O c_1 \wedge \cdots \wedge c_n$ *is defeasibly provable in D iff $c_1 \wedge \cdots \wedge c_n \in \partial_\wedge^-(D)$.*

As far as the computational complexity, [6] proves that the complexity of computing the extension of a defeasible theory without conjunctive obligation is linear in the size of the theory, where the size of the theory is determined by the number of symbols in the theory, and hence if n and r stand for, respectively, the number of atoms and the number of rules in the theory, the complexity is in $O(n * r)$. For the complexity of computing the conjunctive extension of a defeasible theory we have to take into account the complexity of the Evaluate Conjunctive Obligation algorithm and the number of times we have to compute it. This can be determined as follows: let m be the number of conjunctive obligations in the theory, and k the number of conjuncts in the longest conjunctive obligation. For each of them we have to compute the extension of $pot(D, C)$, thus we have to perform $O(m * k * O(n * r))$ computations on top of the computation of the extension (i.e., $O((m+n) * r)$).

Proposition 4.4 *The conjunctive extension of a theory can be computed in polynomial time.*

Notice that the algorithm Evaluate Conjunctive Obligation can be use the evaluate any conjunctive obligation not only the conjunctive obligations occurring in a theory. All we have to do is to compute the conjunctive extension of the theory and then evaluate the single conjunctive obligation, and as we have just seen this can be computed in polynomial time.

5 Summary

We have proposed an extension of Defeasible Deontic Logic able to handle the so called Pragmatic Oddity paradox. The mechanism we used to achieve this result was to provide a schema that allows us to give a guard to the derivation of conjunctive obligations ensuring that each individual obligation does not depend on the violation of the other obligation. The mechanism is given by the proof theory of defeasible logic.

While the complexity of the logic is polynomial and hence feasible the algorithm we propose is not optimal. Nonetheless, this is practical for most real life applications, in which it is likely there will be few conjunctive obligations, each with only a small number of conjuncts; however, the next step is to to devise an optimal algorithm to implement the novel proof conditions.

Acknowledgments

A preliminary version of the paper presenting the idea of the logic was presented at Jurix 2019 [9]. We thanks the anonymous reviewers for their valuable comments on an earlier version of the paper.

References

[1] Grigoris Antoniou, David Billington, Guido Governatori, and Michael J. Maher. Representation results for defeasible logic. *ACM Transactions on Computational Logic*, 2(2):255–287, 2001.

[2] José Carmo and Andrew JI Jones. Deontic logic and contrary-to-duties. In *Handbook of philosophical logic*, pages 265–343. Springer, 2002.

[3] Roderick M Chisholm. Contrary-to-duty imperatives and deontic logic. *Analysis*, 24(2):33–36, 1963.

[4] Lou Goble. A logic for deontic dilemmas. *Journal of Applied Logic*, 3(3-4):461–483, 2005.

[5] Guido Governatori. Thou shalt is not you will. In Katie Atkinson, editor, *Proceedings of the Fifteenth International Conference on Artificial Intelligence and Law*, pages 63–68, New York, 2015. ACM.

[6] Guido Governatori, Francesco Olivieri, Antonino Rotolo, and Simone Scannapieco. Computing strong and weak permissions in defeasible logic. *Journal of Philosophical Logic*, 42(6):799–829, 2013.

[7] Guido Governatori, Vineet Padmanabhan, Antonino Rotolo, and Abdul Sattar. A defeasible logic for modelling policy-based intentions and motivational attitudes. *Logic Journal of the IGPL*, 17(3):227–265, 2009.

[8] Guido Governatori and Antonino Rotolo. Logic of violations: A Gentzen system for reasoning with contrary-to-duty obligations. *Australasian Journal of Logic*, 4:193–215, 2006.

[9] Guido Governatori and Antonino Rotolo. A computational model for pragmatic oddity. In Michał Araszkiewicz and Víctor Rodríguez-Doncel, editors, *JURIX 2019: The 32th international conference on Legal Knowledge and Information Systems*, volume 332 of *Frontiers in Artificial Intelligence and Applications*, pages 187–192, Amsterdam, 2019. IOS Press.

[10] Xavier Parent and Leendert van der Torre. "sing and dance!". In Fabrizio Cariani, Davide Grossi, Joke Meheus, and Xavier Parent, editors, *Deontic Logic and Normative Systems*, pages 149–165, Cham, 2014. Springer International Publishing.

[11] Xavier Parent and Leendert van der Torre. The pragmatic oddity in norm-based deontic logics. In Guido Governatori and Jeroen Keppens, editors, *Proceedings of the 16th edition of the International Conference on Artificial Intelligence and Law*, pages 169–178. ACM, 2017.

[12] Henry Prakken and Marek J. Sergot. Contrary-to-duty obligations. *Studia Logica*, 57(1):91–115, 1996.

Is Free Choice Permission Admissible in Classical Deontic Logic?

Guido Governatori

Data61, CSIRO, Australia

Antonino Rotolo

Alma Human AI, University of Bologna, Italy

Abstract

We explore how, and if, free choice permission (**FCP**) can be accepted when we consider deontic conflicts between certain types of permissions and obligations. **FCP** can license, under some minimal conditions, the derivation of an indefinite number of permissions. We discuss this and other drawbacks and present four Hilbert-style classical deontic systems admitting a guarded version of **FCP**. The systems that we present are not too weak from the inferential viewpoint, as far as permission is concerned, and do not commit to weakening any specific logic for obligations.

Keywords: Free Choice Permission, Weak and Strong Permission

1 Introduction and Background

A significant part of the literature in deontic logic revolves around the discussions of puzzles and paradoxes which show that certain logical systems are not acceptable—typically, this happens with deontic **KD**, i.e., Standard Deontic Logic (**SDL**)—or which suggest that obligations and permissions should enjoy some desirable properties.

One well-known puzzle is the the so-called Free Choice Permission paradox, which was originated by the following remark by von Wright in [24, p. 21]:

> "On an ordinary understanding of the phrase 'it is permitted that', the formula '$\mathbf{P}(p \vee q)$' seems to entail '$\mathbf{P}p \wedge \mathbf{P}q$'. If I say to somebody 'you may work or relax' I normally mean that the person addressed has my permission to work and also my permission to relax. It is up to him to choose between the two alternatives."

Usually, this intuition is formalised by the following schema:

$$\mathbf{P}(p \vee q) \to (\mathbf{P}p \wedge \mathbf{P}q) \tag{FCP}$$

Many problems have been discussed in the literature around **FCP**: for a comprehensive overview, discussion, and some solutions, see [15, 11, 22].

Three basic difficulties can be identified, among the others [11, p. 43]:

- **Problem 1: Permission Explosion Problem** – "That if anything is permissible, then everything is, and thus it would also be a theorem that nothing is obligatory," [22], for example "If you may order a soup, then it is not true that you ought to pay the bill" [6];

- **Problem 2: Closure under Logical Equivalence Problem** – "In its classical form **FCP** entails that classically equivalent formulas can be substituted to the scope of a permission operator. This is also implausible: It is permitted to eat an apple or not iff it is permitted to sell a house or not";

- **Problem 3: Resource Sensitivity Problem** – "Many deontic logics become resource-insensitive in the presence of **FCP**. They validate inferences of the form 'if the patient with stomach trouble is allowed to eat one cookie then he is allowed to eat more than one'".

We focus on another basic problem: how, and if, **FCP** can be accepted when we have incompatibilities between certain varieties of permissions and prohibitions/obligations. The issue is that since Problem 1 licenses the derivation that anything is permitted provided that something is permitted, no prohibition/obligation is allowed, otherwise we get an inconsistency [22]. In doing so, we offer simple logics that take two of the three problems above into account.

The layout of the paper is as follows. The remainder of this section briefly comments on the three major problems mentioned above: the Permission Explosion Problem (Section 1.1), the Closure under Logical Equivalence Problem (Section 1.2), and the Resource Sensitivity Problem (1.3). Section 2 illustrates the theoretical intuitions and assumptions that we adopt to analyse free choice permission. In particular, we assume the distinction between norms and obligations/permissions, and we study the role of deontic incompatibilities, the duality principle, and why free choice permission is strong permission. Section 3 quickly reviews some work that have direct implications for our proposal. Finally, Section 4 presents some minimal deontic systems, four Hilbert-style deontic systems admitting guarded variants of **FCP**: the systems that we present are not too weak from the inferential viewpoint, as far as permission is concerned, and do not commit to weakening any specific logic for obligations. Some conclusions end the paper. An appendix offers proofs of the formal properties of the proposed systems presented in Section 4.

1.1 Problem 1: Permission Explosion Problem

One of the most acute problems springing from **FCP** is obtained in **SDL**, where, if at least one obligation $\mathbf{O}p$ is true, then by necessitation and propositional logic, we get $\mathbf{O}(p \vee q)$. Since axiom **D** is in **SDL**, i.e $\mathbf{O}p \rightarrow \neg\mathbf{O}\neg p$ is valid, we trivially obtain $\neg\mathbf{O}\neg(p \vee q)$, thus, assuming the Duality principle

$$\mathbf{P} =_{def} \neg\mathbf{O}\neg \qquad \text{(Duality)}$$

we derive through **FCP** that $\mathbf{P}q$. Hence, **SDL** licenses that, if something is obligatory, then everything is permitted.

However, a careful analysis shows that this undesired result is not strictly due to **SDL** as such, but to adopting any monotonic modal deontic logic [10], i.e. any system just equipped with inference rule **RM**:

$$\frac{\vdash p \to q}{\vdash \mathbf{O}p \to \mathbf{O}q} \tag{RM}$$

or, alternatively with

$$\frac{\vdash p \equiv q}{\vdash \mathbf{O}p \equiv \mathbf{O}q} \tag{RE}$$

plus the following axiom schema

$$\mathbf{O}(a \wedge b) \to (\mathbf{O}a \wedge \mathbf{O}b). \tag{M}$$

Indeed, assume Classical Propositional Logic (**CPL**), **FCP**, and **RM** for **P**[1] and consider the following derivation:

1. $p \to (p \vee q)$ **CPL**
2. $\mathbf{P}p \to \mathbf{P}(p \vee q)$ 1, **RM**
3. $\mathbf{P}p \to (\mathbf{P}p \wedge \mathbf{P}q)$ 2, **FCP, CPL**
4. $\mathbf{P}p \to \mathbf{P}q$ 3, **CPL**.

In this context, it is enough if we have that $\mathbf{P}p$ is true to derive that any other permission $\mathbf{P}q$ is true as well, i.e., $\mathbf{P}p \vdash \mathbf{P}q$ for any p,q. Whenever **FCP** is accepted, such a problem strictly depends on the characteristic schemata and inference rules of monotonic modal logics, as the above derivation—or a simple semantic analysis—shows. Hence, permission explosion is not a problem of **SDL**, but of any weaker modal deontic logic which is at least closed under classical implication or which is closed under logical equivalence and allows for the distribution of **P** over implication. Notice that **Duality** plays no substantial role. Accordingly, we can have that **RM** is valid for permission, if **P** and **O** are duals and the logic for **O** is a monotonic modal logic, or **P** is independent of **O** and **RM** is assumed for **P**.

In conclusion, if we want not to completely reject the intuition behind **FCP**, we have two non-exclusive options to be explored in order to avoid the Permission Explosion Problem:

No-CPL: abandon **CPL** and adopt suitable non-classical logical connectives;

No-RM: abandon inference rule **RM** (or schema **M**) and endorse very weak modal logics (i.e., the classical ones [10, chap. 8]). [2]

Our paper aims at exploring under what conditions **No-CPL** can be avoided by accepting at least a restricted version of **FCP**. Hence, it seems that **No-RM** thesis must be accepted.

---

[1] With **RM-P** we mean the inference rule $\vdash p \to q/ \vdash \mathbf{P}p \to \mathbf{P}q$. Indeed, it is standard result that every system closed under **RM** for an operator is closed under the rule of the dual of the operator [cf. 10, p. 238–239, 243]. We will use **RM** to refer in general to the rule $\vdash p \to q/\vdash \Box p \to \Box q$ for any modal operator $\Box$.

[2] We state in Section 1.2 why it is convenient not to drop **RE**.

1.2 Problem 2: Closure under Logical Equivalence Problem

In the previous section we mentioned that **RM** must be weakened. Hence, we can also drop **RE** and keep axiom schema **M**. This choice could look satisfactory for those who consider problematic the fact that the logic for **P** is closed under logical equivalence.

We take here another route. Incidentally, one can argue that the implausibility of "It is permitted to eat an apple or not iff it is permitted to sell a house or not" does not depend on **RE**, but rather on the fact that "It is permitted to eat an apple or not" is **P**⊤, which looks quite odd. However, besides this problem—which would lead us to commit to specific philosophical views—dropping **RE** has in general two controversial technical side effects:

- it rejects standard semantics for modal logics, since the class of all neighbourhood frames validate **RE**: [10] argued in fact that classical systems (i.e., containing **RE** but not **RM**) are the minimal modal logics;
- it fails to make, for instance, $\mathbf{O}p$ and $\mathbf{P}\neg p$ logically incompatible under the Duality Principle (while $\mathbf{O}p$ and $\neg \mathbf{O}p$ of course are); similarly, $\mathbf{O}\neg p$ and $\mathbf{O}\neg\neg p$, or $\mathbf{O}(p \vee q)$ and $\mathbf{O}(\neg p \wedge \neg q)$, are not incompatible too (while they of course should be).

In conclusion, we standardly assume that **RE** holds both for permissions and obligations, which means that any logic for free choice permission must be a *classical system of deontic logic* in [10]'s sense, i.e., any modal deontic logic closed under logical equivalence and not under logical consequence.

1.3 Problem 3: Resource Sensitivity Problem

It has been noted [19] that from "You may eat an apple or a pear", one can infer "You may eat an apple and that You may eat a pear", but not "You may eat an apple and a pear" [7, p. 2].

We simply observe that the systems proposed in Section 4 do not license in general the inference above. However, a thoughtful treatment of this problem—the Resource Sensitivity Problem—goes beyond the scope of this paper. In fact, it has been widely discussed in the literature that it is strictly related to considerations from action theory, which have often found solutions shifting from **CPL** to non-classical logics such as the substructural ones [see, among others, 7, 4, 11]. In conclusion, we do not commit here to find any suitable solution to such a problem.

2 Three Basic Intuitions

We are going to present some deontic systems that accommodate restricted variants of **FCP**. This is done under some minimal philosophical assumptions, which can in principle be compatible with several deontic theories. In this section, we illustrate such fundamental intuitions and assumptions.

2.1 The Distinction between Norms and Obligations

We assume in the background a conceptual distinction between norms, on one side, and obligations and permissions, on the other side. The general idea of norms is that they describe conditions under which some behaviours are deemed as 'legal'. In the simplest case, a behaviour can be qualified by an obligation (or a prohibition, or a permission), but often norms additionally specify the consequences of not complying with them, and what sanctions follow from violations and whether such sanctions compensate for the violations. The scintilla for this idea is the very influential contribution [1], which is complementary to the (modal) logic-based approaches to deontic logic. The key feature of this approach is that norms are dyadic constructs connecting applicability conditions to a deontic consequence. A large number of such pairs would constitute an interconnected system called a *normative system* [for more recent proposals in this direction, see 20, 21, 14, 12].

To be clear, *this paper does not present any logic of norms, but our proposal for a logic of obligations and permissions—with restricted variants of **FCP**—can be better understood if one keeps in mind some intuitions about how norms should logically behave* and about the relation between the logic of norms and deontic logic. In particular, our assumptions are:

- obligations and permissions exist because norms generate them when applicable;
- once obligations and permissions are generated from norms—which requires us to reason about norms—we can still perform some reasoning with the resulting obligations and permissions—this is the task of *deontic logic in a strict sense*, i.e., the logic of obligations and permissions;
- norms can be in conflict—without being inconsistent— but this does not hold for obligations and permissions.

Hence, we distinguish two levels of analysis: a *norm-logic level* and a resulting *deontic-logic level*. This paper only technically deals with the second level of analysis.

Assume for example that we have two norms $n_1 : p \Rightarrow \mathbf{O}\neg q$ and $n_2 : p \Rightarrow \mathbf{P}q$, where $\Rightarrow$ is any if-then suitable logical relation connecting applicability conditions of norms and their deontic effects. We can indeed have them—for example, in a legal system—but the point is what obligations/permissions we can obtain from them. A rather standard assumption is that in order to correctly derive deontic conclusions we need to solve the conflict between n_1 and n_2. Specifically, our general view is prudent (or skeptical, as one says in non-monotonic logics), because, unless we know how to solve the conflict (typically, by establishing that n_1 is stronger than n_2 or vice versa), we do not know if $\mathbf{O}\neg q$ or $\mathbf{P}q$ holds. Since we do not accept that both can hold, it is pointless to consider at the deontic level that $\mathbf{O}\neg q$ *and* $\mathbf{P}q$ are true—while any logic of norms can have both n_1 and n_2.

In conclusion, we impose deontic consistency at the deontic-logic level, i.e.,

$\mathbf{O}p \wedge \mathbf{O}\neg p \rightarrow \bot$.

2.2 Deontic Incompatibilities, Duality, and FCP

With the above said, the issue is whether **FCP** is an appropriate principle to adopt for normative reasoning. Our view is that this principle in general is not, even when Problem 1 and 2 above are solved. We provide below a simple counterexample to it, which considers the interplay between free choice permissions and prohibitions.

Example 2.1 When you have dinner with guests the etiquette allows you to eat or to have a conversation with your fellow guests. However, it is forbidden to speak while eating.

The full representation of the example is that each choice is permitted when one refrains from exercising the other one. In a situation when one eats, there is the prohibition to speak, while when one speaks, there is the prohibition to eat. Hence, it means that we can detach any single permission only if the content of such permission is not forbidden. Given that Example 2.1 provides a counterexample to **FCP**, the question is whether we want to derive the individual permissions when one of the two disjuncts holds and we already satisfy the disjunctive permission. The reason is that the individual permissions, each on its own, can trigger other obligations or permissions. The following example illustrates this scenario.

Example 2.2 Suppose a shop has the following policy for clothes bought online. If the size of an item is not a perfect fit, then the customer is entitled to either exchange the item for free or to keep the item and receive a 10$ refund. However, customers electing to keep the item are not entitled to the refund, and customers opting for the refund are not entitled to exchange the item for free. Furthermore, customers who elect to exchange the item (when entitled to do so) have to return it with the original package.

The example can be formalised as follows:

$$online \wedge \neg fit \rightarrow \mathbf{P}(exchange \vee refund)$$
$$exchange \rightarrow \mathbf{O}\neg refund$$
$$refund \rightarrow \mathbf{O}\neg exchange$$
$$\mathbf{P}exchange \wedge exchange \rightarrow \mathbf{O}original$$

Suppose that a customer elects to exchange an item bought online that is not a perfect fit instead of asking for the refund. Intuitively, given that we cannot derive that exchanging is not forbidden ($\mathbf{O}\neg exchange$) at least the weak permission of exchanging the item should hold. However, in a deontic logic without **FCP** (or a restricted version of it) we are not able to derive the permission, and then we are not able to derive other obligations or permissions depending on it: in the example, the obligation to return the item with the original package.

We will return in Section 2.3 to the logical import of the above scenarios in a classical system of deontic logic. For the moment, taking stock of the examples we just notice that **FCP** could be reformulated as follows:

$$(\mathbf{P}(p \vee q) \wedge (\neg \mathbf{O}\neg p \wedge \neg \mathbf{O}\neg q)) \rightarrow (\mathbf{P}p \wedge \mathbf{P}q). \quad (1)$$

However, assuming **Duality**, $\neg \mathbf{O}\neg p$ is equivalent to $\mathbf{P}p$, thus (1) reduces to

$$(\mathbf{P}(p \vee q) \wedge (\mathbf{P}p \wedge \mathbf{P}q)) \rightarrow (\mathbf{P}p \wedge \mathbf{P}q). \quad (2)$$

(2) is a propositional tautology. Thus, (1) does not extend the expressive power of the logic unless one assumes a logic where obligation and permission are not the duals.

2.3 Strong Permission, Classical Systems, and FCP

When permission is no longer the dual of obligation, we enter the territory of strong permission [25, 3, 2] [3]. As is well-known, while it is *sufficient* to show that $\mathbf{O}\neg p$ is *not* the case to argue that p is weakly permitted, this does not hold for strong permission, for which the normative system *explicitly* says that there exists at least one norm permitting p [2, p. 353–355].

In order to keep track of these two cases at the deontic-logic level, we can standardly distinguish in the deontic language two permission operators, $\mathbf{P_w}$ for weak permission (such that $\mathbf{P_w}p =_{def} \neg \mathbf{O}\neg p$) and $\mathbf{P_s}$ for strong permission (where **Duality** does not hold).

What is the minimal logic of strong permission at the deontic level in which some reasonable version of free choice permission can be accepted?

We mentioned that **RM** must be rejected. In fact, besides the Permission Explosion Problem, one may also argue that it is reasonable not to derive $\mathbf{P_s}(p \vee q)$ from any $\mathbf{P_s}p$ because we could have in the background that the normative system consists just of an explicit norm $a \Rightarrow \mathbf{P_s}p$. If we have that, in presence of some version of free choice permission, you may also detach $\mathbf{P_s}q$, which is against the above-mentioned intuition that the strong permission should follow from explicit norms, or from combinations of them in normative systems where all disjuncts are explicitly considered [see, e.g., the discussion in 2, p. 354–355].

Second, as said above, deontic consistency should be ensured:

$$\mathbf{O}p \wedge \mathbf{P_s}\neg p \rightarrow \bot \quad (\mathbf{D_s})$$
$$\mathbf{O}p \wedge \mathbf{O}\neg p \rightarrow \bot \quad (\mathbf{D_w})$$

Notice that $\mathbf{D_w}$ is the standard **D** axiom of Standard Deontic Logic establishing the so called *external consistency* of obligations that, in turn, implies consistency among obligations and (weak) permissions. From $\mathbf{D_s}$ we obtain, as expected,

---
[3] Besides von Wright's theory [25], there is another sense in the literature of strong permission [16].

that strong permission entails weak permission [see, e.g., 2, p. 354], but not the other way around:

$$\mathbf{P_s}p \to \mathbf{P_w}p.$$

This is reasonable because the fact that at the norm-level we derive that p is permitted using an explicit permissive norm n means that no prohibitive norm n' (forbidding p) successfully applies or prevails over n.

What about free choice permission? Coupling Assumptions 1 and 2 with the distinction between weak and strong permission allows us to identify a guarded variant of **FCP** for strong permission, consisting of two schemata:

$$(\mathbf{P_s}(p \vee q) \wedge \mathbf{O}\neg p) \to \mathbf{P_s}q \qquad \text{(FCP}_\mathbf{O}\text{)}$$
$$(\mathbf{P_s}(p \vee q) \wedge \mathbf{P_w}p \wedge \mathbf{P_w}q) \to (\mathbf{P_s}p \wedge \mathbf{P_s}q) \qquad \text{(FCP}_\mathbf{P}\text{)}$$

These schemata take stock of what we said: you can detach from a disjunctive strong permission any single strong permission only if this last is weakly permitted.

The idea of the combination of the two axioms is that from repeated applications of **FCP$_\mathbf{O}$** and from a disjunctive permission, we can obtain the maximal sub-disjunction such that no element is forbidden, and then, the application of the **FCP$_\mathbf{P}$** allows us to derive the individual strong permissions that are not forbidden. Notice that we cannot assume the following formula as the axiom for free choice permission.

$$\mathbf{P_s}\left(\bigvee_{i=1}^{n} p_i\right) \wedge \left(\bigwedge_{j=1}^{m<n} \mathbf{O}\neg p_j\right) \to \bigwedge_{k=m+1}^{n} \mathbf{P_s}p_k$$

The problem is that we do not know in advance how many elements of the disjunctive permission are (individually) forbidden. Consider for example, a theory consisting of the following formulas:

$$\mathbf{P_s}(p \vee q \vee r \vee s \vee t) \qquad \mathbf{O}\neg p \qquad \mathbf{O}\neg q \qquad \mathbf{O}\neg r$$

Here, one could use the conjunction $\mathbf{O}\neg p \wedge \mathbf{O}\neg q$ to obtain $\mathbf{P_s}r$, $\mathbf{P_s}s$ and $\mathbf{P_s}t$, but then we have a contradiction from $\mathbf{P_s}r$ and $\mathbf{O}\neg r$ (from axiom $\mathbf{D_s}$). Notice, that in general, we are not able to use **FCP$_\mathbf{O}$** to detach a single (strong) permission, but a disjunction corresponding to the "remainder" of the disjunction, that is, in the case above, $\mathbf{P_s}(s \vee t)$. Then, we can use the **FCP$_\mathbf{P}$** to "lift" the remaining elements from weak permissions to strong permissions. The only case when we can obtain an individual strong permission from a permissive disjunction is when the remainder is a singleton; but this means, that all the other elements of the permissive disjunction were forbidden. This further means that a disjunctive strong permission holds if at least one of its elements can be legally exercised. Going back to the example, if one extends the theory with $\mathbf{O}\neg s$, then we can derive $\mathbf{P_s}t$.

Consider the situation described in Example 2.1. The scenario can be formalised as follows (where e and s stand for "to eat" and "to speak"):

$$\mathbf{P_s}(e \vee s)$$
$$s \to \mathbf{O}\neg e$$
$$e \to \mathbf{O}\neg s$$

In a logic endorsing the unrestricted version of free choice permission, we have $\mathbf{P_s}e$ and $\mathbf{P_s}s$. This means that as soon as one exercises one of the choices, we get that the other choice is at the same time permitted and forbidden, a situation that is either paradoxical or contradictory. Thus, the only way to avoid this kind of conflict is to refrain from exercising any of the two choices. However, this means that one is not really free to choose between the two options. Accordingly, either one has to adopt a restricted version of the free choice permission or abandon it. Notice, that axiom $\mathbf{FCP_O}$ allows us to conclude that given e, s is forbidden ($\mathbf{O}\neg s$), and thus that e is permitted ($\mathbf{P_s}e$); similarly, one gets $\mathbf{P_s}s$ from s, which implies $\mathbf{O}\neg e$. Similarly, for Example 2.2 when we formalise it using strong permission $\mathbf{P_s}$ instead of $\mathbf{P}$, Axiom $\mathbf{FCP_O}$ allows us to derive $\mathbf{P_s}$*exchange* from which we can conclude $\mathbf{O}$*original*.

Consider $\mathbf{FCP_P}$. One may argue why, in symmetry with $\mathbf{FCP_O}$, we cannot rather have

$$(\mathbf{P_s}(p \vee q) \wedge \mathbf{P_w}p) \to \mathbf{P_s}p \qquad (\mathbf{FCP2_P})$$

Technically, it is obvious that $\mathbf{FCP2_P}$ implies $\mathbf{FCP_P}$ but not the other way around, so both options are available. The variant $\mathbf{FCP_P}$ is more prudent in that it licenses the detachment of an individual strong permission *only if* the normative system explicitly deals with that specific disjunct, while the second allows for the derivation in a slightly more relaxed way. So, if one wants to strictly reframe the structure of standard $\mathbf{FCP}$ in a guarded version but does not want $\mathbf{FCP2_P}$, then $\mathbf{FCP_P}$ is the right option.

We should notice that the above schemata for free choice permission do not necessarily require the technical idea of deontic consistency, unless we assume—but we don't—that obligation implies strong permission, and despite the fact that the consistency problem can occur if we endorse $\mathbf{D_s}$—as we do— and so that strong permission implies weak permission.

3 Related Works

Most of the work on the development of logical systems related to the problem of Free Choice Permission concentrate on logics accepting the $\mathbf{FCP}$ principle. Some work focus on the resource aspects and propose the use of substructural logics to address the problem, see for example [7]. Similarly to our work, in the sense of a non-normal deontic logic, is the proposal in [6, 5]. In fact, even though they have a different philosophical backgrounds based on the of open reading of permissions [17, 9], they propose *simple non-normal axiomatisations for obligation and permission*—as we do—which avoid, e.g., Problem 1 and which are based on the concept of free choice permission as strong permission or, anyway, as a type of permission without **Duality**.

The scenario in Example 2.2 indicates that Deontic Logic should accept **FCP**, but at the same time Example 2.1 points out that it cannot accept in an unrestricted form. In this regard, the proposal by Asher and Bonevac [6] shares with us the idea of limiting the applicability of **FCP**. Their solution is based on a deontic logic taking a non-monotonic logic as the underlying reasoning mechanism instead classical propositional logic as we do. Accordingly, in their system instances of **FCP** are derivable unless they are defeated. In addition, their logic is not closed under logical equivalence.

4 Four Minimal Deontic Axiomatisations with Guarded Free Choice Permission

Finally, we present some minimal deontic systems, four Hilbert-style deontic systems admitting a guarded version of **FCP**. The systems that we present are not too weak from the inferential viewpoint, as far as permission is concerned, and do not commit to weakening any specific logic for obligations.

4.1 Language, Axioms and Inference Rules

The modal language and the concept of well formed formula are defined as usual [see 10, 8]. We just recall that we have three modal operators, two $\Box$ operators, **O** for obligations and $\mathbf{P_s}$ for strong permissions, and $\mathbf{P_w}$ for weak permission. As usual, we assume $\mathbf{P_w}$ to be an abbreviation for $\neg \mathbf{O} \neg$.

For convenience, let us synoptically recall below all relevant schemata and inference rules, where $\Box \in \{\mathbf{O}, \mathbf{P_s}\}$.

Inference Rules:

$\mathbf{RE} := \vdash A \equiv B \Rightarrow \vdash \Box A \leftrightarrow \Box B$

$\mathbf{RM} := \vdash A \to B \Rightarrow \vdash \Box A \to \Box B$

Schemata:

$\mathbf{M} := \Box(p \wedge q) \to (\Box p \wedge \Box q)$

$\mathbf{FCP_O} := (\mathbf{P_s}(p \vee q) \wedge \mathbf{O}\neg p) \to \mathbf{P_s}q$

$\mathbf{FCP_P} := (\mathbf{P_s}(p \vee q) \wedge \mathbf{P_w}p \wedge \mathbf{P_w}q) \to (\mathbf{P_s}p \wedge \mathbf{P_s}q)$

$\mathbf{FCP2_P} := (\mathbf{P_s}(p \vee q) \wedge \mathbf{P_w}p) \to \mathbf{P_s}p$

$\mathbf{D_s} := \mathbf{O}p \wedge \mathbf{P_s}\neg p \to \bot$

$\mathbf{D_w} := \mathbf{O}p \wedge \mathbf{P_w}\neg p \to \bot$

$\mathbf{P_sP_w} := \mathbf{P_s}p \to \mathbf{P_w}p$.

Given the discussion of Section 2, we can identify some deontic systems, as specified in Table 1. Notice that we consider also systems $\mathbf{FCP_2}$ and $\mathbf{FCP_4}$, which are monotonic, so they contain **RM**. Strictly speaking, this is the limit which we cannot trespass, since we have restricted forms of Permission Explosion. We will return on this in the concluding section of the paper.

Deontic System	Properties	Derivable
$\mathbf{E} := \mathbf{RE}$		
$\mathbf{Min} := \mathbf{RE} \oplus \mathbf{D_s} \oplus \mathbf{D_w}$		$\mathbf{P_sP_w}$
$\mathbf{FCP_1} := \mathbf{Min} \oplus \mathbf{FCP_O} \oplus \mathbf{FCP_P}$		$\mathbf{P_sP_w}$
$\mathbf{FCP_2} := \mathbf{FCP_1} \oplus \mathbf{M}$	$\mathbf{FCP_1} \subset \mathbf{FCP_2}$	$\mathbf{P_sP_w}$
$\mathbf{FCP_3} := \mathbf{Min} \oplus \mathbf{FCP_O} \oplus \mathbf{FCP2_P}$	$\mathbf{FCP_1} \subset \mathbf{FCP_3}$	$\mathbf{P_sP_w}, \mathbf{FCP_P}$
$\mathbf{FCP_4} := \mathbf{FCP_3} \oplus \mathbf{M}$	$\mathbf{FCP_2} \subset \mathbf{FCP_4}$ $\mathbf{FCP_3} \subset \mathbf{FCP_4}$	$\mathbf{P_sP_w}, \mathbf{FCP_O}$ $\mathbf{FCP_P}, \mathbf{FCP2_P}$

Table 1
Deontic Systems

4.2 Semantics and System Properties

Let us begin with standard concepts. Assume that PROP is the set of atomic sentences.

Definition 4.1 A *deontic neighbourhood frame* $\mathcal{F}$ is a structure $\langle W, \mathcal{N_O}, \mathcal{N_P} \rangle$ where

- W is a non-empty set of possible worlds;
- $\mathcal{N_O}$ and $\mathcal{N_P}$ are functions $W \mapsto 2^{2^W}$.

Definition 4.2 A *deontic neighbourhood model* $\mathcal{M}$ is a structure $\langle W, \mathcal{N_O}, \mathcal{N_P}, V \rangle$ where $\langle W, \mathcal{N_O}, \mathcal{N_P} \rangle$ is a deontic neighbourhood frame and V is an evaluation function PROP $\mapsto 2^W$.

Definition 4.3 [Truth in a model] Let $\mathcal{M}$ be a model $\langle W, \mathcal{N_O}, \mathcal{N_P}, V \rangle$ and $w \in W$. The truth of any formula p in $\mathcal{M}$ is defined inductively as follows:

(i) standard valuation conditions for the boolean connectives;
(ii) $\mathcal{M}, w \models \mathbf{O}p$ iff $||p||_\mathcal{M} \in \mathcal{N_O}(w)$,
(iii) $\mathcal{M}, w \models \mathbf{P_s}p$ iff $||p||_\mathcal{M} \in \mathcal{N_P}(w)$,
(iv) $\mathcal{M}, w \models \mathbf{P_w}p$ iff $W - ||p||_\mathcal{M} \notin \mathcal{N_O}(w)$,

where, as usual, $||p||_\mathcal{M}$ is the truth set of p wrt to $\mathcal{M}$

$$||p||_\mathcal{M} = \{w \in W : \mathcal{M}, w \models p\}.$$

A formula p is *true at a world* in a model iff $\mathcal{M}, w \models p$; *true in a model* $\mathcal{M}$, written $\mathcal{M} \models p$ iff for all worlds $w \in W$, $\mathcal{M}, w \models p$; *valid in a frame* $\mathcal{F}$, written $\mathcal{F} \models p$ iff it is true in all models based on that frame; *valid in a class* $\mathcal{C}$ *of frames*, written $\mathcal{C} \models p$, iff it is valid in all frames in the class. An inference rule $P_1, \ldots P_n \Rightarrow C$ (where $P_1, \ldots P_n$ are the premises and C the conclusion) is valid in a class $\mathcal{C}$ of frames iff, for any $\mathcal{F} \in \mathcal{C}$, if $\mathcal{F} \models P_1, \ldots, \mathcal{F} \models P_n$ then $\mathcal{F} \models C$[4].

We can now characterise different classes of deontic neighbourhood frames that are adequate of the deontic systems in Table 1.

---
[4] Of course, if any P_k has the form $\vdash p$ then $\mathcal{F} \models P_1$ trivially means $\mathcal{F} \models p$.

Definition 4.4 [Frame Properties] Let $\mathcal{F} = \langle W, \mathcal{N}_{\mathbf{O}}, \mathcal{N}_{\mathbf{P}} \rangle$ be a deontic neighbourhood frame.

- **□-supplementation**: $\mathcal{F}$ is □-supplemented, $\square \in \{\mathbf{O}, \mathbf{P}\}$, iff for any $w \in W$ and $X, Y \subseteq W$, $X \cap Y \in \mathcal{N}_\square(w) \Rightarrow X \in \mathcal{N}_\square(w) \& Y \in \mathcal{N}_\square(w)$;
- **$\mathbf{P_w}$-coherence**: $\mathcal{F}$ is $\mathbf{P_w}$-coherent iff for any $w \in W$ and $X \subseteq W$, $X \in \mathcal{N}_{\mathbf{O}}(w) \Rightarrow W - X \notin \mathcal{N}_{\mathbf{O}}(w)$;
- **$\mathbf{P_s}$-coherence**: $\mathcal{F}$ is $\mathbf{P_s}$-coherent iff for any $w \in W$ and $X \subseteq W$, $X \in \mathcal{N}_{\mathbf{P}}(w) \Rightarrow W - X \notin \mathcal{N}_{\mathbf{O}}(w)$;
- **$\mathbf{FCP_O}$-permission**: $\mathcal{F}$ is $\mathbf{FCP_O}$-permitted iff for any $w \in W$ and $X, Y \subseteq W$, $X \cup Y \in \mathcal{N}_{\mathbf{P}}(w) \& W - Y \in \mathcal{N}_{\mathbf{O}}(w) \Rightarrow X \in \mathcal{N}_{\mathbf{P}}(w)$;
- **$\mathbf{FCP_P}$-permission**: $\mathcal{F}$ is $\mathbf{FCP_P}$-permitted iff for any $w \in W$ and $X, Y \subseteq W$, $X \cup Y \in \mathcal{N}_{\mathbf{P}}(w) \& W - X \notin \mathcal{N}_{\mathbf{O}}(w) \& W - Y \notin \mathcal{N}_{\mathbf{O}}(w) \Rightarrow X \in \mathcal{N}_{\mathbf{P}}(w) \& X \in \mathcal{N}_{\mathbf{P}}(w)$;
- **$\mathbf{FCP2_P}$-permission**: $\mathcal{F}$ is $\mathbf{FCP2_P}$-permitted iff for any $w \in W$ and $X, Y \subseteq W$, $X \cup Y \in \mathcal{N}_{\mathbf{P}}(w) \& W - X \notin \mathcal{N}_{\mathbf{O}}(w) \Rightarrow X \in \mathcal{N}_{\mathbf{P}}(w)$;

Below are some relevant characterisation results. All the proofs for this section are in the Appendix.

Lemma 4.5 *For any deontic neighbourhood frame $\mathcal{F}$,*

(i) $\mathbf{D_s}$ *is valid in the class of $\mathbf{P_s}$-coherent frames;*

(ii) $\mathbf{D_w}$ *is valid in the class of $\mathbf{P_w}$-coherent frames;*

(iii) $\mathbf{FCP_O}$ *is valid in the class of $\mathbf{FCP_O}$-permitted frames;*

(iv) $\mathbf{FCP_P}$ *is valid in the class of $\mathbf{FCP_P}$-permitted frames;*

(v) $\mathbf{FCP2_P}$ *is valid in the class of $\mathbf{FCP2_P}$-permitted frames;*

Completeness results for the four deontic systems are ensured.

Theorem 4.6

(i) $\mathbf{E}$ *is sound and complete w.r.t. the class of deontic neighbourhood frames;*

(ii) $\mathbf{Min}$ *is sound and complete w.r.t. the class of $\mathbf{P_s}$- and $\mathbf{P_w}$-coherent frames;*

(iii) $\mathbf{FCP_1}$ *is sound and complete w.r.t. the class of $\mathbf{FCP_O}$- and $\mathbf{FCP_P}$-permitted frames;*

(iv) $\mathbf{FCP_2}$ *is sound and complete w.r.t. the class of $\mathbf{P}$-supplemented, $\mathbf{FCP_O}$- and $\mathbf{FCP_P}$-permitted frames;*

(v) $\mathbf{FCP_3}$ *is sound and complete w.r.t. the class of $\mathbf{FCP_O}$- and $\mathbf{FCP2_P}$-permitted frames;*

(vi) $\mathbf{FCP_4}$ *is sound and complete w.r.t. the class of $\mathbf{P}$-supplemented, $\mathbf{FCP_O}$- and $\mathbf{FCP2_P}$-permitted frames.*

Next, a corollary showing the relative strength of the four deontic systems.

Corollary 4.7

(i) $\mathbf{FCP_1} \subset \mathbf{FCP_2} \subset \mathbf{FCP_4}$ and
$\mathbf{FCP_1} \subset \mathbf{FCP_3} \subset \mathbf{FCP_4}$.

(ii) Let $\mathbf{L_1}, \mathbf{L_2} \in \{\mathbf{FCP_i}, 1 \leq i \leq 4\}$, and let $\mathcal{C}_1$ and $\mathcal{C}_2$ be classes of frames adequate for $\mathbf{L_1}$ and $\mathbf{L_2}$. If $\mathbf{L_1} \subset \mathbf{L_2}$ then $\mathcal{C}_2 \subset \mathcal{C}_1$.

Finally, we are going to examine the issue of decidability. To this end we recall the result by Lewis [18], who proved that every intensional logic that is axiomatisable by axioms that do not contain iterative operators (non-iterative axioms) has the finite model property; A formula (axiom) A is non-iterative iff for every subformula $\Box_i B/\Diamond_i B$ of A, B does not contain a modal operator. It is immediate to verify that the axioms $\mathbf{D_s}$, $\mathbf{D_w}$, $\mathbf{M}$, $\mathbf{FCP_O}$, $\mathbf{FCP_P}$ and $\mathbf{FCP2_P}$ are non-iterative, hence we have the following theorem.

Theorem 4.8 *The logics* $\mathbf{FCP_1}$, $\mathbf{FCP_2}$, $\mathbf{FCP_3}$ *and* $\mathbf{FCP_4}$ *have the finite model property, and hence are decidable.*

5 Conclusions

In this paper we have investigated how, and if the notion of free choice permission is admissible in modal deontic logic. As is well known, several problems can be put forward in regard to this notion, the most fundamental of them being the so-called Permission Explosion Problem, according to which all systems containing **FCP** and closed under **RM** and **RM-P** license the derivation of any arbitrary permission whenever at least one specific permission is true.

We argued (Section 1.1) that a plausible solution to this problem is to jump from monotonic into classical deontic logics, i.e., systems closed under **RE** but not **RM**. This solution does not necessarily mean that the resulting deontic system is very weak, as far as permission is concerned, if further schemata are added (Sections 2.3 and 4.2).

The basic intuitions for extending classical deontic logics are the following:

(i) We assume in background the distinction between norms and obligations/permissions. While we conceptually accept that the normative system may contain conflicting norms, it is logically inadmissible that such norms generate actual conflicting obligations/permissions since conflicts must be rationally solved, otherwise no obligation/permission can be obtained; hence, we validate schemata $\mathbf{D_s}$ and $\mathbf{D_w}$;

(ii) Free choice permission is strong permission, meaning that it is a permission generated by explicit permissive norms;

(iii) The possibility of detaching single strong permissions from disjunctive strong permissions, i.e., $\mathbf{P_s}q$ from $\mathbf{P_s}(p \lor q)$ strictly depends on the fact that $\mathbf{O}\neg p$ is not the case.

Taking the above points into account, we thus proposed different guarded variants of **FCP** that significantly increase the inferential power of the logic. In particular, four Hilbert-style classical deontic systems were presented.

We observed that two of these systems are classical modal systems, while we can have other two acceptable systems which are monotonic. In fact, the fact that those two systems are closed under **RM** does not lead to full Permission Explosion, but only to a "controlled" version of it: indeed, in systems like **FCP$_2$** any permission is obtainable via free choice permission *only if* it is not incompatible with existing prohibitions.

Some directions for future work can be identified. In particular:

- It is still an open issue to fully discuss the Resource Sensitivity Problem in our setting. In fact, while we argued that this problem goes beyond our paper, there are scenarios where our intuitions are relevant for this problem as well. For example, suppose that there is a fruit basket in the kitchen containing a banana and an apple. Bob and Alice are permitted to eat the banana or the apple and Alice first eats the former. Bob cannot do anything but take the apple. However, if Bob is allergic of apples, so no permission can be reasonably derived because it is forbidden for him to eat the apple.

- Our idea of free choice permission relies on the fact that no strong permission can be detached from a disjunctive permissive expression if another norm allows for deriving a conflicting obligation. Hence a full understanding of schemata such as **FCP$_O$** or **FCP$_P$** may benefit for an explicit logical treatment of the logic of norms adopting defeasible reasoning [11]; we plan to investigate how to integrate the approach presented in the paper and the computationally oriented approach offered by Defeasible Deontic Logic [13].

Acknowledgments

This work was partially supported by the EU H2020 research and innovation programme under the Marie Skłodowska-Curie grant agreement No. 690974 for the project MIREL: *MIning and REasoning with Legal texts*.

References

[1] Alchourrón, C.E. and E. Bulygin, *Normative Systems*, Springer Verlag, 1971.
[2] Alchourrón, C.E. and E. Bulygin, *Permission and permissive norms*, in: W.K. et al., editor, *Theorie der Normen*, Duncker & Humblot, 1984.
[3] Alchourrón, C.E. and E. Bulygin, *The expressive conception of norms*, in: R. Hilpinen, editor, *New Studies in Deontic Logic*, D. Reidel, 1981 pp. 95–125.
[4] Anglberger, A.J.J., H. Dong and O. Roy, *Open reading without free choice*, in: F. Cariani, D. Grossi, J. Meheus and X. Parent, editors, *Deontic Logic and Normative Systems*, (2014), pp. 19–32.
[5] Anglberger, A.J., N. Gratzl and O. Roy, *Obligation, free choice, and the logic of weakest permissions*, The Review of Symbolic Logic **8** (2015), pp. 807–827.
[6] Asher, N. and D. Bonevac, *Free choice permission is strong permission*, Synthese **145** (2005), pp. 303–323.
[7] Barker, C., *Free choice permission as resource-sensitive reasoning*, Semantics and Pragmatics **3** (2010), pp. 1–38.
[8] Blackburn, P., M. de Rijke and Y. Venema, *Modal Logic*, Cambridge University Press, 2001.
[9] Broersen, J., *Action negation and alternative reductions for dynamic deontic logics*, Journal of Applied Logic **2** (2004), pp. 153–168.
[10] Chellas, B.F., *Modal logic: an introduction*, Cambridge University Press, 1980.
[11] Dong, H., "Permission in Non-Monotonic Normative Reasoning", PhD thesis, University of Bayreuth, 2017.

[12] Governatori, G., F. Olivieri, E. Calardo and A. Rotolo, *Sequence semantics for norms and obligations*, in: O. Roy, A. Tamminga and M. Willer, editors, *Deontic Logic and Normative Systems*, (2016), pp. 93–108.

[13] Governatori, G., F. Olivieri, A. Rotolo and S. Scannapieco, *Computing strong and weak permissions in defeasible logic*, Journal of Philosophical Logic **42** (2013), pp. 799–829.

[14] Governatori, G. and A. Rotolo, *Logic of violations: a Gentzen system for reasoning with contrary-to-duty obligations*, Australasian Journal of Logic **4** (2006), pp. 193–215.

[15] Hansson, S.O., *The varieties of permissions*, in: D. Gabbay, J. Horty, X. Parent, R. van der Meyden and L. van der Torre, editors, *Handbook of Deontic Logic and Normative Systems*, College Publications, 2013.

[16] Kamp, H., *Free choice permission*, Proceedings of the Aristotelian Society **74** (1973), pp. 57–74.

[17] Lewis, D., "A Problem About Permission", in: *Essays in Honour of Jaakko Hintikka: On the Occasion of His Fiftieth Birthday on January 12, 1979*, ed. by E. Saarinen, R. Hilpinen, I. Niiniluoto and M.P. Hintikka, Dordrecht: Springer Netherlands, 1979, pp. 163–175.

[18] Lewis, D., *Intensional logics without iterative axioms*, Jounrnal of Philosophical Logic **3** (1974), pp. 457–466.

[19] Lokhorst, G.-J.C., *Deontic Linear Logic with Petri Net Semantics*, tech. rep., FICT, Center for the Philosophy of Information and Communication Technology. Rotterdam, 1997.

[20] Makinson, D., *On a fundamental problem of deontic logic*, in: P. McNamara and H. Prakken, editors, *Norms, Logics and Information Systems. New Studies in Deontic Logic and Computer Science*, IOS Press, 1999 pp. 29–54.

[21] Makinson, D. and L. van der Torre, *Permission from an input/output perspective*, Journal of Philosophical Logic **32** (2003), pp. 391–416.

[22] McNamara, P., *Deontic logic*, in: E.N. Zalta, editor, *The Stanford Encyclopedia of Philosophy*, Metaphysics Research Lab, Stanford University, 2018.

[23] Pacuit, E., *Neighborhood Semantics for Modal Logic*, Springer, 2017.

[24] von Wright, G.H., *An Essay in Deontic Logic and the General Theory of Action with a Bibliography of Deontic and Imperative Logic*, North-Holland Pub. Co, 1968.

[25] von Wright, G.H., *Norm and action: A logical inquiry*, Routledge and Kegan Paul, 1963.

A Basic Properties of the Deontic Systems

Let us start by proving Lemma 4.5.

Lemma 4.5 *For any deontic neighbourhood frame $\mathcal{F}$,*

(i) $\mathbf{D_s}$ *is valid in the class of $\mathbf{P_s}$-coherent frames;*

(ii) $\mathbf{D_w}$ *is valid in the class of $\mathbf{P_w}$-coherent frames;*

(iii) $\mathbf{FCP_O}$ *is valid in the class of $\mathbf{FCP_O}$-permitted frames;*

(iv) $\mathbf{FCP_P}$ *is valid in the class of $\mathbf{FCP_P}$-permitted frames;*

(v) $\mathbf{FCP2_P}$ *is valid in the class of $\mathbf{FCP2_P}$-permitted frames;*

Proof. The proof for case (i) is straightforward. The proof of (ii) is trivial and standard. Both are omitted.

Case (iii) – Consider any frame $\mathcal{F}$ that is $\mathbf{FCP_O}$-*permitted* but such that $\mathcal{F} \not\models \mathbf{FCP_O}$. This means that there exists a model $\mathcal{M} = \langle W, \mathcal{N_O}, \mathcal{N_P}, V \rangle$ based on $\mathcal{F}$ such that $\mathcal{M} \not\models \mathbf{FCP_O}$, i.e., there is a world $w \in W$ where

$$\mathcal{M}, w \models \mathbf{P_s}(p \vee q) \wedge \mathbf{O}\neg p \tag{A.1}$$

$$\mathcal{M}, w \not\models \mathbf{P_s}q \tag{A.2}$$

By construction, from (A.2) we have $||q||_\mathcal{M} \notin \mathcal{N}_\mathbf{P}(w)$, while from (A.1) we have $||p||_\mathcal{M} \cup ||q||_\mathcal{M} \in \mathcal{N}_\mathbf{P}(w)$ and $W - ||p||_\mathcal{M} \in \mathcal{N}_\mathbf{O}(w)$, so $\mathcal{F}$ is not $\mathbf{FCP_O}$-permitted.

Cases (iv) and (v) – The proofs are similar to the one for Case (iii) and are omitted. □

The definitions of some basic notions and of canonical model for the classical bimodal logic **E** (just consisting of **RE** for **O** and $\mathbf{P_s}$) are standard.

In the rest of this section when we refer to a Deontic System **S** we mean one the logic axiomatised in Section 4.

Definition A.1 [**S**-maximality] A set w is maximal iff it is **S**-consistent and for any formula p, either $p \in w$, or $\neg p \in w$.

Lemma A.2 (Lindenbaum's Lemma) *For any Deontic System **S**, any consistent set w of formulae can be extended to an **S**-maximal set w^+.*

Definition A.3 [Canonical Model [10, 23]] A canonical neighbourhood model $\mathcal{M} = \langle W, \mathcal{N}_\mathbf{O}, \mathcal{N}_\mathbf{P}, V \rangle$ for any system **S** in our language $\mathcal{L}$ (where $\mathbf{S} \supseteq \mathbf{E}$) is defined as follows:

(i) W is the set of all the **S**-maximal sets.

(ii) For any propositional letter p, $||p||_\mathcal{M} := |p|_\mathbf{S}$, where $|p|_\mathbf{S} := \{w \in W \mid p \in w\}$.

(iii) If $\square \in \{\mathbf{O}, \mathbf{P_s}\}$, let $\mathcal{N}_\square := \bigcup_{w \in W} \mathcal{N}_\square(w)$ where for each world w, $\mathcal{N}_\square(w) := \{||a_i||_\mathcal{M} \mid \square a_i \in w\}$.

Lemma A.4 (Truth Lemma [10, 23]) *If $\mathcal{M} = \langle W, \mathcal{N}_\mathbf{O}, \mathcal{N}_\mathbf{P}, V \rangle$ is canonical for **E**, then for any $w \in W$ and for any formula p, $p \in w$ iff $\mathcal{M}, w \models p$.*

Thus, we have as usual basic completeness result for **E**. To cover the other systems, it is enough to prove that all frame properties for the relevant schemata and rules are canonical.

Lemma A.5 *The frame properties of Definition 4.4 are canonical.*

Proof. The proofs for $\square$-*supplementation*, $\mathbf{P_w}$-*coherence*, and $\mathbf{P_s}$-*coherence* are standard.

$\mathbf{FCP_P}$-*permission* – Let us consider a canonical model $\mathcal{M}$ for $\mathbf{FCP_P}$, any world w in it, and any truth sets such that $||p||_\mathcal{M} \cup ||q||_\mathcal{M} \in \mathcal{N}_\mathbf{P}(w)$ and $W - ||q||_\mathcal{M} \in \mathcal{N}_\mathbf{O}(w)$. Clearly, $||p \vee q||_\mathcal{M} \in \mathcal{N}_\mathbf{P}(w)$. Since $\mathbf{FCP_P}$ is valid (Lemma 4.5), then $\mathbf{P_s}p \in w$. By construction, this means that $||p||_\mathcal{M} \in \mathcal{N}_\mathbf{P}(w)$, thus the model is $\mathbf{FCP_P}$-permitted.

$\mathbf{FCP_O}$-*permission* and $\mathbf{FCP2_P}$-*permission*– Similar to the case above. □

Hence, the following result is ensured.

Theorem 4.6

(i) **E** *is sound and complete w.r.t. the class of deontic neighbourhood frames;*

(ii) **Min** *is sound and complete w.r.t. the class of* $\mathbf{P_s}$- *and* $\mathbf{P_w}$-*coherent frames;*

(iii) $\mathbf{FCP_1}$ *is sound and complete w.r.t. the class of* $\mathbf{FCP_O}$- *and* $\mathbf{FCP_P}$-*permitted frames;*

(iv) $\mathbf{FCP_2}$ *is sound and complete w.r.t. the class of* **P**-*supplemented,* $\mathbf{FCP_O}$- *and* $\mathbf{FCP_P}$-*permitted frames;*

(v) $\mathbf{FCP_3}$ *is sound and complete w.r.t. the class of* $\mathbf{FCP_O}$- *and* $\mathbf{FCP2_P}$-*permitted frames;*

(vi) $\mathbf{FCP_4}$ *is sound and complete w.r.t. the class of* **P**-*supplemented,* $\mathbf{FCP_O}$- *and* $\mathbf{FCP2_P}$-*permitted frames.*

Finally, let us prove Corollary 4.7.

Corollary 4.7

(i) $\mathbf{FCP_1} \subset \mathbf{FCP_2} \subset \mathbf{FCP_4}$ *and*
    $\mathbf{FCP_1} \subset \mathbf{FCP_3} \subset \mathbf{FCP_4}$.

(ii) *Let* $\mathbf{L_1}, \mathbf{L_2} \in \{\mathbf{FCP_i}, 1 \leq i \leq 4\}$, *and let* $\mathcal{C}_1$ *and* $\mathcal{C}_2$ *be classes of frames adequate for* $\mathbf{L_1}$ *and* $\mathbf{L_2}$. *If* $\mathbf{L_1} \subset \mathbf{L_2}$ *then* $\mathcal{C}_2 \subset \mathcal{C}_1$.

Proof. *Case* (i) – For $\mathbf{FCP_1} \subset \mathbf{FCP_2}$ the inclusion is trivial given that every axiom of $\mathbf{FCP_1}$ is also an axiom of $\mathbf{FCP_2}$. To show that the inlcusion is strict consider the model $\mathcal{M} = \langle W, \mathcal{N_O}, \mathcal{N_P}, V \rangle$, where:

- $W = \{w_1, w_2, w_3, w_4, w_5\}$;
- $V(a) = \{w_1, w_4, w_5\}$, $V(b) = \{w_2, w_3, w_4\}$ and $V(c) = \{w_1, w_2\}$;
- $\mathcal{N_O}(w_1) = \{\{w_4\}\}$; and
- $\mathcal{N_P}(w_1) = \{\{w_1, w_2, w_3\}, \{w_1, w_2\}\}$.

It is easy to verify that the model is $\mathbf{FCP_O}$-permitted, $\mathbf{P_s}(\neg a \vee c)$ and $\mathbf{O}(a \wedge c)$ are true in w_1: $||\neg a \vee c||_\mathcal{M} = \{w_1, w_2, w_3\} \in \mathcal{N_P}(w_1)$ and $||a \wedge c||_\mathcal{M} = \{w_4\} \in \mathcal{N_O}(w_1)$. However, the model is not **O**-supplemented: $||a \wedge c||_\mathcal{M} = \{w_4\} \in \mathcal{N_O}(w_1)$, $\{w_4\} = ||a||_\mathcal{M} \cap ||c||_\mathcal{M}$, but $||a||_\mathcal{M}, ||c||_\mathcal{M} \notin \mathcal{N_O}(w_1)$, falsifying the following instance of **M**: $\mathbf{O}(a \wedge c) \rightarrow \mathbf{O}a \wedge \mathbf{O}c$.

For $\mathbf{FCP_1} \subset \mathbf{FCP_3}$ it is immediate to verify that $\mathbf{FCP2_P}$ implies $\mathbf{FCP_P}$ in **CPL** but not the other way around, and the same relationship holds for the corresponding semantic conditions.

For $\mathbf{FCP_2} \subset \mathbf{FCP_4}$, the result follows from $\mathbf{FCP_1} \subset \mathbf{FCP_3}$.

For $\mathbf{FCP_3} \subset \mathbf{FCP_4}$, the inclusion is trivial and we can reuse the model to show the strictness of the inclusion between $\mathbf{FCP_1}$ and $\mathbf{FCP_2}$.

Case (*ii*) – The result follows from Case (i) above and Theorem 4.6. □

Moral Principles: Hedged, Contributory, Mixed

Aleks Knoks [1]

University of Zurich
Philosophisches Seminar, Zürichbergstrasse 43
CH-8044 Zürich, Switzerland

Abstract

It's natural to think that the principles expressed by the statements "Promises ought to be kept" and "We ought to help those in need" are defeasible. But how are we to make sense of this defeasibility? On one proposal, moral principles have *hedges* or built-in unless clauses specifying the conditions under which the principle doesn't apply. On another, such principles are *contributory* and, thus, do not specify which actions ought to be carried out, but only what counts in favor or against them. Drawing on a defeasible logic framework, this paper sets up three models: one model for each proposal, as well as a third model capturing a mixed view on principles that combines them. It then explores the structural connections between the three models and establishes some equivalence results, suggesting that the seemingly different views captured by the models are closer than standardly thought.

Keywords: moral principles, defeasibility, reasons, defeasible logic

1 Introduction

Consider the following moral principles:

(Promise-keeping) If an agent has promised to X, then she ought to X.

(Beneficence) If an agent can help someone in need by X-ing, then she ought to X. [2]

It's natural to think that these principles—or something in their vicinity—are getting at important truths, and that they should have some role to play in our accounts of morality. However, anyone who accepts them—in fact, anyone who thinks that there are *some* principles like them—faces a challenge: They must explain what happens in cases where such principles come into conflict, such as:

Drowning Child. You have promised a friend to meet her for dinner. En route to the restaurant, you come across a child who has fallen in a shallow pond. The child is crying in distress, and all the evidence suggests that she is going

---

[1] aleks.knoks@uni.lu
[2] Both principles are mentioned in W. D. Ross' list of "basic duties"—see [25, pp. 21–2].

to drown, unless you help her. However, if you rescue the child, you will get your clothes wet and muddy, and won't make it to the dinner

Applying the two rules to this case leads to the conclusion that you ought to both have dinner with your friend and save the child. Taking this at face value means classifying the scenario as a tragic dilemma, or a situation where the agent can't do what she ought to no matter how she acts.[3] And this in spite of the strong intuition that the right thing for you to do is to save the child.

There seem to be two plausible things to say about cases involving conflicts between Promise-keeping and Beneficence, and other principles like them. Both imply that these principles are defeasible.[4] First, one could hold that moral principles are *contributory* or that they do not (by themselves) specify which actions ought to be carried out, but only what counts in favor or against them. Applying this view to the scenario, one could say that, even though there's a genuine conflict between the two principles, it's not a dilemma, because Beneficence outweighs or overrides Promise-keeping. So what you ought to do all-things-considered is save the child. Alternatively, one could hold that moral principles have implicit *hedges* or unless clauses that specify the circumstances under which they don't apply. Applying this view to the Drowning Child, one could say that the conflict between Beneficence and Promise-keeping is only apparent because, say, Promise-keeping doesn't apply when helping those in need means saving their lives. So, on this view too, what you ought to do all-things-considered is save the child.[5]

We will state these views on principles more precisely in later sections. For now simply note that they are naturally thought of and usually presented as distinct, even rival.[6] My aim in this paper is to contribute to a systematic theory of moral principles by exploring the relations between these two views and a mixed view combining them. I will devise a formal model of each, drawing on a simple defeasible logic framework, and establish some results, suggesting that the views modeled are closer than one may think.

The remainder of this paper is structured as follows. Section 2 sets up the stage by formally stating the problem that conflicts between principles give rise to. Sections 3–5 present the models of, respectively, the view on which rules are contributory, the view on which they are hedged, and the mixed view. Section 6 establishes the main results of this paper: The model of the view on which rules are contributory turns out to be equivalent to a fragment of the model on which rules are hedged, and the latter turns out to be equivalent to a restricted fragment of the model of the mixed view. Section 7 concludes and discusses

---

[3] This is how dilemmas are usually characterized—see, e.g., [5].

[4] I'm using the term *defeasible* loosely here, meaning that a principle can engender an ought in a situation and then fail to engender it in a slightly different situation.

[5] For views on which moral principles have hedges see [6], [27], see also [1], [3], and [31] for kindred views in epistemology. For a classical defense of contributory moral principles see the work of W. D. Ross [25]. Views that are naturally thought of as ones on which principles are contributory, but can also have hedges include [11], [15], and [32].

[6] See, for instance, [2, Sec. 1.2] in ethics and [1] in epistemology.

some directions for future research.

2 Preliminaries and the naive view

As our background, we assume the language of propositional logic with the standard connectives. The turnstile $\vdash$ will stand for classical logical consequence. To avoid unnecessary clutter when formalizing particular cases, we assume that our background language allows for materially inconsistent atomic formulas that can't jointly be true, representing such statements as "It's Friday" and "It's Monday." All the formulas we'll encounter should be thought of as relativized to an agent in a situation. Also, we make use of the customary deontic operator. A formula of the form $\bigcirc X$ should be read as saying that it ought to be the case that X, or that morality requires that X. Also, the *ought* here is all-things-considered and not *pro tanto*.

As a first stab, we represent moral principles as (vertically ordered) pairs of formulas of the form $\dfrac{X}{\bigcirc Y}$, where X and Y are formulas of propositional logic.[7] The first expresses a descriptive feature of the situation, the second a normative one, an ought. Now think back to the Drowning Child. Letting p and d stand for the propositions, respectively, that you've made a promise to your friend to dine with her, and that you dine with her, we could represent the relevant instance of Promise-keeping as $\dfrac{p}{\bigcirc d}$. Think of this pair of formulas by analogy with (indefeasible) inference rules of logical systems. The idea is that it lets you infer $\bigcirc d$ whenever p obtains. From now on, then, we'll often refer to our formal representations of principles as *rules*. We denote them with the Greek letter δ, with subscripts, and also introduce two functions $Premise[\cdot]$ and $Conclusion[\cdot]$ for selecting their elements: Where δ stands for $\dfrac{X}{Y}$, the expression $Premise[\delta]$ will stand for the proposition X and $Conclusion[\delta]$ for Y. Also, where $\mathcal{D}$ is a set of rules, we let $Conclusion[\mathcal{D}]$ stand for $\{Conclusion[\delta] : \delta \in \mathcal{D}\}$.

We represent particular cases with the help of the notion of a context.

Definition 2.1 [Contexts] A *context* c is a structure of the form $\langle \mathcal{W}, \mathcal{D} \rangle$, where $\mathcal{W}$ is a set of propositional formulas—capturing the normatively-relevant descriptive features of the scenario and called the *hard information*—and $\mathcal{D}$ is a set of rules of the form $\dfrac{X}{\bigcirc Y}$.

To see the notion in play, let's use it to formalize our running example: Let p and d be as before, and let c and s stand for the propositions that a child is drowning, and that you save the child. (We assume that d and s are materially inconsistent.) The scenario can then be captured in the context $c_1 = \langle \mathcal{W}, \mathcal{D} \rangle$ where $\mathcal{W}$ is the set $\{c, p\}$ and $\mathcal{D}$ contains the familiar rule $\delta_1 = \dfrac{p}{\bigcirc d}$, as well as the relevant instance of Beneficence, namely, the rule $\delta_2 = \dfrac{c}{\bigcirc s}$, which says

---

[7] Principles are often formalized as pairs of formulas—see, e.g., [7,8,13,14].

that you ought to save the child in case she is drowning.

Why does it seem like Promise-keeping and Beneficence reveal important truths about morality? Well, one possible answer is that their instances—together with instances of all other principles—are what link the descriptive features of situations to the normative ones, or, roughly, what happens to what ought to happen.[8] It seems natural to explicate this intuitive idea in the present framework as follows: There's a context standing for every situation, and the (infinite) set of all contexts shares a common set of rules $\mathcal{D}$, containing every instance of Promise-keeping, Beneficence, and other schemas capturing moral principles. Now, one might hope that the logic governing the interaction between these principles is just the good old classical logic.[9] We can capture this view—the naive view alluded to in this section's title—in our framework in two steps. The first is to introduce the notion of triggered rules:

Definition 2.2 [Triggered rules] Let $c = \langle \mathcal{W}, \mathcal{D} \rangle$ be a context. The rules from $\mathcal{D}$ that are *triggered* in c are those that belong to the set $Triggered(c) = \{\delta \in \mathcal{D} : \mathcal{W} \vdash Premise[\delta]\}$.

And the second is to specify which ought formulas follow from a context:

Definition 2.3 [Consequence, first pass] Let $c = \langle \mathcal{W}, \mathcal{D} \rangle$ be a context. Then $\bigcirc X$ follows from c just in case $\bigcirc X$ follow from $Conclusion[Triggered(c)]$ by standard deontic logic.[10]

Now let's apply these definitions to the context capturing the Drowning Child. It's easy to see that both δ_1 and δ_2 are triggered in c_1, and that both $\bigcirc d$ and $\bigcirc s$ follow from it. And in light of the fact that d and s are materially inconsistent, a formula of the form $\bigcirc X$ follows from c_1 for any X whatsoever. This, of course, is no good. So we have to abandon the naive view and change either the way we think about moral principles, or the logic governing their interaction, or both.

3 Contributory principles

According to one prominent view, there is no real problem here because the moral principles in play in the Drowning Child and other scenarios like it are contributory: They do not—not by themselves anyway—specify which actions ought to be carried out, but only what speaks in favor or against carrying them out.[11] This section sets up a simple model of this view, drawing on the work of Horty [7,8].[12]

We represent contributory principles as default rules of the form $\frac{X}{Y}$. Intuitively, a rule of this form can be thought of as saying that X exerts some sort

---

[8] This idea is widely shared among ethicists.
[9] Again, the idea that moral principles (whatever their shape) are governed by classical logic is widespread among ethicists—see, e.g., the remarks in [6].
[10] For a nice presentation of standard deontic logic, see [16, Sec. 2].
[11] See footnote 5 for references.
[12] The model presented here is a fragment of the model set up in [8].

of normative pressure that Y obtains. Functionally, it will let us infer Y from X by default.

Contributory rules are usually associated with *relative weights*, and it's standard to represent these weights formally by means of a priority relation. So where δ and δ' are (contributory) rules, a statement of the form $\delta \leq \delta'$ will mean that δ' has at least as much weight as δ, or that δ' is at least as strong as δ. Following standard practice, we assume that the relation $\leq$ is reflexive and transitive, as well as write $\delta < \delta'$ when $\delta \leq \delta'$ and not $\delta' \leq \delta$.

The next natural step would be to adapt the notion of a context to the idea that rules are contributory. Before taking it, however, we need to introduce the notion of *contrary rules*. Our definition will draw on the concept of minimal inconsistency:

Definition 3.1 [Minimally inconsistent subsets of contrary rules] Let $\mathcal{D}$ be a set of contributory rules and $\mathcal{D}' \subseteq \mathcal{D}$. Then $\mathcal{D}'$ is a minimally inconsistent subset of $\mathcal{D}$ just in case $Conclusion[\mathcal{D}'] \vdash \bot$ and there's no $\mathcal{D}'' \subset \mathcal{D}'$ with $Conclusion[\mathcal{D}''] \vdash \bot$.

Definition 3.2 [Contrary contributory rules] Let $\mathcal{D}$ be a set of contributory rules and δ, δ' two rules from $\mathcal{D}$. Then δ and δ' are *contrary* against the background of $\mathcal{D}$, written as $contrary_\mathcal{D}(\delta, \delta')$, if and only if there's a minimally inconsistent subset $\mathcal{D}'$ of $\mathcal{D}$ with $\delta, \delta' \in \mathcal{D}'$.

Notice that Definitions 3.1 and 3.2 capture and generalize the intuitive idea that *two* rules are contrary when their conclusions are inconsistent. The recourse to minimally inconsistent subsets is needed to account for cases where there's a set of rules the conclusions of which are pairwise consistent, but jointly inconsistent. As an example, consider the rules $\frac{\top}{a}$, $\frac{\top}{b}$, and $\frac{\top}{\neg(a\&b)}$. Were we to say that two rules are contrary just in case their conclusions are inconsistent, the inconsistency of these three rules would slip through. And we don't want that to happen.

Now we can adjust our notion of a context, and we call contexts that represent moral principles as contributory rules *weighted*:

Definition 3.3 [Weighted contexts] A weighted context c is a structure of the form $\langle \mathcal{W}, \mathcal{D}, \leq \rangle$, where $\mathcal{W}$ is a set of propositional formulas, $\mathcal{D}$ is a set of contributory rules, and $\leq$ is a reflexive and transitive relation (a preorder) on $\mathcal{D}$. We assume that weighted contexts are subject to the following constraint:

No Dilemmas: For any $\delta, \delta' \in \mathcal{D}$ with $contrary_\mathcal{D}(\delta, \delta')$, either $\delta \leq \delta'$, or $\delta' \leq \delta$.

The Drowning Child can, then, be captured by the weighted context $c_2 = \langle \mathcal{W}, \mathcal{D}, \leq \rangle$ where $\mathcal{W}$ is the set $\{p, c\}$, where $\mathcal{D}$ contains the rules $\delta_3 = \frac{p}{d}$ and $\delta_4 = \frac{c}{s}$, and where $\delta_3 < \delta_4$. Intuitively, δ_3 says that your having made a promise to dine with your friend speaks in favor of you dining with her, while δ_4 says that the child's drowning (and needing your help) speaks in favor of you saving the child. The relation between the rules, $\delta_3 < \delta_4$, expresses the idea that the latter has strictly more weight, or is strictly stronger, than the former. Notice that this doesn't mean that Beneficence always has more weight than Promise-keeping—there will be many other contexts in which instances of the latter take precedence over the instances of the former. The No Dilemmas constraint captures the following assumption: Moving to contributory principles in response to the problem that conflicts between moral principles give rise to suffices to show that such conflicts aren't tragic dilemmas. The assumption seems to me to be fully justified in the context of this paper.

Now we specify which ought statements follow from a context, relying on three simple definitions.

Definition 3.4 [Outweighed rules] Let $c = \langle \mathcal{W}, \mathcal{D}, \leq \rangle$ be a weighted context. The rules from $\mathcal{D}$ that are *outweighed* in c are those that belong to the set

$$Outweighed(c) = \{\delta \in \mathcal{D} : \text{ there is some } \delta' \in Triggered(c) \text{ such that} \\ (1) \; \delta \leq \delta' \text{ and } (2) \; contrary_\mathcal{D}(\delta, \delta')\}.$$

So a rule is outweighed in a context just in case there's another rule that's triggered, contrary to it, and at least as strong. Notice that this formal notion doesn't match the intuitive sense of *outweighed* perfectly, since it qualifies a rule δ as outweighed if there's another rule that's strictly stronger than δ, as well as if there's another rule that's only as strong as δ. In the latter case, it'd be more fitting to say that δ is counterbalanced. If there was a word in English covering the senses of both *outweighed* and *counterbalanced*, it'd be perfect for our purposes. But given that there isn't one, we work with what we have.

Our next definition combines the notions of triggered and outweighed rules:

Definition 3.5 [Binding rules] Let $c = \langle \mathcal{W}, \mathcal{D}, \leq \rangle$ be a weighted context. The rules from $\mathcal{D}$ that are *binding* in c are those that belong to the set

$$Binding(c) = \{\delta \in \mathcal{D} : \; \delta \in Triggered(c) \text{ and} \\ \delta \notin Outweighed(c)\}.$$

So a rule is binding just in case it is triggered and *not* outweighed. Such binding rules are just what will give us the ought statements:

Definition 3.6 [Consequence, weighted] Let $c = \langle \mathcal{W}, \mathcal{D}, \leq \rangle$ be a weighted context. Then $\bigcirc X$ follows from c just in case $Conclusion[Binding(c)] \vdash X$.

Returning to the Drowning Child scenario, it's easy to see that, on Definition 3.6, $\bigcirc s$ follows from c_2. Both δ_3 and δ_4 get triggered, but only the latter qualifies as binding. Given that δ_4 is strictly stronger than δ_3, $\delta_3 < \delta_4$, and that the two are contrary, $contrary_\mathcal{D}(\delta_3, \delta_4)$, the rule δ_3 comes out outweighed in c_2. Since $Binding(c_2) = \{\delta_4\}$, we have it that $Conclusion[Binding(c_2)] = \{s\}$ and, therefore, that $\bigcirc s$ follows from c_2. Thus, we get the intuitive result that you ought to save the child.

It's worth noting that our model of the view on which moral principles are contributory gives rise to consistent oughts:[13]

---

[13] Many thanks to an anonymous reviewer for pressing me to prove this fact. Among other

Fact 3.7 Let $c = \langle \mathcal{W}, \mathcal{D}, \leq \rangle$ be a weighted context. Then $\bigcirc \bot$ follows from c only if there's a rule δ in $\mathcal{D}$ with $Conclusion[\delta] = \bot$.

Proof. Suppose that $\bigcirc \bot$ follows from the context c. This means that $Conclusion[Binding(c)] \vdash \bot$. Now let's zoom in on some minimally inconsistent subset $\mathcal{D}'$ of $Binding(c)$. (It is guaranteed to exist.) If $\mathcal{D}'$ is a singleton, we're done. So suppose that it isn't. Since $\mathcal{D}'$ is a minimally inconsistent subset of $Binding(c)$ and $Binding(c) \subseteq \mathcal{D}$, Definition 3.2 entails that, for any $\delta, \delta' \in \mathcal{D}'$, we have $contrary_{\mathcal{D}}(\delta, \delta')$. Now zoom in on two such rules δ, δ'. Since $\delta, \delta' \in Binding(c)$, we have $\delta, \delta' \in Triggered(c)$ and $\delta, \delta' \notin Outweighed(c)$. In light of the No Dilemmas constraint, we can be sure that either $\delta \leq \delta'$, or $\delta' \leq \delta$. Without loss of generality, we assume that $\delta \leq \delta'$. Now notice that we have $\delta' \in Triggered(c)$, $\delta \leq \delta'$, and $contrary_{\mathcal{D}}(\delta, \delta')$. This entails $\delta \in Outweighed(c)$, giving us a contradiction. □

Even though the model we just set up is very simple, it's expressive enough to capture some of the *reasons talk* that's so pervasive in contemporary ethical and meta-normative debates.[14] More specifically, the notion of a normative reason can be specified as follows: X is a normative reason for Y in the weighted context c if and only if there's a rule δ of the form $\frac{X}{Y}$ that's triggered in c. When there is such a rule, we say that X's being a reason for Y depends on it.[15] We say that X is outweighed as a reason for Y in c just in case the rule δ that X's being a reason for Y is triggered, but not binding in c. And we say that X is outweighed—or, more precisely, outweighed or counterbalanced—as a reason for Y *by the consideration* Z in c just in case the rule δ that X's being a reason for Y depends on is outweighed by some contrary rule δ' that has Z as a premise.

4 Hedged principles

On an alternative view, situations where moral principles seem to support conflicting recommendations are cases of only apparent conflicts, because principles have built-in hedges guaranteeing that at most one principle applies in any such situation.[16] This section sets up a model of this view.

In Section 2, we expressed principles as rules of the form $\frac{X}{\bigcirc Y}$. Now we do so using the slightly more complex $\frac{X : \{\neg Z_1, \neg Z_2, \ldots\}}{\bigcirc Y}$. The new element $\{\neg Z_1, \neg Z_2, \ldots\}$ is a set of negated propositional formulas standing for the rule's hedge—note that we will often abbreviate it as $\mathcal{Z}$. A hedged principle of this form should be read as, "If X obtains, then it ought to be the case that Y, unless either Z_1 obtains, or Z_2 obtains, or ...." Alternatively, it can be read as, "If X

---

things, this saved me from an embarrassing mistake.
[14] See, e.g., [20,23,26,28,29,30], and [7,8,18] for modeling reasons talk using defeasible logics.
[15] Compare to [8, Section 2.1].
[16] See footnote 5 for references.

obtains, and not-Z_1, not-Z_2, ..., then it ought to be the case that Y."[17] We retain the functions for selecting rule premises and conclusions. Additionally, we introduce a function for selecting a given rule's hedge: If δ is a rule of the form $\frac{X : \mathcal{Z}}{Y}$, let $Hedge[\delta] = \mathcal{Z}$, and if δ is of the form $\frac{X}{Y}$, let $Hedge[\delta] = \emptyset$.

Like we did in the previous section, here too we make use of the idea of contrary rules:[18]

Definition 4.1 [Minimally inconsistent subsets of hedged rules] Let $\mathcal{D}$ be a set of hedged rules and $\mathcal{D}' \subseteq \mathcal{D}$. Then $\mathcal{D}'$ is a *minimally inconsistent subset* of $\mathcal{D}$ just in case $\bigcirc \bot$ follows from $Conclusion[\mathcal{D}']$ in standard deontic logic and there's no $\mathcal{D}'' \subset \mathcal{D}'$ such that $\bigcirc \bot$ follows from $Conclusion[\mathcal{D}'']$ in standard deontic logic.

Definition 4.2 [Contrary hedged rules] Let $\mathcal{D}$ be a set of hedged rules and δ, δ' two rules from $\mathcal{D}$. Then δ and δ' are *contrary* against the background of $\mathcal{D}$, written as $contrary_\mathcal{D}(\delta, \delta')$, if and only if there's a minimally inconsistent subset $\mathcal{D}'$ of $\mathcal{D}$ with $\delta, \delta' \in \mathcal{D}'$.

With the notion of contrary rules in hand, we can adjust the definition of a context from Section 2 to the idea that rules expressing principles can have hedges:

Definition 4.3 [Hedged contexts] A *hedged context* is a structure of the form $\langle \mathcal{W}, \mathcal{D} \rangle$, where $\mathcal{W}$ is a set of propositional formulas and $\mathcal{D}$ is a set of rules, possibly hedged. We assume throughout that hedged contexts are subject to two constraints:

No Dilemmas: For any $\delta, \delta' \in \mathcal{D}$ with $contrary_\mathcal{D}(\delta, \delta')$, either $\neg Premise[\delta] \in Hedge[\delta']$, or $\neg Premise[\delta'] \in Hedge[\delta]$.

No Deviant Pairs of Rules: For any $\delta, \delta' \in \mathcal{D}$, in case $Premise[\delta] = Premise[\delta']$ and $Conclusion[\delta] = Conclusion[\delta']$, then $\delta = \delta'$.

As an illustration, the Drowning Child can be represented as the hedged context $c_3 = \langle \mathcal{W}, \mathcal{D} \rangle$ where $\mathcal{W} = \{p, c\}$ and where $\mathcal{D}$ is comprised of the rules $\delta_5 = \frac{p : \neg c}{\bigcirc d}$ and $\delta_6 = \frac{c}{\bigcirc s}$. The first rule says that you ought to dine with your friend if you've promised to dine with her, unless a child needs help; the second says that you ought to save the child, if the child needs help.[19] Now for the two constraints: No Dilemmas amounts to, again, the assumption that

---

[17] Compare to Reiter's default rules [24].
[18] You may wonder if I shouldn't use indices to keep track of the difference between Definitions 3.2 and 4.2. My reason for not using indices here and elsewhere is to avoid notational clutter. I think that the context always makes it clear whether we're discussing the view on which principles are hedged or the one on which they are contributory, helping disambiguate between the two notions of contrary rules. Parallel considerations apply to the notions of outweighed rules, Definitions 3.4 and 5.4, and undercut rules, Definitions 5.2 and 5.3.
[19] It's natural to wonder if the hedge of the rule δ_5 shouldn't also list the other circumstances in which this rule wouldn't apply—and similarly for δ_6. While I do think that this is a natural consequence of the view, nothing hinges on us working with simplified examples here. For more on the worry that hedged rules might end up being incredibly complex see, e.g., [1].

appealing to hedges succeeds as a response to the problem of conflicting moral principles. No Deviant Pairs of Rules, in turn, rules out pairs of principles that apply in the same circumstances and prescribe the same course of action, but have different hedges: It seems natural to think that any pair of such principles is deviant and should be substituted by one principle with a single hedge.

Our next two definitions determine which ought statements follow from hedged contexts.

Definition 4.4 [Admissible rules] Let $c = \langle \mathcal{W}, \mathcal{D} \rangle$ be a hedged context. The rules from $\mathcal{D}$ that are *admissible* in c are those that belong to the set

$$Admissible(c) = \{\delta \in \mathcal{D} : \delta \in Triggered(c) \text{ and, for no} \\ \neg Z \in Hedge[\delta], \mathcal{W} \vdash Z\}.$$

Definition 4.5 [Consequence, hedged] Let $c = \langle \mathcal{W}, \mathcal{D} \rangle$ be a hedged context. Then $\bigcirc X$ follows from c just in case $\bigcirc X$ follows from $Conclusion[Admissible(c)]$ by standard deontic logic.

Notice that δ_6 does, while δ_5 does not qualify as admissible in c_3. The latter rule gets triggered, but the fact that $\neg c \in Hedge[\delta_5]$ and $\mathcal{W} \vdash c$ precludes it from being added to $Admissible(c_3)$. And given that $Conclusion[Admissible(c_3)] = \{\bigcirc s\}$, we get the intuitive result that $\bigcirc s$ does, while $\bigcirc d$ does not follow from c_3: What you ought to do is save the child.

Let's also note that this model too gives rise to consistent oughts:

Fact 4.6 *Let $c = \langle \mathcal{W}, \mathcal{D} \rangle$ be a hedged context. Then $\bigcirc \bot$ follows from c only if there's a rule δ in $\mathcal{D}$ with $Conclusion[\delta] = \bigcirc \bot$.*

Proof. Suppose that $\bigcirc \bot$ follows from the hedged context c. This means that $\bigcirc \bot$ follow from $Conclusion[Admissible(c)]$ in standard deontic logic. Now consider some minimally inconsistent subset $\mathcal{D}'$ of $Admissible(c)$. In case $\mathcal{D}'$ is a singleton, we have a $\delta \in \mathcal{D}' \subseteq Admissible(c) \subseteq \mathcal{D}$ with $Conclusion[\delta] = \bigcirc \bot$, which would establish the fact. So suppose, toward a contradiction, that $\mathcal{D}'$ is not a singleton set. Given that $\mathcal{D}'$ is a minimally inconsistent subset of $Admissible(c)$ and $Admissible(c) \subseteq \mathcal{D}$, Definition 4.2 tells us that, for any $\delta, \delta' \in \mathcal{D}'$, we have $contrary_\mathcal{D}(\delta, \delta')$. Now consider two such rules δ, δ'. Since $\delta, \delta' \in Admissible(c)$, we know that $\delta, \delta' \in Triggered(c)$, that there's no $\neg Z \in Hedge[\delta]$ with $\mathcal{W} \vdash Z$, and that there's no $\neg Z \in Hedge[\delta']$ with $\mathcal{W} \vdash Z$. Given the No Dilemmas constraint, we can be sure that either $\neg Premise[\delta] \in Hedge[\delta']$ or $\neg Premise[\delta'] \in Hedge[\delta]$ holds true. Without loss of generality, we suppose it is the former. Since $\delta \in Triggered(c)$, we have it that $\mathcal{W} \vdash Premise[\delta]$. Then, however, we have $\neg Premise[\delta] \in Hedge[\delta']$ and $\mathcal{W} \vdash Premise[\delta]$, which contradicts the claim that there's no $\neg Z \in Hedge[\delta']$ with $\mathcal{W} \vdash Z$. □

Unlike the view on which moral principles are contributory, the view on which they are hedged is usually taken to be less congenial to reasons talk. [20]

---

[20] See, e.g., Dancy's [2, pp. 22–9] where it's argued that the view has no hope to account for the phenomena of residual reasons and reason aggregation. If the results established below

However, nothing stands in the way of using the model we just set up to talk and reason about reasons. Thus, we say that X is a normative reason for Y in the hedged context c if and only if there's a rule δ of the form $\dfrac{X\,:\,\mathcal{Z}}{\bigcirc Y}$ that's triggered in c. When there is such a rule δ, we say that X's being a reason for Y depend on δ. We say that X is *defeated* as a reason for Y in c just in case the rule δ that X's being a reason for Y in c depends on is triggered, but not admissible in c. And we say that X is defeated as a reason for Y *by the consideration* Z in $c = \langle \mathcal{W}, \mathcal{D} \rangle$, or, alternatively, that Z is a defeater of X as a reason for Y in c, if and only if the rule δ that X's being a reason for Y depends on is triggered in c, has $\neg Z$ in its hedge, and $\mathcal{W} \vdash Z$. So our some reason talk can be captured in our model of the hedged-principles view too.

Now we have two models of two seemingly very different views on moral principles and conflicts between them. But it turns out to be possible to establish a type of equivalence result between these models. More concretely, there's a many-one equivalence between weighted contexts and a special class of hedged contexts I call *simple*. The special character of these simple contexts has to do with the shape of the hedges of their rules:

Definition 4.7 [Simple hedged contexts] Let $c = \langle \mathcal{W}, \mathcal{D} \rangle$ be a hedged context. We say that c is *simple* just in case, for any rule δ in $\mathcal{D}$, the hedge of δ is the set $\{\neg Premise[\delta'] : \delta' \in \mathcal{D}'\}$ where $\mathcal{D}' \subseteq \{\delta' \in \mathcal{D} : contrary_\mathcal{D}(\delta, \delta')\}$.

On the intuitive level, this simplicity condition can be thought of as a restriction on what rule hedges can do: They can refer *only* to the premises of contrary rules, and, thus, all they can do is resolve conflicts between rules supporting conflicting recommendations.

And I call a weighted context c and a hedged one c' *equivalent* if and only if X is a reason for Y in c just in case X is a reason for Y in c'; X is outweighed as a reason for Y by Z in c just in case X is defeated as a reason for Y by Z in c'; and $\bigcirc X$ follows from c just in case $\bigcirc X$ follows from c'. So the connection between weighted and simple hedged contexts is very close. We'll state the result more precisely in Section 6, along with some other equivalences. Now we turn to a combination of the two views on principles we have discussed.

5 Mixed principles

There's a well-known objection to the views on which principles are contributory: They entail that the moral valences of features are fixed, or, roughly, that if a feature constitutes a reason for some type of action in one situation, then it's *always* a reason for that type of action. This, however, appears to be too strong. Suppose that you've made a promise to meet someone for dinner, but they extracted this promise from you by threat of force. A proponent of the contributory view is forced to think of this case as one where Promise-keeping yields to a weightier principle saying something along the lines of, "If a promise to X was extracted by threat of force, then that speaks against X-ing." And

---

are correct, Dancy's arguments can't work—see [9, Ch. 2] for more on this.

this implies that there's some reason for you to keep the promise. But, intuitively, the moral force that normally comes with promise-making is invalidated or voided in this situation, and there's absolutely no reason for you to keep the promise.[21]

Some philosophers take this objection to be fatal to the contributory view, and it does seem that this view can't handle cases involving what we could call *principle-invalidating conditions* adequately. However, there's a view that's naturally thought of as its extension that can, namely, the *mixed view* on which principles are contributory, but can also have hedges.[22] In the remainder of this section, we set up a model of this view.

We use rules of the form $\dfrac{X : \{\neg Z_1, \ldots, \neg Z_n\}}{Y}$ to express mixed principles. A rule having this form should be read as, "If X obtains, then there's normative pressure that Y obtains, unless either Z_1, or Z_2, or, ..., Z_n obtain." Adapting the by now familiar notion of contrary rules to the mixed setting is a trivial exercise: All we need to do is substitute *mixed rules* for *contributory rules* in Definitions 3.1 and 3.2 from Section 3. We won't do this explicitly, proceeding directly to the next step, that is, to defining mixed contexts:

Definition 5.1 [Mixed contexts] A *mixed context* c is a structure of the form $\langle \mathcal{W}, \mathcal{D}, \leq \rangle$, where $\mathcal{W}$ is a set of propositional formulas, $\mathcal{D}$ is a set of mixed rules, and $\leq$ is a preorder on $\mathcal{D}$. Yet again, we assume that mixed contexts are subject to two familiar constraints:

No Dilemmas: For any $\delta, \delta' \in \mathcal{D}$ with $contrary_{\mathcal{D}}(\delta, \delta')$, either $\delta \leq \delta'$, or $\delta' \leq \delta$.

No Deviant Pairs of Rules: For any $\delta, \delta' \in \mathcal{D}$, in case $Premise[\delta] = Premise[\delta']$ and $Conclusion[\delta] = Conclusion[\delta']$, then $\delta = \delta'$.

To see the notion at work, we formalize the case sketched above. Let p and d stand for the propositions that they stood for before, namely, that you've made a promise to dine with your friend, and that you dine with her. And let t express the proposition that the promise was extracted from you by threat of force. We can then express the case as the context $c_4 = \langle \mathcal{W}, \mathcal{D}, \leq \rangle$, where $\mathcal{W} = \{p, t\}$, where $\mathcal{D}$ is a singleton set containing the rule $\delta_7 = \dfrac{p : \neg t}{d}$, and where $\leq$ is empty.

Next, we introduce a new notion that operates on the information stored in the hedges of mixed rules:[23]

Definition 5.2 [Undercut rules] Let $c = \langle \mathcal{W}, \mathcal{D}, \leq \rangle$ be a mixed context. The rules from $\mathcal{D}$ that are *undercut* in it are those that belong to the set

---

[21] Particularists are particularly fond of raising this objection, in both moral and epistemic domains—see, e.g., [1,2,10]. Also, the ideas that duress and deceit invalidates promises, and that a promise's getting invalidated is very different from it being outweighed is widespread among moral philosophers—see, e.g., [4,19].

[22] Again, see footnote 5 for references.

[23] The term *undercut* comes from epistemology where it's customary to distinguish between rebutting and undercutting defeat—see, e.g., [21] and [22].

$$Undercut(c) \;=\; \{\delta \in \mathcal{D}: \quad \text{there's a } \neg Z \in Hedge[\delta] \text{ such that } \mathcal{W} \vdash Z\}.$$

This definition might remind you of the way we got to consequences of hedged contexts. However, the notion of undercutting is meant to serve a more specific function here, namely, to get certain rules completely out of play in determining which ought statements follow from a context. What's more, it can be defined in our model of the hedged-principles view too, as follows:

Definition 5.3 [Undercut rules, hedged] Let $c = \langle \mathcal{W}, \mathcal{D} \rangle$ be a hedged context. The rules from $\mathcal{D}$ that are *undercut* in it are those that belong to the set

$$Undercut(c) \;=\; \{\delta \in \mathcal{D}: \quad \text{there's some } \neg Z \in Hedge[\delta] \text{ with } \mathcal{W} \vdash Z \text{ and} \\ \text{there is no } \delta' \in \mathcal{D} \text{ such that} \\ (1)\ Z = Premise[\delta'] \text{ and } (2)\ contrary_\mathcal{D}(\delta, \delta')\}.$$

Notice that, on this definition, a hedge rule can be undercut only by a consideration that is not a premise of some rule that's contrary to it.

The addition of undercut rules means that we need to modify the notion of outweighed rules slightly. More specifically, we need to make sure that the rule that's responsible for outweighing is not itself undercut:

Definition 5.4 [Outweighed rules, mixed] Let $c = \langle \mathcal{W}, \mathcal{D}, \leq \rangle$ be a mixed context. The rules from $\mathcal{D}$ that are *outweighed* in it belong to the set

$$Outweighed(c) \;=\; \{\delta \in \mathcal{D}: \quad \text{there is some } \delta' \in Triggered(c) \text{ such that} \\ (1)\ \delta \leq \delta',\ (2)\ contrary_\mathcal{D}(\delta, \delta'),\ \text{and} \\ (3)\ \delta' \notin Undercut(c)\}.$$

With this, the model is pretty much set up. We only need to combine the above definitions in the analogue of admissible/binding rules for mixed context and define consequence.

Definition 5.5 [Optimal rules] Let $c = \langle \mathcal{W}, \mathcal{D}, \leq \rangle$ be a mixed context. The rules from $\mathcal{D}$ that are *optimal* in it are those that belong to the set

$$Optimal(c) \;=\; \{\delta \in \mathcal{D}: \quad \delta \in Triggered(c), \\ \delta \notin Outweighed(c), \\ \delta \notin Undercut(c)\}.$$

Definition 5.6 [Consequence, mixed] Let $c = \langle \mathcal{W}, \mathcal{D}, \leq \rangle$ be a mixed context. Then $\bigcirc X$ follows from c just in case $Conclusion[Optimal(c)] \vdash X$.

Returning to the context c_4, it's easy to see that the statement $\bigcirc d$ doesn't follow from it. This is as it should be. What's more, we also get the intuitive result that there's no reason for you to meet with the person for dinner—that is, as soon as we specify how reasons are to be identified in the model. Here we reuse the idea from Section 3, with a small twist to it: X is a reason for Y in c if and only if there's a rule of the form $\dfrac{X:Z}{Y}$ that's triggered in c *and* not undercut. And X is outweighed as a reason for Y by the consideration Z in c just in case the rule δ that X's being a reason for Y depends on is outweighed by a contrary rule δ' that has $\mathcal{Z}$ as a premise. With this, p, that you've made a promise to meet your friend for dinner, doesn't qualify as a reason for d, to have dinner with her, in c_4. So the mixed view can account for the case adequately.

We close the section by noting that the oughts that the model of the mixed view gives rise to are also consistent:

Fact 5.7 *Let $c = \langle \mathcal{W}, \mathcal{D}, \leq \rangle$ be a mixed context. Then $\bigcirc \perp$ follows from c only if there's a rule δ in $\mathcal{D}$ with $Conclusion[\delta] = \perp$.*

Proof. Similar to the proof of Fact 3.7. □

6 Relations

Having laid out the models of the three views on principles, we can explore the relations between them. This is the goal of this section.

First off, it should be clear that our model of the view on which principles are contributory corresponds to a fragment of the model of the mixed view: It's easy to see that there's a one-one correspondence between weighted contexts and a special class of mixed contexts, namely, those the hedges of all rules of which are empty.

More surprisingly, it's possible to establish a many-one correspondence between a special class of mixed contexts and hedged contexts. It's natural to think of this correspondence as the main result of this paper.[24] As a first step, we introduce an auxiliary notion that will help us avoid clutter in proofs:

Definition 6.1 [Rule counterparts] Let δ be of the form $\dfrac{X : \mathcal{Z}}{\bigcirc Y}$ and $c = \langle \mathcal{W}, \mathcal{D}, \leq \rangle$ some mixed context. If there's a rule $\delta' \in \mathcal{D}$ with $Premise[\delta'] = X$ and $Conclusion[\delta'] = Y$, we say that δ' is the (mixed) *counterpart* of δ in the context c, written as $counterpart_c(\delta) = \delta'$. Similarly, in case δ is of the form $\dfrac{X : \mathcal{Z}}{Y}$ and the set of rules of some hedged context $c = \langle \mathcal{W}, \mathcal{D} \rangle$ contains some rule δ' with $Premise[\delta'] = X$ and $Conclusion[\delta'] = \bigcirc Y$, we say that δ' is the (hedged) *counterpart* of δ in c, written as $counterpart_c(\delta) = \delta'$.

Now notice that there's a natural procedure for transforming mixed contexts into hedged ones:

Definition 6.2 [Derived hedged contexts] Let $c = \langle \mathcal{W}, \mathcal{D}, \leq \rangle$ be a mixed context. We construct a hedged context c' from it as follows. Let $c' = \langle \mathcal{W}', \mathcal{D}' \rangle$, where $\mathcal{W}' = \mathcal{W}$ and $\mathcal{D}'$ is acquired from $\langle \mathcal{D}, \leq \rangle$ by the following procedure: For every rule $\delta \in \mathcal{D}$,

A. Let $\mathcal{D}_\delta = \{\delta' \in \mathcal{D} : \delta \leq \delta'$ and $contrary_\mathcal{D}(\delta, \delta')\}$;

B. set $\mathcal{Z} = \{\neg Premise[\delta'] : \delta' \in \mathcal{D}_\delta\}$;

C. finally, replace $\delta \in \mathcal{D}$ with the rule $\dfrac{Premise[\delta] : Hedge[\delta] \cup \mathcal{Z}}{\bigcirc Conclusion[\delta]}$.

We call the class of mixed contexts that have hedged counterparts *regular*:

---

[24] I structure the presentation of the result after Lewis' [12, Sec. 4]. Thanks to Paolo Santorio for pointing me to Lewis' paper.

Definition 6.3 [Regular mixed contexts] Let $c = \langle \mathcal{W}, \mathcal{D}, \leq \rangle$ some mixed context. We say that c is *regular* if and only if, for any $\delta, \delta' \in \mathcal{D}$ with $contrary_\mathcal{D}(\delta, \delta')$, neither $\neg Premise[\delta'] \in Hedge[\delta]$, nor $\neg Premise[\delta] \in Hedge[\delta']$.

What does this regularity condition do? Well, first, notice that the No Dilemmas constraint on mixed contexts guarantees that any two contrary rules are related by $\leq$, and, thus, that all conflicts between them get resolved. The regularity condition, then, ensures that rule hedges do not interact with this in any way. So it can be thought of as enforcing continuity between the view on which principles are contributory and the mixed one: What accounts for the resolution of conflicts between conflicting principles is their contributory character, not their hedges.

Either way, if a hedged context is acquired from a regular mixed context by Definition 6.2, they are equivalent:

Theorem 6.4 (Mixed-to-Hedged) *Let $c = \langle \mathcal{W}, \mathcal{D}, \leq \rangle$ be some regular mixed context and $c' = \langle \mathcal{W}', \mathcal{D}' \rangle$ a (hedged) context derived from c by the procedure specified in Definition 6.2. Then c and c' are equivalent, that is:*

1. X is a reason for Y in c if and only if X is a reason for Y in c';

2. X is outweighed as a reason for Y by a consideration Z in c if and only if X is defeated as a reason for Y by Z in c'; and

3. $\bigcirc X$ follows from c if and only if $\bigcirc X$ follows from c'.

Proof. As a first step, we establish the following claim: $\delta \in Undercut(c)$ if and only if $counterpart_{c'}(\delta) \in Undercut(c')$. $\Rightarrow$ Consider an arbitrary $\delta \in Undercut(c)$. There must be some Z with $\mathcal{W} \vdash Z$ and $\neg Z \in Hedge[\delta]$. Since c is regular, there's no $\delta^* \in \mathcal{D}$ with both $contrary_\mathcal{D}(\delta, \delta^*)$ and $Z = Premise[\delta^*]$. Now consider $counterpart_{c'}(\delta) = \delta'$. In light of Definition 6.2, there's no $\delta'' \in \mathcal{D}'$ with both $contrary_{\mathcal{D}'}(\delta', \delta'')$ and $\neg Premise[\delta'']$ in $Hedge[\delta']$. Yet $\neg Z \in Hedge[\delta']$ and $\mathcal{W}' \vdash Z$ (since $\mathcal{W}' = \mathcal{W}$). So $\delta' \in Undercut(c')$. $\Leftarrow$ Consider an arbitrary $\delta \in Undercut(c')$. This means that there's some $\neg Z \in Hedge[\delta]$ such that $\mathcal{W}' \vdash Z$ and there's no $\delta^* \in \mathcal{D}'$ with both $Z = Premise[\delta^*]$ and $contrary_{\mathcal{D}'}(\delta, \delta^*)$. Now refocus on $counterpart_c(\delta) = \delta'$. Either $\neg Z \in Hedge[\delta']$ or not. If yes, we're done. So suppose not. Then, in light of Definition 6.2, there has to be some $\delta'' \in \mathcal{D}$ such that $Z = Premise[\delta'']$, $contrary_\mathcal{D}(\delta', \delta'')$, and $\delta' \leq \delta''$. Then, however—again, by Definition 6.2—there must be a $\delta^* \in \mathcal{D}'$ with $Z = Premise[\delta^*]$ and $contrary_{\mathcal{D}'}(\delta, \delta^*)$, which gives us a contradiction.

1. The first clause follows directly from the above claim and the definitions of reasons in the two models.

2. Without loss of generality, we prove only the the left-to-right direction of the second clause: Suppose X is outweighed as a reason for Y by Z in c. This entails that there are rules $\delta = \dfrac{X : Z}{Y}$ and $\delta^* = \dfrac{Z : Z'}{W}$ in $\mathcal{D}$ such that $contrary_\mathcal{D}(\delta, \delta^*)$, $\delta \leq \delta^*$, $\delta, \delta^* \in Triggered(c)$, and $\delta, \delta^* \notin Undercut(c)$. By the construction of c', there are rules $\delta', \delta'' \in \mathcal{D}'$ such that $counterpart_c(\delta') = \delta$, $counterpart_c(\delta'') = \delta^*$, and $\neg Premise[\delta''] \in Hedge[\delta']$. Since $Premise[\delta''] = Z$ and $\mathcal{W}' \vdash Z$, the rule δ' gets defeated by Z in c'. Since $\delta \notin Undercut(c)$, we

can be sure that $\delta' \notin Undercut(c')$. Thus, X is defeated as a reason for Y by Z in c'.

3. Instead of proving the third clause directly, we prove a claim from which it quickly follows, namely, $\{\bigcirc X : X \in Conclusion[Optimal(c)]\} = Conclusion[Admissible(c')]$.

$\subseteq$ Consider some formula $\bigcirc X$ in the set on the left hand side. Clearly, $X \in Conclusion[Optimal(c)]$, and so there's a rule δ in $Optimal(c)$ such that $Conclusion[\delta] = X$. Definitions 6.2 and 5.5 (Optimal rules) entail that there's a δ' in $\mathcal{D}'$ such that $counterpart_c(\delta') = \delta$, $\delta' \in Triggered(c')$ and there's no $\delta'' \in Triggered(c')$ with $\neg Premise[\delta''] \in Hedge[\delta']$. The fact that $\delta \in Optimal(c)$ also entails that $\delta \notin Undercut(c)$, whence $\delta' \notin Undercut(c')$. This is actually enough to conclude that $\delta' \in Admissible(c')$. And given that $Conclusion[\delta'] = \bigcirc Conclusion[\delta] = \bigcirc X$, we're done. $\supseteq$ Take an $\bigcirc X \in Conclusion[Admissible(c')]$. Clearly, there's a rule $\delta \in Admissible(c')$ such that $Concluson[\delta] = \bigcirc X$. Now, $\delta \in Triggered(c')$ and for no $\neg Z \in Hedge[\delta]$ do we have $\mathcal{W} \vdash Z$. By Definition 6.2, there must be $\delta' \in \mathcal{D}$ with $counterpart_{c'}(\delta') = \delta$. It's easy to see that $\delta' \in Triggered(c)$, $\delta' \notin Undercut(c)$, and that $\delta' \notin Outweighed(c')$—otherwise, δ wouldn't be admissible. From here, $\delta' \in Optimal(c)$, and so $\bigcirc Conclusion[\delta'] = Conclusion[\delta] = \bigcirc X$ is in the set $\{\bigcirc X : X \in Conclusion[Optimal(c)]\}$. □

Although every regular mixed context is equivalent to some hedged context, two regular mixed contexts can be equivalent to the same hedged context. As an example, take the toy contexts $c_5 = \langle \mathcal{W}, \mathcal{D}, \leq \rangle$ where $\mathcal{W} = \{a, b\}$, $\mathcal{D} = \{\delta_8 = \dfrac{a : \emptyset}{c}, \delta_9 = \dfrac{b : \emptyset}{d}\}$, and $\delta_8 < \delta_9$, and c_6 which is like c_5, except for, in its case, the ordering on rules is $\delta_9 < \delta_8$. Since c and d here are consistent (by assumption), the ordering on rules has no bearing on which ought statements follow from these contexts. So it seems natural to think of it as providing surplus information. It's not difficult to see that applying Definition 6.2 to c_5 and c_6 results in the same hedged contexts, from which the surplus information is absent. (Notice that a similar pair of contexts shows that the correspondence between weighted and simple hedged contexts is many-one.)

For the other direction, we can, again, define a procedure for transforming hedged contexts into mixed ones:

Definition 6.5 [Derived mixed contexts] Let $c = \langle \mathcal{W}, \mathcal{D} \rangle$ be a hedged context. We construct a mixed context c' from it as follows. First, we define an ordering on the rules from $\mathcal{D}$, using their hedges: For any two rules $\delta, \delta' \in \mathcal{D}$, let

$\delta \preceq \delta'$ if and only if $\neg Premise[\delta'] \in Hedge[\delta]$ and $contrary_\mathcal{D}(\delta, \delta')$.

Now let c' be the mixed context $\langle \mathcal{W}', \mathcal{D}', \leq \rangle$, where
1. $\mathcal{W}' = \mathcal{W}$;
2. $\mathcal{D}'$ is the set of rules δ' obtained thus: For every rule $\delta = \dfrac{X : Z_{Old}}{\bigcirc Y}$ in $\mathcal{D}$,
   A. Let $\mathcal{D}_\delta = \{\delta'' \in \mathcal{D} : \delta \preceq \delta''\}$;
   B. let $\mathcal{P} = \{\neg Premise[\delta''] : \delta'' \in \mathcal{D}_\delta\}$;

C. set δ' to be $\dfrac{X : \mathcal{Z}_{Old}\backslash \mathcal{P}}{Y}$;

3. $\delta \leq \delta'$ if and only if $\dfrac{Premise[\delta] : \mathcal{Z}}{\bigcirc Conclusion[\delta]} \preceq \dfrac{Premise[\delta'] : \mathcal{Z}'}{\bigcirc Conclusion[\delta']}$.

Any mixed context constructed by this procedure turns out to be equivalent to the hedged context it's constructed from:

Theorem 6.6 (Hedged-to-Mixed) *Let the context $c' = \langle \mathcal{W}', \mathcal{D}', \leq' \rangle$ be derived from some hedged context $c = \langle \mathcal{W}, \mathcal{D} \rangle$ by the procedure specified in Definition 6.5. Then c' and c are equivalent.*

Proof. Let $c'' = \langle \mathcal{W}'', \mathcal{D}'' \rangle$ be the result of applying Definition 6.2 to c'. By Theorem 6.4, the contexts c' and c'' are equivalent. What we need to do is show that $c = c''$. It's obvious that $\mathcal{W} = \mathcal{W}''$. So it remains to show that $\mathcal{D} = \mathcal{D}''$:

Without loss of generality, we establish only $\mathcal{D} \subseteq \mathcal{D}''$: Consider an arbitrary rule δ from $\mathcal{D}$. By Definition 6.5, there must be some rule δ' in $\mathcal{D}'$ such that $counterpart_c(\delta') = \delta$. By Definition 6.2, there must be a rule δ'' in $\mathcal{D}''$ such that $counterpart_{c'}(\delta'') = \delta'$. Clearly, $Premise[\delta''] = Premise[\delta]$ and $Conclusion[\delta''] = Conclusion[\delta]$. Now we need to show that $Hedge[\delta''] = Hedge[\delta]$.

$Hedge[\delta] \subseteq Hedge[\delta'']$: Consider an arbitrary $\neg Z \in Hedge[\delta]$. Either there's some rule δ^* such that $contrary_\mathcal{D}(\delta, \delta^*)$ and $Z = Premise[\delta^*]$, or there isn't. If yes, then, by Definition 6.5, there's a rule $\delta^\dagger \in \mathcal{D}'$ such that $counterpart_c(\delta^\dagger) = \delta^*$ and $\delta' \leq \delta^\dagger$. Also, notice that $contrary_{\mathcal{D}'}(\delta', \delta^\dagger)$. By Definition 6.2, $\delta^\dagger$ is in $\mathcal{D}'_{\delta'}$, and so $\neg Premise[\delta^\dagger] \in Hedge[\delta'']$. But $Premise[\delta^\dagger] = Premise[\delta^*] = Z$. So, $\neg Z \in Hedge[\delta'']$. If there's no rule δ^* with $contrary_\mathcal{D}(\delta, \delta^*)$ and $Z = Premise[\delta^*]$, then, by Definition 6.5, $\neg Z \in Hedge[\delta']$, and, from this and Definition 6.2, $\neg Z \in Hedge[\delta'']$.

$Hedge[\delta''] \subseteq Hedge[\delta]$: Take some $\neg Z \in Hedge[\delta'']$. Either there is a rule $\delta^* \in \mathcal{D}''$ with $contrary_{\mathcal{D}''}(\delta'', \delta^*)$ and $Premise[\delta^*] = Z$, or not. If yes, then, by Definition 6.2, there's a rule $\delta^\dagger \in \mathcal{D}'$ such that $counterpart_{c''}(\delta^\dagger) = \delta^*$, $contrary_{\mathcal{D}'}(\delta', \delta^\dagger)$, and $\delta' \leq \delta^\dagger$. In light of Definition 6.5, $\delta' \leq \delta^\dagger$ is enough to conclude that $\delta^\dagger \in \mathcal{D}'_{\delta'}$, and so that $counterpart_c(\delta') = \delta \preceq counterpart_c(\delta^*)$. This latter fact means that $\neg Premise[counterpart_c(\delta^*)]$ is in $Hedge[\delta]$. But $Premise[counterpart_c(\delta^*)] = Premise[\delta^*]$, and so $\neg Z \in Hedge[\delta]$. In case there's no rule $\delta^* \in \mathcal{D}''$ with $contrary_{\mathcal{D}''}(\delta'', \delta^*)$ and $Premise[\delta^*] = Z$, Definition 6.5 entails that $\neg Z \in Hedge[\delta']$, and this fact together with Definition 6.2 entail that $\neg Z \in Hedge[\delta]$. □

So there's a close connection between regular mixed and hedged contexts. But what about irregular mixed contexts that have no hedged counterparts? Notice that any such context will contain at least one pair of contrary rules δ, δ' with $\delta \leq \delta'$ and either $\neg Premise[\delta] \in Hedge[\delta']$, or $\neg Premise[\delta'] \in Hedge[\delta]$. The former type of pattern, that is, one where $\neg Premise[\delta] \in Hedge[\delta']$, strikes me as borderline incoherent: The principle instantiated by δ' is stronger than the one instantiated by δ, and yet the feature that makes the weaker principle apply is also what undercuts the stronger one. But the latter type of pattern

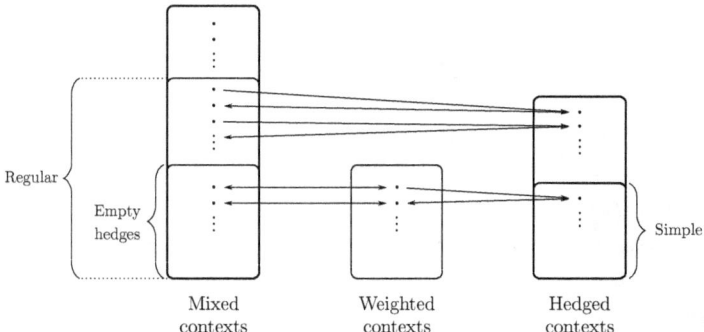

Fig. 1. Relations between various types of contexts

is a different matter; and if there are clear cases exhibiting it, the mixed view has an upper hand over the view on which principles are hedged. Why? Well, because only the mixed view has the resources to distinguish such cases from the mundane ones where the stronger principle outweighs the weaker one. [25]

In Section 4, we already discussed the relation between the model on which principles are contributory and the one on which they are hedged. Here we state it in the form of a corollary to Theorems 6.4 and 6.6: [26]

Corollary 6.7 (Contributory is simple hedged) *For every weighted context c, there's a (simple) hedged context c' such that c and c' are equivalent, and, for every simple hedged context c', there's a weighted context c such that c and c' are equivalent.*

The relations between all three types of contexts—mixed, weighted, and hedged—are summarized in Figure 1.

7 Conclusion and outlook

The main goal of this paper was to explore the connections between three different takes on moral principles. On the first, they are contributory; on the second, they are hedged; on the third, they are contributory, but can also have hedges. We saw that there are close connections between the models of these views. This strongly suggests that the views themselves are much closer than standardly thought. And while it might be obvious that the mixed view extends the contributory-principles view, it would be quite surprising

---

[25] Unfortunately, all the examples I could think of here are quite controversial. Here's one such, coming from the epistemic domain. First, notice that the rules "If an agent perceives that X, then that perception speaks in favor of believing that X" and "If an agent has outstanding testimony that X, then it speaks in favor of believing that X" are the epistemic analogues of contributory/mixed moral principles. Epistemologists sometimes invoke an infallible Epistemology Oracle reporting the truth to the agent—see, e.g., [34]—and we might imagine a situation where some agent sees a red-looking object, but is also told, by this Oracle, that the object is blue. One could hold that, in this case, the Oracle's testimony doesn't simply outweigh, but also undercuts the Perception rule.

[26] The proof runs parallel to those of the theorems. It's also simpler.

if it turned out that the contributory-principles view was only as expressive as the hedged-principles view, or if the differences between the mixed view and the hedged-principles view were only cosmetic. It may be too early to claim that the correspondence results established here show that, since the models we set up here are very simple. So one direction for future research is to extend the correspondence results established here to more expressive models of the views. Another one is to explore the ramifications of this result for claims about views on principles advanced in the philosophical literature.[27] Yet another is to explore the connections between hedged rules and "exclusionary rules" discussed in [8, Sec. 6], or, roughly, rules that take other rules out of consideration. On the face of it, having an exclusionary rule δ that gets triggered when X obtains, taking some other rule δ' out of consideration, isn't all that different from thinking of δ' as having $\neg X$ listed in its hedge. Finally, it would be interesting to explore how the models and results presented here might be relevant in the context of the debate between generalists and specificationists about rights.[28]

References

[1] Bradley, D., *Are there indefeasible epistemic rules?*, Philosopher's Imprint **19** (2019), pp. 1–19.
[2] Dancy, J., "Ethics without Principles," Oxford University Press, 2004.
[3] Elga, A., *How to disagree about how to disagree*, in: R. Feldman and T. Warfield, editors, *Disagreement*, Oxford University Press, 2010 pp. 175–86.
[4] Frederick, D., *Pro-tanto obligations and ceteris-paribus rules*, Journal of Moral Philosophy **12** (2015), pp. 255–66.
[5] Goble, L., *Normative conflicts and the logic of 'ought'*, Noûs **43** (2009), pp. 450–89.
[6] Holton, R., *Principles and particularisms*, in: *Proceedings of the Aristotelian Society, Suppl. Volume 76*, 2002, pp. 191–210.
[7] Horty, J., *Reasons as defaults*, Philosophers' Imprint **7** (2007).
[8] Horty, J., "Reasons as Defaults," Oxford University Press, 2012.
[9] Knoks, A., "Defeasibility in Epistemology," Ph.D. thesis, University of Maryland, College Park (2020).
[10] Lance, M. and M. Little, *Particularism and antitheory*, in: D. Copp, editor, *The Oxford Handbook of Ethical Theory*, Oxford University Press, 2006 pp. 567–94.
[11] Lance, M. and M. Little, *Where the laws are*, Oxford Studies in Metaethics **2** (2007), pp. 149–71.
[12] Lewis, D., *Ordering semantics and premise semantics for counterfactuals*, Journal of philosophical logic **10** (1981), pp. 217–34.
[13] Makinson, D. and L. van der Torre, *Input/output logics*, Journal of Philosophical Logic **29** (2000), pp. 383–408.
[14] Makinson, D. and L. van der Torre, *Constraints for input/output logics*, Journal of Philosophical Logic **30** (2001), pp. 155–85.
[15] McKeever, S. and M. Ridge, "Principled Ethics: Generalism as a Regulative Ideal," Oxford University Press, 2006.

---

[27] In [9, Ch. 2] I make some steps in this direction.
[28] See [17] and [33, Sec. 5.2] for an introduction to the debate and further pointers to the literature. Also, since this is the final footnote, I thank the anonymous reviewers for their feedback and insightful comments.

[16] McNamara, P., *Deontic logic*, in: E. N. Zalta, editor, *The Stanford Encyclopedia of Philosophy*, Metaphysics Research Lab, Stanford University, 2019 .
[17] Mullins, R., *Moral conflict and the logic of rights*, Philosophical Studies **117** (2020), pp. 633–51.
[18] Nair, S. and J. Horty, *The logic of reasons*, in: D. Star, editor, *The Oxford Handbook of Reasons and Normativity*, Oxford University Press, 2018 pp. 67–84.
[19] Owens, D., *Duress, deception and the validity of a promise*, Mind **116** (2007), pp. 293–315.
[20] Parfit, D., "On What Matters: Volume One," Oxford University Press, 2011.
[21] Pollock, J., "Knowledge and Justification," Princeton: Princeton University Press, 1974.
[22] Pollock, J. and J. Cruz, "Contemporary Theories of Knowledge," Rowman & Littlefield Publishers, 1999.
[23] Raz, J., "Engaging Reasons: On the Theory of Value and Action," Oxford University Press, 1999.
[24] Reiter, R., *A logic for default reasoning*, Artificial Intelligece **13** (1980), pp. 81–132.
[25] Ross, W. D., "The Right and the Good," Oxford University Press, 1930.
[26] Scanlon, T. M., "What We Owe to Each Other," Cambridge, MA: Harvard University Press, 1998.
[27] Scanlon, T. M., *Principles and particularisms*, in: *Proceedings of the Aristotelian Society, Suppl. Volume 74*, 2000, pp. 301–17.
[28] Schroeder, M., "Slaves of the Passions," New York: Oxford University Press, 2007.
[29] Schroeder, M., "Reasons First," Oxford University Press, 2021, (forthcoming).
[30] Star, D., *Introduction*, in: D. Star, editor, *The Oxford Handbook of Reasons and Normativity*, Oxford University Press, 2018 pp. 1–21.
[31] Titelbaum, M., *Rationality's fixed point (or: in defense of right reason)*, Oxford Studies in Epistemology **5** (2015), pp. 253–294.
[32] Väyrenen, P., *A theory of hedged moral principles*, Oxford Studies in Metaethics **4** (2009), pp. 91–132.
[33] Wenar, L., *Rights*, in: E. N. Zalta, editor, *The Stanford Encyclopedia of Philosophy*, Metaphysics Research Lab, Stanford University, 2021 .
[34] White, R., *Epistemic permissiveness*, Philosophical Perspectives **19** (2005), pp. 445–59.

A Reduction in Violation Logic

Timo Lang [1]

TU Vienna
Vienna, Austria

Abstract

Using proof theoretic methods, we show that a substantial fragment of violation logic as developed by Governatori, Rotolo et al. can be translated into classical modal logic. A number of consequences of this result are discussed. Furthermore, we present a new criterion for axiomatizations of violation logic and comment on the definability of the $\otimes$-operator.

Keywords: deontic logic, contrary to duty reasoning.

1 Introduction

In a series of works [7,2,5,6,3] Governatori, Rotolo et al. introduced a family of logics intended to model contrary-to-duty reasoning. To this end they extend classical modal logic E (which features the operator O, for 'obligation') by an additional operator $\otimes$ with the intended meaning that [5]

> [t]he interpretation of a chain like $a \otimes b \otimes c$ is that a is obligatory, but if it is violated (i.e., $\neg a$ holds), then b is the new obligation (and b compensates for the violation of a); again, if the obligation of b is violated as well, then c is obligatory [...]

For these so-called $\otimes$-chains a variety of rules and axioms are proposed, resulting in a number of different systems of *violation logic*. One therefore has two levels of obligations, one stemming from the $\otimes$-chains, and the other one from the O modality of the underlying logic E. As the authors put it in [5] regarding their semantics for the $\otimes$-operator,

> We [...] split the treatment of $\otimes$-chains and obligations; the intuition is that chains are the generators of obligations and permissions [...]

In the present paper, we investigate this role of $\otimes$-chains as generators of obligations using proof theoretic methods. Our main result is that $\otimes$-chains can be replaced by formulas in the underlying logic E which generate exactly the same obligations. This yields a translation of a large fragment of violation logic into the base logic E. As a consequence, tools available for E – such

---

[1] Research supported by FWF Project W1255-N23.

as neighbourhood semantics on the model theoretic side, or cutfree Gentzen systems on the proof theoretic side – can be used to study violation logics. We establish coNP-completeness of the 'translatable' fragment of violation logic, and close with some remarks on the choice of axioms for $\otimes$-chains.

2 Preliminaries

Classical Modal Logic

The deontic logic underlying the treatment of $\otimes$-chains is given by axiomatic extensions of the classical non-normal modal logic E (see [4]). We have a language with a countably infinite set Var of propositional variables (denoted $a, b, c, \ldots$), a constant $\bot$ (falsum) and the following connectives:

$$\wedge, \rightarrow \text{ (binary) and O (unary)}$$

Any formula built from variables, constants and the above connectives will be called a *deontic formula* and denoted by uppercase letters A, B. Additional connectives are defined as abbreviations: $\neg A := A \rightarrow \bot$ (negation), $\top := \neg \bot$ (verum), $A \equiv B := (A \rightarrow B) \wedge (B \rightarrow A)$ (equivalence). For a set Γ of formulas, $\bigwedge \Gamma$ denotes the conjunction of all formulas in Γ, with the convention that $\bigwedge \emptyset := \top$. We call *classical* any formula not containing O, and CL denotes the sets of those classical formulas which are theorems of classical logic.

The logic E is defined to be the smallest logic of deontic formulas containing CL and closed under the rules

$$\frac{A \quad A \rightarrow B}{B} \text{ (MP)} \quad \text{and} \quad \frac{A \equiv B}{OA \equiv OB} \text{ (O-RE)}.$$

It will be convenient for our purposes to have a notion of *derivations from assumptions* in an axiomatic extension of E. Here a set Γ will play the role of *local assumptions*, whereas a set Δ plays the role of additional *axioms*.

Definition 2.1 Let $\Delta \cup \Gamma \cup \{A\}$ be a set of deontic formulas. A $(E+\Delta)$-*proof of A from Γ* is a tree of deontic formulas built from rules (MP) and (O-RE), and which is rooted in A. Its leaves are either substitution instances of formulas from $\text{CL} \cup \Delta$, or formulas from Γ. The latter type of leaves are called *local assumptions*.[2] We impose the following locality condition: No instance of (O-RE) appears below a local assumption in the proof.

We write $\Gamma \vdash_{E+\Delta} A$, and say that A is derivable from Γ in $E + \Delta$, if there is a $(E + \Delta)$-proof of A from Γ. Finally, we identify the *logic* $E + \Delta$ with its derivability relation $\vdash_{E+\Delta}$.

The locality condition reflects the well-known fact that modal rules such as (O-RE) should not be applied to local assumptions in modal logic, cf. the chapter on proof theory in [1]. The following Deduction Theorem holds:

Fact 2.2 (Deduction Theorem) $\Gamma \cup \{B\} \vdash_{E+\Delta} A \iff \Gamma \vdash_{E+\Delta} B \rightarrow A$.

Proof. See [1]. □

---

[2] More precisely, we call local assumptions only those leaves which are not at the same time instances of formulas from Δ or classical theorems.

We write $C(A)$ for a formula in which some occurences of a subformula A are distinguished, and subsequently $C(B)$ for the result of replacing in C all these distinguished occurences of A by B. Then:

Fact 2.3 (Uniform Substitution) *For any formula $C(A)$, the following rule is admissible in $E + \Delta$:*

$$\frac{A \equiv B}{C(A) \equiv C(B)} \text{ (O-RE')}$$

Proof. See [1]. □

Neighbourhood semantics We review the notion of neighbourhood models, which form the standard semantics of classical modal logics. A *neighbourhood model* $\mathcal{W} = \langle W, \mathcal{N}, V \rangle$ is composed of the following elements:

- a nonempty set W of worlds
- a neighbourhood function $\mathcal{N} : W \to \mathcal{P}(\mathcal{P}(W))$
- a valuation function $V : Var \to \mathcal{P}(W)$

By abuse of notation, we write $w \in \mathcal{W}$ for worlds w instead of $w \in W$. Given e neighbourhood model $\mathcal{W}$, we can define the notion $\langle \mathcal{W}, w \rangle \models A$ of *truth at a world* $w \in W$ by induction on the deontic formula A: $\langle \mathcal{W}, w \rangle \not\models \bot$, $\langle \mathcal{W}, w \rangle \models a :\Leftrightarrow w \in V(a)$, $\langle \mathcal{W}, w \rangle \models A \wedge B :\Leftrightarrow \langle \mathcal{W}, w \rangle \models A$ and $\langle \mathcal{W}, w \rangle \models B$, $\langle \mathcal{W}, w \rangle \models A \to B :\Leftrightarrow \langle \mathcal{W}, w \rangle \not\models A$ or $\langle \mathcal{W}, w \rangle \models B$, and finally

$$\langle \mathcal{W}, w \rangle \models OA \quad :\Leftrightarrow \quad [A]_\mathcal{W} \in \mathcal{N}(w)$$

where $[A]_\mathcal{W} := \{w \in \mathcal{W} \mid \langle \mathcal{W}, w \rangle \models A\}$. The part $\mathcal{F}_\mathcal{W} = \langle W, \mathcal{N} \rangle$ of a neighbourhood model $\mathcal{W}$ is called a *neighbourhood frame*, and conversely $\langle W, \mathcal{N}, V \rangle$ is called a neighbourhood model *based on* $\mathcal{F}$. *Truth on a frame* is defined as follows: $\mathcal{F} \models A$ if for all models $\mathcal{W}$ based on $\mathcal{F}$ and all worlds $w \in \mathcal{W}$, $\langle \mathcal{W}, w \rangle \models A$. For a set $\Gamma \cup \Delta \cup \{A\}$ of deontic formulas, we define the following semantic consequence relation:

$\Gamma \models_\Delta A$ iff for all neighbourhood models $\mathcal{W}$ and $w \in \mathcal{W}$, if $\mathcal{F}_\mathcal{W} \models \bigwedge \Delta$ and $\langle \mathcal{W}, w \rangle \models \bigwedge \Gamma$, then $\langle \mathcal{W}, w \rangle \models A$.

Fact 2.4 (Soundness and Completeness) $\Gamma \models_\Delta A \iff \Gamma \vdash_{E+\Delta} A$.

Proof. This follows from the strong completeness theorem for E with respect to neighbourhood models (see [4]) and the Deduction Theorem. □

Local assumptions in $E + \Delta$ therefore correspond to truths at a certain world.

Violation Logics

We now discuss a family of logics which were originally introduced in [7], and then developed in a series of subsequent article (e.g., [2,5,6]). On the syntactic level, they extend classical modal logics by an operator $\otimes$, which comes in any arity $n > 0$. A formula

$$A_1 \otimes A_2 \otimes A_3 \otimes \ldots \otimes A_n$$

is meant to model a chain of obligations and corresponding compensations: A_1 is obligatory, but if A_1 is violated, then the new (secondary) obligation is A_2; the fulfillment of A_2 compensates the violation of A_1; if however A_2 is violated as well, then there is a new (ternary) obligation A_3, and so on.

Example 2.5 Consider three propositional variables w, p and f with meaning $w=$'it is the weekend', $p=$'parking downtown' and $f=$'paying a fine'. Then the intended meaning of the formula

$$A_{Ex} = w \to (\neg p) \otimes f$$

taken from [6] is: *On weekends it is forbidden to park downtown; but if one does so, one has to pay a fine.* The formula A_{Ex} will serve as a running example throughout this article.

We will call various systems for logics with $\otimes$ *violation logics*, a term coined in [7]. A formula of violation logic (henceforth just called a formula) is any expression A built from $\bot, \wedge, \to, O$ and $\otimes$ obeying the following *nesting condition*: No pair of operators from $\{O, \otimes\}$ appears nested in A. For example, $\neg(Oa \wedge (b \otimes c \otimes d))$ is a formula of violation logic, whereas $\neg O(a \wedge (b \otimes c))$ is not. A formula of the form $A_1 \otimes \ldots \otimes A_n$ $(n > 0)$ is called a $\otimes$-*chain*. Due to the nesting condition, every formula A_i occuring in a $\otimes$-chain is classical.

Concerning rules and axioms for $\otimes$, the literature contains a large variety of different systems, with no optimal candidate singled out. For the sake of the present article, we pick a system which is close to the one described in [5]; But we remark already here that our results apply to different systems as well, an observation which will be made precise later (Corollary 4.4). That being said, we will have the following two rules for $\otimes$:

$$\frac{A \equiv B}{\nu \otimes A \otimes \nu' \equiv \nu \otimes B \otimes \nu'} \quad (\otimes\text{-RE})$$

$$\frac{A \equiv B}{\nu \otimes A \otimes \nu' \otimes B \otimes \nu'' \equiv \nu \otimes A \otimes \nu' \otimes \nu''} \quad (\otimes\text{-contraction})$$

Here, a string such as $\nu \otimes A \otimes \nu'$ stands symbolically for a $\otimes$-chain containing the (classical) formula A at some position. It is allowed that ν or ν' are empty, so that A is the first or last element of the chain. The rule ($\otimes$-RE) is the generalization of (O-RE) to the language of violation logic, and ($\otimes$-contraction) is a principle of redundancy elimination.

As axioms, we take the following set Σ of formulas:

$$a_1 \otimes \ldots \otimes a_n \wedge \bigwedge_{i=1}^{k} \neg a_i \to Oa_{k+1} \quad \text{(O-detachment)}$$

$$a_1 \otimes \ldots \otimes a_n \otimes a_{n+1} \to a_1 \otimes \ldots \otimes a_n \quad (\otimes\text{-shortening})$$

$$a_1 \otimes \ldots \otimes a_{n+1} \wedge \neg a_1 \to a_2 \otimes \ldots \otimes a_{n+1} \quad (\otimes\text{-detachment})$$

Here, $n \geq 1$ and $0 \leq k < n$.[3] The axiom (O-detachment) captures the intended meaning of $\otimes$-chains as descriptions of compensatory obligations: If the first k obligations expressed in a $\otimes$-chain $a_1 \otimes \ldots \otimes a_k \otimes a_{k+1} \otimes \ldots \otimes a_n$ have been violated, then the next obligation a_{k+1} comes into effect. We refer the reader to [5] for an extensive discussion of the system.

We again define a notion of derivations from assumptions.

Definition 2.6 Let $\Delta \cup \Gamma \cup \{A\}$ be a set of formulas. A $(V_\Sigma + \Delta)$-*proof of A from* Γ is a tree of formulas built from the rules (MP), (O-RE), ($\otimes$-RE) and ($\otimes$-contraction), and which is rooted in A. Its leaves are either substitution instances of formulas from $CL \cup \Delta \cup \Sigma$, or formulas from Γ. The latter type of leaves are called *local assumptions*. We impose the following locality condition: No instance of (O-RE), ($\otimes$-RE) or ($\otimes$-contraction) appears below a local assumption.

We write $\Gamma \vdash_{V_\Sigma + \Delta} A$, and say that A is derivable from Γ in $V_\Sigma + \Delta$, if there is a $(V_\Sigma + \Delta)$-proof of A from Γ. Finally, we identify the *violation logic* $V_\Sigma + \Delta$ with its derivability relation $\vdash_{V_\Sigma + \Delta}$.

Fact 2.7 (Deduction Theorem) $\Gamma \cup \{B\} \vdash_{V_\Sigma + \Delta} A \iff \Gamma \vdash_{V_\Sigma + \Delta} B \to A$.

Proof. By induction on the height of proofs. □

The Deduction Theorem equips us with the following mode of inference in violation logic: If we can prove A from assumption B *without using rules* (O-RE), ($\otimes$-RE) *or* ($\otimes$-contraction) *below the assumption B*, then we can infer $B \to A$.

Example 2.8 We look again at the formula A_{Ex} from Example 2.5. The following proof shows that $\{A_{Ex}, w, p\} \vdash_{V_\Sigma} Of$, which means that parking downtown on a weekend leads to the obligation of paying a fine:

$$\frac{\dfrac{w \quad w \to (\neg p) \otimes f}{(\neg p) \otimes f}(MP) \quad \dfrac{p}{\neg\neg p}}{\dfrac{\neg\neg p \wedge ((\neg p) \otimes f) \quad \neg\neg p \wedge ((\neg p) \otimes f) \to Of}{Of}(MP)}$$

(local assumption) w; (local assumption) $w \to (\neg p) \otimes f$; (local assumption) p; (instance of O-detachment)

A double line abbreviates some steps of 'classical reasoning', i.e. the use of classical theorems and (MP). Since none of the rules (O-RE), ($\otimes$-RE) or ($\otimes$-contraction) are applied in the proof above, we can also conclude, e.g., $\{A_{Ex}, w\} \vdash_{V_\Sigma} p \to Of$.

3 A Reduction Theorem

Throughout this section, we work in violation logics $V_\Sigma + \Delta$ where Δ consists of deontic axioms only, and hence the meaning of the $\otimes$-chains is given by the axiom set Σ. The set Δ might for example consist of the single axiom

$$Oa \to \neg O(\neg a) \tag{D}$$

---

[3] By the convention on empty conjunctions, it follows that $a_1 \otimes \ldots \otimes a_n \wedge \top \to Oa_1$ is an instance of (O-detachment).

in which case $V_\Sigma + \Delta$ is the logic $D^\otimes$ from [5].

The technical results we are going to present apply to a fragment of violation logic that we will call the *chain negative fragment*.

Definition 3.1 (Chain Negative Fragment) An occurence of a $\otimes$-chain in a formula A is called positive if there is an even number (including zero) of implicational subformulas $B \to C$ of A such that the chain appears in B.[4] Otherwise, the occurence is called negative. We call a formula *chain negative* (resp. chain positive) if all occurences of $\otimes$-chains in it are negative (resp. positive).

For example, the chain $a \otimes b$ appears positively in the formulas $a \otimes b$, $\neg\neg(c \wedge a \otimes b)$ and $c \to a \otimes b$, and negatively in $\neg(a \otimes b)$, $(a \otimes b) \to Oc$ and $(a \otimes b) \wedge c \to Od$.[5] The simplest nontrivial example of a chain positive formula is a $\otimes$-chain. Intuitively, a chain negative formula is a formula in which $\otimes$-chains appear *only as assumptions, but not as conclusions*.

As our main result, we will now show that questions about the chain negative fragment of violation logic can be answered without using the machinery of violation logic, but with a suitable reduction to the underlying deontic logic $E + \Delta$ instead. To this end, we first give a meaning to $\otimes$-chains as deontic formulas.

Definition 3.2 (π-translation) The translation π from $\otimes$-chains to deontic formulas is inductively defined as follows:

$$\pi(\otimes A) := OA\,^6$$

$$\pi(A_1 \otimes \ldots \otimes A_n \otimes A_{n+1}) := \pi(A_1 \otimes \ldots \otimes A_n) \wedge \left((\bigwedge_{i=1}^{n} \neg A_i) \to OA_{n+1} \right)$$

As an example, we have $\pi(a \otimes b \otimes c) = Oa \wedge (\neg a \to Ob) \wedge (\neg a \wedge \neg b \to Oc)$. In the following we will write π in closed form as

$$\pi(A_1 \otimes \ldots \otimes A_n) = \bigwedge_{i=1}^{n} \left((\bigwedge_{j=1}^{i-1} \neg A_j) \to OA_i \right)$$

where by a harmless abuse of notation, we identify the conjunct $\top \to OA_1$, corresponding to the index $i = 1$, with the formula OA_1. We extend π to arbitrary formulas by letting it commute with $\wedge$, $\to$ and O, so that for example

$$\pi(A_{Ex}) = \pi(w \to (\neg p) \otimes f) = w \to O(\neg p) \wedge (\neg\neg p \to Of).$$

Given a set Γ of formulas, $\pi(\Gamma)$ denotes $\{\pi(A) \mid A \in \Gamma\}$.

---

[4] This is the standard notion of a positive/negative occurence of a subformula, see e.g. Definition 24.18 in [9].
[5] Recall that $\neg A = A \to \bot$ by definition.
[6] $\otimes A$ denotes a $\otimes$-chain of length 1.

We point out that the meaning given to ⊗-chains by the translation π is quite close to the intuitive interpretation of ⊗-chain from [5], which was already quoted in the introduction:

[t]he interpretation of a chain like $a \otimes b \otimes c$ is that a is obligatory, but if it is violated (i.e., $\neg a$ holds), then b is the new obligation (and b compensates for the violation of a); again, if the obligation of b is violated as well, then c is obligatory [...]

As a first observation, the axioms for ⊗-chains remain true if translated via π:

Lemma 3.3 (Axiom Soundness) *For any axiom $A \in \Sigma$, $\vdash_E \pi(A)$.*

Proof. Below are the three axioms schemes and their respective π-translations:

(O-detachment) $\quad a_1 \otimes \ldots \otimes a_n \wedge \bigwedge_{i=1}^{k} \neg a_i \to O a_{k+1}$

$\qquad \bigwedge_{i=1}^{n} \left((\bigwedge_{j=1}^{i-1} \neg a_j) \to O a_i \right) \wedge (\bigwedge_{i=1}^{k} \neg a_i) \to O a_{k+1}$

(⊗-shortening) $\quad a_1 \otimes \ldots \otimes a_n \otimes a_{n+1} \to a_1 \otimes \ldots \otimes a_n$

$\qquad \bigwedge_{i=1}^{n+1} \left((\bigwedge_{j=1}^{i-1} \neg a_j) \to O a_i \right) \to \bigwedge_{i=1}^{n} \left((\bigwedge_{j=1}^{i-1} \neg a_j) \to O a_i \right)$

(⊗-detachment) $\quad a_1 \otimes \ldots \otimes a_{n+1} \wedge \neg a_1 \to a_2 \otimes \ldots \otimes a_{n+1}$

$\qquad \bigwedge_{i=1}^{n+1} \left((\bigwedge_{j=1}^{i-1} \neg a_j) \to O a_i \right) \wedge \neg a_1 \to \bigwedge_{i=2}^{n+1} \left((\bigwedge_{j=2}^{i-1} \neg a_j) \to O a_i \right)$

It is cumbersome but easy to check that the translations are provable in E. In fact, they are all instances of classical theorems. $\square$

We now want to argue that in some sense, $A_1 \otimes \ldots \otimes A_n$ and its translation $\pi(A_1 \otimes \ldots \otimes A_n)$ are equivalent. One half of this claim holds in the literal sense:

Lemma 3.4 (Chain Soundness) $\vdash_{V_\Sigma} A_1 \otimes \ldots \otimes A_n \to \pi(A_1 \otimes \ldots \otimes A_n)$.

Proof. Let $1 \leq i \leq n$. From local assumptions $A_1 \otimes \ldots \otimes A_n$ and $\bigwedge_{j=1}^{i-1} \neg A_j$, we can infer OA_i using the axiom (O-detachment). So by the Deduction Theorem, we can infer $(\bigwedge_{j=1}^{i-1} \neg A_j) \to OA_i$ for each $1 \leq i \leq n$, and by further classical reasoning we obtain

$$\bigwedge_{i=1}^{n} \left((\bigwedge_{j=1}^{i-1} \neg A_j) \to OA_i \right)$$

which is precisely $\pi(A_1 \otimes \ldots \otimes A_n)$. $\square$

Corollary 3.5 *For every chain negative formula N, $\vdash_{V_\Sigma} \pi(N) \to N$.*

Proof. By induction on the structure of N. Simultaneously, one has to prove that $\vdash_{V_\Sigma} P \to \pi(P)$ for chain positive P. Both statements are trivially true if the formula does not contain ⊗. Furthermore, if P is a ⊗-chain we can use the Chain Soundness Lemma.

As an example for the inductive step, assume that a chain negative formula N is of the form $A \to B$. Then A is chain positive and B is chain negative. By induction hypothesis, we therefore have $\vdash_{V_\Sigma} A \to \pi(A)$ and $\vdash_{V_\Sigma} \pi(B) \to B$. From this and classical reasoning we obtain

$$\vdash_{V_\Sigma} (\pi(A) \to \pi(B)) \to (A \to B)$$

which is what we need since $\pi(A \to B) = \pi(A) \to \pi(B)$.

The other cases are similar. We note that the induction step for formulas beginning with O is trivial, since by the nesting condition, such formulas do not contain the $\otimes$-operator. □

Remark 3.6 The converse of Lemma 3.4 does not hold, i.e. in V_Σ we cannot prove $A_1 \otimes \ldots \otimes A_n$ from its π-translation. The intuitive reason for this is that in Σ, we do not have any axiom at hand which *creates* $\otimes$-chains from deontic formulas. For a formal argument, consider an alternative translation τ of formulas which replaces all $\otimes$-chains in a formula by $\bot$. For any axiom $A \in \Sigma$, an easy inspection shows that $\vdash_{V_\Sigma} \tau(A)$. In words: The axioms of violation logic remain true if $\otimes$-chains are interpreted as contradictions.

By a simple induction on proof length it follows that $\vdash_{V_\Sigma} \tau(A)$ for any theorem A of V_Σ. Hence if $\pi(A_1 \otimes \ldots \otimes A_n) \to A_1 \otimes \ldots \otimes A_n$ was provable for all $\otimes$-chains $A_1 \otimes \ldots \otimes A_n$, then so would be its τ-translation $\pi(A_1 \otimes \ldots \otimes A_n) \to \bot$, which cannot be the case.

Nevertheless, we will see that the deontic formula $\pi(A_1 \otimes \ldots \otimes A_n)$ is as strong as the $\otimes$-chain $A_1 \otimes \ldots \otimes A_n$ *when it comes to the derivation of deontic formulas*: In particular, the obligations arising from $A_1 \otimes \ldots \otimes A_n$ are exactly the obligations arising from $\pi(A_1 \otimes \ldots \otimes A_n)$.

This follows from the Reduction Theorem below, which is our main technical result in this article. We first state and prove the theorem and then discuss its technical and conceptual ramifications.

Theorem 3.7 (Reduction Theorem for the chain negative fragment)
For any chain negative formula N, the following holds:

$$\vdash_{V_\Sigma + \Delta} N \quad \text{if and only if} \quad \vdash_{E + \Delta} \pi(N).$$

Proof. The direction from right to left is easy: If $\vdash_{E+\Delta} \pi(N)$, then obviously also $\vdash_{V_\Sigma + \Delta} \pi(N)$ since violation logic has all the axioms and rules of E. But then $\vdash_{V_\Sigma + \Delta} N$ follows from Corollary 3.5, since N is chain negative.

For the direction from left to right, we argue by induction on the length of a proof δ witnessing $\vdash_{V_\Sigma + \Delta} N$.

(i) Assume first that δ has height 1, i.e. N is an axiom of $V_\Sigma + \Delta$.
   (a) If N is a substitution instance of a classical theorem, then $\pi(N)$ is again a substitution instance of the the same classical theorem, since π commutes with boolean connectives. Hence $\vdash_{E+\Delta} \pi(N)$.
   (b) Similarly, if N is a substitution instance of a formula in Δ, then $\pi(N)$ is again a substitution instance of the the same formula in Δ, since π commutes with boolean connectives and O. Hence $\vdash_{E+\Delta} \pi(N)$.
   (c) If N is a substitution instance of a formula in Σ, then $\vdash_{E+\Delta} \pi(N)$ by the Axiom Soundness Lemma (Lemma 3.3).

(ii) If the last step in δ is an instance of (MP) $A, A \to B / B$, then by induction hypothesis $\vdash_{E+\Delta} \pi(B)$ and $\vdash_{E+\Delta} \pi(A \to B)$. Since $\pi(A \to B)$ equals $\pi(A) \to \pi(B)$, we can conclude $\vdash_{E+\Delta} \pi(B)$ by applying (MP) in E.

(iii) If the last step in δ is an instance of (O-RE) $A \equiv B / OA \equiv OB$, then by induction hypothesis $\vdash_{E+\Delta} \pi(A \equiv B)$. Since $\pi(A \equiv B)$ equals $\pi(A) \equiv \pi(B)$, we can conclude $\vdash_{E+\Delta} O\pi(A) \equiv O\pi(A)$ by applying (O-RE) in E, and $O\pi(A) \equiv O\pi(B)$ equals $\pi(OA \equiv OB)$.

(iv) Assume that the last step in δ is an inference

$$\frac{A \equiv B}{\nu \otimes A \otimes \nu' \equiv \nu \otimes B \otimes \nu'} \quad (\otimes\text{-RE}).$$

By induction hypothesis $\vdash_{E+\Delta} \pi(A \equiv B)$. Since A, B occur in a $\otimes$-chain, they must be classical formulas by the nesting condition, and so the premise $\pi(A \equiv B)$ equals $A \equiv B$. Now the deontic formula $\pi(\nu \otimes A \otimes \nu')$ arises from replacing some occurences of B in $\pi(\nu \otimes B \otimes \nu')$ by the formula A. Hence

$$\frac{A \equiv B}{\pi(\nu \otimes A \otimes \nu') \equiv \pi(\nu \otimes B \otimes \nu')}$$

is an instance of (O-RE') (cf. Lemma 2.3), and so $\vdash_{E+\Delta} \pi(\nu \otimes A \otimes \nu' \equiv \nu \otimes B \otimes \nu')$ as desired.

(v) Assume that the last step in δ is an inference ($\otimes$-contraction). We only consider a characteristic case:

$$\frac{A \equiv B}{X \otimes A \otimes Y \otimes B \otimes Z \equiv X \otimes A \otimes Y \otimes Z} \quad (\otimes\text{-contraction})$$

Again A and B must be classical, and so we have $\vdash_{E+\Delta} A \equiv B$ by induction hypothesis. Now arguing in $E + \Delta$, we can use (O-RE') to derive from $A \equiv B$ the equivalence

$$\pi(X \otimes A \otimes Y \otimes B \otimes Z) \equiv \pi(X \otimes A \otimes Y \otimes A \otimes Z)$$

Written verbosely, the formula $\pi(X \otimes A \otimes Y \otimes A \otimes Z)$ equals

$$OX \wedge (\neg X \to OA) \wedge (\neg X \wedge \neg A \to OY) \wedge (\neg X \wedge \neg A \wedge \neg Y \to OA)$$
$$\wedge (\neg X \wedge \neg A \wedge \neg Y \wedge \neg A \to OZ).$$

By using classical reasoning we see that the fourth conjunct can be omitted since it is implied by the second conjunct. Furthermore, the second $\neg A$ in the last conjunct can be removed. The above formula is therefore equivalent to

$$OX \wedge (\neg X \to OA) \wedge (\neg X \wedge \neg A \to OY) \wedge (\neg X \wedge \neg A \wedge \neg Y \to OZ)$$

which is precisely $\pi(X \otimes A \otimes Y \otimes Z)$. Hence we have $\vdash_{E+\Delta} \pi(X \otimes A \otimes Y \otimes B \otimes Z \equiv X \otimes A \otimes Y \otimes Z)$ as desired.

This concludes the proof of the Reduction Theorem. $\square$

It is instructive to single out a special case of Theorem 3.7.

Theorem 3.8 (Reduction Theorem, Special Case) *Let $\Gamma \cup \{D\}$ be a set of deontic formulas. Then for any chain positive formula P, the following are equivalent:*

(i) $\Gamma \cup \{P\} \vdash_{V_\Sigma + \Delta} D$

(ii) $\Gamma \cup \{\pi(P)\} \vdash_{E + \Delta} D$

(iii) $\Gamma \cup \{\pi(P)\} \vdash_{V_\Sigma + \Delta} D$

In particular, this holds if P is a $\otimes$-chain.

Proof. $\Gamma \cup \{P\} \vdash_{V_\Sigma + \Delta} D$ is equivalent to $\vdash_{V_\Sigma + \Delta} \bigwedge (\Gamma \cup \{P\}) \to D$ by the Deduction Theorem. Since $\bigwedge (\Gamma \cup \{P\}) \to D$ is chain negative, its provability is equivalent to $\vdash_{E + \Delta} \pi(\bigwedge (\Gamma \cup \{P\}) \to D)$ by the Reduction Theorem. Now $\pi(\bigwedge(\Gamma \cup \{P\}) \to D)$ equals $\bigwedge(\Gamma \cup \{\pi(P)\}) \to D)$ since neither Γ nor D contain $\otimes$-chains by assumption. So by the Deduction Theorem, we obtain equivalence with $\Gamma \cup \{\pi(P)\} \vdash_{E + \Delta} D$. We have thus established (i)↔(ii), and applying (i)↔(ii) to $\pi(P)$ instead of P yields (ii)↔(iii). □

Conceptually, of most importance is the equivalence (i)↔(iii) in the case that P is a $\otimes$-chain C, and its meaning can then be described as follows:

Within a context of deontic formulas, using a $\otimes$-chain C as an assumption has exactly the same effect as using its translation $\pi(C)$.

In other words, as long as we are only interested in the role of $\otimes$-chains as generators of obligations (under some circumstances described by deontic formulas), then we may as well replace all chains by their π-translations.

The questions which are not covered by the Reduction Theorem are those about the *generation of $\otimes$-chains from deontic assumptions* as well as those about *relations between different $\otimes$-chains*, such as the question when one $\otimes$-chain implies another one. We will come back to this in Section 5.

Example 3.9 Recall the formula $A_{Ex} = w \to (\neg p) \otimes f$ from Example 2.5. For any set Γ of deontic formulas, we may ask whether

$$\{A_{Ex}\} \cup \Gamma \vdash_{V_\Sigma + \Delta} Of$$

holds, i.e. whether under the assumption of A_{Ex}, the deontic circumstances expressed in Γ lead to the obligation of paying a fine. By the (special case of the) Reduction Theorem, this question is equivalent to asking whether

$$\{\pi(A_{Ex})\} \cup \Gamma \vdash_{E + \Delta} Of$$

holds, where $\pi(A_{Ex}) = w \to (O(\neg p) \land (\neg \neg p \to Of))$.

Remark 3.10 The Reduction Theorem is formulated relative to violation logics $V_\Sigma + \Delta$ with a fixed axiomatization

$$\Sigma = \{(\text{O-detachment}), (\otimes\text{-contraction}), (\otimes\text{-shortening})\}$$

of $\otimes$-chains (whereas the deontic axioms Δ can be anything). Nevertheless, the proof is modular and can be adapted to violation logics $V_\Pi + \Delta$ where Π is a different axiomatization of chains: We only have to check that the Axiom

Soundness Lemma (Lemma 3.3) and the Chain Soundness Lemma (Lemma 3.4) hold for the axiomatization Π, and then the proof of the Reduction Theorem goes through. Note in particular that the Chain Soundness Lemma holds for any Π which contains (O-detachment).

Remark 3.11 An easy example demonstrating that the Reduction Theorem does not hold for the full language of violation logic is the following. Consider the (chain positive!) formula $P = \pi(a \otimes b) \to (a \otimes b)$. P is not provable in V_Σ: Recall Remark 3.6, where it is argued that if P was provable in V, then so would be $\tau(P) = \pi(a \otimes b) \to \bot = \neg(Oa \wedge (\neg a \to Ob))$. But this is not a theorem of V_Σ, since it is easily seen to be falsifiable in E. On the other hand $\pi(P) = \pi(a \otimes b) \to \pi(a \otimes b)$ is obviously a theorem of E.

4 Applications of the Reduction Theorem

Throughout this section, Δ denotes a set of deontic formulas.

Corollary 4.1 *The violation logic $V_\Sigma + \Delta$ is conservative over $E + \Delta$.*

Proof. Let D be a formula without $\otimes$-chains. Then D is in the chain negative fragment and furthermore $\pi(D) = D$, and so we have $\vdash_{V_\Sigma + \Delta} D$ iff $\vdash_{E + \Delta} D$ by the Reduction Theorem. □

This conservativity result also follows from the *sequence semantics* for violation logic, see e.g. [5].

The main point of a reduction as expressed in Theorem 3.7 is that the logic $E + \Delta$ one reduces to is well studied, and one can transfer results about it back to the 'new' logic $V_\Sigma + \Delta$. Let us see some examples.

Corollary 4.2 *The validity problem for the chain negative fragment of the violation logic V_Σ is coNP-complete.*

Proof. By the Reduction Theorem, $\vdash_{V_\Sigma} D$ is equivalent to $\vdash_E \pi(D)$ for a chain negative D, and the mapping $D \mapsto \pi(D)$ is computable in polynomial (in fact, quadratic) time. Since theoremhood in E is coNP-decidable ([10], Theorem 3.3), the same therefore holds for V_Σ. On the other hand the chain negative fragment of V_Σ is a conservative extension of CL, which is coNP-hard. □

By the same argument, complexity (or just decidability) results can be obtained for other violation logics $V_\Sigma + \Delta$: We only have to know the complexity of the underlying deontic logic $E + \Delta$. As far as we know, no decidability results for violation logics have been established so far.

It also follows from the Reduction Theorem that the neighbourhood semantics of classical modal logics provides a complete semantics for the chain negative fragment of violation logic. This semantics is simpler than the sequence semantics proposed in [5,6].

Corollary 4.3 *Let $\Gamma \cup \{D\}$ be a set of deontic formulas. Then for any chain positive formula P, $\Gamma \cup \{P\} \vdash_{V_\Sigma + \Delta} D$ iff for every neighbourhood model $\mathcal{W}$*

with $\mathcal{F}_\mathcal{W} \models \Delta$ the following is true: For any world $w \in \mathcal{W}$, if $\langle \mathcal{W}, w \rangle \models \bigwedge \Gamma$ and $\langle \mathcal{W}, w \rangle \models \pi(P)$, then $\langle \mathcal{W}, w \rangle \models D$.

Proof. By the Reduction Theorem, $\Gamma \cup \{P\} \vdash_{V_\Sigma + \Delta} D$ is equivalent to $\Gamma \cup \{\pi(P)\} \vdash_{E+\Delta} D$, which in turn is equivalent to $\Gamma \cup \{\pi(P)\} \models_\Delta D$ by Fact 2.4. □

So within a context of deontic formulas, having a $\otimes$-chain $C = a \otimes b \otimes c$ as a local assumption amounts to assuming the truth of

$$\pi(a \otimes b \otimes c) = Oa \land (\neg a \to Ob) \land (\neg a \land \neg b \to Oc)$$

at a world of a neighbourhood model $\mathcal{W}$.

Corollary 4.4 *Let $\Pi \neq \Sigma$ be any alternative axiomatization of $\otimes$-chains containing at least (O-detachment), and such that $\vdash_E \pi(A)$ for every $A \in \Pi$. Then for any set of deontic formulas Δ, the chain negative fragments of $V_\Sigma + \Delta$ and $V_\Pi + \Delta$ coincide.*

Proof. By Remark 3.10, the proof of the Reduction Theorem goes through for $V_\Pi + \Delta$ under the given assumptions. But then $V_\Sigma + \Delta$ and $V_\Pi + \Delta$ have the same characterization of their chain negative fragment (which does not depend on Σ or Π), namely

$$\vdash_{V_\Sigma + \Delta} N \quad \text{iff} \quad \vdash_{E+\Delta} \pi(N) \quad \text{iff} \quad \vdash_{V_\Pi + \Delta} N.$$

□

An immediate consequence of Corollary 4.4 is that the axioms ($\otimes$-shortening) and ($\otimes$-detachment) are never needed for proving formulas in the chain negative fragment of $V_\Sigma + \Delta$. As another consequence, consider the axiom ($\otimes$-I)

$$\left(a_1 \otimes \ldots \otimes a_n \land \left((\bigwedge_{i=1}^n \neg a_i) \to b_1 \otimes \ldots \otimes b_m \right) \right) \to a_1 \otimes \ldots \otimes a_n \otimes b_1 \otimes \ldots \otimes b_m$$

for creating $\otimes$-chains which is considered in [7,3], but not in [5,6]. It is easy to see that its π-translation is a theorem of E, and so by Corollary 4.4 its inclusion as an additional axiom has no effect on the chain negative fragment.

An axiomatization of $\otimes$-chains to which the Reduction Theorem does *not* apply is the one given in [2], where axioms such as $a \otimes (\neg a) \equiv \top$ are included. Indeed, the π-translation of the latter axiom is $Oa \land (\neg a \to O\neg a) \equiv \top$, which does not hold in E.

Another consequence of the Reduction Theorem is that questions in violation logic can be tackled using the proof theory of classical modal logics. For example, [8] presents cutfree Gentzen systems for the logics

$$E, \quad EC = E + Oa \land Ob \to O(a \land b) \quad \text{and} \quad M = E + O(a \land b) \to Oa \land Ob$$

which are called **Eseq**, **ECseq** and **Mseq** respectively.

Corollary 4.5 *Let Δ be $\emptyset$ (or $\{Oa \land Ob \to O(a \land b)\}$, or $\{O(a \land b) \to Oa \land Ob\}$). Then for any chain negative formula N, $\vdash_{V_\Sigma + \Delta} N$ iff there is a cutfree proof of $\pi(N)$ in **Eseq** (or **ECseq**, or **Mseq**).*

Example 4.6 Here is a Gentzen-style proof establishing $\{A_{Ex}, w, p\} \vdash_{V_\Sigma} Of$ by means of the π-translation (cf. Example 2.8):

$$\cfrac{w \Rightarrow w \quad \cfrac{\cfrac{\cfrac{\cfrac{p \Rightarrow p}{p, \neg p \Rightarrow}(\neg L)}{p \Rightarrow \neg\neg p}(\neg R) \quad Of \Rightarrow Of}{\neg\neg p \to Of, p \Rightarrow Of}(\to L)}{O(\neg p) \land (\neg\neg p \to Of), p \Rightarrow Of}(\land L)}{w \to (O(\neg p) \land (\neg\neg p \to Of)), w, p \Rightarrow Of}(\to L)$$

5 More on the interpretation of ⊗-chains

Arguably, the formalization of many contrary-to-duty reasoning scenarios in the framework of violation logic remains in the chain negative fragment. Recall that in particular all questions of the form

Given some (deontic) circumstances, which obligations arise from a ⊗-chain?

are expressible. The Reduction Theorem then suggests that in the chain negative fragment, the 'meaning' of a ⊗-chain can be identified with its π-translation (assuming, of course, one believes that the meaning of ⊗-chains is given by their proof-theoretic behaviour). Furthermore, we have seen (Corollary 4.4) that this identification is to some extent independend of the exact axiomatization Σ of ⊗-chains.

If we move beyond the chain negative fragment, the precise axiomatization of ⊗-chains matters more. So let us now consider an arbitrary violation logic $V_\Pi + \Delta$ where Π satisfies the premises of Corollary 4.4, and for which therefore the Reduction Theorem holds (Δ is again any set of deontic axioms). A typical question outside the chain negative fragment is: When does a ⊗-chain C imply another ⊗-chain C', i.e. when does $\vdash_{V_\Pi + \Delta} C \to C'$ hold? A good axiomatization Π should give a tangible meaning to the notion of implication between chains. Hence, the question we have to ask is:

When *should* a ⊗-chain C imply another ⊗-chain C'?

Here is one possible proposal. We say that a chain C *deontically subsumes* another chain C' over $V_\Pi + \Delta$ if for every deontic formula D, $\vdash_{V_\Pi + \Delta} C' \to D$ implies $\vdash_{V_\Sigma\Pi + \Delta} C \to D$. In words: C deontically subsumes C' if every obligation arising from C' already arises from C.

Definition 5.1 The violation logic $V_\Pi + \Delta$ is *faithful* if it proves $C \to C'$ for every pair C, C' of chains where C deontically subsumes C'.

So in a faithful violation logic, the meaning of an implication $C \to C'$ between chains is that of deontic subsumption. From the Reduction Theorem arises a simple characterization of deontic subsumption:

Lemma 5.2 C *deontically subsumes* C' *iff* $\vdash_{E+\Delta} \pi(C) \to \pi(C')$.

Proof. Assume that C deontically subsumes C'. Since $\vdash_{V_\Pi + \Delta} C' \to \pi(C')$ (Lemma 3.4), we also have $\vdash_{V_\Pi + \Delta} C \to \pi(C')$ by deontic subsumption. But then $\vdash_{E+\Delta} \pi(C) \to \pi(C')$ by the Reduction Theorem. Conversely, if $\vdash_{E+\Delta} \pi(C) \to \pi(C')$ and D is a deontic formula implied by C', then $\vdash_{E+\Delta} \pi(C') \to D$ by the Reduction Theorem, and so $\vdash_{E+\Delta} \pi(C) \to D$. Then again by Lemma 3.4, $\vdash_{V_\Pi + \Delta} C \to D$ follows. □

For our basic violation logic V_Σ, we can show the following:

Theorem 5.3 V_Σ *is not faithful.*

Proof. (Sketch) Let a, b be two distinct variables. The counterexample will be the two chains
$$C = a \otimes (\neg a) \quad \text{and} \quad C' = a \otimes (\neg a) \otimes b.$$
Their respective π-translations are $\pi(C) = Oa \wedge (\neg a \to O(\neg a))$ and $\pi(C') = Oa \wedge (\neg a \to O(\neg a)) \wedge (\neg a \wedge \neg\neg a \to Ob)$. Since $\pi(C)$ implies $\pi(C')$, we know by Lemma 5.2 that C deontically subsumes C'. However, while $C' \to C$ is an instance of ($\otimes$-shortening), V_Σ fails to prove $C \to C'$. We show this by providing a countermodel in the *sequence semantics* of [5]. A sequence model extends a neighbourhood model $\mathcal{W} = \langle W, \mathcal{N}, V \rangle$ by a function $\mathcal{C}$ which maps each world w to a set $\mathcal{C}_w$ of finite nonempty sequences $\langle X_1, \ldots, X_n \rangle$ of sets of worlds, and which obeys the following closure conditions:

(i) If $\langle X_1, \ldots, X_n \rangle \in \mathcal{C}_w$ and $n > 1$, then $\langle X_1, \ldots, X_{n-1} \rangle \in \mathcal{C}_w$

(ii) Let $L \in \mathcal{C}_w$ be a list in which a set of worlds X occurs at a certain position. Then $\mathcal{C}_w$ must contain also all lists arising from removing or introducing copies of X at a later position in L.

(iii) If $\langle X_1, \ldots, X_n \rangle \in \mathcal{C}_w$ and for some $0 \leq k < n$, $w \notin X_1 \cup \ldots \cup X_k$, then $X_{k+1} \in \mathcal{N}(w)$ and $\langle X_{k+1}, \ldots, X_n \rangle \in \mathcal{C}_w$

The satisfaction clauses of the standard neighbourhood semantics are then extended by setting $\langle \mathcal{W}, \mathcal{C}, w \rangle \models A_1 \otimes \ldots \otimes A_n :\Leftrightarrow \langle [A_1]_\mathcal{W}, \ldots, [A_n]_\mathcal{W} \rangle \in \mathcal{C}_w$. It is proved in [5] that $\vdash_{V_\Sigma} A$ iff A holds in all sequence models. So our task is to construct a sequence model in which C holds, but C' fails. It will suffice to have two worlds w, v. Assume that $V(a) = \{w, v\}$ and $V(b) = \{w\}$. We let $\mathcal{N}(w) = \{\{w, v\}\}$. The value of $\mathcal{N}$ on other worlds is not relevant. Neither is the choice of $\mathcal{C}_v$, which can be set to $\emptyset$ to trivially satisfy the closure conditions. We let $\mathcal{C}_w$ consist of all sequences of the form
$$\langle \{w, v\}, \ldots, \{w, v\} \rangle \quad \text{or} \quad \langle \{w, v\}, \emptyset, X_1, \ldots, X_n \rangle$$
where $n \geq 0$ and each X_i is either $\{w, v\}$ or $\emptyset$. Then $\mathcal{C}_w$ satisfies the closure conditions, and $\langle [a]_\mathcal{W}, [\neg a]_\mathcal{W} \rangle = \langle \{w, v\}, \emptyset \rangle \in \mathcal{C}_w$, whereas $\langle [a]_\mathcal{W}, [\neg a]_\mathcal{W}, [b]_\mathcal{W} \rangle = \langle \{w, v\}, \emptyset, \{w\} \rangle \notin \mathcal{C}_w$, and so $\langle \mathcal{W}, \mathcal{N}, w \rangle \not\models a \otimes (\neg a) \to a \otimes (\neg a) \otimes b$. □

We have already seen in Remark 3.6 that $\otimes$-chains are not equivalent to their π-translation over V_Σ. From the above theorem, we can conclude that no translation with that property exists:

Corollary 5.4 (Undefinability of $\otimes$-chains over V_Σ) *There is no translation π^* from $\otimes$-chains to deontic formulas such that*

$$\vdash_{V_\Sigma} A_1 \otimes \ldots \otimes A_n \equiv \pi^*(A_1 \otimes \ldots \otimes A_n)$$

for all $\otimes$-chains $A_1 \otimes \ldots \otimes A_n$.

Proof. Assume that such a translation exists, and let C, C' be the two $\otimes$-chains from the proof of Theorem 5.3. Since C deontically subsumes C' and $\vdash_{V_\Sigma} C' \to \pi^*(C')$, we have $\vdash_{V_\Sigma} C \to \pi^*(C')$. Now because $\vdash_{V_\Sigma} \pi^*(C') \to C'$, we can conclude $\vdash_{V_\Sigma} C \to C'$, contradiction. □

From the proof of Corollary 5.4, we can extract the following observation: If in a violation logic every $\otimes$-chain is definable by a deontic formula, then the violation logic is faithful. However it is not so clear if definability of $\otimes$-chains is desirable. On a technical level, it trivializes the treatment of $\otimes$-chains, and in some sense deprives the $\otimes$-chains of their status as logical entities in their own right. If on the other hand definability does not hold, one has the burden of finding an intuition about $\otimes$-chains which is robust enough to allow for the acceptance and rejection of the principles proposed for them (such as the principle of faithfulness).

We remark that it is possible to have faithfulness without having definability of $\otimes$-chains: We obtain such a logic by formally adding to V_Σ the rule

$$\frac{\pi(C) \to \pi(C')}{C \to C'} \ .$$

(To show that $\otimes$-chains are not definable in the resulting logic, the argument in Remark 3.6 can be applied.)

Earlier on, we already mentioned the axiom ($\otimes$-I)

$$\left(a_1 \otimes \ldots \otimes a_n \wedge \left((\bigwedge_{i=1}^n \neg a_i) \to b_1 \otimes \ldots \otimes b_m \right) \right) \to a_1 \otimes \ldots \otimes a_n \otimes b_1 \otimes \ldots \otimes b_m$$

which appears in [3]. From ($\otimes$-I) we can prove $a \otimes (\neg a) \to a \otimes (\neg a) \otimes b$, the implication which was used as a counterexample to faithfulness in Theorem 5.3. This suggests the following question, to which we do not know the answer:

Is the extension of V_Σ by ($\otimes$-I) a faithful violation logic?

Finally, let us comment on the definability of $\otimes$-chains again. The easiest, but also the least illuminating way of achieving this is to add a scheme like $A_1 \otimes \ldots \otimes A_n \equiv \pi(A_1 \otimes \ldots \otimes A_n)$ to the violation logic at hand. It might also be of interest to have a 'natural' axiomatization of $\otimes$-chains which implies definability. For example, consider the following axiomatization of $\otimes$-chains:

$$\Sigma^* = \text{(O-detachment)} + (\otimes\text{-I}) + (\text{O}\otimes): Oa \to \otimes a$$

Theorem 5.5 *The violation logics $V_{\Sigma^*} + \Delta$ and $V_\Sigma + \Delta$ coincide on the chain negative fragment, and in $V_{\Sigma^*} + \Delta$ every $\otimes$-chain is definable via*

$$A_1 \otimes \ldots \otimes A_n \equiv \pi(A_1 \otimes \ldots \otimes A_n).$$

In particular, $V_{\Sigma^*} + \Delta$ is faithful.

Proof. $V_{\Sigma^*} + \Delta$ satisfies the premises of Corollary 4.4, and so its chain negative fragment coincides with that of $V_\Sigma + \Delta$. The Chain Soundness Lemma is satisfied in V_{Σ^*} because Σ^* contains (O-detachment). Hence for definability, it suffices to show by induction on n that

$$\vdash_{V_{\Sigma^*} + \Delta} \pi(A_1 \otimes \ldots \otimes A_n) \to A_1 \otimes \ldots \otimes A_n.$$

The base case $n = 1$ is precisely the axiom (O$\otimes$). For the induction step, we first note that the assumption $\pi(A_1 \otimes \ldots \otimes A_n \otimes A_{n+1})$ equals

$$\pi(A_1 \otimes \ldots \otimes A_n) \wedge \left((\bigwedge_{i=1}^{n} \neg A_i) \to OA_{n+1} \right)$$

by the definition of π. Now by the induction hypothesis, we can replace $\pi(A_1 \otimes \ldots \otimes A_n)$ by $A_1 \otimes \ldots \otimes A_n$ and OA_{n+1} by $\otimes A_{n+1}$. The axiom ($\otimes$-I) then yields $A_1 \otimes \ldots \otimes A_n \otimes A_{n+1}$ as desired. □

Hence if one accepts (O-detachment) and ($\otimes$-I) as true principles for $\otimes$-chains but rejects their definability, one must argue against the validity of the axiom (O$\otimes$).

6 Conclusion

We have isolated the 'chain negative fragment' of violation logic, and showed how questions in this fragment can be systematically reduced to questions in the underlying classical modal logic. This made it possible to use results about classical modal logic to reason in violation logic. On top of that, we have seen that truth in the chain negative fragment is to some extent independent of the axiomatization of $\otimes$-chains. Concerning future work, we believe that the main challenge for violation logic lies in the search for intuititive, yet sufficiently formal criteria which discriminate between different possible axiomatizations of $\otimes$-chains. One such criterion called 'faithfulness' was suggested here.

7 Acknowledgements

We are indebted to Guido Governatori for bringing the subjection of violation logics to our attention during a lecture at the Technical University of Vienna. Chris Fermüller read various versions of the draft and made helpful suggestions. Finally, we want to thank the two anonymous referees for their valuable comments.

References

[1] Blackburn, P., J. F. van Benthem and F. Wolter, "Handbook of modal logic," Elsevier, 2006.
[2] Calardo, E., G. Governatori and A. Rotolo, *A preference-based semantics for ctd reasoning*, in: F. Cariani, D. Grossi, J. Meheus and X. Parent, editors, *Deontic Logic and Normative Systems* (2014), pp. 49–64.

[3] Calardo, E., G. Governatori and A. Rotolo, *Sequence semantics for modelling reason-based preferences*, Fundam. Inform. **158** (2018), pp. 217–238.
    URL https://doi.org/10.3233/FI-2018-1647
[4] Chellas, B. F., "Modal logic: an introduction," Cambridge university press, 1980.
[5] Governatori, G., F. Olivieri, E. Calardo and A. Rotolo, *Sequence semantics for norms and obligations*, in: O. Roy, A. M. Tamminga and M. Willer, editors, *Deontic Logic and Normative Systems - 13th International Conference, DEON 2016, Bayreuth, Germany, July 18-21, 2018* (2016), pp. 93–108.
[6] Governatori, G., F. Olivieri, E. Calardo, A. Rotolo and M. Cristani, *Sequence semantics for normative agents*, in: M. Baldoni, A. K. Chopra, T. C. Son, K. Hirayama and P. Torroni, editors, *PRIMA 2016: Princiles and Practice of Multi-Agent Systems - 19th International Conference, Phuket, Thailand, August 22-26, 2016, Proceedings*, Lecture Notes in Computer Science **9862** (2016), pp. 230–246.
    URL https://doi.org/10.1007/978-3-319-44832-9_14
[7] Governatori, G. and A. Rotolo, *Logic of violations: A gentzen system for reasoning with contrary-to-duty obligations*, Australasian Journal of Logic **4** (2006), pp. 193–215.
[8] Lavendhomme, R. and T. Lucas, *Sequent calculi and decision procedures for weak modal systems*, Studia Logica **66** (2000), pp. 121–145.
[9] Takeuti, G., "Proof theory," Courier Corporation, 2013.
[10] Vardi, M. Y., *On the complexity of epistemic reasoning*, in: *[1989] Proceedings. Fourth Annual Symposium on Logic in Computer Science*, 1989, pp. 243–252.

Interpretive Normative Systems

Juliano Maranhão [1]

University of São Paulo Law School and Center for AI- C4AI/USP
Largo São Francisco, 92- São Paulo- Brasil

Giovanni Sartor [2]

CIRSFID-Alma AI, University of Bologna, Via Galliera 3, I-40121 Bologna, Italy
European University Institute, Law Department. Villa Salviati.Via Bolognese 156, 50139 Firenze, Italy

Abstract

We provide a formal definition of normative systems, which is compatible with different conceptions of the relation between law and morality. We embed a model for balancing values into an architecture of i/o logics representing conceptual, deontological and axiological rules. In particular, we provide a formal representation of three versions of the so-called Radbruch's formula, according to which legal obligations hold unless they reach a certain degree of immorality. Accordingly, we define eight different entailment relations, which correspond to eight different legal theories concerning the relation between law and morality.

Keywords: normative systems, input/output logics, balancing values.

1 Introduction

The regulation of human action has two sides. On the one side it aims to achieve certain values, i.e., goals that are socially desirable. Such values may consist in individual entitlements or rights (e.g. freedom of speech, property, privacy) or collective/social objectives (e.g. public health, national security, etc.). On the other side, the regulation specifies that certain actions may or may not be accomplished under certain antecedent conditions. The first is the dimension of consequentialism (also called teleology or axiology), according to which actions are evaluated according to their future impact on the relevant values: they are prohibited if they have a negative aggregated impact on the relevant values and they are permitted otherwise. The second is the dimension of deontology,

---

[1] Juliano Maranhão acknowledges the support by the Fundação de Apoio à Pesquisa do Estado de São Paulo (FAPESP 2019/07665-4) and the IBM Corporation to the Center for Artificial Intelligence (C4AI/USP).

[2] Giovanni Sartor has been supported by the H2020 European Research Council (ERC) Project "CompuLaw" (G.A. 833647)

according to which actions are evaluated according to the context in which they were accomplished: they are impermissible (or respectively permissible) if they are accomplished under conditions that trigger, through a rule, their prohibition (respectively permission).

The two dimensions should ideally be aligned, since the deontological rules in the regulation are meant to serve the values aimed at by the regulation. The alignment is successfully achieved when the circumstances under which rules prohibit (permit) an action correspond to the circumstances under which the action would be detrimental (favourable) to the relevant values. However, a mismatch is also possible: what is deontologically prohibited may be axiologically required (having a positive impact on the relevant values) and what is deontologically permitted may be axiologically prohibited. For simplicity's sake, we assume that axiological components only pertain to political morality, while deontological components only pertain to positively enacted law. However, as we shall remark later, our approach can also deal with the incorporation of axiological components in the positively enacted law.

A long standing problem in legal theory concerns exactly the criteria for the identification of valid law in case of mismatch between law and morality. In the contemporary debate, the difficulty rests on how to sustain the authority of legal rules while excepting their application when it would lead to morally unacceptable results.

Non-positivist theories, such as those put forward by R. Dworkin [5] and R. Alexy [2], affirm a necessary but nuanced relation between law and morality: on the one hand legal interpretation and argumentation may include evaluative efforts meant to align deontology and axiology; on the other hand, in some cases, immorality may entail legal invalidity. In particular, Alexy refers to the formula originally proposed by Gustav Radbruch [13] to determine the (in)validity of Nazi's laws: *laws enacted by proper authority and power are legally valid unless they reach an unbearable degree of immorality or injustice.*

Positivist theories on the other hand, reject the view that the identification of law is necessarily dependant on moral considerations, while accepting that the immorality of a law may justify its modification or even the refusal to apply it, when this would lead to morally unacceptable consequences [8].

In our framework, the identification of the obligations and permissions derived from the normative system vary according to the version of the Radbruch's formula assumed, which, in its turn, reflects a particular conception about the morality of Law.

Our effort has not only a theoretical import for legal philosophy, but also a practical import for the design of intelligent normative agents. In a human-centered AI, artificial agents must not blindly apply predefined rules, but also be able to determine how best to apply such rules and even refrain from complying with them when that might offend the underlying social values and individual rights.

In Section 2 we introduce the normative sets of conceptual, modulation, deontological and axiological rules. In Section 3 we introduce a logical architecture of i/o logics based on our concept of Normative Systems. In Section 4 we define an operator of axiological entailment. In Section 5 we introduce the concept of normative theories and we identify eight different legal theories regarding the relations between law and morality, based on three versions of Radbruch's formula.

2 Normative Sets

We shall use the term *"normative set"* to refer to sets of different kinds of rules: (i) a set of *conceptual rules*; (ii) a set of *modulation rules*; (iii) a set of *deontological rules* (iv) a set of *axiological rules*.

Conceptual rules consist in the ascription of a legal meaning or concept, *i.e.* they state that the entities described by certain factors *count as* (are to be classified as) instances of the ascribed concept (see [7]). We represent conceptual rules in the form (a, c) where a is the triggering factor (or conjunction of factors) and c is the ascribed concept. For instance, a conceptual rule stating that a message exchange stored in a mobile phone (sms) counts as "data" can be represented as (sms, dat).

Modulation rules specify the extent to which the presence of a factor affects the impact of actions on values. Such modulations reflect both causal connections (that the action, given the factor is likely to produce a certain individual or social outcome) and evaluative assessments (that the outcome of the action will count as an impact on the value). The values may consist in individual or social rights, moral principles, or collective goals.

We distinguish three kinds of modulation rules: baseline, intensifier and attenuator rules (following [4]). A baseline rule specifies that an action has a certain impact on a value, in the absence of relevant circumstances. That is, baseline modulation rules are those pairs where the body is a tautology, while intensifiers and attenuators are rules where the body is non-tautological. An intensifier rule specifies that the presence of a factor (the intensifier) increases the action's impact on the value (its index is positive). An attenuator rule specifies that the presence of a factor (the attenuator) decreases the action's impact on the value (its index is negative). We represent modulation rules in the form $(a, V^x)_i$, where a is the triggering factor, V is the affected value, x is the action at stake, and i is the extent of the modulation. For a baseline example, consider the rule specifying that the action consisting in the access to any item in a search by the police demotes the value of Privacy to the extent 0.2, which we model as $(\top, Privacy^{acc})_{.2}$. For an intensifier, consider the rule that the impact of this action on privacy is increased if the item is a mobile phone $(mob, Priv^{acc})_{.8}$. For an attenuator, consider that the impact is decreased if the mobile phone is not personal $(\neg pers, Priv^{acc})_{-.4}$.

We distinguish two kinds of rules establishing obligations or permssions, deontological and axiological ones. Such rules lead to deontic conclusions which may be in conflict.

Deontological rules link the (deontological) prohibition or permission of a given action to the presence of certain antecedent conditions. We model deontological rules in the form (a, x), where a is the triggering factor (or concept) and x is the obligatory or permitted action. For instance, we represent as $(\neg sord, \neg acc)$ the rule prohibiting police officers from accessing personal documents without a search & seizure order.

Axiological rules make the (axiological) obligation or permission of an action dependant on on the action's impact on a value. They are partitioned into two sets: those linking the prohibition of an action to a value demoted by that action; and those linking the permission of an action to a value promoted by that action. We represent axiological prohibitions in the form $(V^x, \neg x)_i$, where V is the value demoted by action x, which is consequently prohibited, and i is the weight of the value. We represent axiological permissions as $(V^x, x)_i$ where V is the promoted value, and x is the consequently permitted action. For instance, let us assume that access to a mobile phone by police officers demotes privacy, which is a reason for prohibiting it, while it promotes public safety, which is consequently a reason to permit it. We can model these rules as $(Priv^{acc}, \neg acc)_{.4}$ and $(Saf^{acc}, acc)_{.6}$.

3 Normative systems

Reasoning with each kind of rules (conceptual, modulation, deontological, or axiological) has different logical properties and therefore requires a different output operator in an architecture of i/o logics (for an introduction to i/o logics see [9] and [12]).

Let L be a standard propositional language with propositional variables and logical connectives: $\neg, \wedge, \vee, \rightarrow, \bot, \top$. Let $Val = \{V_1^x, V_2^x, ...V_1^y, V_2^y, ...\}$ be a set of values. We say that $N \subseteq G \times G$, where $G \in \{L, Val\}$ is a *normative set* and that each $r \in N$ is a *rule*. For any $A \subseteq G$, $N(G)$ is the image of N under G, that is $N(G) = \{x : (a, x) \in N, \text{ for some } a \in G\}$. We write simply $N(a)$ to abbreviate $N(\{a\})$. To state that x is the output of input a to normative set N, we may write $x \in out_i(N, a)$, or $(a, x) \in out_i(N)$. For any normative set N we define $body(N) = \{a : (a, x) \in N\}$.

Therefore, normative sets contain pairs of propositions or pairs linking a proposition or value to another proposition or value. In order to simplify the exposition, we shall consider actions as propositions (action-propositions), which will be the scope of deontic operators. We shall employ the classical consequence operator Cl. In this paper, it is possible that an action impacts different values. However possible combinations of values will be assessed in the balancing model, so we do not need to consider conjunctions of values or logical inferences among them. Given that we consider values as primitive entities, which are logically independent of each other and that no consequence relations among them are of interest, we shall consider weaker versions of i/o logics, where no consequence operator is applied to the output (of the set of modulation rules). We shall use the following i/o operators:

Definition 3.1 Let N be a normative set, $A \subseteq L$ and $\mathscr{V}$ the set of all maximal consistent sets v in classical propositional logic. Then, we define the following output operators:

(i) *simple minded*: $out_1(N, A) = Cl(N(Cl(A)))$
(ii) *weak*: $out_{1^-}(N, A) = N(Cl(A))$
(iii) *basic*: $out_2(N, A) = \bigcap \{out_1(N, v) : A \subseteq v, \text{ for } v \in \mathscr{V} \text{ or } v = L \}$
(iv) *weak basic*: $out_{2^-}(N, A) = \bigcap \{out_{1^-}(N, v) : A \subseteq v, \text{ for } v \in \mathscr{V} \text{ or } v = L\}$
(v) *basic reusable*: $out_4(N, A) = \bigcap \{out_1(N, v) : A \subseteq v \text{ and } out_1(N, v) \subseteq v, \text{ for } v \in \mathscr{V} \text{ or } v = L\}$

Definition 3.2 Let N be a normative set and $P \subseteq (L \times L)$ a set of explicit permissions. Then, $(a, x) \in perm_i(P, N)$ iff $(a, x) \in out_i(N \cup Q)$, for some singleton or empty $Q \subseteq P$.

One may combine normative sets N_1 and N_2 and output operators out_i, out_j, by making the output of a normative set (possibly joined with the input set) the input of the output operation on the other normative set, that is $out_{i,j}(N_1, N_2, A) = out_i(N_1, out_j(N_2, A) \cup I)$, where $I \in \{A, \emptyset\}$. We call *sequence* a chain of combinations of normative sets.

Definition 3.3 (*Normative System*) Let $A, I \subseteq L$. Let $N_1, ..., N_n, N$ be normative sets and $r \in \{0, 1\}$.
Then $(N_1^{out_1, r_1}, ..., N_n^{out_n, r_n})$, where out_j is the output operator associated to set N_j is a *sequence of normative sets* iff for all N_j, $1 \leq j \leq n$, it holds that $out_j, ... out_n(N_j, N_{j+1}, ..., N_n, A) = out_j(N_j, out_{j+1}, ..., out_n(N_{j+1}, ..., N_n, A) \cup I)$, where $N_j \subseteq N$ and $I = A$, if $r_i = 1$, or $I = \emptyset$, if $r_i = 0$. A *normative system* is a class of sequences of normative sets.

We shall write N^{i,r_1} as an abbreviation of N^{out_i, r_1} and $out_{k,l}(N, M, A)$ to abbreviate $out_k, out_l(N, M, A)$.

Our model constructs a particular structure or architecture of normative systems, where the set of conceptual rules (box C) contributes to the determination of which deontological rules and which value assessments are triggered. Following [10] we assume that the set of conceptual rules is governed by a basic reusable output operator and the set of deontological rules is governed by a basic output operator. Their combination is given by the identities: $out_{2,4}(O_d, C, A) = out_2(O_d, out_4(C, A) \cup A)$ and $perm_{2,4}(P_d, C, A) = perm_2(P_d, out_4(C, A) \cup A)$.

From now on, we may write O/P_d or O/P_v for referring to both obligation and permission rules, $O_{d/v}$ and $P_{d/v}$ for both deontological and axiological rules and $O/P_{d/v}$ to include all modalities. The value assessment employs the set of modulation rules and two sets of axiological rule. The set of modulation rules (M) links facts and concepts to the value-impacts of the action in the presence of such fact and concepts. It is governed by a weakened basic output operator. One set of axiological rules (P_v) links each value to the permission of the action that promotes it, and the other (O_v) links each value to the prohibitions of the action that demotes it. Both are governed by the axiological output operator $out_\succ$, defined in Section 4.

The combination of these normative sets is given by the following identity:

$out_{\succ,2^-,4}(O/P_v, M, C, A) = out_\succ(O/P_v, out_{2^-}(M, out_4(C, A) \cup A))$

Hence our discussion shall involve the following structures:

$\langle O/P_d, C \rangle = \{(O_d^{2,0}, C^{4,1}), (P_d^{2,0}, C^{4,1})\}$
$\langle O/P_v, M, C \rangle = \{(O_v^{\succ,0}, M^{2^-,0}, C^{4,1}), (P_v^{\succ,0}, M^{2^-,0}, C^{4,1})\}$
$\langle O/P_{d/v}, M, C \rangle = \langle O/P_d, C \rangle \cup \langle O/P_v, M, C \rangle$

The normative system is specified by indicating the rules of each normative set in the corresponding structure. The structure $\langle O/P_{d/v}, M, C \rangle$ of normative systems is represented in the figure below. The arrows indicate the direction of the outputs and inputs of each normative set.

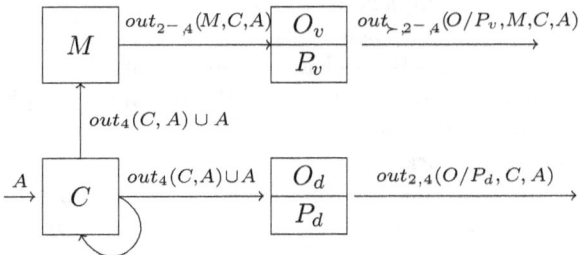

4 Axiological entailment

An axiological entailment presupposes a determination of the comparative moral merits of the choice of performing an action rather than abstaining from it. The action may consist in any behaviour, e.g., having an abortion rather that continuing the pregnancy or accessing an *sms* message, rather than respecting its confidentiality.

The comparison depends on the evaluations expressed by the quantitative indexes of modulation rules (for influence on impact on values) and axiological rules (for weighs of values). For generality's sake we assume that such indexes can take arbitrary numerical assignments within given ranges. These numbers can be restricted to any scales that may be convenient for the chosen domain of application. Here we shall use the positions (0,.2,.4,.6,.8,1) in the examples. What matters is that the numerical assignments reflect some relative importance of the elements at stake, as part of a reasoning with dimensions and magnitudes, and how such assessment of relative importance affects the outputs of the systems and its overall coherence.

4.1 Evaluation of axiological rules

The entailment of axiological rules may involve three kinds of rules –conceptual, modulation, and axiological ones–, so that their evaluation depends on the intensity of factors and the weights of values.

The evaluation model basically compares, for each given action, its impact on the set of values it promotes against its impact on the set of values it demotes, given the constellation of factors, i.e the context in which the action is performed. Two clarifications are of central importance to understand the model here proposed.

First, we only consider the assessment of impact of a single action on values and therefore we only compare the values promoted against the values demoted by that specific action, so that a decision takes place whether that action should or should not be performed on moral grounds. There is no room in this model to compare and decide among different and logically independent actions in terms of their impacts on values. Typically, a claim before a court questions the legality of a particular action and the court must decide whether that action under evaluation should be performed or not (should be forbidden or permitted, should be punished or not be punished). So we keep the same structure regarding its axiological evaluation. We acknowledge that there may be contexts where a judicial decision compares and chooses among alternative courses of action, for instance, between the consumer's right to receive a new product or to have his money back. However we shall leave this kind of value assessment to future work.

Second, we assume that the direction of impact of an action on a value –i.e., whether the action promotes or demotes the value– is invariant, although the extent of the promotion or demotion may be *intensified* or *attenuated* by the presence of factors in the context of performance. By saying that the direction of impact is invariant, we mean that irrespective of how many attenuating factors are taken into account, the impact of an action in the promotion of a particular value never shifts to the demotion of that value. And vice-versa the impact of the action at stake on the demotion of a value never shifts to its promotion.

Let us illustrate the rationality behind the model with an example. Suppose the rules of a condominium forbid people to take the elevator during the pandemics. Suppose now that one inhabitant has a medical emergency. Then one could evaluate whether following the rule would lead to immoral results. The factor "medical emergency" is an intensifier w.r.t the promotion of the value of the patient's health, which would lead to a permission to use the elevator. But now consider that the emergency does not hinder the patient's ability to walk (for instance, it is a toothache) and that she lives in the second floor. So the proportional influence of the set of factors on the promotion of the patient's health may become null or negative, but one would not say that the action of taking the elevator would now demote her health in that particular context. Actually the action still promotes health even in presence of those attenuating factors. But in such cases the proportional impact of the action is so low that it becomes morally irrelevant to legal considerations, that is, it will not play a role in a consideration whether to follow the rule or not. Hence, in the model here proposed, attenuating factors only affects the degree of moral impact of the action on a value.

Considering that the direction of impact of the action on a value is invariant, then for a given an action x, the set Val of values may be partitioned into the set of values Val^x_{Dem} which are demoted by the action and a set of values

Val^x_{Prom} which are promoted by the action. The relative importance or weight of each value, denoted by w_V, is given by a *weight function* $w : Val \longrightarrow [0, 1]$.

Both features, i.e. the direction of impact of the action on a value and the weight of the value may be directly represented in our architecture by defining the O_v box and P_v box respectively as $O_v = \{(V^x, \neg x)_{w_V} : V \in Val^x_{Dem}\}$ and $P_v = \{(V^x, x)_{w_V} : V \in Val^x_{Prom}\}$.

Let us now move to modulation rules. As noted in Section 2, the extent to which an action promotes or demotes the relevant values is determined by the baseline impact of the action and by the context (the constellation of factors) in which the action takes place. The influence of a factor on the action's impact on a value is given by the *modulation function* $\Delta : L^2 \times Val \longrightarrow [-1, 1]$. We denote by $\Delta^m_V(x)$ the influence of the modulating factor $m \in body(M)$ on the impact of the action $x \in L$ on the value $V^x \in Val$.

Considering that we shall not model the evaluation of sets of different actions, but only the impact of a single action on the promotion against the demotion of given values, we shall omit the reference to the action at stake in the indication of its impact on a value, i.e., we shall indicate such impact with $(V, \neg x)_w \in O_v$ and $(V, x)_w \in P_v$, rather than $(V^x, \neg x)_w \in O_v$ and $(V^x, x)_w \in P_v$.

If the influence of a factor m on a value V is positive ($\Delta^m_V > 0$), m is an *intensifier* of the impact of the action at stake on value V. If the influence is negative ($\Delta^m_V < 0$), then m is an *attenuator* of its impact on V. If there is no influence ($\Delta^m_V = 0$), m is *neutral*.

By a λ-evaluation we mean an evaluation assignment $\lambda_i = [\Delta_i, w_i]$, where Δ_i is a modulation function and w_i is a weight function, and we denote by Λ the set of all λ-evaluations. The proportional influence of a modulating factor m on value V, denoted by ϕ^m_V, is the product of the index of the modulation rule (indicating the intensification or attenuation due to the factor) and of the index of the axiological rule (indicating the weight of the value), that is:

Definition 4.1 (*Proportional influence of a modulating factor on a value*) Let $(m, V)_{\Delta^m_V} \in M$ and $(V, x)_{w_V} \in O/P_v$. Then: $\phi^m_V = \Delta^m_V \times w_V$

Now we extend the definition of proportional influence to cover the impact of a set of factors B on a set of values W, such an impact being the sum of the proportional influences of each factor.

Definition 4.2 (*Proportional influence of factors on values*) Let $Q = \{(m_1, V_1)_{i_1}, ..., (m_n, V_k)_{i_n}\} \subseteq M$ and $U \subseteq O/P_v$ such that $U = \{(V_1, x)_{j_1}, ..., (V_k, x)_{j_k}\}$. Then, for factors $B = body(Q)$ and values $W = body(U)$ we have:

$$\Phi^B_W = \sum_{\substack{1 \leq i \leq n \\ 1 \leq j \leq k}} \phi^{m_i}_{V_j}$$

4.2 Axiological Output

Given the above definitions, we are able to define the axiological output ($out_\succ$) operator. The idea is to compare the proportional impact of an action on the values it demotes *vis-à-vis* its impact on the values it promotes, considering only those values which are triggered by the input. The set $M(A)$ of the modulating factors involved in the comparison is the subset of $body(M)$, which is triggered by the input $A \subseteq L$, that is, $M(A) = Cl(A) \cap body(M)$. Since the normative system used in our model also includes conceptual rules in the sequence, we have $M(A) = (out_4(C, A) \cup A) \cap body(M)$. In their turn, the sets of values involved in the comparison are those subsets of $body(O_v)$ and of $body(P_v)$, which are triggered by the output of the set of modulation rules. That is, we are going to compare set of demoted values triggered by input A, i.e., $O_v(A) = out_i(M, A) \cap body(O_v)$, against the set of promoted values triggered by input A, i.e., $P_v(A) = out_i(M, A) \cap body(P_v)$. In our architecture, where the sequence includes conceptual rules, we compare $O_v(A) = out_{2-,4}(M, C, A) \cap body(O_v)$ against $P_v(A) = (out_{2-,4}(M, C, A) \cap body(P_v)$.

If, for a given constellation of factors, the proportional impact of the action on the *demoted* values is positive and stronger than its proportional impact on the promoted values, then there is an overall axiological prohibition to do it. On the other hand, if the proportional impact of the action on the *promoted* values is positive and stronger that its proportional impact on the demoted values, then there is an axiological explicit permission to do it. Hence, we have the following definition of the value output operator $out_\succ$. We may write simply $O/P_v(A)$ to abbreviate $\Phi^{M(A)}_{O/P_v(A)}$.

Definition 4.3 [*Axiological output*] Consider $NS = \langle O/P_v, M \rangle$, $A \subseteq L$ and $x \in L$. Then $x \in out_\succ(O_v, M, A)$ iff: (i) $y \in O_v(out(M, A)))$, (ii) $x \in Cl(y)$ and (iii) $0 < O_v(A) > P_v(A)$. The same holds, *mutatis mutandis*, for $out_\succ(P_v, M, A)$.

It is worth mentioning that, contrary to all the other output operators discussed so far, the axiological output is defeasible, *i.e.* it does not satisfy the property of Strengthening the Input, according to which if $b \vdash a$, and $(a, x) \in out(N)$, then $(b, x) \in out(N)$ (see example 4.4).

When assessing whether there is convergence of axiological and deontological outputs we need to compare the intensity of the action's impact on each value, relatively to given contexts (constellations of input factors).

A modulating factor may trigger more than one value (directly or indirectly, *i.e.* by detaching other modulating factors) and the impact on a single value may be affected by different modulating factors. Therefore it is interesting to compare modulating factors in terms of the influence of each on the impact on the aggregate of values, as well as to observe how much each value is impacted by the action in a given context.

In order to compare modulating factors, we call the quantity $\Phi^{\{m\}}_{O/P_v(m)}$, where $O/P_v(m) = \{V : V \in out_{2-}(M, m)\}$, the *strength* of the modulating factor $m \in body(M)$. It represents the sum of the all impacts of the action (on the promoted values or on the demoted values), which are triggered by the modulating factor m. We are going to abbreviate by $m_1 > m_2$ the comparison of strengths of different modulating factors $\Phi^{\{m_1\}}_{O/P_v(m_1)} > \Phi^{\{m_2\}}_{O/P_v(m_2)}$.

In order to compare how much different values are impacted by a given input, we shall use $M(a)$ to denote the set of modulating factors triggered by input a, that is $M(a) = out_4(C,a) \cap body(M)$, and we are going to abbreviate the quantity $\Phi_{\{V\}}^{M(a)}$ by $V(a)$, which represents the extent of the action's impact on a single value V, given input a. So, given $a,b \in L$, the expression $V_1(a) > V_2(b)$ denotes $\Phi_{\{V_1\}}^{M(a)} > \Phi_{\{V_2\}}^{M(b)}$.

Let us illustrate these notations with a hypothetical λ- evaluation, which represents the *Riley vs California* case, where the values of privacy, public safety, and property rights were affected. The US case law before that decision included a rule according to which an officer could access personal property when arresting an individual due to a criminal offense. This rule could be explained by the following considerations on the underlying value impacts: the modulating factor "arrest" intensifies the promotion of public safety (through the action search) so as to outweigh the extent to which the factors property and "personal data" intensify the demotion (through the same action) of property rights and privacy respectively. However, as considered by the court, if the item collected is a mobile phone, then the negative impact on privacy is intensified to the extent that the promotion of public safety is outweighed. This led the court to introduce an exception for searches involving mobile phones.

Example 4.4 [Riley vs California] Consider $NS = \langle O/P_v, M, C \rangle$:
$O_v = \{(Priv, \neg acc)_{.4}, (Pright, \neg acc)_{.4}\}$, $P_v = \{(Saf, acc)_{.6}\}$
$M = \{(\top, Priv)_{.2}, (\top, Pright)_{.0}, (\top, Saf)_{.2}, (dat, Priv)_{.6}, (prop, Pright)_{.4},$
$(arrest, Saf)_{.8}, (mob, Priv)_1\}$, $C = \{(mob, data), (mob, prop)\}$
We have that $Saf(arrest) = 0.6$, $Pright(prop) = 0.16$, $Priv(data) = 0.32$ and the factor mobile played a strong intensifying role with $Priv(mob) = 0.72$. The strength of factors each factor is $mob = 0.88$, $arrest = 0.6$, $dat = 0.32$ and $prop = 0.16$. So, we have $mob > arrest > dat > prop$ and, comparing the values, it holds that $Saf(arrest) > Pright(prop) + Priv(dat)$, but $Priv(mob) + Pright(mob) > Saf(arrest)$.

Hence, the balancing above explains the shift in the U.S case law given the factor "mobile phone", as we have that $acc \in out_\succ(P_v, M, C, \{arrest, prop, data\})$, but it also holds that $\neg acc \in out_\succ(O_v, M, C, \{arrest, mob\})$. That is, it is morally admissible for the police to access property items and personal data in an arrest, but it is immoral to access the content of a mobile phone, for the impact on privacy, in that case, is severely intensified (a mobile phone is conceptually both property and data).

Based on the strength of the impacts on the values triggered by an input, we define the proportional impact of an entailed axiological rule as the difference between the values promoted (demoted) and demoted (promoted) in the entailment.

Definition 4.5 *(Proportional impact of a rule)* Consider a normative system $NS = \langle O/P_v, M, C \rangle$, and $(a,x) \in out_\succ(O_v, M, C)$. Then, $\sigma(a,x) = O_v(a) - P_v(a)$ is the proportional impact of the rule (a,x). The same holds, *mutatis* mutandis, for $(a, x) \in out_\succ(P_v, M, C)$.

In Example 4.4, the proportional impact of prohibiting access to the content of a mobile phone in an arrest is $\sigma(mob \wedge arrest, \neg acc) = 0.28$.

5 Normative theories

A normative system is the object of assertions by jurists (legal doctrine) who describe the systems through normative propositions, i.e., statements that certain obligations and permissions hold given certain factors, according to a normative system.

Normative propositions, while being descriptive of a given normative system (as viewed by the interpreter), also reflect the evaluative aspects of the described system, namely, the ascription of intensities of influence (to modulation rules) or the ascription of weights of values (to axiological rules). Such λ-evaluations contribute to determine the axiological obligations/permissions delivered by the system, and consequently, what normative propositions would be true about it.

Definition 5.1 Let $NS = \langle O/P_{d/v}, M, C \rangle$ be a normative system and $b, x \in L$. The for a given λ evaluation:
$NS \models^\lambda \mathbb{O}_d(x/b)$ iff $x \in out_{2,4}(O_d, C, b)$
$NS \models^\lambda \mathbb{P}^-_d(x/b)$ iff $\neg x \notin out_{2,4}(O_d, C, b)$
$NS \models^\lambda \mathbb{P}^+_d(x/b)$ iff $x \in perm_{2,4}(O_d, P_d, C, b)$
$NS \models^\lambda \mathbb{O}_v(x/b)$ iff $x \in out_{>,2^-,4}(O_v, M, C, b)$
$NS \models^\lambda \mathbb{P}^-_v(x/b)$ iff $\neg x \notin out_{>,2^-,4}(O_v, M, C, b)$
$NS \models^\lambda \mathbb{P}^+_v(x/b)$ iff $x \in out_{>,2^-,4}(P_v, M, C, b)$

Each normative proposition describes an entailed deontological or axiological rule, with the exception of negative permissive propositions, which describe the non-derivability of such a rule. Thus, following Alchourrón [1], we distinguish a negative sense of permission $\mathbb{P}^-_{d/v}(x/b)$, as the absence of prohibition, from a positive sense of permission as an entailed deontological or axiological permission $\mathbb{P}^+_{d/v}(x/b)$.

A normative theory Th^λ_{NS} about a normative system NS is the set of all normative propositions describing the rules entailed by that normative system based on the λ-evaluation, that is on given modulation and weight functions: $Th^\lambda_{NS} = \{\alpha : NS \models^\lambda \alpha\}$. We say that a normative system leads to a conflict, relatively to a certain input factors when, given that input, the systems delivers the prohibition and the permission of the same action. We distinguish conflicts of normative propositions according to the kind of rules which contribute to produce the conflict:

Definition 5.2 (*Consistency, Coherence and Stability of normative theories*)
For any given $b \in L$, a normative theory is:
b-inconsistent iff $\bot \in out_2(O/P_d, b)$; b-incoherent iff $\bot \in out_{2,4}(O/P_d, C, b)$;
b-λ-unstable iff there is $x \in L$ for which $\{\mathbb{O}_v(\neg x/b), \mathbb{P}_d(x/b)\} \subseteq Th^\lambda_{NS}$ or $\{\mathbb{O}_d(\neg x/b), \mathbb{P}_v(x/b)\} \subseteq Th^\lambda_{NS}$

In other words, inconsistency captures cases in which deontological rules directly deliver incompatible conclusions, proper incoherence the case in which the conflict of deontological rules is triggered by a conceptual classification, and proper instability the case in which deontological rules are in conflict with axiological rules. We also may say that a normative theory is strongly stable, relatively to an input, if the corresponding deontological normative propositions are matched by corresponding axiological proposition, and that it is weakly stable, if the deontological propositions are not conflicted by axiological propositions.

We propose here an interpretation of Radbruch's formula, based on the concept of "proportional impact" of a rule, as the key to define different entailment relations and, accordingly, different legal theories.

Definition 5.3 (*Negative Radbruch's Formula*) Let $NS = \langle O/P_{d/v}, M, C \rangle$ be a normative system, $b, x \in L$ and λ an evaluation, then:
(i) $NS \models^\lambda_{rad-} \mathbb{P}^+(x/b)$ iff $NS \models^\lambda \mathbb{P}^+_d(x/b)$ and it is not the case that $NS \models^\lambda \mathbb{O}_v(\neg x/b)$ and $\sigma(b,x) \geq r$, where r is a treshold index;
(ii) $NS \models^\lambda_{rad-} \mathbb{O}(x/b)$ iff $NS \models^\lambda \mathbb{O}_d(x/b)$ and it is not the case that $NS \models^\lambda \mathbb{P}^+_v(\neg x/b)$ and $\sigma(b,x) \geq r$.

According to Definition 5.3, morality only has a censorial role: it produces no legal conclusions and only excludes the application of highly immoral deontological rules.

Definition 5.4 (*Positive Radbruch's Formula*) Let $NS = \langle O/P_{d/v}, M, C \rangle$ be a normative system, $b, x \in L$ and λ an evaluation, then:
(i) $NS \models^\lambda_{rad+} \mathbb{P}^{+/-}(x/b)$ iff $NS \models^\lambda \mathbb{P}^{+/-}_d(x/b)$ and it is not the case that both $NS \models^\lambda \mathbb{O}_v(\neg x/b)$ and $\sigma(b,x) \geq r$; otherwise $NS \models^\lambda \mathbb{O}(\neg x/b)$
(ii) $NS \models^\lambda_{rad+} \mathbb{O}(x/b)$ iff $NS \models^\lambda \mathbb{O}_d(x/b)$ and it is not the case that both $NS \models^\lambda \mathbb{P}^+_v(\neg x/b)$ and $\sigma(b,x) \geq r$; otherwise $NS \models^\lambda \mathbb{P}^+(\neg x/b)$

According to Definition 5.4, morality has both a censorial role and a generative one, delivering outputs with high moral merit (proportional impact above threshold).

Definition 5.5 (*Dual Radbruch's Formula*) Let $NS = \langle O/P_{d/v}, M, C \rangle$ be a normative system, $b, x \in L$ and λ an evaluation, then:
(i) $NS \models^\lambda_{dual} \mathbb{P}^{+/-}(x/b)$ iff $NS \models^\lambda \mathbb{P}^{+/-}{}_v(x/b)$ and it is not the case that both $NS \models^\lambda \mathbb{O}_d(\neg x/b)$, and $\sigma(b,x) \leq r$; otherwise $NS \models^\lambda \mathbb{O}(\neg x/b)$
(ii)$NS \models^\lambda_{dual} \mathbb{O}(x/b)$ iff $NS \models^\lambda \mathbb{O}_v(x/b)$ and it is not the case that both $NS \models^\lambda \mathbb{P}^+{}_d(\neg x/b)$ and $\sigma(b,x) \leq r$; otherwise $NS \models^\lambda \mathbb{P}^+(\neg x/b)$

According to Definition 5.5, morality has both a censorial role and a generative one. The difference from Definition 5.4 lies in those cases where axiological outputs are not conflicted by deontological rules. By the Dual Radbruch's formula all such axiological outputs are delivered by the legal system, while in the Positive Radbruch Formula an axiological output is only delivered when it exceeds the moral threshold.

Our definitions of the Radruch formulas also cover the limit cases when the threshold is null ($r = 0$) or infinite ($r = \infty$). This allows us to capture eight different legal theories, which differs with respect to the specific question whether external considerations of morality may generate valid law.

Definition 5.6 Let $NS = \langle O/P_{d/v}, M, C \rangle$ be a normative system, $b, x \in L$, λ an evaluation and r a given threshold in a Radbruch's formula. Then:

- Closed Positivism[3] : $NS \models^\lambda_{cpos} \mathbb{O}/\mathbb{P}(x/b)$ iff $NS \models^\lambda_{rad} \mathbb{O}/\mathbb{P}(x/b)$ and $r = \infty$
- Open Positivism: $NS \models^\lambda_{opos} \mathbb{O}/\mathbb{P}(x/b)$ iff $NS \models^\lambda_{dual} \mathbb{O}/\mathbb{P}(x/b)$ and $r = \infty$
- Strong Censorial Non-Positivism: $NS \models^\lambda_{scnp} \mathbb{O}/\mathbb{P}(x/b)$ iff $NS \models^\lambda_{rad^-} \mathbb{O}/\mathbb{P}(x/b)$ and $r = 0$
- Weak Censorial Non-Positivism: $NS \models^\lambda_{wcnp} \mathbb{O}/\mathbb{P}(x/b)$ iff $NS \models^\lambda_{rad^-} \mathbb{O}/\mathbb{P}(x/b)$ and $0 < r < \infty$
- Strong Generative Non-Positivism: $NS \models^\lambda_{sgnp} \mathbb{O}/\mathbb{P}(x/b)$ iff $NS \models^\lambda_{rad^+} \mathbb{O}/\mathbb{P}(x/b)$ and $r = 0$
- Weak Generative Non-Positivism: $NS \models^\lambda_{wgnp} \mathbb{O}/\mathbb{P}(x/b)$ iff $NS \models^\lambda_{rad^+} \mathbb{O}/\mathbb{P}(x/b)$ and $0 < r < \infty$
- Absolute Natural Law: $NS \models^\lambda_{anl} \mathbb{O}/\mathbb{P}(x/b)$ iff $NS \models^\lambda_{dual} \mathbb{O}/\mathbb{P}(x/b)$ and $r = 0$
- Relative Natural Law: $NS \models^\lambda_{rnl} \mathbb{O}/\mathbb{P}(x/b)$ iff $NS \models^\lambda_{dual} \mathbb{O}/\mathbb{P}(x/b)$ and $0 < r < \infty$

For *Closed Positivism* only deontological outputs are delivered, while axiological outputs are irrelevant to legal validity. For *Open Positivism* all deontological outputs are delivered together with the axiological outputs that are consistent (not conflicting) with them. For *Strong Censorial Non-Positivism*, only those deontological outputs are valid, which are consistent with all axiological outputs. For *Weak Censorial Non-Positivism*, the deontological outputs are delivered, which are not inconsistent with those highly ranked axiological outputs above the assumed threshold. For *Strong Generative Non-Positivism*, all axiological outputs are delivered plus those deontological outputs that are consistent with them. For *Weak Generative Non-Positivism*, those axiological outputs with high proportional impact (above the threshold) are delivered together with those deontological outputs which are consistent with them. For *Absolute Natural Law*, only axiological outputs are delivered. For *Relative Natural Law*, axiological outputs of high proportional impact are delivered independently of consistency with deontological outputs, while axiological outputs of lesser impact are delivered only if consistent with delivered deontological outputs. One could say, in a theory resembling Finnis' [6], that Relative Natural Law would contend that matters with low moral significance (e.g. coordination problems) could be left to discretionary choices by authorities, while sensitive matters should be ruled by moral reasoning.

In the table below we present the output of the normative systems for

---

[3] using the rad^- or the rad^+ entailment relation results in the same positivist theory of validity.

Riley v California	Output of the normative system	Explanation of court decision
Closed Positivism	Permitted	Change the Law
Open Positivism	Permitted	Change the Law
Strong Censorial Non-Positivism	Gap	Fill
Weak Censorial Non-Positivism	Gap/Permitted	Fill/Change the Law
Strong Generative Non-Positivism	Forbidden	Apply the Law
Weak Generative Non-Positivism	Forbidden / Permitted	Apply/Change the Law
Absolute Natural Law	Forbidden	Apply the Law
Relatve Natural Law	Forbidden/ Permitted	Apply/Change the Law

each legal theory concerning the *Riley* case and the corresponding theoretical explanation for the court's decision to prohibit access to the mobile phone. Notice that the positively enacted law provides the deontological output that $\mathbb{P}^+{}_d(acc/mob \land arrest)$, while the axiological output is $\mathbb{O}_v(\neg acc/mob \land arrest)$ with a proportional impact $\sigma(mob \land arrest, \neg acc) = 0.28$, according to the assumed λ-evaluation. The normative propositions describing the content of the normative system would be either a positive permission, a prohibition or a negative permission (a gap). According to these theories the decision of the U.S court –forbidding access to the content of the mobile phone– would have different explanations: that the court changed the existing law (*contra legem*), that it applied the existing law (*secundum legem*), or that it filled a gap by discretion creating new law (*extra legem*).

With respect to the weak versions of non-postivism, the outcome of the normative systems and the corresponding explanations would depend on whether the theory assumes a Radbruch's threshold above or below 0.28. Suppose the theory assumed a threshold $r = 0.6$. Then the weak non-positivist theory would maintain that it is permitted to access the content of a mobile phone in an arrest, since the reached level of immorality is below the threshold. But now suppose that that positive law authorized the search of an individual's mobile phone independently of any arrest. Then the axiological output would be $\mathbb{O}_v(\neg acc/mob)$ with a proportional impact $\sigma = 0.76$ (thus above the 0.6 threshold). Therefore the final outcome, for accessing the content of the mobile phone independently of an arrest would be either a gap (weak non-positivism) or a prohibition to access (strong non-positivism).

6 Final Remarks

By combining an architecture of i/o logics and a model of balancing values, we have proposed a formal concept of normative system, where obligations and permissions may be assessed in terms of their impact on the promotion or demotion of moral values. Based on this concept and on three interpretations of the so-called Radbruch's formula, we have formally defined eight different conceptions of the connection between law and morality. The above analysis assumes that axiological consideration are external to the positively enacted law, pertaining to political morality. However, our approach is also compatible with the assumption that axiological considerations are internal to the positively enacted law, as legal principle o fundamental rights, in particular those enshrined in a Constitution. In future investigations, following the latter approach, we may define corresponding versions of Constitutionalism from classical negative (censorial) constitutionalism to different generative forms of neo-constitutional moralism and principialism.

The framework here proposed brings together two parallel lines of research in AI & law: on the one hand the study of the role of values in case-based legal argumentation ([14] and [15]), and on the other hand the study of statutory interpretation as the dynamical modification of combined normative sets, including conceptual qualification, conditional rules and values ([3], [10] and [11]). One of the difficulties in the last approach is how to set up and formalize criteria for the choice between alternative normative systems that satisfy a revision function. This paper offers a conceptual and formal basis for the balancing of values that may be used as criteria both to trigger and to choose between possible results of revisions of normative systems. The modelling of constructive legal interpretation by revision functions based on the framework here proposed will be left to future work, as well as the effort to characterize the axiological output operator here advanced.

References

[1] Alchourrón, C., *Logic of norms and logic of normative propositions*, Logique et analyse **12** (1969), pp. 242–68.
[2] Alexy, R., "The Argument from Injustice. A Reply to legal positivism," Oxford University Press, 2003.
[3] Boella, G. and L. van der Torre, *A logical architecture of a normative system*, in: *Deontic Logic and Artificial Normative Systems - DEON 2006* (2006), pp. 24–35.
[4] Dancy, J., "Ethics Without Principles," Oxford University Press, 2004.
[5] Dworkin, R. M., "Law's Empire," Kermode, 1986.
[6] Finnis, J. M., "Natural Law and Natural Rights," Clarendon, 1980.
[7] Grossi, D., J.-C. Meyer and F. Dignum, *The many faces of counts-as: A formal analysis of constitutive rules*, Journal of Applied Logic **6** (2008), pp. 192–217.
[8] Hart, H. L. A., *Positivism and the separation of law and morals*, in: *Essays in Jurisprudence and Philosophy*, Clarendon, 1983 pp. 49–87, (1st ed. 1958.).
[9] Makinson, D. and L. van der Torre, *Input/output logics*, Journal of Philosophical Logic **29** (2000), pp. 383–408.
[10] Maranhão, J., *A logical architecture for dynamic legal interpretation*, in: *Proceedings of the Eight International Conference on AI and Law ICAIL '17,*, ACM Press, 2017 pp. 129–38.
[11] Maranhão, J. and E. G. de Souza, *Contraction of combined normative sets*, in: *Deontic Logic and Normative Systems: 14th International Conference, DEON 2018*, Springer, 2018 pp. 247–261.
[12] Parent, X. and L. van der Torre, *Input/output logic*, in: J. Horty, D. Gabbay, X. Parent, R. van der Meyden and L. van der Torre, editors, *Handbook of Deontic Logic and Normative Systems, Volume 1*, College Publications, 2013 pp. 499–544.
[13] Radbruch, G., *Statutory lawlessness and supra-statutory law*, Oxford Journal of Legal Studies **6** (2006), pp. 1–11, (1st ed. 1946.).
[14] Sartor, G., *The logic of proportionality: Reasoning with non-numerical magnitudes*, German Law Journal (2013), pp. 1419–57.
[15] Sartor, G., *Consistency in balancing: from value assessments to factor-based rules*, in: D. Duarte and S. Sampaio, editors, *Proportionality in Law: An Analytical Perspective*, Springer, 2018 pp. 121–36.

A Logical Analysis of Freedom of Thought

Réka Markovich [1]

University of Luxembourg

Olivier Roy

University of Bayreuth

Abstract

This paper studies the logical form and properties of one prominent category of epistemic rights: the freedom of thought and belief. We do so in the broadly Hohfeldian formalization of rights developed by [Markovich, 2020,Markovich, 2019], but extended with tools from doxastic logic. The resulting analysis reveals subtle differences in the way freedom of thought can be analyzed, and how these differences affect the logical properties of this doxastic right and the normative positions it incorporates.

Keywords: rights and duties, normative positions, epistemic rights, doxastic logic, action logic, multi-modal logic, conceptual analysis

1 Introduction

The freedom of thought and belief is one of the most fundamental and intimate human rights declared not only by the United Nations' Universal Declaration of Human Rights, but also by the European Convention of Human Rights and the vast majority of western constitutions. This paper studies freedom of thought and belief from a logical point of view.

We investigate freedom of thought as an epistemic right. According to a new approach in philosophy, an epistemic right is one that protects and governs the distribution and accessibility of epistemic goods [Watson, 2021]. For the development of a formal analysis within the theory of the normative (or Hohfeldian) positions (see Section 2), we assume that an epistemic right, in the narrow sense, is a right pertaining to a certain state of knowledge or belief of the right holder. [2] Next to the freedom of thought, such rights include someone's right to know—or to not know—her medical test's result, the citizens' right

---

[1] This work was supported by the Fonds National de la Recherche Luxembourg through the project Deontic Logic for Epistemic Rights (OPEN O20/14776480).

[2] In a broader sense, rights, where not the right-holder's, but the duty-bearer's epistemic state is concerned by the right, can also be considered as epistemic rights, such as the right to be forgotten, or the right to privacy, studied in e.g. [Aucher et al., 2011,Aucher et al., 2010,Cuppens and Demolombe, 1996].

to know the declaration of assets of members of parliament, the consumers' right to not be misled by advertisements, or, for example, the right to truth: the right, in the case of grave violations of human rights, of the victims and their families or societies to have access to the truth of what happened. In this paper, we focus on freedom of thought and belief, and investigate its content and logical properties through an analysis using Hohfeldian conceptions.

While many systems at the intersection of epistemic logic, deontic logic, and the logic of agency have been developed e.g. [Broersen, 2011], the notion of epistemic rights *as normative positions* has not yet been investigated in logic. The deontic logic literature has been mostly concerned with epistemic obligation, Åqvist's paradox [Åqvist, 1967,Hulstijn, 2008], and the theory of knowledge-based obligations [Pacuit et al., 2006]. In their work on privacy policies, Aucher et al. [Aucher et al., 2011,Aucher et al., 2010] investigated both the obligation and the permission to know something, differentiating between obligatory and permitted knowledge and obligatory and permitted messages.

The notion of epistemic rights *per se* is not completely new to the philosophical literature, but so far it has been restricted to the right to believe when discussing justification in epistemology, see for instance [Dretske, 2000], or focused on epistemic obligation [Feldman, 1988,Stapleford, 2012]. It is a recent development that epistemic rights are discussed *as* a group of legal rights by Watson [2018, 2019, 2020], a categorization with which we agree. Before, if epistemic rights were discussed together with normative positions, it was always in comparison or contrast with them [Wenar, 2003,Altschul, 2021]. Wenar even claims that the "epistemic (...) realms contain no claims, powers, or immunities" [Wenar, 2015]. This paper can be seen as challenging that view, by showing that analyzing the epistemic rights using Hohfeldian categories yields interesting insights.

The paper is structured as follows. In section 2, we briefly introduce the theory of normative positions that we will use, and section 3 presents the details of the language and semantics, including the doxastic operators. In section 4 we turn to freedom of thought proper, and section 5 concludes by pointing to directions for future work. The contribution of this paper is conceptual, and the mathematical observations we make are elementary. The proofs are thus omitted, and we leave aside the study of the meta-logical properties (completeness, decidability and tracktability) of the systems that we are using.

2 The Theory of Normative Positions

The theory of normative positions goes back to Hohfeld's typology [Hohfeld, 1923] and its seminal formalizations in [Kanger, 1971] and [Lindahl, 1994]. See [Makinson, 1986] and [Sergot, 2013] for a critical assessment of this tradition. New formal approaches to Hohfeldian rights have been developed and presented by several authors, for instance [Sartor, 2005], [Gelati et al., 2004,Gelati et al., 2002], [Governatori and Rotolo, 2008], and more recently in [Dong and Roy, 2017], and [Markovich, 2020,Markovich, 2019]. The work in this paper builds on the latter, but we change the semantics somewhat and extend it with tools from epistemic logic.

Hohfeld proposed to distinguish between four types of atomic right-positions (Claim, Privilege, Power, Immunity) and their correlative duty positions (Duty, No-claim, Liability, Disability). See Figure 1, taken from [Markovich, 2020]. The right-positions in the left square are claim-right and privilege. A claim-

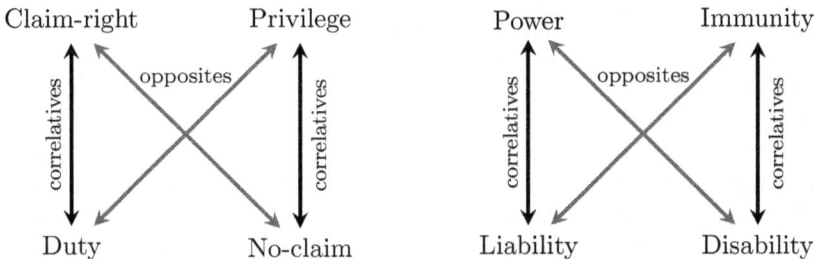

Fig. 1. The Hohfeldian atomic types of rights, and their correlative

right of an agent concerns the counter-party's actions. The counter-party has an obligation to do the certain thing, and this obligation is directed to the right-holder. Hohfeld calls this a duty, in the narrow sense. The seller's right against the buyer that the latter pays the purchase price, for instance, is a claim-right. The freedom or privilege [3] to do something, on the other hand, is understood as not being the subject of a claim-right coming from the counter-party. A land owner's right to use her own land refers to her privilege in the sense that the counter-party does not have the claim-right against her to refrain from that use. [4] Privilege can thus be seen as a directed version of the standard (weak) permission in deontic logic.

The normative positions in the right square capture the agent's ability to change an (other) agent's normative positions. For that reason, they have been called "higher order" or *capacitative* [Fitch, 1967]. They thus capture the norm-changing potential, or lack thereof, of an agent [Dong and Roy, 2017, Markovich, 2020]. A land owner, for instance, has a right—here a power—to sell her land and the other agent(s), for instance, the one who so far has rented a house on it, is (are) exposed, that is, liable (in the Hohfeldian sense), to this change: his relevant normative positions will change. The land owner has immunity, though, regarding her neighbor selling the land: the neighbor is unable to change the owner's normative positions concerning the land. This counts as a disability, meaning that he does not have a power to do that.

---

[3] 'Freedom' is an often used alternative for 'privilege' in the literature dealing with Hohfeld.
[4] In this particular case the owner has, in fact, such a privilege against any counter-party. The owner's position is a so-called absolute position or a multital (vs paucital) right. Her property rights, thus this privilege of hers too, are to be considered against every other agent. [Simmonds, 2001] and [Markovich, 2020].

While it can be argued that the type of deontic actions—changing someone's normative positions—that are involved in the capacitative square is of a different kind than the actions that claim-rights and privileges concern [Jones and Sergot, 1996,Dong and Roy, 2017,Markovich, 2020], here we analyze them using a simple combination of alethic and agentive modalities. We do so because these are actually actions whose execution is *possible* if and only if the actor has the power to do so. This simplification allows us to focus on the formalization of freedom of thought while keeping our logical language and its interpretation relatively simple. In the Conclusion we briefly discuss the consequences of adopting a more dynamic modeling of actions for some of the results presented below.

The main characteristic of Hohfeldian theory is that the normative positions are inherently relational. Not only duties are directed, but also all the other positions: for instance, the dual[5] of a duty, a privilege (or freedom) is to be interpreted as being free from a given other agent's, the counterparty's claim-right. One can, however, also express the idea of an absolute duty in this relational framework. Hohfeld himself differentiated between paucital and multital positions: in the former we consider one given relation between two parties, while in the latter one agent is a party in a series (conjunction) of such relations [Simmonds, 2001,Markovich, 2020]. We will formalize freedom of thought as such a multital right: we have it against/with regard to everyone else.[6]

Hohfeld's theory identifies four atomic types of rights to resolve the terminological confusion arising from (over)using the word 'right' while meaning different concepts. Legal language, though, still uses the word 'right' or sometimes, as in our case, 'freedom', to refer to different positions, or, often, their complex combinations. As we argue below, this is indeed the case for freedom of thought. It consists of a combination of at least three atomic types: a multital privilege, a multital claim-right, and a multital immunity.

3 Language and Semantics

We analyze freedom of thought using a combination of standard deontic logic augmented with directed operators [Markovich, 2020], and epistemic/doxastic logic.

Definition 3.1 Let A be a finite set of agents and Φ a set of propositional letters. The language $\mathcal{L}$ is defined as follows:

$$p \in \Phi \mid \varphi \wedge \psi \mid \neg \varphi \mid \{E_a \varphi \mid O_{a \to b} \varphi \mid B_a \varphi\}_{a,b \in A} \mid \Box \varphi$$

$\mathcal{L}$ thus extends the propositional logic with four modalities. E_a is the agency modality and should be read as "agent a sees to it that...". $O_{a \to b}$

---

[5] Or "opposite" in the less precise Hohfeldian term.
[6] As a matter of fact, *everyone* has it against everyone else (which would be a complete directed graph from the graph-theoretical point of view [Markovich, 2019]), but for the sake of simplicity, here we analyze *one's* freedom of thought.

is a directed obligation modality, and should be read as "agent a has a duty towards b that...". B_a, on the other hand, is a doxastic modality, to be read as "agent a believes that...". The $\Box$ modality is the universal, alethic modality "it is necessary that." All these modalities have duals: the weak permissions operator, i.e. $P_{a \to b}...$, which stands for $\neg O_{a \to b} \neg...$; $\langle B_a \rangle...$ which stands for $\neg B_a \neg...$; and $\Diamond...$, which stands for $\neg \Box \neg...$.

We make the following assumptions regarding the logical behavior of these modalities. We take the deontic modalities $O_{a \to b}$ to be normal modalities validating the D axiom, i.e. $O_{a \to b} \varphi \to P_{a \to b} \varphi$. So the deontic fragment of our language is standard deontic logic. For the agentive modalities E_a, we take them to be non-normal, validating only substitution under logical equivalence and the T axiom ($E_a \varphi \to \varphi$). As it turns out the logical behavior of freedom of thought will be strongly influenced by what additional assumptions are made about the logic of E_a, for instance that the agents always see to it that necessarily true formulas hold ($\Box \varphi \to E_a \varphi$), or that the operator is regular (if $\Box(\varphi \to \psi)$ then $E_a \varphi \to E_a \psi$). Observe that from the T axiom it also follows that no agent can see to it that a contradiction holds. The doxastic modalities B_a are assumed to be normal modalities validating D ($B_a \varphi \to \langle B_a \rangle \varphi$). We do not, in particular, assume that the belief modalities are either positively or negatively introspective.

Given these assumptions, the language $\mathcal{L}$ is interpreted over frames containing a neighborhood function for each E_a, a deontic ideality relation for each $O_{a \to b}$, and a doxastic accessibility relation for each B_a.

Definition 3.2 Let A be a finite set of agents. A frame $\mathfrak{F}$ is a tuple of the following form:
$$\mathfrak{F} = \langle W, \{f_a, R_a^B, R_{a,b}^O\}_{a,b \in A} \rangle$$

Here W is set of possible worlds. The function $f_a : W \to \wp\wp(W)$ is a neighborhood function such that, for all $w \in W$ and $X \in f_a(w)$, we have $w \in X$. Both $R_a^B \subseteq W^2$ and $R_{a,b}^O \subseteq W^2$ are serial, binary relations. A model $\mathcal{M}$ is a frame $\mathfrak{F}$ together with a valuation function $V : \Phi \to \wp(W)$.

With this in hand the truth conditions of formula of our language is defined in the standard way. We have only defined explicitly the case for the modalities.

Definition 3.3 Let $||\varphi|| = \{w : \mathcal{M}, w \models \varphi\}$. Then:

- $\mathcal{M}, w \models E_a \varphi \Leftrightarrow ||\varphi|| \in f_a(w)$
- $\mathcal{M}, w \models \mathbf{O}_{a \to b} \varphi \Leftrightarrow \forall w'(w R_{a,b}^O w' \Rightarrow \mathcal{M}, w' \models \varphi)$
- $\mathcal{M}, w \models \mathbf{B}_a \varphi \Leftrightarrow \forall w'(w R_a^B w' \Rightarrow \mathcal{M}, w' \models \varphi)$
- $\mathcal{M}, w \models \Box \varphi \Leftrightarrow \forall w', \mathcal{M}, w' \models \varphi$

When a formula of the form $\Box \varphi$ is true in a model we will say that φ is necessarily true, and similarly for $\Box \neg \varphi$ and "necessarily false". Otherwise we will say that φ is contingent in a model. Validity in models, frames, and classes thereof, are defined as usual.

Since we do not make any specific assumptions regarding the interaction between these modalities, the set of validities over our intended class of frames is completely axiomatized by all propositional tautologies, the logic ET for the agentive modality E_a, KD for $O_{a \to b}$ and B_a, S5 for $\Box$, and the standard inclusion axioms relating the universal modality $\Box$ to the other operators in the language.

4 Freedom of Thought

We are now ready to address freedom of thought. We first provide some legal foundation for our analysis, and then move to the formalization itself.

4.1 Legal Foundations

The United Nation's document, The Universal Declaration of Human Rights [7], as its name says, declares what are the human rights. [8] Article 18 is about freedom of thought: "Everyone has the right to freedom of thought, conscience and religion; this right includes freedom to change his religion or belief (...)." Article 19 says: "Everyone has the right to freedom of opinion (...); this right includes freedom to hold opinions without interference (...)." [9] [10] The United Nation's Office of the High Commissioner for Human Rights' (OHCHR) general Comment Nr. 22 interpreting Article 18 and 19 says: [11]

> The right to freedom of thought, conscience and religion (which includes the freedom to hold beliefs) in article 18.1 is far-reaching and profound; it encompasses freedom of thought on all matters, personal conviction and the commitment to religion or belief (...). [Article 18] does not permit any limitations whatsoever on the freedom of thought and conscience or on the freedom to have or adopt a religion or belief of one's choice. These freedoms are protected unconditionally, as is the right of everyone to hold opinions

---

[7] The document was proclaimed by the United Nations General Assembly in Paris on 10 December 1948 (General Assembly resolution 217 A) as a common standard of achievements for all peoples and all nations. https://www.un.org/en/universal-declaration-human-rights/

[8] We do not go into the philosophical discussion on where these rights come from (natural law vs. legal positivism), nor on what political or legal legitimacy the UN has. We only analyze formally what the declared human rights' content and implications are based on the official interpretation. The reader unwilling to accept the Declaration as a legal source because of the legitimacy questions regarding international law can instead consider a national constitution's relevant article, the wording of which is very much alike the Declaration.

[9] While Article 19 is about freedom of opinion and its expression, we believe that its internal part—freedom of opinion—is to be discussed together with the internal part of freedom of thought referring to the same thing as far as the formalization is concerned.

[10] The freedom of thought, conscience and religion has an internal and an external realm: the internal concerns the beliefs without concerning their expression, while the external concerns the manifestation of beliefs, such as religious practices—just like the separation between freedom of opinion and its expression. We intentionally cite only the parts of the Declaration concerning the internal realm, as our current investigation is only concerned with this.

[11] General Comment No. 22: The right to freedom of thought, conscience and religion (Art. 18): . 30/07/93. CCPR/C/21/Rev.1/Add.4, General Comment No. 22. (General Comments) https://bit.ly/37T15Uc

without interference in article 19 (...) The Committee observes that the freedom to "have or to adopt" a religion or belief necessarily entails the freedom to choose a religion or belief, including the right to replace one's current religion or belief with another or to adopt atheistic views, as well as the right to retain one's religion or belief. Article 18 bars coercion that would impair the right to have or adopt a religion or belief, including the use of threat of physical force or penal sanctions to compel believers or non-believers to adhere to their religious beliefs and congregations, to recant their religion or belief or to convert.

In the terminology of normative positions, freedom of thought, as its name suggests, includes a freedom (or privilege), but a multital one. There is no duty of ours toward anyone regarding our beliefs' content. But this privilege in itself would be a rather weak position, so freedom of thought also involves protections, in two ways. On one hand, it means a claim-right against everyone else not to interfere with it. What interfering with practicing a freedom of thought would be is, of course, debatable. One might raise the question whether it is really possible at all, for example, to force someone to believe in something. The common reference in this regard is Orwell's dystopia, 1984, and its thought police and thought crime concepts.[12] Whether forcing someone to adopt or change a belief is possible in reality is rather a question that psychology or neuroscience could answer; we do not need to commit ourselves in this matter, we only need to represent that it is forbidden. A freedom and a joint claim-right against everyone else is a frequent combination: these are what are usually called civil liberties.[13] On the other hand, the Declaration, according to its official interpretation, is also ruling out the possibility of changing this freedom (for example, by a country introducing penal sanctions, that is, duties to accept some specific view). This is what Hohfeld identifies as immunity: the other party's inability to change our normative positions (for example, in this case, imposing a duty as to what to believe in).

4.2 Formalization

Three components have, thus, to be analyzed: the freedom, the claim-right, and the immunity components of freedom of thought. We look at them in turn. Throughout we assume that the right-holder is a given agent a, and that the right bears on a's doxastic attitudes towards a given proposition φ.

---

[12] In [Hulstijn, 2008], when combining deontic and epistemic logics, the possible need for the—as Hulstijn refers to it—'freedom of thought' axiom $K_i\varphi \to PK_i$ is raised "to exclude the definition of 'thought crimes' in Orwell's 1984" (the author then recants this as it would go against the purpose of access control policies). We don't think knowledge would be a good description of thought, especially in the context of this freedom. We, therefore, will use a belief operator.

[13] In the reception of Hohfeld, it was raised that he missed identifying this kind of liberty as an atomic type of right, but as is shown in [Markovich, 2020], this combination of Hohfeldian notions expressing what a civil liberty is rejects these opinions: a civil liberty (that [Bentham, 1843] calls vested liberty, while [Wright, 1963] calls simply as 'right'), is not an atomic, but a compositional or molecular type of right.

4.2.1 Freedom

As observed in the previous section, freedom of thought consists partly in being, indeed, a freedom in the Hohfeldian sense. One has no duty towards anybody else not to believe φ:

$$\bigwedge_{b\in A} \neg O_{a\to b}\neg \mathbf{B}_a\varphi$$

For readability we will use the dual $P_{a\to b}...$ instead of $\neg O_{a\to b}\neg...$:

$$\bigwedge_{b\in A} P_{a\to b}\mathbf{B}_a\varphi \qquad \text{(FoT-F-}B_a\varphi\text{)}$$

This formulation does not rule out that a is in fact under the obligation to believe φ, i.e. that she is not permitted not to believe it. This privilege is indeed consistent with an(y) agent b having a claim-right against a that she (a) believes φ. To rule that out one can instead require that a has a multital privilege to hold any of the three possible attitudes towards φ: belief, disbelief, or suspending judgment.

$$\bigwedge_{b\in A}(P_{a\to b}\mathbf{B}_a\varphi \land P_{a\to b}\mathbf{B}_a\neg\varphi \land P_{a\to b}(\langle\mathbf{B}_a\rangle\varphi \land \langle\mathbf{B}_a\rangle\neg\varphi)) \qquad \text{(FoT-F-}a\text{-}\varphi\text{)}$$

Semantically, this condition restricts the application of freedom of thought to propositions that are contingent in a particular model.[14] We have indeed assumed that the doxastic modalities B_a are normal and that beliefs are consistent, i.e. they validate the D axiom. Under these assumptions about the belief modalities, (FoT-F-a-φ) predicts that freedom of thought does not apply to necessary truths or necessary falsities.

This restriction to contingent formulas is a consequence of the idealizations that we have made regarding the belief operators. The standard logical models of belief assume that the logic of modalities is either KD45 or K4.3 [Fagin et al., 2003,Stalnaker, 2006,Pacuit, 2013]. Here the modalities B_a are weaker. Yet, they still represent the agents as consistent and logically omniscient. This has the direct consequence that agents do not have the freedom to, for instance, suspend judgment about necessary truth or necessary falsities. A number of solutions to the logical omniscience problem have been proposed (c.f. [Hawke et al., 2019] and references therein), but introducing them here would go beyond the scope of this paper. This is a question of the adequate model of belief, not primarily of the logical form of freedom of thought. Even if we were to decide to adopt a weaker doxastic logic, this would arguably not affect the logical form of (FoT-F-a-φ). What this would change is its logical behavior.

---

[14] It furthermore imposes a richness condition on the set of states that are normatively ideal from the perspective of a towards b. This richness assumption appears less controversial than the restriction to the contingent formula that we discuss this the main text.

4.2.2 Claim-right

Freedom of thought also consists of a multital claim-right to refrain from interfering with us practicing our freedom, that is, forcing us to not hold certain beliefs. As with freedom, we will first consider this claim-right as bearing on simply believing φ, and consider later on the consequences of extending it simultaneously to disbelief and suspending judgment.

A first attempt at capturing this claim-right is in terms of the others' correlated duty to refrain:

$$\bigwedge_{b\in A} \mathbf{O}_{b\to a}\neg E_b \neg B_a \varphi \qquad \text{(FoT-C1)}$$

This first attempt is, perhaps, overly strong. It rules out any attempt to convince someone, or simply teaching or instructing. This is not what freedom of thought forbids. It is rather the *forceful* intervention into someone's beliefs. The idea that someone is forced or prevented, against her will, to hold or form certain beliefs is beyond the scope of the language and the models that we are working with. Something which *is*, however, within the expressive power of our language is the idea that interventions which not only result in a not believing something, but rather make this impossible, are forbidden.

$$\bigwedge_{b\in A} \mathbf{O}_{b\to a}\neg E_b \neg \Diamond B_a \varphi \qquad \text{(FoT-C-}a\text{-}\Diamond\text{)}$$

The logical behavior of this formalization of the claim-right turns out to be not completely satisfactory, and it depends heavily on the assumptions that one makes regarding the logic of E_a. Recall that the T axiom for E_a rules out that a does the impossible. On the other hand, since $\Box$ is the universal modality, we get $\mathcal{M}, w \models \Diamond B_a \varphi$ implies $\mathcal{M}, v \models \Diamond B_a \varphi$ for any v in W. In other words, $\mathcal{M}, w \models \Diamond B_a \varphi$ implies that $\neg \Diamond B_a \varphi$ is true nowhere in $\mathcal{M}$. But then it is also impossible for b to actively rule out that possibility, which in turn entails that a has a (trivial) claim-right against b regarding a's belief in φ.

Observation 1 *For any model $\mathcal{M}$ and state w, if $\mathcal{M}, w \models \Diamond B_a \varphi$, then for all v we have $\mathcal{M}, v \models \bigwedge_{b\in A} \mathbf{O}_{b\to a}\neg E_b \neg \Diamond B_a \varphi$.*

Perhaps surprisingly, the other direction of the implication is also valid, provided that one makes the additional assumption that in any state there is at least something trivial that a does. Recall that this additional assumption translates syntactically to $\Box\varphi \to E_a\varphi$, which semantically corresponds to the fact that W is an element of $f_a(w)$ for any a and w.[15] This is a property that some agency operators satisfy, notably any normal ones like the "Chellas stit" [Belnap et al., 2001]. We do not need full normality, though. It is sufficient that f_a "contains the unit" [Pacuit, 2017].

---

[15] In standard neighborhood semantics without the universal modality the syntactic correspondent of this condition is $E_a \top$.

Observation 2 *For any model $\mathcal{M}$ where all f_a contain the unit, and state w, if $\mathcal{M}, w \models \bigwedge_{b \in A} \mathbf{O}_{b \to a} \neg E_b \neg \Diamond B_a \varphi$ then $\mathcal{M}, w \models \Diamond B_a \varphi$.*

FoT-C-a-$\Diamond$ and $\Diamond B_a \varphi$ thus become equivalent when we assume that agents can always see to it that necessary truths hold. This is a form of deontic collapse: the claim-right component of freedom of thought, an "ought", collapses to a modal fact about a's belief, an "is". To the extent that one sees this as an undesirable consequence of this particular model, this can be used as yet another argument against assuming that $\Box \varphi \to E_a \varphi$. Classical agency operators, e.g. in [Kanger and Kanger, 1966] or the *dstit* and the *astit* [Belnap et al., 2001] also invalidate this principle. On the other hand, the culprit is not *only* the assumption that necessary truths are always (trivially) seen to it that. For one thing the direction from $B_a \varphi$ to FoT-C-a-$\Diamond$ follows just from assuming that agents are not doing impossible things, which is a plausible and in any case much more common assumption. Furthermore, the full equivalence follows rather from the *combination* of assuming $\Box \varphi \to E_a \varphi$ and the fact that $\Diamond B_a \varphi$, given a model, is never contingent. This suggests that FoT-C-a-$\Diamond$ might not be quite the right analysis of this claim-right.

As an alternative to (FoT-C-a-$\Diamond$) we could instead express the claim-right not as bearing on the sheer possibility of holding a particular belief, but instead on being forced to *adopt* a particular belief, here viewed as something that the agent actively does.[16] This would give the following:

$$\bigwedge_{b \in A} \mathbf{O}_{b \to a} \neg E_b \neg E_a B_a \varphi \qquad \text{(FoT-C-}a\text{-}E\text{)}$$

FoT-C-a-E is logically independent both from FoT-C1 and FoT-C-a-$\Diamond$. It furthermore avoids the ought-is collapse that we observed for the latter. As before, however, we gain some logical interactions between these different formalizations of the claim-right by assuming $\Box \varphi \to E_a \varphi$. Indeed, with that additional assumption FoT-C-a-E implies FoT-C-a-$\Diamond$:

Observation 3 *For any model $\mathcal{M}$ where all f_a contain the unit, and state w, if $\mathcal{M}, w \models \bigwedge_{b \in A} \mathbf{O}_{b \to a} \neg E_b \neg E_a B_a \varphi$ then $\mathcal{M}, w \models \bigwedge_{b \in A} \mathbf{O}_{b \to a} \neg E_b \neg \Diamond B_a \varphi$.*

The converse direction still fails, however, even when all f_a contains the unit. So even in this case FoT-C-a-E avoids the ought-is collapse.

As we did for the privilege to believe, the claim-right component of freedom of thought can of course be expanded to the three possible attitudes that an

---

[16] Using an action operator in front of the belief operator to refer to some kind of agency regarding one's own belief, on the one hand, accords well with the phrasing of the OHCHR comment on the Declaration talking about to *have* or *adopt* a belief; and, on the other hand, has its epistemological foundations in the view of doxastic voluntarism [Chignell, 2018].

agent can hold with respect to a particular formula φ.

$$\bigwedge_{b \in A} \mathbf{O}_{b \to a}(\neg E_b \neg E_a B_a \varphi \wedge \neg E_b \neg E_a B_a \neg \varphi \wedge \neg E_b \neg E_a(\langle B_a \rangle \varphi \wedge \langle B_a \rangle \neg \varphi))$$

(FoT-C-a-φ)

As before, this expansion also restricts the claim-right to consistent formulas, even without the assumption that agents always see to it that necessary propositions hold. Indeed, if φ is necessary in a particular model then the second and the third conjuncts become false because $\neg \varphi$ becomes necessarily false. If φ is instead necessarily false, then it is the first and the third conjuncts that become false.

Another notable fact regarding FoT-C-a-φ, is that this formula is consistent even in the case that a and b are the same agent. So imposing the claim-right in that case boils down to saying that everyone has a duty towards herself not to force herself to hold a particular belief regarding φ. From the legal point of view, this is rather questionable. This conclusion follows from the fact that we do not make any assumptions regarding the iteration of agency operators, so it could of course be avoided by, for instance, assuming that refraining from refraining, i.e. $\neg E_a \neg E_a...$, is equivalent to doing $E_a...$—c.f. again the discussion in [Belnap et al., 2001]. On the other hand, this conclusion could be avoided without entering this substantial debate about doing and refraining, by simply restricting the multitality to all agents $b \neq a$.

4.2.3 Immunity

The last component of freedom of thought is the immunity it involves. As we observed above, freedom of thought is viewed as inalienable and indispensable. No legal statement nor legislative act could take that freedom away. Any act entailing the negation of freedom of thought would turn out to be an invalid law. In other words, such action is not possible, which translates into an Hohfeldian immunity.

$$\bigwedge_{b \in A} \neg \Diamond (E_b(\neg(\text{FoT-F-}a\text{-}\varphi)) \vee E_b(\neg(\text{FoT-C-}a\text{-}\varphi)))$$

(FoT-I-a-φ)

Unlike our first formulation of the freedom and the claim-right constituents of freedom of thought, (FoT-I-a-φ) implicitly covers the three possible attitudes that a can take towards φ (believing, disbelieving, and suspending judgment).

This formulation of immunity behaves differently from its freedom and claim-right components when it comes to φ being necessarily true or necessarily false. Recall that if φ is necessarily true or necessarily false in a particular model then both the freedom and the claim-right become necessarily false. This is not the case for (FoT-I-a-φ). This formula is satisfiable even when φ is not contingent. If, however, we assume $\Box \varphi \to E_a \varphi$, then, as before, (FoT-I-a-φ) becomes always false when φ is not contingent.

There is one important aspect which is not covered, however, by (FoT-I-a-φ), namely that this immunity might apply recursively, so to speak, to limit everyone's powers to change that very immunity. One could indeed interpret the wide-ranging character of freedom of thought in that sense, i.e. that it also protects itself.[17] The expressive power of our language is too weak to capture such cases of self-reference, so we limit ourselves to some general observations.

The self-referential character of this immunity could be captured by more expressive languages containing fixed-points operators, for instance the modal mu-calculus [Pratt, 1981,Kozen, 1983]. This language extends basic modal ones like ours with the smallest and largest fixpoint operators μ and ν, as well as propositional variables, here simply x. One can use these additional resources to capture the self-reference in the immunity as follows:

$$\mu x. \bigwedge_{b \in A} \neg \Diamond (E_b \neg (\text{FoT-F-}a\text{-}\varphi) \vee E_b \neg (\text{FoT-C-}a\text{-}\varphi) \vee E_b \neg x)$$

This formula should be interpreted as saying that it is impossible for any agent b to either negate the freedom, the claim-right, or *this very immunity*, expressed by the variable x, here bound by the operator μ. The scope of this operator indicates what the variable x implicitly refers to. So once unpacked, this self-reference through x contains another self-reference, which needs to be unpacked, and so on, creating an ascending hierarchy of higher-order immunities regarding lower-order ones.

Without going into the details of this potential fixed-point extension of our language, we can already observe that the variable x is in the scope of an even number of negations, which means that the formula itself semantically corresponds to a monotone operator, which in turns guarantees the existence of smallest and largest fixpoints. So this formulation would be recursive, but not viciously circular.

4.2.4 Logical properties of freedom of thought

We are now in position to put together the three components of freedom of thought, and highlight some of their logical properties. First, to recast, our proposal for the formalization of a's freedom of thought regarding proposition φ is the following.

$$(\text{FoT-F-}a\text{-}\varphi) \wedge (\text{FoT-C-}a\text{-}\varphi) \wedge (\text{FoT-I-}a\text{-}\varphi) \qquad (\text{FoT-}a\text{-}\varphi)$$

---

[17] While we intentionally refrained from becoming involved in the natural law–positive law debate, here we need to mention that considering the UN Declaration's article to be self-referential might greatly depend on what philosophical assumptions one has: according to a natural law approach, the whole Declaration can be considered as merely descriptive; it might be said, therefore, that the impossibility of changing or taking away freedom of thought doesn't come from this very article, it comes because of people's inability to intervene with what is there by nature (which is often referred to as the inalienability of human rights). This way, the point of regarding the immunity not to be self-referential is that whatever is written in the Declaration does not change people's immunity concerning any change in their human rights, with their freedom of thought among them. We leave the discussion of this approach and its logical consequences, though, to future work.

That is, we analyze freedom of thought as a conjunction of three different multital, complex normative positions. These constituting normative positions are themselves complex because they cover all three possible doxastic attitudes that a can hold towards φ.

This complex right can be reformulated as a conjunction of three corresponding statements of freedom of thought regarding believing φ, disbelieving φ, and suspending judgment. In other words, (FoT-a-φ) is equivalent to the conjunction of a freedom, a claim-right and an immunity regarding a believing φ, a freedom, a claim-right and an immunity regarding disbelieving φ, and similarly for suspending judgment. Indeed, the freedom part of this right is simply a conjunction of three freedoms, one for each doxastic attitude. One obtains a similar conjunction for the second part, the claim-right, by observing that these are conjunctions of normal obligation operators, which of course distribute over conjunctions. Finally, the immunity can be rewritten with a $\Box$ operator scoping over a conjunction of negated agency statements, and this $\Box$ is also a normal modality. This observation is important to the extent that some of the atypical logical behavior that we have observed in the previous sections, for instance regarding necessary truths or falsities, can be avoided, if one wishes, by formulating restricted versions of freedom of thought, applicable for instance only to one of the three possible doxastic attitudes.

The logical behavior of (FoT-a-φ) is otherwise very limited. Necessary truth can be substituted in its scope, but otherwise all three properties that constitute normal modalities fail for this formula.

Observation 4

(i) *Substitution under logical equivalence holds:* $\Box(\varphi \leftrightarrow \psi)$ *and (FoT-a-φ) together imply (FoT-a-ψ).*

(ii) *Closure under Conjunction fails: (FoT-a-φ) and (FoT-a-ψ) together do not imply (FoT-a-$\varphi \wedge \psi$).*

(iii) *Regularity fails:* $\Box(\varphi \to \psi)$ *and (FoT-a-φ) together do not imply (FoT-a-ψ)*

(iv) *Necessitation fails:* φ *being valid in our class of models does not entail (FoT-a-φ).*

5 Conclusion

We analyzed the logical structure and properties of freedom of thought. We argued, *contra* [Wenar, 2015], that there is a theoretical basis for viewing this right as a particular case of Hohfeldian positions, i.e. a combination of freedom, claim-right and immunity, and that these can be analyzed using a combination of deontic logic, doxastic logic, and logic of agency. We then proceeded to study the logical behavior of these rights, and showed how this behavior depends on making particular assumptions on its constitutive belief, obligation, and agency operators. This logical analysis has also allowed to show an ought-is collapse for one formalization of the claim-right, and highlighted the potential recursivity of so-called inalienable immunities. This last point constitutes a natural next step to extend our analysis, both from a legal and a logical point of view.

One important modeling choice that should be revisited in future work is to capture the agency operator in a "static" way, following the tradition in the theory of normative positions [Kanger, 1971,Sergot, 2001] or in stit theory [Belnap et al., 2001], for instance. This choice turns out to be crucial in explaining the ought-is collapse for (FoT-C-a-$\Diamond$). We conjecture that this collapse would have not occurred had we used dynamic modalities in the style of Dynamic Epistemic Logic [Pacuit, 2013], as we do in [Markovich and Roy, 2021] for the right to know, or update semantics [Klein and Marra, 2020], to study the effect of the agents' actions.

Another important point that we have not touched on is the relation between freedom of thought and so-called conscientious objections. The OHCHR comment writes:

> The Covenant does not explicitly refer to a right to conscientious objection, but the Committee believes that such a right can be derived from article 18, inasmuch as the obligation to use lethal force may seriously conflict with the freedom of conscience and the right to manifest one's religion or belief.

The possibility to decline an otherwise existing duty is a power in Hohfeldian terms. We already showed that a (multital) claim-right, a (multital) freedom and a (multital) immunity are all part of the freedom of thought, so having this power would mean that all the four atomic right positions are incorporated in this one human right. Aiming at the formal representation of this power's derivability from the freedom brings up some considerations that we haven't made, though. On the one hand, the OHCHR interpretation talks about a very special duty that can be refused: using lethal force, but there are other cases in the legal literature where the notion of conscientious objection comes up, for instance medical practitioners not providing certain treatments to their patients [Shanawani, 2016], so the crucial point in formalization should be the reason of a "serious conflict" with beliefs, which refers to a supposed possibility of incompatibility between given beliefs and given actions. This brings us—as the comment also refers to it—to the external realm of the freedom of thought that we have intentionally omitted from our investigation so far: the manifestation, the freedom of choosing one's actions accordingly. There the immunity, for instance, is not absolute: duties can be imposed regarding our actions even if these have something to do with the (otherwise) free manifestation of our beliefs. Also the power we discuss here is not absolute: while one could make up a religion declaring paying taxes incompatible with one's innermost beliefs, courts would hardly accept rejection of paying taxes counting as a conscientious objection. This, of course, can be interpreted also not as defeasibility of the power to reject actions being in serious conflict with our conscience, but a need to define the notions of conscience and serious conflict more precisely. We leave this investigation to later papers.

References

[Altschul, 2021] Altschul, J. (2021). Epistemic entitlement. *Internet Encyclopedia of Philosophy*.

[Aucher et al., 2011] Aucher, G., Boella, G., and Torre, L. (2011). A dynamic logic for privacy compliance. *Artificial Intelligence & Law*, 19(2/3):187 – 231.

[Aucher et al., 2010] Aucher, G., Boella, G., and Torre, L. V. D. (2010). Privacy policies with modal logic: the dynamic turn. In *Deontic Logic in Computer Science (DEON 2010)*, pages 196–213.

[Belnap et al., 2001] Belnap, N. D., Perloff, M., and Xu, M. (2001). *Facing the future: agents and choices in our indeterminist world*. Oxford University Press, Oxford.

[Bentham, 1843] Bentham, J. (1843). *The works of Jeremy Bentham*. W. Tait ; Simkin, Marshall, and Co., Edinburgh ; London.

[Broersen, 2011] Broersen, J. M. (2011). Deontic epistemic stit logic distinguishing modes of mens rea. *J. Appl. Log.*, 9(2):137–152.

[Chignell, 2018] Chignell, A. (2018). The Ethics of Belief. In Zalta, E. N., editor, *The Stanford Encyclopedia of Philosophy*. Metaphysics Research Lab, Stanford University, spring 2018 edition.

[Cuppens and Demolombe, 1996] Cuppens, F. and Demolombe, R. (1996). A deontic logic for reasoning about confidentiality. In Brown, M. A. and Carmo, J., editors, *Deontic Logic, Agency and Normative Systems, DEON '96: Third International Workshop on Deontic Logic in Computer Science, Sesimbra, Portugal, 11-13 January 1996*, Workshops in Computing, pages 66–79. Springer.

[Dong and Roy, 2017] Dong, H. and Roy, O. (2017). Dynamic logic of power and immunity. In *Logic, Rationality, and Interaction : 6th International Workshop, LORI 2017, Sapporo, Japan, September 11-14, 2017, Proceedings*, volume 10455 of *Lecture Notes in Computer Science*, pages 123–136. Springer, Berlin, Heidelberg.

[Dretske, 2000] Dretske, F. (2000). Entitlement: Epistemic rights without epistemic duties? *Philosophy and Phenomenological Research*, 60:591?606.

[Fagin et al., 2003] Fagin, R., Moses, Y., Halpern, J. Y., and Vardi, M. Y. (2003). *Reasoning about knowledge*. MIT press.

[Feldman, 1988] Feldman, R. (1988). Epistemic obligations. *Philosophical Perspectives*, 2:235–256.

[Fitch, 1967] Fitch, F. B. (1967). A Revision of Hohfeld's Theory of Legal Concepts. *Logique et Analyse*, 10(39/40):269–276.

[Gelati et al., 2002] Gelati, J., Governatori, G., Rotolo, A., and Sartor, G. (2002). Actions, institutions, powers. Preliminary notes. In *International Workshop on regulated Agent-Based Social Systems: Theories and Applications*, pages 131–147.

[Gelati et al., 2004] Gelati, J., Rotolo, A., Sartor, G., and Governatori, G. (2004). Normative autonomy and normative co-ordination: Declarative power, representation, and mandate. *Artificial Intelligence and Law*, 12(1):53–81.

[Governatori and Rotolo, 2008] Governatori, G. and Rotolo, A. (2008). A computational framework for institutional agency. *Artificial Intelligence and Law*, 16(1):25–52.

[Hawke et al., 2019] Hawke, P., Özgün, A., and Berto, F. (2019). The fundamental problem of logical omniscience. *Journal of Philosophical Logic*, pages 1–40.

[Hohfeld, 1923] Hohfeld, W. N. (1923). Fundamental legal conceptions applied in judicial reasoning. In Cook, W. W., editor, *Fundamental Legal Conceptions Applied in Judicial Reasoning and Other Legal Essays*, pages 23–64. New Haven : Yale University Press.

[Hulstijn, 2008] Hulstijn, J. (2008). Need to know: Questions and the paradox of epistemic obligation. In van der Meyden, R. and van der Torre, L. W. N., editors, *Deontic Logic in Computer Science, 9th International Conference, DEON 2008, Luxembourg, Luxembourg, July 15-18, 2008. Proceedings*, volume 5076 of *Lecture Notes in Computer Science*, pages 125–139. Springer.

[Jones and Sergot, 1996] Jones, A. and Sergot, M. (1996). A formal characterization of institutionalised power. *Journal of the IPGL*, 3:417–443.

[Kanger, 1971] Kanger, S. (1971). New foundations of ethical theory. In Hilpinen, R., editor, *Deontic Logic: Introductory and Systematic Readings*, pages 36–58. D. Reidel, Dordrecht.

[Kanger and Kanger, 1966] Kanger, S. and Kanger, H. (1966). Rights and parlamentalism. *Theoria*, 32:85–115.

[Klein and Marra, 2020] Klein, D. and Marra, A. (2020). From oughts to goals: A logic for enkrasia. *Studia Logica*, 108(1):85–128.

[Kozen, 1983] Kozen, D. (1983). Results on the propositional μ-calculus. *Theoretical computer science*, 27(3):333–354.

[Lindahl, 1994] Lindahl, L. (1994). Stig Kanger's Theory of Rights. In D. Prawitz, B. Skyrms, D. W., editor, *Logic, Methodology and Philosophy of Science IX*, pages 889–911. Elsevier Science Publsiher, New York.

[Makinson, 1986] Makinson, D. (1986). On the formal representation of rights relations: Remarks on the work of Stig Kanger and Lars Lindahl. *Journal of Philosophical Logic*, 15(4):403–425.

[Markovich, 2019] Markovich, R. (2019). Rights and Punishment: The Hohfeldian Theory's Applicability and Morals in Understanding Criminal Law. *IFCoLog Journal of Logics and their Applications*, 6(5):847–864.

[Markovich, 2020] Markovich, R. (2020). Understanding Hohfeld and Formalizing Legal Rights: the Hohfeldian Conceptions and Their Conditional Consequences. *Studia Logica*, 108.

[Markovich and Roy, 2021] Markovich, R. and Roy, O. (2021). Formalizing the Right to Know: Epistemic Rights as Normative Positions. accepted for LNGAI2021.

[Pacuit, 2013] Pacuit, E. (2013). Dynamic epistemic logic i: Modeling knowledge and belief. *Philosophy Compass*, 8(9):798–814.

[Pacuit, 2017] Pacuit, E. (2017). *Neighborhood semantics for modal logic*. Springer.

[Pacuit et al., 2006] Pacuit, E., Parikh, R., and Cogan, E. (2006). The logic of knowledge based obligation. *Synthese*, 149(2):311–341.

[Pratt, 1981] Pratt, V. R. (1981). A decidable mu-calculus: Preliminary report. In *22nd Annual Symposium on Foundations of Computer Science (sfcs 1981)*, pages 421–427. IEEE.

[Åqvist, 1967] Åqvist, L. (1967). Good samaritans, contrary-to-duty imperatives, and epistemic obligations. *Nous*, 1(4):361–379.

[Sartor, 2005] Sartor, G. (2005). *Legal Reasoning. A Treatise of Legal Philosophy and General Jurisprudence*. Springer.

[Sergot, 2001] Sergot, M. (2001). A computational theory of normative positions. 2(4):581–622.

[Sergot, 2013] Sergot, M. (2013). Normative Positions. In Gabbay, D., Horty, J., Parent, X., van der Meyden, R., and van der Torre, L., editors, *Handbook of Deontic Logic and Normative Systems*, pages 353–406. College Publications.

[Shanawani, 2016] Shanawani, H. (2016). The challenges of conscientious objection in health care. *Journal of Religion and Health*, 55:384–393.

[Simmonds, 2001] Simmonds, N. (2001). Introduction. In *Hohfeld: Fundamental legal conceptions as applied in judicial reasoning*, Classical Jurisprudence series. Ashgate, Aldershot, new ed. / edited by David Campbell and Philip Thomas. edition.

[Stalnaker, 2006] Stalnaker, R. (2006). On logics of knowledge and belief. *Philosophical studies*, 128(1):169–199.

[Stapleford, 2012] Stapleford, S. (2012). Epistemic duties and failure to understand one's evidence. *Principia: an international journal of epistemology*, 16(1):147–177.

[Watson, 2021] Watson, L. (2021). *The Right to Know: Epistemic Rights and Why We Need Them*. Routledge.

[Wenar, 2003] Wenar, L. (2003). Epistemic rights and legal rights. *Analysis*, 63(2):142–146.

[Wenar, 2015] Wenar, L. (2015). Rights. In Zalta, E. N., editor, *The Stanford Encyclopedia of Philosophy*. Metaphysics Research Lab, Stanford University, fall 2015 edition.

[Wright, 1963] Wright, G. H. v. G. H. (1963). *Norm and Action : a Logical Enquiry*. International library of philosophy and scientific method. Routledge and Kegan Paul, London.

Deontic Logic Based on Inquisitive Semantics

Karl Nygren[1]

Department of Philosophy, Stockholm University

Abstract

This paper introduces deontic logic based on inquisitive semantics. A semantics for action formulas is introduced where each action formula is associated with a set of alternatives. Deontic operators are then interpreted as quantifying over all alternatives associated with the action formulas within their scope. It is shown how this construction provides solutions to problems related to free choice permissions and obligations, including issues concerning Hurford disjunctions. The main technical result is a complete axiomatization of the logic.

Keywords: Alternatives, free choice, inquisitive semantics, permission.

1 Introduction

The aim of this paper is to introduce deontic logic based on inquisitive semantics. I define deontic operators that take actions, rather than propositions, as arguments; actions are then given an interpretation based on inquisitive semantics. I will show that this provides solutions to various problems concerning free choice inferences, including problems related to Hurford disjunctions.

Inquisitive semantics originated as a framework aiming to provide a uniform account of both statements and questions. In recent years, the framework of inquisitive semantics has evolved into a full-fledged theory with associated logic; for an overview, see [10] and [12]. Inquisitive semantics has been applied in several different areas moving beyond the original motivation, e.g. epistemic logic [14], dependency [9], and semantics for imperatives [2,11]. Inquisitive semantics moves beyond standard truth-conditional semantics by interpreting sentences as sets of sets of information states, where an information state is modeled as a set of possible worlds. The idea is that the meaning of a question can be given in terms of the information needed to resolve the question, whereas the meaning of a statement is given in terms of the information needed to establish the statement.

The main source of inspiration for this paper is the semantics of imperatives presented by Ciardelli and Aloni [11] (an earlier account is presented in [2]). They provide an action-theoretic interpretation of inquisitive semantics, and

---

[1] karl.nygren@philosophy.su.se

show how it can be used to give a semantics for imperatives. The logic introduced in this paper is naturally seen as the deontic counterpart of Ciardelli and Aloni's account of imperatives [11], as their notion of imperative entailment can be simulated in it.

The paper is structured as follows. Section 2 introduces the problems of free choice inferences, discuss solutions suggested in the literature, and discuss the role of Hurford disjunctions. In Section 3 I introduce the logic IDL, discuss how it handles free choice problems, and show that Ciardelli and Aloni's [11] notion of imperative entailment can be simulated in IDL. A complete axiomatization of IDL is introduced in Section 4. Section 5 concludes the paper.

2 Free choice inferences

Free choice inferences are inferences involving *choice-offering* permissions and obligations, i.e. permissions and obligations that offer a choice between two or more possible options. Typically, free choice is introduced by the use of disjunction in the scope of deontic operators. Consider the following sentences:

(i) Jane may buy the green car or the blue car.

(ii) Jane may buy the green car and Jane may buy the blue car.

Arguably, the most natural interpretation of sentence (i) is that it conveys an offer to choose between two individually permitted options: Jane is both permitted to buy the green car, and permitted to buy the blue car. Hence, one expects sentence (ii) to be entailed by (i). In general, one expects the following *Free Choice Principle* (FCP) to be valid, where P is a modal operator expressing permission:

$$P(\varphi \vee \psi) \models P\varphi \wedge P\psi$$

The standard account of deontic modalities interprets sentences like the ones above in a modal framework consisting of a set of possible worlds together with a relation associating each world w with those worlds that are considered ideal from the point of view of w (the 'ideal-at-w' worlds). In this framework, permission has the following truth conditions: $P\varphi$ is true in w if and only if φ is true in some ideal-at-w world. The standard account fails to validate FCP, since $P(\varphi \vee \psi)$ simply says that there is some ideal world where φ or ψ is true; not that there is an ideal world where φ is true *and* an ideal world where ψ is true.

Even more problematic is the fact that adding FCP as an axiom to an otherwise normal modal logic with permission analyzed as a diamond modality trivializes permission in the sense that any permission is equivalent to any other permission [23]. If one assumes that P is upwards monotonic (i.e. $\varphi \models \psi$ implies $P\varphi \models P\psi$) and that disjunction introduction holds (i.e. $\varphi \models \varphi \vee \psi$), then FCP together with the transitivity of entailment implies that $P\varphi \models P\psi$.

Problems with free choice inferences also arise in the analysis of obligation. On the standard account, obligation is analyzed as a normal box operator O with the following truth conditions: $O\varphi$ is true in w if and only if φ is true in all ideal-at-w worlds. This makes the O operator upwards monotonic. This

property, together with disjunction introduction, allows for the inference of an
obligation over a disjunction from the obligation of one of the disjuncts, an
inference known as Ross' paradox [26]:

$$O\varphi \models O(\varphi \vee \psi).$$

This property allows for several problematic instances, for example the inference of (iv) from (iii):

(iii) Jane is obliged to post the letter.

(iv) Jane is obliged to post the letter or burn it.

The problem here is that sentence (iv) seems to offer *more* choice than (iii). If Jane is obliged to post the letter or burn it, it clearly seems that she can fulfil her obligation by burning the letter; an option which is not offered to her in (iii). The standard approach to obligation fails to account for the *choice-offering* aspect of certain obligations. In particular, an obligation applied to a disjunction seem to imply that each disjunct is permitted:

$$O(\varphi \vee \psi) \models P\varphi \wedge P\psi.$$

2.1 Proposed solutions

There are many proposed solutions to the problems of free choice inferences in the literature; here, I review some of them.

Several approaches adopt variants of the 'open reading' of permission, where an action is permitted if every way of performing it is normatively okay [5,6,25,27,30,31]. FCP comes out as valid under this interpretation of permission if disjunctive actions are interpreted as the sum of the ways to perform each disjunct. However, this approach is problematic as an explication of free choice permission. For one thing, it is very strong to require that every way of performing an action is normatively okay for it to be permitted. As Giordani and Canavotto point out: "ordinary choices can be risky: we are ordinarily allowed to choose between alternative actions even if there are ways of performing such actions that lead to a violation of the law." [18, p. 89]. In addition, given a classical underlying logic, the approach suffers from what is known as the 'vegetarian free lunch problem' [21]. When the underlying propositional logic is classical and the permission operator allows for substitution of classical logical equivalents within its scope, FCP makes the following a theorem:

$$P\varphi \models P(\varphi \wedge \psi) \wedge P(\varphi \wedge \neg \psi).$$

To see why this is problematic, consider the following example [21, p. 208]: 'If you may order a vegetarian meal, then you may order a vegetarian meal and pay for it, and you may order a vegetarian meal and not pay for it'. Hansson takes the free lunch problem, together with other problems that crop up when FCP is combined with substitution of classical equivalents within the scope of deontic operators (see [21, pp. 214–217]), to indicate that the *single sentence assumption* is incorrect. According to the single sentence assumption, "[f]free choice between a and b can be represented as a property of a single sentence, namely $a \vee b$." [21, p. 218]. Instead, Hansson suggests,

> ... (free choice) permission to perform either a or b is not a function of a single sentence $a \vee b$ but a function of the two sentences a and b. It is a function of two variables, not one. Similarly, (free choice) permission to perform either a, b, or c is a function of three variables, etc. [21, p. 218].

Another way to react to free lunch problems is to give up substitution of classical logical equivalents in the scope of deontic operators. In the formal semantics literature, *alternative semantics* has been put to use in the analysis of deontic modalities and imperatives [1,2,3,29]. In alternative semantics, disjunctions are interpreted as sets containing the *alternatives* of the disjunction. Different versions of alternative semantics differ in how they construe alternatives, but the standard approach is to use an *exact* semantics, where alternatives of a disjunction are identified with the classical propositions expressed by each disjunct. Thus, a sentence such as $p \vee q$ is interpreted as the set $\{|p|, |q|\}$, where $|p|$ is the classical proposition expressed by p and $|q|$ is the classical proposition expressed by q. Deontic operators are then taken to quantify over all alternatives associated with their arguments: $P(p \vee q)$ is true in a world w if some ideal-at-w world is contained in $|p|$, and some ideal-at-w world is contained in $|q|$; $O(p \vee q)$ is true in a world w if every ideal-at-w world is contained in either $|p|$ or $|q|$, and there are ideal-at-w worlds in both $|p|$ and $|q|$.

Other approaches in the same style as the alternative semantics ones are based on truthmaker semantics [4,16,17]. In exact truthmaker semantics, the exact truthmakers of a sentence are sensitive to the syntactic structure of the sentence. The interpretation of a disjunctive sentence $p \vee q$ is the sum of the exact truthmakers of p and the exact truthmakers of q, and the interpretation of a conjunctive sentence $p \wedge q$ generally has a different set of exact truthmakers than those associated with p and q. As shown by Anglberger, Faroldi and Korbmacher [4], giving truth conditions to permissions and obligations in terms of properties of exact truthmakers makes it possible to validate FCP while avoiding free lunch problems.

The approach considered in this paper is similar to the alternative semantics approach and the truthmaker approach. An important difference, however, is that the inquisitive semantics used in this paper is not exact in the above sense. In inquisitive semantics, there is not in general a one-to-one connection between the disjuncts and the alternatives of a disjunctive sentence. As I will argue in the next section, exact approaches to alternative semantics fail to give a satisfactory account of certain types of disjunctions where one disjunct entails another.

2.2 Hurford's constraint

Hurford disjunctions are disjunctive sentences where one of the disjuncts entails the other, either logically or locally in relation to a context. Consider the following sentences [22, p. 410]:

(v) John is an American or a Californian.

(vi) The painting is of a man or a bachelor.

(vii) The value of x is different from 6 or greater than 6.

Hurford's constraint is the following principle, stating that sentences such as the ones above are always infelicitous:

> The joining of two sentences by *or* is unacceptable if one sentence entails the other; otherwise the use of *or* is acceptable. [22, p. 410]

Hurford style disjunctions have also been observed to be present in interrogative contexts [15], as well as in deontic and imperative contexts [11,15,29]. Simons notes the infelicity of the following sentence [29, p. 303]:

(viii) Jane may/must wear a dress or a red dress.

In the context of imperatives, Ciardelli and Aloni [11] argue that the following sentence is infelicitous:

(ix) Get an American or a Californian to do this job!

A corresponding equally infelicitous obligation sentence can be formed:

(x) Jane is obliged to hire an American or a Californian.

A standard explanation of the infelicity of Hurford disjunctions is that they involve *redundancy* in the sense that the whole disjunction is equivalent to one of the disjuncts [15,24,28]. A speaker uttering a Hurford disjunction could have conveyed the same information by simply using the weaker disjunct. The general idea behind the redundancy explanation is to derive Hurford's constraint from a more general ban on redundant operations. For example, Katzir and Singh [24, p. 210] propose a *local redundancy* principle, here quoted from Ciardelli, Groenendijk and Roelofsen [12, p. 171]:

> **Local redundancy:** A sentence is deviant if its logical form contains a binary operator $\circ$ applying to two arguments A and B, and the outcome $A \circ B$ is semantically equivalent to one of the arguments.

It has been pointed out that there are several apparent counterexamples to Hurford's constraint [15]. Consider for example the following sentences, which do not appear infelicitous:

(xi) Jane is having dinner with John, with James, or with both.

(xii) Jane is obliged to have dinner with John, with James, or with both.

The redundancy based explanation is therefore extended with the notion of *exhaustive strengthening* [7,8,15]. The idea is that the weaker disjuncts in (xi) and (xii) receive an exhaustive interpretation. Thus, 'Jane is having dinner with John' and 'Jane is having dinner with James' are interpreted, in the relevant context, as 'Jane is having dinner with John, and not with James' respectively 'Jane is having dinner with James, and not with John'. Under this interpretation, neither disjunct entails the other. This kind of exhaustive interpretation is not available for the weaker disjuncts in e.g. (v) and (x): 'American' cannot be strengthened to 'American and not Californian'.

Ciardelli and Roelofsen [15] argue that in approaches to alternative semantics where the alternatives of a disjunction are the classical propositions expressed by each disjunct, the explanation of Hurford disjunctions using a general ban on redundancy fails. The reason for this is that alternative semantics construes the meaning of a disjunction as the set containing exactly the propositions corresponding to each disjunct. Thus, the meaning of $p \vee q$ is construed as the set $\{|p|, |q|\}$, which is distinct from the meaning of each of the two disjuncts. Hence, even when one of p or q entails the other, the logical form of $(p \vee q)$ does not exhibit any redundancy in the relevant sense. This failure of the redundancy explanation carries over to deontic Hurford disjunctions. Under the alternative semantics analysis of deontic modalities, the meaning of $D(p \vee q)$, where D is a deontic operator, crucially depends on properties of both $|p|$ and $|q|$, even when $|p| \subset |q|$, i.e. p strictly entails q. Hence, no redundancy is involved in the derivation of the meaning of $D(p \vee q)$. Without any further stipulations, the redundancy explanation of deontic Hurford disjunctions does not work for the standard alternative semantics analysis. A similar argument can be made against the exact truthmaker approach as well. In the approach of Anglberger, Faroldi and Korbmacher [4] or Fine [16,17], the truth conditions of $D(p \vee q)$ depend on the exact truthmakers of $p \vee q$, which equals the sum of the exact truthmakers of p and the exact truthmakers of q. Even if one of p and q entails the other, the truth of $D(p \vee q)$ depends on both the exact truthmakers of p and the exact truthmakers of q. Hence, no redundancy is involved when the meaning of $D(p \vee q)$ is derived.

Thus, while the exact alternative semantics approach solves the problems with free choice permissions, it does not provide a satisfactory account of deontic Hurford disjunctions.

3 The deontic logic IDL

In this section, I will define the logic IDL, show how it provides solutions to the problems of free choice inferences, and argue that it provides a better account of deontic Hurford disjunctions than the approaches based on exact alternative semantics or truthmaker semantics. I will also show that imperative entailment of [11] can be simulated in IDL.

3.1 Inquisitive semantics for action formulas

In this section, I will introduce inquisitive semantics for a language of action expressions. This action language is in fact a fragment of the language of *basic inquisitive logic* InqB [10], and the semantics I will use is derived from the semantics of InqB. For a review of InqB, see Chapters 2 and 3 of [10]. The action-theoretic interpretation of the semantics is largely based on the interpretation of [11] (see also [2]).

Let Π be a set of *atomic action formulas*. The set of *action formulas* $\mathcal{L}_A$ is defined by the following grammar, where a ranges over Π:

$$\alpha ::= a \mid \neg \alpha \mid \alpha \wedge \alpha \mid \alpha \mathbin{\mathpalette\makebbW\relax} \alpha.$$

I use α, β, δ as symbols ranging over action formulas. The intended reading of action formulas are as follows: $\neg \alpha$ denotes the action of not doing α, $\alpha \wedge \beta$ denotes the action of doing both α and β, and $\alpha \mathbin{\mathpalette\makebbW\relax} \beta$ denotes the action of doing either α or β.[2] An action formula without any occurrences of $\mathbin{\mathpalette\makebbW\relax}$ is called a *classical action formula*. The set of classical action formulas is denoted $\mathcal{L}_A^{cl}$ and I use γ, τ as symbols ranging over elements of $\mathcal{L}_A^{cl}$. Note that $\mathcal{L}_A^{cl} \subset \mathcal{L}_A$.

The semantics of action formulas is given in terms of *action models* of the form $\mathcal{M} = (W, V)$, where W is taken to be a set of *conducts*, and $V : \Pi \to \mathcal{P}(W)$ is a valuation function. Intuitively, a conduct represents the actions performed by an implicit agent over some relevant stretch of time. The valuation function V assigns sets of conducts to atomic action formulas, so that if a conduct w is in $V(p)$, it means that the conduct w encodes the performance of a p-action, and if w is not in $V(p)$, then the conduct w encodes that no p-action is performed. On this picture, a model (W, V) can be seen as a decision situation for the implicit agent. Following Aloni and Ciardelli [2], a conduct w is said to *execute* a basic action formula a if $w \in V(a)$. *Execution conditions* for general action formulas are recursively defined as follows, where $\mathcal{M}$ is an action model:

$$|a|_\mathcal{M} = V(a)$$
$$|\neg \alpha|_\mathcal{M} = W \setminus |\alpha|_\mathcal{M}$$
$$|\alpha \wedge \beta|_\mathcal{M} = |\alpha|_\mathcal{M} \cap |\beta|_\mathcal{M}$$
$$|\alpha \mathbin{\mathpalette\makebbW\relax} \beta|_\mathcal{M} = |\alpha|_\mathcal{M} \cup |\beta|_\mathcal{M}.$$

Note that $|\alpha|$ is the equivalent of the standard truth set of propositional logic.

A key observation is that some action formulas can be executed in different ways. Typically, an action formula of the form $a \mathbin{\mathpalette\makebbW\relax} b$ can be executed either by executing a or by executing b. Now, to arrive at a semantics for action formulas that allows distinguishing different ways to execute them, I will use the notion of alternatives as defined in inquisitive semantics. Say that a set $X \subseteq W$ of conducts is an *option*. As a first step, action formulas are given an interpretation in terms of *choice sets*, which are collections of options. For each action formula α, its associated choice set in a model $\mathcal{M}$, denoted $[\![\alpha]\!]_\mathcal{M}$, is recursively defined by the following clauses (cf. [2]):

$$[\![a]\!]_\mathcal{M} = \{V(a)\}$$
$$[\![\neg \alpha]\!]_\mathcal{M} = \{W \setminus \bigcup [\![\alpha]\!]_\mathcal{M}\}$$
$$[\![\alpha \wedge \beta]\!]_\mathcal{M} = \{X \cap Y \mid X \in [\![\alpha]\!]_\mathcal{M}, Y \in [\![\beta]\!]_\mathcal{M}\}$$
$$[\![\alpha \mathbin{\mathpalette\makebbW\relax} \beta]\!]_\mathcal{M} = [\![\alpha]\!]_\mathcal{M} \cup [\![\beta]\!]_\mathcal{M}.$$

---

[2] I use the symbol $\mathbin{\mathpalette\makebbW\relax}$ to denote disjunction, since it will be given an interpretation that differs from the interpretation of classical disjunction $\vee$.

The *alternatives* of a choice set are the maximal options included in it. Formally, the *alternative closure* of a choice set A, denoted $\text{ALT}(A)$, is defined by
$$\text{ALT}(A) = \{X \in A \mid \text{ for all } Y \in A, \text{ if } X \subseteq Y \text{ then } X = Y\}.$$
Given an action formula α, its *alternative set* relative to a model $\mathcal{M}$ is denoted $\text{ALT}_{\mathcal{M}}(\alpha)$, and defined by $\text{ALT}(\llbracket \alpha \rrbracket_{\mathcal{M}})$.[3] Intuitively, one may understand $X \in \text{ALT}_{\mathcal{M}}(\alpha)$ as saying that α is executed in a uniform way across the conducts in X. I will refer to the options in $\text{ALT}_{\mathcal{M}}(\alpha)$ as the *alternatives for α in $\mathcal{M}$*.

The execution conditions of an action formula α can be recovered from the set of alternatives for α in the sense that a conduct w executes α if and only if w is included in one of the ways of executing α. For any model $\mathcal{M}$ and any action formula α,
$$w \in |\alpha|_{\mathcal{M}} \quad \text{if and only if} \quad w \in X \text{ for some } X \in \text{ALT}_{\mathcal{M}}(\alpha).$$
In other words, $|\alpha|_{\mathcal{M}} = \bigcup \text{ALT}_{\mathcal{M}}(\alpha)$ for all models $\mathcal{M}$ and action formulas α.

Two action formulas α and β are said to be *equivalent* if for all models $\mathcal{M}$, $\text{ALT}_{\mathcal{M}}(\alpha) = \text{ALT}_{\mathcal{M}}(\beta)$.

Say that an action formula α is *basic* if $\text{ALT}_{\mathcal{M}}(\alpha)$ is a singleton for any $\mathcal{M}$. If α is not basic, it is said to be *choice-offering*. If α is a basic action formula, then the only alternative for α is $|\alpha|_{\mathcal{M}}$: that is, $\text{ALT}_{\mathcal{M}}(\alpha) = \{|\alpha|_{\mathcal{M}}\}$. It is easily verified that all classical action formulas are basic. On the other hand, not all basic action formulas are classical; for example, every negated action formula is basic, but not necessarily classical. However, every basic formula is *equivalent* to a classical one [10, p. 54]. Hence, basic action formulas in $\mathcal{L}_A$ can be characterized as precisely those action formulas that are equivalent to a classical action formula.

An important feature of the inquisitive semantics for action formulas is that for any action formula α, one can recursively compute a set of formulas called the *resolutions* of α, which can roughly be taken to denote the alternatives for α. Formally, the set of resolutions of α, denoted $\mathcal{R}(\alpha)$, is defined as follows (cf. [10]):

- $\mathcal{R}(\alpha) = \{\alpha\}$, if α is an atomic action formula or a negated action formula;
- $\mathcal{R}(\alpha \wedge \beta) = \{\alpha' \wedge \beta' \mid \alpha' \in \mathcal{R}(\alpha), \beta' \in \mathcal{R}(\beta)\}$;
- $\mathcal{R}(\alpha \vee\!\!\!\vee \beta) = \mathcal{R}(\alpha) \cup \mathcal{R}(\beta)$.

By definition, each resolution in $\mathcal{R}(\alpha)$ is a basic action formula, although not necessarily classical. As already noted, every basic action formula is equivalent to a classical formula, and there is a translation cl from $\mathcal{L}_A$ to $\mathcal{L}_A^{cl}$ such that if α is a basic action formula, then α is equivalent to $cl(\alpha)$.[4] Since every resolution

---

[3] By Proposition 2.2.7. in [10, p. 52], every action expression has a non-empty set of alternatives. This property does not hold in first-order inquisitive semantics.

[4] cl commutes with $\wedge$, and $\neg$, and replaces $\vee\!\!\!\vee$ with the classical disjunction $\vee$, which can be defined in terms of $\neg$ and $\wedge$.

is a basic formula, it is natural to define the set of *classical resolutions* of a formula φ as
$$\mathcal{R}^{cl}(\alpha) = \{cl(\alpha') \mid \alpha' \in \mathcal{R}(\alpha)\}.$$
Classical resolutions provide a normal form for action formulas [10, p. 54]:

Proposition 3.1 (Normal form) *For any action formula α of $\mathcal{L}_A$, α is equivalent ot $\bigvee \mathcal{R}^{cl}(\alpha)$.*

In other words, any action formula α is equivalent to a disjunction of classical action formulas. In particular, this observation provides a handle on the alternatives for action formulas, in the sense that any alternative for an action formula α equals the set $|\gamma|$ for some $\gamma \in \mathcal{R}^{cl}(\alpha)$. For any model $\mathcal{M}$ and any action formula α, assuming $\mathcal{R}^{cl}(\alpha) = \{\gamma_1, \ldots, \gamma_k\}$,
$$\text{ALT}_\mathcal{M}(\alpha) = \text{ALT}_\mathcal{M}(\gamma_1 \mathbin{\!\!\vee\!\!} \cdots \mathbin{\!\!\vee\!\!} \gamma_k) = \text{ALT}(\{|\gamma_1|_\mathcal{M}, \ldots, |\gamma_k|_\mathcal{M}\}).$$

3.2 Language and semantics of IDL

The language $\mathcal{L}_{\mathsf{IDL}}$ of IDL is defined by the following grammar, where a ranges over Π and α ranges over $\mathcal{L}_A$:
$$\varphi ::= a \mid P\alpha \mid \neg\varphi \mid \varphi \land \varphi \mid \Box\varphi.$$

Formulas of the form $P\alpha$ are used to express that α is permitted. $\Box$ expresses 'it is settled that...' and is included in order to talk about properties that hold regardless of what conduct is executed. The propositional connectives $\lor$, $\rightarrow$ and $\leftrightarrow$ are defined in the usual way, as is the dual $\Diamond$ of $\Box$. Note that iteration of deontic operators is not allowed: for example, formulas of the form $PP\alpha$ are not well-formed. This is motivated by the idea that deontic operators attach to action formulas to form *deontic sentences*, which express true or false facts about the normative status of actions. Note also that $\mathcal{L}_A^{cl} \subset \mathcal{L}_{\mathsf{IDL}}$, whereas $\mathbin{\!\!\vee\!\!}$ is only allowed within the scope of P.

Formulas are interpreted on *deontic action models* $\mathcal{M} = (W, D, V)$, where (W, V) is an action model, and $D \subseteq W \times W$ is a binary relation over W. By $D(w)$ I mean the set of all conducts v such that $(w, v) \in D$ Intuitively, D associates each conduct w with the set $D(w)$ of conducts that are *legal* given that w is performed.[5] Formulas are evaluated at pairs consisting of a deontic action model and a conduct in that model:

$\mathcal{M}, w \models a$ iff $w \in V(a)$;
$\mathcal{M}, w \models P\alpha$ iff   for all $X \in \text{ALT}_\mathcal{M}(\alpha)$, $X \cap D(w) \neq \emptyset$;
$\mathcal{M}, w \models \neg\varphi$ iff   not $\mathcal{M}, w \models \varphi$;
$\mathcal{M}, w \models \varphi \land \psi$ iff $\mathcal{M}, w \models \varphi$ and $\mathcal{M}, w \models \psi$;
$\mathcal{M}, w \models \Box\varphi$ iff   for all $v \in W$, $\mathcal{M}, v \models \varphi$.

---

[5] It is natural to assume that the set of legal conducts is constant in the model: for all $w, v \in W$, $D(w) = D(v)$. Say that models satisfying this property are *uniform*. In uniform models, what is legal to do for the agent does not depend on what the agent actually does. While uniformity is a natural property in the present context, one might still argue that some decision situations are non-uniform. For example, Anglberger, Gratzl and Roy mention cases of moral hazards or state-act-dependence [5].

Validity in a model (notation $\mathcal{M} \models_{\mathsf{IDL}} \varphi$) is defined as truth at all conducts in the model. Validity (notation $\models_{\mathsf{IDL}} \varphi$) and semantic consequence (notation $\Phi \models_{\mathsf{IDL}} \varphi$) are defined as truth respectively preservation of truth at all conducts in all models. I will drop the subscript IDL whenever there is no risk for confusion.

The most interesting semantic clause is of course the clause for formulas of the form $P\alpha$, which states that all alternatives for α, i.e. every possible way to execute α, contains some legal conduct. This operator is essentially the same as the one suggested by Ciardelli [10, pp. 247–249] in the context of modal inquisitive logic. The clause for $\square$ makes it a universal modality [19]. Note that for any classical action formula γ, the set $\{w \in W \mid \mathcal{M}, w \models \gamma\}$ coincides with the execution conditions $|\gamma|_\mathcal{M}$. Intuitively, if γ is a classical action formula, then $\mathcal{M}, w \models \gamma$ means that α is executed at w.

An operator for expressing obligation is defined from the permission operator as follows:
$$O\alpha \stackrel{\text{def}}{=} \neg P\neg\alpha \wedge P\alpha.$$

Definitions of obligation along these lines are suggested by Gustafsson [20] and Castro and Maibaum [6]. The following clause can be derived:

$$\mathcal{M}, w \models O\alpha \quad \text{iff} \quad D(w) \subseteq |\alpha|_\mathcal{M}, \text{ and}$$
$$\text{for all } X \in \text{ALT}_\mathcal{M}(\alpha), X \cap D(w) \neq \emptyset.$$

That is, α is obligatory if each legal conduct executes α, and for each possible way to execute α, there is a corresponding legal conduct.[6]

3.3 Reasoning with permissions and obligations

If γ is a classical action formula, then the following clause can be derived:

$$\mathcal{M}, w \models P\gamma \quad \text{iff} \quad |\gamma|_\mathcal{M} \cap D(w) \neq \emptyset.$$

Hence, P behaves like a normal modal diamond when there is a classical action formula within its scope. Things become more interesting when the formula within the scope of the permission operator is choice-offering.

Let γ and τ be two classical action formulas. If γ and τ are independent in the model $\mathcal{M}$, in the sense that neither $|\gamma|_\mathcal{M}$ nor $|\tau|_\mathcal{M}$ is properly included in the other, then the following implication is valid in $\mathcal{M}$ (cf. [10, p. 249]):

$$\mathcal{M} \models P(\gamma \mathbin{\mathpalette\make@circled W} \tau) \rightarrow P\gamma \wedge P\tau.$$

In the general case of two arbitrary action formulas α and β, it may happen that some alternative for α is properly contained in some alternative for β, or vice versa. Hence, in general it cannot be assumed that $\text{ALT}(\alpha) \cup \text{ALT}(\beta) \subseteq \text{ALT}(\alpha \mathbin{\mathpalette\make@circled W} \beta)$, and quantifying over the alternatives for α and the alternatives

---

[6] One may argue that obligation should not imply permission, for example by appealing to conditional obligations: 'If Smith murders Jones he ought to do so gently' intuitively does not imply 'If Smith murders Jones he may do so gently'. In the present approach, such derivations can be blocked by introducing O as a primitive operator, and extending the models with an additional relation associating each conduct with a *required* option.

for β cannot always be reduced to quantifying over the alternatives for $\alpha \mathbin{\!\!\vee\!\!} \beta$. Hence, FCP does not hold in general:

$$\not\models P(\alpha \mathbin{\!\!\vee\!\!} \beta) \to P\alpha \land P\beta.$$

Similarly, it can be shown that the definition of obligation does not allow inferring the permission of each disjunct from an obligatory disjunction. In general,

$$\not\models O(\alpha \mathbin{\!\!\vee\!\!} \beta) \to P\alpha \land P\beta.$$

However, FCP does hold under suitable assumptions. Let α and β be any action formulas, and let $\mathcal{M}$ be any model. α is *alternative to* β *in* $\mathcal{M}$, denoted $\alpha \bowtie_\mathcal{M} \beta$, if for all $X \in \text{ALT}_\mathcal{M}(\alpha)$, for all $Y \in \text{ALT}_\mathcal{M}(\beta)$, $X \not\subset Y$.

Proposition 3.2 *Let α and β be any action formulas and $\mathcal{M}$ any model. Then $\alpha \bowtie_\mathcal{M} \beta$ if and only if* $\text{ALT}_\mathcal{M}(\alpha) \subseteq \text{ALT}_\mathcal{M}(\alpha \mathbin{\!\!\vee\!\!} \beta)$.

Proof. Suppose that $\alpha \bowtie_\mathcal{M} \beta$ is not the case. Then there are $X \in \text{ALT}_\mathcal{M}(\alpha)$ and $Y \in \text{ALT}_\mathcal{M}(\beta)$ such that $X \subset Y$. Since $X, Y \in [\![\alpha]\!]_\mathcal{M} \cup [\![\beta]\!]_\mathcal{M} = [\![\alpha \mathbin{\!\!\vee\!\!} \beta]\!]_\mathcal{M}$, it must hold that $X \notin \text{ALT}_\mathcal{M}(\alpha \mathbin{\!\!\vee\!\!} \beta)$. Hence, $\text{ALT}_\mathcal{M}(\alpha) \not\subseteq \text{ALT}_\mathcal{M}(\alpha \mathbin{\!\!\vee\!\!} \beta)$.

Suppose that $\text{ALT}_\mathcal{M}(\alpha) \not\subseteq \text{ALT}_\mathcal{M}(\alpha \mathbin{\!\!\vee\!\!} \beta)$. Then there is $X \in \text{ALT}_\mathcal{M}(\alpha)$ such that $X \notin \text{ALT}_\mathcal{M}(\alpha \mathbin{\!\!\vee\!\!} \beta)$. Then there is $Y \in [\![\alpha \mathbin{\!\!\vee\!\!} \beta]\!]_\mathcal{M} = [\![\alpha]\!]_\mathcal{M} \cup [\![\beta]\!]_\mathcal{M}$ such that $X \subset Y$. Since $X \in \text{ALT}_\mathcal{M}(\alpha)$, it must hold that $Y \in [\![\beta]\!]_\mathcal{M}$. But then there is $Z \in \text{ALT}_\mathcal{M}(\beta)$ such that $X \subset Y \subseteq Z$, and so it is not the case that $\alpha \bowtie_\mathcal{M} \beta$. □

Using Lemma 3.2, it is readily verified that for all action formulas α and β, the following *guarded* version of FCP is valid for any model $\mathcal{M}$:

$$\alpha \bowtie_\mathcal{M} \beta \quad \text{implies} \quad \mathcal{M} \models P(\alpha \mathbin{\!\!\vee\!\!} \beta) \to P\alpha.$$

To see this, assume $\mathcal{M}, w \models P(\alpha \mathbin{\!\!\vee\!\!} \beta)$; then any member of $\text{ALT}_\mathcal{M}(\alpha \mathbin{\!\!\vee\!\!} \beta)$ contains some legal conduct. If $\alpha \bowtie_\mathcal{M} \beta$, then Lemma 3.2 implies that $\text{ALT}_\mathcal{M}(\alpha) \subseteq \text{ALT}_\mathcal{M}(\alpha \mathbin{\!\!\vee\!\!} \beta)$. Hence, each member of $\text{ALT}_\mathcal{M}(\alpha)$ must contain some legal conduct, and so $\mathcal{M}, w \models P\alpha$.

By the definition of obligation, it also holds that for all action formulas α and β and all models $\mathcal{M}$,

$$\alpha \bowtie_\mathcal{M} \beta \quad \text{implies} \quad \mathcal{M} \models O(\alpha \mathbin{\!\!\vee\!\!} \beta) \to P\alpha.$$

As argued by Ciardelli and Roelofsen [15], the way alternatives are construed in inquisitive semantics provides a nice way of deriving the infelicity of Hurford disjunctions using the redundancy explanation, both in declarative and interrogative contexts. Basically, a disjunction $a \mathbin{\!\!\vee\!\!} b$, where e.g. $|a|$ is properly included in $|b|$, is semantically equivalent to b. Since $\text{ALT}(a \mathbin{\!\!\vee\!\!} b) = \{|b|\}$, the option set $|b|$ is the only alternative, and so the disjunction $a \mathbin{\!\!\vee\!\!} b$ is redundant in the relevant sense. Ciardelli and Aloni [11] extend this account to imperative Hurford disjunctions. It is straightforward to adapt the account to deontic Hurford disjunctions as well. Let $D(a \mathbin{\!\!\vee\!\!} b)$ be a deontic disjunction, where D expresses either permission or obligation, and let $|a|$ be properly included in $|b|$. Since the logical form of $D(a \lor b)$ involves $a \lor b$, which is semantically equivalent to b, the whole deontic disjunction is redundant in the relevant sense. These considerations provides a justification for the conditional version of FCP: one is only allowed to infer the permission of each disjunct from a permitted disjunction $P(\alpha \lor \beta)$ or an obligatory disjunction $O(\alpha \lor \beta)$ when both disjuncts are independent from each other in the relevant sense, i.e. when $\alpha \bowtie \beta$ and $\beta \bowtie \alpha$.

The inclusion of the universal modality $\Box$ in the language of IDL in combination with the fact that any action formula is equivalent to the disjunction of its classical resolutions, makes it possible to capture variants of the above guarded versions of FCP as IDL-validities. Recall the definition of $\mathcal{R}^{cl}(\alpha)$ from Section 3.1. Let γ be a classical action formula, let α be any action formula and let $\mathcal{M}$ be any model. Since $\text{ALT}_{\mathcal{M}}(\gamma) = \{|\gamma|_{\mathcal{M}}\}$ and $\text{ALT}_{\mathcal{M}}(\alpha) = \text{ALT}(\{|\tau_1|_{\mathcal{M}}, \ldots, |\tau_k|_{\mathcal{M}}\})$, assuming $\mathcal{R}^{cl}(\alpha) = \{\tau_1, \ldots, \tau_k\}$, it follows that $\gamma \bowtie_{\mathcal{M}} \alpha$ if and only if for all $1 \leq i \leq k$, if $|\gamma|_{\mathcal{M}} \subseteq |\tau_i|_{\mathcal{M}}$ then $|\gamma|_{\mathcal{M}} = |\tau_i|_{\mathcal{M}}$. Thus, if γ is a classical action formula, α is any action formula and $\mathcal{M}$ any model, then

$$\gamma \bowtie_{\mathcal{M}} \alpha \quad \text{if and only if} \quad \mathcal{M} \models \bigwedge_{\tau \in \mathcal{R}^{cl}(\alpha)} (\Box(\gamma \to \tau) \to \Box(\gamma \leftrightarrow \tau)).$$

As a consequence, the following validities hold, where γ is a classical action formula and α is any action formula:

$$\models \left(P(\gamma \lor \alpha) \land \bigwedge_{\tau \in \mathcal{R}^{cl}(\alpha)} (\Box(\gamma \to \tau) \to \Box(\gamma \leftrightarrow \tau)) \right) \to P\gamma,$$

and

$$\models \left(O(\gamma \lor \alpha) \land \bigwedge_{\tau \in \mathcal{R}^{cl}(\alpha)} (\Box(\gamma \to \tau) \to \Box(\gamma \leftrightarrow \tau)) \right) \to P\gamma.$$

Hence, if γ is alternative to α, it is possible to detach the permission $P\gamma$ from the permission $P(\gamma \lor \alpha)$ or the obligation $O(\gamma \lor \alpha)$.

While FCP only holds in a guarded version since the alternatives for α and β are not necessarily included in the alternatives for $\alpha \lor \beta$, it is readily verified that $\text{ALT}_{\mathcal{M}}(\alpha \lor \beta) \subseteq \text{ALT}_{\mathcal{M}}(\alpha) \cup \text{ALT}_{\mathcal{M}}(\beta)$ for any model $\mathcal{M}$. Based on this observation, it follows that

$$\models P\alpha \land P\beta \to P(\alpha \lor \beta).$$

It can also be shown that the vegetarian free lunch problem is avoided:

$$\not\models P\alpha \to P(\alpha \land \beta) \land P(\alpha \land \neg \beta).$$

Let a express the action of ordering a vegetarian meal, and let b express the action of paying for the meal. Consider the model $\mathcal{M} = (W, D, V)$ where

$W = \{w_1, w_2, w_3\}$, $D(w_i) = \{w_1, w_3\}$ for $i = 1, 2, 3$, $V(a) = \{w_1, w_2\}$ and $V(b) = \{w_1\}$. In this model, the legal conducts are those where a vegetarian meal is not ordered at all (conduct w_3) or where a vegetarian meal is ordered and payed for (conduct w_1). The conduct where a vegetarian meal is ordered but not payed for (conduct w_2) is not legal. Clearly $\mathcal{M}, w \models Pa$, but $\mathcal{M}, w \not\models P(a \wedge \neg b)$ for any conduct w.

Using the definition of obligation, it is easily verified that Ross' paradox is blocked:

$$\not\models O\alpha \to O(\alpha \mathbin{\!\vee\!\!\vee} \beta).$$

Let a mean that the letter is posted, and let b mean that the letter is burnt. Let $\mathcal{M} = (W, D, V)$ be a model such that $W = \{w_1, w_2\}$, $D(w_i) = \{w_1\}$ for $i = 1, 2$, $V(a) = \{w_1\}$, and $V(b) = \{w_2\}$. In this model, the only legal conduct is the one where the letter is posted. Then clearly $\mathcal{M}, w \models Oa$ for any w. However, one alternative for $a \mathbin{\!\vee\!\!\vee} b$ contains a non-legal conduct, and so $\mathcal{M}, w \not\models O(a \mathbin{\!\vee\!\!\vee} b)$ for all w. Hence, $\mathcal{M}$ is a countermodel to the Ross formula. In fact, burning the letter is clearly incompatible with posting the letter, in the sense that one cannot do both; hence, one would expect that the obligation to post the letter contradicts the obligation to post or burn the letter. The present account makes the correct prediction in this case. Consider any model where a and b are incompatible, i.e. $|a| \cap |b| = \emptyset$. If Oa holds at a conduct w in the model, then $D(w) \subseteq |a|$, and so $D(w) \cap |b| = \emptyset$. Hence, $O(a \mathbin{\!\vee\!\!\vee} b)$ fails at w in the model.

3.4 Imperative entailment

Ciardelli and Aloni [11] suggest an account of imperative entailment based on inquisitive semantics. In their framework, an *imperative* is a sentence of the form $!\alpha$, where α is an action formula. Action formulas are defined and interpreted as in Section 3.1. Imperatives are then given a semantics in terms of *compliance conditions*. Intuitively, an option X (i.e. a set of conducts) complies with an imperative $!\alpha$ if all conducts in X execute α, and for each possible way to execute α, there is a corresponding conduct in X. Formally, the following clause is defined, where $\mathcal{M} = (W, V)$ is an action model and $X \subseteq W$:

$$\mathcal{M}, X \models !\alpha \quad \text{iff} \quad X \subseteq |\alpha|_\mathcal{M}, \text{ and for all } Y \in \text{ALT}_\mathcal{M}(\alpha), X \cap Y \neq \emptyset.$$

Imperative entailment is then defined as a relation on the set of imperatives as follows. Let $!\alpha$ and $!\beta$ be two imperatives. Then $!\alpha$ *entails* $!\beta$, notation $!\alpha \models_I !\beta$, if for all action models $\mathcal{M}$ and all options X of $\mathcal{M}$, $\mathcal{M}, X \models !\alpha$ implies $\mathcal{M}, X \models !\beta$.

Clearly, the semantic clause for imperatives is very similar to the semantic clause for the obligation operator O in IDL. Let $\mathcal{M} = (W, V)$ be an action model and let $\mathcal{M}_D = (W, D, V)$ be a deontic action model based on $\mathcal{M}$. The following equivalence is easily verified:

$$\mathcal{M}_D, w \models O\alpha \quad \text{iff} \quad \mathcal{M}, D(w) \models !\alpha.$$

With this equivalence at hand, the following result can be proved:[7]

Proposition 3.3 $!\alpha \models_I !\beta$ if and only if $O\alpha \models_{\mathsf{IDL}} O\beta$.

4 Axiomatization of IDL

In this section, I propose an axiomatization of IDL, and prove soundness and completeness. The axiomatization utilizes the fact that the language for action formulas $\mathcal{L}_A$ is a fragment of the language of basic inquisitive logic InqB [10,13], which has a complete axiomatization.

4.1 The logic InqB

The language $\mathcal{L}_{\mathsf{InqB}}$ of InqB is defined by the following grammar:

$$\alpha ::= a \mid \bot \mid \alpha \to \alpha \mid \alpha \wedge \alpha \mid \alpha \mathbin{\!\!\vee\!\!} \alpha,$$

where p ranges over a set of atomic formulas. Negation is defined by $\neg \alpha \stackrel{\text{def}}{=} \alpha \to \bot$. Note that by letting $\mathcal{L}_{\mathsf{InqB}}$ be constructed from the set Π of atomic action formulas, one can consider $\mathcal{L}_A$ as a fragment of $\mathcal{L}_{\mathsf{InqB}}$.

The language of InqB is interpreted on *information models* $\mathcal{M} = (W, V)$, where W is a set of possible worlds and V is a valuation function for atomic formulas. Formulas are interpreted at a pair consisting of a model $\mathcal{M} = (W, V)$ and a set $X \subseteq W$, according to the following clauses:

$\mathcal{M}, X \models a$ iff $X \subseteq V(a)$;
$\mathcal{M}, X \models \bot$ iff $X = \emptyset$;
$\mathcal{M}, X \models \alpha \to \beta$ iff for all $Y \subseteq X$, $\mathcal{M}, Y \models \alpha$ implies $\mathcal{M}, Y \models \beta$;
$\mathcal{M}, X \models \alpha \wedge \beta$ iff $\mathcal{M}, X \models \alpha$ and $\mathcal{M}, X \models \beta$;
$\mathcal{M}, X \models \alpha \mathbin{\!\!\vee\!\!} \beta$ iff $\mathcal{M}, X \models \alpha$ or $\mathcal{M}, X \models \beta$.

Let α and β be two formulas. α *entails* β (notation $\alpha \models_{\mathsf{InqB}} \beta$) if for all models $\mathcal{M} = (W, V)$ and all $X \subseteq W$, if $\mathcal{M}, X \models \alpha$, then $\mathcal{M}, X \models \beta$.

The axiomatization of InqB from [13] extends intuitionistic logic with the following axiom schemas:

(i) $(\neg \alpha \to \beta \mathbin{\!\!\vee\!\!} \delta) \to (\neg \alpha \to \beta) \mathbin{\!\!\vee\!\!} (\neg \alpha \to \delta)$

(ii) $\neg\neg a \to a$, for any atomic formula a.

Derivability in this system is denoted by $\vdash_{\mathsf{InqB}}$, and $\alpha \dashv\vdash_{\mathsf{InqB}} \beta$ is shorthand for $\alpha \vdash_{\mathsf{InqB}} \beta$ and $\beta \vdash_{\mathsf{InqB}} \alpha$. The following result is proved by Ciardelli and Roelofsen [13] (see also [10, Chapter 3]):

Theorem 4.1 (Soundness and completeness of InqB) *For any α and β of $\mathcal{L}_{\mathsf{InqB}}$, $\alpha \models_{\mathsf{InqB}} \beta$ if and only if $\alpha \vdash_{\mathsf{InqB}} \beta$.*

For the purposes of this paper, it suffices to note the following corollary:

Corollary 4.2 *For any action formulas α and β in $\mathcal{L}_A$, α and β are equivalent if and only if $\alpha \dashv\vdash_{\mathsf{InqB}} \beta$.*

---

[7] Fine [17, p. 641] proves a similar link between his logic of imperatives and his logic of 'free choice obligation'.

In particular, it holds that for any action formula α, $\alpha \dashv\vdash_{\mathsf{InqB}} \mathbb{W}\, \mathcal{R}^{cl}(\alpha)$.

4.2 Axiomatization of IDL

I will use the completeness of InqB to define a sound and complete axiom system for IDL. This axiom system consists of axioms for propositional logic, S5 axioms for $\Box$, and all instances of the below axiom schemas, where α, β range over $\mathcal{L}_A$ and γ, τ range over $\mathcal{L}_A^{cl}$:

A1 $\neg P \neg (\gamma \to \tau) \to (\neg P \neg \gamma \to \neg P \neg \tau)$

A2 $P\gamma \to \Diamond \gamma$

A3 $P(\gamma \mathbb{W} \alpha) \wedge \Box(\tau \to \gamma) \to P(\tau \mathbb{W} \gamma \mathbb{W} \alpha)$

A4 $P(\alpha) \wedge P(\beta) \to P(\alpha \mathbb{W} \beta)$

A5 $\left(P(\gamma_1 \mathbb{W} \cdots \mathbb{W} \gamma_k) \wedge \bigwedge_{1 \le i \le k}(\Box(\gamma_1 \to \gamma_i) \to \Box(\gamma_1 \leftrightarrow \gamma_i)) \right) \to P\gamma_1$, for any $k \ge 2$

The rules of inference are modus ponens, necessitation for $\Box$, and replacement of equivalent action formulas within the scope of P:

RE From $\alpha \dashv\vdash_{\mathsf{InqB}} \beta$, infer $P\alpha \leftrightarrow P\beta$.

Theorem 4.3 (Soundness) *The axiom system for IDL is sound.*

The proof of the soundness result is straightforward and omitted. Here, I sketch the completeness proof.

Let α be any action formula in $\mathcal{L}_A$. Let $\mathcal{M}$ be a model. Define the $\mathcal{M}$-*filtering* of α as follows:

$$\mathcal{F}_{\mathcal{M}}(\alpha) = \{\gamma \in \mathcal{R}^{cl}(\alpha) \mid \text{ for all } \tau \in \mathcal{R}^{cl}(\alpha), \text{ if } |\gamma|_{\mathcal{M}} \subseteq |\tau|_{\mathcal{M}} \text{ then } |\gamma|_{\mathcal{M}} = |\tau|_{\mathcal{M}}\}.$$

Let Γ be a maximally consistent set. Define the Γ-*filtering* of α as follows:

$$\mathcal{F}_{\Gamma}(\alpha) = \{\gamma \in \mathcal{R}^{cl}(\alpha) \mid \text{ for all } \tau \in \mathcal{R}^{cl}(\alpha), \text{ if } \Box(\gamma \to \tau) \in \Gamma \text{ then } \Box(\gamma \leftrightarrow \tau) \in \Gamma\}.$$

The proof of the following lemma is straightforward using the definitions of $\text{ALT}_{\mathcal{M}}(\alpha)$ and $\mathcal{F}_{\mathcal{M}}(\alpha)$.

Lemma 4.4 *Let α be any action formula and let $\mathcal{M}$ be a model. Then $\text{ALT}_{\mathcal{M}}(\alpha) = \{|\gamma|_{\mathcal{M}} \mid \gamma \in \mathcal{F}_{\mathcal{M}}(\alpha)\}$.*

Lemma 4.5 *Let α be any action formula and let Γ be a maximally consistent set. Then $P\alpha \in \Gamma$ if and only if $P\gamma \in \Gamma$ for all $\gamma \in \mathcal{F}_{\Gamma}(\alpha)$.*

Proof. ($\Rightarrow$). Suppose that there is $\gamma \in \mathcal{F}_{\Gamma}(\alpha)$ such that $P\gamma \notin \Gamma$. By construction of $\mathcal{F}_{\Gamma}(\alpha)$, for all $\tau \in \mathcal{R}^{cl}(\alpha)$, if $\Box(\gamma \to \tau) \in \Gamma$ then $\Box(\gamma \leftrightarrow \tau) \in \Gamma$. Since Γ is maximally consistent, $\Box(\gamma \to \tau) \to \Box(\gamma \leftrightarrow \tau) \in \Gamma$ for all $\tau \in \mathcal{R}^{cl}(\alpha)$. Assume $\mathcal{R}^{cl}(\alpha) = \{\gamma_1, \ldots, \gamma_k\}$ with $\gamma = \gamma_1$. Since Γ is maximally consistent, $\bigwedge_{1 \le i \le k}(\Box(\gamma \to \gamma_i) \to \Box(\gamma \leftrightarrow \gamma_i)) \in \Gamma$. Hence, by Axiom **A5**, $P(\gamma_1 \mathbb{W} \cdots \mathbb{W} \gamma_k) \notin \Gamma$. Since $\alpha \dashv\vdash_{\mathsf{InqB}} \gamma_1 \mathbb{W} \cdots \mathbb{W} \gamma_k$, it follows by **RE** that $P\alpha \leftrightarrow P(\gamma_1 \mathbb{W} \cdots \mathbb{W} \gamma_k) \in \Gamma$, and so $P\alpha \notin \Gamma$.

($\Leftarrow$). Suppose $P\gamma \in \Gamma$ for all $\gamma \in \mathcal{F}_\Gamma(\alpha)$. Let $\mathcal{F}_\Gamma(\alpha) = \{\gamma_1, \ldots, \gamma_k\}$. By repeated use of Axiom **A4**, $P(\gamma_1 \mathbin{\mathpalette\make@circled\wedge} \cdots \mathbin{\mathpalette\make@circled\wedge} \gamma_k) \in \Gamma$. Let A be the set of all classical action formulas occurring in $\mathcal{R}^{cl}(\alpha)$ that are not members of $\mathcal{F}_\Gamma(\alpha)$. Assume $A = \emptyset$. Then $\alpha \dashv\vdash_{\mathsf{InqB}} \gamma_1 \mathbin{\mathpalette\make@circled\wedge} \cdots \mathbin{\mathpalette\make@circled\wedge} \gamma_k$, and so by **RE**, $P\alpha \leftrightarrow P(\gamma_1 \mathbin{\mathpalette\make@circled\wedge} \cdots \mathbin{\mathpalette\make@circled\wedge} \gamma_k) \in \Gamma$. Hence, $P\alpha \in \Gamma$. Assume $A = \{\tau_1, \ldots, \tau_l\} \neq \emptyset$. Then by construction of $\mathcal{F}_\Gamma(\alpha)$, for each $\tau \in A$, there is $\gamma \in \mathcal{F}_\Gamma(\alpha)$ such that $\Box(\tau \to \gamma) \in \Gamma$. Using Axiom **A3** and **RE**, it can be shown that $P(\gamma_1 \mathbin{\mathpalette\make@circled\wedge} \cdots \mathbin{\mathpalette\make@circled\wedge} \gamma_k \mathbin{\mathpalette\make@circled\wedge} \tau_1 \mathbin{\mathpalette\make@circled\wedge} \cdots \mathbin{\mathpalette\make@circled\wedge} \tau_l) \in \Gamma$. Since $\mathcal{R}^{cl}(\alpha) = \mathcal{F}_\mathcal{M}(\alpha) \cup A$ and $\alpha \dashv\vdash_{\mathsf{InqB}} \bigvvee \mathcal{R}^{cl}(\alpha)$, **RE** implies that $P\alpha \leftrightarrow P(\gamma_1 \mathbin{\mathpalette\make@circled\wedge} \cdots \mathbin{\mathpalette\make@circled\wedge} \gamma_k \mathbin{\mathpalette\make@circled\wedge} \tau_1 \mathbin{\mathpalette\make@circled\wedge} \cdots \mathbin{\mathpalette\make@circled\wedge} \tau_l) \in \Gamma$. Hence, $P\alpha \in \Gamma$. $\square$

Define the *modal depth* of a formula φ, denoted $md(\varphi)$, as follows:

- $md(a) = 0$
- $md(\neg\varphi) = md(\varphi)$
- $md(\varphi \wedge \psi) = \max(\varphi, \psi)$
- $md(P\alpha) = 1$
- $md(\Box\varphi) = md(\varphi) + 1$

Say that a formula φ is *strictly less complex* than ψ, denoted $\varphi \prec \psi$, if either φ is a proper subformula of ψ, or $md(\varphi) < md(\psi)$. It can be verified that $\prec$ is a well-founded strict partial order on $\mathcal{L}_{\mathsf{IDL}}$, and hence an ordering that is suitable for inductive proofs.

If Γ is a maximally consistent set, then $\Gamma^\Box = \{\varphi \mid \Box\varphi \in \Gamma\}$. Fix a maximally consistent set Γ_0. Define the *canonical model* $\mathcal{M} = (W, D, V)$ for Γ_0 as follows:

- W is the set of maximally consistent sets Γ such that $\Gamma^\Box = \Gamma_0^\Box$;
- $w \in D(v)$ if and only if $\{P\gamma \mid \gamma \in \mathcal{L}_A^{cl} \text{ and } \gamma \in w\} \subseteq v$;
- $V(a) = \{w \in W \mid a \in w\}$.

Lemma 4.6 (Truth lemma) *For any formula φ, $\mathcal{M}, w \models \varphi$ if and only if $\varphi \in w$.*

Proof. The proof is by induction on the complexity of φ in terms of $\prec$. The cases for the propositional connectives and formulas of the form $\Box\psi$ are omitted. I will here sketch the proof for the case where $\varphi = P\alpha$.

($\Rightarrow$). Suppose $\mathcal{M}, w \models P\alpha$.

Suppose that $\gamma \notin \mathcal{F}_\mathcal{M}(\alpha)$. Then there is $\tau \in \mathcal{R}^{cl}(\alpha)$ such that $|\gamma|_\mathcal{M} \subset |\tau|_\mathcal{M}$. This implies that $\mathcal{M}, v \models \gamma \to \tau$ for all $v \in W$ and there is $u \in W$ such that $\mathcal{M}, u \not\models \gamma \leftrightarrow \tau$. Since $\gamma \to \tau \prec P\alpha$ and $\gamma \leftrightarrow \tau \prec P\alpha$, by the induction hypothesis $\gamma \to \tau \in v$ for all $v \in W$ and there is $u \in W$ such that $\gamma \leftrightarrow \tau \notin u$. By construction of the canonical model, $\Box(\gamma \to \tau) \in w$ and $\Box(\gamma \leftrightarrow \tau) \notin w$. Hence, $\gamma \notin \mathcal{F}_w(\alpha)$. Hence, $\mathcal{F}_w(\alpha) \subseteq \mathcal{F}_\mathcal{M}(\alpha)$.

Suppose that $\gamma \in \mathcal{F}_w(\alpha)$. Then $\gamma \in \mathcal{F}_\mathcal{M}(\alpha)$. By Lemma 4.4, $|\gamma|_\mathcal{M} \in \mathrm{ALT}_\mathcal{M}(\alpha)$. By the initial assumption, for all $X \in \mathrm{ALT}_\mathcal{M}(\alpha)$, $X \cap D(w) \neq \emptyset$, and so $|\gamma|_\mathcal{M} \cap D(w) \neq \emptyset$. Then there is $u \in W$ such that $\mathcal{M}, u \models \gamma$ and $u \in D(w)$. By the induction hypothesis $\gamma \in u$. By construction of the canonical model, $P\gamma \in w$.

Since $\gamma \in \mathcal{F}_w(\alpha)$ was chosen arbitrarily, $P\gamma \in w$ for all $\gamma \in \mathcal{F}_w(\alpha)$. By Lemma 4.5, $P\alpha \in w$.

($\Leftarrow$). Suppose $P\alpha \in w$.

Suppose that $\gamma \notin \mathcal{F}_w(\alpha)$. Then there is $\tau \in \mathcal{R}^{cl}(\alpha)$ such that $\Box(\gamma \to \tau) \in w$ and $\Box(\gamma \leftrightarrow \tau) \notin w$. By construction of the canonical model, $\gamma \to \tau \in v$ for all $v \in W$ and there is $u \in W$ such that $\gamma \leftrightarrow \tau \notin u$. Since $\gamma \to \tau \prec P\alpha$ and $\gamma \leftrightarrow \tau \prec P\alpha$, by the induction hypothesis it follows that $\mathcal{M}, v \models \gamma \to \tau$ for all $v \in W$ and there is $u \in W$ such that $\mathcal{M}, u \not\models \gamma \leftrightarrow \tau$. Hence, $|\gamma|_{\mathcal{M}} \subset |\tau|_{\mathcal{M}}$, and so $\gamma \notin \mathcal{F}_{\mathcal{M}}(\alpha)$. It follows that $\mathcal{F}_{\mathcal{M}}(\alpha) \subseteq \mathcal{F}_w(\alpha)$.

Assume $X \in \text{ALT}_{\mathcal{M}}(\alpha)$. By Lemma 4.4, $X = |\gamma|_{\mathcal{M}}$ for some $\gamma \in \mathcal{F}_{\mathcal{M}}(\alpha)$. Then $\gamma \in \mathcal{F}_w(\alpha)$. By Lemma 4.5 and the initial assumption, $P\gamma \in w$. By construction of the canonical model, there is $u \in W$ such that $\gamma \in u$ and $u \in D(w)$. By the induction hypothesis, $\mathcal{M}, u \models \gamma$, and so $u \in |\gamma|_{\mathcal{M}}$. Hence, $X \cap D(w) \neq \emptyset$. Since X was chosen arbitrarily, $\mathcal{M}, w \models P\alpha$. □

Having established the truth lemma, the completeness follows.

Theorem 4.7 (Completeness) *The axiom system for IDL is complete.*

5 Conclusion

In this paper, I introduced and studied the logic IDL. In this logic, deontic operators attach to action formulas, and action formulas are given interpretations based on inquisitive semantics. I showed that IDL solves many of the various problems related to free choice inferences, including Hurford style disjunctions. In particular, the logic validates guarded versions of the free choice principle, which I argued is a natural consequence of how Hurford disjunctions are handled. I also showed that the notion of imperative entailment defined by Ciardelli and Aloni [11] can be simulated in IDL. Finally, I supplied IDL with a complete axiomatization.

The approach presented in this paper is only one step towards solving the problems of free choice inferences. For example, the approach in this paper cannot account for the intuition that free choice effects disappear under negation (see e.g. [1,3]). A sentence such as 'Jane may not have soup or salad' typically entails 'Jane may not have soup' and 'Jane may not have salad'. However, in the present approach, $\neg P(\alpha \vee\!\!\!\vee \beta)$ does not imply $\neg P\alpha \wedge \neg P\beta$. This is a natural topic for further research. Another topic for further research is to consider the permission operator of IDL in a general modal inquisitive logic setting (cf. [10, pp. 247–249]).

References

[1] Aloni, M., *Free choice, modals and imperatives*, Natural Language Semantics **15** (2007), pp. 65–94.
[2] Aloni, M. and I. Ciardelli, *A logical account of free-choice imperatives*, in: M. Aloni, M. Franke and F. Roelofsen, editors, *The dynamic, inquisitive, and visionary life of φ, $?\varphi$, and $\Diamond\varphi$: A festschrift for Jeroen Groenendijk, Martin Stokhof, and Frank Veltman*, ILLC Publications, Amsterdam, 2013 pp. 1–17.

[3] Alonso-Ovalle, L., "Disjunction in Alternative Semantics," Ph.D. thesis, University of Massachusetts, Amherst (2006).
[4] Anglberger, A., F. L. G. Faroldi and J. Korbmacher, *An exact truthmaker semantics for permission and obligation*, in: O. Roy, A. Tamminga and M. Willer, editors, *Deontic Logic and Normative Systems: 13th International Conference, DEON 2016*, College Publications, London, 2016 pp. 16–31.
[5] Anglberger, A., N. Gratzl and O. Roy, *Obligation, free choice, and the logic of weakest permissions*, The Review of Symbolic Logic **8** (2015), pp. 807–827.
[6] Castro, P. F. and T. S. E. Maibaum, *Deontic action logic, atomic boolean algebras and fault-tolerance*, Journal of Applied Logic **7** (2009), pp. 441–466.
[7] Chiercia, G., D. Fox and B. Spector, *Hurford's constraint and the theory of scalar implicatures: Evidence for embedded implicatures*, in: P. Egré and G. Magri, editors, *Presuppositions and Implicatures*, MIT Press, Cambridge, MA, 2009 pp. 47–62.
[8] Chiercia, G., D. Fox and B. Spector, *Scalar implicatures as grammatical phenomenon*, in: P. Portner, C. Maienborn and K. von Heusinger, editors, *Semantics: An international handbook of natural langauge meaning, Vol. 3*, Mouton de Gruyter, Berlin, 2012 pp. 2297–2331.
[9] Ciardelli, I., *Dependency as question entailment*, in: S. Abramsky, J. Kontinen, J. Väänänen and H. Vollmer, editors, *Dependence Logic: theory and applications*, Springer International Publishing Switzerland, 2016 pp. 129–181.
[10] Ciardelli, I., "Questions in Logic," Ph.D. thesis, ILLC University of Amsterdam, the Netherlands (2016).
[11] Ciardelli, I. and M. Aloni, *Choice-offering imperatives in inquisitive and truth-maker semantics*, Presented at 'Imperatives: worlds and beyond', Hamburg University (2016).
[12] Ciardelli, I., J. Groenendijk and F. Roelofsen, "Inquisitive Semantics," Oxford University Press, Oxford, 2018.
[13] Ciardelli, I. and F. Roelofsen, *Inquisitive logic*, Journal of Philosophical Logic **40** (2011), pp. 55–94.
[14] Ciardelli, I. and F. Roelofsen, *Inquisitive dynamic epistemic logic*, Synthese **192** (2015), pp. 1643–1687.
[15] Ciardelli, I. and F. Roelofsen, *Hurford's constraint, the semantics of disjunction, and the nature of alternatives*, Natural Language Semantics **25** (2017), pp. 199–222.
[16] Fine, K., *Compliance and command I – categorical imperatives*, The Review of Symbolic Logic **11** (2018), pp. 609–633.
[17] Fine, K., *Compliance and command II, imperatives and deontics*, The Review of Symbolic Logic **11** (2018), pp. 634–664.
[18] Giordani, A. and I. Canavotto, *Basic action deontic logic*, in: O. Roy, A. Tamminga and M. Willer, editors, *Deontic Logic and Normative Systems*, College Publications, London, 2016 pp. 80–92.
[19] Goranko, V. and S. Passy, *Using the universal modality: Gains and questions*, Journal of Logic and Computation **2** (1992), pp. 5–30.
[20] Gustafsson, J. E., *Permissibility is the only feasible deontic primitive*, Philosophical Perspectives **34** (2020), pp. 117–133.
[21] Hansson, S. O., *The varieties of permission*, in: D. Gabbay, J. Horty, X. Parent, R. van der Meyden and L. van der Torre, editors, *Handbook of Deontic Logic and Normative Systems*, College Publications, 2013 pp. 195–240.
[22] Hurford, J., *Exclusive or inclusive disjunction*, Foundations of Language **11** (1974), pp. 409–411.
[23] Kamp, H., *Free choice permission*, Proceedings of the Aristotelian Society **74** (1973), pp. 57–74.
[24] Katzir, R. and R. Singh, *Hurford disjunctions: Embedded exhaustification and structural economy*, in: U. Etzeberria, A. Fălăuş, A. Irurtzun and B. Leferman, editors, *Proceedings of Sinn und Bedeutung 18*, 2013 pp. 201–216.
[25] Meyer, J.-J. C., *A different approach to deontic logic: Deontic logic viewed as a variant of dynamic logic*, Notre Dame Journal of Formal Logic **29** (1988), pp. 109–136.
[26] Ross, A., *Imperatives and logic*, Theoria **7** (1941), pp. 53–71.
[27] Segerberg, K., *A deontic logic of action*, Studia Logica **41** (1982), pp. 269–282.

[28] Simons, M., *Disjunction and alternativeness*, Linguistics and Philosophy **24** (2001), pp. 597–619.
[29] Simons, M., *Dividing things up: The semantics of* or *and the modal/*or *interaction*, Natural Language Semantics **13** (2005), pp. 271–316.
[30] Trypuz, R. and P. Kulicki, *On deontic action logics based on Boolean algebra*, Journal of Logic and Computation **25** (2015), pp. 1241–1260.
[31] van der Meyden, R., *The dynamic logic of permission*, Journal of Logic and Computation **6** (1996), pp. 465–479.

Input/Output Logic With a Consistency Check–The Case of Permission

Maya Olszewski [1]

Technische Universität Wien
Institute of Logic & Computation
Vienna, Austria

Xavier Parent [2]

Technische Universität Wien
Institute of Logic & Computation
Vienna, Austria

Leendert van der Torre [3]

University of Luxembourg
Computer Science
Esch-sur-Alzette, Luxembourg

Abstract

In Input/output (I/O) logic, one makes a distinction between three kinds of permission, called negative, positive static and positive dynamic permission. They have been studied semantically and axiomatically by Makinson and van der Torre in the particular case where the underlying I/O operation for obligation is one of the standard systems. In this paper, we investigate what happens when the underlying I/O operation is one of the constrained I/O operations recently introduced by Parent and van der Torre. Their distinctive feature is two-fold. First, they are not closed under logical consequence. Second they have a built-in consistency check, which filters out excess outputs and allows them to properly deal with contrary-to-duty reasoning. The main contribution of this paper is the characterization of the positive static permission with a set of rules, called subverse rules. Due to the fact that the studied logics are different from the original framework, although the proof of the characterization result is similar to the original one, it still includes novel arguments. This is the definition of a first positive permission proof system for constrained output.

Keywords: Deontic logic, input/output logic, permission, normative system.

---

[1] Supported by WWTF MA16-028 and FWF W1255-N23. E-mail: maya@logic.at
[2] Supported by WWTF MA16-028. E-mail: xavier@logic.at
[3] E-mail: leon.vandertorre@uni.lu

1 Introduction

The aim of this paper is to analyse three kinds of permission operations, derived from the Input/output (I/O) logics O_1 and O_3, introduced by Parent and van der Torre [17]. The analysis looks at negative permission, positive static permission and positive dynamic permission, such as Makinson and van der Torre have done in 2003 for the unconstrained I/O logics $out_1 - out_4$ [13].[4] There are two main differences between the older and the newer systems. The newer systems are augmented with a consistency check. It is this property which puts these logics at the level of constrained output, which can deal with contrary-to-duty (CTD) reasoning. They also lack the weakening of the output (WO) rule. (WO allows to infer (a, y) from (a, x) and $x \vdash y$, where (a, x) represents the norm that if a, then x ought to be the case, and $x \vdash y$ means that y logically follows from x). We analyse the differences that these changes cause to the different kinds of permissions and try to get rule-sets that fully characterize the permission operations. This leads us to introduce the first proof systems for positive permission in terms of constrained output.

With permission being far less studied than obligation, we see it as important to give it its fair share of spotlight. In practice, normative codes such as traffic rules often include both obligatory and permissive norms, and so it is vital when modeling such rules to have a good understanding of the choice of permission at hand as well as of the underlying (input/output) logic. As they lack WO, we argue that O_1 and O_3 can be better suited for modelling normative reasoning compared to the *out* logics. We briefly recall below the argument given in [14,15] against WO.

WO yields as a special case the principle of conjunction elimination, warranting the move from $(a, x \wedge y)$ to (a, x). As suggested for example by Hamblin [7], Goble [6] and Hansen [8, p. 91], such a principle is counter-intuitive in those cases where x and y are not separable, so that (to quote Hansen) "failing a part [of the order] means that satisfying the remainder no longer makes sense. E.g. if I am to satisfy the imperative 'buy apples and walnuts', and the walnuts [...] and the apples [are meant to] land in a Waldorf salad, then it might be unwanted and a waste of money to buy the walnuts if I cannot get the apples" [8, p. 91].

WO is also undesirable with respect to the issue of deontic detachment. Deontic detachment (DD) is the law : from $(\top, x)$ and (x, y) infer $(\top, y)$, where $\top$ denotes a tautology. It is a special case of the law known as cumulative transitivity (CT): from (a, x) and $(a \wedge x, y)$ infer (a, y). Counter-examples have been given to deontic detachment (see, e.g. [11,3]). They can be blocked by replacing CT with the following variant rule— we call it "aggregative cumulative transitivity" (ACT): from (a, x) and $(a \wedge x, y)$ infer $(a, x \wedge y)$. This substitute rule makes sense only in a system without WO. Here is an example. The Luxembourgish traffic laws [1] say that if one wants to park one's car at a

---

[4] Hansson [10] provides an enlightening overview of the major issues in deontic logic that are specific for permission.

parking spot having a park meter during the times specified on the street sign, then one should buy a ticket. They also say that, if a parking ticket is purchased, then it should be put on display inside the vehicle. The obligation to put the ticket on display no longer holds, if the obligation to pay is violated (for instance the ticket has been forged). Thus, the correct conclusion is: one should pay-and-display the ticket.

We believe that the permission operations defined in terms of O_1 and O_3 are worth studying as well. Similarly to the logics O_1 and O_3, the permission operations underlying those new logics also lack the WO rule. They also have a consistency proviso restraining the application of two rules, one of them being AND and the other being ACT. This is needed to block the well-known pragmatic oddity from [18] among other things.

Regarding WO, the same situation can be expected to arise with "may" as with "must". And indeed it does. For illustration's sake, consider the following example. Restaurants often have a lunch-menu (l), and typically they have the option to order a starter (s), a main course (m) and a dessert (d), a starter and a main course or a main course and a desert. However it is not generally allowed to order a starter and a desert, without there being a formal prohibition, but there is a lack of a positive permission. Let $(a, x)^p$ denote the conditional permission to do x given a. We have $(l, s \wedge m \wedge d)^p$ but not $(l, s \wedge d)^p$. As a second case, consider a modified version of Feldman's medication example [5, p. 87]. Let a and b be two medicines such that medicine a needs medicine b in order to be safe for use in the treatment of disease d. In that case we have that $(d, a \wedge b)^p$ but not $(d, a)^p$.

There is another class of I/O logics to compare to O_1 and O_3, namely constrained I/O logics [12]. They are better suited for normative reasoning than unconstrained I/O logics, as they are capable of handling CTD reasoning. We could, in principle, define the three kinds of permission using constrained I/O logic as the underlying logic for obligation, similarly as they are defined for unconstrained I/O logic. The main downside to this approach is that (to our knowledge) there is no axiomatic characterization of constrained I/O logic that is "intrinsic". Straßer et al. [22] provide a dynamic proof theory of constrained I/O logics—it is that of the adaptive logic (AL) framework (see e.g. [21] for a general introduction). First, unconstrained and constrained I/O logics are embedded within some suitable modal logics. Next, the adaptive counterparts of all the constrained I/O operations are given. Representation results are provided for the modal characterizations in both the unconstrained setting and the constrained setting. It would be interesting to investigate the relationship between their account and ours. We leave this issue as a topic for future research. One would need to go beyond their framework in its current form, which does not cover the new I/O logics from [14,16,17] yet, and has no apparatus for handling positive permissions.

As mentioned above, there are important differences between the classical I/O logics and the operations O_1 and O_3. Because of these differences, the proofs given by Makinson and van der Torre [13] do not always go through.

The formal challenge thus consists in finding alternative proofs to the ones Makinson and van der Torre give, taking into consideration the nature of the new logics. We prove the characterization of the positive static permission operation by its subverse rule-set by showing that a result called the non-repetition property holds, such as Makinson and van der Torre did. However since intermediary results do not hold, the proof of the non-repetition property for O_1 is different. For O_3, the result is somewhat similar, as it also uses phasing of the derivation. In the present case the whole derivation cannot be phased, and one can phase only certain sub-parts of derivations, which is enough to prove the non-repetition property.

This paper is structured as follows. Section 2 gives the required background on I/O logic, section 3 outlines the differences between the classical and the new I/O logics, sections 4, 5 and 6 present respectively the negative permission, the static positive permission and the static dynamic permission. Finally, in section 7 we outline a few directions for future research.

2 Background

This section gives a brief review of the basic notions of I/O logic that are used throughout this work.

2.1 Semantics

I/O logic uses *conditional norms*, which are pairs of the form (a, x), where a is called the *body* of the norm and x the *head* of the norm. The norm (a, x) can be read as *if a, then x is obligatory*. For a set of norms G, $h(G)$ is the set of all heads of elements of G and $b(G)$ the set of all bodies of elements of G. $G(A)$ is defined as $\{x : (a, x) \in G \text{ for some } a \in A\}$.

The four unconstrained output operations of I/O logic that have first been introduced are the following, where G is a set of norms, A a set of formulae of a propositional language, Cn the consequence operation of classical propositional logic and $\mathcal{L}$ the set of all boolean formulae:

Definition 2.1 (Classical unconstrained I/O operations [12])

- *Simple-minded output*: $out_1(G, A) = Cn(G(Cn(A)))$
- *Basic output*: $out_2(G, A) = \cap \{Cn(G(V)) : A \subseteq V, V \text{ complete}\}$
  $= \cap \{out_1(V) : A \subseteq V, V \text{ complete}\}$

  A set V is *complete* iff $V = \mathcal{L}$ or $V \subseteq \mathcal{L}$ is maximally consistent.
- *Reusable simple-minded output*:
  $out_3(G, A) = \cap \{Cn(G(B)) : A \subseteq B = Cn(B) \supseteq G(B)\}$
- *Reusable basic output*:
  $out_4(G, A) = \cap \{Cn(G(V)) : A \subseteq V \supseteq G(V), V \text{ complete}\}$

Parent and van der Torre have introduced new logics O_1 and O_3 corresponding to out_1 and out_3 with an additional consistency check and without the rule WO [17]. They solve a problem that was present in the earlier systems: How to prevent the pragmatic oddity and the drowning problem? The pragmatic oddity [18] arises from the possibility of detaching a CTD obligation in a violation context, and aggregating it with its associated primary obligation. The following is a typical example: "you should keep your promise and apologize for not keeping it" can be derived from "you should keep your promise", "if you do not keep your promise you should apologize" and "you do not keep your promise" [16]. The drowning problem arises when a primary obligation no longer holds after a violation has occurred.[5]

Let $x \dashv\vdash y$ stand for $(x \vdash y)$ and $(y \vdash x)$. Then the systems Parent and van der Torre present are defined in the following way:

Definition 2.2 (New I/O logics [17])

- *Single-step detachment*: $x \in O_1(G, A)$ iff there exists some finite $M \subseteq G$ and a set $B \subseteq Cn(A)$ such that $M \neq \emptyset$, $B = b(M)$, $x \dashv\vdash \wedge h(M)$ and $\{x\} \cup B$ is consistent. $O_1(G) = \{(A, x) : x \in O_1(G, A)\}$.

- *Iterated detachment*: $x \in O_3(G, A)$ iff there exists some finite $M \subseteq G$ and a set $B \subseteq Cn(A)$ such that $M(B) \neq \emptyset$, $x \dashv\vdash \wedge h(M)$ and
  · $\forall B'(B \subseteq B' = Cn(B') \supseteq M(B') \Rightarrow b(M) \subseteq B')$
  · $\{x\} \cup B$ is consistent.
  $O_3(G) = \{(A, x) : x \in O_3(G, A)\}$.

M is called the *witness* of (A, x).

2.2 Proof Theory

Each of the previously defined output operations have their associated proof system, called $deriv_i$, for $i \in \{1, ..., 4\}$ for the classical I/O logics and D_i for $i \in \{1, 3\}$ for the new ones, each of which consists of the following sets of rules:

- $deriv_1 = \{\text{TAUT, SI, WO, AND}\}$
- $deriv_2 = \{\text{TAUT, SI, WO, AND, OR}\}$
- $deriv_3 = \{\text{TAUT, SI, WO, AND, CT}\}$
- $deriv_4 = \{\text{TAUT, SI, WO, AND, OR, CT}\}$
- $D_1 = \{\text{EQ, SI, R-AND}\}$
- $D_3 = \{\text{EQ, SI, R-ACT}\}$

Where the rule names have the following meaning:

- TAUT - tautology
- SI - strengthening of the input
- WO - weakening of the output
- AND - conjunction of the output
- OR - disjunction of the input
- CT - cumulative transitivity
- EQ - equivalence
- R-AND - restricted AND
- R-ACT - restricted aggregative cumulative transitivity

---
[5] Other approaches are possible. It is often thought that the CTD scenarios involve two kinds of obligations, prima facie (ideal, etc) obligations vs. all-things-considered (actual, etc) obligations. [4,20] are two examples of a formal setting articulating such a distinction. Our take is different. We are interested in obligations which still hold even if violated, as opposed to obligations satisfying ought-implies-can.

Those rules are the following:

$$\frac{-}{(\top, \top)} \text{TAUT} \qquad \frac{(a,x) \quad b \vdash a}{(b,x)} \text{SI}$$

$$\frac{(a,x) \quad x \vdash y}{(a,y)} \text{WO} \qquad \frac{(a,x) \quad (a,y)}{(a, x \wedge y)} \text{AND}$$

$$\frac{(a,x) \quad (b,x)}{(a \vee b, x)} \text{OR} \qquad \frac{(a,x) \quad (a \wedge x, y)}{(a,y)} \text{CT}$$

$$\frac{(a,x) \quad (a,y) \quad a \wedge x \wedge y \not\vdash \bot}{(a, x \wedge y)} \text{R-AND} \qquad \frac{(a,x) \quad x \dashv\vdash y}{(a,y)} \text{EQ}$$

$$\frac{(a,x) \quad (a \wedge x, y) \quad a \wedge x \wedge y \not\vdash \bot}{(a, x \wedge y)} \text{R-ACT}$$

We say that $(a,x) \in deriv_i(G)$ (or $D_i(G)$) iff (a,x) is derivable from G using the rules of $deriv_i$ (or D_i). We say that $(A,x) \in deriv_i(G)$ (or $D_i(G)$) iff $(a,x) \in deriv_i(G)$ (or $D_i(G)$), where a is a conjunction of formulas in A. Equivalently, we say that $x \in deriv_i(G, A)$ (or $D_i(G, A)$).

Looking at the proof systems another difference between the classical and the new systems becomes apparent: the latter lack WO, whereas it is present in the former ones.

For simplifying derivation representations, let us define a generalized version of R-AND:

$$\frac{(a, x_1) \quad \ldots \quad (a, x_n) \quad a \wedge x_1 \wedge \ldots \wedge x_n \not\vdash \bot}{(a, x_1 \wedge \ldots \wedge x_n)} \text{G-R-AND}$$

which is a short version of n consecutive R-AND applications.

D_1 and D_3 are sound and complete w.r.t. the semantics [17], i.e. $(A,x) \in O_i(G)$ iff $(A,x) \in D_i(G)$ and so O_i and D_i can be interchanged when needed for $i \in \{1,3\}$.

We use the notation of O and D when we talk about the output operations with the consistency check O_1 and O_3, and out and $deriv$ for the classical output operations out_1-out_4.

Parent et al. [14] define the notion of derivation as follows.

Definition 2.3 (Derivation)
Let D be a proof system. A *derivation* of (a, x) from a set of norms G is a finite sequence of pairs ending with (a, x), each of which is either an element of G or follows from earlier pairs in the sequence using the rules of D. The elements of G being used in a derivation are called the *leaves* of the derivation, and it is required that all leaves have a consistent fulfilment, i.e. for all leaves (a, x), $a \wedge x$ is consistent. The length of a derivation is the length of the sequence.

In this work we mostly represent derivations graphically using proof trees.

3 O versus *out*

Already at this point there is one significant difference when it comes to O versus *out*: whereas *out* is a closure operation [13], O is not, as it does not satisfy inclusion: take $G = \{(x, \neg x)\}$, $\neg x \notin O(G, x)$ so $G \not\subseteq O(G)$. However monotony ($G \subseteq H \Rightarrow O(G) \subseteq O(H)$) and idempotence ($O(O(G)) = O(G)$) both hold, as shown below. (Note that one half of idempotence is established for O_1 only.)

Proposition 3.1 *(Monotony)*
Let $O = O_1, O_3$ be an output operation, G, H be sets of norms with $G \subseteq H$ and let $(a, x) \in O(G)$. Then $(a, x) \in O(H)$.

Proof. Assume $(a, x) \in O(G)$. By the definitions of O_1 and O_3, there exists a witness M for (a, x), with $M \subseteq G$. As $G \subseteq H$, one can take the same M as witness to get that $(a, x) \in O(H)$. □

The following sequence of results leads to showing that the left-in-right direction of idempotence holds for O_1:

Lemma 3.2 Let $O = O_1$ be an output operation, G be a set of norms. Let M be the witness for (a, x). Then M does not contain a pair of the form (a_i, x_i) with $a_i \wedge x_i \vdash \bot$.

Proof. Suppose M contains a pair of the form (a_i, x_i) with $a_i \wedge x_i \vdash \bot$. We know, by definition of O_1 that $x \dashv\vdash \wedge h(M)$, so $x \vdash x_i$, thus $a_i \wedge x \vdash \bot$. But $a_i \in b(M)$, so $b(M) \cup \{x\} \vdash \bot$ by monotony for $\vdash$, which contradicts the definition of the witness M. □

Lemma 3.3 Let $O = O_1$ be an output operation, G be a set of norms. Let $(a, x) \in O(G)$ and M be the witness for (a, x). Then $M \subseteq O(G)$.

Proof. Let $O = O_1$, $(a_i, x_i) \in M$. $\{(a_i, x_i)\}$ is finite and non-empty, $a_i \vdash a_i$, $x_i \vdash x_i$ and $\{x_i, a_i\} \not\vdash \bot$ by Lemma 3.2. So $(a_i, x_i) \in O(G)$. □

Proposition 3.4 *(Idempotence, left-to-right)*
Let $O = O_1$ be an output operation, G be a set of norms. Then $(a, x) \in O(G) \Rightarrow (a, x) \in O(O(G))$.

Proof. Let $(a, x) \in O(G)$ and $M = \{(a_1, x_1), ..., (a_n, x_n)\}$ be the witness for (a, x). By Lemma 3.3, $M \subseteq O(G)$. So $(a, x) \in O(O(G))$. □

Proposition 3.5 *(Idempotence, right-to-left)*
Let $O = O_1, O_3$ be an output operation, G be a set of norms.
Then $(a, x) \in O(O(G)) \Rightarrow (a, x) \in O(G)$.

Proof. Take $(a, x) \in O(O(G))$. By completeness, there exists a derivation of (a, x) from $O(G)$ in the corresponding proof system D. We have that every leaf $(a_i, x_i) \in O(G)$. Let $\{(a_1, x_1), ..., (a_n, x_n)\} \subseteq O(G)$ be the enumeration of the leaves of that derivation. Then there also exists a derivation of (a_i, x_i) from G in the corresponding proof system D. Let $\{(a_{i_1}, x_{i_1}), ..., (a_{i_m}, x_{i_m})\} \subseteq G$ be the enumeration of the leaves of that derivation. We have that every leaf $(a_{i_j}, x_{i_j}) \in G$ for j such that $1 \leq j \leq m$. Putting those derivations together, we can get a derivation of (a,x) from G where the leaves are $\{(a_{1_1}, x_{1_1}), ..., (a_{n_m}, x_{n_m})\} \subseteq G$. By soundness, $(a, x) \in O(G)$. □

4 Negative Permission

Negative permission is the most straightforward permission of the three kinds we are going to discuss. Something is said to negatively permitted if it is not prohibited.

Definition 4.1 (Negative permission [13])
Let G be a set of norms and O an output operation. Then $(a, x) \in negperm(G)$ iff $(a, \neg x) \notin O(G)$.

We will now discuss if the results on negative permission from Makinson and van der Torre's work [13] still hold in this new setting. Let us first look at what Horn rules the negative permission operation satisfies. In Makinson and van der Torres's fashion let us call the premises of the rules of the form $(\alpha, \varphi) \in O(G)$ a *substantive premise* and the premises of the form $\theta \in Cn(\gamma)$ and $\bigwedge(\alpha \wedge \varphi) \not\vdash \bot$ *auxiliary premise*. The idea behind the inverse of a Horn rule is the following: having one or more substantive premises, one takes one of them, negates its head and puts it as permitted in the conclusion. In retribution one takes the conclusion, negates its head and puts it as permitted in the premises. The other premises are left unchanged. Intuitively it says that if a group of conditional obligations imply some conclusion, which is also a conditional obligation, then taking all the premises in this group with the exception of one and combining it with the permission to not do the conclusion, then this implies that we also have the permission to not do what the excluded obligation stated (otherwise we would have the obligation of the conclusion). The updated Horn rules fit rules such as R-AND and R-ACT and their inverses. A Horn rule has the form:

$$(\text{HR}): (\alpha_i, \varphi_i) \in O(G) \ (i \leq n) \ \& \ \theta_j \in Cn(\gamma_j) \ (j \leq m)$$
$$\& \ \bigwedge_{k=0}^{n} (\alpha_k \wedge \varphi_k) \not\vdash \bot \Rightarrow (\beta, \psi) \in O(G)$$

Its inverse has the form:

$$(\text{HR})^{-1}: (\alpha_i, \varphi_i) \in O(G) \ (i < n) \ \& \ (\beta, \neg\psi) \in negperm(G)$$
$$\& \ \theta_j \in Cn(\gamma_j) \ (j \leq m) \ \& \ \bigwedge_{k=0}^{n} (\alpha_k \wedge \varphi_k) \not\vdash \bot$$
$$\Rightarrow (\alpha_n, \neg\varphi_n) \in negperm(G)$$

The inverses of each rule are given in Table 1.

Proposition 4.2 *Let $O = O_1, O_3$ be an output operation. If O satisfies a rule of the form* (HR), *then the corresponding negperm operation satisfies the inverse(s)* $(\text{HR})^{-1}$.

Rule	(HR)	$(HR)^{-1}$	$(HR)^{\perp}$
EQ	$\dfrac{(a,x) \quad x \dashv\vdash y}{(a,y)}$	$\dfrac{(a,x)^p \quad x \dashv\vdash y}{(a,y)^p}$	$\dfrac{(a,x)^p \quad x \dashv\vdash y}{(a,y)^p}$
SI	$\dfrac{(a,x) \quad b \vdash a}{(b,x)}$	$\dfrac{(a,x)^p \quad a \vdash b}{(b,x)^p}$	$\dfrac{(a,x)^p \quad b \vdash a}{(b,x)^p}$
R-AND	$\dfrac{(a,x) \quad (a,y)}{(a, x \wedge y)}$	$\dfrac{(a,(\beta \vee x))^p \quad (a,(\beta \vee y))^p}{(a,(\beta \vee x \wedge y))^p}\ \top A \beta \vee x \vee a$	$\dfrac{(a,x)^p \quad (a,y)^p}{(a, x \wedge y)^p}\ \top A \beta \vee x \vee a$
R-ACT	$\dfrac{(x,a)\quad (\beta,y)}{(a, x\vee y)}\ \top A \beta \vee x \vee a$	$\dfrac{(a, \neg(x\vee y))^p}{(a,\neg x)^p}\ (a, x\vee y)^o\ \top A \beta \vee x \vee a$	$(a, x\vee y)^o\ (a,x)^p\ \top A \beta \vee x \vee a$

Table 1: An enumeration of all the rules satisfied by the proof systems corresponding to O_1 and O_3 as well as their inverse and subverse rules. Superscript o indicates an obligatory norm, superscript p a permissive one.

Proof. The proof of EQ is trivial and the proof of SI is similar to the original paper by Makinson and van der Torre, so we omit them here.
Let G be a set of norms.

- Let O satisfy R-AND, $(a,x) \in O(G)$, $(a, \neg(x \wedge y)) \in negperm(G)$ and $a \wedge x \wedge y \not\vdash \bot$. Then $(a, x \wedge y) \notin O(G)$ by definition of $negperm$. As $(a,x) \in O(G)$ and $a \wedge x \wedge y \not\vdash \bot$, by R-AND for O we have $(a,y) \notin O(G)$. So $(a, \neg y) \in negperm(G)$.

- (i) Let O satisfy R-ACT, $(a,x) \in O(G)$, $(a, \neg(x \wedge y)) \in negperm(G)$ and $a \wedge x \wedge y \not\vdash \bot$. Then $(a, x \wedge y) \notin O(G)$ by definition of $negperm$. As $(a,x) \in O(G)$ and $a \wedge x \wedge y \not\vdash \bot$, by R-ACT for O we have that $(a \wedge x, y) \notin O(G)$, so $(a \wedge x, \neg y) \in negperm(G)$.

  (ii) Let O satisfy R-ACT, $(a \wedge x, y) \in O(G)$, $(a, \neg(x \wedge y)) \in negperm(G)$ and $a \wedge x \wedge y \not\vdash \bot$. Then $(a, x \wedge y) \notin O(G)$ by definition of $negperm$. As $(a \wedge x, y) \in O(G)$ and $a \wedge x \wedge y \not\vdash \bot$, by R-ACT for O we have that $(a, x) \notin O(G)$, so $(a, \neg x) \in negperm(G)$. □

5 Static Positive Permission

The static positive permission takes into account two explicit sets of norms. A set G of explicit obligations and a set P of explicit permissions. Something is said to be statically permitted if one can get it as output from the obligation set together with a single permission.

Definition 5.1 (Static positive permission [13])
Let G be a set of explicit obligations and P a set of explicit permissions and O an output operation. Then $(a,x) \in statperm(P,G)$ iff $(a,x) \in O(G \cup Q)$ for some $Q = \{(c,z)\} \subseteq P$ or $Q = \emptyset$.

For static permission, the definition yields that $O(G) \subseteq statperm(P,G)$ as O is monotone. What is different with O than with out is that $statperm$ is no longer a closure operation in its argument P as inclusion does not hold: take $P = \{(x, \neg x)\}, G = \emptyset$. Then $(x, \neg x) \in P$ but $(x, \neg x) \notin statperm(P,G)$, so $P \not\subseteq statperm(P,G)$. However, monotony holds ($P \subseteq Q$ implies $statperm(P,G) \subseteq statperm(Q,G)$) as O is monotonous and idempotence ($statperm(P,G) = statperm(statperm(P,G),G)$) also holds.

Proposition 5.2 *(Idempotence)*
Let $O = O_1, O_3$ be an output operation.
Then $statperm(P,G) = statperm(statperm(P,G),G)$.

Proof. To show the inclusion from right to left, one can take the same approach as for Proposition 3.4, using proof theory.

For the other way, assume $(a,x) \in statperm(P,G)$, let M be the witness for (a,x), and $B = b(M)$. By definition, since $\{x\} \cup B$ is consistent, M is also a witness for (B,x) and so $(B,x) \in statperm(P,G)$. Now one can take $M' = \{(B,x)\}$ to be the witness for (a,x) in $statperm(statperm(P,G),G)$, and thus $(a,x) \in statperm(statperm(P,G),G)$. □

statperm also is not a closure operation in its argument G, as inclusion does not hold: take $G = \{(x, \neg x)\}, P = \emptyset$. Then $(x, \neg x) \in G$ but $(x, \neg x) \notin statperm(P, G)$, so $G \not\subseteq statperm(P, G)$. Monotony holds as O is monotonous, but here idempotence fails: take $G = \emptyset, P = \{(a, x), (a, y)\}$ such that $a \wedge x \wedge y \not\vdash \bot$. Then $(a, x \wedge y) \notin statperm(P, G) = O(\{(a, x)\}) \cup O(\{(b, y)\})$ but $(a, x \wedge y) \in statperm(P, statperm(P, G)) = O(\{(a, x), (a, y)\})$.

Let us define the subverse of Horn rules, which are the rules satisfied by *statperm*. Here, one of the substantive premises as well as the conclusion of the Horn rule are changed from being an obligatory norm to being a permissive norm. This simply says that if we have a set of obligations that imply another obligation, then having the same set of obligation with the exception of one premise, which now is a permission, will change the conclusion from an obligation into a permission:

$$(\text{HR})^{\downarrow}: (\alpha_i, \varphi_i) \in O(G) \ (i < n) \ \& \ (\alpha_n, \varphi_n) \in statperm(P, G)$$

$$\& \ \theta_j \in Cn(\gamma_j) \ (j \leq m) \ \& \ \bigwedge_{k=0}^{n} (\alpha_k \wedge \varphi_k) \not\vdash \bot$$

$$\Rightarrow (\beta, \psi) \in statperm(P, G)$$

The subverses of each Horn rule for O_1 and O_3 are given in Table 1. We will now prove a series of results leading up to the proof that the subverse set is sufficient to characterize the static permission operation *statperm*. The way to get there mimics the way Makinon and van der Torre took in 2003 [13].

Proposition 5.3 *Let O be O_1 or O_3. If O satisfies a rule of the form* (HR), *then the corresponding statperm operation satisfies the subverse(s)* $(\text{HR})^{\downarrow}$.

We omit the proof, as it is virtually the same as the original one [13].

Makinson and van der Torre have shown that for O the subverse set of a Horn rule is sufficient to characterize the corresponding static permission operation [13]. They have established that the problem reduces to showing that the non-repetition property holds. The *non-repetition property* is satisfied if for any $(b, y) \in O(G \cup \{(c, z)\})$ there exists a derivation of (b, y) from $G \cup \{(c, z)\}$ using the rules of the corresponding proof system, such that (c, z) is attached to at most one leaf node.

Proposition 5.4 *Consider O_1 and D_1. Let D be a derivation of (b, y) with a leaf-set L, in which some leaves are used more than once. Then there exists a derivation D' of (b, y) from a leaf-set $L' \subseteq L$ where every leaf is used at most once.*

The proof given by Makinson and van der Torre [13] in the original framework does not work in the new setting, because of the consistency proviso restraining the application of AND. We provide an alternative proof, which also would have worked for the original framework.

Proof. D is a derivation from L to (b, y), so by soundness and completeness, it holds that $y \in O_1(N, b)$ for N consisting of the norms present in the leaf-set

L. By definition of O_1, $\exists M \subseteq N$ and $B \subseteq Cn(b)$ with $B = b(M)$ and $M \neq \emptyset$, $y \dashv\vdash \bigwedge h(M)$ and $\{y\} \cup B \not\vdash \bot$.

Let $\{(a_1, x_1), ..., (a_n, x_n)\} = M$.

Then $a_1 \wedge ... \wedge a_n \wedge x_1 \wedge ... \wedge x_n \dashv\vdash \bigwedge B \wedge y \not\vdash \bot$.

As $B \subseteq Cn(b)$, $b \vdash a_1 \wedge ... \wedge a_n = \bigwedge B$. We can thus build the following derivation. For visual effect we omit the auxiliary premises in the proof tree.

$$\cfrac{\cfrac{\cfrac{(a_1, x_1)}{(a_1 \wedge ... \wedge a_n, x_1)}\text{SI} \quad ... \quad \cfrac{(a_n, x_n)}{(a_1 \wedge ... \wedge a_n, x_n)}\text{SI}}{\cfrac{(a_1 \wedge ... \wedge a_n, x_1 \wedge ... \wedge x_n)}{\cfrac{(b, x_1, ... x_n)}{(b, y)}\text{EQ}}\text{SI}}\text{G-R-AND}}$$

Put $L' = M \subseteq N$. This derivation uses all elements of L' only once, so all norms of the initial leaf-set L are used at most once. □

Corollary 5.5 O_1 *satisfies the non-repetition property.*

Corollary 5.6 *The subverse set of EQ, SI, R-AND suffices to characterize the static permission operation based on O_1.*

Let us look at O_3 now. The following result has been adapted from the original framework [13] to fit O_3. This proof is inspired by the work of Makinson and van der Torre. Similarly to their proof, we are phasing the derivation a certain way. The difference is that, given the nature of the output operations we are considering, we are not able to phase the full derivation, and restrict ourselves to certain sub-derivations.

Lemma 5.7 *Let D be a derivation using the rules EQ, SI, R-ACT. Then at any line $l : (a, x)$ of the derivation:*

- *the head of l, x, classically implies the head of any line above it in D that l is based on*

- *the conjunction of body and head of l, $a \wedge x$, both classically implies the body and the head of any line above it in D that l is based on*

The proof is a straightforward proof by induction on the length of the derivation, and is omitted here.

Lemma 5.8 *Let $D = \{l_1, ..., l_n\}$ be a derivation of (b, y) from leaf-set L using the rules EQ, SI, R-ACT, and let l_i be a line where R-ACT is applied. Then there exists a derivation D' of (b, y) from leaf-set L, which is alike D, except for the fact that the two sub-derivation above line l_i follow the order SI, R-ACT, EQ.*

Proof. Consider the derivation D.

$$\cfrac{\cdots\ (a_i, x_i)\ \cdots \qquad \cdots\ (a_j, x_j)\ \cdots}{\cfrac{(a, x') \qquad (a \wedge x', x'') \qquad a \wedge x' \wedge x'' \not\vdash \bot}{l_i : (a, x' \wedge x'')}\ \text{R-ACT}}$$

$$(b, y)$$

Let l_i be a line of the conclusion of an R-ACT rule. Let d_1 be the left sub-derivation, and d_2 the right sub-derivation, with (a, x') and $(a \wedge x', x'')$ as their respective roots and $L(d_1), L(d_2)$ as leaf-sets. The rule EQ is invertible both with SI and R-ACT, it can be applied at any point in the derivation. Without loss of generality assume that EQ is applied at the bottom of the derivations d_1 and d_2. This leaves rules SI and R-ACT above in the upper parts of d_1 and d_2. By Lemma 5.7, it holds that $a \wedge x' \wedge x'' \vdash a_k$ and $a \wedge x' \wedge x'' \vdash x_k$ for every norm (a_k, x_k) from which $(a, x' \wedge x'')$ follows in the derivation D, so for all the lines in d_1 and d_2. This gives that in d_1 and d_2, R-ACT followed by SI can be inverted to SI followed by R-ACT; the following derivation

$$\cfrac{\cfrac{(b, y_1) \qquad (b \wedge y_1, y_2) \qquad b \wedge y_1 \wedge y_2 \not\vdash \bot}{(b, y_1 \wedge y_2)}\ \text{R-ACT} \qquad c \vdash b}{(c, y_1 \wedge y_2)}\ \text{SI}$$

can be transformed into:

$$\cfrac{\cfrac{(b, y_1) \quad c \vdash b}{(c, y_1)}\ \text{SI} \qquad \cfrac{(b \wedge y_1, y_2) \quad c \wedge y_1 \vdash b \wedge y_1}{(c \wedge y_1, y_2)}\ \text{SI} \qquad c \wedge y_1 \wedge y_2 \not\vdash \bot}{(c, y_1 \wedge y_2)}$$

The fact that $c \wedge y_1 \wedge y_2 \not\vdash \bot$ follows from
- $a \wedge x' \wedge x'' \vdash c$
- $a \wedge x' \wedge x'' \vdash y_1$
- $a \wedge x' \wedge x'' \vdash y_2$
- $a \wedge x' \wedge x'' \not\vdash \bot$

So the derivations d_1 and d_2 can be phased to SI, R-ACT, EQ. □

Theorem 5.9 *In a derivation using at most the rules EQ, SI, R-ACT and having two leaves (a, x) and root (b, y), one of those leaves can be eliminated.*

Proof. Looking at derivations having two leaves labelled with (a, x), we are in the following scenario, with the depicted R-ACT node n being the meeting point of two sub-derivations both containing (a, x):

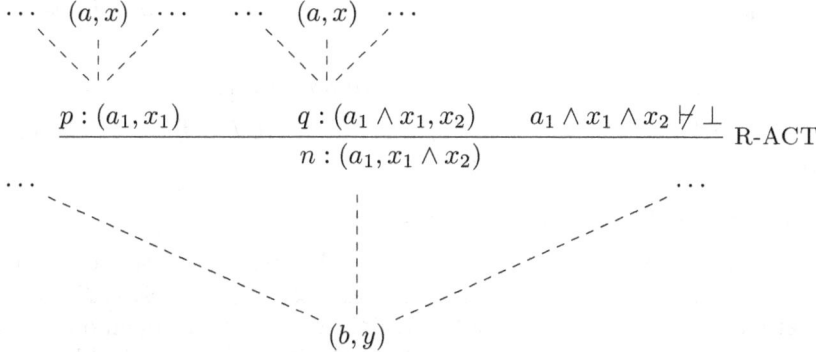

By Lemma 5.8, we know that those two sub-derivations can be replaced by derivations where the order of the rules is SI, R-ACT, EQ. Call n the meeting node of the two sub-trees d_1 and d_2 containing (a,x), and such that the sub-tree with n as root is phased SI, R-ACT, EQ.

The rest of the proof is very similar to the proof Observation 3 (c) done by Makinson and van der Torre [13]. The rule R-ACT goes from (a,x), $(a \wedge x, y)$ with $a \wedge x \wedge y \not\vdash \bot$ to $(a, x \wedge y)$. We call (a,x) the minor premise and $(a \wedge x, y)$ the major premise. The succession of R-ACT can be written in a way where no major premise of an application of R-ACT is the conclusion of another application of R-ACT. This has been shown for ACT [13] (from (a,x) and $(a \wedge x, y)$ to $(a, x \wedge y)$), and it still holds for its restricted version.

As node q is a major premise of R-ACT, it is not the conclusion of another R-ACT application, which means that the sub-tree d_2 has as only leaf (a,x) and root q and uses only one SI. The sub-tree d_1 which has p as root uses SI and R-ACT. By this and by Lemma 5.7, it holds that $x_2 \dashv\vdash x$ and $x_1 \vdash x$. So $x_1 \dashv\vdash x_1 \wedge x$, and one can delete from the tree the sub-tree d_2 with root q as well as node n, leaving a derivation with a single (a,x) node:

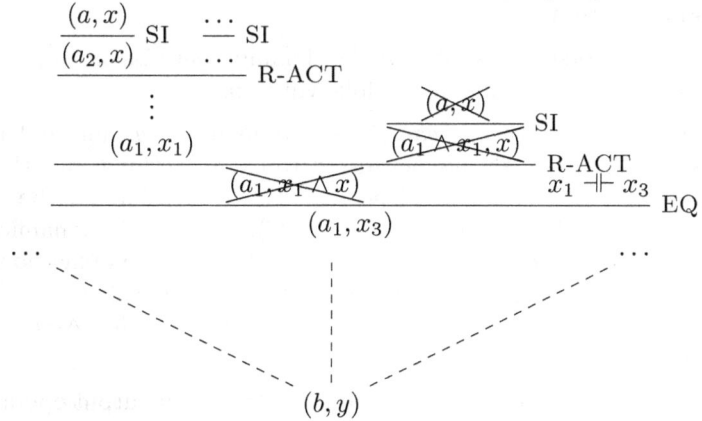

Corollary 5.10 *In a derivation using at most the rules EQ, SI, R-ACT and*

having multiple leaves (a, x) and root (b, y), all but one of those leaves can be eliminated.

Corollary 5.11 O_3 satisfies the non-repetition property.

Corollary 5.12 The subverse set of EQ, SI, R-ACT suffices to characterize the static permission operation based on O_3.

6 Dynamic Positive Permission

Similarly to the static positive permission, the dynamic positive permission takes into account a set of obligations G and a set of permissions P. However, the static positive permission is not a straightforward result of an output operation. The main idea is that (a, x) is dynamically permitted if adding $(a, \neg x)$ to the set of obligations causes a conditional prohibition of something that is permitted under that same condition for the set G. This can be understood as a form of conflict resolution; one allows something, if allowing the opposite causes conflicts with the already existing permissions. The definition as found in the original framework [13] had to be adapted slightly, as there, exact opposites are used in order to detect conflicts ((a, x) and $(a, \neg x)$). In the old systems $out_1 - out_4$ this was not a problem, as WO was always present. The WO rule allowed to include any norm derivable from the exact opposite in this conflict resolution. In the new systems this can no longer be used as is, because of the lack of WO. Instead we have to work with collectively inconsistent pairs so that all those norms that are no longer derivable via WO are still considered explicitly.

Definition 6.1 (Dynamic positive permission)

Let G be a set of explicit obligations and P a set of explicit permissions and O an output operation. Then $(a, x) \in dynperm(P, G)$ iff $\exists c, u, v$ s.t. $(c, u) \in O(G \cup \{(a, \neg x)\})$ and a pair $(c, v) \in statperm(P, G)$ with $u \wedge v$ inconsistent and c consistent.

To get a better understanding of how the dynamic permission works and how it detects conflicts, let us look at the following example.

Example 6.2 Let f denote *eating with fingers*. c denote *clean* and e denote *eat*. Let us assume that it is always permitted to eat something, but that if something is not clean, then we should not eat it. So let G and P be such that $(\neg c, \neg e) \in O(G)$ and $(\top, e) \in statperm(P, G)$. Then it is dynamically permitted to not eat with fingers, because adding $(\top, f)$ to the obligation set would allow us to derive $(\neg c, f)$ which is in conflict with $(\neg c, \neg e)$:

$(\neg c, f) \in O(G \cup \{(\top, f)\})$ for $(\neg c, \neg e) \in statperm(P, G)$ with $f \wedge \neg e \vdash \bot$. So $(\top, \neg f) \in dynperm(P, G)$.

Makinson and van der Torre give a general proof that if an output operation satisfies a Horn rule, then the dynamic permission operation satisfies its inverse. We cannot follow the same path as they did, as they use certain properties that do not hold, such as inclusion. We do not give a general proof, but we show that for the systems O_1 and O_3 specifically this holds.

Proposition 6.3 *Let $O = O_1, O_3$ be an output operation. If O satisfies a Horn rule of the form (HR), then the corresponding dynamic permission satisfies the inverse of the Horn rule $(HR)^{-1}$.*

Proof. This proof makes use of the proof theory, as D_1 and D_3 are sound and complete w.r.t. O_1 and O_3 respectively. The relevant Horn rules and their inverses are given in Table 1.

- EQ is straightforward. Details are omitted.
- Suppose O satisfies SI. Let P and G be such that $(a, x) \in dynperm(P, G)$ and $a \vdash b$. We have $(c, v) \in statperm(P, G)$ and $(c, u) \in O(G \cup \{(a, \neg x)\})$ for some u, v and c such that c consistent but $u \wedge v \vdash \bot$. By completeness, (c, u) is derivable from $G \cup \{(a, \neg x)\}$ in the corresponding proof system D. Given SI, (c, u) is derivable from $G \cup \{(b, \neg x)\}$. So by soundness $(c, u) \in O(G \cup \{(b, \neg x)\})$, and hence $(b, x) \in statperm(P, G)$. Hence $dynperm$ satisfies $(SI)^{-1}$.
- Suppose O satisfies R-AND. Take P and G such that $(a, x) \in O(G)$, $(a, \neg(x \wedge y)) \in dynperm(P, G)$ and $a \wedge x \wedge y \not\vdash \bot$. We have $(c, v) \in statperm(P, G)$ and $(c, u) \in O(G \cup \{(a, x \wedge y)\})$ for u, v and c such that $u \wedge v \vdash \bot$ and c consistent. By completeness, (c, u) is derivable from $G \cup \{(a, x \wedge y)\}$ and (a, x) is derivable from G in the corresponding proof system D. D has R-AND and $a \wedge x \wedge y \not\vdash \bot$. So one can combine the two derivations to obtain a derivation of (c, u) from $G \cup \{(a, y)\}$. By soundness $(c, u) \in O(G \cup \{(a, y)\})$. This implies $(a, \neg y) \in statperm(P, G)$, and shows that $dynperm$ satisfies $(R\text{-}AND)^{-1}$.
- Suppose O satisfies R-ACT.
  - Take P and G such that $(a, x) \in O(G)$, $(a, \neg(x \wedge y)) \in dynperm(P, G)$ and $a \wedge x \wedge y \not\vdash \bot$. Then $(c, v) \in statperm(P, G)$ and $(c, u) \in O(G \cup \{(a, x \wedge y)\})$ for c, v and u such that c consistent and $u \wedge v \vdash \bot$. By completeness, (a, x) is derivable from G, and (c, u) is derivable from $G \cup \{(a, x \wedge y)\}$ in the corresponding proof system D. D has R-ACT as rule and $a \wedge x \wedge y \not\vdash \bot$. One can combine the two derivations to obtain a derivation of (c, u) from $G \cup \{(a \wedge x, y)\}$. By soundness $(c, u) \in O(G \cup \{(a \wedge x, y)\})$. It follows that $(a \wedge x, \neg y) \in dynperm(P, G)$. Hence $dynperm$ satisfies the first version of $(R\text{-}ACT)^{-1}$.
  - Take P and G such that $(a \wedge x, y) \in O(G)$, $(a, \neg(x \wedge y)) \in dynperm(P, G)$ and $a \wedge x \wedge y \not\vdash \bot$. Then $(c, v) \in statperm(P, G)$ and $(c, u) \in O(G \cup \{(a, x \wedge y)\})$ for such that there is some c, v and u such c consistent and $u \wedge v \vdash \bot$. By completeness, $(a \wedge x, y)$ is derivable from G and (c, u) is derivable from $G \cup \{(a, x \wedge y)\}$ in the corresponding proof system D. D has R-ACT and $a \wedge x \wedge y \not\vdash \bot$. So one can combine the two derivations to obtain a derivation of (c, u) from $G \cup \{(a, x)\}$. By soundness, $(c, u) \in O(G \cup \{a, x\})$, and hence $(a, \neg x) \in dynperm(P, G)$. This shows that $dynperm$ satisfies the second version of $(R\text{-}ACT)^{-1}$.

$\square$

7 Conclusion and Future Work

In this paper we introduce the first proof systems for permission in terms of constrained output. We use the two logics of constrained output with a consistency check. The proofs are generalizations of the proofs of Makinson and van der Torre for unconstrained output [13]. Only constrained output can handle CTD reasoning, so O_1/O_3 together with the permissive norms defined in this paper is the first approach satisfying the following minimal requirements:

- detachment semantics for obligation and permissive norms (negative, static, dynamic) which can reason about CTD and dilemmas in a consistent way
- proof systems for these semantics both for obligation and one kind of permission (static)

As topics for future research, we firstly would like to find out whether the inverse rule-set is enough to fully characterize the negative permission and the positive dynamic permission operations. Furthermore, it would be desirable to find general proofs, as some of the proofs we provided are tailored to O_1 and O_3 specifically. This would allow to include any future systems in the analysis.

There are several papers about permission as exception/derogation [2,9,19]. We leave it as a topic for future research to investigate if the account studied in this paper yield any new insight on this notion.

Finally, we only consider two operations O_1 and O_3, whereas there are four classical I/O operations out_1-out_4. Indeed, only two operations with a consistency check have been defined so far. The definition of O_2/O_4 such that they satisfy all the desired properties remains an open problem.

References

[1] *Recueil de Législation Routière*, http://legilux.public.lu/eli/etat/leg/code/route/20190531, accessed: 2019-06-12.
[2] Boella, G. and L. van der Torre, *Permissions and Obligations in Hierarchical Normative Systems*, in: *Proceedings of the 9th International Conference on Artificial Intelligence and Law*, ICAIL '03 (2003), p. 109–118.
[3] Cariani, F., *Deontic logic and natural language*, in: D. Gabbay, J. Horty, X. Parent, R. van der Meyden and L. van der Torre, editors, *Handbook of Deontic Logic and Normative Systems*, College Publications, London, 2020 Volume 2. To appear.
[4] Carmo, J. and A. J. I. Jones, *Deontic logic and contrary-to-duties*, in: D. M. Gabbay and F. Guenthner, editors, *Handbook of Philosophical Logic: Volume 8* (2002), pp. 265–343.
[5] Feldman, F., "Doing the Best We Can: An Essay in Informal Deontic Logic," D. Reidel Publishing Company, Dordrecht, 1986.
[6] Goble, L., *A logic of good, should, and would: Part I*, Journal of Philosophical Logic **19** (1990), pp. 169–199.
[7] Hamblin, C., "Imperatives," Blackwell, Oxford, 1987.
[8] Hansen, J., "Imperatives and Deontic logic," Ph.D. thesis, University of Leipzig (2008).
[9] Hansen, J., *Reasoning about permission and obligation*, in: S. O. Hansson, editor, *David Makinson on Classical Methods for Non-Classical Problems*, Springer Netherlands, Dordrecht, 2014 pp. 287–333.
[10] Hansson, S. O., *The varieties of permission*, in: D. Gabbay, J. Horty, X. Parent, R. van der Meyden and L. van der Torre, editors, *Handbook of Deontic Logic and Normative Systems*, College Publications, London, 2013 pp. 195–240, volume 1.

[11] Lassiter, D., "Graded Modality," Oxford University Press, 2017.
[12] Makinson, D. and L. van der Torre, *Input/output logics*, Journal of Philosophical Logic **29** (2000), pp. 383–408.
[13] Makinson, D. and L. van der Torre, *Permission from an input/output perspective*, Journal of Philosophical Logic **32** (2003), pp. 391–416.
[14] Parent, X., D. Gabbay and L. an der Torre, *Intuitionistic basis for input/output logic*, in: S. O. Hansson, editor, *David Makinson on Classical Methods for Non-Classical Problems*, Springer Netherlands, Dordrecht, 2014 pp. 263–286.
[15] Parent, X. and L. van der Torre, *I/O logics without weakening*, To appear in *Filosofiska Notiser*.
[16] Parent, X. and L. van der Torre, *The pragmatic oddity in norm-based deontic logics*, in: G. Governatori, editor, *Proceedings of the 16th Edition of the International Conference on Articial Intelligence and Law*, ICAIL '17 (2017), p. 169–178.
[17] Parent, X. and L. van der Torre, *I/O logics with a consistency check*, in: J. Broersen, C. Condoravdi, S. Nair and G. Pigozzi, editors, *Deontic Logic and Normative Systems* (2018), pp. 285–299.
[18] Prakken, H. and M. Sergot, *Dyadic deontic logic and contrary-to-duty obligations*, in: D. Nute, editor, *Defeasible Deontic Logic*, Kluwer Academic Publishers, Dordrecht, 1997 pp. 223–262.
[19] Stolpe, A., *A theory of permission based on the notion of derogation*, Journal of Applied Logic **8** (2010), pp. 97 – 113.
[20] Strasser, C., *A deontic logic framework allowing for factual detachment*, Journal of Applied Logic **9** (2010), pp. 61–80.
[21] Straßer, C., "Adaptive Logics for Defeasible Reasoning. Applications in Argumentation, Normative Reasoning and Default Reasoning," Trends in logic **38**, Springer, Berlin, 2014.
[22] Straßer, C., M. Beirlaen and F. V. D. Putte, *Adaptive logic characterizations of input/output logic*, Studia Logica **104** (2016), pp. 869–916.

Term-Sequence-Dyadic Deontic Logic

Takahiro Sawasaki[1,2,4]

Graduate School of Letters,
Hokkaido University, Sapporo, Hokkaido, Japan

Katsuhiko Sano[1,3,5]

Faculty of Humanities and Human Sciences,
Hokkaido University, Sapporo, Hokkaido, Japan

Abstract

We often talk about general obligations of someone towards someone given that a certain condition holds, such as "Anyone has an obligation towards one's mother to be with her, given that she is lonely." To formalize such obligations with quantifiers, it is necessary to relativize obligations to agents and a given condition. The paper thus proposes a quantified dyadic deontic logic, which can not only index each dyadic deontic operator by a pair of terms but also quantify variables in the pair. The proposed logic can accommodate some class of normative conflicts without contradictions nor deontic explosions. Hilbert system of the logic is proposed and it is shown to be sound and strongly complete for an intended Kripke semantics. Moreover, a cut-free sequent calculus for the logic is provided.

Keywords: term-modal logic, conditional obligation, normative conflict.

1 Introduction

How can we formalize conditional obligations with quantifiers of someone towards someone, such as "Anyone has an obligation towards one's mother to be with her given that she is old and lonely"? The idea of making obligations relativized to something like agents or conditions is often found in the literature (e.g. [4,7,16,46,49,24,25,1,43,15]), but few deontic logics along with the idea have been developed over first order logic.

---

[1] We would like to thank two reviewers of DEON2020/2021 for their helpful comments and suggestions. Both authors were partially supported by JSPS KAKENHI Grant-in-Aid for Scientific Research (B) Grant Number 17H02258.

[2] The work of the first author was partially supported by The Uehiro Foundation on Ethics and Education.

[3] The work of the second author was partially supported by JSPS KAKENHI Grant-in-Aid for Scientific Research (C) Grant Number 19K12113.

[4] Email: taka.sawasaki562@gmail.com

[5] Email: v-sano@let.hokudai.ac.jp

To formalize such conditional obligations, term-modal logic developed by [44,12] is useful since it allows us not only index each modal operator by a single term, but also quantify variables in the term. The logic thus enables us to formalize an obligation *of someone* like "Anyone has an obligation to be with one's mother" by $\forall x \forall y (Myx \to O_x Wxy)$, where Myx, Wxy, and O_x stand for "y is x's mother", "x is with y", and "x has an obligation", respectively. Moreover, [41,15,14] have developed expansions of term-modal logic such that each modal operator is indexed by a finite sequence of terms.[6] In these expansions, we can naturally formalize not only the obligation above, but also an obligation of someone *towards someone* like "Anyone has an obligation towards one's mother to be with her" with $\forall x \forall y (Myx \to O_{xy} Wxy)$, where O_{xy} stands for "x has an obligation towards y".

The paper thus proposes a combination of term-modal logic and conditional logic which we call *term-sequence-dyadic deontic logic* (**TDDL**). The proposed logic enables us to index each dyadic deontic operator by a pair of terms and even to quantify variables in the pair. Van Fraassen's and Chellas' conditional deontic logics [47,6] are well known in the literature, but for the sake of simplicity, term-sequence-dyadic deontic logic is based on the conditional logic **CK** introduced in [5,6]. It nevertheless makes it possible to formalize "Anyone has an obligation towards one's mother to be and talk with her given that she is old and lonely" as $\forall x \forall y (Myx \to O_{xy}(Wxy \wedge Txy | Oy \wedge Ly))$, where Oy, Ly, Txy, and $O_{xy}(\cdot | Oy \wedge Ly)$ stand for "y is old", "y is lonely", "x talks with y", and "x has an obligation towards y given that y is old and lonely", respectively.

Furthermore, the paper also presents a sequent calculus for **TDDL** to show the cut elimination and Craig interpolation theorems via purely proof-theoretic methods. The proof-theoretic studies for conditional logics have been done in [9,17,37,26,2,18,33,51,39], but much of their results are for *propositional* conditional logics. On the other hand, the results of [2] for first order conditional logic are obtained via a *semantic* method. Thus our sequent calculus has a novelty in this regard.

In addition to these formal results, we argue that our **TDDL** can be applied to accommodate two kinds of normative conflicts, i.e., situations in which incompatible obligations are directed towards *different* agents and situations in which incompatible obligations are directed towards the same agent under *different* conditions. We also use our cut-free sequent calculus to prove *in a purely proof-theoretic way* that the normative conflicts of the first and second kinds do not give rise to contradictions or arbitrary obligations. This proof-theoretic result is desirable particularly in deontic logic, since it does not invoke any semantic notions so that it can make **TDDL** to some extent compatible with a philosophical thesis found e.g. in [30,21] that norms are neither true nor false.

Our **TDDL** should be compared with Frijters' term-modal dyadic deontic logic **NCL** given in [14, pp. 130–133]. His logic **NCL** is a combination of a term-modal logic and **CK** with the universal modal operator such that each modal operator is indexed by a formula and a pair of agents. Thus, except that the universal modal operator is supplemented to it, it is quite similar to **TDDL**. However, our **TDDL** still

---

[6] For other directions of developments of term-modal logic, see e.g. [23,35,48,28].

has a novel formal aspect, i.e., a cut-free sequent calculus.

The paper proceeds as follows. Sect. 2 introduces the syntax and semantics for TDDL, and Sect. 3 provides an sound and strongly complete axiomatization of it (Theorems 3.1, 3.10). Sects. 4 presents a sequent calculus for TDDL to show the cut elimination and Craig interpolation theorems (Theorems 4.4, 4.7). Sect. 5 demonstrates that TDDL can accommodate some class of normative conflicts. The section also proves in a purely proof-theoretic way that the accommodated normative conflicts do not give rise to contradictions nor arbitrary obligations (Proposition 5.1).

2 Syntax and Semantics for TDDL

The *language L* of *term-sequence-dyadic deontic logic* TDDL consists of a countably infinite set Var = $\{x, y, \ldots\}$ of *variables*, a countably infinite set Con = $\{c, d, \ldots\}$ of *constants*, a countably infinite set Pred = $\{P, Q, \ldots\}$ of *predicate symbols* each of which has a fixed finite arity, and the set of *logical constants*: $\bot, \rightarrow, \forall$, and $\mathbf{O}$. The set Term of *terms* is given by Var $\cup$ Con and a *formula* φ is defined recursively as follows:

$$\varphi ::= Pt_1 \ldots t_n \mid \bot \mid (\varphi \rightarrow \varphi) \mid \forall x \varphi \mid \mathbf{O}_{tt}(\varphi|\varphi),$$

where P is a predicate symbol with arity n, x is a variable, and $t, t_1, \ldots, t_n$ are terms. The boolean connectives $\neg, \wedge, \vee, \leftrightarrow, \top$ and the existential quantifier $\exists$ are defined as usual, i.e., $\neg\varphi := \varphi \rightarrow \bot$, $\varphi \wedge \psi := \neg(\varphi \rightarrow \neg\psi)$, $\varphi \vee \psi := \neg\varphi \rightarrow \psi$, $\varphi \leftrightarrow \psi := (\varphi \rightarrow \psi) \wedge (\psi \rightarrow \varphi)$, $\top := \bot \rightarrow \bot$ and $\exists x \varphi := \neg\forall x \neg\varphi$. The sets FV($T$), FV($\Gamma$) of free variables in sets T, Γ of terms and formulas are defined respectively as usual, except that FV($\mathbf{O}_{ts}(\varphi|\psi)) := $ FV(t, s) $\cup$ FV(φ, ψ). Substitutions $t[s/x]$, $\varphi[s/x]$ of a term s for a variable x in a term t and a formula φ are also defined as usual, where any bound variables in φ are relabelled, if necessary, to avoid clashes.

The intended reading of $\mathbf{O}_{ts}(\varphi|\psi)$ is "given that ψ, agent t has an obligation towards agent s to see to it that φ." This means that, when formalizing obligatory sentences, we often somehow insert the expression "see to it that" in the very sentences which enables us to interpret the ought-to-do sentences as the ought-to-be sentences. For example, when formalizing "Adam has an obligation towards Barbara to be and talk with her," we first interpret it as "Adam has an obligation towards Barbara to see to it that he is and talks with her" and then formalize the latter. In addition, the monadic operator $\mathbf{O}_{ts}$ for obligations, the dyadic and monadic operators $\mathbf{P}_{ts}(\cdot|\cdot)$, $\mathbf{P}_{ts}$ for permissions are defined as $\mathbf{O}_{ts}\varphi := \mathbf{O}_{ts}(\varphi|\top)$, $\mathbf{P}_{ts}(\varphi|\psi) := \neg\mathbf{O}_{ts}(\neg\varphi|\psi)$ and $\mathbf{P}_{ts}\varphi := \mathbf{P}_{ts}(\varphi|\top)$, respectively. Analogously, we read $\mathbf{P}_{ts}(\varphi|\psi)$ as "agent t is permitted towards agent s to see to it that φ given that ψ."

The Kripke semantics for TDDL is given as follows. A *frame* is a tuple $F = (W, D, f)$, where W is a nonempty set of *worlds*; D is a function which maps each world w to a nonempty set D_w of *agents*; f is an *indexed selection function* that maps each pair $(d, e) \in \bigcup_{w \in W}(D_w \times D_w)$ to a selection function, i.e., $f_{de} : W \times 2^W \rightarrow 2^W$;

f also satisfies the *cumulative domain condition*[7] :

for all $w, v \in W$, $X \subseteq W$ and $(d,e) \in D_w \times D_w$, if $v \in f_{de}(w, X)$ then $D_w \subseteq D_v$.

A *model* is a tuple $M = (F, I)$, where F is a frame and I is an *interpretation* that maps each constant c to an agent $I(c) \in \bigcap_{w \in W} D_w$, and each world w and each predicate symbol P with arity n to a subset $I(P, w)$ of D_w^n. An *assignment* α is a function from Var to $\bigcup_{w \in W} D_w$ and its domain is extended to Term by $\alpha(c) := I(c)$ for any constant c. The assignment $\alpha(x|d)$ stands for the same assignment as α except for assigning d to x. The satisfaction relation and validity are given as follows.

Definition 2.1 Let $M = (W, D, f)$ be a model, α an assignment, w a world in W, and φ a formula such that $\alpha(x) \in D_w$ for all $x \in \mathsf{FV}(\varphi)$. We sometimes write $(\alpha(t), \alpha(s))$ as $\alpha(t,s)$ for short. The *satisfaction relation* $M, \alpha, w \models \varphi$ between M, α, w and φ is defined as follows.

$M, \alpha, w \models P t_1 \ldots t_n$ iff $(\alpha(t_1), \ldots, \alpha(t_n)) \in I(P, w)$
$M, \alpha, w \not\models \bot$
$M, \alpha, w \models \psi \to \gamma$ iff $M, \alpha, w \models \psi$ implies $M, \alpha, w \models \gamma$
$M, \alpha, w \models \forall x \psi$ iff   for all agents $d \in D_w$, $M, \alpha(x|d), w \models \psi$
$M, \alpha, w \models \mathbf{O}_{ts}(\gamma|\psi)$ iff   for all worlds $v \in W$,
                   $v \in f_{\alpha(t,s)}(w, [\![\psi]\!]_\alpha^M)$ implies $M, \alpha, v \models \gamma$

where $[\![\psi]\!]_\alpha^M = \{ v \in W \mid M, \alpha, v \models \psi \}$.

Definition 2.2 A formula φ is *valid* if, for all models M, assignments α, and worlds w such that $\alpha(x) \in D_w$ for all $x \in \mathsf{FV}(\varphi)$, it holds that $M, \alpha, w \models \varphi$. A set Γ of formulas is *satisfiable* if there exists a model M, an assignment α and a world w such that $M, \alpha, w \models \varphi$ for all $\varphi \in \Gamma$.

Two remarks should be noted here. The first is that the terms t, s in the satisfaction relation of $\mathbf{O}_{ts}(\cdot|\cdot)$ are not the agents but their *names*, contrary to a usual treatment in multiagents systems based on propositional logic. The second is involved in the interpretation of the clause "$v \in f_{\alpha(t,s)}(w, [\![\psi]\!]_\alpha^M)$." In the paper we interpret this clause as "v is made acceptable at w to t by s given that ψ." This interpretation is intuitive to some extent in the sense that it can be regarded as a generalization of a usual interpretation that v is acceptable at w [31].

As expected from the semantics given, converse Barcan-like formulas are valid whereas Barcan-like formulas are not.

Proposition 2.3 $\mathbf{O}_{ts}(\forall x \varphi | \psi) \to \forall x \mathbf{O}_{ts}(\varphi | \psi)$ is valid if $x \notin \mathsf{FV}(t,s) \cup \mathsf{FV}(\psi)$.

Proof. Suppose $x \notin \mathsf{FV}(t,s) \cup \mathsf{FV}(\psi)$ and fix any model M, α, w such that $\alpha(z) \in D_w$ for all $z \in \mathsf{FV}(\mathbf{O}_{ts}(\forall x \varphi | \psi) \to \forall x \mathbf{O}_{ts}(\varphi | \psi))$. Assume also $M, \alpha, w \models \mathbf{O}_{ts}(\forall x \varphi | \psi)$. Take any agent $d \in D_w$ and any world v such that $v \in f_{\alpha(x|d)(t,s)}(w, [\![\psi]\!]_{\alpha(x|d)}^M)$. We show $M, \alpha(x|d), v \models \varphi$. Since $x \notin \mathsf{FV}(t,s) \cup \mathsf{FV}(\psi)$, we find that $\alpha(x|d)(t,s) =$

---

[7] As argued e.g. in [3,14], as for deontic logic, constant domain semantics seems a tenable semantics. However, when a constant domain semantics was adopted for **TDDL**, we have to present a cut-free sequent calculus for **TDDL** with Barcan-like formulas, which seems not so simple. To obtain a simple cut-free sequent calculus for **TDDL**, we have taken a cumulative domain semantics in this paper.

$\alpha(t, s)$ and $[\![\psi]\!]^M_{\alpha(x|d)} = [\![\psi]\!]^M_\alpha$. Thus we have $v \in f_{\alpha(t,s)}(w, [\![\psi]\!]^M_\alpha)$. Together with $M, \alpha, w \models O_{ts}(\forall x \varphi | \psi)$, we deduce that $M, \alpha, v \models \forall x \varphi$. The cumulative domain condition also guarantees $d \in D_v$ from $v \in f_{\alpha(t,s)}(w, [\![\psi]\!]^M_\alpha)$ and $d \in D_w$. Hence we obtain $M, \alpha(x|d), v \models \varphi$. □

Proposition 2.4 $\forall x O_{ts}(\varphi | \psi) \to O_{ts}(\forall x \varphi | \psi)$ *is not valid even if* $x \notin FV(t, s) \cup FV(\psi)$.

Proof. Consider a constant c, a formula $\forall x O_{cc}(Px | \top) \to O_{cc}(\forall x Px | \top)$ and a model $M = (W, D, f, I)$ such that $W = \{w, v\}$, $D_w = \{c\}$, $D_v = \{c, d\}$, $f_{cc}(w, W) = \{v\}$, $I(P, v) = \{c\}$ and $c^I = c$. It is easy to see that $M, \alpha, w \not\models \forall x O_{cc}(Px | \top) \to O_{cc}(\forall x Px | \top)$ and $x \notin FV(c) \cup FV(\top)$, no matter how the assignment α is given. □

Both formulas are natural counterparts to converse Barcan and Barcan formulas, but, as far as we know, they have rarely been considered in the literature on quantified conditional logics. For example, no mentions on converse Barcan or Barcan formulas are found in [13,32,2]. Only a few mentions on both formulas are found in [10, p. 121].

3 Axiomatization of TDDL

This section provides a sound and strongly complete axiomatization for TDDL. The Hilbert system $\mathcal{H}(\text{TDDL})$ for TDDL consists of axioms PC1, PC2, PC3 and U; inference rules MP, G, RCEA and RCK, displayed in Table 1. When $n = 0$ in RCK, RCK essentially means that we may infer $O_{ts}(\varphi | \psi)$ from φ. We define the notion of *proof* in $\mathcal{H}(\text{TDDL})$ as usual and sometimes write $\vdash_\mathcal{H} \varphi$ to mean that φ is provable in $\mathcal{H}(\text{TDDL})$.

PC1	$\varphi \to (\psi \to \varphi)$
PC2	$(\varphi \to (\psi \to \gamma)) \to ((\varphi \to \psi) \to (\varphi \to \gamma))$
PC3	$(\neg \psi \to \neg \varphi) \to (\varphi \to \psi)$
U	$\forall x \varphi \to \varphi[t/x]$
MP	From $\varphi \to \psi$ and φ, infer ψ
G	From $\varphi \to \psi[y/x]$, infer $\varphi \to \forall x \psi$ if $y \notin FV(\varphi, \forall x \psi)$
RCEA	From $\psi \leftrightarrow \psi'$, infer $O_{ts}(\varphi | \psi) \leftrightarrow O_{ts}(\varphi | \psi')$
RCK	From $(\varphi_1 \land \cdots \land \varphi_n) \to \varphi$, infer $(O_{ts}(\varphi_1 | \psi) \land \cdots \land O_{ts}(\varphi_n | \psi)) \to O_{ts}(\varphi | \psi)$

Table 1
Hilbert system $\mathcal{H}(\text{TDDL})$

It is straightforward to establish soundness.

Theorem 3.1 (Soundness) *If a formula φ is provable in $\mathcal{H}(\text{TDDL})$ then φ is valid.*

For strong completeness of $\mathcal{H}(\text{TDDL})$, we employ the canonical model construction. The canonical models for propositional and first order conditional logics are introduced in [27,5] and [10,11], respectively. As far as we know, only [11] and [2] consider varying domain semantics for first order conditional logics, but none of them are designed so as to have cumulative domain. The canonical models for term-modal logics and their analogues are introduced in [41,15,48,28]. Amongst them, [48] and [41] construct the canonical models to have cumulative domain.[8] The canonical model we provide in the paper can be seen as an adaptation of them for conditional logics.

For a set V of variables, we define $L(V)$ as the language obtained from L by replacing Var with V. We also denote $L(FV(\Gamma))$ by $L(\Gamma)$.

Definition 3.2 Let V, V' be sets of variables. The notation $L(V) \sqsubset L(V')$ means that $V \subseteq V'$ and $V' \setminus V$ is countably infinite.

Let us now say that

- a formula φ is *provable from* Γ in $\mathcal{H}(\text{TDDL})$, denoted by $\Gamma \vdash_{\mathcal{H}} \varphi$, if there exists a finite subset Δ of Γ such that $\vdash_{\mathcal{H}} \bigwedge \Delta \to \varphi$, where $\bigwedge \Delta$ denotes the conjunction of all formulas in Δ ($\bigwedge \varnothing := \top$);
- Γ is *inconsistent* if $\Gamma \vdash_{\mathcal{H}} \bot$;
- Γ is *consistent* if Γ is not inconsistent;
- Γ is a *maximally consistent set* (*MCS* for short) if Γ is consistent and for all formulas φ in $L(\Gamma)$, $\varphi \in \Gamma$ or $\neg \varphi \in \Gamma$;
- Γ has $\forall$-*property* if, for all $\forall x \varphi$ in $L(\Gamma)$, there exists a term $t \in \text{Term}(\Gamma)$ such that $\varphi[t/x] \to \forall x \varphi \in \Gamma$.

In addition to the above, we put $\text{Var}^+ := \text{Var} \cup \text{Var}'$ throughout the paper, where Var' is a "fresh" countably infinite set of variables disjoint from Var.

Proposition 3.3 *Let Γ be an MCS in $L(\text{Var}^+)$. If $\Gamma \vdash_{\mathcal{H}} \varphi$ then $\varphi \in \Gamma$ for all φ in $L(\Gamma)$.*

Lemma 3.4 *Let Γ be a consistent set in $L(\text{Var}^+)$ such that $L(\Gamma) \sqsubset L(\text{Var}^+)$. There exists an MCS Γ^+ with $\forall$-property in $L(\text{Var}^+)$ such that $\Gamma \subseteq \Gamma^+$ and $L(\Gamma^+) \sqsubset L(\text{Var}^+)$.*

Definition 3.5 Define

$$W^c := \{ \Gamma \mid \Gamma \text{ is an MCS with } \forall\text{-property in } L(\text{Var}^+) \text{ such that } L(\Gamma) \sqsubset L(\text{Var}^+) \};$$
$$|\varphi| := \{ \Theta \in W^c \mid \varphi \in \Theta \}.$$

The *canonical model* M^c is the tuple (W^c, D^c, f^c, I^c) where

- $D^c_\Gamma = \text{Term}(\Gamma)$;
- for all $(t, s) \in \bigcup_{\Gamma \in W^c}(D^c_\Gamma \times D^c_\Gamma)$, f^c_{ts} is defined as follows:
  · if $X = |\psi|$ for some formula ψ in $L(\Gamma)$,

$$\Delta \in f^c_{ts}(\Gamma, |\psi|) \quad \text{iff} \quad \{ \gamma \mid O_{ts}(\gamma|\psi) \in \Gamma \} \subseteq \Delta;$$

  · otherwise, $f^c_{ts}(\Gamma, X) := \varnothing$;
- $(t_1, \ldots, t_n) \in I^c(P, \Gamma) \quad \text{iff} \quad P t_1 \ldots t_n \in \Gamma$;
- $I^c(c) = c$.

---

[8] [48] considers a constant domain semantics, but for the lack of Barcan-like formulas, constructs the canonical pseudo model having cumulative domain via the techniques inspired by [8].

Proposition 3.6 *For all $(t, s) \in \bigcup_{\Gamma \in W^c}(D_\Gamma^c \times D_\Gamma^c)$, the canonical model's f_{ts}^c is well-defined.*

Proof. Suppose $|\psi| = |\psi'|$. We show the following equivalence

$$O_{ts}(\gamma|\psi) \in \Gamma \quad \text{iff} \quad O_{ts}(\gamma|\psi') \in \Gamma$$

holds for any formula ψ, ψ' in $L(\Gamma)$. We first establish $\vdash_\mathcal{H} \psi \to \psi'$. Suppose not. Since the set $\{\psi, \neg\psi'\}$ is then consistent, by Lemma 3.4 we have an MCS $\Delta \in W^c$ such that $\{\psi, \neg\psi'\} \subseteq \Delta$. Thus, we can deduce by $|\psi| = |\psi'|$ that $\psi', \neg\psi' \in \Delta$, which contradicts the consistency of Δ. Hence $\vdash_\mathcal{H} \psi \to \psi'$. Similarly, $\vdash_\mathcal{H} \psi' \to \psi$. As $\vdash_\mathcal{H} \psi \leftrightarrow \psi'$ holds, it follows from RCEA that $\vdash_\mathcal{H} O_{ts}(\gamma|\psi) \leftrightarrow O_{ts}(\gamma|\psi')$. This gives the above equivalence. □

Proposition 3.7 *The canonical model is a model.*

Proof. We need to confirm that $I^c(c) \in \bigcap_{\Gamma \in W^c} D_\Gamma^c$ and that the frame of the canonical model satisfies the cumulative domain condition, i.e., for all $\Gamma, \Delta \in W^c$, $X \subseteq W^c$ and $(t, s) \in (D_\Gamma^c)^2$, if $\Delta \in f_{ts}(\Gamma, X)$ then $D_\Gamma \subseteq D_\Delta$. The former is immediate so we confirm only the latter. Take any $\Gamma, \Delta \in W^c$, $X \subseteq W^c$ and $(t, s) \in (D_\Gamma^c)^2$. Suppose $\Delta \in f_{ts}^c(\Gamma, X)$ and $u \in D_\Gamma^c$. By $f_{ts}^c(\Gamma, X) \neq \emptyset$, it holds that $\{\gamma \mid O_{ts}(\gamma|\psi) \in \Gamma\} \subseteq \Delta$ for some formula ψ in $L(\Gamma)$ such that $X = |\psi|$. Also, $O_{ts}(Pu \to Pu|\psi) \in \Gamma$ holds. Therefore $Pu \to Pu \in \Delta$ hence $u \in D_\Delta^c$, as required. □

Lemma 3.8 *Let Γ be an MCS with $\forall$-property in $L(\text{Var}^+)$ such that $L(\Gamma) \sqsubset L(\text{Var}^+)$. Given that $\neg O_{ts}(\varphi|\psi) \in \Gamma$, there exists an MCS Δ with $\forall$-property in $L(\text{Var}^+)$ such that $L(\Delta) \sqsubset L(\text{Var}^+)$, $\neg\varphi \in \Delta$ and $\Delta \in f_{ts}^c(\Gamma, |\psi|)$.*

Proof. Suppose $\neg O_{ts}(\varphi|\psi) \in \Gamma$. To establish that $\Delta_0 := \{\neg\varphi\} \cup \{\gamma \mid O_{ts}(\gamma|\psi) \in \Gamma\}$ is consistent, suppose for contradiction that Δ_0 is inconsistent. Then, where $O_{ts}(\gamma_i|\psi) \in \Gamma$, $\vdash_\mathcal{H} \gamma_1 \wedge \cdots \wedge \gamma_n \to \varphi$ for some n. We obtain $\Gamma \vdash_\mathcal{H} \bot$ as follows.

1. $\vdash_\mathcal{H} \gamma_1 \wedge \cdots \wedge \gamma_n \to \varphi$ Supposition for contradiction
2. $\vdash_\mathcal{H} O_{ts}(\gamma_1|\psi) \wedge \cdots \wedge O_{ts}(\gamma_n|\psi) \to O_{ts}(\varphi|\psi)$ 1, RCK
3. $\Gamma \vdash_\mathcal{H} O_{ts}(\varphi|\psi)$ 2, $O_{ts}(\gamma_i|\psi) \in \Gamma$, PC
4. $\Gamma \vdash_\mathcal{H} \bot$ 3, $\neg O_{ts}(\varphi|\psi) \in \Gamma$, PC

However, Γ should be consistent so a contradiction occurs. Thus Δ_0 is consistent. Also, $L(\Delta_0) \sqsubset L(\text{Var}^+)$ from the definition of Δ_0 and $L(\Gamma) \sqsubset L(\text{Var}^+)$.

By applying Lemma 3.4 to Δ_0, we obtain an MCS Δ^+ with $\forall$-property in $L(\text{Var}^+)$ such that $\Delta_0 \subseteq \Delta^+$ and $L(\Delta^+) \sqsubset L(\text{Var}^+)$. We need to show that $\neg\varphi \in \Delta^+$ and $\Delta^+ \in f_{ts}^c(\Gamma, |\psi|)$. The former is obvious. For the latter, note that $FV(\psi) \subseteq D_\Gamma^c$ and $t, s \in D_\Gamma^c$ from $\neg O_{ts}(\varphi|\psi) \in \Gamma$. Then $f_{ts}^c(\Gamma, |\psi|) = \{\Delta \in W^c \mid \{\gamma \mid O_{ts}(\gamma|\psi) \in \Gamma\} \subseteq \Delta\}$. Therefore $\Delta^+ \in f_{ts}^c(\Gamma, |\psi|)$ from $\{\gamma \mid O_{ts}(\gamma|\psi) \in \Gamma\} \subseteq \Delta_0 \subseteq \Delta^+$. □

Lemma 3.9 *Let $M^c = (W^c, D^c, f^c, I^c)$ be the canonical model and α^c the assignment defined by $\alpha^c(x) = x$ for all $x \in \text{Var}^+$. For all $\varphi, \Gamma \in W^c$ such that φ is in $L(\Gamma)$,*

$$M^c, \alpha^c, \Gamma \models \varphi \quad \text{iff} \quad \varphi \in \Gamma.$$

Proof. We prove only when φ is $\mathbf{O}_{ts}(\gamma|\psi)$. For the right-to-left direction, suppose $\mathbf{O}_{ts}(\gamma|\psi) \in \Gamma$. To show $M^c, \alpha^c, \Gamma \models \mathbf{O}_{ts}(\gamma|\psi)$, fix any world $\Delta \in W^c$ such that $\Delta \in f_{ts}^c(\Gamma, [\![\psi]\!]_{\alpha^c}^{M^c})$. We show $M^c, \alpha^c, \Delta \models \gamma$. Note first that $FV(\psi) \subseteq D_\Gamma^c$ and $t, s \in D_\Gamma^c$ since $\mathbf{O}_{ts}(\gamma|\psi)$ is in $L(\Gamma)$; that $[\![\psi]\!]_{\alpha^c}^{M^c} = |\psi|$ by inductive hypothesis, where recall $|\psi| = \{ \Delta' \in W^c \,|\, \psi \in \Delta' \}$. It then follows from $\Delta \in f_{ts}^c(\Gamma, [\![\psi]\!]_{\alpha^c}^{M^c})$ that $\{ \gamma' \,|\, \mathbf{O}_{ts}(\gamma'|\psi) \in \Gamma \} \subseteq \Delta$. Since $\gamma \in \Delta$ from $\mathbf{O}_{ts}(\gamma|\psi) \in \Gamma$, by inductive hypothesis we obtain $M^c, \alpha^c, \Delta \models \gamma$, as required. For the other direction, suppose $\mathbf{O}_{ts}(\gamma|\psi) \notin \Gamma$. We show $M^c, \alpha^c, \Gamma \not\models \mathbf{O}_{ts}(\gamma|\psi)$. In this case, $\neg \mathbf{O}_{ts}(\gamma|\psi) \in \Gamma$ since $\mathbf{O}_{ts}(\gamma|\psi)$ is in $L(\Gamma)$. Hence by Lemma 3.8 we obtain an MCS Δ with $\forall$-property in $L(\text{Var}^+)$ such that $L(\Delta) \sqsubset L(\text{Var}^+)$, $\Delta \in f_{ts}^c(\Gamma, |\psi|)$ and $\neg \gamma \in \Delta$, which implies $\gamma \notin \Delta$. From $\Delta \in f_{ts}^c(\Gamma, |\psi|)$ and $\gamma \notin \Delta$, we deduce by inductive hypothesis that $\Delta \in f_{ts}^c(\Gamma, [\![\psi]\!]_{\alpha^c}^{M^c})$ but $M^c, \alpha^c, \Delta \not\models \gamma$. This implies $M^c, \alpha^c, \Gamma \not\models \mathbf{O}_{ts}(\gamma|\psi)$. □

Strong completeness of $\mathcal{H}(\text{TDDL})$ now follows from Lemmas 3.4 and 3.9.

Theorem 3.10 (Strong completeness) *Any consistent set of formulas in L is satisfiable.*

4 Sequent Calculus for TDDL

This section presents a sequent calculus $\mathcal{G}(\text{TDDL})$ for TDDL to show the cut elimination and Craig interpolation theorems.

Let us first illustrate a notion of sequents following [34, p. 7]. Given finite *multisets* Γ, Δ of formulas, an expression $\Gamma \Rightarrow \Delta$ we call a *sequent*. Intuitively, it means that some formulas in Δ follow from all the formulas in Γ. When $\Gamma = \varnothing$, the sequent $\Rightarrow \Delta$ means that some formulas in Δ follow without any assumptions. On the other hand, when $\Delta = \varnothing$, the sequent $\Gamma \Rightarrow$ means that a contradiction follows from all the formulas in Γ.

The sequent calculus $\mathcal{G}(\text{TDDL})$ consists of initial sequents (id) and ($\bot$); structural rules ($\Rightarrow w$), ($w \Rightarrow$), ($\Rightarrow c$), ($c \Rightarrow$) and (Cut); logical rules ($\Rightarrow \rightarrow$), ($\rightarrow \Rightarrow$), ($\Rightarrow \forall$), ($\forall \Rightarrow$) and (O), displayed in Table 2.
For short, we often write the rule (O) as

$$\frac{(\psi_i \Leftrightarrow \psi)_{1 \leq i \leq n} \quad \varphi_1, \ldots, \varphi_n \Rightarrow \varphi}{\mathbf{O}_{ts}(\varphi_1|\psi_1), \ldots, \mathbf{O}_{ts}(\varphi_n|\psi_n) \Rightarrow \mathbf{O}_{ts}(\varphi|\psi)} \quad (O)$$

where $\psi_i \Leftrightarrow \psi$ is a pair of two sequents $\psi_i \Rightarrow \psi$ and $\psi \Rightarrow \psi_i$. The rule (O) is imported in the form of the two-sided rule from the one-sided rule (CK_g) introduced in [37, p. 14]. The reader may regard (O) as "applying RCK together with RCEA." We define the notion of *derivation* in $\mathcal{G}(\text{TDDL})$ as usual and write $\vdash_\mathcal{G} \varphi$ to mean that φ is derivable in $\mathcal{G}(\text{TDDL})$.

Example 4.1 The sequent

$$\forall x \forall y (Mxy \rightarrow \mathbf{O}_{xy}(Wxy \wedge Txy|Oy \wedge Ly)), Mab \Rightarrow \mathbf{O}_{ab}(Wab|Ob \wedge Lb)$$

is derivable in $\mathcal{G}(\text{TDDL})$ as follows, where ($\wedge \Rightarrow$) is a derived rule in $\mathcal{G}(\text{TDDL})$.

$$\varphi \Rightarrow \varphi \quad (id) \qquad\qquad \bot \Rightarrow \quad (\bot)$$

$$\frac{\Gamma \Rightarrow \Delta}{\Gamma \Rightarrow \Delta, \varphi} \ (\Rightarrow w) \qquad\qquad \frac{\Gamma \Rightarrow \Delta}{\varphi, \Gamma \Rightarrow \Delta} \ (w \Rightarrow)$$

$$\frac{\Gamma \Rightarrow \Delta, \varphi, \varphi}{\Gamma \Rightarrow \Delta, \varphi} \ (\Rightarrow c) \qquad\qquad \frac{\varphi, \varphi, \Gamma \Rightarrow \Delta}{\varphi, \Gamma \Rightarrow \Delta} \ (c \Rightarrow)$$

$$\frac{\Gamma \Rightarrow \Delta, \varphi \quad \varphi, \Theta \Rightarrow \Sigma}{\Gamma, \Theta \Rightarrow \Delta, \Sigma} \ (Cut)$$

$$\frac{\varphi, \Gamma \Rightarrow \Delta, \psi}{\Gamma \Rightarrow \Delta, \varphi \to \psi} \ (\Rightarrow \to) \qquad \frac{\Gamma \Rightarrow \Delta, \varphi \quad \psi, \Theta \Rightarrow \Sigma}{\varphi \to \psi, \Gamma, \Theta \Rightarrow \Delta, \Sigma} \ (\to \Rightarrow)$$

$$\frac{\Gamma \Rightarrow \Delta, \varphi[y/x]}{\Gamma \Rightarrow \Delta, \forall x \varphi} \ (\Rightarrow \forall)^{\dagger} \qquad \frac{\varphi[t/x], \Gamma \Rightarrow \Delta}{\forall x \varphi, \Gamma \Rightarrow \Delta} \ (\forall \Rightarrow)$$

$$\frac{\psi_1 \Rightarrow \psi \quad \psi \Rightarrow \psi_1 \quad \cdots \quad \psi_n \Rightarrow \psi \quad \psi \Rightarrow \psi_n \quad \varphi_1, \ldots, \varphi_n \Rightarrow \varphi}{\mathbf{O}_{ts}(\varphi_1|\psi_1), \ldots, \mathbf{O}_{ts}(\varphi_n|\psi_n) \Rightarrow \mathbf{O}_{ts}(\varphi|\psi)} \ (O)$$

†: y is not a free variable in $\Gamma, \Delta, \forall x \varphi$.

Table 2
Sequent calculus $\mathcal{G}(\text{TDDL})$

$$\frac{Mab \Rightarrow Mab \quad \dfrac{Ob \wedge Lb \Rightarrow Ob \wedge Lb \quad \dfrac{Wab \Rightarrow Wab}{Wab \wedge Tab \Rightarrow Wab} (\wedge \Rightarrow)}{\mathbf{O}_{ab}(Wab \wedge Tab|Ob \wedge Lb) \Rightarrow \mathbf{O}_{ab}(Wab|Ob \wedge Lb)} (O)}{\dfrac{Mab \to \mathbf{O}_{ab}(Wab \wedge Tab|Ob \wedge Lb), Mab \Rightarrow \mathbf{O}_{ab}(Wab|Ob \wedge Lb)}{\forall x \forall y (Mxy \to \mathbf{O}_{xy}(Bxy \wedge Txy|Ox \wedge Lx)), Mab \Rightarrow \mathbf{O}_{ab}(Wab|Ob \wedge Lb)} (\forall \Rightarrow)} (\to \Rightarrow)$$

Proposition 4.2 (Equipollence) *A formula φ is provable in $\mathcal{H}(\text{TDDL})$ iff $\Rightarrow \varphi$ is derivable in $\mathcal{G}(\text{TDDL})$.*

Proof. The left-to-right direction is established by induction on length of a proof of φ. The other direction immediately follows from the claim that $\vdash_{\mathcal{G}} \Gamma \Rightarrow \Delta$ implies $\vdash_{\mathcal{H}} \bigwedge \Gamma \to \bigvee \Delta$, where $\bigwedge \Gamma$ is the conjunction of all formulas in Γ ($\bigwedge \emptyset := \top$) and $\bigvee \Delta$ is the disjunction of all formulas in Δ ($\bigvee \emptyset := \bot$). The claim is shown by induction on height of a derivation of $\Gamma \Rightarrow \Delta$. In the case that the derivation ends with an application of (O), we obtain a proof in $\mathcal{H}(\text{TDDL})$ of the corresponding formula to $\Gamma \Rightarrow \Delta$ by applying RCK together with RCEA. □

To show the cut elimination theorem for $\mathcal{G}(\text{TDDL})$, let us make a few preliminaries. To begin, we define two alternatives of $\mathcal{G}(\text{TDDL})$.

Definition 4.3 The sequent calculus $\mathcal{G}^{-}(\text{TDDL})$ is the calculus obtained from $\mathcal{G}(\text{TDDL})$ by removing (Cut). On the one hand, the sequent calculus $\mathcal{G}^{*}(\text{TDDL})$ is the calculus obtained from $\mathcal{G}(\text{TDDL})$ by replacing (Cut) with the extended rule (Cut^*) which is introduced in [34],:

$$\frac{\Gamma \Rightarrow \Delta, \varphi^m \quad \varphi^n, \Theta \Rightarrow \Sigma}{\Gamma, \Theta \Rightarrow \Delta, \Sigma} \ (Cut^*),$$

where φ is called a *cut-formula* and $m, n \geq 0$ denote the numbers of occurrences of φ. The notions of derivation in $\mathcal{G}^-(\text{TDDL})$ and $\mathcal{G}^*(\text{TDDL})$ are defined in the same way as in $\mathcal{G}(\text{TDDL})$.

In addition, we say that a derivation $\mathfrak{D}$ in $\mathcal{G}^*(\text{TDDL})$ is of the (Cut^*)-*bottom form* if the last applied rule in $\mathfrak{D}$ is (Cut^*) and there are no other applications of (Cut^*) in $\mathfrak{D}$. We also define the *complexity* of cut-formula φ and the *weight* of a derivation $\mathfrak{D}$ by the number of logical constants other than $\bot$ occurring in φ and the number of sequents occurring in $\mathfrak{D}$ except for its root, respectively. Finally, we assume in what follows that free variables and bound variables in derivations are thoroughly separated.

Theorem 4.4 (Cut elimination) *If* $\Gamma \Rightarrow \Delta$ *is derivable in* $\mathcal{G}(\text{TDDL})$ *then it is derivable in* $\mathcal{G}^-(\text{TDDL})$.

Proof. Since (Cut) is an instance of (Cut^*), it suffices to show that $\vdash_{\mathcal{G}^*} \Gamma \Rightarrow \Delta$ implies $\vdash_{\mathcal{G}^-} \Gamma \Rightarrow \Delta$. It is not difficult to see that this is obtained from the following claim:

Given a derivation $\mathfrak{D}$ of (Cut^*)-bottom form of a sequent $\Gamma \Rightarrow \Delta$ in $\mathcal{G}^*(\text{TDDL})$, there is a derivation of $\Gamma \Rightarrow \Delta$ in $\mathcal{G}^-(\text{TDDL})$.

We establish the claim by double induction on complexity of cut-formula and weight of such a derivation $\mathfrak{D}$. Since the other cases are dealt with in the same way as in first order logic, we confine ourselves to the case that both of the left and right upper sequents (of the only one application) of (Cut^*) in $\mathfrak{D}$ are obtained by (O). We may also assume that both of the numbers m, n of cut-formula are more than zero, because otherwise a derivation of $\Gamma \Rightarrow \Delta$ in $\mathcal{G}^-(\text{TDDL})$ is immediately obtained from one of the upper sequents of (Cut^*) by applying $(\Rightarrow w)$ and $(w \Rightarrow)$ repeatedly. Thus, $\mathfrak{D}$ is now of the form

$$\frac{\dfrac{(\psi_i \Leftrightarrow \psi)_{1 \leq i \leq k} \quad \overline{\varphi_k} \Rightarrow \varphi}{\overline{O_{ts}(\varphi_k|\psi_k)} \Rightarrow O_{ts}(\varphi|\psi)}(O) \quad \dfrac{(\psi \Leftrightarrow \psi')^n \quad (\psi'_j \Leftrightarrow \psi')_{1 \leq j \leq l} \quad \varphi^n, \overline{\varphi'_l} \Rightarrow \varphi'}{(O_{ts}(\varphi|\psi))^n, \overline{O_{ts}(\varphi'_l|\psi'_l)} \Rightarrow O_{ts}(\varphi'|\psi')}(O)}{\overline{O_{ts}(\varphi_k|\psi_k)}, \overline{O_{ts}(\varphi'_l|\psi'_l)}, \Rightarrow O_{ts}(\varphi'|\psi')} (Cut)^*$$

where $\overline{O_{ts}(\varphi_k|\psi_k)} = O_{ts}(\varphi_1|\psi_1), \ldots, O_{ts}(\varphi_k|\psi_k)$, $\overline{O_{ts}(\varphi'_l|\psi'_l)} = O_{ts}(\varphi'_1|\psi'_1), \ldots, O_{ts}(\varphi'_l|\psi'_l)$, $\overline{\varphi_k} = \varphi_1, \ldots, \varphi_k$, and $\overline{\varphi'_l} = \varphi'_1, \ldots, \varphi'_l$. Put derivations $\mathfrak{D}^i_1, \mathfrak{D}^i_2$ and $\mathfrak{D}_3$ as

$$\mathfrak{D}^i_1 = \frac{\psi_i \Rightarrow \psi \quad \psi \Rightarrow \psi'}{\psi_i \Rightarrow \psi'}(Cut^*) \quad \mathfrak{D}^i_2 = \frac{\psi' \Rightarrow \psi \quad \psi \Rightarrow \psi_i}{\psi' \Rightarrow \psi_i}(Cut^*)$$

$$\mathfrak{D}_3 = \frac{\overline{\varphi_k} \Rightarrow \varphi \quad \varphi^n, \overline{\varphi'_l} \Rightarrow \varphi'}{\overline{\varphi_k}, \overline{\varphi'_l} \Rightarrow \varphi'}(Cut^*)$$

respectively. Note that each (Cut^*) is eliminable by inductive hypothesis since the complexities of $\mathfrak{D}_1^i, \mathfrak{D}_2^i$ and $\mathfrak{D}_3$ are reduced. By using $\mathfrak{D}_1^i, \mathfrak{D}_2^i$ and $\mathfrak{D}_3$, we can construct a derivation of the same sequent in $\mathcal{G}^-(\mathbf{TDDL})$ as follows.

$$\dfrac{\begin{array}{ccc} \mathfrak{D}_1^i, \mathfrak{D}_2^i & \vdots & \mathfrak{D}_3 \\ (\psi_i \Leftrightarrow \psi')_{1 \leq i \leq k} & (\psi'_j \Leftrightarrow \psi')_{1 \leq i \leq l} & \overline{\varphi_k}, \overline{\varphi'_l} \Rightarrow \varphi' \end{array}}{\overline{\mathbf{O}_{ts}(\varphi_k|\psi_k)}, \overline{\mathbf{O}_{ts}(\varphi'_l|\psi'_l)} \Rightarrow \mathbf{O}_{ts}(\varphi'|\psi')} \text{ (O)}$$

Hence the claim was established. □

It is remarked that the Craig interpolation theorem for conditional logic **CK** [6] was established in [36, Theorem 6.11]. We can still preserve the theorem in our setting. That is, we show that $\mathcal{G}(\mathbf{TDDL})$ enjoys the Craig interpolation theorem by Maehara method [29] as a corollary of the cut elimination theorem.

Definition 4.5 A *partition* of a sequent $\Gamma \Rightarrow \Delta$ is a pair $((\Gamma_1, \Delta_1), (\Gamma_2, \Gamma_2))$ of pairs of finite multisets of formulas such that $\Gamma = \Gamma_1, \Gamma_2$ and $\Delta = \Delta_1, \Delta_2$. In what follows, a partition is denoted by $(\Gamma_1 : \Delta_1), (\Gamma_1 : \Delta_2)$.

Lemma 4.6 *Let $\Gamma \Rightarrow \Delta$ be a sequent derivable in $\mathcal{G}(\mathbf{TDDL})$. If $(\Gamma_1 : \Delta_1), (\Gamma_2 : \Delta_2)$ is a partition of the sequent $\Gamma \Rightarrow \Delta$, then there is an* interpolant *φ of it, i.e., φ satisfies the following: the sequents $\Gamma_1 \Rightarrow \Delta_1, \varphi$ and $\varphi, \Gamma_2 \Rightarrow \Delta_2$ are provable in $\mathcal{G}(\mathbf{TDDL})$, $\mathrm{FV}(\varphi) \subseteq \mathrm{FV}(\Gamma_1, \Delta_1) \cap \mathrm{FV}(\Gamma_2, \Delta_2)$, and $\mathrm{Pred}(\varphi) \subseteq \mathrm{Pred}(\Gamma_1, \Delta_1) \cap \mathrm{Pred}(\Gamma_2, \Delta_2)$.*

Proof. Suppose that $\Gamma \Rightarrow \Delta$ be a sequent derivable in $\mathcal{G}(\mathbf{TDDL})$ hence also in $\mathcal{G}^-(\mathbf{TDDL})$. Our proof is done by induction on the height of the derivation $\mathcal{D}$ of $\Gamma \Rightarrow \Delta$ in $\mathcal{G}^-(\mathbf{TDDL})$. In what follows, we exclude partitions of the form $(\varnothing : \varnothing)$, $(\Gamma : \Delta)$ or $(\Gamma : \Delta), (\varnothing : \varnothing)$, since $\top$ and $\bot$ are easily seen to be interpolants for these partitions, respectively. We skip the base case and directly move to the inductive step. Here we only show the cases where $\Gamma \Rightarrow \Delta$ is obtained by (O).

Let us deal with the case where a sequent is obtained by (O). Let

$$\dfrac{(\varphi_i \Leftrightarrow \varphi)_{1 \leq i \leq n} \quad (\Sigma_{i1}, \Sigma_{i2})_{1 \leq i \leq n} \Rightarrow \psi}{(\mathbf{O}_{xy}(\Sigma_{i1}|\varphi_i), \mathbf{O}_{xy}(\Sigma_{i2}|\varphi_i))_{1 \leq i \leq n} \Rightarrow \mathbf{O}_{xy}(\psi|\varphi)} \text{ O}$$

be the last step of a derivation where $\Sigma_i := \Sigma_{i1}, \Sigma_{i2}$ and $\mathbf{O}_{xy}(\Sigma|\varphi)$ means $\{\mathbf{O}_{xy}(\gamma|\varphi) \mid \gamma \in \Sigma\}$. There are two cases depending on which side of a partition of the lower sequent contains $\mathbf{O}_{xy}(\psi_0|\varphi_0)$. We show only the case that $\mathbf{O}_{xy}(\psi_0|\varphi_0)$ lies in the right side of a partition of the lower sequent. Let $((\mathbf{O}_{xy}(\Sigma_{i1}|\varphi_i))_{1 \leq i \leq n} : \varnothing), (\mathbf{O}_{xy}(\Sigma_{i2}|\varphi_i))_{1 \leq i \leq n} : \mathbf{O}_{xy}(\psi_0|\varphi_0))$ be a partition. It is remarked that $(\mathbf{O}_{xy}(\Sigma_{i1}|\varphi_i))_{1 \leq i \leq n}$ is not empty. Fix such k. Since $\varphi_0 \Rightarrow \varphi_k$ is derivable by assumption, our inductive hypothesis implies there is an interpolant ρ_k for $(\varphi_0 : \varnothing), (\varnothing : \varphi_k)$, i.e., both of $\varphi_0 \Rightarrow \rho_k$ and $\rho_k \Rightarrow \varphi_k$ are derivable in $\mathcal{G}(\mathbf{TDDL})$ and $\mathrm{X}(\rho_k) \subseteq \mathrm{X}(\varphi_0) \cap \mathrm{X}(\varphi_k)$ where X is FV or Pred. Since $\varphi_k \Rightarrow \varphi_0$ is derivable in $\mathcal{G}(\mathbf{TDDL})$ (by assumption), we can also obtain that $\rho_k \Rightarrow \varphi_0$ is derivable in $\mathcal{G}(\mathbf{TDDL})$. By a similar argument we can show that all sequents in $(\varphi_i \Leftrightarrow \rho_k)_{1 \leq i \leq n}$, i.e., $\{\rho_k \Rightarrow \varphi_i, \varphi_i \Rightarrow \rho_k \mid 1 \leq i \leq n\}$ are derivable in $\mathcal{G}(\mathbf{TDDL})$.

Moreover, by inductive hypothesis to the other premises of the above last application of (O), we obtain:

- both $(\Sigma_{i1})_{1 \leq i \leq n} \Rightarrow \chi$ and $\chi, (\Sigma_{i2})_{1 \leq i \leq n} \Rightarrow \psi_0$ are derivable in $\mathcal{G}(\mathbf{TDDL})$,

- $X(\chi) \subseteq X((\Sigma_{i1})_{1\leqslant i\leqslant n}) \cap X((\Sigma_{i2})_{1\leqslant i\leqslant n}, \psi_0)$ where X is FV or Pred.

Then, it is easy to see that $\mathbf{O}_{xy}(\chi|\rho_k)$ is an interpolant for our partition. This finishes to show the case where a sequent is derived by the rule (O). □

Theorem 4.7 (Craig interpolation theorem) *If* $\Rightarrow \varphi \to \psi$ *is derivable in* $\mathcal{G}(\text{TDDL})$, *then there is a formula* χ *such that both* $\Rightarrow \varphi \to \chi$ *and* $\Rightarrow \chi \to \psi$ *are derivable in* $\mathcal{G}(\text{TDDL})$, $\text{Var}(\chi) \subseteq \text{Var}(\varphi) \cap \text{Var}(\psi)$, *and* $\text{Pred}(\chi) \subseteq \text{Pred}(\varphi) \cap \text{Pred}(\psi)$.

Proof. Suppose that $\Rightarrow \varphi \to \psi$ is derivable in $\mathcal{G}(\text{TDDL})$. By the following derivation:

$$\dfrac{\Rightarrow \varphi \to \psi \qquad \dfrac{\varphi \Rightarrow \varphi \qquad \psi \Rightarrow \psi}{\varphi, \varphi \to \psi \Rightarrow \psi} \to\Rightarrow}{\varphi \Rightarrow \psi} \, (Cut)$$

the sequent $\varphi \Rightarrow \psi$ is also derivable in $\mathcal{G}(\text{TDDL})$, i.e., also in $\mathcal{G}^-(\text{TDDL})$. Consider a partition $(\varphi : \varnothing), (\varnothing : \psi)$ of $\varphi \Rightarrow \psi$. It follows from Lemma 4.6 that there is an interpolant χ of the partition. Then, both $\Rightarrow \varphi \to \chi$ and $\Rightarrow \chi \to \psi$ are derivable in $\mathcal{G}(\text{TDDL})$ and $\text{Var}(\chi) \subseteq \text{Var}(\varphi) \cap \text{Var}(\psi)$ and $\text{Pred}(\chi) \subseteq \text{Pred}(\varphi) \cap \text{Pred}(\psi)$, as required. □

5 Application to Normative Conflicts

In this final section, we argue that TDDL can accommodate some class of normative conflicts. We also show in a purely proof-theoretic way that the accommodated normative conflicts do not give rise to contradictions or arbitrary obligations. This proof-theoretic result is desirable particularly in deontic logic since it makes TDDL to some extent compatible with the thesis that norms are neither true nor false.

It is widely accepted in the literature [21,20,31] that the standard deontic logic SDL (i.e., the normal modal propositional logic KD) cannot properly formalize situations called *normative conflicts* in which an agent has obligations to do incompatible things. In SDL, the only natural formalization of such situations is $OP \land O\neg P$, which contradicts an axiom D ($\neg(O\varphi \land O\neg\varphi)$) in SDL. As is often pointed out, the logic obtained from SDL by removing D, i.e., the smallest normal modal propositional logic K, still fails to formalize normative situations properly. It is because an arbitrary obligation $O\psi$ follow from a normative conflict $OP \land O\neg P$ via a formula $(OP \land O\neg P) \to O\psi$ provable in K (e.g. [20, pp. 297–8]). Therefore, it has been explored by many deontic logicians how normative conflicts can be accommodated, i.e., how they can be formalized without giving rise to a contradiction nor an arbitrary obligation.

Our approach to accommodate normative conflicts is to relativize obligations to a pair of agents and a condition. It is a familiar strategy in the literature and in fact found e.g. in [24,25,19,50]. However, TDDL can accommodate a larger class of normative conflicts than their logics can.

Normative conflicts of the first kind that TDDL can accommodate are situations in which incompatible obligations are directed towards different agents. Consider a situation in which Adam (a) has obligations towards Barbara (b) and Charles (c) to be with her and him, respectively, but cannot be with both. Let Wab and Wac represent "Adam is with Barbara" and "Adam is with Charles". Since we may identify Wac with $\neg Wab$ in this specific example, the current situation is simply formalized in **TDDL** by

$$O_{ab}Wab \land O_{ac}\neg Wab.$$

We can confirm that the formula does not give rise to a contradiction or an arbitrary obligation because Barbara and Charles are different persons. More precisely, we can falsify the following two formulas by considering an interpretation I such that $I(b) \neq I(c)$:

$$(O_{ab}Wab \land O_{ac}\neg Wab) \to \bot$$
$$(O_{ab}Wab \land O_{ac}\neg Wab) \to O_{ts}(\varphi|\psi)$$

Deontic logics with no indices are difficult to accommodate normative conflicts of the first kind with a succinct formulation.

Normative conflicts of the second kind that **TDDL** can accommodate are situations in which incompatible obligations are directed towards the same agent under different conditions. For example, consider a situation in which Adam has a conditional obligation towards Barbara to be with her given that she is old. Assume also that in the situation he has another conditional obligation towards her not to be with her given that COVID-19 is still spreading. Let Wab, Ob and C represent "Adam is with Barbara", "Barbara is old" and "COVID-19 is spreading", respectively. Then the current situation is formalized in **TDDL** by

$$O_{ab}(Wab|Ob) \land O_{ab}(\neg Wab|C).$$

We can confirm that the formula does not also give rise to a contradiction or an arbitrary obligation because the conditions Ob and C are not logically equivalent. More precisely, we can falsify the following two formulas considering a model M and an assignment α such that $[\![Ob]\!]_\alpha^M \neq [\![C]\!]_\alpha^M$:

$$(O_{ab}(Wab|Ob) \land O_{ab}(\neg Wab|C)) \to \bot$$
$$(O_{ab}(Wab|Ob) \land O_{ab}(\neg Wab|C)) \to O_{ts}(\varphi|\psi)$$

Deontic logics with no conditional obligations face difficulties in accommodating normative conflicts of the second kind. The most well-known difficulty is that such logics cannot express non-monotonicity of obligations. For example, consider the above situation in which Adam has a conditional obligation towards Barbara to be with her given that she is old. Consider also another situation in which Adam has a conditional obligation towards Barbara to be with her given that she is old but not lonely. Let also Lb represent "b is lonely". Then, the most natural formalizations in deontic logics with no conditional obligations would be $Ob \to O_{ab}Wab$ and $Ob \land \neg Lb \to O_{ab}Wab$. However, since the former implies the latter, these formalizations do not reflect our intuition that the latter may fail even if the former holds. Thus, deontic logics with no conditional obligations are difficult to accommodate normative conflicts of the second kind while keeping non-monotonicity of obligations.

Normative conflicts of the second kind need more fine-grained formalizations when they follow from another situations involved in quantification. Consider a situation in which Adam has an obligation towards Barbara not to be with her given that COVID-19 is still spreading but in which she is his mother. Suppose also that we accept that anyone has an obligation towards one's mother to be with her given that she is old. Then, from the current situation, a normative conflict of the second kind follows in which Adam has an obligation towards Barbara not to be with her given that COVID-19 is still spreading, as well as another obligation towards her to be with her given that she is old. We shall call such a normative conflict *derived normative conflict*.

Our **TDDL** can formalize derived normative conflicts as well. Let Wab, Ob, C and Mba represent "Adam is with Barbara", "Barbara is old", "COVID-19 is spreading" and "Barbara is Adam's mother", respectively. Then, the first situation is formalized in **TDDL** by

$$\mathbf{O}_{ab}(\neg Wab|C) \wedge Mba \wedge \forall x \forall y (Myx \rightarrow \mathbf{O}_{xy}(Wxy|Oy)),$$

from which the derived normative conflict above is obtained in the following form:

$$\mathbf{O}_{ab}(\neg Wab|C) \wedge \mathbf{O}_{ab}(Wab|Ob).$$

As before, we can confirm that the formula does not give rise to a contradiction $\bot$ or an arbitrary obligation $\mathbf{O}_{ts}(\varphi|\psi)$.

Against our claim requiring conditionals for the second kind, there might be the following objection: no additional machineries are necessary to accommodate the second kind, since we can formalize even such conditions just by adding the third index to modal operators. For example, consider the aforementioned situation in which Adam has an obligation towards Barbara to be with her given that she is old and in which he has another obligation towards her not to be with her given that COVID-19 is still spreading. According to the objection, it can be formalized as $\mathbf{O}_{abo}Wab \wedge \mathbf{O}_{abc}Wab$, where o and c denotes the conditions that Barbara is old and that COVID-19 is spreading. Clearly, it does not give rise to contradictions or arbitrary obligations. For ease of reference, we shall refer to an approach formalizing conditions in such a way as the indexing approach.

However, the use of indices representing conditions should be avoided to keep the same interpretation of logical constants in a given context. Consider a situation in which Adam has an obligation to be *and* talk with Barbara given that she is old *and* lonely. Whatever formalization is given to it, we would require that two logical constants "and" in it be formalized in the same way. The indexing approach basically faces a difficulty when formalizing logical constants appearing in both of the antecedent and the consequent of a conditional obligation. Amongst those which might fall into the indexing approach, Gabbay [16] and Tamminga [43] can avoid the difficulty. It is because the former allows indices to be formulas and the latter represents conditionals by actions of some group which have no logical forms. However, they are based on propositional logic, so difficult to accommodate derived normative conflicts requiring quantifiers to formalize.

Against our claim that derived normative conflicts require quantifiers for formalization, one might think that finite conjunctions or disjunctions are enough and that quantifiers are not necessary in deontic logic, since we may assume the number of agents to be finite. We can answer this criticism by putting forth two points.

First, as pointed out e.g. in Hilpinen and McNamara [22, p. 53] and Frijters [14, p. 70], there are deontic sentences in which the *de re/de dicto* distinction should be made. For example, it may be the case that Adam has an obligation towards Barbara to take someone's class in the university (since she is sending him money for class), but not that there is someone such that he has an obligation towards her to take the person's class in the university. Given a formula Cax representing "Adam takes x's class in the university", the intuition can be captured in TDDL with formulas $O_{ab}\exists xCax$ and $\exists xO_{ab}Cax$ since the former does not imply the latter. It seems difficult to capture this intuition in the framework of propositional deontic logic. See Frijters [14, p. 70] for a similar example.

Second, even when the domain of quantification is a finite set $\{a_1,\ldots,a_n\}$ of agents, it is not a decisive proposition in deontic logic that a finite conjunction $Pa_1 \wedge \cdots \wedge Pa_n$ and a universally quantified sentence $\forall xPx$ have the same truth value. It is because we sometimes consider a universally quantified sentence like "everyone should be honest" to be true without knowing how the domain of quantification is like. We can find this interesting idea in F. Ramsey's *Philosophical Papers* [38, p. 145], in which he calls such universal sentences *variable hypotheticals*. Of course our current semantics is not suitable for this idea. However, with this idea we can claim the need of quantifiers for deontic logic.

To sum up, TDDL can accommodate the class of normative conflicts of the first and second kinds. As is often said, SDL and K cannot accommodate the first kind, nor the second kind. Kooi and Tamminga [24,25], Glavanicova[19] and Yamada [50] can accommodate the first kind, but not the second kind due to the lack of conditional obligations. Gabbay [16] and Tamminga [43] can be applied to accommodate some of the second kind, but not all. They cannot be applied to accommodate derived normative conflicts due to the lack of quantifiers.

Finally, we show in a purely proof-theoretic way that the normative conflicts of the first and second kinds do not give rise to contradictions nor arbitrary obligations.

Proposition 5.1 *Let x, y, z, w be pairwise distinct variables and P, Q, R, S, T pairwise distinct nullary predicate symbols.*

1. *A sequent $O_{xy}P \wedge O_{xz}\neg P \Rightarrow \bot$ is not derivable in $\mathcal{G}(\text{TDDL})$.*
2. *A sequent $O_{xy}P \wedge O_{xz}\neg P \Rightarrow O_{xw}S$ is not derivable in $\mathcal{G}(\text{TDDL})$.*
3. *A sequent $O_{xy}(P|Q) \wedge O_{xy}(\neg P|R) \Rightarrow \bot$ is not derivable in $\mathcal{G}(\text{TDDL})$.*
4. *A sequent $O_{xy}(P|Q) \wedge O_{xy}(\neg P|R) \Rightarrow O_{xw}(S|T)$ is not derivable in $\mathcal{G}(\text{TDDL})$.*

Proof. We first give proofs of items 1, 2. Recall that $\neg\varphi$, $\varphi \wedge \psi$ and $O_{ts}\varphi$ are abbreviations of $\varphi \to \bot$, $\neg(\varphi \to \neg\psi)$ and $O_{ts}(\varphi|\top)$, respectively. Recall also that $\mathcal{G}^-(\text{TDDL})$ is the sequent calculus obtained from $\mathcal{G}(\text{TDDL})$ by removing (Cut). By induction on $i \in \mathbb{N}$, we can establish the fact that a sequent

$$O_{xy}P^j, O_{xz}\neg P^k, (O_{xy}P \wedge O_{xz}\neg P)^l \Rightarrow (O_{xy}P \to \neg O_{xz}\neg P)^m, \neg O_{xz}\neg P^n, \bot^o, O_{xw}S^p, S^q$$

is not derivable in $\mathcal{G}^-(\text{TDDL})$ with at most height i for any $j, k, l, m, n, o, p, q \geq 0$, where each superscript denotes the number of occurrences of each formula. We can use the fact to show items 1, 2. We first show item 1, i.e., that $O_{xy}P \wedge O_{xy}\neg P \Rightarrow$

$\perp$ is not derivable in $\mathcal{G}(\mathbf{TDDL})$. Suppose not. By the cut elimination theorem of $\mathcal{G}(\mathbf{TDDL})$ (Theorem 4.4), $\mathbf{O}_{xy}P \wedge \mathbf{O}_{xy}\neg P \Rightarrow \perp$ is derivable in $\mathcal{G}^-(\mathbf{TDDL})$. On the other hand, the above fact implies that it is not derivable in $\mathcal{G}^-(\mathbf{TDDL})$. This is a contradiction so item 1 holds. By the same argument, item 2 also holds.

We then give proofs of items 3, 4. By induction on $i \in \mathbb{N}$, we can establish the fact that a sequent

$$\mathbf{O}_{xy}(P|Q)^j, \mathbf{O}_{xy}(\neg P|R)^k, (\mathbf{O}_{xy}(P|Q) \wedge \mathbf{O}_{xy}(\neg P|R))^l \Rightarrow$$
$$(\mathbf{O}_{xy}(P|Q) \to \neg \mathbf{O}_{xy}(\neg P|R))^m, \neg \mathbf{O}_{xy}(\neg P|R)^n, \perp^o, \mathbf{O}_{xw}(S|T)^p, S^q$$

is not derivable in $\mathcal{G}^-(\mathbf{TDDL})$ with at most height i for any $j, k, l, m, n, o, p, q \geq 0$, where each superscript denotes the number of occurrences of each formula. Thus, by the same argument as for items 1, 2, we can prove items 3, 4. □

Therefore, we can claim that **TDDL** is to some extent compatible with the thesis that norms are neither true nor false.

Conclusion

We shall close the paper by listing four directions for further studies.

The first direction to be pursued is to change the framework of **TDDL** on conditionals. For example, the transition of the current framework **CK** to the influential framework **CD** introduced in [6] probably makes **TDDL** more acceptable as deontic logic.

The second direction is to add function symbols and the equality symbol. As done in Fitting et al. [12], we may include function symbols into our syntax without any technical difficulty, i.e., we can still keep all the technical results in this paper. It is interesting to consider if we may further include the equality symbol. (It is remarked that Fitting et al. [12] did not consider the equality symbol in their syntax.)

The third direction to be studied is to change the framework of **TDDL** on domain. The semantics of **TDDL** is a varying domain semantics but induces the cumulative domain condition on it. The condition, however, is hard to accept from philosophical viewpoints. This challenge might be overcome by combining **TDDL** with common sense modal predicate logics recently developed by [45,42,40], which do not require such any conditions on domain.

Finally, the fourth possible direction is to find how to accommodate normative conflicts in which incompatible obligations are directed towards the *same* agent under the *same* conditions. However, as for this final direction, we can also take a philosophical position denying that such normative conflicts actually exist. For example, see Yamada [50].

References

[1] Beirlaen, M. and C. Straßer, *A Paraconsistent Multi-agent Framework for Dealing with Normative Conflicts*, in: Lecture Notes in Computer Science, Springer Berlin Heidelberg, 2011 pp. 312–329.

[2] Benzmüller, C., *Cut-Elimination for Quantified Conditional Logic*, Journal of Philosophical Logic **46** (2016), pp. 333–353.

[3] Calardo, E., "Non-normal Modal Logics, Quantification, and Deontic Dilemmas: A Study in Multi-relational Semantics," phdthesis, University of Bologna (2013).
[4] Castañeda, H.-N., *The Paradoxes of Deontic Logic: The Simplest Solution to all of them in one Fell Swoop*, in: *New Studies in Deontic Logic*, Springer Netherlands, 1981 pp. 37–85.
[5] Chellas, B. F., *Basic Conditional Logic*, Journal of Philosophical Logic **4** (1975), pp. 133–153.
[6] Chellas, B. F., "Modal Logic: An Introduction," Cambridge University Press, 1980.
[7] Cholvy, L. and F. Cuppens, *Reasoning about Norms Provided by Conflicting Regulations*, Norms, Logics and Information Systems: New Studies in Deontic Logic and Computer Science (1998), pp. 247–264.
[8] Corsi, G., *A Unified Completeness Theorem for Quantified Modal Logics*, The Journal of Symbolic Logic **67** (2002), pp. 1483–1510.
[9] de Swart, H. C. M., *A Gentzen- or Beth-type System, a Practical Decision Procedure and a Constructive Completeness Proof for the Counterfactual Logics VC and VCS*, Journal of Symbolic Logic **48** (1983), pp. 1–20.
[10] Delgrande, J. P., *A First-order Conditional Logic for Prototypical Properties*, Artificial Intelligence **33** (1987), pp. 105–130.
[11] Delgrande, J. P., *On First-order Conditional Logics*, Artificial Intelligence **105** (1998), pp. 105–137.
[12] Fitting, M., L. Thalmann and A. Voronkov, *Term-Modal Logics*, Studia Logica **69** (2001), pp. 133–169.
[13] Friedman, N., J. Y. Halpern and D. Koller, *First-order Conditional Logic for Default Reasoning Revisited*, ACM Transactions on Computational Logic **1** (2000), pp. 175–207.
[14] Frijters, S., "All Doctors Have an Obligation to Care for Their Patients: Term-modal Logics for Ethical Reasoning with Quantified Deontic Statements," Ph.D. thesis, Ghent University (2021).
[15] Frijters, S. and F. V. D. Putte, *Classical Term-modal Logics*, Journal of Logic and Computation (2020).
[16] Gabbay, D. M. and G. Governatori, *Dealing with Label Dependent Deontic Modalities*, Norms, Logics and Information Systems. New Studies in Deontic Logic (1999), pp. 311–330.
[17] Gent, I. P. et al., *A Sequent-or Tableau-style System for Lewis's Counterfactual Logic V*, Notre Dame Journal of Formal Logic **33** (1992), pp. 369–382.
[18] Girlando, M., B. Lellmann, N. Olivetti and G. L. Pozzato, *Standard Sequent Calculi for Lewis' Logics of Counterfactuals*, in: *Logics in Artificial Intelligence*, Springer International Publishing, 2016 pp. 272–287.
[19] Glavanicova, D., Δ *-TIL and Normative Systems*, Organon F **23** (2016), pp. 204–221.
[20] Goble, L., *Prima Facie Norms, Normative Conflicts, and Dilemmas*, in: D. Gabbay, J. Horty, X. Parent, R. van der Meyden and L. van der Torre, editors, *Handbook of Deontic Logic and Normative Systems* (2013), pp. 241–352.
[21] Hansen, J., G. Pigozzi and L. van der Torre, *Ten Philosophical Problems in Deontic Logic*, Normative Multi-agent Systems (2007), pp. 1–26.
[22] Hilpinen, R. and P. McNamara, *Deontic Logic: A Historical Survey and Introduction*, in: D. Gabbay, J. Horty, X. Parent, R. van der Meyden and L. van der Torre, editors, *Handbook of Deontic Logic and Normative Systems*, College Publications, Milton Keynes, UK, 2013, 1st edition pp. 3–136.
[23] Kooi, B., *Dynamic Term-modal Logic*, in: J. van Benthem, S. Ju and F. Veltman, editors, *A Meeting of the Minds. Proceedings of the Workshop on Logic, Rationality and Interaction, Beijing, 2007*, Texts in Computing Computer Science 8 (2008), pp. 173–185.
[24] Kooi, B. and A. Tamminga, *Conflicting Obligations in Multi-agent Deontic Logic*, in: *Deontic Logic and Artificial Normative Systems*, Springer Berlin Heidelberg, 2006 pp. 175–186.
[25] Kooi, B. and A. Tamminga, *Moral Conflicts Between Groups of Agents*, Journal of Philosophical Logic **37** (2008), pp. 1–21.
[26] Lellmann, B. and D. Pattinson, *Sequent Systems for Lewis' Conditional Logics*, in: L. F. del Cerro, A. Herzig and J. Mengin, editors, *Logics in Artificial Intelligence* (2012), pp. 320–332.
[27] Lewis, D., "Counterfactuals," Blackwell Publishers, 1973.
[28] Liberman, A. O., A. Achen and R. K. Rendsvig, *Dynamic Term-modal Logics for First-order Epistemic Planning*, Artificial Intelligence **286** (2020), p. 103305.
[29] Maehara, S., *On Interpolation Theorem of Craig*, Sugaku **12** (1961), pp. 235–237, in Japanese.
[30] Makinson, D., *On a Fundamental Problem of Deontic Logic*, in: P. McNamara and H. Prakken, editors, *Norms, Logics and Information Systems: New Studies in Deontic Logic and Computer Science*, January, Cambridge University Press, Cambridge, 1999 pp. 29–53.
URL http://ebooks.cambridge.org/ref/id/CBO9781107415324A009
[31] McNamara, P., *Deontic Logic*, in: E. N. Zalta, editor, *The Stanford Encyclopedia of Philosophy*, Metaphysics Research Lab, Stanford University, 2019, summer 2019 edition .

[32] Milošević, M. and Z. Ognjanović, *A First-order Conditional Probability Logic*, Logic Journal of IGPL **20** (2011), pp. 235–253.
[33] Negri, S. and G. Sbardolini, *Proof Analysis for Lewis Counterfactuals*, The Review of Symbolic Logic **9** (2016), p. 44–75.
[34] Ono, H., "Proof Theory and Algebra in Logic," Springer Nature Singapore Pte Ltd., 2019.
[35] Orlandelli, E. and G. Corsi, *Decidable Term-Modal Logics*, in: F. Belardinelli and E. Argente, editors, *Multi-Agent Systems and Agreement Technologies*, Lecture Notes in Artificial Intelligence **10767** (2018), pp. 147–162, eUMAS 2017, AT 2017, Lecture Notes in Computer Science.
[36] Pattinson, D. and L. Schröder, *Cut Elimination in Coalgebraic Logics*, Information and Computation **208** (2010), pp. 1447–1468.
[37] Pattinson, D. and L. Schröder, *Generic Modal Cut Elimination Applied to Conditional Logics*, Logical Methods in Computer Science **7** (2011), pp. 1–28.
[38] Ramsey, F. P., "Philosophical Papers," Cambridge University Press, 1990.
[39] Sano, K. and M. Ma, *Sequent Calculi for Normal Update Logics*, in: *Logic and Its Applications*, Springer Berlin Heidelberg, 2019 pp. 132–143.
[40] Sawasaki, T. and K. Sano, *Frame Definability, Canonicity and Cut Elimination in Common Sense Modal Predicate Logics*, Journal of Logic and Computation (2020), exaa067.
URL https://doi.org/10.1093/logcom/exaa067
[41] Sawasaki, T., K. Sano and T. Yamada, *Term-Sequence-Modal Logics*, in: P. Blackburn, E. Lorini and M. Guo, editors, *Logic, Rationality, and Interaction: 7th International Workshop, LORI 2019, Chongqing, China, October 18–21, 2019, Proceedings*, Theoretical Computer Science and General Issues **11813** (2019), pp. 244–258.
[42] Seligman, J., *Common Sense Modal Predicate Logic*, Presentation (2017), Non-classical Modal and Predicate Logics: The 9th International Workshop on Logic and Cognition, Guangzhou, China, 4 December 2017.
[43] Tamminga, A., *Deontic Logic for Strategic Games*, Erkenntnis **78** (2011), pp. 183–200.
[44] Thalmann, L., "Term-Modal Logic and Quantifier-free Dynamic Assignment Logic," Ph.D. thesis, Uppsala University (2000).
[45] van Benthem, J., "Modal Logic for Open Minds," CSLI Publications, Stanford, 2010.
[46] van der Torre, L., *Contextual Deontic Logic: Normative Agents, Violations and Independence*, Annals of Mathematics and Artificial Intelligence **37** (2003), pp. 33–63.
[47] Van Fraassen, B. C., *The logic of conditional obligation*, Journal of Philosophical Logic **1** (1972), pp. 417–438.
[48] Wang, Y. and J. Seligman, *When Names are not Commonly Known: Epistemic Logic with Assignments*, in: G. Bezhanishvili, G. D'Agostino, G. Metcalfe and T. Studer, editors, *Advances in Modal Logic*, Proceedings of the12th Conference on Advances in Modal Logic, Held in Bern, Switzerland, August 27-31, 2018 **12** (2018), pp. 611–628.
[49] Yamada, T., *Logical Dynamics of Some Speech Acts that Affect Obligations and Preferences*, Synthese (2008).
[50] Yamada, T., *Moral Dilemmas and the Contrary-to-duty Scenarios in Dynamic Logic of Acts of Commanding — the Significance of Moral Considerations behind Moral*, in: byunghan Kim, J. Brendle, G. Lee, F. Liu, R. Ramanujam, S. M. Srivastava, A. Tsuboi and L. Yu, editors, *Proceedings of the 14th and 15th Asian Logic Conferences*, 2019, pp. 249–269, pDF's page numbers are not correct.
[51] Zach, R., *Non-analytic Tableaux for Chellas's Conditional Logic CK and Lewis's Logic of Counterfactuals VC*, Australasian Journal of Logic (2018).

Forms and Norms of Indecision in Argumentation Theory

Daniela Schuster [1]

Universität Konstanz
78464 Konstanz

Abstract

One main goal of argumentation theory is to evaluate arguments and to determine whether they should be accepted or rejected. When there is no clear answer, a third option, being undecided, has to be taken into account. Indecision is often not considered explicitly, but rather taken to be a collection of all unclear or troubling cases. However, current philosophy makes a strong point for taking indecision itself to be a proper object of consideration. This paper aims at revealing parallels between the findings concerning indecision in philosophy and the treatment of indecision in argumentation theory. By investigating what philosophical forms and norms of indecision are involved in argumentation theory, we can improve our understanding of the different uncertain evidential situations in argumentation theory.

Keywords: Labelling Approach, Statement Labelling, Indecision, Doxastic Indecision, Suspension of Judgement

1 Introduction

Abstract argumentation theory is concerned with modelling arguments and their relation of defeat. One main goal of argumentation theory is to determine of a given set of arguments which of those should be accepted and which of those should be rejected. This yields an extension – a set of accepted arguments – and an antiextension – a set of of rejected arguments. Additionally, it is possible to distil a third set of arguments that is sometimes not characterised explicitly, which is the set of arguments one is undecided about. The involvement of indecision is made explicit in the labelling-approach of abstract argumentation theory, where arguments are labelled according to three different labels: in, out and undecided.

Very often indecision is only seen as a quite useful but rather unimportant byproduct when characterising the acceptability and rejectability of arguments. From a philosophical point of view, however, besides acceptance and rejection, indecision is the third main doxastic response, and it should be taken to be equally important. Especially in the last couple of years, researchers in

---

[1] daniela.schuster@uni-konstanz.de

the philosophical disciplines of epistemology and philosophy of mind started to focus more on this third, neutral stance. They try to investigate the norms of indecision (which is done in epistemology) and the different phenomena involved in this broad concept (which is done in philosophy of mind), see for example [18] for the distinction. The philosophical investigations in both of these areas can be very useful when applied to argumentation theory. Investigating the notion of indecision more precisely can help us to observe the different options of how to use indecision as a tool to describe uncertain, doubtful, or conflicting information. This will allow us to find ways to improve the representation of the given knowledge and to make certain decisions less bold and more understandable and trust-worthy.

In this paper, I aim at taking the first steps in transferring the mentioned philosophical considerations to argumentation theory and in revealing important parallels. I will shed some light on the different forms of indecision that can be found in abstract argumentation theory and in what way they correspond to particular philosophical phenomena. Moreover, I want to illustrate by what means the various semantics of abstract argumentation theory treat indecision differently and how this relates to the epistemological debate about rationality norms.

In a second step, I want to dive deeper into the more structured level of statement-labelling, by considering not only the argument as such, but also its conclusion. I will attempt to reveal some of the philosophical norms and forms of indecision in these conclusions. Although statement evaluation has barely been considered explicitly in argumentation theory, this enterprise is even more important from a philosophical perspective, as arguably what determines our actions, is more our beliefs in the form of statements and less the arguments we have in mind themselves.

The paper is structured in the following way. In Section 2, some philosophical findings about indecision will be presented, first from the sub-field of epistemology, which is concerned with rationality norms (Section 2.1), and second from the sub-field of philosophy of mind, which can provide insights into the different forms of indecision (Section 2.2). In Section 3 possible transfers from the philosophical findings to the area of argumentation theory will be suggested. In Section 3.1, parallels will be drawn between indecision in philosophy and indecision in *abstract argumentation theory*. Here we are operating on the level of arguments. For this, some formal backgrounds will be introduced (Section 3.1.1), such that the different forms of indecision (Section 3.1.2) and the norms of indecision involved in abstract argumentation theory (Section 3.1.3) can be revealed. In a second step, in Section 3.2, parallels between indecision in philosophy and indecision on the statement level of argumentation theory will be considered. Finally, in the outlook (Section 4), I will conclude these considerations and point to some further areas where the philosophical notion of indecision can be applied fruitfully.

2 Indecision in Philosophy

In Philosophy, the concept of indecision generally refers to a neutral stance that a person can take towards something. This stance lies somewhere in between believing and disbelieving. Philosophers take the "something" one is undecided about (and which is also the object of belief and disbelief) [2] to be a proposition. Although there is quite some dispute about the nature of propositions and their representation, in this paper I will assume that propositions can simply be represented by sentences. The philosophical investigations concerning indecision can broadly be divided into belonging to one of the two philosophical areas: philosophy of mind and epistemology. In Section 2.1, I will focus on the epistemological part, discussing philosophical norms of indecision. In Section 2.2, the investigations of philosophers of mind will help us to understand different forms of indecision.

The terminology concerning indecision is not uniform in philosophy. There are different terms (such as non-belief, indecision, suspension, agnosticism, withholding belief) that are used differently in the literature, see [9, p. 166]. I will comment on this briefly in Section 2.2, when describing different forms of indecision. Generally, I will use the term "indecision" (in philosophical terminology) to refer to any neutral position that a subject can adopt towards a topic, besides believing or disbelieving it.

2.1 Norms about Indecision in Philosophy

Epistemologists focus on providing rules (or norms) for when to adopt a certain doxastic attitude. Until recently, modern epistemologists were mostly only concerned with formulating norms about when a proposition should be believed or disbelieved, see [9, p. 165]. The third option of being neutral was not investigated as such, but merely considered a byproduct. At best, the third option of being undecided has been seen as a base case or fall-back position for not-considered propositions before "proper" reasoning yields to a decision about the truth of a proposition, i.e., believing or disbelieving it.

One can find such philosophical attitudes in the history of philosophy. For example, Descartes formulated in his *Meditationes* [6] very strict necessary conditions of when one should believe a proposition. He thought that he should only believe the things he was absolutely certain about. In order to fulfil these very strict rules for belief, he used his approach of methodological doubt and considered every possible statement as undecided. [3] In his approach, he laid no restrictions on when something can or cannot be undecided, hence disregarding any norms concerning indecision.

---

[2] For reasons of treating the the three doxastic stances: belief, disbelief and indecision equally, we will assume here that in the philosophical considerations indecision has propositional content, although this has been questioned, for example, in [7].

[3] In his strategic doubt, Descartes actually first tried to *disbelieve* every proposition he used to belief, so e.g., disbelieve that he has two hands, yielding to his desired point of departure of being in doubt about everything.

Only recently philosophers became more interested in epistemological concepts tied to this third neutral stance. They started to investigate such phenomena as indecision and formulated norms about when one should or should not be undecided about a proposition. Friedman's ignorance norm in [10], for example, suggests a necessary condition for indecision,[4] stating that a person is only allowed to be undecided about a certain proposition, if she does not know the proposition or its negation already. This requirement forbids being undecided when the proposition in question is already known by the subject. In [11] she even suggests a norm that forbids to be undecided on a matter, if one already entertained a belief on that matter. Thereby Friedman formulates a necessary condition for indecision: not knowing or respectively not believing the relevant proposition already. Other norms, in the form of principles, are, for example, discussed in [16]. Two very basic norms that one can find in the philosophical literature on indecision are the *Absence of Evidence Norm*, stating that you should be undecided about a proposition p, when there is no evidence speaking in favour or against p, see [8, p. 60], and what one could call the *Balanced Evidence Norm*, stating that you should be undecided about p, when the evidence for and against p is equally balanced, see for example [17].

In Sections 3.1.3 and 3.2 I will discuss how these norms about indecision find parallels in argumentation theory. First, however, I want to present some findings from philosophy of mind that show the diversity of the phenomenon.

2.2 Forms of Indecision in Philosophy

Recent investigations in philosophy of mind suggest that there is not just one way in which a person can take a neutral stance or be undecided about a proposition. Most of the philosophers in the current discussion aim at describing what one does if one "suspends judgement" or "suspends belief" about a matter. Accordingly, these philosophers argue that this is the third *attitude*, besides believing and disbelieving, that one can take towards a proposition. Although current scholars strongly disagree on how to use the term of suspension of judgement, most of them do agree that there are different ways in which a person can be neutral about a proposition.

Example 2.1 Let us consider the following example situations of indecision:

1) A caveman, who is undecided about the proposition that quarks exist.

2) Me, being undecided about the proposition that the guava fruit has a lot of vitamin C.

3) Me, being undecided whether the covid pandemic will be over next year.

4) Me, being undecided about the proposition that a God exists.

Furthermore, let us assume, first, that I only briefly heard about the guava

---

[4] In fact, Friedman is here talking about suspension and not about indecision. As mentioned before, we will use the term "indecision" instead and come back to the discussion about the differences in the next sub-section.

fruit and I never thought about its nutritional values. Secondly, that I do think about the covid pandemic a lot and whether it will be over soon, but that I cannot decide the question by looking at my evidence. Thirdly, let us assume that I also thought about the existence of God, but decided that there is no way to find out, so I will stay undecided.

In the few different situations, different forms of indecision are involved. Situation 1 is a famous example going back to Hájek [12, p. 205, footnote], that describes a typical case of mere *non-belief*. With the term *non-belief* philosophers describe situations in which a person neither believes nor disbelieves a certain proposition, i.e., a lack of both belief and disbelief. Most philosophers agree that phenomena like suspension of judgement or indecision as such should not be identified with mere non-belief. Wedgwood [19, p. 272] for example notices that non-belief is in fact not even an attitude, a position one can take, or a "mental state" at all, since, for instance stones can be in this situation of non-belief as well and we would not say that stones can suspend judgement about something. Situation 1 is somehow similar to this, as the caveman could not even understand the concepts involved in the proposition. Clearly, situations 2, 3 and 4 already all involve a more sophisticated form of indecision than situation 1 does.

Even so, philosophy of mind also makes a point for distinguishing situation 2, 3 and 4 from each other. What we want to describe with suspension of judgement is a *higher* or *more sophisticated* form of indecision; a situation in which we cognitively consider the proposition but are still neutral about its truth. Some sort of "considering" or "thinking about" condition has to be fulfilled, when we want to say that someone suspends judgement. Although the form of indecision involved in situation 2 already comes closer to suspension of judgement (because I do understand the relevant sentence), I am still lacking the so-called cognitive contact to the proposition, as I have never even thought about it. Most scholars would not describe me as suspending about the proposition that the Guava fruit has a lot of vitamin C, if I never even thought about it.
This "considering" element is given both in situation 3 and 4. In both cases, I had the respective proposition at least temporally in mind. Still, scholars like Friedman [9], Wagner [18], or Raleigh [15] argue that there is still some important and somehow categorical difference between the two situations 3 and 4. In situation 3, although I am undecided, I still continue to think about the question whether the covid pandemic will be over next year. In fact, almost every day I get new evidence speaking for and new evidence speaking against the proposition. Moreover, I might be strongly motivated to find out about the matter, because I might want to plan a trip for next year. This is what distinguishes situation 3 from situation 4, according to [9] or [18]. In situation 4, I "settled" the indecision and thereby closed the question about God's existence. Friedman argues that this settlement comes with forming

a *sui generis* attitude of suspension, that is not reducible to non-belief in any sense, while Wagner takes the settling element, that is necessary for proper suspension, to be an endorsement of my own indecision. According to this account, non-belief is constitutive but not sufficient for suspension.[5] Regardless of what the specific accounts take this settling element to be, it can be noticed that in situation 3 this kind of settlement or "closing the question" is missing.

To illustrate the differences, I think it can be helpful to set the different forms in relation to each other by visualising them along an axis of "engaging" or "being concerned" or "being involved" with a proposition.

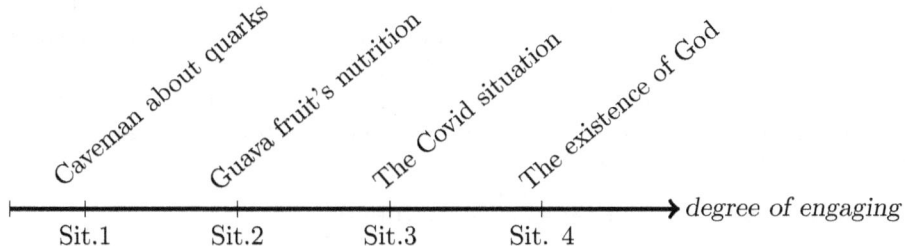

The caveman does not engage with the proposition about the existence of quarks at all, as he does not even understand it. While I do understand the proposition about the nutrition of the Guava fruit, I still do not really engage with it, as I never even entertained it in thought. Compared to this, I engage with the proposition about the covid pandemic a lot. In fact, I constantly think about it. Situation 4, though, falls a bit out of the pattern. One the one hand one could argue that I, in fact, engage less with the proposition about God's existence than with the proposition about the Covid pandemic, in the sense that I do not think about the proposition constantly anymore. On the other hand, however, I am still in some sense more *bound* to the proposition in situation 4, as I definitely include the proposition into (the third category of) my belief sets. In this sense, although I do not think about it a lot anymore, I am yet more identified and involved with the proposition and I am further in the evaluating process than I am with the proposition about the covid pandemic. I closed the question about God's existence and I integrated the proposition into my doxastic household. In this sense situation 4 has to be placed on the very right of the axis.

In the next section, I will relate these different philosophical forms and the epistemological considerations from Section 2.1 to the field of argumentation theory.

---

[5] For scholars like Raleigh [15] or Wedgwood [19] it is sufficient that I form some sort of meta-belief about my own doxastic situation. This can be regarded as a "committing" element, too.

3 Indecision in Argumentation Theory

As stated in the introduction, the main goal of this paper is to connect the usage of indecision in argumentation theory with the philosophical concept of indecision. We will do this on two levels: first on the argument level (Section 3.1) and second on the statement level (Section 3.2). The doxastic trisection of philosophy (believing, disbelieving and indecision) finds correspondence both on the argument level of abstract argumentation theory and on the statement level of structured argumentation theory. As described above, philosophers take propositions, that are structurally equal to statements or sentences, to be the object of the doxastic stances. Hence, parallels between philosophy and the statement level of argumentation theory are easier to draw than the parallels with the argument level. At the argument level, in abstract argumentation theory, arguments are the relevant objects for evaluation, which are structurally very different from propositions. Nevertheless, a correspondence between the three doxastic responses one can take towards a proposition in philosophy and three possible reactions one can show towards an argument in abstract argumentation theory can still be found.

3.1 Indecision on the Level of Arguments

3.1.1 Formal Background

Before I can shed some light on the parallels between indecision in abstract argumentation theory and philosophy, some basic notions of abstract argumentation theory have to be introduced. This sub-section does not aim at providing a complete overview of the definitions and theorems of abstract argumentation theory. Only the parts that are directly relevant for the considerations concerning indecision are presented. The following definitions can for example be found in [1].

Abstract Argumentation Theory focuses on modelling arguments and their relation of attack on an abstract level. This is modelled in argumentation frameworks, which allow for an illustration of arguments (nodes of the graph) and their attack (edges of the graph). More precise features, such as the internal structure of the arguments and how they attack each other is not representable at this abstract level.

Definition 3.1 [Argumentation Framework] An Abstract Argumentation Framework is a pair (Ar, att), where Ar is a set of arguments and $att \subset Ar \times Ar$ is a relation of attack between the arguments.

Besides the modelling challenge, abstract argumentation theory also deals with evaluating arguments in order to choose a proper subset of acceptable arguments among the modelled ones.

In general, there are two approaches for this: the extension approach and the labelling approach. The labelling approach provides a function that maps each argument to a label, that is either in, out or undec (undecided).

Definition 3.2 [Labelling] Given a set of labels, $\{\text{in}, \text{out}, \text{undec}\}$, a labelling is a function $Lab : Ar \to \{\text{in}, \text{out}, \text{undec}\}$ which maps each argument to one of the three possible labels.

In contrast to this, there is the extension approach. The extension approach simply yields a subset of the considered arguments (the extension) which consists of the accepted arguments. The arguments that are rejected or undecided are then defined in terms of the extension. Although the two approaches are convertible into each other, see [1], the labelling approach provides a straight-forward way to distinguish the three possible states of an argument and speaks about an undecided evaluation explicitly. Therefore, I will focus on the labelling approach in this paper.

The challenge then is to decide which argument should get which label. This decision is not to be made arbitrarily, but has to follow certain rules. These rules are given by different so-called semantics. Quite basic rules are provided by admissible semantics.

Definition 3.3 [Admissible Semantics] Let (Ar, att) be an argumentation framework. A labelling Lab is called an admissible labelling (or a labelling according to admissible semantics), iff the following two conditions hold:

- every in-labelled argument $A \in Ar$ is legally in, i.e., $\forall B \in Ar$: if $(B, A) \in att$ then $Lab(B) = $ out.
- every out-labelled argument $A \in Ar$ is legally out, i.e., $\exists B \in Ar$, such that $(B, A) \in att$ and $Lab(B) = $ in.

Admissible semantics demand that only arguments that are *legally* in or out, should get the respective label. Informally speaking, an argument is legally in, if all its attackers (if there even are any) are labelled out. An argument is legally out, if it has at least one attacker that is labelled in.

Another semantics, that we we will introduce here and need in Section 3.1.3, as it differs from admissible semantics in some important manner, is complete semantics.

Definition 3.4 [Complete Semantics] Let (Ar, att) be an argumentation framework. A labelling Lab is called a complete labelling (or a labelling according to complete semantics), iff the following three conditions hold:

- every in-labelled argument $A \in Ar$ is legally in, i.e., $\forall B \in Ar$: if $(B, A) \in att$ then $Lab(B) = $ out.
- every out-labelled argument $A \in Ar$ is legally out, i.e., $\exists B \in Ar$, such that $(B, A) \in att$ and $Lab(B) = $ in.
- every undec-labelled argument $A \in Ar$ is legally undec, i.e., $\exists B \in Ar$, such that $(B, A) \in att$ and $Lab(B) \neq $ out and $\forall B \in Ar$: if $(B, A) \in att$ then $Lab(B) \neq $ in

Complete semantics fulfil the requirements of admissible semantics and fulfil on top of that the requirement that every undec-labelled argument has to be also legally undec. An argument is legally undec, iff it is neither legally in nor legally out. This means that an argument is legally undec iff *none* of its attackers are labelled in and it has at least one attacker that is *not* labelled out.

Admissible and complete semantics are only two of the many possible semantics that have been investigated in argumentation theory. All semantics make demands on the labellings of the arguments. Some semantics (e.g., grounded semantics) only yield one allowed labelling. Most semantics, like admissible and complete semantics, however, usually allow for a plurality of possible labellings. In such situations the question of the justification status of an argument is raised. If an argument is, for example, labelled in according to one complete labelling and labelled out in another one, how should we evaluate that argument after all?

A rather detailed analysis of different justification statuses is provided in [20]. Wu et. al. consider in particular the case of complete semantics and note that the labellings produced by complete semantics can be interpreted as the different reasonable stances one can take towards an argument. They suggest a function that maps each argument to its justification status, which they take to be the set of all possible labels each argument gets from the different complete labellings.

Definition 3.5 [Justification Status] Let (Ar, att) be an argumentation framework. For $A \in Ar$ let $J(A)$ be the justification status of A, given by the function $J : Ar \to \mathcal{P}(\{\text{in}, \text{out}, \text{undec}\})$.[20, p. 16]

J maps every $A \in Ar$ to a subset of $\{\text{in}, \text{out}, \text{undec}\}$ consisting of all the labels A gets by one or more complete labelling.

Considering complete semantics, in [20, p. 16] 6 possible justification statuses are obtained:[6] $\{\text{in}, \text{out}, \text{undec}\}, \{\text{in}, \text{undec}\}, \{\text{out}, \text{undec}\}, \{\text{in}\}, \{\text{out}\}, \{\text{undec}\}$.

The basic idea behind justification statuses and about abstract argumentation theory in general can be visualised in the following example:

Example 3.6 Imagine you get introduced to a group of friends and you are trying to evaluate the different people and their relationships. You get the following information:[7]

A: Alice says that Carole is a Liar.
C: Carole says that David is a Liar.
D: David says that Alice is a Liar.
B: Everybody agrees that Bob is really trustworthy.
E: Emily says that Fred is a funny guy.
F: Fred says that Emily does not know him.

---

[6] Note that the two remaining options $\{\emptyset\}$ and $\{\text{in}, \text{out}\}$ are not considered, as [20] only take non-empty labelling-sets into account and complete semantics have the characteristic of being "abstention allowing", meaning that if there is a complete labelling that labels A in and there is another complete labelling that labels A out, there has to be a third one that labels A undec, [1, p. 27].

[7] The example is an adapted version from the example in [20, p. 21].

The argumentation graph of the argument can be visualised like this:

The admissible labellings of the given graph are the following:

$$\{\{B,F\},\{E\},\{A,C,D\}\}, \ \{\{B,E\},\{F\},\{A,C,D\}\}, \ \{\{B\},\emptyset,\{A,C,D,E,F\}\},$$
$$\{\{F\},\{E\},\{B,A,C,D\}\}, \ \{\{E\},\{F\},\{B,A,C,D\}\}, \ \{\emptyset,\emptyset,\{B,A,C,D,E,F\}\},$$

while only the first three are also complete labellings.[8]

The justification status of the arguments (according to complete semantics) are:

$J(A) = \{\text{undec}\}$, $J(C) = \{\text{undec}\}$, $J(D) = \{\text{undec}\}$,
$J(B) = \{\text{in}\}$, $J(E) = \{\text{in, out, undec}\}$, $J(F) = \{\text{in, out, undec}\}$.

3.1.2 Forms of Indecision on the Argument Level

As we have seen, in abstract argumentation theory arguments are generally evaluated by labelling them according to the three labels: in, out, undec. Although philosophy is concerned with propositions rather than arguments, this trisection still finds some correspondence. Philosophers are concerned with determining when a person does (or should) believe, disbelieve or be undecided about a proposition. Believing a proposition can be said to correspond to labelling an argument in, disbelieving to labelling it out and being undecided to labelling it undec. If a subject believes a proposition, she takes this proposition to be true, if we label an argument in, we accept it to be valid and possibly even sound and thereby take its conclusion (and its premises) to be true.

Within this correspondence, parallels between the different forms of indecision in philosophy and the different forms of indecision in argumentation theory can be drawn. These parallels can be useful to transfer important insights from philosophy to argumentation theory and vice versa. The parallels can then be used to also apply further developments in the philosophical considerations about indecision to argumentation theory.

The authors in [20] interpret arguments with justification status {out} to be clearly (or strongly) rejected and arguments with the justification status {in} to be strongly accepted. In comparison to that, they interpret

---

[8] Note that the notation $\{\{A\},\{B\},\{C\}\}$ corresponds to a labelling in which A is labelled in, B is out and C is undec.

arguments with justification status {in, undec} to be only weakly accepted (and arguments with justification status {out, undec} respectively only weakly rejected). So, two of the six possible statuses represent (two forms of) acceptance and another two represent rejection. Finally, the remaining two justification statuses {in, out, undec} and {undec} represent indecision. In [20] {in, out, undec} is called an undetermined borderline case, while {undec} is taken to be a determined borderline case.

The difference between determined and undetermined resides exactly in the difference between committed and uncommitted indecision. In the case of the justification status {in, out, undec} all options are still available, i.e., the argument may be labelled in, it may be labelled out and it may be labelled undec. In the determined borderline case of the justification status {undec}, it is decided that the argument has to be labelled undec. In all possible (in this case complete) labellings the argument is labelled undec. Hence, the question "What label does the argument get?" is no longer open. The only plausible label in this case is undec. In comparison to this, in the undetermined borderline case of the justification status {in, out, undec}, the different labellings disagree on how to label the argument. This can be interpreted as a situation of a vote for which there are some voices voting for in, some for out and some for undec. It is also possible to interpret this situation within one reasoning subject. The different voters would then correspond to different points of evidence *one* subject has, that are pointing in different directions, i.e., towards in, out or undec. Both interpretations take the question of what label the argument should get to be not settled, i.e., to be still an open question. This shows us that the difference between suspension of judgement as a settled form of indecision and other non-committing forms of "mere indecision", investigated by philosophers, is reflected in the two different undecided justification statuses an argument can get. In the example situations from section 2.2, we would, give the proposition [9] that the covid pandemic will be over next year, the justification status {in, out, undec}, as it is still an open question whether we will accept, reject or stay undecided about the matter, whereas the proposition that a God exists will get the committing justification status {undec}, as it is a closed question in the sense that I committed myself to be undecided about it.

This can also be visualised in Example 3.6. While argument A, C, and D get the justification status {undec}, argument F and E get the justification status {in, out, undec}. In the case of A, C and D each argument attacks one of the other three, yielding a circle of attack. Given the information of the example, there is no way to decide who is a Liar and who is not.[10] Hence, the only reasonable thing to do is to label all three arguments undec. The

---

[9] Again, note the different object of the justification statueses: arguments in the case of abstract argumentation theory vs. propositions in the philosophical example.

[10] Of course this is a reformulation of the famous Liar Paradox.

justification status expresses this commitment to the indecision. On the other hand, for the arguments F and E we have more options. We could label both undec, or we could, for example, label F in and E out. In this scenario, we might believe that Fred has a good insight on who of the group actually knows him and, therefore, we should believe him when he says that Emily does not know him. Then, however, we should not trust any of Emily's statements about Fred, and argument E is labelled out. The situation for argument E and F is more open and flexible, and less committed to indecision, than the situation for A, C and D.

One can observe that the different justification statuses of E and F compared to A, C and D do not stem from the different arguments as such, but simply from the fact that E and F are involved in an *even* cycle of attack, while A, C and D form an *odd* cycle of attack. This can be seen more clearly when we change the argument E to "Emily says that Fred is a Liar" and F to "Fred says that Emily is a Liar". Although, in terms of content, we seem to have the same situation with E and F now as we have with A, C and D, the arguments E and F will still get the justification status $\{\text{in}, \text{out}, \text{undec}\}$.[11] The situation of A, C and D, i.e., the situation that all involved arguments *have to* be labelled undec in fact only occurs in odd-length attacking cycles. This was observed in [14, p. 242] and also revisited later, e.g., in [1, p. 21]. Many scholars regard the unequal treatment of odd and even cycles in semantics like complete semantics as problematic. Therefore, other not-admissible based semantics have been developed that allow for an equal treatment of the cycles, see [2].

For the purpose of this paper, however, it is only relevant that there are cases, in which the justification status is purely $\{\text{undec}\}$ and that those cases can be seen to represent a committed form of indecision, as in the example above and that there also are cases of arguments with the non-committing justification status $\{\text{in}, \text{out}, \text{undec}\}$.

Although I think that the example as it is presented in 3.6 is better suited than the adapted example to describe, why the situation for E and F is less committed, than the situation for A, C and D, there is still some difference between the situations, even when changing the content of argument E and F to the respective Liar sentences. The difference consists in how "easy" it is for a person that is presented with the arguments to give the arguments a label that is different from *undec*. Although in the adapted example of E and F, it still might seem to be the most reasonable action to label both arguments *undec*, you can still do otherwise. If you, for example, happen to believe women more than men, you might think that Emily is more trustworthy. In believing what Emily says, you thereby believe that Fred is a Liar and take argument E to defeat argument F. Then you will reject argument F and thereby all of the attackers of E are labelled *out*, allowing

---

[11] I thank an anonymous reviewer for stressing this point.

you to label E in. This is not possible in the same way for the threefold attacking cycle of A, C and D. Say, that you, for some external reason, believe Alice. Hence, you will believe what she says and argument A will successfully defeat argument C. You take Alice' word and believe that Carole is a Liar and hence you will reject argument C. Then, argument D has no (not-rejected) attacker anymore; so that you should accept it. As argument D, however, tells you that Alice is a Liar, you will have to question your starting point, which is that you belief Alice, again. One can argue that in such odd cycles of attack, you will not manage to break out of the cycle and hence you will not be able to react otherwise than to take each argument to be undecided.

In the next section, I will move from considerations about different forms of indecision in abstract argumentation theory to the involved norms concerning indecision.

3.1.3 Norms of Indecision on the Argument Level

In Section 2.1, I described the development in epistemology that started with not considering explicit norms for indecision and treating indecision merely as a lack of belief and disbelief and as a fallback position for propositions that do not fulfil the requirements of belief and disbelief or are simply not investigated yet. This was typified by the story of Descartes. Only in the last couple of years, epistemologists started to investigate norms about indecision specifically. This development finds some correspondence in the different semantics of argumentation theory.

One can note that admissible semantics only formulate rules (or necessary conditions) about when an argument can be accepted (labelled in) or rejected (labelled out). There is no such requirement for when an argument is labelled undec. A labelling that simply labels all arguments undec always fulfils the requirements of admissible semantics. This means that admissible semantics only provide *norms* (in the form of restrictions) for acceptance and rejection, but no norms about when an argument should or should not be regarded as undecided. Hence, the label undec is always suitable for any argument.

Complete semantics, on the other hand, expand admissible semantics (as they adopt the necessary conditions on when an argument is legally in or out), but also provide restrictions on when an argument can be labelled undec. The treatment of the undecided-label is what distinguishes complete semantics from admissible semantics. While in admissible semantics, the set of undecided arguments is basically only a collection of whatever is left over, (because it has not been determined yet or because it does not manage to be labelled in or out,) in complete semantics there are rules about when an argument can legally be called undecided.[12]

---

[12] The attitude that the undec label represents arguments for which there is no decision yet, although there should be a clear decision is even better illustrated in preferred or stable

The difference is also visible in Example 3.6. While the last labelling, in which every argument is labelled undec, is admissible, this labelling is not complete. In particular, complete labellings forbid to label the argument B that is not attacked by any other argument undec. This rule can be interpreted in philosophical terminology as a norm stating that, if there is good evidence for a proposition, then one should not be undecided about it, but believe it, and if there is good evidence against a proposition, then one should not be undecided either, but disbelieve it.

Moreover, the different justification statuses also show some parallels with the different norms of indecision from epistemology. In [20, p. 17], it is proven that an argument A gets the justification status {undec}, iff A is not labelled in in any admissible set *and* A is not labelled out in any admissible set. Moreover, an argument A gets the justification status {in, out, undec}, iff A is labelled in in at least one admissible labelling *and* A is labelled out in at least one admissible labelling, see [20, p. 19-20]. The first case can be interpreted as having no evidence for or against the argument (or having no one voting pro or con the argument), while the second case can be interpreted as having evidence for and evidence against the argument (or having both voices that vote for the argument and against it). Hence, the first case corresponds to the absence of evidence norm from philosophy and the second case to the balanced evidence norm, that were introduced in Section 2.1.

It can be concluded that different philosophical norms concerning indecision as well as the epistemological development are represented on the argument level of argumentation theory. Next, I want to consider statements, instead of arguments and reveal parallels with philosophical investigations on this level, too.

3.2 Indecision on the Level of Statements

Up to now, I have only addressed the evaluation of arguments as such. It can be argued, however, that it is not so much the arguments themselves, but sentences or propositions that form the doxastic situation of an agent, on which the agent ultimately bases her actions on. In order to reveal this doxastic situation, one has to dive deeper and extract the inner structure of the arguments, i.e., revealing the premises and conclusions involved. By doing this, we end up on the level of statements. In the following sub-section, I will present some transfers of the philosophical findings concerning forms and norms of indecision to the statement level of argumentation theory. When philosophical considerations come into play, the concentration on the statement level is particularly desirable, because statements are structurally similar to propositions, which are, as described, the main objects of investigation in philosophy. The parallels can therefore be drawn more straight-forwardly.

---

semantics, that try to maximise the number of accepted and rejected arguments or even try to bring the set of undecided arguments to be the empty set, see [1] for an overview

Considering all this, it is quite astonishing that assessing justification statuses to statements has gotten, in comparison to arguments, only little attention. Just recently, Baroni and Riveret describe in [4] possibilities to transfer the justification of arguments to justification of statements by evaluating the conclusions of the arguments in question.[13] In general, they distinguish two approaches, which they call an argument-focused and a statement-focused approach, to determine the status of statements. In both approaches the status of a statement is eventually determined by the statuses of the arguments speaking for or against that statement. For the purpose of this paper, the different approaches are not directly relevant. What I will focus on here, is the different types of statement labellings that are considered in [4]. They distinguish basically between three types of statement labellings.

- Bivalent Labellings: Bivalent statement labellings only allow for two possible labels a statement can obtain. [4, p. 839] take the possible labels to be {yes, no}.
- Doubt-Tolerant Labellings: doubt-tolerant labellings allow for a third label between the definite answers of yes and no of the bivalent labellings. This third, intermediate label expresses some sort of doubt. Baroni et. al. call in [4, p. 848] the three labels {yes, fal, ni} where yes means that a statement is accepted (or verified), fal means that a statement is falsified and ni means that there is doubt about the status of the statement.
- Ignorance-Aware Labllings: ignorance-aware labellings further divide the group of undetermined statements. Besides yes and fal there are two intermediate labels: unk and ni, yielding the set {yes, fal, unk, ni}, where unk stands for unknown statements and is meant to capture statements for which there is no evidence or lack of knowledge, and ni captures the statements for which the evidence indicates indecision.

The differences of the three labelling-types can be understood best, when one takes a look at the following example from [4], which was introduced in [3, p. 489]:
"Suppose that Dr. Smith says to you: 'Given your clinical data I conclude you are affected by disease D1'. Suppose then that another equally competent physician Dr. Jones says to you: 'Given your clinical data I conclude you are not affected by disease D1'. Your view on the justification of the statements s1='I am affected by disease D1' and ¬s1='I am not affected by disease D1' may become quite uncertain. In a different situation, at home, you use an off-the-shelf test kit suggesting you have caught disease D2. You then undertake a serious and reliable clinical test, which excludes disease D2. Would you consider the same status for the statement s2='I am affected by disease D2' and the statement s1? [....] Consider [as well the] statement s3='I

---

[13] In fact Wu et. al. consider in [20] also briefly how to transfer the justification statuses to statements.

am affected by D3', where D3 is a poorly studied and initially asymptomatic disease you only know by name." [4, p. 793–794]

The authors in [4] appeal to the intuition that there should be a different justification status for the statement s1 and the statement s2, although in both cases there are arguments for *and* arguments against the statement. Moreover, we want to say that the status of statement s3 should be different from the statuses of s1 and s2, too. s3 should intuitively get a justification status of "full ignorance" as there is no evidence that concerns the third disease (and thereby statement s3).
In fact, the evaluation of the different labellings [14] by means of the example shows that only the ignorance-aware labelling can account for this intuition. The different labellings provide the following results:

	s1	¬s1	s2	¬s2	s3
Bivalent Labellings	no	no	no	yes	no
Doubt-Tolerant Labellings	ni	ni	fal	yes	ni
Ignorance-Aware Labellings	ni	ni	fal	yes	unk

One can see that the bivalent labellings cannot even illustrate the different intuitive justification statuses of statement s1 and s2. The doubt-tolerant labellings already do a better job, recognising that the question about s1 is, in comparison to s2, not decided, and hence taking both s1 and its negation to be ni, while s2 is labelled fal and its negation is labelled yes. However, there is no way to distinguish the justification status of s1 (and ¬s1) from the justification status of s3. This is only achievable in the ignorance-aware labellings. Recall that ignorance-aware labellings distinguish between two forms of the middle, undetermined status: the label ni represents "conflicting support", while unk represents the "absence of support", [4, p. 848]. This fits our intuitions of the example, as s3 is a statement, for which there is no support at all, while s1 is a statement for which the support is conflicting.

With the two different labels from ignorance-aware labellings that represent indecision, the authors in [4] want to distinguish cases where the evidence is absent from cases where the evidence is equally balanced. This distinction is exactly reflected in the two basic norms for indecision from epistemology: the absence of evidence norm and the balanced evidence norm.
Moreover, the different forms of indecision, found in our philosophical range are represented in the example and covered (at least to some extent) by the ignorance-aware labellings. As we have seen, there is no evidence at all,

---

[14] They first investigate different formalism of structured argumentation and how the different types of labellings can be implemented there. They then investigate which outcomes the different labellings yield for the provided example. For some formalisms they distinguish between a sceptical and a credulous approach. This distinction only yields a different treatment of statement s1 and its negation, but is silent on the other statements. For reason of simplicity, we will only consider the results of the sceptical labellings.

speaking for or against s3, so s3 is labelled unk. The subject has not even considered s3, and hence it is comparable to the statement "The guava fruit has a lot of vitamin C", that I have never considered. Thus, in the range of indecision, s3 similarly has to be placed quite at the beginning of the axis of engaging. It is clear that we have some form of "not engaged" indecision, as the subject is not concerned with the statement at all. On the other hand, s1 is a statement the subject already collected evidence for. There is one physician arguing for s1 and another, equally competent physician, arguing against s1. The subject has engaged with the statement in question and came to the conclusion that she cannot tell, whether she has the disease. This is clearly a form of more engaged indecision, as the subject not only considered the statement s2, but actually collected quite some evidence for (and against) it. This can be compared to the example of me wondering whether the covid pandemic will be over next year. I, too, collected evidence for and against the proposition (and in fact new evidence gets through to me every day), but I still cannot decide, because the evidence seems more or less balanced.

We see that the ignorance-aware labellings allow us to distinguish two distinct forms of indecision. It might be valuable, however, to represent even more than two forms of indecision in this framework. When we look back to the axis that represented the example situations of 2.1, we already find four distinct forms of indecision. As said above, the situation of me being undecided about the Guava fruit's nutritional value can be compared to the indecision concerning statement s3, saying that I have disease D3 that I never heard of. Both cases represent a rather passive and not engaged form of indecision as I have not considered the respective statement or proposition. In the philosophical example, however, the cavemen's case exemplifies another form, that is even more left on the axis and hence even less engaged. In argumentation theory, it can also be eligible, to make a more fine-grained distinction between such little-engaged forms. Although s3 is a statement that has not been considered yet, the reasoning subject or system at least understands the statement. However, there can be cases in which the system cannot grasp or process the statement, when, for example, the statement contains words or phrases which do not belong to the language of the respective argumentation theory system. Such cases would correspond to the cavemen being undecided [15] about whether quarks exist. In such a case, the system should be able to reply a different form of indecision than in the case of s3. While s3 is a statement with respect to which the system is not equipped with any argument for or against it, it still could in principle evaluate the statement with a decisive label if, for example, at some later point new arguments would come into play. With a statement that is not even included in the system's language, this is different. The system here should give a reply that shows full ignorance about what to to with that statement.

---

[15] As said above, philosophers would not use the term indecision, but mere non-belief here.

On the other side of the axis, there is room for finer distinctions, too. Philosophers distinguish between rather open, but yet reflective forms of indecision (the Covid case) from settled forms of indecision (the question about God's existence). As we have seen, the example of s1 and ¬s1 represent a similar situation as the Covid case, as the subject will be undecided whether she has disease D1 but will still look for further evidence, possibly changing the label of the statement with the help of further arguments. There might be other examples, however, where the argumentation framework is build in a way in which it is clear that a statement *cannot* get a different label than that of being undecided. The cycles of attack that have been regarded in section 3.1.2 provide an outline of a possible example of such situations. To distinguish these two forms with two different labels can be helpful for a system to recognise when there is no need for further deliberation or further arguments concerning a certain statement, and when it is worth to allow further arguments in order to reevaluate.

4 Conclusion and Outlook

In this paper I laid out some first steps toward a better understanding of the forms of indecision involved in argumentation theory. For this end, I presented philosophical considerations concerning different phenomena related to indecision and norms about when a subject should or should not be undecided about a matter. I tried to show that a lot of the philosophical investigations find some correspondence both on the argument level of abstract argumentation theory and on the level of statement evaluation. By having revealed parallels of this kind, we took the first steps toward using philosophical arguments to better evaluate how indecision is used in argumentation theory. The philosophical considerations can, for example, be used to evaluate the status of indecision in different semantics. Once we have drawn these parallels, and understood the use of indecision and the different forms involved in argumentation theory, we are able to apply further developments in philosophy concerning indecision or other neutral stances to argumentation theory, too. This is interesting, because indecision can be a useful tool for indicating unclear or critical situations.

For this enterprise, investigating indecision on a statement level seems crucial. For further research, it is, hence, desirable to concentrate on the different ways to evaluate statements and on the different ways to transfer justification of arguments to justification of statements, yielding statement-labellings. Although recent work already allows for a distinction between two kinds of indecision-labels on a statement level, I argued that it is necessary to distinguish even more forms of indecision. It is desirable that an argumentation system is capable of distinguishing cases in which it cannot deal with the input, from cases in which it has no evidence for or against a statement; and cases for which the evidential situation seems balanced, so it might need further arguments, from cases that clearly cannot be decided. Additionally, the system should be capable of storing the information and reporting these different situations. These skills can help the system to signal when it reaches its limits of application or when it needs further evidence or arguments to decide. This can help to reduce random decisions and make the systems choices more grounded and traceable.

Another challenge in the field of statement evaluation will be to not only consider the conclusions of an argument but also the premises. This is especially relevant when considering the different forms of defeat, such as undercutting or rebutting defeat, see [13].

On the other hand, it can also be interesting to take a look at an even higher level than arguments. When non-unique semantics are considered, the question about which labelling (or extension) is to choose suggests itself immediately. For example, Dauphin et. al. investigate in [5] certain principles that decision graphs that are used for choosing among multiple extensions should satisfy. Concerning such choosing strategies, a better understanding of the underlying concepts of indecision can also be very useful for cases in which the available evidence does clearly point to one direction.

References

[1] Baroni, P., M. Caminada and M. Giacomin, *An introduction to argumentation semantics*, Knowledge Engineering Review **26** (2011), p. 365.
[2] Baroni, P., M. Giacomin and G. Guida, *Scc-recursiveness: a general schema for argumentation semantics*, Artificial Intelligence **168** (2005), pp. 162–210.
[3] Baroni, P., G. Governatori, H.-P. Lam and R. Riveret, *On the justification of statements in argumentation-based reasoning*, in: KR, Citeseer, 2016, pp. 521–524.
[4] Baroni, P. and R. Riveret, *Enhancing statement evaluation in argumentation via multi-labelling systems*, Journal of Artificial Intelligence Research **66** (2019), pp. 793–860.
[5] Dauphin, J., M. Cramer and L. van der Torre, *Abstract and concrete decision graphs for choosing extensions of argumentation frameworks*, Computational Models of Argument (2018).
[6] Descartes, R., "Meditationes de prima philosophia in Oeuvres VII," Edited by Charles Adam and Paul Tannery, Paris, 1996.
[7] Friedman, J., *Question-directed attitudes*, Philosophical Perspectives **27** (2013), pp. 145–174.
[8] Friedman, J., *Rational agnosticism and degrees of belief*, Oxford studies in epistemology **4** (2013), pp. 57–81.
[9] Friedman, J., *Suspended judgment*, Philosophical studies **162** (2013), pp. 165–181.
[10] Friedman, J., *Why suspend judging?*, Noûs **51** (2017), pp. 302–326.
[11] Friedman, J., *Inquiry and belief*, Noûs **53** (2019), pp. 296–315.
[12] Hájek, A., *Agnosticism meets bayesianism*, Analysis **58** (1998), pp. 199–206.
[13] Pollock, J. L., "Cognitive carpentry: A blueprint for how to build a person," MIT Press, 1995.
[14] Pollock, J. L., *Defeasible reasoning with variable degrees of justification*, Artificial intelligence **133** (2001), pp. 233–282.
[15] Raleigh, T., *Suspending is believing*, Synthese (2019), pp. 1–26.
[16] Rosa, L., *Logical principles of agnosticism*, Erkenntnis **84** (2019), pp. 1263–1283.
[17] Schroeder, M., *Stakes, withholding, and pragmatic encroachment on knowledge*, Philosophical Studies **160** (2012), pp. 265–285.
[18] Wagner, V., *Agnosticism as settled indecision*, Philosophical Studies (forthcoming).
[19] Wedgwood, R., *The aim of belief*, Philosophical perspectives **16** (2002), pp. 267–297.

[20] Wu, Y., M. Caminada and M. Podlaszewski, *A labelling-based justification status of arguments*, Studies in Logic **3** (2010), pp. 12–29.

Goal-Directed Decision Procedures for Input/Output Logics

Alexander Steen[1]

University of Luxembourg, FSTM
6, Avenue de la Fonte
L-4364 Esch-sur-Alzette, Luxembourg

Abstract

Input/Output (I/O) logics address the abstract study of conditional norms. Here, norms are represented as pairs of formulas instead of statements that themselves carry truth-values. I/O logics have been studied thoroughly in the past, including further applications and refinements. In this paper, a class of automated reasoning procedures is presented that, given a set of norms and a concrete situation, decide whether a specific state of affairs is obligatory according to the output operations of I/O logics. The procedures are parametric in the underlying logical formalism and can be instantiated with different classical objects logics, such as propositional logic or first-order logic. The procedures are shown to be correct, and a proof-of-concept implementation for propositional I/O logics is surveyed.

Keywords: Deontic logic, I/O logics, Automated reasoning, Normative reasoning.

1 Introduction

Input/Output (I/O) logics have been devised by Makinson and van der Torre [8] as a class of formal systems for norm-based deontic reasoning. Intuitively, they formalize the question which obligations can be detached from a given set of conditional norms and a specific situation. I/O logics differ from other deontic logics, such as Standard Deontic Logic (SDL, a modal logic of type **KD**) and Dyadic Deontic Logic (DDL) [1], in the sense that the norms themselves are not part of the object logic and hence do not carry truth values. Furthermore, in SDL and DDL the deontic operators are evaluated with respect to a set of possible words, whereas in I/O logics they are evaluated with respect to a set of norms. An overview of deontic logic formalisms can be found in the literature, see e.g. [7].

The field of automated reasoning studies the conception, implementation, application and evaluation of methods for automating logical inferences on the computer [14]. This includes, among others, methods for deciding satisfiability and tautology, for model generation, and for computer algebra systems. The

---

[1] E-Mail: `alexander.steen@uni.lu`; ORCID ID: 0000-0001-8781-9462

study of automated deduction systems denotes one of the earliest concerns of artificial intelligence, and is today often referred to as symbolic AI; in contrast to recently successful approaches using statistical and learning-based approaches. One of the core applications is automated theorem proving (ATP): ATP systems are computer programs that, given a set A of axioms and a conjecture C as input, try to prove that C is a logical consequence of A, i.e., that C is true whenever every formula in A holds. In this context, the search for a proof is conducted autonomously so that no intervention or advice from human users is necessary. Unfortunately, most ATP systems focus on classical logics only and hence there are only few systems available for automating logics relevant to deontic reasoning. Notable exceptions are ATP systems for (normal) modal logics, but since these logics suffer from various theoretical drawbacks, their application to normative reasoning in, e.g., legal contexts [5] is limited.

In this paper, a first structured step is taken towards automation of I/O formalisms: Decision procedures for four different deontic operators of (unconstrained) I/O logic are presented that decide whether a formula x can be detached as an obligation given a set of norms and a situation (put in I/O logic terms: the procedures decide whether a formula x is in the output given a certain input). They are shown to be sound and complete, and to be decidable if the underlying logical language is decidable. Furthermore, a prototype implementation of the procedures is presented. This implementation is freely available as a web application and can be used to conduct own experiments.

Related work. I/O logics have also been employed in the context of studying conditional permissions [10]. Also, there exist extensions of I/O logics, called *constrained I/O logics*, that address the classical deontic paradoxes [9] such as contrary-to-duty scenarios. Recent work furthermore addresses weaker notions of I/O logic that allow for a fined-grained control over employed inference principles [12].

From a computational perspective, there are comparably few related approaches available. Complexity aspects of I/O logics have been studied [17]. However, the methods used in the analysis do not yield means for implementing respective decision procedures. Quite recent work focuses on automating other deontic logics via shallow semantical embeddings into classical higher-order logic [3]. However, such an approach is not yet available for all unconstrained I/O logic operations [2], and indeed seems more complex than for other logical systems in the context of deontic reasoning [4]. There are representation results available for expressing I/O logics in modal logic; and there exists an alternative proof-theoretic (dynamic) characterization of I/O logic [16]. However, these results have, up to the author's knowledge, not yet been utilized in the context of automated reasoning systems.

2 I/O Logics

I/O logic is used for studying conditional norms, e.g., obligations under some legal code. Here, conditional codes are represented as pairs of formulas and therefore do not carry truth values themselves, whereas declarative statements are usual Boolean formulas that come from some logical language L.

Let L be a logical language that is closed under the truth-functional connectives such as conjunction ($\wedge$) and disjunction ($\vee$). From a semantical perspective, it is assumed that the eligible logical languages considered in the following come with a derivation relation $\vdash$ for which the operation $Cn(A)$, given by $Cn(A) = \{x \in L \mid A \vdash x\}$, is a Tarskian closure operator. A prominent example for L in the context of I/O logics is the language of classical propositional logic with the usual consequence relation (often assumed in the literature). However, also first-order logic or even higher-order logic languages are possible. In the following, it is assumed that L comes from a classical logic.

A normative system $N \subseteq L \times L$ is a set of pairs (a, x) of formulas. The pair (a, x) represents the conditional obligation that *given a, it ought to be x*. By convention, given a norm (a, x) the first element a is also referred to as the body and the second element x is referred to as the head. The image of N, denoted $N(A)$, where A is a set of formulas, is given by $N(A) = \{x \in L \mid (a, x) \in N \text{ for some } a \in A\}$. Given a normative system N and a set of formulas A (the input set), $out(N, A)$ denotes the output of A under N where out is the respective output operator.

The semantics of I/O logics is operational in the sense that the meaning of normative concepts is given by generated outputs given a set of norms and an input. The four output operators out_i, $i \in \{1, 2, 3, 4\}$, studied in the literature are defined as follows [8]:

$out_1(N, A) = Cn(N(Cn(A)))$
$out_2(N, A) = \bigcap \{Cn(N(V)) \mid V \supseteq A, V \text{ complete}\}$
$out_3(N, A) = \bigcap \{Cn(N(B)) \mid A \subseteq B = Cn(B) \supseteq N(B)\}$
$out_4(N, A) = \bigcap \{Cn(N(V)) \mid A \subseteq V \supseteq N(V), V \text{ complete}\}$

where a set $V \subseteq L$ is called complete iff $V = L$ or V is a maximally consistent set.

A proof-theoretic characterization of the different output operations can be achieved by putting $out_i(N) = \{(A, x) \mid x \in out_i(N, A) \text{ for some } A \subseteq L\}$, $i \in \{1, 2, 3, 4\}$. The specific inference rules are the following (the first component of each pair are assumed to be singleton sets and the curly braces are omitted):

SI: From (a, x) to (b, x) if $b \vdash a$
WO: From (a, x) to (a, y) if $x \vdash y$
AND: From $(a, x), (a, y)$ to $(a, x \wedge y)$
OR: From $(a, x), (b, x)$ to $(a \vee b, x)$
CT: From $(a, x), (a \wedge x, y)$ to (a, y)

In earlier work [8] it is shown that (a, x) is in the respective set $out_i(N)$ if and only if it is contained in the least superset of $N \cup (\top, \top)$ that is closed under the inference rules as follows: $\{SI, WO, AND\}$ for out_1, $\{SI, WO, AND, OR\}$ for out_2, $\{SI, WO, AND, CT\}$ for out_3, and $\{SI, WO, AND, OR, CT\}$ for out_4. For non-singleton sets A, derivability of (A, x) from N is reduced to derivability of (a, x), where $a = a_1 \wedge \ldots \wedge a_n$ is some conjunction of the elements of A.

The empty input is interpreted as an empty conjunction and assumed to be a tautology; hence $(\emptyset, x)$ is reduced to derivability of $(\top, x)$.

Although there is an adequate syntactic characterization for I/O logics that can be used to derive outputs, the above calculi are not *machine-oriented* in the sense that they can be implemented as-is in an effective manner on a computer. This is, in particular, because the rules *SI* and *WO* allow to derive an infinite number of outputs, and it is not immediately clear what intermediate derivations actually contribute to the ultimate proof goal. However, without effective means of automation it is challenging, or even impossible, to apply I/O logics to practical scenarios, e.g., in the context of multi-agent systems or legal reasoning use cases [13].

3 Decision Procedures for I/O logic

In this section, four different decision procedures are presented, one for each output operation, that allow automated reasoning within I/O logics in an effective way. It is implicitly assumed that $\vdash$ is a sound and complete derivation relation for L. Furthermore, it is assumed that inputs to the decision procedures are finite, i.e. N is a finite set of norms and A is a finite set of formulae.

Let IN-OUT$_i$, $i \in \{1, 2, 3, 4\}$, denote the following decision problem: *Given a formula $x \in L$, a set of norms N and an input A, is it the case that $x \in out_i(N, A)$?* The decision procedure that addresses the respective decision problem IN-OUT$_i$ is denoted $\mathsf{IO}_i^\vdash$. By convention, $(N, A, x) \in \mathsf{IO}_i^\vdash$ is written iff $\mathsf{IO}_i^\vdash$ gives "yes" for a set of norms N, input A and prospective output x; i.e., it is identified with the subset of parameter tuples for which it decides positively.

Note that because it is not fully specified what logical formalism L is employed, the procedures $\mathsf{IO}_i^\vdash$ presented in the following in fact describe a class of parametric decision procedures that can be used in different contexts. As an example, if L is taken as classical propositional logic the implementation of the $\mathsf{IO}_i^\vdash$ is straight-forward, and they are guaranteed to terminate and to yield correct results.

It is also possible to apply each $\mathsf{IO}_i^\vdash$ to logics that are not decidable: Already existing automated theorem provers can be utilized as oracle for $\vdash$, e.g., in the context of first-order logic or higher-order logic. This way a wide range of input-output logic reasoners can be implemented with comparably low effort. Of course, for logics that are not decidable the procedures $\mathsf{IO}_i^\vdash$ might never terminate; as usual in automated reasoning in expressive logics.

The decision procedures are given in pseudo-code in the following. Apart from the usual set-theoretic functions, a function for calculating the disjunctive normal form (DNF) of a formula is used: For a given formula x, the function DNF(x) gives the disjunctive normal form of x, represented as a set $\{x_1, x_2, \ldots, x_n\}$ of formulas where each x_i does not contain any disjunction.

3.1 Simple-minded output

Listing 1 shows the decision procedure $\mathsf{IO}_1^\vdash$ for deciding IN-OUT$_1$ with respect to a logical language L with underlying derivation relation $\vdash$.

```
1  Input:   N = {(b_1, h_1), (b_2, h_2), ...} set of norms
2           A = {a_1, ..., a_m} set of formulae
3           x formula
4  Output:  Yes or No
5
6  N' := {(b, h) ∈ N | A ⊢ b}
7  if {h | (b, h) ∈ N'} ⊢ x then
8     return Yes
9  else
10    return No
11 endif
```

Listing 1: Decision procedure $\mathsf{IO}_1^\vdash$ for IN-OUT$_1$.

The general idea of $\mathsf{IO}_1^\vdash$ is the following: First, the subset N' of norms whose body is satisfied by the input A is calculated. Then, the procedure returns Yes if and only if the heads of N' satisfy x. Note that this approach allows to effectively handle the infinite sets $Cn(A)$ and $Cn(N(Cn(A)))$ that cannot be computed exhaustively in an explicit way. The procedure $\vdash$ used here acts as oracle. Depending on the logic L that is assumed, the effort for implementing such a procedure may vary. An example implementation is described in §4.

Adequateness of $\mathsf{IO}_1^\vdash$ is ensured as shown in the following:

Theorem 3.1 (Partial correctness of $\mathsf{IO}_1^\vdash$) *$\mathsf{IO}_1^\vdash$ is sound and complete for* IN-OUT$_1$; *in particular,* $x \in out_1(N, A)$ *if and only if* $(N, A, x) \in \mathsf{IO}_1^\vdash$.

Proof For the first direction, assume $x \in out_1(N, A)$ for some formula $x \in L$. By definition, $N(Cn(A)) \vdash x$ hence there exists some subset $\{(b_1, h_1), \ldots, (b_m, h_m)\} \subseteq N$ of norms such that $A \vdash \bigwedge_{i=1}^m b_i$ and $\bigwedge_{i=1}^m h_i \vdash x$. Since N' as computed in line 6 is the largest subset of N for which each body of $(b, h) \in N'$ it holds that $A \vdash b$, by monotony of $\vdash$ it also holds that $\{h \mid (b, h) \in N'\} \vdash x$. As a consequence, the if condition in line 7 is true and thus $(N, A, x) \in \mathsf{IO}_1^\vdash$.

For the second part, assume $(N, A, x) \in \mathsf{IO}_1^\vdash$. Then, by definition, there is a subset N' of norms such that $\{h \mid (b, h) \in N'\} \vdash x$. It furthermore holds that $A \vdash b$ for each $(b, h) \in N'$ by construction. As a consequence, it holds that $x \in Cn(N'(Cn(A)))$. Since $N' \subseteq N$ it follows that $x \in Cn(N(Cn(A)))$ and hence $x \in out_1(N, A)$. □

Theorem 3.2 (Total correctness of $\mathsf{IO}_1^\vdash$) *Let $\vdash$ be a decidable derivation relation for L. $\mathsf{IO}_1^\vdash$ terminates and is sound and complete for* IN-OUT$_1$.

Proof Theorem 3.1 already yields soundness and completeness. Termination is straight-forward: As all input is finite and since $\vdash$ is decidable by assumption the set N' can be constructed in finite time. Also, the if-condition in line 7 can be evaluated in finite time as $\vdash$ is decidable by assumption. □

3.2 Basic output

Listing 2 presents the decision procedure $\mathsf{IO}_2^\vdash$ for deciding IN-OUT$_2$.

```
1  Input:    N = {(b₁,h₁),(b₂,h₂),...} set of norms
2            A = {a₁,...,aₘ} set of formulae
3            x formula
4  Output: Yes or No
5
6  D := DNF(⋀ A)
7  for all d ∈ D do
8     N' := ∅
9     for all (b,h) ∈ N do
10       C := {(b',h') ∈ N | h' ⊢ h}
11       if d ⊢ ⋁{b' | (b',h') ∈ C} then
12          N' := N' ∪ {(b,h)}
13       endif
14    endfor
15    if {h | (b,h) ∈ N'} ⊬ x
16       return No
17    endif
18 endfor
19
20 return Yes
```

Listing 2: Decision procedure $\mathsf{IO}_2^\vdash$ for IN-OUT$_2$.

$\mathsf{IO}_2^\vdash$ is a modified version of the respective procedure for IN-OUT$_1$, adapted to incorporate the validity of the OR rule. If the output of a formula can be established for different inputs, then it is also in the output set for the disjunction of these inputs. From a semantical perspective this amounts to incorporating reasoning by cases (cf. definition of out_2 in §2): If a norm n is triggered by every complete extension of the input, then the head of n is contained in the output set. In order to reflect this in $\mathsf{IO}_2^\vdash$, the DNF of input A is computed first and the following steps are done for each clause $d \in \text{DNF}(A)$: The subset of triggered norms N' (with respect to d) is generated. If $n \in N$ is a norm, then let $n' \in N$ denote a n-compatible norm if and only if the body of n' is at least as strong as the body of n, i.e., it holds that $h' \vdash h$ where h and h' are the heads of n and n', respectively. Intuitively, every n-compatible norm can be considered a conditional case in which n's head is triggered. In order to check whether the head of a given norm n is triggered in every complete extension of d, the set of n-compatible norms is first collected in a set C (cf. line 10). Subsequently, it is checked whether the disjunction of every body in C is entailed by d (if-condition in line 11). If this is the case, the norm n is added to the set of triggered norms N'. This is iteratively conducted for every norm on N; as soon as the for-loop (cf. lines 9–14) has terminated, the set N' contains every norm that is triggered by d in the basic setting. Finally, similar to $\mathsf{IO}_1^\vdash$, is it checked whether the prospective output x is entailed by the heads of N' (cf. line 15). If this is not the case, No is returned prematurely. Conversely, if this condition holds for every clause d, $\mathsf{IO}_2^\vdash$ ultimately returns Yes (cf. line 20).

The following results establish adequateness for IN-OUT$_2$:

Theorem 3.3 (Partial correctness of $IO_2^\vdash$) $IO_2^\vdash$ *is sound and complete for* IN-OUT$_2$*; in particular, $x \in out_2(N, A)$ if and only if $(N, A, x) \in IO_2^\vdash$.*

Proof For the left-to-right direction, the contrapositive is shown. Assume that $(N, A, x) \notin IO_2^\vdash$. By definition, it follows that $\{h \mid (b, h) \in N'\} \nvdash x$, where N' is the set generated in lines 9-14 with respect to some $d \in \text{DNF}(A)$. It remains to be shown that there is no norm $n \in N$ that was incorrectly not included in N'. Let $n = (b', h') \in N \setminus N'$ be a norm such that $\{h \mid (b, h) \in N'\} \nvdash h'$. By construction it holds that $d \nvdash \bigvee\{b \mid (b, h) \in N \text{ and } h \vdash h'\}$, and hence there exists at least one complete extension $V \supseteq A \cup \{d\} \supseteq A$ such that $V \vdash \neg b$ for every (b, h) with $h \vdash h'$. It follows that $N(V) \nvdash h'$ for some complete $V \supseteq A$. By generalization, it holds that $N(V) \nvdash x$ and thus $x \notin out_2(N, A)$.

For the second part, assume $(N, A, x) \in IO_2^\vdash$. Then, by construction, for every clause $d \in \text{DNF}(A)$ there exists a set $N' \subseteq N$ such that $\{h \mid (b, h) \in N'\} \vdash x$. Let $d \in \text{DNF}(A)$ and let $(b, h) \in N'$ be some norm from the respective set N'. It follows that there exists some set of norms $\{(b'_1, h'_1), \ldots, (b'_m, h'_m)\} \subseteq N$ such that $h'_i \vdash h$, for each $1 \le i \le m$, and $d \vdash b'_1 \vee \ldots \vee b'_m$. By monotony, it also holds that $V \vdash b'_1 \vee \ldots \vee b'_m$ for every complete set $V \supseteq \{d\}$. Since V is complete, it follows that $V \vdash b'_i$ for some $1 \le i \le m$ and hence $N(V) \vdash h$. By generalization, it holds that $N(V) \vdash \bigwedge\{h \mid (b, h) \in N'\}$ and thus $N(V) \vdash x$. Since this is the case for every $d \in \text{DNF}(A)$, it follows that $N(V) \vdash x$ for every complete $V \supseteq A$ and hence $x \in out_2(N, A)$.
□

Theorem 3.4 (Total correctness of $IO_2^\vdash$) *Let $\vdash$ be a decidable derivation relation for L. $IO_2^\vdash$ terminates and is sound and complete for* IN-OUT$_2$.

Proof Theorem 3.3 already yields soundness and completeness. The termination argument is analogous to the the $IO_1^\vdash$ case if $\vdash$ is decidable. □

3.3 Reusable output

Listing 3 shows the decision procedure $IO_3^\vdash$ for deciding IN-OUT$_3$ with respect to a logical language L with underlying derivation relation $\vdash$.

In $IO_3^\vdash$ a more complex *proof search* is conducted in order to accommodate the interdependence between the operators $Cn(.)$ and $N(.)$ in the semantics of out_3. The basic approach is quite similar to the $IO_1^\vdash$ procedure for out_1: A set N' of norms is calculated that is triggered by the input A; then, it is checked whether the heads of these norms N' satisfy the output x. However, since out_3 validates the CT rule, it is also possible that x it not satisfied directly by the heads of N' but rather by some superset of N' which is triggered by $A \cup \{h \mid (b, h) \in N'\}$. Put differently, the output is reused to further strengthen the input and, in turn, to possibly trigger more outputs. As this can be done repeatedly, the $IO_3^\vdash$ procedure iteratively updates the input (called A') by the heads of the triggered norms, and subsequently collects all newly triggered norms (by A') in an updated set N'. If in some iteration N' satisfies the output x, the proof search succeeds; if, however, x is not satisfied and there are no new norms triggered, the process is terminated and a negative answer

```
1  Input:    N = {(b₁,h₁),(b₂,h₂),...} set of norms
2           A = {a₁,...,aₘ} set of formulae
3           x formula
4  Output:  Yes or No
5
6  A' := A
7  N' := {(b,h) ∈ N | A' ⊢ b}
8  N̄  := N \ N'
9
10 while not {h | (b,h) ∈ N'} ⊢ x do
11    A' := A' ∪ {h | (b,h) ∈ N'}
12    M  := {(b,h) ∈ N̄ | A' ⊢ b}
13    if M = ∅ then
14       return No
15    else
16       N' := N' ∪ M
17       N̄  := N̄ \ M
18    endif
19 endwhile
20 return Yes
```

Listing 3: Decision procedure $IO_3^\vdash$ for IN-OUT3.

is returned. The termination condition intuitively reflects that N' is a fixed point with respect to $Cn(.)$ and $N(.)$ and moreover does not satisfy x.

Let A^* be the least superset of A that is closed both under Cn and N. The following results establish adequateness for IN-OUT3, using so-called *bulk increments* [15]:

Theorem 3.5 (Partial correctness of $IO_3^\vdash$) $IO_3^\vdash$ *is sound and complete for* IN-OUT3*; in particular,* $x \in out_3(N, A)$ *if and only if* $(N, A, x) \in IO_3^\vdash$.

Proof Assume $x \in out_3(N, A)$ for some formula $x \in L$. Then, it holds that $x \in Cn(N(A^*))$. This implies that there exists a subset $N' = \{(b_1, h_1), \ldots, (b_m, h_m)\} \subseteq N$ of norms such that $A^* \vdash \bigwedge_{i=1}^{m} b_i$ and $\bigwedge_{i=1}^{m} h_i \vdash x$. By construction, in every iteration it holds that $A \subseteq A'$ and either $N(A') \vdash x$ in which case already Yes is returned, or A' will eventually reach a fixed point that is A^*. In the latter case $N(A') \vdash x$ iff $N(A^*) \vdash x$, which holds by assumption, and Yes is returned. In either case, $(N, A, x) \in IO_3^\vdash$.

For the second part, assume $(N, A, x) \in IO_3^\vdash$. Then, by construction there exists some A' such that $A \subseteq A' \subseteq A^*$ and $N(A') \vdash x$. Since N is monotone, it also holds that $N(A^*) \vdash x$ and hence $x \in out_3(N, A)$. □

Theorem 3.6 (Total correctness of $IO_3^\vdash$) *Let* $\vdash$ *be a decidable derivation relation for* L. $IO_3^\vdash$ *terminates and is sound and complete for* IN-OUT3.

Proof Theorem 3.5 already yields soundness and completeness. As $\vdash$ is decidable and every input is finite, every loop iteration itself terminates. There are only a finite number of loop iterations, as the set $\overline{N}$ is monotonously decreasing and $Cn(.)$ is monotone and idempotent. If no new norms can be triggered, the loop is terminated. □

```
1  Input:    N = {(b₁,h₁),(b₂,h₂),...} set of norms
2            A = {a₁,...,aₘ} set of formulae
3            x formula
4  Output:   Yes or No
5
6  N' := ∅
7  N̄  := N \ N'
8
9  while {h | (b,h) ∈ N'} ⊬ x do
10     D := DNF(a₁ ∧ ... ∧ aₘ ∧ ⋀{h | (b,h) ∈ N'})
11     M := ∅
12     for all d ∈ D do
13        M' := ∅
14        for all (b,h) ∈ N̄ do
15           C := {(b',h') ∈ N | h' ⊢ h}
16           if d ⊢ ⋁{b' | (b',h') ∈ C} then
17              M' := M' ∪ {(b,h)}
18           endif
19        endfor
20        M := M ∩ M'
21     endfor
22     if M = ∅ then
23        return No
24     else
25        N' := N' ∪ M
26        N̄  := N̄ \ M
27     endif
28  endwhile
29
30  return Yes
```

Listing 4: Decision procedure $\mathsf{IO}_4^\vdash$ for IN-OUT₄.

3.4 Basic reusable output

Listing 4 shows the decision procedure $\mathsf{IO}_4^\vdash$ for deciding IN-OUT₄ with respect to a logical language L with underlying derivation relation $\vdash$.

The procedure $\mathsf{IO}_4^\vdash$ combines the incremental proof search approach of $\mathsf{IO}_3^\vdash$ with the method for calculating the triggered norms in the basic output scenario as incorporated by $\mathsf{IO}_2^\vdash$: In the resulting procedure, the set of (basic) triggered norms is collected by N', initially empty. The set $\overline{N}$ will store the subset of norms from N not (yet) triggered. As long as the heads contained in N' do not entail the prospective output x (cf. while condition in line 9) the while loop will incrementally augment the set N' with norms triggered by the input A and the heads of N'. To that end, the disjunctive normal form D of this expression is calculated first (cf. line 10). In contrast to $\mathsf{IO}_3^\vdash$, the set M of newly triggered norms in each iteration cannot be calculated directly, but rather proceeds in a similar fashion to $\mathsf{IO}_2^\vdash$: After the inspection of each clause $d \in D$, the set M invariantly contains all triggered norms by all earlier clauses including the current clause d (cf. line 20). If no new norms can be triggered in the basic setting, i.e., after termination of the for-loop in lines 14–19, the procedure

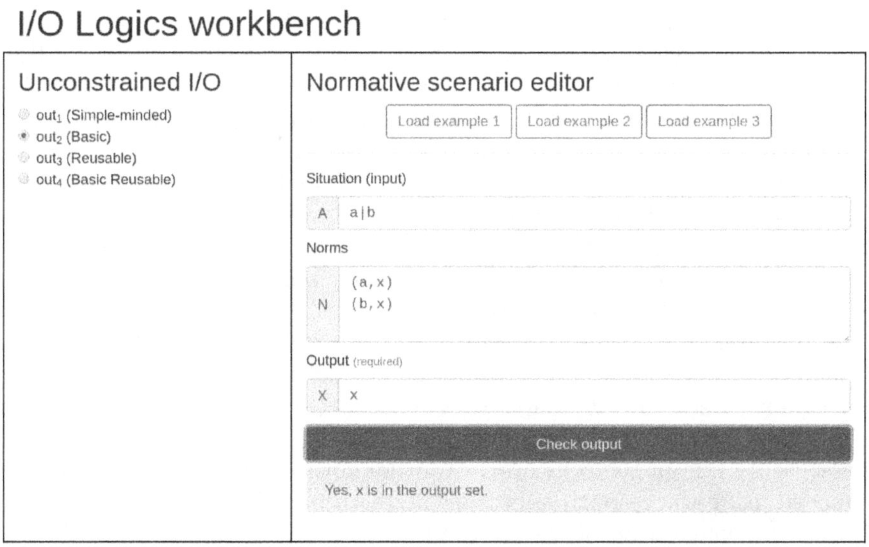

I/O Logics workbench

Unconstrained I/O	Normative scenario editor
○ out$_1$ (Simple-minded) ● out$_2$ (Basic) ○ out$_3$ (Reusable) ○ out$_4$ (Basic Reusable)	[Load example 1] [Load example 2] [Load example 3] Situation (input) A : a | b Norms N : (a, x) (b, x) Output (required) X : x [Check output] Yes, x is in the output set.

Figure 1. The I/O Logic Workbench: An open-source implementation of the $IO_i^\vdash$ procedures as a browser-based application.

returns No. Otherwise, the outer loop is continued with the augmented set of triggered norms N'. If the while-loop terminates, the output x is entailed by the heads in N' and $IO_4^\vdash$ returns Yes.

The following results establish adequateness for IN-OUT$_4$:

Theorem 3.7 (Partial correctness of $IO_4^\vdash$) $IO_4^\vdash$ *is sound and complete for* IN-OUT$_4$; *in particular,* $x \in out_4(N, A)$ *if and only if* $(N, A, x) \in IO_4^\vdash$.

Proof The argument is analogous to the proof of Theorem 3.5. However, in every incremental step, the set of triggered norms corresponds to the basic output, hence accommodating the principle of reasoning by cases analogously to Theorem 3.3. □

Theorem 3.8 (Total correctness of $IO_4^\vdash$) *Let* $\vdash$ *be a decidable derivation relation for* L. $IO_4^\vdash$ *terminates and is sound and complete for* IN-OUT$_4$.

Proof Theorem 3.7 already yields soundness and completeness. As $\vdash$ is decidable and every input is finite, every loop iteration itself terminates. There is only a finite number of while loop iterations as the set $\overline{N}$ is monotonously decreasing and $Cn(.)$ is monotone and idempotent. If no new norms can be triggered, the loop is terminated directly. □

4 Implementation

A prototype implementation of the decision procedures presented in this paper is freely available as an open-source software library at GitHub.[2]

---
[2] See github.com/I-O-Logic for the source code files and further information.

This above library constitutes the basis for the I/O Logics Workbench (IOLW) that provides graphical means for reasoning in I/O logics. IOLW is a browser-based application and is implemented in JavaScript. There is no need for any backend server infrastructure, as IOLW is implemented purely as a client-side application. Hence, it runs in every reasonably current browser, ready-to-use for conducting own experiments without any installation or set-up. An instance of IOLW is hosted at the author's personal web site. [3]

The user interface of IOLW is presented in Fig. 1. In the left menu panel, a user can choose which *out* operation should be used for the reasoning process. On the right side, the input A, the set of norms N and a prospective output x can be entered. The input language is an ASCII representation of propositional logic, where |, & and ~ denote disjunction, conjunction and negation, respectively. The input A is a comma separated list of formulas, whereas the set of norms N is, as usual, represented as a set of pairs. Every norm is entered as a separate line in the text area. Additionally, some example scenarios can be loaded using the respective buttons at the top.

The implementation of the decision procedure library and the IOLW will be extended with further I/O operations and input logics, cf. further work in §5 below for more details.

5 Conclusion

In this paper four decision procedures are presented, one of each out_i operation, $i \in \{1, 2, 3, 4\}$, that abstract from the underlying classical logical language L. These procedures are designed to decide whether a given formula $x \in L$ is in the output $out_i(N, A)$, given a set of norms N and an input A. They are shown to be correct (sound and complete) and to be decidable if the derivation relation $\vdash$ of the underlying logic L is decidable.

Instead of deciding for every prospective output $x \in L$ individually, the ideas underlying the decision procedures can also be used to calculate a finite base $\{x_1, \ldots, x_n\} \subset L$ of the output set $out_i(N, A)$ itself. Such a set can be constructed by modifying the presented procedures in such a way that all triggered norms are collected in a result set. Deciding IN-OUT$_i$ can then be reduced to checking entailment with respect to $\{x_1, \ldots, x_n\}$.

The output operators with so-called *throughput* [8], denoted out_i^+ for $i \in \{1, 2, 3, 4\}$, can easily be covered by the procedures presented in this paper: Intuitively, these operators behave similar to the respective operators without throughput with the exception that the input A is incorporated into the output set (in addition to the generated output). It is known that out_2^+ and out_4^+ coincide and that they collapse to classical consequence (cf. [9] for details). Moreover, the operators out_1^+ and out_3^+ can be expressed in terms of their non-throughput counterpart [9]. As a consequence, the decision procedures for all the out_i^+ operators can simply be reduced to the underlying routines for $\vdash$ (in case of out_2^+ and out_4^+) and to the routines for out_1 and out_3 (in case of out_1^+

---

[3] See alexandersteen.de/iol for details.

and out_3^+, respectively).

On the practical side the procedures are quite simple to implement, since already existing implementations of decision procedures for $\vdash$ can be used as a black box. A prototypical browser-based implementation for classical propositional logic as underlying logical language is presented. The implementation is open-source, publicly available at GitHub, and can be used for conducting (small) independent normative experiments. The browser-based graphical user interface is primarily intended to serve as a pedagogical tool, e.g., to be used in university teaching for a more interactive exposure to logical reasoning. However, the decision procedures themselves can easily be used as general components in larger software systems.

Future work. The presented procedures only address *unconstrained* input/output operations. While they are interesting operations for different applications, it is pointed out in the literature that they are not fully fit for usage in normative and deontic context [9]; e.g. due to lack of robustness to the usual deontic paradoxes. Further work thus focuses on generalizing the procedures to *constrained* input/output logics [9] that address these aspects.

It is planned to investigate whether the presented decision procedures may contribute to the practical employment of so-called *logical input/output nets* [6], *lions* for short, which combine different normative systems and output operators in a graph structure. Each node of a lion could be implemented by an independent instance of some appropriate $\mathsf{IO}_i^\vdash$ procedure.

Furthermore, a prototypical implementation for first-order logic as an underlying formalism is ongoing work. Also, empirical studies have to be conducted for assessing the practical effectiveness of the proposed approach for larger normative systems, e.g., in the context of reasoning with large legal knowledge bases [13].

Finally, the computational approach presented in this paper may be generalized to allow for the employment of non-classical logics as underlying logical formalism, e.g., for intuitionistic I/O logics [11] and further variants.

Acknowledgements

The author would like to thank the anonymous reviewers for their valuable feedback that led to significant improvements of the paper. The author acknowledges financial support from the Luxembourg National Research Fund (FNR) under grant CORE AuReLeE (C20/IS/14616644).

References

[1] Åqvist, L., *Deontic logic*, in: *Handbook of philosophical logic*, Springer, 2002 pp. 147–264.

[2] Benzmüller, C., A. Farjami, P. Meder and X. Parent, *I/O logic in HOL*, FLAP **6** (2019), pp. 715–732.

[3] Benzmüller, C., A. Farjami and X. Parent, *A dyadic deontic logic in HOL*, in: J. M. Broersen, C. Condoravdi, N. Shyam and G. Pigozzi, editors, *DEON* (2018), pp. 33–49.

[4] Benzmüller, C. and X. Parent, *I/O logic in HOL – First Steps*, CoRR **abs/1803.09681** (2018).
    URL http://arxiv.org/abs/1803.09681
[5] Boella, G. and L. W. N. van der Torre, *Regulative and constitutive norms in normative multiagent systems*, in: *KR* (2004), pp. 255–266.
[6] Boella, G. and L. W. N. van der Torre, *A logical architecture of a normative system*, in: *DEON*, Lecture Notes in Computer Science **4048** (2006), pp. 24–35.
[7] Gabbay, D. et al., editors, "Handbook of Deontic Logic and Normative Systems," College Publications, 2013.
[8] Makinson, D. and L. W. N. van der Torre, *Input/Output Logics*, J. Philosophical Logic **29** (2000), pp. 383–408.
[9] Makinson, D. and L. W. N. van der Torre, *Constraints for Input/Output Logics*, J. Philosophical Logic **30** (2001), pp. 155–185.
[10] Makinson, D. and L. W. N. van der Torre, *Permission from an input/output perspective*, J. Philosophical Logic **32** (2003), pp. 391–416.
[11] Parent, X., D. Gabbay and L. v. d. Torre, *Intuitionistic basis for input/output logic*, in: S. O. Hansson, editor, *David Makinson on Classical Methods for Non-Classical Problems*, Springer Netherlands, Dordrecht, 2014 pp. 263–286.
[12] Parent, X. and L. W. N. van der Torre, *I/O logics with a consistency check*, in: J. M. Broersen, C. Condoravdi, N. Shyam and G. Pigozzi, editors, *DEON* (2018), pp. 285–299.
[13] Robaldo, L. et al., *Formalizing GDPR provisions in reified I/O logic: the DAPRECO knowledge base*, Journal of Logic, Language and Information (2019).
[14] Robinson, J. A. and A. Voronkov, editors, "Handbook of Automated Reasoning (in 2 volumes)," Elsevier and MIT Press, 2001.
[15] Stolpe, A., "Norms and Norm-System Dynamics," Ph.D. thesis, University of Bergen, Norway (2008).
[16] Straßer, C., M. Beirlaen and F. V. D. Putte, *Adaptive logic characterizations of input/output logic*, Studia Logica **104** (2016), pp. 869–916.
[17] Sun, X. and L. Robaldo, *On the complexity of input/output logic*, Journal of Applied Logic **25** (2017), pp. 69 – 88.

Prioritized Defaults and Formal Argumentation

Christian Straßer

Institute Philosophy II, Ruhr-University Bochum
Bochum, Germany

Pere Pardo

Philosophy Dept., University of Milan
Milano, Italy

Abstract

Default logic and formal argumentation are paradigmatic methods in the study of nonmonotonic inference. Defeasible information often comes in different strengths stemming from different degrees of reliability in epistemic applications or from varying strengths of authorities issuing norms in deontic applications. In both paradigms methods have been developed to deal with prioritized knowledge bases. Questions of comparability of these methods therefore naturally arise. Argumentation theory has been developed with a strong emphasis on unification. It is therefore a desideratum to obtain natural representations of various approaches to (prioritized) default logic within frameworks of structured argumentation, such as ASPIC. Important steps in this direction have been presented in Liao et al. (2016, 2018). In this work we identify and address some problems in earlier translations, we broaden the focus from total to modular orderings of defaults, and we consider non-normal defaults.

Keywords: Argumentation, Defaults, Priorities, Argument Strength, ASPIC$^+$, ASP, Hypothetical reasoning,

---

1 Introduction

In [11] Dung proposed formal argumentation as an abstract unifying framework for nonmononotonic inference types. He demonstrated how default logic and other formal methods can be embedded in it. By now, several frameworks of structured argumentation have been proposed such as ASPIC$^+$ [22], ABA [7] and sequent-based argumentation [2] that share the same ambition but with a stronger emphasis on providing a logical form to arguments and attacks. Their unifying nature has been demonstrated by embedding nonmonotonic logics and other logical systems in them (see [14] for a recent survey).

Default logic has been considered, e.g. in assumption-based argumentation [7]. More recently the focus turned on prioritized forms of default logic, in particular we mention [19,18] and [26]. In this paper we continue these investigations. The advantages of such characterizations are manifold (see below for applications in deontic logic). First, they substantiate the unifying status of argumentation theory as an overarching framework for nonmonotonic reasoning; second, they allow for a dialectical perspective on default reasoning; third, the highly parametrizable argumentative characterizations may give rise to intuitive variants of existing approaches; fourth, such characterizations may inspire new variants for measuring argument strength in structured argumentation (see [3] for a survey) and highlight shortcomings; and fifth, (variants of) the translations may give rise to argumentative characterizations of other nonmonotonic logics. For an example of the latter, in [18] a new *disjoint* variant of weakest link lifting has been proposed to characterize greedy default reasoning.

Such considerations are of particular interest in the context of deontic logic. Clearly, depending on the source of a norm or its role in a normative framework (such as an ethical or legal system), norms come in different strengths. Both default logic [16] and argumentation theory [6,25] allow to aggregate deontic reasons in the form of arguments, albeit argumentation theory allows for a more direct model of their interplay, i.e. the way they support and defeat each other. Methods of dealing with prioritzed defaults have been shown to be based on different intuitions that give rise to different results when applied to deontic scenarios such as the order puzzle [16,18] (Ex. 6.5 below).

In this paper we study whether formal argumentation is able to make these intuitions formally precise, i.e., *whether* and *how* they can be explicated as an interplay of arguments of different strengths. This gives rise to a new perspective on and understanding of these intuitions and so may help in demarcating their use for different applications of defeasible reasoning. In [23] logical studies of the interplay of reasons have been motivated to investigate the way reasons of various strengths defeat and accrue. Formal argumentation offers a very natural formal framework for representing this type of dynamics. Moreover, answering the *how* question leads to conceptual insights of interest to argumentation theorists. For instance, in our study, we delineate a role for hypothetical arguments when reasoning with prioritized default information (see e.g., Section 6).

In this paper we identify and solve some problems in previous characterizations of default logics and generalize some of them from total to modular orders. After a brief presentation of default logics (Section 2) and elements of structured argumentation (Section 3), we sketch how these elements induce a translation from defaults to arguments, and define what does it mean that an argumentation semantics characterizes a default logic (for an order type) under a given translation. Then we proceed with characterizations of the default logics: Greedy (Section 4), Hansen (Section 5) and Brewka-Eiter (Section 6). After this, we move to non-normal default theories as proposed by Reiter [24] and Lukaszewicz [20] (Section 7) and offer characterizations for them as well (Section 8). This way we also obtain an argumentative characterization of answer set programming. The paper concludes with Section 9.

2 Prioritized Normal Defaults

In this section we define three central approaches to reasoning with prioritized default information: the greedy, Hansen [13] and Brewka-Eiter [9,10] approaches.[1] Following [19,18], we work with a simplified language $\text{Lit}_\top = \text{Lit} \cup \{\top\}$ consisting of a set Lit of literals $\ell_i \in \{p_i, \neg p_i\}$ and a true proposition $\top$. We also define negation for literals $-\ell$ by: $-p = \neg p$ and $-\neg p = p$.

Definition 2.1 A *default theory* is a triple $\mathsf{N} = (\mathcal{F}, \mathcal{N}, \preceq)$ consisting of:

- a consistent[2] set of *facts* $\mathcal{F} \subseteq \text{Lit}_\top$ with $\top \in \mathcal{F}$,
- a set of defaults $\mathcal{N} \subseteq \text{Lit}_\top \times \text{Lit}$, written $\ell \Rightarrow \ell'$, with $\ell' \neq -\ell$; and
- an order $\preceq \subseteq \mathcal{N} \times \mathcal{N}$.

We define $\text{head}(\ell \Rightarrow \ell') = \ell'$, $\text{body}(\ell \Rightarrow \ell') = \ell$ and $r \prec r'$ iff $r \preceq r'$ and $r' \not\preceq r$.

Throughout the paper, the reader may interpret a default theory in terms of a *normative system*, consisting of prioritized conditional norms $(\mathcal{N}, \preceq)$ and factual information $\mathcal{F}$ in view of which some norms are triggered and give rise to obligations. We also assume that $\mathsf{N} = (\mathcal{F}, \mathcal{N}, \preceq)$ is a default theory for which $\preceq$ is a partial, total, modular or flat order on a finite set $\mathcal{N}$. A reflexive and transitive relation $\preceq$ on $\mathcal{N}$ is called a *modular order* [17] if it admits a ranking: an order-preserving function $f : \mathcal{N} \to \mathbb{N}$ satisfying $f(r) \leq f(r')$ iff $r \preceq r'$. A *flat order* is a modular order with $\preceq = \mathcal{N} \times \mathcal{N}$.[3] For total or modular orders, moreover, $\preceq$ will be defined from an assignment of strengths to defaults, denoted $\ell \Rightarrow^k \ell'$, with smaller numbers meaning lower strength. The same convention will apply later on to defeasible rules, arguments and non-normal defaults. Finally, for any function f applied to a set X, we denote the resulting set of values by $f[X] = \{f(x) : x \in X\}$.

Definition 2.2 Let $\mathcal{N}' \subseteq \mathcal{N}$ be a set of defaults. $\text{out}_\mathsf{N}(\mathcal{N}')$ is the $\subseteq$-smallest set X satisfying: $\ell \in \mathcal{F} \cup X$ and $\ell \Rightarrow \ell' \in \mathcal{N}'$ implies $\ell' \in X$.[4] $\text{Trig}_\mathsf{N}(\mathcal{N}')$ is the set of rules in $\mathcal{N}$ whose bodies are contained in $\text{out}_\mathsf{N}(\mathcal{N}') \cup \mathcal{F}$. $\text{Con}_\mathsf{N}(\mathcal{N}')$ is the set of rules in $\mathcal{N}$ whose heads are consistent with $\text{out}_\mathsf{N}(\mathcal{N}')$. And $\text{TrigCon}_\mathsf{N}(\mathcal{N}') = \text{Trig}_\mathsf{N}(\mathcal{N}') \cap \text{Con}_\mathsf{N}(\mathcal{N}')$.

We now discuss three central approaches for reasoning with prioritized default theories. Typically one proceeds by iteratively building a *scenario* (a set of defaults) so that its *extension* (the set of propositional commitments)

---

[1] In [19] the Hansen approach is called *Optimization* and Brewka-Eiter is called *Reduction*.
[2] A set $X \subseteq \text{Lit}_\top$ is *consistent*, denoted $X \not\vdash \bot$, iff $\neg\top \notin X$ and for all atoms p, $\{p, \neg p\} \not\subseteq X$.
[3] A *partial order* $\preceq$ on $\mathcal{N}$ is a reflexive, transitive and antisymmetric relation; $\preceq$ is *total* in case $r \preceq r'$ or $r' \preceq r$ for each $r, r' \in \mathcal{N}$. Note that partial (unlike modular) orders allow for incomparable defaults while modular (unlike partial) orders allow for defaults of the same strength. Although a flat order $\preceq = \mathcal{N} \times \mathcal{N}$ is obviously a total relation, in this paper *total* only refers to the antisymmetric case, that is a total (partial) order.
[4] This is the definition for deontic applications. While in epistemic applications facts are part of the output, in the deontic case literals are only part of the output if they are detached from defaults, read as conditional norms.

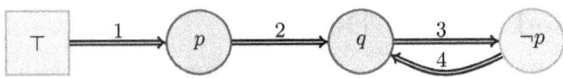

Fig. 1. The greedy extension $\{p, q\}$ for Ex. 2.4 is highlighted.

remains consistent; that is, the set of literals detachable from the given facts and the defaults collected. The first approach is greedy reasoning. A greedy reasoner will in each iteration select and detach a maximally strong default that is triggered by her propositional commitments.

Definition 2.3 A *greedy scenario* for N is given by $\mathbf{gr} = \bigcup_{i \geq 0} \mathbf{gr}_i$ where

$$\mathbf{gr}_0 = \emptyset \quad \text{and} \quad \mathbf{gr}_{i+1} = \begin{cases} \mathbf{gr}_i & \text{TrigCon}_\mathsf{N}(\mathbf{gr}_i) \setminus \mathbf{gr}_i = \emptyset \\ \mathbf{gr}_i \cup \{r\} & r \in \max_{\preceq}(\text{TrigCon}_\mathsf{N}(\mathbf{gr}_i) \setminus \mathbf{gr}_i). \end{cases}$$[5]

X is a *greedy extension* of N, denoted $X \in \mathbf{greedy}(\mathsf{N})$, iff there is a greedy scenario $\mathbf{gr}$ for which $X = \text{out}_\mathsf{N}(\mathbf{gr})$ ($= \text{head}[\mathbf{gr}]$).

Example 2.4 Consider the default theory N based on fact $\mathcal{F} = \{\top\}$ and defaults $\mathcal{N} = \{\top \Rightarrow^1 p, \ p \Rightarrow^2 q, \ q \Rightarrow^3 \neg p, \ \neg p \Rightarrow^4 q\}$. Fig. 1 depicts the only greedy scenario $\mathbf{gr} = \{\top \Rightarrow^1 p, \ p \Rightarrow^2 q\}$, with extension $X = \{p, q\}$.

The Hansen approach, designed for deontic reasoning, always selects a strongest default preserving output consistency —be it triggered or not.

Definition 2.5 A *Hansen scenario* $\mathbf{ha}$ for N is obtained by the following construction: $\mathbf{ha} = \bigcup_{i \geq 0} \mathbf{ha}_i$ where $\mathbf{ha}_0 = \emptyset$ and

$$\mathbf{ha}_{i+1} = \begin{cases} \mathbf{ha}_i & \{r \in \mathcal{N} \setminus \mathbf{ha}_i \mid \text{out}_\mathsf{N}(\mathbf{ha}_i \cup \{r\}) \not\vdash \bot\} = \emptyset \\ \mathbf{ha}_i \cup \{r\} & r \in \max_{\preceq}(\{r \in \mathcal{N} \setminus \mathbf{ha}_i \mid \text{out}_\mathsf{N}(\mathbf{ha}_i \cup \{r\}) \not\vdash \bot\}) \end{cases}$$

X is a *Hansen extension* of N, denoted $X \in \mathbf{hansen}(\mathsf{N})$, iff there is a Hansen scenario $\mathbf{ha}$ for which $X = \text{out}_\mathsf{N}(\mathbf{ha})$ ($\subseteq \text{head}[\mathbf{ha}]$).

Example 2.6 (Ex. 2.4, cont'd.) The only Hansen scenario is $\mathbf{ha} = \{\neg p \Rightarrow^4 q, \ q \Rightarrow^3 \neg p, \ p \Rightarrow^2 q\}$, with extension $X = \emptyset$. While the propositional commitments did not increase, the inferential commitments (scenario) did.

In [19,18] Brewka-Eiter (in short: BE) extensions X have been characterized as fixpoints of greedy reasoning, in the sense of being greedy extensions for the defaults triggered by X itself.

Definition 2.7 X is a *BE extension* of N, denoted $X \in \mathbf{be}(\mathsf{N})$, iff $X \in \mathbf{greedy}(\mathsf{N}^X)$, where the default theory $\mathsf{N}^X = (\mathcal{F}, \mathcal{N}^X, \preceq^X)$, is defined by: $\mathcal{N}^X = \{\top \Rightarrow \ell \mid \ell' \Rightarrow \ell \in \mathcal{N}, \ell' \in X \cup \mathcal{F}\}$ and $\top \Rightarrow \ell \preceq^X \top \Rightarrow \ell'$ iff $\max_{\preceq}\{\ell_0 \Rightarrow \ell \in \mathcal{N}\}_{\ell_0 \in X} \preceq \max_{\preceq}\{\ell_0 \Rightarrow \ell' \in \mathcal{N}\}_{\ell_0 \in X}$.

---

[5] In case $\preceq$ is a partial order, $\max_{\preceq} X$ must be read as the set of $\preceq$-*maximal* elements of X. In a modular order $\preceq$, all $\preceq$-maximal elements are $\leq$-maximum in the ranking function, in which case we abusively let $\max_{\preceq} X$ denote an arbitrary element of this set (e.g. in Def. 2.7).

Example 2.8 (Ex. 2.4 cont'd.) The unique BE extension is $X = \{\neg p, q\}$. Note that $\mathcal{N}^X = \{\top \Rightarrow^1 p, \top \Rightarrow^3 \neg p, \top \Rightarrow^4 q\}$ and $\{\neg p, q\}$ is the only greedy extension of N^X. Also observe that $X' = \{p, q\}$ is *not* a BE extension, since the greedy extension for $\mathcal{N}^{X'} = \{\top \Rightarrow^1 p, \top \Rightarrow^2 q, \top \Rightarrow^3 \neg p\}$ is X, not X'.

We note that the original characterization [10] is more stringent in that it demands that X is both a greedy extension of N and of N^X.

Definition 2.9 X is a *BEo extension* (original BE extension) of N, denoted $X \in \mathtt{beo}(\mathsf{N})$, iff $X \in \mathtt{greedy}(\mathsf{N}) \cap \mathtt{greedy}(\mathsf{N}^X)$, where N^X is defined as in Def. 2.7. [6]

Example 2.10 (Ex. 2.4 cont'd.) N has no BEo extension since its unique BE extension $\{\neg p, q\}$ is not a greedy extension of N.

As shown in [10], the BEo extensions have an alternative characterization. [7]

Proposition 2.11 *Let* $\mathsf{N} = \langle \mathcal{F}, \mathcal{N}, \preceq \rangle$ *be a default theory based on a total order* $\preceq$. *Then,* $X \in \mathtt{BEo}(\mathsf{N})$ *iff (i)* $X \in \mathtt{greedy}(\mathsf{N})$ *and (ii) for every* $\ell_1 \Rightarrow^k \ell' \in \mathcal{N}$ *for which* $\ell_1 \in X$ *and* $\ell' \notin X$ *there is a* $\ell_2 \Rightarrow^{k'} -\ell' \in \mathcal{N}$ *for which* $\ell_2, -\ell' \in X$ *and* $k' > k$.

3 Argumentative Characterizations

In the following sections, we translate prioritized default logics into structured argumentation. Our characterizations are phrased in a simple ASPIC$^+$ setting [18]. We denote defeasible rules by double arrows $\Rightarrow$ and strict rules by single arrows $\rightarrow$. For each default logic approach and order type, we translate a default theory $\mathsf{N} = (\mathcal{F}, \mathcal{N}, \preceq)$ into $\mathsf{N}_{\mathtt{arg}} = (\mathtt{Arg}, \mathtt{defeat})$, an abstract argumentation framework [11] (AF, in short). Arg is the set of arguments obtained as in Fig. 2 by specifying:

(i) a *formal language* $\mathcal{L}$ based on $\mathsf{Lit}_\top$,

(ii) a map $\mathcal{N} \mapsto \mathcal{D}$ from defaults to sets of *defeasible rules* $\mathcal{D} \subseteq \wp(\mathcal{L}) \times \mathcal{L}$,

(iii) a set of *strict rules* $\mathcal{S} \subseteq \wp(\mathcal{L}) \times \mathcal{L}$.

defeat is a relation over Arg obtained from defining (iv)–(vii):

(iv) a (partial) *contrariness function* $\overline{\cdot} : \mathcal{L} \rightarrow \wp(\mathcal{L})$,

(v) a *lifting* of the order $\preceq \mapsto \preceq_{\mathtt{arg}}$ from defeasible rules to arguments, [8]

(vi) an *attack* relation over Arg: $(a, b) \in \mathtt{attack}$ iff there is a $b' \in \mathtt{Sub}(b)$ with a last rule of the form $r = \mathtt{head}(b'') \Rightarrow \phi$ and such that

---

[6] We thank an anonymous reviewer for pointing out this difference between [19,18] and the original Brewka-Eiter definiton found in [10]. We also note that the original definition is phrased in a richer language.

[7] This is a simplified version of Prop. 2 in [10], adapted to the present setting.

[8] All of the proposed translations map each default $\ell \Rightarrow^k \ell'$ into an identical defeasible rule $\ell \Rightarrow^k \ell'$ and (possibly) additional defeasible rules. The ordering $\preceq$ on $\mathcal{N}$ is uniformly expanded to all defeasible rules in $\mathcal{D}$ and subsequently lifted to the level of arguments.

$a =$	$\ell \Rightarrow \phi$	$a_1 \ldots a_m \Rightarrow \phi$	$a_1 \ldots a_m \to \phi$
condition	$\ell \in \mathcal{F}$	$a_1, \ldots, a_m \in \mathsf{Arg}$	$a_1, \ldots, a_m \in \mathsf{Arg}$
	$\ell \Rightarrow \phi \in \mathcal{D}$	$r = \phi_1 \ldots \phi_m \Rightarrow \phi \in \mathcal{D}$	$\phi_1 \ldots \phi_m \to \phi \in \mathcal{S}$
$\mathsf{head}(a) =$	ϕ	ϕ	ϕ
$\mathcal{D}(a) =$	$\{\ell \Rightarrow \phi\}$	$\{r\} \cup \bigcup_{i=1}^{m} \mathcal{D}(a_i)$	$\bigcup_{i=1}^{m} \mathcal{D}(a_i)$
$\mathsf{Sub}(a) =$	$\{a\}$	$\{a\} \cup \bigcup_{i=1}^{m} \mathsf{Sub}(a_i)$	$\to idem$
$\mathsf{PSub}(a) =$	$\emptyset$	$\bigcup_{i=1}^{m} \mathsf{Sub}(a_i)$	$\to idem$

Fig. 2. From facts $\mathcal{F}$ and rules $\mathcal{D} \cup \mathcal{S}$ one inductively defines the set Arg of arguments (top), and the functions (bottom) mapping each argument to: its conclusion head : Arg $\to$ Lit, its set of defeasible rules $\mathcal{D}$: Arg $\to \wp(\mathcal{D})$, and its (proper) subarguments (PSub) Sub : Arg $\to \wp(\mathsf{Arg})$. ϕ_i denotes the conclusion $\mathsf{head}(a_i) = \phi_i$.

rebut: $\mathsf{head}(a) \in \overline{\phi}$ or *undercut* (when needed): $\mathsf{head}(a) \in \overline{r}$,[9]

and in either case we say *a attacks b in b'*. Items (iv)–(vi) determine *defeat*:

$(a, b) \in$ defeat iff a attacks b in some b' and, in case of a rebut, $a \not\prec b'$.

Finally, the argumentation semantics (stable, in our case) also depends on:

(vii) a notion of *conflict-freeness*: a set $\mathcal{X} \subseteq \mathsf{Arg}$ is *conflict-free* [21] iff for no $a, b \in \mathcal{X}$, $(a, b) \in$ defeat.

(Only in Sections 6.2 and 8, item (vii) will consist of an attack-based notion of conflict-freeness, where $\mathcal{X}$ is *conflict-free* iff for no $a, b \in \mathcal{X}$, $(a, b) \in$ attack). A simple illustration of a translation defined by these components can be found at the beginning of Section 4.

Definition 3.1 Where (Arg, defeat) is an AF, a set $\mathcal{X} \subseteq \mathsf{Arg}$ is *stable* iff it is conflict-free and for each $a \in \mathsf{Arg} \setminus \mathcal{X}$ there is a $b \in \mathcal{X}$ such that $(b, a) \in$ defeat.

The set of stable sets of $\mathsf{N}_{\mathsf{arg}}$, the *stable extensions*, is written $\mathsf{stb}(\mathsf{N}_{\mathsf{arg}})$. Where $\mathcal{D}' \subseteq \mathcal{D}$, $\mathsf{Arg}(\mathcal{D}')$ denotes the set of $a \in \mathsf{Arg}$ for which $\mathcal{D}(a) \subseteq \mathcal{D}'$.

Definition 3.2 Let $\mathsf{head}_{\mathsf{lit}}[\mathcal{X}] =_{\mathsf{df}} \mathsf{head}[\mathcal{X}] \cap \mathsf{Lit}$.[10] A translation $\mathsf{N} \mapsto \mathsf{N}_{\mathsf{arg}}$ *characterizes* $\mathsf{method} \in \{\mathsf{greedy}, \mathsf{hansen}, \mathsf{be}, \ldots\}$ for a class $\mathcal{C}$ of orders (total, modular, ...) if for each default theory $\mathsf{N} = (\mathcal{F}, \mathcal{N}, \preceq)$ with $\preceq \in \mathcal{C}$:

(1) $X \in \mathsf{method}(\mathsf{N})$ implies $X = \mathsf{head}_{\mathsf{lit}}[\mathcal{X}]$ for some $\mathcal{X} \in \mathsf{stb}(\mathsf{N}_{\mathsf{arg}})$, and
(2) $\mathcal{X} \in \mathsf{stb}(\mathsf{N}_{\mathsf{arg}})$ implies $\mathsf{head}_{\mathsf{lit}}[\mathcal{X}] \in \mathsf{method}(\mathsf{N})$.

While the *weakest link* lifting is the predominant notion of argument strength, in this paper we will often work with disjoint weakest link [18].

Definition 3.3 *Weakest link* $\succ_{\mathsf{w}} \subseteq \mathsf{Arg} \times \mathsf{Arg}$ is defined by: $a \succ_{\mathsf{w}} b$ iff there is an $r \in \mathcal{D}(b)$ such that for all $r' \in \mathcal{D}(a)$, $r' \succ r$.[11]

---

[9] Here we suppose that the names of rules are part of $\mathcal{L}$. Note that in line with [22] for undercuts the strengths of arguments don't matter.
[10] For some of the translations presented and their stable extensions $\mathcal{X}$, $\mathsf{head}_{\mathsf{lit}}[\mathcal{X}] \neq \mathsf{head}[\mathcal{X}]$.
[11] Since we focus on modular and total orders we do not consider more fine-grained variants such as democratic or elitist weakest link [22].

Definition 3.4 *Disjoint weakest link* $\succ\ \subseteq\ \mathsf{Arg} \times \mathsf{Arg}$ *is defined by:* $a \succ b$ *iff there is a* $r \in \mathcal{D}(b) \setminus \mathcal{D}(a)$ *such that for all* $r' \in \mathcal{D}(a) \setminus \mathcal{D}(b)$, $r' \succ r$.

Given that our focus is on flat, total and modular orders, we can define $a \succeq b$ iff $b \not\succ a$. Our characterizations make use of the following results about disjoint weakest link $\succ$.

Fact 3.5 $a \succ b$ *iff* $\mathcal{D}(b) \setminus \mathcal{D}(a) \neq \emptyset$ *and (i)* $\mathcal{D}(a) \setminus \mathcal{D}(b) = \emptyset$ *or (ii) there is a* $r' \in \min_{\preceq}(\mathcal{D}(b) \setminus \mathcal{D}(a))$ *such that for all* $r \in \min_{\preceq}(\mathcal{D}(a) \setminus \mathcal{D}(b))$, $r \succ r'$.

Fact 3.6 *(i)* $a \succeq a$ *and* $a \not\succ a$ *and (ii) if* $a \succ b$ *and* $\mathcal{D}(c) \subseteq \mathcal{D}(a)$, *then* $c \succ b$.

Lemma 3.7 *Where* $\preceq\ \subseteq \mathcal{N} \times \mathcal{N}$ *is modular, if* $a_1 \succ a_2 \succ a_3$ *then* $a_1 \succ a_3$.

Proof Sketch. Let $\mathcal{D}_i = \mathcal{D}(a_i)$. By Fact 3.5, the claim $a_i \succ a_j$ splits into cases (α_{ij}) $\mathcal{D}_i \subseteq \mathcal{D}_j$ and $\mathcal{D}_j \setminus \mathcal{D}_i \neq \emptyset$ and (β_{ij}) $r_{ji} \prec \mathcal{D}_i \setminus \mathcal{D}_j \neq \emptyset$ for some $r_{ji} \in \mathcal{D}_j \setminus \mathcal{D}_i$. Thus, $a_1 \succ a_2$ and $a_2 \succ a_3$ give four cases to prove $a_1 \succ a_3$, but only the proof from (β_{12}, β_{23}) is not immediate. For this, the subcase $r_{21} \notin \mathcal{D}_3 \setminus \mathcal{D}_1$ gives $r_{32} \prec r_{21}$ which can be shown to make r_{32} a witness for (β_{13}), i.e. $r_{32} = r_{31}$; for the subcase $r_{21} \in \mathcal{D}_3 \setminus \mathcal{D}_1$ the witness r_{31} for (β_{13}) is either a $\preceq$-minimum of $\{r_{21}, r_{32}\}$ (if $r_{32} \notin \mathcal{D}_1$) or r_{21} (if $r_{32} \in \mathcal{D}_1$, as this implies $r_{21} \prec r_{32}$). In all cases (β_{13}) follows, which implies $a_1 \succ a_3$. □

Lemma 3.8 *Where* $\preceq\ \subseteq \mathcal{N} \times \mathcal{N}$ *is total, (i)* $(a \succeq b$ *and* $b \succeq a)$ *iff* $\mathcal{D}(a) = \mathcal{D}(b)$ *iff* $(a \not\succ b$ *and* $b \not\succ a)$, *and (ii)* $\preceq\ \subseteq \mathsf{Arg} \times \mathsf{Arg}$ *is also total.*

Proof Sketch. Similar to the proof of Lemma 3.7. □

Fact 3.9 *Let* $a_0 \in \mathsf{Sub}(a)$. *(i) If* $a \succ b$, $a_0 \succ b$. *(ii) If* $b \succ a_0$, $b \succ a$.

4 Greedy extensions

Before generalizing the greedy approach to non-normal default theories in Sec. 8 we report on previous results and demonstrate why it is difficult to go beyond total or flat orders. For the greedy approach the translation $\mathsf{N}_{\mathsf{arg}}$ of a default theory $\mathsf{N} = (\mathcal{F}, \mathcal{N}, \preceq)$ with a total order $\preceq$ is very simple [19,18,26]:

(i) $\mathcal{L} = \mathsf{Lit}_\top$ (ii) $\mathcal{D} = \mathcal{N}$ (iii) $\mathcal{S} = \emptyset$
(iv) $\overline{\ell} = \{-\ell\}$ (v) disjoint weakest link   (vi) only rebut.

Proposition 4.1 *The translation above characterizes* $\mathsf{greedy}(\mathsf{N})$ *for default theories* $\mathsf{N} = (\mathcal{F}, \mathcal{N}, \preceq)$ *where* $\preceq$ *is a total or flat order.*

Example 4.2 Prop. 4.1 does not generalize to modular or partial orders. For the modular case, Fig. 3 (left) depicts the ranked defaults in $\mathsf{N} = \langle \{\top\}, \{\top \Rightarrow^1 p, \top \Rightarrow^1 q, p \Rightarrow^2 \neg q, q \Rightarrow^2 \neg p\}, \preceq \rangle$. Note that $\{\top \Rightarrow^1 p, \top \Rightarrow^1 q\}$ is not a greedy scenario, but it is a stable extension of $\mathsf{N}_{\mathsf{arg}}$ in view of defeats.[12]

Fig. 3 (right) shows the partial order $\preceq\ = \{(\top \Rightarrow p, \top \Rightarrow \neg q), (\top \Rightarrow q, \top \Rightarrow \neg p)\}$. This renders $\top \Rightarrow p, \top \Rightarrow \neg p$ incomparable, as well as $\top \Rightarrow q, \top \Rightarrow \neg q$. Now, a greedy reasoner will first choose between $\top \Rightarrow \neg p$ and $\top \Rightarrow \neg q$. However, in the argumentative setting $\{\top \Rightarrow p, \top \Rightarrow q\}$ is a stable extension.

---

[12] We thank Leon van der Torre for proposing this example.

Fig. 3. Greedy vs. stable semantics in non-flat or non-total orders. (Left) A modular order for which no greedy scenario matches the stable extension (in green); red arrows represent defeat under disjoint weakest link. (Right) A similar case with a partial order; ordered pairs are represented by dotted arrows.

Despite such direct counterexamples, an indirect characterization exists via linear extensions of modular orders $\preceq$, i.e. total orders $\sqsubseteq$ satisfying $\preceq \subseteq \sqsubseteq$, denoted $\sqsubseteq \in \text{lin}(\preceq)$.[13]

Proposition 4.3 *Let* $\mathsf{N}^{\preceq} = (\mathcal{F}, \mathcal{N}, \preceq)$ *be based on a modular order* $\preceq$. *Then, X is a greedy extension of* $\mathsf{N}^{\preceq}$ *iff X is a greedy extension of* $\mathsf{N}^{\sqsubseteq} = (\mathcal{F}, \mathcal{N}, \sqsubseteq)$ *for some linear extension* $\sqsubseteq$ *of* $\preceq$. *That is,* $\text{greedy}(\mathsf{N}^{\preceq}) = \bigcup_{\sqsubseteq \in \text{lin}(\preceq)} \text{greedy}(\mathsf{N}^{\sqsubseteq})$.

Proof Sketch. ($\subseteq$) Let $X \in \text{greedy}(\mathsf{N}^{\preceq})$ be based on a scenario $\mathbf{gr} = \bigcup_{k=1}^{n} \mathbf{gr}_k$. We find a $\sqsubseteq \in \text{lin}(\preceq)$ such that $X \in \text{greedy}(\mathsf{N}^{\sqsubseteq})$. Enumerate $\mathcal{N}$ by $\langle r^i \rangle_{i \leq |\mathcal{N}|}$ and define $\sqsubseteq$ by: $r^i \sqsubset r^j$ iff (1) $r^i \prec r^j$ or (2) $\text{rank}(r^i) = \text{rank}(r^j)$ and (2a) $r^j \in \mathbf{gr}_k$ and $r^i \notin \mathbf{gr}_k$ for some $k \geq 1$, or else, (2b) $i < j$. An induction over the rounds $k = 1, \ldots, |\mathcal{N}|$ shows that $\mathbf{gr}$ is a greedy scenario for $\mathsf{N}^{\sqsubseteq}$. ($\supseteq$) Let now $\sqsubseteq \in \text{lin}(\preceq)$ and $X \in \text{greedy}(\mathsf{N}^{\sqsubseteq})$ be based on $\mathbf{gr} = \bigcup_{k=1}^{n} \mathbf{gr}_k$. An induction over k shows that $\mathbf{gr}$ is also a greedy scenario for $\mathsf{N}^{\preceq}$. □

So, in order to characterize a default theory N based on a modular order $\preceq$, we need to translate all linear extensions of $\prec$ and apply Prop. 4.1. The next example shows that this strategy does not work for partial orders.

Example 4.4 Consider again $\mathcal{N} = \{\top \Rightarrow q, \top \Rightarrow p, p \Rightarrow \neg q, q \Rightarrow \neg p\}$ (Fig. 3, left) but now with the partial order $\prec = \{(\top \Rightarrow p, p \Rightarrow \neg q), (\top \Rightarrow q, q \Rightarrow \neg p)\}$. Note first that $\mathbf{gr} = \{\top \Rightarrow p, \top \Rightarrow q\}$ is a greedy scenario. The set of linear extensions $<$ of $\prec$ splits into two classes: (1) $\top \Rightarrow q < \top \Rightarrow p$ and (2) $\top \Rightarrow p < \top \Rightarrow q$. For linear extensions satisfying (1), any greedy scenario will first choose $\top \Rightarrow p$ and then $p \Rightarrow \neg q$. All remaining cases satisfy (2), in which case any greedy scenario will first choose $\top \Rightarrow q$ and then $q \Rightarrow \neg p$.

5 Hansen extensions

We now present a translation for Hansen extensions that is adequate for modular orderings. But first we show that the translations offered in the literature so far are not adequate for this purpose.

---

[13] In view of the order-extension principle such an extension always exists.

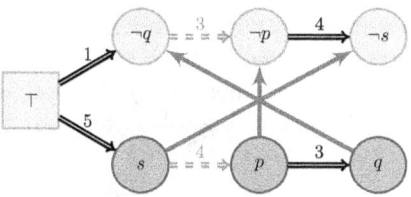 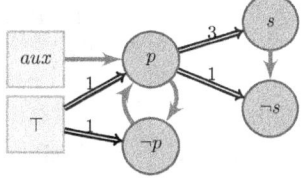

Fig. 4. Left: Counter-example for the Hansen translation in [18]; dashed arrows represent $\Rightarrow_r$ rules, and red arrows defeat. Right: Counter-example with a modular order for the Hansen translation in [19]; defeats on super-arguments are omitted.

5.1 Problems with previous characterizations

In [18] the proposed translation of Hansen for linear orders introduces a second type of defeasible rule, denoted by $\Rightarrow_r$. The $\Rightarrow_r$ rules read as conditional permissions (resp. possibilities, in the epistemic case), and so they are not part of the propositional commitments for obligations (resp. beliefs). Arguments can chain $\Rightarrow$ and $\Rightarrow_r$ rules all the same, but conclusions inherit the permissive reading of any $\Rightarrow_r$ rule occurring in them. Permission (p-)arguments and the previous obligation (o-)arguments attack each other, and again defeat is based on strength. The translation of $\mathsf{N} = (\mathcal{F}, \mathcal{N}, \preceq)$ in [18] defines $\mathcal{D}$ as the closure of $\mathcal{N}$ under *weak contraposition* and using the same strength:

(i) $\mathcal{L} = \mathsf{Lit}_\top$ (ii) $\mathcal{D} = \mathcal{N} \cup \{-\ell' \Rightarrow_r^k -\ell\}_{\ell \Rightarrow^k \ell' \in \mathcal{N}}$ (iii) $\mathcal{S} = \emptyset$
(iv) $\overline{\ell} = \{-\ell\}$ (v) disjoint weakest link (vi) oo, po, op rebuts.

Example 5.1 Let $\mathsf{N} = (\mathcal{F}, \mathcal{N}, \preceq)$ contain the sets $\mathcal{F} = \{\top\}$ and $\mathcal{N} = \{\top \Rightarrow^5 s, \neg p \Rightarrow^4 \neg s, p \Rightarrow^3 q, \top \Rightarrow^1 \neg q\}$. Fig. 4 (left) shows how weak contraposition gives inter alia $s \Rightarrow_r^4 p$ and so in [18] the p-argument $a : \top \Rightarrow^5 s \Rightarrow_r^4 p \Rightarrow^3 q$ defeats the argument $b : \top \Rightarrow^1 \neg q$. In this case, the unique stable extension does not conclude $\neg q$. In contrast, the unique Hansen scenario ha is built through $\mathsf{ha}_3 = \{\top \Rightarrow^5 s, \neg p \Rightarrow^4 \neg s, p \Rightarrow^3 q\}$, detaching only s, and then $\mathsf{ha} = \mathsf{ha}_4 = \mathsf{ha}_3 \cup \{\top \Rightarrow^1 \neg q\}$, so now we conclude $\neg q$.

In [19] another construction was proposed to tackle Hansen extensions based on total orders. It employs an auxiliary argument aux and a function $\mathsf{warg} : \mathsf{Arg} \to \wp(\mathsf{Arg})$ given by $\mathsf{warg}(a) = \uparrow b$ where $b \in \mathsf{sub}(a)$ is the subargument of a with the weakest defeasible top-rule, and $\uparrow b$ is the set of all super-arguments of b. The translation is then:

(i) $\mathcal{L} = \mathsf{Lit}_\top$, (ii) $\mathcal{D} = \mathcal{N}$, (iii) $\mathcal{S} = \emptyset$,
(iv) $\overline{\ell} = \{-\ell\}$ (v) weakest link (vi) rebut based on **defeat**,

where now **defeat** is extended into $\mathsf{Arg} \cup \{\mathsf{aux}\}$ as follows:

if a defeats b then $\begin{cases} a \text{ defeats each } c \in \mathsf{warg}(b) & \text{if } a \notin \mathsf{warg}(b) \\ \mathsf{aux} \text{ defeats each } c \in \mathsf{warg}(b) & \text{if } a \in \mathsf{warg}(b). \end{cases}$

Unfortunately this translation does not work for rankings.

Example 5.2 Consider the modularly ordered $\mathsf{N} = (\mathcal{F}, \mathcal{N}, \preceq)$ in Fig. 4 (right) with $\mathcal{F} = \{\top\}$ and $\mathcal{N} = \{\top \Rightarrow^1 \neg p, \top \Rightarrow^1 p, p \Rightarrow^3 s, p \Rightarrow^1 \neg s\}$. Note that $\mathsf{ha} = \{\top \Rightarrow^1 p,\ p \Rightarrow^3 s\}$ forms a Hansen extension. Moving to the argumentative setting of [19] let $a' = \top \Rightarrow^1 p$, $a = a' \Rightarrow^3 s$ and $b = a' \Rightarrow^1 \neg s$. Since $\mathsf{warg}(b) = \uparrow a' = \{a', a, b\}$ and a defeats b, the argument aux defeats a', a and b. Hence, the set $\{a, a'\}$ that would match ha is not a stable extension.

5.2 A new translation

For a default theory $\mathsf{N} = (\mathcal{F}, \mathcal{N}, \preceq)$ with a modular order $\preceq$, we define its translation into an AF $\mathsf{N}_{\mathsf{arg}}$. To this end, we express a commitment to a default $\ell \Rightarrow \ell'$ in the object language by $\ell \mapsto \ell'$ and encode weak contraposition ($\Rightarrow_r$ rules in [18]) with the help of strict rules. (This maneuver restricts the chaining of rules, which was problematic in Ex. 5.1). We also make explicit the above permissive reading with an operator $!\ell$ denoting $-\ell$ *is permissible* (or $-\ell$ *is possible*, in the epistemic case). Commitments to defaults $\ell \mapsto \ell'$ can be attacked by stating ℓ in conjunction with $!\ell'$ or $-\ell'$. Altogether our translation looks as follows:

(i) $\mathcal{L} = \ell_\top \mid \ell_\top \mapsto \ell \mid !\ell \mid !\ell \vee !\ell' \mid \ell_\top \wedge \ell'_\top \mid \ell_\top \wedge !\ell'$
where $\ell_\top, \ell'_\top \in \mathsf{Lit}_\top$ and $\ell, \ell' \in \mathsf{Lit}$

(ii) the set $\mathcal{D} = \tau[\mathcal{N}]$ is defined by $\tau(\ell \Rightarrow^k \ell') = \{(\ell \Rightarrow^k \ell'), (\top \Rightarrow^k (\ell \mapsto \ell'))\}$

(iii) strict rules in $\mathcal{S}$ consist of all instances of the form (for literals $\neq \top$):[14]

$\quad$ R!1 $(\ell_1 \mapsto \ell),\ -\ell \to !\ell_1 \qquad$ R!V $(\ell_1 \mapsto \ell),\ (\ell_2 \mapsto -\ell) \to !\ell_1 \vee !\ell_2$
$\quad$ R!2 $(\ell_1 \mapsto \ell),\ !\ell \to !\ell_1 \qquad$ R2V $!\ell_1 \vee !\ell_2,\ (\ell_3 \mapsto \ell_2) \to !\ell_1 \vee !\ell_3$
$\quad$ RAG1 $\ell, \ell' \to \ell \wedge \ell' \qquad\qquad$ ROR $!\ell \vee !\ell \to !\ell$
$\quad$ RAG2 $\ell, !\ell' \to \ell \wedge !\ell'$

(iv) contraries are: $\overline{\ell} = \{-\ell, !\ell\}$ and $\overline{(\ell \mapsto \ell')} = \{\ell \wedge \phi \mid \phi \in \overline{\ell'}\}$

(v) disjoint weakest link, and $\quad$ (vi) rebut.

Example 5.3 (Ex. 5.1, cont'd) The translation handles the problems in Ex. 5.1. Now we have, inter alia, the following (six) arguments:

$$\left.\begin{array}{l}\top \Rightarrow^5 s \\ \top \Rightarrow^4 (\neg p \mapsto \neg s)\end{array}\right] \to !\neg p \qquad \left.\begin{array}{l}\top \Rightarrow^1 \neg q \\ \top \Rightarrow^3 (p \mapsto q)\end{array}\right] \to !p$$

Since there are no defeats, Arg is a stable extension. Its conclusions in Lit correspond to the Hansen extension: $\mathsf{head}_{\mathsf{lit}}[\mathsf{Arg}] = \{s, \neg q\} = \mathsf{head}[\mathsf{ha}]$. The conclusions not in Lit provide additional information: e.g., $!p$ tells us that our conclusion $\neg q \in \mathsf{head}[\mathsf{ha}]$ is non-robust under learning that p. I.e., if we were to learn that p is true or obligatory, then $\neg q$ would cease to be obligatory.

Proposition 5.4 *The translation above characterizes* $\mathsf{hansen}(\mathsf{N})$ *for default theories* $\mathsf{N} = (\mathcal{F}, \mathcal{N}, \preceq)$ *where* $\preceq$ *is a modular order.*

Proof Sketch. (1) Let $X \in \mathsf{hansen}(\mathsf{N})$ be based on the scenario ha. We show that $\mathcal{X} = \mathsf{Arg}(\bigcup \tau[\mathsf{ha}]) \in \mathsf{stb}(\mathsf{N}_{\mathsf{arg}})$. ($\mathcal{X}$ is conflict-free.) This follows from the !-operator tracking of conflicts: if $a \in \mathcal{X}$ and $\mathsf{head}(a) \in \{!\ell_1 \vee !\ell_2, !\ell_1\}$ then

---

[14] In order to keep things simple we don't consider $\wedge$-elimination rules.

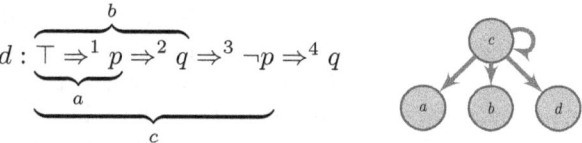

Fig. 5. Counter-example for the translation of Brewka-Eiter in [19,18]. (Left) Arguments obtained under last link. (Right) Defeats prevent any stable extension.

$\{\ell_1, \ell_2\} \not\subseteq \text{head}[\mathcal{X}]$. ($\mathcal{X}$ defeats every $a \in \text{Arg} \setminus \mathcal{X}$.) Note that any commitment to a default in $\mathcal{N} \setminus \text{ha}$ is defeated by $\mathcal{X}$: (†) if $\ell \Rightarrow^k \ell' \notin \text{ha}$, there are $\{b_1, b_2\} \subseteq \mathcal{X}$ stronger than k and with heads $\{\ell, !\ell'\}$. Let, without loss of generality, $a \notin \mathcal{X}$ satisfy $\text{PSub}(a) \subseteq \mathcal{X}$. It is easy to see that the top-rule of a is defeasible, leaving us with two cases: $a = a_1 \Rightarrow^k \ell'$ where $\text{head}(a_1) = \ell$; and $a = \top \Rightarrow^k (\ell \mapsto \ell')$. In either case, $\ell \Rightarrow^k \ell' \notin \text{ha}$. By (†), there is a $b \in \mathcal{X}$ that defeats a.

(2) Let $\mathcal{X} \in \text{stb}(\mathsf{N}_{\text{arg}})$ and $\text{ha}^{\mathcal{X}} = \{\ell \Rightarrow^k \ell' \mid \top \Rightarrow^k (\ell \mapsto \ell') \in \mathcal{X}\}$. One can show that (i) $\mathcal{D}[\mathcal{X}] \subseteq \text{ha}^{\mathcal{X}}$ and (ii) $\text{Arg}(\text{ha}^{\mathcal{X}}) \subseteq \mathcal{X}$. We enumerate $\mathcal{N}$ by $\langle r^i = \ell_i \Rightarrow^{k_i} \ell'_i \rangle_{i=1}^n$ so that (a) $k_i \geq k_{i+1}$ and (b) if $k_i = k_j$ and $r^i \in \text{ha}^{\mathcal{X}}$ while $r^j \notin \text{ha}'$, then $i < j$. Consider a Hansen-style construction but based on this enumeration, i.e. $\text{ha}_{i+1} = \text{ha}_i \cup \{r^{i+1}\}$ in case $\text{out}_N(\text{ha}_i \cup \{r^{i+1}\}) \not\vdash \bot$, and let $\text{ha} = \bigcup_i \text{ha}_i$. One can prove by induction: that ha always selects an available $\max_{\preceq}$-element (and so ha is a Hansen scenario); and, using (i)–(ii), that $\text{ha}^{\mathcal{X}} = \text{ha}$. These claims imply: $\text{head}_{\text{lit}}[\mathcal{X}] = \text{head}[\text{ha}^{\mathcal{X}}] = \text{head}[\text{ha}] \in \text{hansen}(\mathsf{N})$. □

6 Brewka-Eiter extensions

6.1 Brewka-Eiter extensions as defined in Liao et al.

Let us start by pointing out some problems with the translation proposed in [18,19] for total orders. There, the translation of be is defined as the translation for greedy except for the lifting, now using *last link*. That is,

(i) $\mathcal{L} = \text{Lit}_\top$ (ii) $\mathcal{D} = \mathcal{N}$ (iii) $\mathcal{S} = \emptyset$
(iv) $\overline{\ell} = \{-\ell\}$ (v) last link (vi) rebut.

Definition 6.1 *Last link* $\succ_{\text{la}} \subseteq \text{Arg} \times \text{Arg}$ is defined by: $a \succeq_{\text{la}} b$ iff the last defeasible link of a is greater than or equal to ($\succeq$) the last defeasible link of b.

Example 6.2 (Ex. 2.4, cont'd) Consider again the theory N based on the defaults $\mathcal{N} = \{\top \Rightarrow^1 p, p \Rightarrow^2 q, q \Rightarrow^3 \neg p, \neg p \Rightarrow^4 q\}$, see Fig. 1. Now, if we choose $X = \{q, \neg p\}$, we have $\mathcal{N}^X = \{\top \Rightarrow^1 p, \top \Rightarrow^3 \neg p, \top \Rightarrow^4 q\}$. The only greedy extension for $\mathcal{N}^X$ will collect the heads $\{\neg p, q\} = X$, so X is a greedy fixpoint of N and so is a BE extension. Now, for the stable semantics with last link [18], things look differently: no stable extension exists in such argumentation framework, depicted in Fig. 5. If, moreover, inconsistent arguments such as c are filtered out [19], then the stable extension concludes $X' = \{p, q\}$ which does not correspond to the BE extension $X = \{\neg p, q\}$ (recall Ex. 2.8).

Given a default theory N, its BE extensions result from a sort of hypothetical reasoning from candidate theories N^X. Thus, the function $\ell \Rightarrow \ell' \in \mathcal{N} \mapsto \top \Rightarrow \ell' \in \mathcal{N}^X$ turns each $\ell \in X$ into a hypothesis and asks us to reason about its direct consequences ℓ' and beyond. What is more, BE extensions allow for hypothetical bootstrapping: in Fig. 5, $X = \{q, \neg p\}$ is not *factually* grounded, as any chaining of $\mathcal{N}$ rules from $\top$ to q leads to p; instead, a BE reasoner hypothesizes q and $\neg p$ and justifies one in terms of the other. Let us capture the BE method by adding hypothetical arguments in structured argumentation. To this end, we expand the language $\mathsf{Lit}_\top$ with: (1) hypotheses $[\ell]$, introduced via defeasible rules of maximal strength $r_\ell = \top \Rightarrow^\omega [\ell]$, and (2) undercuts $\neg r_\ell$ against hypotheses, via rules of minimal strength $\top \Rightarrow^0 \neg r_\ell$. In turn, any (possibly hypothetical) argument for ℓ can defeat such undercutter $\neg r_\ell$. Altogether the translation looks as follows:

(i) $\mathcal{L} = \ell_\top \mid [\ell] \mid r_\ell \mid \neg r_\ell$ for arbitrary $\ell_\top \in \mathsf{Lit}_\top$ and $\ell \in \mathsf{Lit}$

(ii) $\mathcal{D} = \mathcal{N} \cup \{[\ell] \Rightarrow^k \ell'\}_{\ell \Rightarrow^k \ell' \in \mathcal{N}} \cup \{r_\ell = \top \Rightarrow^\omega [\ell]\}_{\ell \in \mathsf{Lit}} \cup \{\top \Rightarrow^0 \neg r_\ell\}_{\ell \in \mathsf{Lit}}$

(iii) an empty set of strict rules $\mathcal{S} = \emptyset$

(iv) contraries defined by: $\overline{\ell} = \{-\ell\}$, $\overline{\neg r_\ell} = \{\ell\}$ and $\overline{r_\ell} = \{\neg r_\ell\}$

(v) weakest link lifting, and (vi) rebut and undercut defeats.

Henceforth we let $\mathsf{Arg}(\mathsf{N}_{\mathsf{arg}})$ denote the set of all arguments for the theory $\mathsf{N}_{\mathsf{arg}}$. Our translation generates arguments of the following types:

- *factual arguments:* $\mathsf{FArg}(\mathsf{N}_{\mathsf{arg}})$ $\begin{cases} \ell_1 \Rightarrow^{k_2} \ell_2 \Rightarrow \ldots \Rightarrow^{k_n} \ell_n \\ \top \Rightarrow^0 \neg r_\ell \end{cases}$

- *hypothetical arguments:* $\mathsf{HArg}(\mathsf{N}_{\mathsf{arg}})$ $\top \Rightarrow^\omega [\ell_1] \Rightarrow^{k_2} \ell_2 \Rightarrow \ldots \Rightarrow^{k_n} \ell_n$.

where the *undercutting arguments* $\top \Rightarrow^0 \neg r_\ell$ are further collected in a set $\mathsf{UArg}(\mathsf{N}_{\mathsf{arg}})$. The attack dynamics plays out as follows. A reasoner is always free to introduce hypotheses via $\top \Rightarrow^\omega [\ell]$, but these arguments are by default undercut by arguments of the type $\top \Rightarrow^0 \neg r_\ell$. The hypothesis $[\ell]$ thus needs to be backed up by an argument with conclusion ℓ that defeats the undercutter (note that $\overline{\neg r_\ell} = \{\ell\}$). Such an argument can rest upon hypotheses, e.g. $\top \Rightarrow^\omega [\ell_1] \Rightarrow \ldots \Rightarrow \ell$.[15] As we will see in Section 6.2 this is exactly responsible for the bootstrapping occurring in Examples like Ex. 2.4 for BE extensions. This bootstrapping does not occur in the original BE extensions, and so the characterization of BEo will disallow attacks from hypothetical arguments on factual ones. But more on that in Sec. 6.2.

Example 6.3 Consider the default theory $\mathsf{N} = (\mathcal{N}, \mathcal{F}, \preceq)$ given by $\mathcal{F} = \{\top\}$ and $\mathcal{N} = \{\top \Rightarrow^2 \neg p, \top \Rightarrow^1 p, p \Rightarrow^3 p\}$. Our translation gives, among others, the arguments depicted in Fig. 6. The dark arguments form one stable extension, the light ones the other. They correspond to the BE extensions.

Proposition 6.4 *The translation above characterizes* $\mathsf{be}(\mathsf{N})$ *for any default theory* $\mathsf{N} = (\mathcal{F}, \mathcal{N}, \preceq)$ *with a modular order* $\preceq$.

---

[15] The contrariness function avoids a particular type of circularity: $[\ell]$ cannot reinstate itself (or arguments based on it). Still, arguments based on $[\ell]$ can reinstate $[\ell]$ (see Ex. 6.3).

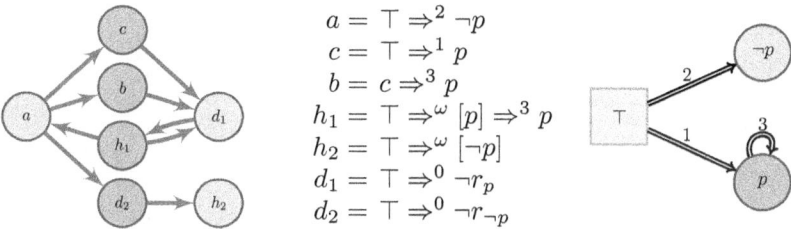

Fig. 6. Stable sets and BE extensions in Ex. 6.3. (Left) The two stable extensions of N_{arg}. (Center) Arguments in N_{arg}. (Right). The two BE extensions of N.

Proof Sketch. (1) Let $X \in \mathsf{greedy}(\mathsf{N}^X)$. Let $\mathcal{X}$ be the set of all arguments a such that for all $b \in \mathsf{Sub}(a)$, $\mathsf{head}(b) \in \{\ell, [\ell]\}$ for some $\ell \in X$. When proving that $\mathcal{X} \in \mathsf{stb}(\mathsf{N}_{\mathsf{arg}})$, we have to show that it is (i) conflict-free and it (ii) attacks every $a \in \mathsf{Arg} \setminus \mathcal{X}$. We show (ii). As in the proof of Prop. 5.4, we can assume that the top-rule of a is defeasible. So $\mathsf{head}(a)$ has the form (a) ℓ or (b) $\neg r_\ell$, or (c) $[\ell]$ for some $\ell \in \mathsf{Lit}$. In case (a), the top-rule of a is $\phi \Rightarrow^l \ell$ with $\phi \in \{\ell', [\ell']\}$ for some $\ell' \in X$. So, $\top \Rightarrow^{l'} \ell \in \mathsf{N}^X$ for some $l' \geq l$. Since $\ell \notin X$, $-\ell \in X$ and so there is a $l'' \geq l'$ for which $\top \Rightarrow^{l''} -\ell \in \mathcal{N}^X$. Thus, $b = [\ell''] \Rightarrow^{l''} -\ell \in \mathcal{X}$ defeats a. Cases (b) and (c) are shown similarly.

(2) Let $\mathcal{X} \in \mathsf{stb}(\mathsf{N}_{\mathsf{arg}})$, $X = \mathsf{head}_{\mathsf{lit}}[\mathcal{X}]$, and enumerate $\mathcal{N}^X$ by $\langle r_i = \top \Rightarrow^{k_i} \ell'_i \rangle_{i=1}^n$ in accordance with: (1) if $k_i > k_j$ then $i < j$ and (2) if $k_i = k_j$ and some $\ell_i \Rightarrow \ell'_i \in \mathcal{D}[\mathcal{X}]$ but there is no $\ell_j \Rightarrow \ell'_j \in \mathcal{D}[\mathcal{X}]$, then $i < j$. Define:

$$\mathsf{gr}_0 = \emptyset \quad \text{and} \quad \mathsf{gr}_{i+1} = \begin{cases} \mathsf{gr}_i \cup \{r_{i+1}\} & \text{if } -\ell'_{i+1} \notin \mathsf{head}[\mathsf{gr}_i] \\ \mathsf{gr}_{i+1} = \mathsf{gr}_i & \text{else} \end{cases}$$

and let $\mathsf{gr} = \bigcup_{i \geq 0} \mathsf{gr}_i$. We then prove that gr is a greedy scenario of N^X satisfying $\mathsf{head}[\mathsf{gr}] = X$, by inductively showing that (a) $\mathsf{gr}_{i+1} \setminus \mathsf{gr}_i \subseteq \max_{\preceq}(\mathsf{TrigCon}_{\mathsf{N}^X}(\mathsf{gr}_{i+1}) \setminus \mathsf{gr}_i)$; and (b) $r_i \in \mathsf{gr}_i$ iff $\ell'_i \in X$. □

6.2 The original Brewka-Eiter extensions

We now move to a characterization of the original BE extensions. These are also not characterized by last link. A way to see this is the order puzzle ([18,16]).

Example 6.5 Let $\mathcal{N} = \{\top \Rightarrow^1 q, q \Rightarrow^3 p, \top \Rightarrow^2 \neg p\}$. Let $a_1 = \top \Rightarrow^1 q$. According to last link the argument $a_2 = a_1 \Rightarrow^3 p$ defeats $b = \top \Rightarrow^2 \neg p$ and $\{a_1, a_2\}$ will form the only stable extension in the characterization by Liao et al. However, as can easily be seen, there is no BEo extension of this theory.[16]

Moreover, in theories that do have BEo extensions the characterization in terms of last link lifting is also not adequate.

Example 6.6 Let $\mathcal{N} = \{\top \Rightarrow^1 p, p \Rightarrow^4 \neg q, \top \Rightarrow^2 q, q \Rightarrow^3 \neg p\}$. We consider the arguments and the argumentation framework based on last link in Fig. 7.

---

[16] Given the importance of the Order puzzle in the discussion of prioritized obligations, the existence of a BE extension justifies [18,19] in their removal of the Greedy condition from the original Brewka-Eiter method, at least in the context of deontic applications.

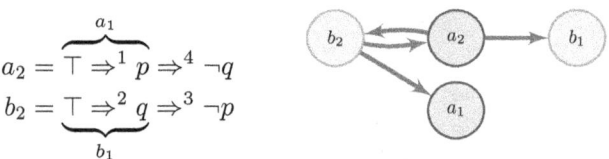

Fig. 7. Example 6.6. The highlighted arguments form a stable extension with last link that does not represent a BEo extension.

The only BEo extension of this theory is $\{q, \neg p\}$, while $\{a_1, a_2\}$ form a stable extension in last link with conclusions $\{p, \neg q\}$.

We now show that our translation for BE extensions presented above can be adjusted with a minor twist to also account for the original BE extensions. In Section 6.1 we have allowed for attacks without type restrictions. In particular, hypothetical arguments could attack factual arguments. In order to characterize the original BE extensions we use the same translation, but restrict the range of the attack relation to $\mathsf{Arg}(\mathsf{N}_{\mathsf{arg}}) \times \mathsf{Arg}(\mathsf{N}_{\mathsf{arg}})) \setminus (\mathsf{HArg}(\mathsf{N}_{\mathsf{arg}}) \times \mathsf{FArg}(\mathsf{N}_{\mathsf{arg}}))$. That is, we don't allow for attacks from hypothetical to factual arguments. This may create situations in which factual arguments attack but do not defeat hypothetical arguments, which is why conflict-freeness has now to be considered on the basis of attacks, not defeats.

Let in the remainder of the section $\mathsf{N}_{\mathsf{arg}}$ stand for the translation of a given normative theory $\mathsf{N} = (\mathcal{F}, \mathcal{N}, \preceq)$ based on a total order $\preceq$.

Stable extensions $\mathcal{X}$ of $\mathsf{N}_{\mathsf{arg}}$ have the property that the (literal) conclusions of the factual arguments in $\mathcal{X}$ stand in a 1:1 relation to the conclusions of the hypothetical arguments in $\mathcal{X}$. Formally, where for $a \in \mathsf{Arg}(\mathsf{N}_{\mathsf{arg}}) \setminus \mathsf{UArg}(\mathsf{N}_{\mathsf{arg}})$, $\mathsf{head}^\star(a) = \ell$ iff $\mathsf{head}(a) \in \{\ell, [\ell]\}$, we have:

Lemma 6.7 *Let $\mathcal{X} \in \mathsf{stb}(\mathsf{N}_{\mathsf{arg}})$ and $X = \mathsf{head}_{\mathsf{lit}}[\mathcal{X}]$. Then, $\mathsf{head}_{\mathsf{lit}}[\mathcal{X} \cap \mathsf{FArg}(\mathsf{N}_{\mathsf{arg}})] = \mathsf{head}^\star[\mathcal{X} \cap \mathsf{HArg}(\mathsf{N}_{\mathsf{arg}})]\ (= X)$.*

Proof Sketch. We show ($\Rightarrow$) by an induction over the length of a factual argument for ℓ in $\mathcal{X}$. The other direction is shown analogously. *Base.* Suppose $\ell \in \mathcal{F} \cap \mathcal{X}$. The only attacker of $h = \top \Rightarrow^\omega [\ell]$ is $\top \Rightarrow^0 \neg r_\ell$. The latter is attacked by ℓ. So, $h \in \mathcal{X}$ and $[\ell] \in X$. *Inductive step.* Consider $a = \ell_1 \Rightarrow^1 \ell_2 \Rightarrow \ldots \Rightarrow^{n-1} \ell_n \in \mathcal{X}$. By the inductive hypothesis $\ell_{n-1} \in X$. Consider $h = \top \Rightarrow^\omega [\ell_{n-1}] \Rightarrow^{n-1} \ell_n$. The only attack in $\top \Rightarrow^\omega [\ell_{n-1}]$ is by $\top \Rightarrow^0 \neg r_{\ell_{n-1}}$. The latter is defeated by $\ell_1 \Rightarrow^1 \ell_2 \Rightarrow \ldots \Rightarrow^{n-2} \ell_{n-1} \in \mathcal{X}$ and thus defended by $\mathcal{X}$. So, $\top \Rightarrow^\omega [\ell_{n-1}] \in \mathcal{X}$. The only way to attack h is by an argument with conclusion $-\ell_n$. Since such an argument is attacked by a and since $\mathcal{X}$ is stable, it is defeated by $\mathcal{X}$. So h is defended by $\mathcal{X}$ and thus $[\ell_n] \in X$. □

We are now in a position to prove the adequacy of our translation.

Proposition 6.8 *The translation above characterizes $\mathsf{BEo}(\mathsf{N})$ for default theories $\mathsf{N} = (\mathcal{F}, \mathcal{N}, \preceq)$ where $\preceq$ is total.*

Proof Sketch. For Item (1) let $X \in \mathsf{BEo}(\mathsf{N})$. Let $\mathsf{N}'_{\mathsf{arg}}$ be the translation

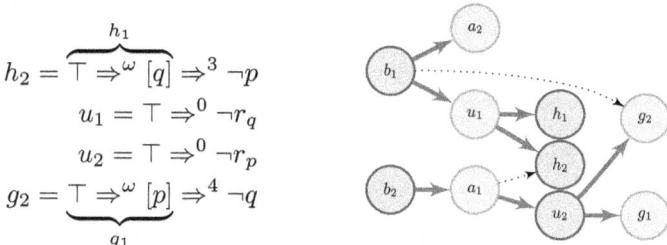

Fig. 8. Example 6.9. Dotted arrows present attacks that are not defeats.

of N adequate for greedy extensions from Sec. 4. By Prop. 4.1, there is an $\mathcal{X}' \in \mathsf{stb}(\mathsf{N}'_{\mathsf{alg}})$ for which $\mathsf{head}_{\mathsf{lit}}[\mathcal{X}'] = X$. Let $\mathcal{X} = \mathcal{X}' \cup \{h \in \mathsf{HArg}(\mathsf{N}_{\mathsf{arg}}) \mid \mathsf{head}^\star[\mathsf{Sub}(h)] \subseteq X\} \cup \{\top \Rightarrow^0 \neg r_\ell \mid \ell \in \mathsf{Lit} \setminus X\}$. Note that, $(\star)$, $\mathsf{head}_{\mathsf{lit}}[\mathcal{X}] = X$. So, we only have to show that $\mathcal{X} \in \mathsf{stb}(\mathsf{N}_{\mathsf{arg}})$.

The conflict-freeness of $\mathcal{X}$ follows immediately in view of $(\star)$ and the conflict-freeness of $\mathcal{X}'$. Consider now an $a \in \mathsf{Arg}(\mathsf{N}_{\mathsf{arg}}) \setminus \mathcal{X}$. If a is in $\mathsf{Arg}(\mathsf{N}'_{\mathsf{arg}})$, by the stability of $\mathcal{X}'$ it is defeated by $\mathcal{X}'$. If a is of the form $\top \Rightarrow^0 \neg r_\ell$, by the definition of $\mathcal{X}$, $\ell \in X$. So, there is a $b \in \mathcal{X}'$ with conclusion ℓ that defeats a. Suppose now that $a \in \mathsf{HArg}(\mathsf{N}_{\mathsf{arg}})$. Without loss of generality we assume that $\mathsf{PSub}(a) \subseteq \mathcal{X}$. If a is of the form $\top \Rightarrow^\omega [\ell]$, $\ell \notin X$ and so $\top \Rightarrow^0 \neg r_\ell \in \mathcal{X}$ which defeats a. Suppose a is of the form $\top \Rightarrow^\omega [\ell_1] \Rightarrow^{k_1} \ldots \Rightarrow^{k_{n-1}} \ell_n$. Since $\top \Rightarrow^\omega [\ell_1] \Rightarrow^{k_1} \ldots \Rightarrow^{k_{n-2}} \ell_{n-1} \in \mathcal{X}$, $\ell_{n-1} \in X$. Since $a \notin \mathcal{X}$, $\ell_n \notin X$. By Prop. 2.11, there is a $\ell' \Rightarrow^k -\ell_n \in \mathcal{N}$ for which $\ell', -\ell_n \in X$ and $k > k_{n-1}$. Then, $\top \Rightarrow^\omega [\ell'] \Rightarrow^k -\ell_n \in \mathcal{X}$ defeats a. In sum, we have shown $\mathcal{X} \in \mathsf{stb}(\mathsf{N}_{\mathsf{arg}})$.

For Item (2) let $\mathcal{X} \in \mathsf{stb}(\mathsf{N}_{\mathsf{arg}})$ and $X = \mathsf{head}_{\mathsf{lit}}[\mathcal{X}]$. We use Prop. 2.11 and show that (i) $X \in \mathtt{greedy}(\mathsf{N})$ and that X also satisfies Prop. 2.11 (ii).

Concerning (i), we first notice that the fragment of our translation for BEo without hypothetical and undercutting arguments is identical to the translation for greedy. Let $\mathcal{X}' = \mathcal{X} \setminus (\mathsf{HArg}(\mathsf{N}_{\mathsf{arg}}) \cup \mathsf{UArg}(\mathsf{N}_{\mathsf{arg}}))$. Let $\mathsf{N}'_{\mathsf{arg}}$ be the translation for greedy of N. Then, $\mathcal{X}' \in \mathsf{stb}(\mathsf{N}'_{\mathsf{arg}})$ due to the fact that there are no attacks from hypothetical and from undercutting arguments on factual arguments. By Lemma 6.7, $X = \mathsf{head}_{\mathsf{lit}}[\mathcal{X}'] \in \mathtt{greedy}(\mathsf{N})$. Concerning (ii), suppose $\ell_1 \Rightarrow^l \ell' \in \mathcal{N}$, $\ell_1 \in X$ and $\ell' \notin X$. Consider $h = \top \Rightarrow^\omega [\ell_1] \Rightarrow^k \ell'$. By Lemma 6.7, $h' = \top \Rightarrow^\omega [\ell_1] \in \mathcal{X}$ and $h \notin \mathcal{X}$. Since $h \notin \mathcal{X}$ and by the stability of $\mathcal{X}$ there is a defeater a of h in $\mathcal{X}$. Since $h' \in \mathcal{X}$, the defeat is in ℓ' and so $\mathsf{head}^\star(a) = -\ell'$. So, a is of the form $\ldots \Rightarrow \ell_2 \Rightarrow^{k'} -\ell'$. Since $a > h$, $k' > k$, and $\ell_2 \Rightarrow^{k'} -\ell' \in \mathcal{N}$. □

Example 6.9 We take another look at the theory from Example 6.6. Fig. 8 shows the argumentation framework and the additional arguments for our translation of the original BE construction. As desired, the only stable extension $\{b_1, b_2, h_1, h_2, u_2\}$ corresponds to the unique original BE extension $\{q, \neg p\}$.

7 Non-Normal Defaults

We now move to a more general setting, in which defaults include a list of literals, called *justifications*. Defaults are now of the form:

$$r = \frac{\ell_0 \quad \ell_1, \ldots, \ell_n}{\ell}$$

where $\text{justif}(r) = \{\ell_1, \ldots, \ell_n\}$ are the justifications, and as before $\text{head}(r) = \ell$ and $\text{body}(r) = \ell_0$. A default is *non-normal* in case $\text{justif}(r) \neq \{\text{head}(r)\}$.

Definition 7.1 A *default theory* $\mathsf{N} = (\mathcal{F}, \mathcal{N}, \preceq)$ is as in Def. 2.1 except that now $\mathcal{N}$ is a set of (possibly) non-normal defaults. For $X, X' \subseteq \text{Lit}$ we define:

$$\text{Con}_\mathsf{N}(X) = \{r \in \mathcal{N} \mid \text{each } \ell' \in \text{justif}(r) \text{ is consistent with } X\}$$
$$\text{TrigCon}^\star_\mathsf{N}(\mathcal{N}', X') = (\text{Trig}_\mathsf{N}(\mathcal{N}') \setminus \mathcal{N}') \cap \text{Con}_\mathsf{N}(X').$$

Definition 7.2 ([24]) A set $X \subseteq \text{Lit}$ is a *Reiter extension* of N if it is obtained by the following procedure: [17]

---
guess $X \subseteq \text{Lit}$; $X' \leftarrow \emptyset$; $\mathcal{N}' \leftarrow \emptyset$
loop until $\max_\preceq(\text{TrigCon}^\star_\mathsf{N}(\mathcal{N}', X')) \cap \text{Con}_\mathsf{N}(X) = \emptyset$
    $r \leftarrow \max_\preceq(\text{TrigCon}^\star_\mathsf{N}(\mathcal{N}', X')) \cap \text{Con}_\mathsf{N}(X)$
    $X' \leftarrow X' \cup \{\text{head}(r)\}$; $\mathcal{N}' \leftarrow \mathcal{N}' \cup \{r\}$
if $(X' = X)$ return X

We write $\texttt{reiter}(\mathsf{N})$ for the set of Reiter extensions of N.

Being dissatisfied with the pseudo-inductive Def. 7.2, the fact that for some default theories no Reiter extensions exists, and some possibly counter-intuitive outputs, Lukaszewicz proposed the following variant.

Definition 7.3 ([20]) Where $\mathsf{N} = (\mathcal{F}, \mathcal{N}, \preceq)$ has a total or flat order $\preceq$, let

$$\text{Con}_\mathsf{N}(X, J) = \{r \in \mathcal{N} \mid \text{each } \ell' \in J \cup \text{justif}(r) \text{ is consistent with } X \cup \{\ell\}\}.$$

We say that X is a *Lukaszewicz extension* if it is obtained as follows: [18]

$X \leftarrow \emptyset$; $J \leftarrow \emptyset$; $\mathcal{N}' \leftarrow \emptyset$
loop until $(\text{Trig}_\mathsf{N}(\mathcal{N}') \setminus \mathcal{N}') \cap \text{Con}_\mathsf{N}(X, J) = \emptyset$
    $r \leftarrow \max_\preceq((\text{Trig}_\mathsf{N}(\mathcal{N}') \setminus \mathcal{N}') \cap \text{Con}_\mathsf{N}(X, J))$
    $X \leftarrow X \cup \{\text{head}(r)\}$; $J \leftarrow J \cup \text{justif}(r)$; $\mathcal{N}' \leftarrow \mathcal{N}' \cup \{r\}$
return X

We write $\texttt{luk}(\mathsf{N})$ for the set of all Lukaszewicz extensions of N.

Fact 7.4 *Every Lukaszewicz extension X produced by the algorithm in Def. 7.3 is consistent with every justification ℓ collected in J.*

Example 7.5 (Adapted from [20]) Let the flat default theory N consist of: [19]

---

[17] Our definitions Defs. 7.2–7.3 generalize the deontic versions of Reiter and Lukaszewicz (featuring no priorities). The original definitions correspond to flat orders.

[18] See Section 5 of [20] for this characterization of what are there called *m-extensions*.

[19] Read e = *evening*, d = *doing the dishes*, t = *before test* and s = *study*. The defaults then read: in the evening you ought to do the dishes (if possible and unless you have to study); and before a test, you have to study (if possible).

$a_0 = e \Rightarrow d \wedge \langle d \rangle \wedge \langle \neg s \rangle$
$b_0 = t \Rightarrow s \wedge \langle s \rangle$
$a_1 = a_0 \rightarrow d$
$a_2 = a_0 \rightarrow \langle d \rangle$
$a_3 = a_0 \rightarrow \langle \neg s \rangle$
$b_1 = b_0 \rightarrow s$
$b_2 = b_0 \rightarrow \langle s \rangle$

Fig. 9. Examples 8.1 and 8.5. (Left) The arguments. (Center) The argumentation framework for Reiter. (Right) The argumentation framework for Lukaszewicz.

$$\mathcal{F} = \{\top, e, t\} \quad \text{and} \quad \mathcal{N} = \left\{ \frac{e \quad d, \neg s}{d}, \frac{t \quad s}{s} \right\}.$$

While the only Reiter extension is $X_1 = \{s\}$, besides X_1 we have the additional Lukaszewicz extension $X_2 = \{d\}$.

8 Translating Non-Normal Defaults

Let us then characterize the Reiter and Lukaszewicz methods argumentatively. Since both methods are generalizations of Greedy reasoning, they face the same problems for partial and modular orders (Ex. 4.2). Hence, our translations only concern total and flat orders. In both cases we translate a default r with a consistency requirement $\ell \in \mathsf{justif}(r)$ into a defeasible rule $\tau(r)$ containing an expression $\langle \ell \rangle$ that represents this *consistency of* ℓ. The two translations only differ in the contrariness function: justifications $\langle \ell \rangle$ are allowed to attack in Lukasiewicz, while in Reiter they are not.

8.1 Prioritized Reiter

We provide a translation of a default theory $\mathsf{N} = \langle \mathcal{F}, \mathcal{N}, \preceq \rangle$ based on a total or flat order $\preceq$ (induced by a ranking f) to $\mathsf{N}_{\mathsf{arg}}$ by:

(i) $\mathcal{L} = \ell_\top \mid \langle \ell \rangle \mid \phi \wedge \ldots \wedge \phi'$ \hfill where $\ell_\top \in \mathsf{Lit}_\top$ and $\ell \in \mathsf{Lit}$

(ii) $\mathcal{D} = \tau[\mathcal{N}]$ with $\tau(r) = \mathsf{body}(r) \Rightarrow^k \mathsf{head}(r) \wedge \bigwedge_{\ell \in \mathsf{justif}(r)} \langle \ell \rangle$ \hfill for $f(r) = k$

(iii) $\mathcal{S}$ consists of all instances of the form (AGG) $(\ldots \wedge \phi \wedge \ldots) \rightarrow \phi$

(iv) contraries are: $\overline{\ell} = \{-\ell\}$, $\overline{\langle \ell \rangle} = \{-\ell\}$, and $\overline{\phi_1 \wedge \ldots \wedge \phi_n} = \bigcup_{i=1}^n \overline{\phi_i}$

(v) disjoint weakest link

(vi) defeat based on rebut, and \quad (vii) conflict-freeness based on attacks.

Example 8.1 (Ex. 7.5 cont'd) In this example we have, inter alia, the arguments and the argumentation framework presented in Fig. 9. The unique stable extension including b_1 corresponds to the unique Reiter extension $\{s\}$.

Proposition 8.2 *The above translation characterizes* $\mathsf{reiter}(\mathsf{N})$ *for any non-normal default theory* $\mathsf{N} = (\mathcal{F}, \mathcal{N}, \preceq)$ *with a total or flat order* $\preceq$.

Proof Sketch. (1) We sketch the case for a total order $\preceq$. Let $X = \bigcup_{i=1}^n X_i \in \mathsf{reiter}(\mathsf{N})$ be based on the scenario $\mathsf{rei} = \bigcup_{i=1}^n \mathsf{rei}_i = \{r_1, \ldots, r_n\}$. Let $\mathcal{X}$

consist of all arguments a for which $\mathcal{D}(a) \subseteq \tau[\mathbf{rei}]$. We have to show that $\mathcal{X}$ is (a) conflict-free and (b) $\mathcal{X}$ defeats every $b \in \mathsf{Arg} \setminus \mathcal{X}$. One can show that $(\star)$, where $b \in \mathcal{X}$ with $\tau(r_{i+1}) \in \mathcal{D}(b)$, for each $\ell \in X_i$ there is an $a \in \mathcal{X}$ with $\mathsf{head}(a) = \ell$ and $a \succ b$. We show (b). Without loss of generality we assume $\mathsf{PSub}(b) \subseteq \mathcal{X}$. Let $\tau(r)$ be the defeasible top-rule of $b = b_0 \Rightarrow \ell' \wedge \langle \ell'_1 \rangle \wedge \ldots \wedge \langle \ell'_m \rangle$ and l be minimal such that $b_0 \in \mathsf{Arg}(\tau[\bigcup_{i=1}^{l} \mathbf{rei}_i])$. Since $r \notin \mathbf{rei}$, one can show that there is a minimal $k \geq 0$ such that $r \notin \mathsf{Con}_{\mathsf{N}}(X_{l+k})$ and $r_{l+k'} \succ r$ for all $1 \leq k' < k$. So, $-\ell'_i \in X_{l+k}$. In view of $(\star)$ we can then construct an $a \in \mathcal{X}$ that defeats b.

(2) Let $\mathcal{X} \in \mathsf{stb}(\mathsf{N}_{\mathsf{arg}})$. We show that $X = \mathsf{head}_{\mathsf{lit}}[\mathcal{X}] \in \mathsf{reiter}(\mathsf{N})$. We build the scenario $\mathbf{rei} = \bigcup_{i=0}^{n} \mathbf{rei}_i$ by performing a greedy search over the arguments of $\mathcal{X}$: we let $\mathbf{rei}_0 = \emptyset$ and add to $\mathbf{rei}_i$ in the $(i+1)$-th step the strongest default r with $\tau(r) \in \mathcal{D}[\mathcal{X}] \setminus \tau[\mathbf{rei}_i]$ that extends an argument in $\mathsf{Arg}(\tau[\mathbf{rei}_i])$. It can be shown that $\mathbf{rei}$ is a Reiter scenario with extension X. □

Remark 8.3 Answer set programming is closely related to Reiter's default logic. For an extended logic program Π over Lit (with default negation $\sim\ell$), its clauses translate into non-normal defaults as follows:

$$\tau: \ell \leftarrow \ell_1, \ldots, \ell_n, \sim\ell'_1, \ldots, \sim\ell'_m \qquad \longmapsto \qquad \frac{\ell_1 \wedge \ldots \wedge \ell_n \quad -\ell'_1, \ldots, -\ell'_m}{\ell}.$$

resulting in a default theory $\mathsf{N} = (\mathcal{F}, \mathcal{N}, \cdot)$ with $\mathcal{N} = \tau[\Pi]$ and without facts $\mathcal{F} = \emptyset$ (since in logic programming *facts* are empty-bodied clauses $\ell \leftarrow$). For the case of flat orders, this translation establishes a one-one correspondence between *Reiter extensions* of N and *answer sets* of Π, as shown in [12, Prop. 3].

An immediate corollary of Prop. 8.2 is thus an ASPIC$^+$-based representation of answer set programming.[20] Let us remark that this ASPIC$^+$ representation applies not just to the epistemic setting originally assumed in [12, Prop. 3] but also to the deontic setting. The reason is that a logic program translates into a default theory without facts $\mathcal{F} = \emptyset$, so that both settings give rise to the same Reiter extensions under the above correspondence.[21]

Example 8.4 Conflict-freeness based on **defeat** does not work for the proposed translation, as shown by the prioritized defaults and the generated arguments:

$$\mathcal{N} = \left\{ \frac{\top \quad p, s}{p} \; 2, \frac{p \quad \neg s}{\neg s} \; 1 \right\}$$

$$\mathsf{Args} = \left\{ \begin{array}{l} a = \top \Rightarrow^2 (p \wedge \langle s \rangle \wedge \langle p \rangle), \; a' = a \to p, \\ b = a' \Rightarrow^1 (\neg s \wedge \langle \neg s \rangle), \; b' = b \to \neg s \end{array} \right\}$$

Based on **defeat**, the set $\mathcal{X} = \{a, a', b, b'\}$ is conflict-free since b' only attacks

---

[20] Since in this paper we restricted the attention to single-body rules, the result –strictly speaking– only applied to logic programs with $n = 1$ in the clauses. One would have to add an aggregation rule to our translation: $\ell_1, \ldots, \ell_n \to \ell_1 \wedge \ldots \wedge \ell_n$ to obtain the full result.

[21] Recall that in the epistemic setting facts are part of the output. In a nutshell, this setting redefines Def. 2.2 with the output function $\mathsf{out}_{\mathsf{N}}(\mathcal{N}') = $ the $\subseteq$-smallest superset X of $\mathcal{F}$ satisfying: $\ell \in X$ and $\ell \Rightarrow \ell' \in \mathcal{N}'$ implies $\ell' \in X$. For Reiter extensions, then, Def. 7.2 initializes with $X' \leftarrow \mathcal{F}$ (instead of $X' \leftarrow \emptyset$).

(but does not defeat) the other arguments in $\mathcal{X}$. In contrast, under conflict-freeness based on attack, $\mathcal{X}$ is not anymore conflict-free. Moreover, there is no stable extension and neither is there a Reiter-extension of the default theory.

8.2 Prioritized Lukaszewicz

For total or flat orders $\preceq$ the translation $\mathsf{N} \mapsto \mathsf{N}_{\mathsf{arg}}$ is as in Reiter for (i)–(iii) and (v)–(vii); for contraries, justifications $\langle \ell \rangle$ now attack arguments for $-\ell$:

(iv) contraries are: $\overline{\ell} = \{-\ell, \langle -\ell \rangle\}$, $\overline{\langle \ell \rangle} = \{-\ell\}$, and $\overline{\phi_1 \wedge \ldots \wedge \phi_n} = \bigcup_{i=1}^{n} \overline{\phi_i}$.

Example 8.5 (Ex. 7.5 cont'd) We have the argumentation framework depicted in Fig. 9 (right). We highlight the arguments in the stable extension that represents the Lukaszewicz extension that is not a Reiter extension.

Proposition 8.6 *The above translation characterizes* $\mathtt{luk}(\mathsf{N})$ *for any non-normal default theory with a total or flat order* $\preceq$.

Proof Sketch. The proof is very similar to that of Prop. 8.2. □

9 Conclusion

In this paper we have advanced the state of the art concerning argumentative representations of (prioritized) default logic by (a) identifying and fixing problems in previously proposed embeddings, (b) generalizing some of them from total to modular orders, and (c) by considering non-normal defaults. In future work we will investigate: (1) multiple-body rules, (2) more liberal orderings (e.g., partial and pre-orders), (3) fully propositional languages, and (4) embeddings in other argumentation systems such as ABA and (assumptive [8] and prioritized [1]) sequent-based argumentation. Of course, also the inverse question may be asked of how much of structured argumentation can be expressed in default logic. [22] Finally, making use of hypothetical reasoning within structured argumentation theory deserves more future investigations in its own right (see e.g., [5,4]). In this article we have demonstrated that hypothetical reasoning helps to adequately characterize reasoning with priorities that has been proposed in the context of default logic and logic programming [9,10].

Acknowledgments The authors would like to thank the anonymous reviewers for their helpful suggestions. Research for this article was sponsored by the Department of Philosophy Piero Martinetti of the University of Milan under the Project Departments of Excellence 2018–2022 awarded by the Ministry of Education, University and Research (MIUR).

References

[1] Arieli, O., A. Borg and C. Straßer, *Prioritized sequent-based argumentation*, in: *Proceedings of the 17th International Conference on Autonomous Agents and MultiAgent Systems*, International Foundation for Autonomous Agents and Multiagent Systems, 2018, pp. 1105–1113.

---

[22] In [15] ASPIC$^+$ without undercut is characterized in terms of maxicon sets of defaults.

[2] Arieli, O. and C. Straßer, *Sequent-based logical argumentation*, Argument and Computation. **6** (2015), pp. 73–99.
[3] Beirlaen, M., J. Heyninck, P. Pardo and C. Straßer, *Argument strength in formal argumentation*, Journal of Applied Logics-IfCoLog Journal of Logics and their Applications **5** (2018), pp. 629–675.
[4] Beirlaen, M., J. Heyninck and C. Straßer, *Reasoning by cases in structured argumentation.*, in: *Proceedings of the Symposium on Applied Computing*, ACM, 2017, pp. 989–994.
[5] Beirlaen, M., J. Heyninck and C. Straßer, *A critical assessment of pollock's work on logic-based argumentation with suppositions*, in: *Proceedings of Argumentation and Philosophy (sub-workshop of COMMA 2018)*, 2018, p. xx.
[6] Beirlaen, M., C. Straßer and J. Heyninck, *Structured argumentation with prioritized conditional obligations and permissions*, Journal of Logic and Computation **29** (2018), pp. 187–214.
[7] Bondarenko, A., P. M. Dung, R. A. Kowalski and F. Toni, *An abstract, argumentation-theoretic approach to default reasoning*, Artifical Intelligence **93** (1997), pp. 63–101.
[8] Borg, A., *Assumptive sequent-based argumentation*, Journal of Applied Logics **2631** (2020), p. 227.
[9] Brewka, G. and T. Eiter, *Preferred answer sets for extended logic programs*, Artificial Intelligence **109** (1999), pp. 297–356.
[10] Brewka, G. and T. Eiter, *Prioritizing default logic*, in: *Intellectics and computational logic*, Springer, 2000 pp. 27–45.
[11] Dung, P. M., *On the acceptability of arguments and its fundamental role in nonmonotonic reasoning, logic programming and n-person games*, Artifical Intelligence **77** (1995), pp. 321–358.
[12] Gelfond, M. and V. Lifschitz, *Classical negation in logic programs and disjunctive databases*, New Generation Computing **9** (1991), p. 365–385.
[13] Hansen, J., *Prioritized conditional imperatives: problems and a new proposal*, Autonomous Agents and Multi-Agent Systems **17** (2008), pp. 11–35.
[14] Heyninck, J., "Investigations into the logical foundations of defeasible reasoning: an argumentative perspective," Ph.D. thesis, Ruhr-University Bochum (2018).
[15] Heyninck, J. and C. Straßer, *Rationality and maximal consistent sets for a fragment of aspic+ without undercut*, Argument & Computation (2020), pp. 1–45.
[16] Horty, J. F., "Reasons as defaults," Oxford University Press, 2012.
[17] Lehmann, D. J. and M. Magidor, *What does a conditional knowledge base entail?*, Artificial Intelligence **55** (1992), pp. 1–60.
[18] Liao, B., N. Oren, L. van der Torre and S. Villata, *Prioritized norms and defaults in formal argumentation*, Deontic Logic and Normative Systems (2016).
[19] Liao, B., N. Oren, L. van der Torre and S. Villata, *Prioritized norms in formal argumentation*, Journal of Logic and Computation **29** (2018), pp. 215–240.
[20] Łukaszewicz, W., *Considerations on default logic: an alternative approach*, Computational intelligence **4** (1988), pp. 1–16.
[21] Modgil, S. and H. Prakken, *A general account of argumentation with preferences*, Artificial Intelligence **195** (2013), pp. 361–397.
[22] Modgil, S. and H. Prakken, *The aspic+ framework for structured argumentation: a tutorial*, Argument & Computation **5** (2014), pp. 31–62.
[23] Nair, S. and J. Horty, *The logic of reasons*, Oxford Handbooks Online (2018). URL http://dx.doi.org/10.1093/oxfordhb/9780199657889.013.4
[24] Reiter, R., *A logic for default reasoning*, Artifical Intelligence **1–2** (1980).
[25] Straßer, C. and O. Arieli, *Normative reasoning by sequent-based argumentation*, Journal of Logic and Computation **29** (2015), pp. 387–415.
[26] Young, A. P., S. Modgil and O. Rodrigues, *Prioritised default logic as rational argumentation*, in: *Proceedings of the 2016 International Conference on Autonomous Agents & Multiagent Systems*, International Foundation for Autonomous Agents and Multiagent Systems, 2016, pp. 626–634.

www.ingramcontent.com/pod-product-compliance
Lightning Source LLC
Chambersburg PA
CBHW050118170426
43197CB00011B/1629